SPOT TESTS

IN INORGANIC ANALYSIS

SOLE DISTRIBUTORS FOR THE U.S.A. AND CANADA:

D. VAN NOSTRAND COMPANY, INC.

120 Alexander Street, Princeton, N.J.
257 Fourth Avenue, New York 10, N.Y.
25 Hollinger Road, Toronto 16, Canada

FOR THE BRITISH COMMONWEALTH EXCEPT CANADA:

CLEAVER-HUME PRESS, LTD.

31, Wright's Lane, Kensington, London, W.8.

Library of Congress Catalog Card Number 56-13144

SPOT TESTS
IN
INORGANIC ANALYSIS

by

FRITZ FEIGL, Eng., D. Sc.

Laboratório da Produção Mineral, Ministério da Agricultura, Rio de Janeiro; Professor at the University of Brazil; Member of the Austrian and Brazilian Academies of Sciences

Translated by

RALPH E. OESPER, Ph.D.

Professor Emeritus, University of Cincinnati

Fifth, enlarged and revised English Edition

ELSEVIER PUBLISHING COMPANY

AMSTERDAM LONDON NEW YORK PRINCETON

1958

First English Edition, 1937 (translation by J. W. Matthews)
Second English Edition, 1939 (translation by J. W. Matthews)
Third English Edition, 1946 (translation by Ralph E. Oesper, on the basis of the translation by J. W. Matthews)
[under the title *Qualitative Analysis by Spot Tests, Inorganic and Organic Applications*]
Fourth English Edition (in two volumes), 1954
(translation by Ralph E. Oesper)
[under the title *Spot Tests, Vol. I, Inorganic Applications*]
Fifth English Edition, 1958 (translated by Ralph E. Oesper)

PRINTED IN THE NETHERLANDS BY
N.V. DRUKKERIJ G. J. THIEME, NIJMEGEN

FOREWORD TO THE FIFTH EDITION

The last edition of my monograph on the use of Spot Tests in inorganic analysis, published four years ago was practically sold out within a short period; that is why the preparation of this new edition became necessary. A considerable number of new tests, and improvements of many old ones, have been added. Much of the new material has not been published before. Moreover, the text has been brought up to date with regard to experimental details and explanation of the chemistry of the tests on the basis of recent literature.

As explained in the Foreword to the 5th Edition of the volume on the organic applications of spot tests, the publishers felt justified in making the latter a book in its own right, changing the title from *Spot Tests, Vol. II, Organic Applications* to *Spot Tests in Organic Analysis*. A similar change of title has now taken place with the present volume which in its 4th Edition was called *Spot Tests, Vol. I, Inorganic Applications,* and now in its 5th Edition appears under the title *Spot Tests in Inorganic Analysis*. Where in the present text reference is made to "Volume II", the companion volume *Spot Tests in Organic Analysis*, 5th Edition (1956) is meant.

Dr. Ralph E. Oesper, University of Cincinnati, has contributed notably to this new edition not only by his excellent translation of the German manuscript, but also through his sympathetic and understanding assistance in reviewing and extending the text. He has earned the author's heartfelt gratitude for this cooperation.

Rio de Janeiro, February 1958 FRITZ FEIGL

FROM THE FOREWORD TO THE FOURTH EDITION

The text is intended for all who are interested in semimicro and micro methods of qualitative inorganic analysis and its translation to the macro scale. In addition to the requisite manual dexterity, it assumes an understanding and interest in spot tests, which though easily and quickly conducted sometimes involve or are based on rather complicated chemical foundations. Consequently, this text has not been written for the very early stages of chemical instruction; the novices will merely be overwhelmed by the abundance of material. On the other hand, however, the advanced students will find here many important and significant facts of experimental chemistry, and they will gain an insight into the relations between analytical problems and other provinces of chemistry. As before, the author has kept these didactic and most valuable educational features clearly in mind.

It is exactly twenty years since the author, while preparing the second edition of "Spot Tests", came to the conclusion that the pursuit, propagation and extension of research stimulated by spot test analysis must lead to the tripartion which has now been realized in the texts dealing with inorganic spot test analysis, organic spot test analysis, and the chemistry of specific, selective and sensitive reactions. Many enthusiastic coworkers have contributed to this result, and assistance in these three divisions has come even from circles that are remote from analytical chemistry. Above all, however, the author had the good fortune, in 1941, to be invited by the Brazilian government to continue at Rio de Janeiro the researches which he had started in Austria. In the Laboratório da Produção Mineral, unhindered by teaching obligations and routine duties, he was able to concentrate all his efforts on experimental studies and literary work. Since 1949 the Conselho National de Pesquisas has provided assistants and fitted out his laboratory, and so promoted his researches in the field of the chemistry of specific, selective and sensitive reactions. It thereby has done much to advance the cause of spot test analysis. The author's thanks accordingly are properly recorded here, especially to Admiral Alvaro Alberto da Motta e Silva, the president of the National Research Council, and the director of the Laboratório da Produção Mineral, Dr. Alvaro Paiva de Abreu.

Rio de Janeiro, November 1953 FRITZ FEIGL

CONTENTS

Foreword to the Fifth Edition V
From the Foreword to the Fourth Edition VI

CHAPTER 1. DEVELOPMENT, PRESENT STATE AND PROSPECTS OF INORGANIC SPOT TEST ANALYSIS 1

CHAPTER 2. SPOT TEST TECHNIQUES (by PH. W. WEST) 27

1. Introduction 27
2. Laboratory and equipment requirements 28
3. Working methods 31
4. Special techniques 50

References 56

CHAPTER 3. TESTS FOR METALS. CATIONS, AND ANIONS OF METALLO ACIDS 57

A. HYDROGEN SULFIDE GROUP

A1. Basic Sulfide Group

1. Silver 58
2. Mercury 64
3. Lead 72
4. Bismuth 76
5. Copper 80
6. Cadmium 94

A2. Acid Sulfide Group

7. Arsenic 99
8. Antimony 103
9. Tin 107
10. Germanium 112
11. Molybdenum 115
12. Tungsten 119
13. Vanadium 123
14. Gold 127

15–17. Platinum metals 131
15. Platinum, 134 – 16. Palladium, 135 – 17. Osmium and Ruthenium, 141

B. AMMONIUM SULFIDE GROUP

18. Cobalt 143
19. Nickel 149
20. Thallium 154
21. Iron 161
22. Chromium 167
23. Manganese 173
24. Zinc 178
25. Aluminum 182
26. Beryllium 191
27. Titanium 195
28. Zirconium 199
29. Uranium 204
30. Cerium 210
31. Indium and Gallium 213

C. AMMONIUM CARBONATE GROUP

32. Barium 216
33. Strontium 220
34. Calcium 220

D. ALKALI METALS, AMMONIA, AND DERIVATIVES OF AMMONIA

35. Magnesium 223
36. Sodium 229
37. Potassium 230
38. Lithium 233
39. Cesium 234
40. Ammonia (Ammonium salts) 235
41. Hydrazine 239
42. Hydroxylamine 242
References 247

CHAPTER 4. TESTS FOR ACID RADICALS. ANIONS 258

1. Hydrochloric acid 259
2. Hydrobromic acid 262
3. Hydriodic acid 265
4. Hydrofluoric acid 269

5. Hydrocyanic acid 276
6. Mercuric cyanide 280
7. Thiocyanic acid 282
8. Cyanic acid 285
9. Hydrazoic acid 286
10. Ferrocyanic acid (Hydroferrocyanic acid) 287
11. Ferricyanic acid (Hydroferricyanic acid) 291
12. Hypohalogenous acids 294
13. Chloric acid 297
14. Bromic acid 298
15. Iodic acid 299
16. Perchloric acid 300
17. Periodic acid 301
18. Hydrogen sulfide 303
19. Sulfurous acid 307
20. Sulfuric acid 313
21. Thiosulfuric acid 318
22. Hyposulfurous acid (Dithionous acid) 320
23. Persulfuric acid 322
24. Aminosulfonic acid (Sulfamic acid) 325
25. Nitric acid 326
26. Nitrous acid 330
27. Phosphoric acid 333
28. Silicic acid 335
29. Carbonic acid 337
30. Boric acid 339
31. Permanganic acid 344
32. Chromic acid 345
33. Selenious acid and Selenic acid 345
34. Tellurous acid and Telluric acid 349
35. Hydrogen peroxide 353
References 356

CHAPTER 5. TESTS FOR FREE ELEMENTS 361

A. FREE METALS AND ALLOYS

1. General 361
2. Aluminum, Lead, Zinc, Tin 363
3. Mercury (Mercury vapor) 364
4. Arsenic 366
5. Molybdenum, Tungsten, Vanadium 367

B. FREE NON-METALS

6. Free halogens 368
7. Dicyanogen 370

8. Free sulfur . . . 372
9. Free selenium . . . 375
10. Free tellurium . . . 378
11. Free carbon . . . 379
References . . . 382

CHAPTER 6. THE SYSTEMATIC ANALYSIS OF MIXTURES BY SPOT REACTIONS . . . 383

1. Gutzeit's analytical scheme . . . 384
2. Scheme according to Heller and Krumholz . . . 389
3. Analysis scheme according to Krumholz . . . 391
4. Spot analysis of alloys according to Heller . . . 393
5. Ring oven method according to Weisz . . . 394
6. Application of spot tests in preliminary examinations . . . 394
7. Identification of anions in a mixture . . . 404
8. Identification tests for substances insoluble or sparingly soluble in acids ("Insoluble residues") . . . 407
9. Separation by the ring oven method . . . 423
References . . . 426

CHAPTER 7. APPLICATION OF SPOT REACTIONS IN TESTS OF PURITY, EXAMINATION OF TECHNICAL MATERIALS, STUDIES OF MINERALS . . . 428

1. Detection of silver in alloys and plating . . . 429
2. Detection of copper in alloys . . . 429
3. Detection of small amounts of copper in pharmaceutical products or foodstuffs . . . 430
4. Detection of traces of copper in water . . . 431
5. Detection of traces of copper in metallic nickel or nickel salts 432
6. Detection of copper in solutions of alkali cyanides . . . 432
7. Detection of lead in tin platings and enamels . . . 433
8. Detection of lead in alloys and crude metals . . . 433
9. Detection of lead in ores, minerals, and pigments . . . 434
10. Detection of traces of lead in water, alkali salts, and sulfuric acid 435
11. Detection of lead in bismuth compounds . . . 437
12. Detection of traces of metallic lead in criminalistic investigations 437
13. Detection of thallium in minerals, alloys, water, etc. . . . 437
14. Detection of bismuth in alloys . . . 438
15. Detection of tin in minerals, metallic objects, etc. . . . 439
16. Test for metallic tin and tin in alloys . . . 441
17. Simultaneous test for tin and antimony in alloys . . . 441
18. Test for antimony in alloys and mineral products . . . 442

19. Detection of arsenic in minerals 443
20. Detection of molybdenum in tungsten ores and technical materials 444
21. Differentiation between palladium and platinum foil 444
22. Detection of gold in alloys, coating, etc. 445
23. Detection of nickel plating and nickel in alloys 445
24. Detection of traces of nickel in cobalt salts 446
25. Detection of cobalt plating and cobalt in alloys 448
26. Detection of zinc in plated metals 448
27. Detection of traces of iron in fluorides and phosphoric acid . 449
28. Detection of small amounts of iron in mercury salts 450
29. Detection of traces of iron in alumina, pyrolusite, titanium dioxide, silicates, colorless ignition residues 450
30. Detection of slight amounts of iron in metallic copper, and in cobalt- and nickel-alloys 451
31. Detection of traces of iron in metallic zinc and magnesium, and in zinc- and magnesium-salts 452
32. Detection of traces of iron in concentrated nitric acid 453
33. Detection of alumina 454
34. Detection of traces of aluminum in water 454
35. Detection of chromium in rocks and steel 455
36. Detection of manganese in minerals and rocks 456
37. Detection of uranium in minerals 457
38. Detection of beryllium in minerals, ores, and alloys 457
39. Detection of tungsten in minerals 459
40. Detection of titanium in minerals 459
41. Detection of calcium in silicates 460
42. Detection of magnesium in silicate rocks 461
43. Detection of potassium in silicate rocks and glass 461
44. Detection of ferrous iron in silicates and in acid-resistant, silica-free minerals 462
45. Detection of nickel in silicate rocks 463
46. Detection of reducible metals in minerals 463
47. Detection of alkali metals in silicates 464
48. Differentiation of magnesite and dolomite or bräunerite . . 465
49. Identification of calcite (differentiation from dolomite) . . . 466
50. Differentiation of hard- and soft-burned lime 466
51. Differentiation of caustic burned and sintered magnesite . . 467
52. Detection of lime in magnesite 468
53. Detection of calcium in ashes, dry residues, magnesite, dolomite, silicates, etc. 468
54. Differentiation of calcite and aragonite 469
55. Identification of calcium sulfate (gypsum and anhydrite) . . 470
56. Differentiation of gypsum and anhydrite 471
57. Differentiation of minerals containing barium and strontium . 473
58. Orienting tests of glass 473

59. Detection of free metals in oxides, printing on paper, etc. . . . 475
60. Detection of barium sulfate in pigments, paper ash, etc. . . . 476
61. Ferric ferrocyanide 476
62. Procedure for testing woods impregnated with salts . . . 476
63. Mineral tanning agents 477
64. Classification of inks. Age of ink writings 479
65. Detection of fixed alkalis and ammonium salts in alkali cyanides 481
66. Detection of alkali earths and alkali in ash of paper, charcoal, and coal 482
67. Detection of ammonium salts in chemicals 483
68. Detection of ammonium salts in filter papers 483
69. Imprint and developing procedure for chemical detection of heterogeneities in manufactured metals, minerals, etc. 484
70. Electrographic methods 489
71. Detection of fluorine in rocks and in mineral waters 490
72. Detection of free (unbound) sulfur in minerals, inorganic mixtures and polysulfides 491
73. Detection of sulfide sulfur in rocks 491
74. Detection of selenium in minerals, sulfur, and tellurium . . 492
75. Detection of sulfates in hydrofluoric acid and fluorides. . . . 493
76. Detection of sulfate in inorganic fine chemicals 494
77. Detection of traces of nitrate in lead- and manganese-dioxide, antimony pentoxide and acidic metallic oxides 494
78. Detection of traces of nitrate in alkali molybdate, tungstate, and vanadate. 495
79. Detection of traces of chloride in fine chemicals 495
80. Detection of bichromate in the presence of monochromate . . 496
81. Detection of alkali monochromate in the presence of alkali bichromate 497
82. Detection of carbonate in alkali cyanides, sulfites, and sulfides 498
83. Detection of silica in caustic alkalis and alkaline solutions. Vulnerability of glass and other siliceous products to attack by alkalis 499
84. Detection of silicic acid in minerals 499
85. Detection of boron in rocks and enamels 500
86. Detection of phosphates in minerals and rocks 500
87. Detection of chlorine in minerals and rocks 501
88. Detection of free acids and basic compounds in solutions of aluminum salts and of free acids in solutions of copper and cobalt salts 502
89. Detection of traces of hydrogen sulfide in water 503
90. Hydrogen-ion concentration of aqueous solutions 504
91. Orienting reactions for judging samples of water 505
92. Determination of hardness of water; differentiation of distilled and tap water 507

Chapter 1

Development, Present State and Prospects of Inorganic Spot Test Analysis[1]

Analytical chemists have long used single chemical tests carried out in drops of solution on filter paper or on impermeable surfaces. A familiar example is the use of indicator papers to detect rapidly an excess of hydrogen or hydroxyl ions in a drop of a solution. Likewise, the endpoint of certain reactions used in titrimetric analysis, or the completion of electrolytic depositions, can be established by removing a drop of the test solution and bringing it into contact with a suitable reagent on filter paper, on a watch glass, or on a porcelain plate. It has not been definitely established who made the first use of spot reactions for analytical purposes. Probably the earliest published instance was given by F. Runge, who in 1834 used potassium iodide-starch paper to detect free chlorine.[2] In 1859 Schiff [3] employed filter paper impregnated with silver carbonate to reveal uric acid in urine. A drop of the specimen produced a brown fleck of free silver. This appears to be the earliest precise description of a spot test, because the great sensitivity of this reaction was determined in the same manner as at present, i.e. by testing a series of dilute solutions of uric acid. The fundamental work for that division of spot test analysis in which filter paper functions as the reaction medium is a study by Schönbein.[4] He showed that when aqueous solutions rise in strips of filter paper, the water precedes the dissolved material, and the relative height of ascent of solutes can differ enough to make it possible to detect the cosolutes in separate zones. These observations gave the impetus for the classic studies (1861–1907) of Fr. Goppelsroeder, which were compiled in his "*Kapillaranalyse*", published at Dresden in 1910. He made a very extensive study of the capillary rise of solutions and the capillary spreading of drops of solutions in filter paper and investigated the analytical use of these effects, particularly in the examination of organic liquids, dissolved compounds, and dyes. His publications also contain references to the capillary spreading of inorganic salts, effects which later were studied by other chemists, especially Skraup and his associates [5] and Krulla.[6] The findings suggested the problem of discovering whether an

References pp. 25–26

inorganic capillary analysis is possible in which the primary objective is the feasibility of carrying out color reactions in the form of spot tests in the various zones of the paper to detect the materials which had been separated by capillarity. Investigations along this line, which required, above all, the determination of the minimum quantities of substances that can be detected by spot reactions on paper, were conducted (1917–21) by Feigl and Stern [7] with solutions of salts of the metals of the ammonium sulfide group. These studies, and a continuation [8] dealing with the detection of metals of the hydrogen sulfide group, yielded observations which set the course for further studies of spot reactions. Many tests, well known from their accepted use in the classic procedures of qualitative inorganic analysis, where they were executed in test tubes, displayed an unexpected great sensitivity when they were tried as spot reactions on filter paper, since under these latter conditions the picture of a reaction may be quite different from that seen in a test tube. The appearance of the fleck frequently is very different according to the concentration of the reacting partners, the quality of the paper, and other experimental conditions. Not only were sensitive individual tests found to be possible through spot reactions, but sometimes several materials could be detected in a single drop of a solution, provided the reagents were properly chosen. However, the most important finding was that the amounts that can be detected by means of spot reactions on filter paper are so small that microanalytical goals are reached.

In order further to develop the observations gleaned from the first comprehensive studies of spot reactions, with the objective of detecting at least the great majority, if not all, of the anions and cations by means of spot tests, and with the avoidance, wherever possible, of the usual separation procedures, two things were essential: a) the extension and refinement of the technique of working with drops, and b) the employment of familiar and/or new sensitive detection reactions. The attention given to spot reactions not only by the author and his school but subsequently by others (especially N. A. Tananaeff) led in the course of time to great progress in both of these directions and so much material was assembled that there was ample justification for designating this new division of chemical analysis: *spot test analysis*.

It has now become customary to speak of spot reactions or, more correctly, of spot or drop tests, when in a chemical test by the wet method at least one reactant—usually the material being detected or identified—is used in the form of a drop of a solution.* The most usual kind of spot test consists in bringing together drops of the test (sample) solution and the reagent solution

* The usual German designation is *Tüpfelreaktion*. The French use *réaction à la touche*, or *réaction à la goutte*, or *stilliréaction* [9].

on porous substrates such as filter paper, on impermeable media such as spot plates, in microcrucibles, on watch glasses, or in micro test tubes. Another version employs one of the reactants in the solid form, i.e. a little of the material being studied is spotted with a drop of a suitable reagent solution, or a drop of the test solution is brought into contact with a solid reagent. Sometimes, a drop of a solution or a pinch of a solid can be made to evolve a gas, which can then be detected by its action on a reagent paper or on a drop of an appropriate reagent solution. Chapter 2 gives a description of the apparative aids commonly used in these tests and discusses the techniques employed.

It scarcely needs to be stressed that all improvements of the inherently simple technique of spot test analysis were and continue to be of the highest importance in its development. However, it is not its technique alone which has led to systematic spot test analysis. Equally important, in fact even more so in many respects, has been the need of finding appropriate reactions to which the technique can be applied. Even the earliest high quality spot reactions indicated the lines along which spot test analysis would develop. These guiding principles were: a) to use reactions of the highest possible sensitivity and reliability; b) to employ all possibilities of enhancing sensitivity and reliability. With respect to the former, spot test analysis has filled a very useful function. In this effort to adapt chemical reactions which had already been described, many tests that were scattered through the literature and which in part had been forgotten, were revived, tried out again, and improved in certain instances. It frequently became necessary to unravel their chemical basis or to correct erroneous notions. New facts were thus assembled which later proved of value in the search for new analytically useful reagents and reactions. Many reactions to which spot test analysis had recourse, and others which were first used for spot testing, were later employed in qualitative macroanalysis, and sometimes were introduced even into quantitative macro- and microanalysis. It is characteristic of studies intended to extend spot test analysis that they frequently deliver findings of use to analytical chemistry in general, and in many cases they prove to have value also in other provinces of chemistry. The reason for this is the fact that the prime factors in spot test analysis, namely sensitivity and reliability, are important to all provinces of chemical analysis. Therefore experimental chemistry in its widest sense must be pursued in the scientific treatment of these questions.

Soon after the earliest detailed description of spot tests, it was proposed [10], on the basis of a comparative study of the efficiency of spot tests and the tests employed in the classical procedures of qualitative inorganic macro- and microanalysis, that the type of procedure as well as the sensitivity of the tests be expressed in a symbolic form. This seemed all the more necessary since the previous use of the term "sensitivity" had not been uniform and frequently was not even exact. Sometimes sensitivity was taken to mean the smallest quantity of a material that could be detected by a test, and sometimes it was used to indicate the dilution prevailing at a test with no reference to the quantity of material. The expression

$$x[S]y$$

refers to tests conducted in solutions. It contains information of interest to the characterization of tests, to the comparison of the efficiency of tests, and for the statistic of tests. In this expression, the symbols denote:

S = the technique employed, i.e., reactions in micro test tubes, on filter paper, spot plates, etc;
y = the volume (in ml) of the test (sample) solution taken for the test;
x = the quantity of material dissolved in y, which is revealable by the particular technique and expressed in gamma (γ) ($1\ \gamma = 1$ microgram $= 0.001$ mg).

The term *Limit of Identification*, which was first suggested by the author, has found almost universal acceptance as x. When the values of x and y are known, the concentration (or dilution) of the test solution with respect to the material being sought or identified is easily stated. This leads to the *Concentration Limit* [11] or better its reciprocal the *Dilution Limit* [12]. The following simple relation exists between the dilution limit, the volume of the test solution, and the identification limit:

$$\textit{Dilution limit} = 1 : \frac{y \cdot 10^6}{x} = 1 : \frac{\text{Volume of the test solution (in ml)} \cdot 10^6}{\text{Limit of Identification (in } \gamma)}$$

It follows from this expression that the identification limit of a test can be calculated if the dilution (concentration) of a solution and the particular volume of the solution used in the test are known. This computation is used in the experimental determination of the identification limits of tests in which initial solutions of known concentration are systematically diluted.

Hahn (*loc. cit.*) has quite properly pointed out that the numerical value of the identification limit of a test expresses its quantity sensitivity, whereas the dilution (concentration) limit expresses the concentration sensitivity of a test. Therefore, the complete appraisal of a test require a knowledge of both its quantity- and concentration-sensitivity. In the strictest sense, a test is highly sensitive, if, when conducted in the smallest feasible volume of the test solution, the identification limit is small and the dilution limit large. However, cases are known in which a knowledge of the quantity sensitivity of a test (identification limit) has to suffice. This situation is encountered when so-called insoluble materials are to be identified by means of dissolved, or melted, or gaseous reagents. Accordingly, the term *Limit of Identification* is very convenient because it retains its meaning for the characterization of the sensitivity of tests without regard to the solubility or insolubility of the test material.

References pp. 25–26

At the time the symbolic expression $x[S]y$ was proposed for the characterization of a test, there was little or no appreciation of one fact which accumulated experience in spot test analysis has now shown to be of the highest importance. If it is stated that S is a drop reaction, or a precipitation or a color reaction, occurring in a given volume y, this designation of the technique and volume employed is not adequate for the complete characterization of the test. By taking proper measures it is possible to enhance the discernibility of the occurrence of chemical reactions and thus enormously improve the sensitivity of tests. In such cases, we speak of the *conditioning of tests*. However, there is no symbolic designation of this beneficiation which obviously is very important to the certainty of tests, a factor which will be discussed later. Despite this limitation, the symbolic characterization of tests has had a lasting influence. It has brought clarification to the term "sensitivity"; it has led to new and critical reviews of the analytical assessment of chemical reactions.

In this text, the limits of identification and the dilution limits which can be attained by the respective spot tests are given immediately after the outline of the various procedures. A tabular summary (Chapter 8) presents the identification limits of a large number of the more important spot reactions. If the drop volume, which ordinarily is around 0.05 milliliter, is taken into account, the dilution limits corresponding to the identification limits can be readily computed.

However, spot reactions are often conducted with microdrops, i.e. in volumes of 0.03–0.001 ml, and much smaller identification limits are attained then. An inspection of the tabular summary will disclose that few identification limits are greater than 2 gamma per 0.05 ml drop (dilution limit = 1 : 25,000). In most cases this value lies below 0.2 gamma (dilution limit = 1 : 250,000). Far lower identification limits and correspondingly higher dilution limits are reached in many instances. A comparison of the identification limits and dilution limits with those attainable by the classical methods of qualitative microanalysis (particularly precipitations of crystals that are then examined under the microscope) shows that the spot reactions are generally superior with respect to concentration sensitivity (dilution limit). As to quantity sensitivity (identification limit), spot reactions are sometimes as good as micro precipitations, sometimes better, never inferior. Consequently, if it is accepted that the goal of qualitative microanalysis is not the use of a particular technique or sample size, but the detection of absolute and relatively (with respect to the dilution) small quantities of material, then spot reactions should undoubtedly be regarded as "microchemical tests" in the great majority of cases. Sometimes, especially in the Anglo-Saxon countries, spot reactions are called semimicro reactions, and spot test analysis is referred to as semimicroanalysis. In the light of what has just been said, such terms are incorrect and lead to error by such generalization, since logically a spot test which is more sensitive than a classical microchemical test can never be semimicrochemical.* It is permissible and correct to distinguish between semimicro- and micro tests in order to draw the limits between these two classes and to contrast them with macro tests. Semimicro tests may refer to quantities (identification limits) of 10–200 γ; micro tests involve identification limits below 10 γ.**

The numerical values for the sensitivity of a test, i.e. its identification limit and dilution limit, are not constants of the underlying chemical reaction *per se*. This is plainly shown by precipitation reactions in which the numerical value of the solubility product of the solid is a characteristic and genuine reaction constant, and yet the sensitivities are by no means dependent on the smallness of the particular solubility products. The decisive factor for the sensitivity of a reaction is the discernibility of a reaction product. This was first pointed out by Böttger [18] in a study which was of high importance to reaction sensitivity. The fact that the sensitivity of a precipitation reaction does not depend solely on the solubility product of the solid product indicates that other processes besides a reaction which can be represented stoichiometrically must also play a part. This is actually the case. When precipitates appear, quite a few significant process supple-

* In this connection is must be noted that the widely used terms: microchemical identification, determination and the like, cannot withstand close scrutiny. There is no "microchemistry" as a separate branch. All special provinces of chemistry must justify their existence either by dealing with certain groups of material or with special phenomena and regularities. A special field of microchemistry would exist if it were concerned with phenomena and modes of behavior different from those dealt with in "macrochemistry". However, this is not the case, and the term "microchemistry" is just as erroneous as the familiar term "technical chemistry", which has now been replaced by "chemical technology". Of course there is "microtechnique of general chemical experimentation" [13] which is not confined to chemical analysis but is applicable likewise to physico-chemical measurements, preparative studies, demonstration experiments, etc. In the author's opinion, neither the use of a special technique nor the "study of techniques permitting the chemical handling of quantities of material so small that the usual equipment is not longer suitable" [14] can be called microchemistry. The situation is different with respect to the term "microanalysis", which because of its goal (detection and determination of tiny amounts of materials) is actually a special branch of chemical analysis. However, a special microtechnique is not always necessary to reach microanalytical objectives. This fact is clearly demonstrated by spot test analysis and by trace analysis, [15] which both have become important parts of microanalysis, even though they do not require a special microtechnique. See however ref. [16].

** There is as yet no generally accepted nomenclature for and differentiation between macro-, semimicro- and microtests based on values of identification limits. Kirk [17] used the appropriate terms "milligram analysis" and "microgram analysis", and in quantitative work he distinguishes between macro-, micro-, ultramicro-, and submicrogram methods, based on sample sizes of 100 mg, 10–20 mg, 1 mg, 1 gamma, and 0.001 γ respectively. Gillis (*loc. cit.*), has adopted the same classification for qualitative tests. It is obvious that there is no present agreement among chemical analysts as to whether the term "micro" should refer to sample size and the correlative special technique or—as the author believes—to the quantity of the material to be identified or determined, independent of any special technique. In this connection compare ref. [18].

ment the actual chemical change. They include supersaturation, formation of nuclei, nuclear growth, sol formation, coagulation, transformation of amorphous into crystalline forms, recrystallization, adsorption. Such supplementary processes occur to extents that differ according to the prevailing conditions (temperature, concentration, rate of adding the reagent, etc.). Accordingly, the test conditions often determine in great measure the reaction picture, i.e. the time of the visible beginning of the precipitation, the form species and dispersion of the precipitate, the color intensity of precipitates, etc. This is true even when precipitates come out of solutions which contain nothing besides the compound to be detected, and it is still more true in the presence of cosolutes, including even those which do not react with the precipitant being employed. The dependence of a precipitation reaction on partial or accompanying processes of the types just listed is the reason why the same reaction conducted in a volume of several milliliters, in macro- or microdrops, on porous or non-porous substrates, with dissolved or solid reagents, sometimes exhibits wide differences in sensitivity.

With respect to the analytical employment of chemical reactions, it is of greatest importance that the sensitivity of a test is related to the conditions under which it is conducted. The possibility is thus provided of raising the sensitivity of tests by taking appropriate measures, and spot test analysis makes extensive use of such possibilities. For instance, if reactions are conducted by bringing single drops of the test and reagent solution together on filter paper, the reaction seemingly occurs in the plane of the paper. This is an advantage in that colored, either insoluble or soluble, reaction products are held near the site of their production by the capillaries of the paper, and are more readily seen because of the white background provided by the paper. In precipitation reactions, precipitation and filtration take place in the plane of the paper, so to speak. As a consequence, other spot reactions designed to reveal cosolutes can then be applied to the area surrounding the fleck, i.e. to the quasi filtrate. A matter of great importance in spot reactions on paper is that this medium not only brings about a rapid uniform spreading of liquids through its coarse capillaries, but also the fine porosity of the cellulose fibers produces an effective adsorption medium due to the great surface exposed. When a drop of an aqueous solution is placed on filter paper, the dissolved materials, depending on the dilution, are almost always accumulated at the center of the fleck or in certain of its concentric zones. This local enrichment is a very great aid to the improvement of the visibility of a colored reaction product. In case two dissolved materials are present, these may become localized in two distinct zones. This effect constitutes a kind of chromatography, which thus makes it possible to detect,

References pp. 25–26

in the separate zones of the fleck, the materials which have been sufficiently separated by capillary, or more correctly, adsorption processes. Pertinent examples are to be found in this text.[19]

Though a fleck on filter paper is seemingly produced in quite simple fashion, it actually is the net result of a complicated interplay of capillary spreading, diffusion, swelling, adsorption, and chemical reaction. Connected with this is the fact, observed by West and Hamilton [20] and more recently by Rush and Rogers [21] that the identification limits of some sensitive tests conducted on filter paper, and carried out under quite similar conditions, may vary by powers of ten according to the type of paper used. Probably such influence of the type of filter paper holds for all spot tests; consequently the proper choice of paper is highly important when the highest sensitivity is desired. Actually, the influence of the paper on the sensitivity extends much farther. Some chemical changes, especially color reactions, have excellent sensitivity when conducted as spot reactions on paper, whereas they are quite insensitive on a spot plate or in a test tube. Typical examples illustrating this point are provided by the detection of manganese ^{II} through the production of autoxidizable $Mn(OH)_2$ and conversion of benzidine into benzidine blue,[22] and the detection of cadmium [23] by precipitation of the red insoluble $[Fe(\alpha\alpha'dip)_3]$ $[CdI_4]$. These reactions, which are described on pages 175 and 94, respectively, are characteristic instances of the fact, first pointed out by Pavelka,[24] that the surface of filter paper (and interfaces in general) can play an important role in the way a reaction takes place. When spot tests are carried out on filter paper, the influence of the paper is sometimes so great that it may truthfully be said that the paper becomes an active participant in the spot reaction.

An example can be given, however, in which the filter paper functions as a reactant in the strictest sense of the word. It is found in the test (page 344) for traces of permanganate in solutions of alkali chromate.[25] If a drop of the test solution is placed on filter paper, the cellulose is oxidized by the permanganate and the resulting manganese dioxide is deposited in the capillaries of the paper. The chromate remains unaltered and can be removed by washing with water. The residual fleck of manganese dioxide permits the detection of 0.3 γ potassium permanganate in the presence of 20,000 γ potassium chromate, i.e. in a saturated solution of the latter. It would be difficult to conceive of a simpler method of solving this analytical problem, which is a real challenge when conducted by other means.

An important method of raising the sensitivity of spot tests is to use papers impregnated with water-soluble or water-insoluble reagents which yield colored reaction products. (It is best to dry such papers before use.) When a drop of the test solution is placed on the impregnated paper, the reaction products can be distinctly seen in the form of characteristic flecks or rings. In certain instances, spot reactions are carried out on paper impregnated with water-insoluble reagents. Clarke and Hermance [26] showed that the sensitivity of spot tests on filter paper impregnated with an insoluble reagent which acts as an ion donor can be greater by powers of ten than the sensitivity of the same spot test utilizing a water-soluble compound as ion donor. The following comparisons illustrate this fact; the insoluble reagent is given second in each case:

		gamma
Copper with	potassium ferrocyanide	0.5
	zinc ferrocyanide	0.05
Copper with	sodium diethyldithiocarbamate	0.2
	zinc diethyldithiocarbamate	0.002
Molybdenum with	potassium ethylxanthate	0.2
	zinc diethylxanthate	0.01
Cadmium with	sodium sulfide	6.3
	zinc sulfide	0.05

It may be accepted as a general rule that a higher sensitivity can be reached in spot reactions based on the formation of colored insoluble metal compounds if an insoluble compound rather than a soluble reagent serves as the source of the detecting anion. Lower limits of identification can be secured through spot reactions than when the same tests are conducted in a test tube with a solution volume one hundred times as great. The dimethylglyoxime test for nickel and the Prussian blue test for ferrocyanide illustrate this point. As a test tube reaction, the former will reveal 2.8 γ nickel in 2 milliliters of solution; this corresponds to the dilution limit 1 : 700,000. In contrast, if a drop (0.05 ml) of the nickel solution is applied to dimethylglyoxime paper, a definite red appears when as little as 0.015 γ nickel is present (1 : 3,300,000). Analogous figures are obtained in the detection of ferrocyanide. A positive result is just discernible if 3 milliliters of a 1 : 400,000 solution are treated with a ferric solution in a test tube, whereas on ferric chloride paper a blue ring is produced by one drop of potassium ferrocyanide solution even at a dilution 1 : 700,000.

The enhancement of the sensitivity on reagent papers containing insoluble reagents has various reasons. The reaction occurs on the surface of a solid which is finely dispersed throughout the capillaries of the paper, and hence an excess of the reagent is always available when small amounts of material are being detected. Secondly, the diffusion and dilution which invariably accompany the mixing of the reagent and test solutions are avoided. Thirdly, sharply defined flecks result because the colored reaction products remain at the site of their formation. Furthermore, the high degree of

References pp. 25–26

dispersion of the solid reactant is of greatest importance in topochemical reactions of this kind. This latter point is well demonstrated by an observation by Pavelka [27] of the behavior of lead sulfate toward a solution of gallocyanine. When precipitated in a test tube or on a spot plate, the difficultly soluble white salt is only slightly tinted by the dye, whereas when produced on paper by means of a spot reaction the resulting finely divided salt is strongly colored. A sensitive spot test for lead (page 75) has been developed on the basis of this behavior.

It is worth noting that certain reactions have value as tests only when they are conducted as topochemical spot reactions on filter paper impregnated with insoluble reagents, whereas no trace of a reaction can be seen in a test tube. Several typical examples will emphasize this important point. If filter paper is bathed in permanganate solution, the cellulose is partly oxidized and manganese dioxide is deposited in the pores of the paper. The color of the impregnated paper ranges from yellow to brown depending on the concentration of the permanganate solution. If the test paper is then treated with a drop of a dilute solution of a reducing agent, such as oxalic acid or sulfurous acid, the manganese dioxide is reduced to a colorless manganese ^{II} salt and a lighter, or even white fleck appears. Reducing agents in quantities around 1 gamma can be detected in this way.[28] A similar spot test for hydrogen peroxide [29] is based on the oxidation of lead sulfide to sulfate. When a drop of peroxide solution is placed on lead sulfide paper, a white fleck or ring develops on the colored paper. Because of the sharp contrast with the dark background, the test is quite sensitive (identification limit is 0.04 γ hydrogen peroxide). Of course, lead sulfide can be oxidized or manganese dioxide can be reduced in a test tube, but under such circumstances the reactions are absolutely useless for the detection of small amounts of reducing or oxidizing agents. These redox reactions can be effectively employed for qualitative analytical purposes only when they are conducted as spot reactions on paper (see page 353 and 402).

When spot reactions are carried out on paper impregnated with water-insoluble reagents, it is sometimes possible to follow details of the course of the reactions which are impossible to see or which escape notice entirely when the same reactions are conducted in test tubes. Examples are provided by those reactions in which the same precipitating agent brings about a fractional precipitation from a solution if the solubility products of the precipitated materials are far enough apart. A fractional precipitation of this kind can be successfully accomplished in a test tube only when the concentrations of the materials to be precipitated are not too different. A particularly desirable analytical application is sometimes made possible by carrying out fractional precipitations in the course of spot tests on reagent papers.

For instance, rubeanic acid forms dark green, violet, and brown precipitates respectively with copper, nickel, and cobalt salts, and their simultaneous precipitation makes it difficult or impossible to detect these metals in solutions of mixtures of their salts. According to Feigl and Kapulitzas,[30] a fractional precipitation is possible if the following procedure is used. A drop of the solution containing say copper and cobalt is placed on paper impregnated with rubeanic acid, and a dark or olive ring, depending on the copper content, is formed in the center of the test area, and around this is a concentric brown zone of cobalt rubeanate. It is easily possible to detect 0.05 γ copper in the presence of 20,000 times this quantity of cobalt. The same identification limits and limiting proportions can be obtained by the fractional precipitation of copper and nickel rubeanate in separate zones. Small amounts of copper can also be detected in the presence of nickel and cobalt by fractional precipitation on rubeanic acid paper. Velculescu and Cornea [31] described another case of fractional precipitation on filter paper. They found that concentric rings of the corresponding silver halides are produced if a drop of a solution containing iodide, bromide, and chloride is placed on silver nitrate paper. Accordingly, the fractional precipitation on reagent paper is accompanied by a capillary separation of the precipitated products in distinct zones, and thus permits their detection in the presence of each other. (The tests cited are described on pages 88 and 268.)

Another effect, which was first observed and utilized in spot reactions on filter paper impregnated with solid reagents, is the "protective layer" effect. It consists in the enveloping of the surface of a reactive material by a resistant coating produced chemically in such a manner that the underlying parts of the reactive solid are protected against chemical attack. This principle has long been utilized in chemical technology to combat corrosion. In spot test analysis, the protective layer effect can bring about a considerable increase in sensitivity. For example, an excellent test for palladium is possible on paper impregnated with red, water-insoluble nickel dimethylglyoxime.[32] A drop of the neutral or acetic acid solution of the palladium salt is allowed to soak into the paper which is then placed in dilute mineral acid. The nickel dimethylglyoxime of the red paper dissolves almost immediately except at the site of the fleck. A red stain is left, depending on the quantity of palladium. As little as 0.05 γ palladium can be distinctly revealed by this procedure (see page 137). A protective layer effect is active here since the acid-soluble nickel dimethylglyoxime dispersed throughout the paper reacts with Pd^{+2} ions to form the yellow acid-insoluble palladium dimethylglyoxime. The latter remains where it is formed, i.e. on the surface of the red nickel salt and protects the underlying portions from the subsequent brief action of the mineral acid. Quantities of palladium which are too small to be revealed by a drop reaction with dimethylglyoxime on filter paper are adequate to produce this protective effect. The tests for elementary sulfur and selenium described on pages 372 and 375 are based on the formation of protective layers through binding of the sulfur or selenium on the surface of black-brown thallous sulfide, which has been precipitated in the pores of filter paper.[33] The protective layer is formed by spotting the paper with a drop of a solution of sulfur or selenium in an organic liquid or alkali sulfide, and the insoluble product retards or prevents the rapid solution of the underlying thallous sulfide in dilute acid or hydrogen peroxide. Consequently, when the spotted thallous sulfide paper is bathed in dilute mineral acid or hydrogen peroxide solution, the brown paper is completely decolorized except for the flecks on which the protective layer was formed. A test for calcium [34] by means of the osazone of dihydroxy-tartaric acid (page 221) shows that reagents which are difficultly soluble in water can also produce protective layer effects that can be utilized in spot tests conducted on a spot plate. There is no doubt that additional examples of the employment of protecting layers in spot test analysis will be discovered, if use is made of solid reagents whose behavior toward pH changes differs from that of the particular reaction product.

Its capillarity and adsorptive effects, which have contributed so much to sensitivity, have made filter paper the preferred substrate for spot reactions. Other suitable substrates may be used. Przibram [35] recommended impregnated silk threads for spot reactions with microdrops. Winckelmann [36] suggested gelatine * containing certain reagents and Skalos [38] advocated the use of small paraffined porcelain dishes for spot reactions. Tiny drops retain their shape on the paraffine surface and do not coalesce. Filter paper, gelatine foils, and paraffined porcelain dishes are not suitable substrates for spot tests that are to be conducted in strongly acid or strongly basic solutions, or that require considerable elevation of temperature. In such cases it has been found well to use glass or porcelain spot plates (white, black), micro test tubes and microcrucibles. According to Razim,[39] the surface of liquids on spot plates can be heated by infrared radiation.

If drop reactions are conducted in micro test tubes, the sensitivity of tests can often be raised significantly by shaking out with organic liquids which are not miscible with water. Colored products, which are soluble in organic liquids, can be extracted out of the water layer and thus concen-

* In this connection, note should be taken of the somewhat remarkable phenomenon that when chemical reactions are conducted on cellophane impregnated with the reagent or the solution of the metal salt, the product of the reaction frequently remains in the substrate in the colloidal condition. For instance, when metallic sulfide or nickel dimethylglyoxime are formed, the cellophane films are stained brown-black or red, although no solid particles are visible with the naked eye.[37]

trated in a small volume of the organic solvent. In case a colored organic compound, which is soluble in ether, benzene, etc., combines with inorganic ions to yield a salt that is soluble only in water, extraction with an organic solvent may provide a valuable aid in recognizing the reaction in question. A pertinent instance is the specific detection of aluminum by means of the ether-soluble hydroxytriphenylmethane dyes, which yield violet water-soluble aluminum-chelate compounds (compare p. 189). Colored or colorless reaction products, which are insoluble in water and organic liquids, collect in the interface and this analytical flotation will frequently reveal traces of colloidally dispersed precipitates whose presence cannot be established otherwise or only after long standing. The volume of water can be increased beforehand in either extraction or flotation; the separation of the liquid phases then becomes more visible and there is no or only a very slight decrease in sensitivity.

Among the types of reactions which cannot be conducted on filter paper, but which are readily adapted to spot test technique, are fusion and sintering reactions, and also reactions conducted in organic liquids. Fusion and sintering reactions take advantage of the direct reactivity of solids at elevated temperatures, resulting in optically detectable chemical changes which cannot be realized at ordinary temperature or in solution. Striking examples are the formation of hydrogen sulfide when even tiny amounts of elementary sulfur are melted with benzoin (see page 375) and the specific test for lithium, proposed by Stewart and Young [40] which is based on heating the evaporation residue of a drop of a lithium salt solution on soda glass to produce a dull area if the temperature is carried to 300° C. Special note should be given here to the reactions of inorganic solids with colorless low-melting organic compounds to produce colored chelate compounds. See for example, the sensitive tests for Fe_2O_3 and V_2O_5 by means of molten 8-hydroxyquinoline as described on pages 413 and 125. There has been little employment of solid body reactions in the absence of water for spot test purposes, even though the production of colored inner complex salts and color lakes by sintering together the proper reaction partners is extremely notable from the analytical standpoint.[41] The reactions of inorganic compounds dissolved in organic liquids offer much promise in spot test analysis. Excellent examples are the test for antimony ^{III} [42] based on the addition of rhodamine B to a benzene solution of antimony ^{III} iodide (see page 106) and the test for tin with morin (see page 110).

The sensitivity of spot reactions for cations or anions is measured by starting with standard solutions of salts of the highest purity which are then systematically diluted. Drops of known volume are taken for the test in question at each dilution. When this systematic procedure is followed, it is frequently found that a stage is reached, particularly with precipitation reactions, at which tests repeated under seemingly identical conditions, sometimes respond positively, sometimes negatively. What Emich [43] called the "region of uncertain reaction" has been reached. The uncertainty is related to several different physical processes which accompany reactions that can be represented stochiometrically. Another anomaly often encountered should be noted here. In almost all color and precipitation reactions, conducted in the form of spot tests, the intensity of the result of the particular reaction is proportional to the quantity and concentration of the material being detected, a fact that is the basis of spot colorimetry, which will be discussed later. However, it happens frequently that color and precipitation reactions, which are still distinctly visible at a particular dilution, unexpectedly disappear completely at even a slightly greater dilution. Therefore, no judgement should be passed concerning the sensitivity of spot tests without making repeated trials and defining precisely the experimental conditions surrounding the tests. It should also be noted with respect to the determination of sensitivity that different methods of conducting the tests yield different values for the limits of identification and dilution limits. Accordingly, the experimentally determined sensitivity data of a spot test may not be transferred to other methods of carrying out the same underlying test.

As was pointed out, every effort is made to raise the sensitivity of spot tests through the extensive use of sensitive reactions of dissolved, solid, fused, and gaseous materials and by the employment of all possibilities within the limits of certain experimental conditions. However, the sensitivity of tests is not the sole determing factor in spot tests. Since the latter should succeed in the presence of as many accompanying materials as possible, and independent of the quantities of the latter present, the use of unambiguous and unequivocal tests is of prime importance. The criterion for this is the discernibility of the occurrence of the underlying chemical changes in the presence of the accompanying materials and the elimination of the interferences due to them. All experience shows that there are but few absolutely reliable and non-misleading tests. First of all, it should be noted that accompanying materials which do not react with the reagent being used may nevertheless lower the sensitivity of the test. This effect was first pointed out by Schoorl [44] in his studies of the sensitivity of micro-crystalline precipitations. He introduced the term "limiting proportion" to express the ratio of the quantity of material which can just be detected to that of the accompanying material. Limiting proportions are also encountered in spot reactions and therefore it must never be assumed that their limits of identification (determined in pure solutions) are likewise valid in the presence of

References pp. 25–26

cosolutes which do not react with the particular precipitation or color reagent. As a rule, such accompanying materials lower the sensitivity of tests, sometimes to a considerable degree * and independent of the amount present. This is a cogent reason for using tests of the highest sensitivity in spot test analysis, because such tests, despite the deleterious influence of accompanying materials, will still provide values for identification and dilution limits that are adequate for microanalytical demands.

Far more serious than the depression of the sensitivity of tests by the attendant materials are the interferences due to accompanying materials which make their effects evident in regions of medium or even high concentrations of the substance being sought. This category includes the interference arising from the presence of non-reacting colored ions during color reactions. It should also be noted that it is more difficult to see turbidities produced by precipitation reactions in colored solutions than in colorless media. If tests in the vicinity of the particular limit of identification are not involved, some aid may be obtained by conducting such drop reactions on spot plates or in micro test tubes, and by using comparison solutions which contain only interfering colored ionic species in approximately the same concentration as in the test solution.

Analytical flotation, i.e. shaking out with organic liquids, may be used to advantage when colored or colorless precipitations are conducted in a test tube. Minute amounts of a colored precipitate produced in a colored solution can often be made visible if the test is made as a drop reaction on filter paper or, in case this is not feasible because of the need of warming, etc., a drop of the suspension (after dilution if necessary) can be placed on filter paper. The precipitate will be retained in the center of the fleck while the colored liquor diffuses away through the capillaries of the paper and can be effectively separated from the precipitate by repeated treatments with one to three drops of water. However, it must be remembered that not all colored solutes may be washed away in this manner, because certain soluble colored substances (usually with colored anions) are irreversibly adsorbed on filter paper. Examples include the red solutions of complex ferrous salts of α,α'-dipyridyl and phenanthroline, and likewise solutions of acid or amphoteric dyes (for instance rhodamine dyes).

Special attention should be paid to the impairment of tests by those attendant materials which react with the ion being sought or with the precipitating or color reagent. In the former case, only those reactions demand attention which yield soluble products whose formation results in the disappearence of the ion being sought. Usually, stable complex com-

* There are isolated and therefore very important instances of the reverse behavior. Usually these are due to induced reactions. See for example, the test for bismuth described on page 77.

pounds are formed in such cases, but the production of soluble principal valence compounds may also enter this consideration. It is customary to say that the test ion has been masked or sequestered. The extent of the masking depends on the nature and quantity of the materials entering into the masking reaction, the pH, and the particular precipitation or color reagent. A good example is the masking of Fe^{+3} ions by F^- ions; the effect is due to the production of the complex $[FeF_6]^{-3}$ anions. Because of the greatly lowered Fe^{+3} ion concentration, most, but not all tests for iron^{III} fail in acid solutions which contain this complex anion plus an excess of fluoride ions. If the solution of a ferric salt contains a quantity of fluoride ions insufficient for the complete masking of this cation, the residual Fe^{+3} ions will react with the reagent. The same holds for the masking of reactions of cations and anions in general. Accordingly, it is not permissible to declare categorically that all tests for certain ionic species can be prevented by masking agents. It may be taken as a general rule that only large amounts of a masking agent are able to prevent certain tests completely, whereas insufficient amounts can merely accomplish a lowering of the sensitivity of tests. Masking can be annulled by decomposing the compounds produced in the masking action. When this is accomplished in the wet way by adding a material which removes the masking agent from the masked system by forming a stable soluble or insoluble compound, the process is called "demasking".* It has been found that materials which bring about demasking can be detected reliably and with high sensitivity if the masked reaction system is applied as reagent. Pertinent examples are the tests for mercury, fluoride, cyanide, ferrocyanide described on pages 72, 269, 278, 288. In this connection it should be noted that the most rapid and most sensitive test for complex alkalicuprocyanide and likewise the micro detection of insoluble acid-resistant silver halides is based on demasking processes (compare page 277).

If an ionic species is to be detected by means of a particular precipitation or color reaction and the accompanying materials likewise enter into the precipitation or the color production, it is obvious that the analyst is then confronted with the most dangerous type of interference because the interference can operate over wide concentration ranges. Fortunately, a uniform mode of reaction does not always occur over all pH regions, since many

* Both masking and demasking reactions are equilibrium reactions through which certain ionic species are made to disappear or reappear. Accordingly, demasking involves the disturbance of the equilibrium of a masking reaction. The initiation of a reaction through the withdrawal of an ionic species participating in the equilibrium is of course not limited to demasking. A number of sensitive spot reactions are based on disturbances of equilibria, including equilibria of redox reactions. (Compare the detection of basic materials, page 507 and the detection of copper and zinc, pages 84 and 178.)

reactions which lead to the establishment of equilibrium conditions are pH-dependent. Consequently, the reaction of the accompanying materials can be prevented or kept under control in many cases by proper adjustment of the pH. This objective can also be reached in some instances by adding appropriate masking agents in sufficient quantities. Instances are known in which the interferences of even large amounts of accompanying substances are safely avoided by combining an effective adjustment of the pH with masking. Spot test analysis makes extensive use of pH adjustment and particularly of maskings to avert impairments of tests by interfering accompanying materials which otherwise can be removed only by tedious separation procedures. The essential factor in the masking of reactions is to decrease the concentration of certain ionic species to the point at which the ionic concentration necessary to the occurrence of the particular color or precipitation reaction is no longer present. The lowering of an ionic concentration to very low levels can also be achieved by precipitating difficultly soluble compounds. This means that if the reaction theater is transferred from a homogeneous solution to the surfaces of solids which contain the sought ion as a phase component, the same effect can be achieved as by masking in solution. The sensitive test for cobalt with 1-nitroso-2-naphthol (page 144) is an admirable example. This compound, in neutral or acetic acid solution, reacts not only with cobalt ^{II} but also with iron ^{III}, iron ^{II}, copper ^{II}, and uranyl ions to produce the respective colored inner complex salts. However, if these ions are first brought down as tertiary phosphates by means of alkali phosphate, and the suspension is then treated with 1-nitroso-2-naphthol, only red-brown cobalt ^{III} nitrosonaphtholate is formed. This result means that the concentration of metal ions furnished by the other phosphates is not sufficient to exceed the solubility product of the nitrosonaphtholate in question. Therefore, it is quite permissible to speak here of a masking of the iron (uranyl, etc.) nitrosonaphtholate reaction. This type of masking* has no significance in macroanalysis. On the other hand, it can be of advantage in spot test analysis which deals with small amounts of liquid and precipitates.

The Committee of The International Union of Chemistry dealing with the study of new analytical reaction and reagents decided to differentiate between specific and selective reactions and reagents, respectively.[45] Reac-

* Sometimes the term "complexing" is used in place of "masking"; this is not correct. Apart from the fact that a masking may sometimes be due to causes other than the formation of complex compounds, it should be noted that the production of a complex does not necessarily lead to the prevention of a reaction. The pivotal point is not the production of a complex but rather the formation of such soluble complex compounds as are capable of preventing the occurrence of certain reactions. The term "complexing" therefore does not give an exact representation of the meaning of a reaction masking.

References pp. 25–26

tions (reagents) which, under the experimental conditions employed, are indicative of *one* substance are designated as specific, while those which are indicative of a comparatively small number of substances are classified as selective. Accordingly, there are reactions (reagents) that are more or less selective. On the other hand, a reaction (reagent) is either specific or non-specific. Consequently, when these designations are employed, it must be remembered that they apply to analytically useful reactions (reagents) and therefore, strictly speaking, to procedures and not merely to reactions (reagents). A reaction (reagent) may therefore display different degrees of selectivity, or it may even become specific, according to the particular procedure employed, since the latter may bring into play to varying extents such factors as pH adjustment, masking, etc. Accordingly, the same considerations are valid for specificity and selectivity as well as for sensitivity: the underlying chemical change—although characteristic for the test in question—is a necessary but not sufficient condition. As a rule, maximum specificity, selectivity, and sensitivity of tests are attained only by the proper choice and establishment of the circumstances surrounding the reaction, a process known as conditioning.

The importance of characterizing tests by sensitivity, limiting proportions, selectivity, and specificity is now so widely accepted that the pertinent data and statements are given in most of the modern reports of new tests. It is obviously impossible to give exhaustive information regarding the impairment of tests by accompanying materials since the extent of such interferences, under otherwise like conditions, depends on the kind and quantity of the attendant materials. More and more attention is being devoted to the conditioning of tests, in order that the detrimental effects of the accompanying materials may be lessened or even eliminated. These efforts have been successful in a surprisingly large number of cases. Much thanks is due the Committee just mentioned for its valuable services in critically checking numerous tests for anions and cations with respect to their application as spot reactions, including descriptive statements regarding sensitivity and the behavior of other ions.[46]

Spot test analysis has inspired many investigational studies connected directly or indirectly with specificity, selectivity and sensitivity. If explanations can now be supplied for the strikingly changeful morphology of flecks, and if information is now available as to whether the sensitivity and reliability of tests can be enhanced by spot reactions on paper, it is because research along these lines became necessary. The same holds regarding the effective adaptation of known chemical reactions from the macro scale to spot testing, and with respect to the improvement of the actual making of spot tests. However, research of this kind does not go much beyond the

References pp. 25–26

narrow objectives of spot test analysis. Of more general significance is the fact that spot test analysis stimulated the search for new reactions possessing the greatest possible certainty and sensitivity when used for analytical purposes. In the front rank stands the action of organic reagents, whose outstanding usefulness came to light at the very start of the experiments on spot reactions. Efforts to arrive at new organic reagents seemed so important that the author soon began to make special studies concerning the relations between groups in organic compounds and selective actions. Observations were obtained which could not always be utilized directly for analytical purposes but which nonetheless deepened the knowledge of the activity of organic reagents. The correctness of the course chosen was demonstrated by the discovery of new organic reagents, and some of them also proved useful in spot testing. Among these were the first sensitive and selective organic color reagents for silver and barium, namely *p*-dimethylaminobenzylidenerhodanine [47] and rhodizonic acid [48] (see pages 59 and 216). The latter has been found to be an excellent reagent for the identification of insoluble lead salts [49] (page 73). The search for new organic reagents, which is based principally on the concept of group actions and the knowledge of the basic facts of the chemistry of the coordination compounds, has become a special research province in analytical chemistry.[50] The great importance of this field and its rapid development is illustrated best by the fact that as late as 1936 it was still possible to present its essential features in a single paper,[51] whereas now the excellent 4-volume work by Welcher [52] needs to be consulted in all comprehensive studies of organic reagents. No present-day worker who discovers an organic reagent that leads to colored organometallic reaction products would be likely to omit trials to discover the applicability of his new compound as a spot test reagent. This established practice is a good indication of the close ties that now exist between spot test analysis and the search for new organic reagents. It must be pointed out here that studies in the field of organic reagents have values beyond analytical applications. Many new organic and organometallic compounds isolated in such researches have compositions, constitutions, and properties that are of great importance to the chemistry of complexes. Compounds which fit into certain classes of coordination compounds have been prepared but, in addition, members of new classes have appeared [50] including some which represent abnormal valence states of certain metals. Several tests described in this text exemplify such departures from the usual behavior of these metals. The employment of organic reagents has stimulated interest in the other direction, namely the elucidation and correct interpretation of numerous color- and precipitation reactions. Valuable studies of this kind have come from the laboratories of M. Bose,

A. Freiser, S. M. Korenmann, L. Kul'berg, V. J. Kuznetzov, and A. Okáč.

The pioneer service rendered by spot test analysis with respect to the appreciation of indisputability and sensitivity is shown especially by the utilization of catalytic reactions in qualitative analysis. Many of these, including some which play a role in quantitative procedures, have been known for a long time. It has also long been known that catalytic actions can be produced by small quantities of materials and that they limit their effects to definite reaction systems. Nevertheless, it is only within the last twenty years that much consideration has been given to tests which are based on catalyses in homogeneous systems. The stimulus for this new field of interest came from reports on several highly sensitive spot tests based on catalytic reactions and from the indication that in catalytic actions the effects obviously possess specificity and sensitivity.[53] A fine example is the observation published by Raschig [54] in 1915 that the stable brown solution containing sodium azide and iodine is instantly decolorized when crystals of alkali thiosulfate or alkali sulfide are introduced. Although this initiation of the reaction: $2\ NaN_3 + I_2 \rightarrow 2\ NaI + 3\ N_2$ had been cited in the inorganic textbooks as an example of catalysis, the Raschig observation was not used analytically until 1928 and then in the field of spot testing.[55] At present this iodine-azide reaction enjoys wide popularity as a test for sulfide-bound sulfur in both inorganic and organic spot test analysis. Numerous examples of the application of catalysis reactions for the detection of cations and anions are given in this text; they all are characterized by great sensitivity and reliability.[56]

The search for new tests that was stimulated by spot test analysis places in the foreground the question as to the analytical applicability of reactions and the question as to the specificity, selectivity, and sensitivity of chemical methods. These questions go beyond the narrow objectives of spot test analysis since they obviously refer to basic problems of chemical analysis. Satisfactory answers require the assembling, classification, and critical review of the experimental material which is connected directly or indirectly with specificity, selectivity, and sensitivity of methods used in chemical analysis. The pertinent material, based not only on new researches but also on many observations scattered through publications that are often of a much earlier date and that deal with all provinces of chemistry, is so extensive that it is entirely proper to speak of the "chemistry of specific, selective and sensitive reactions". Within the bounds of this special field,[57] spot test analysis presents only a particular method of carrying out and applying specific, selective, and sensitive reactions. On the other hand, the chemistry of specific, selective and sensitive reactions constitutes the scientific background of inorganic and organic spot test analysis, which is

References pp. 25–26

thus brought into close relationship with many special fields of chemistry.

In its early years, spot test analysis was restricted to tests conducted in drops of a test solution. This technique was soon extended through the use of spot reactions carried out directly on solid materials (powders, smooth surfaces, massive fragments). Many identifying reactions which have proved very useful in the examination of technical materials and rocks are in this category. Examples are: the identification of alumina, the differentiation of dolomite and magnesite, and many others given in Chapter 7. Such identification reactions, carried on by spotting, need not be directed toward microchemical goals exclusively. Because of their convenience, they can often replace tedious procedures of qualitative macro- or semi-microanalysis. A fundamental effect of spot test analysis, namely the recognition of colored products at certain points or areas of filter paper or reagent papers, has logically been applied in the so-called off-print process. By this means it is possible to locate and detect inhomogeneities in metals and rocks. The development of this interesting technique is due primarily to the studies by Niessner,[58] Yagoda,[59] and Gutzeit.[60] The localized deposition of colored reaction products is likewise the basis of the methods of electrographic detection originated by Glazunow.[61]

In accord with their limits of identification and the corresponding dilution limits, spot reactions can be applied directly only within certain concentration ranges of the material to be detected. When higher dilutions are presented, preliminary concentration is essential. Evaporation of considerable volumes of liquid was formerly employed almost exclusively, but this has been replaced in many instances by the use of collectors. The latter are suitable solids which have the ability to fix on their surface traces of the sought material. The action of collectors is due to the formation of mixed crystals in some cases, or to mutual precipitation, or adsorption.[62] Traces accumulated from highly dilute solutions by these gathering aids can then be tested directly on the collector by suitable spot reactions, or the gathered material may be put into solution in a small volume before the test is applied. This application of spot tests in the so-called "search for traces" has become of great importance in the examination of metallic samples.

The main field of application of inorganic spot tests is in qualitative analysis. However, spot reactions which lead to colored products can be employed successfully for quantitative determinations also. N. A. Tananaeff [63] was the first (1929) to make "spot colorimetric determinations" by comparing the intensity of color reactions carried out in drops of test and standard solutions on filter paper or on a spot plate. Yagoda [59] improved and refined spot colorimetry considerably through his ingenious idea of conducting spot reactions on "confined" spot test papers. Small quantities

References pp. 25–26

of material can be colorimetrically determined in a single drop with remarkable accuracy by the Yagoda method. Another approach to quantitative results is the extinction method developed by Wenger.[64] This procedure requires the use of several tests of different but known sensitivities. The discovery of which tests fail and which ones respond positively gives an indication of the concentration and therefore of the quantity of the material present.

The foregoing discussion of the development and present state of spot test analysis has indicated that the consideration of the outlook for this branch of analysis will have to be along several lines: (1) development of the technique; (2) additional reactions available for spot testing; (3) applications of spot reactions. There is little likelihood that anything essentially new will develop with regard to the technique of spot test analysis hitherto in use. On the other hand, numerous improvements and refinements may become necessary. Particularly desirable would be an arsenal of stable reagent papers impregnated with water-insoluble compounds to obtain a maximum of sensitivity (compare Clarke and Hermance, *loc. cit.*). The preparation of such reagent papers is not always easy. There are no general procedures that will invariably produce uniformly impregnated papers. When reagent papers are stored for any length of time, the impregnant dusts off, a process clearly due to a growth in the size of the particles (recrystallization). Hence it is important to stabilize, as much as possible, the fine distribution of reagents on and in the capillaries of the paper. Steigmann [65] has described a novel method of preparing papers impregnated with water-insoluble organic salts of water-soluble acid organic reagents. Such reagents are precipitated with "sapamine" (trimethyloleo-aminoethylammonium sulfate), a cationic wetting agent. The dried precipitates are soluble in organic liquids, such as chloroform. If filter paper is bathed in a solution of this kind and the solvent allowed to evaporate, the amorphous organic salt is left as a fine dispersion in the capillaries. Possibly this interesting procedure points the way to the preparation of more stable, strongly impregnated reagent papers. These would be of great importance also for the extension of the technique of utilizing protective layer effects. As yet there are no reports concerning the preparation and use of films (foils) of cellophane or plastics, including those with water-repelling surfaces. Such reagent foils could possibly offer advantages in certain cases.

Flood [66] showed that filter paper impregnated with basic or acidic alumina can accomplish the same chromatographic separations as the alumina columns used in inorganic chromatographic analysis, and first described by Schwab.[67] In other words, cations and anions are fixed in definite separate zones when solutions are allowed to ascend strips of alumina paper. If the latter are then

References pp. 25–26

developed by means of suitable reagents, the adsorptively separated materials can be detected. Iijima [68] and his associates have further elaborated this method and have also experimented with papers impregnated with chromium hydroxide. Hopf [69] carried out spot reactions on filter paper impregnated with basic alumina (starch) and certain reagents. He suggested the term "chromatographic spot tests" for this noteworthy procedure, which probably is capable of further development. It may safely be assumed that sensitive spot tests will play an important, sometimes decisive, role in all kinds of inorganic chromatographic analysis.

The future will certainly bring an added number of spot reactions for the detection of inorganic materials. An important role will be filled by organic reagents which lead to colored products by the dry or wet method. Furthermore, the use of organic reagents as masking agents will offer several advantages with respect to improving the selectivity or even to attain specificity. The production of fluorescing organometallic compounds will receive more attention than heretofore. Gotô [70] has rendered pioneer service in this direction. He showed that the formation of fluorescent metal salts and also the quenching of fluorescence can be used in inorganic spot test analysis The sensitive and strictly specific test for hydrazine (page 239) based on the. formation of salicylaldoxime shows that condensation reactions leading to fluorescent products can also be employed. Less consideration has been given thus far to the utilization of photo-reactions, but the possibilities are excellent. It may be confidently expected that new highly sensitive spot reactions will be developed on the basis of catalytic reactions. Up to now, catalyzable redox reactions have been considered almost exclusively. However, catalytic accelerations of other types of reactions, especially organic, will certainly be employed. A striking example is the detection of cyanide through catalysis of the benzoin rearrangement (see page 280).

The development of spot test analysis has shown that its practical importance resides primarily in the fact that it permits the rapid accomplishment of sensitive identification tests with tiny amounts of materials. Such tests are of interest to all branches of natural science which have need of chemical tests within the bounds of their particular problems. The main fields of application of spot test analysis, which can also give it new stimuli, will reside in the future, as in the present, in trace analysis, in the testing of materials, in factory controls, in criminal investigations, and in paper chromatography.[71] This prediction is based on the finding that the number of publications dealing with applications of previously known spot reactions now exceeds the number of papers concerned with new spot tests. Spot colorimetry is capable of great expansion because additional color reactions can be called on. Further advances may be expected in the detection of inhomogeneities in metals and rocks by the off-print procedure, as well as in the identification of minerals and components of alloys by means of sensitive spot reactions.

A clear knowledge of the underlying chemical changes and of the scientific foundation of all the measures which are required to reach the maximum sensitivity and reliability is indispensable to the understanding of the procedures prescribed in the application of spot reactions. So many phenomena, regularities, and general modes of behavior, drawn from nearly all fields of chemistry, are encountered that, when properly practiced, spot test analysis is anything but routine drudgery; rather it is experimental chemistry applied to the solution of analytical problems. This fact, together with the necessity of working precisely with small quantities of materials, confers a high didactic value on spot test analysis and justifies its inclusion in the chemical curriculum.[72] However, it should be emphasized that a critical sense and the ability to make correct judgements acquired in previous courses of chemical instruction are essential to a proper understanding of the theory and practice of spot test analysis. For this reason, and also because of the growing use of spot reactions in qualitative organic analysis (see Volume II), spot test analysis should not be included at too early stages of chemical training; it should be reserved for the later phases.* The practising professional chemist, who can derive much stimulus and many suggestions from spot testing, will decide for himself when and where the application of spot reactions will be of advantage to his own special field and problems. The rather detailed description and discussion of the individual spot reactions, as well as the many illustrations taken from the testing of commercial materials, included in this text have been composed to meet these aspects of spot test analysis. Furthermore, the presentation is given in such manner as to make possible an adaptation to the macro scale.

The following additional volumes can be consulted with profit:

Reports of the International Committee on New Analytical Reactions and Reagents of the International Union of Chemistry:

First Report, Leipzig, 1938

* The use of spot test analysis as a laboratory practice in institutions of higher learning dates from 1929 when it appeared in the official syllabus of the Institute of Analytical Chemistry at the Technical University of Delft (Prof. C. J. van Nieuwenburg). The author had practical experience with student courses in the laboratories of the University and the Volkshochschule in Vienna from 1921–1938, and later in the Chemical School of the University of Brazil at Rio de Janeiro. At present, prominent centers of instruction and research in spot test analysis are the chemistry department of Louisiana State University at Baton Rouge (Prof. Ph. W. West) and the University of Birmingham (Dr. R. Belcher). The number of papers issued indicates that the Soviet universities are paying considerable attention to spot test analysis.

Second Report (edited by P. E. Wenger and R. Duckert *et al.*), New York and Amsterdam, 1948

Third Report (edited by C. Duval), Paris, 1948

Fourth Report (edited by P. E. Wenger and Y. Rusconi), Paris, 1950

F. Feigl, *Chemistry of Specific, Selective and Sensitive Reactions*, New York, 1949.

REFERENCES

1. Compare F. Feigl, *Amer. Soc. Testing Mat. Spec. Technical Publ.*, 98 (1950) 1; F. Feigl and Ph. W. West, *Mikrochem. ver. Mikrochim. Acta*, 36/37 (1951) 191; F. Feigl, *ibid.*, 39 (1952) 368; Ph. W. West, *Analyst*, 77 (1952) 611.
2. B. Anft, *J. Chem. Educ.*, 32 (1955) 972 gives a comprehensive report of Runge's pioneering studies of spot tests and paper chromatography.
3. H. Schiff, *Ann.*, 109 (1859) 67.
4. C. F. Schönbein, *Ann.*, 114 (1861) 275.
5. Z. Skraup and associates, *Monatsh.*, 31 (1910) 1067.
6. R. Krulla, *Z. Phys. Chem.*, 66 (1909) 307.
7. F. Feigl and R. Stern, *Z. anal. Chem.*, 60 (1921) 28.
8. F. Feigl and F. Neuber, *Z. anal. Chem.*, 62 (1923) 373.
9. See for instance, R. Delaby and J. A. Gautier, *Analyse qualitative minérale à l'aide des Stilliréactions*, Paris, 1940.
10. F. Feigl, *Mikrochemie*, 1 (1923) 4.
11. F. L. Hahn, *Mikrochemie*, 8 (1930) 75.
12. K. Heller, *Mikrochemie*, 8 (1930) 141.
13. A. A. Benedetti-Pichler, *Mikrochemie*, 36/37 (1951) 24.
14. J. Gillis, *Experientia*, 8 (1952) 365.
15. Compare E. B. Sandell, *Colorimetric Determination of Traces of Metals*, New York, 1950.
16. C. L. Wilson, *Mikrochim. Acta*, (1953) 58.
17. P. L. Kirk, *Quantitative Ultramicroanalysis*, New York, 1950.

17a. Committee on Nomenclature, Division of Analytical Chemistry of A.C.S., *Anal. Chem.*, 24 (1952) 1348. Compare C. L. Wilson, *J. Royal Institute Chem.*, Part 4 (1952) 194.

18. W. Böttger, *Chem. Ztg.*, 33 (1904) 1003.
19. Comp. A. Okáč and P. Černý, *Chem. Abstracts*, 46 (1952) 3896.
20. Ph. W. West and W. C. Hamilton, *Mikrochem. ver. Mikrochim. Acta*, 38 (1951) 700.
21. R. M. Rush and L. B. Rogers, *Mikrochim. Acta*, (1955) 821.
22. F. Feigl, *Chem. Ztg.*, 44 (1920) 689.
23. F. Feigl and L. I. Miranda, *Ind. Eng. Chem., Anal. Ed.*, 16 (1944) 141.
24. F. Pavelka, *Mikrochemie*, 23 (1937) 202.
25. F. Feigl and H. A. Suter, *Chemist Analyst*, 32 (1943) 4.
26. B. L. Clarke and H. W. Hermance, *Ind. Eng. Chem., Anal. Ed.*, 9 (1937) 292.
27. F. Pavelka, *Mikrochemie*, 7 (1929) 303.
28. F. Feigl and R. Uzel, *Mikrochemie*, 19 (1936) 136.
29. R. Kempf, *Z. anal. Chem.*, 89 (1933) 88.
30. F. Feigl and H. J. Kapulitzas, *Mikrochemie*, 8 (1930) 239.
31. A. Velculescu, and J. Cornea, *Z. anal. Chem.*, 94 (1933) 225.
32. F. Feigl, *Chemistry & Industry*, 57 (1938) 1161.
33. F. Feigl and N. Braile, *Chemist Analyst*, 32 (1943) 76.
34. F. Feigl, *Rec. trav. chim.*, 58 (1939) 472.
35. E. M. Przibram, *Chem. Zentr.*, I, (1939) 475.
36. J. Winckelmann, *Mikrochim.*, 210 (1931) 437; 122 (193) 17; 14 (1934) 171; 16 (1936) 203
37. G. Hirsch, *Dissertation*, Vienna, 1927.
38. G. Skalos, *Mikrochem. ver. Mikrochim. Acta*, 32 (1944) 233.
39. W. W. Razim, *Ind. Eng. Chem., Anal. Ed.*, 14 (1942) 278.

40. O. J. Stewart and D. W. Young, *J. Am. Chem. Soc.*, 57 (1935) 695.
41. F. Feigl, L. I. Miranda and H. A. Suter, *J. Chem. Educ.*, 21 (1944) 19.
42. Ph. W. West and W. C. Hamilton, *Anal. Chem.*, 24 (1952) 1025.
43. F. Emich, *Ber.*, 43 (1910) 10.
44. N. Schoorl, *Z. anal. Chem.*, 46 (1907) 658.
45. Compare *Mikrochim. Acta*, 1 (1937) 253.
46. Compare the *Second Report* of the International Committee on New Analytical Reactions and Reagents of the International Union of Chemistry, edited by P. E. Wenger, R. Duckert, C. J. van Nieuwenburg and J. Gillis, New York and Amsterdam, 1948.
47. F. Feigl, *Z. anal. Chem.*, 7 (1926) 4, 380.
48. F. Feigl, *Mikrochemie*, 2 (1924) 188.
49. F. Feigl and H. A. Suter, *Ind. Eng. Chem., Anal. Ed.*, 14 (1942) 840.
50. Compare F. Feigl, *Anal. Chem.*, 21 (1949) 1298.
51. F. Feigl, *Ind. Eng. Chem., Anal. Ed.*, 8 (1936) 404.
52. F. J. Welcher, *Organic Analytical Reagents*, New York, 1947/48.
53. F. Feigl, *Z. angew. Chem.*, 44 (1931) 739.
54. F. Raschig, *Ber.*, 48 (1915) 2088.
55. F. Feigl, *Z. anal. Chem.*, 74 (1928) 369.
56. Compare Ph. West, *Anal. Chem.*, 23 (1951) 176.
57. F. Feigl, *Chemistry of Specific, Selective and Sensitive Reactions*, New York, 1949.
58. M. Niessner, *Z. angew. Chem.*, 52 (1949) 721.
59. H. Yagoda, *Ind. Eng. Chem., Anal. Ed.*, 9 (1937) 79.
60. G. Gutzeit, *Eng. Mining J.*, 143 (1942) 57.
61. A. Glazunow, *Oesterr. Chem. Ztg.*, 41 (1938) 217.
62. E. B. Sandell, *Colorimetric Determination of Traces of Metals*, 2nd ed., New York, 1950.
63. N. A. Tananaeff, papers since 1929, especially Russian publications.
64. P. E. Wenger and associates, *Helv. Chim. Acta*, 29 (1946) 1698; 30 (1947) 1636; 31 (1948) 290; 32 (1949) 1865; *Bull. soc. chim. France*, 15 (1948) 517; *Anal. Chim. Acta*, 1 (1947) 190.
65. A. Steigmann, *J. Soc. Chem. Ind.*, 64 (1945) 88.
66. H. Flood, *Z. anal. Chem.*, 129 (1940) 327.
67. G. Schwab and A. N. Gosh, *Z. angew. Chem.*, 53 (1940) 327.
68. Sh. Iijima and coworkers, *Chem. Abstracts*, 42 (1948) 7197.
69. P. P. Hopf, *J. Chem. Soc.*, (1946) 785.
70. H. Gotô, *J. Chem. Soc. Japan*, 59 (1938) 199, 203, 365, 371, 547, 625, 805, 797, 1215, 1357, 1362; 60 (1939) 937, 940.
71. Comp. R. J. Block, R. Le Strange and G. Zweig, *Paper Chromatography*, New York, 1952; J. N. Balston and B. E. Talbot, *A Guide to Filter Paper and Cellulose Powder Chromatography*, London, 1952; F. Cramer, *Papierchromatographie*, 2nd ed., Berlin, 1953; E. Lederer and M. Lederer, *Chromatography, A Review of Principles and Applications*, Amsterdam, 1953.
72. Regarding the execution of spot reaction demonstration experiments see F. Feigl, *J. Chem., Educ.*, 20 (1943) 137, 174, 240, 294, 298, 300; 21 (1944) 347, 479; 22 (1945) 36, 342, 554; 34 (1957) 457; F. Feigl and G. B. Heisig, *ibid.*, 29 (1952) 192.

Chapter 2

Spot Test Techniques

1. Introduction

The term "spot test analysis" is a generic term referring to sensitive and selective tests based on chemical reactions whereby the use of a drop of the test or reagent solution is an essential step. The tests are microanalytical or semi-microanalytical in nature and are applicable for the investigation of both inorganic and organic compounds. An important part in spot test analysis is played by the actual manipulations with drops of unknown substances and reagents, and the method is not dependent on the use of auxiliary optical magnification.

In general, spot test procedures are the ultimate in simplicity. The elegance of the method derives from the nature of the reagents used, together with the advantageous use of reaction conditions, so that the utmost of sensitivity and selectivity can be obtained with a minimum of physical and chemical operations. As much as possible, separations and conditioning reactions are integrated in the test procedure so that the final test becomes a unitized operation that can be applied directly for the identification of the substance in question. The tests are ordinarily run by using one of the following techniques:

1. By bringing together one drop each of the test solution and reagent on porous or non-porous supporting surfaces such as paper, glass, or porcelain.
2. By placing a drop of the test solution on a medium impregnated with appropriate reagents (filter paper, asbestos, gelatin).
3. By placing a drop of reagent solution on a small quantity of the solid specimen (fragments or pulverized particles, evaporation or ignition residues).
4. By subjecting a drop of reagent or a strip of reagent paper to the action of liberated gases from a drop of the test solution or from a minute quantity of the solid specimen.
5. In an extended sense, spot reactions may also include tests accomplished by adding a drop of test solution to a larger volume (0.5 to 2 ml) of reagent solution and then extracting the reaction products with organic solvents.

The choice of procedure to be followed will ordinarily be dictated by the nature of the sample and the reagents available. The equipment and manipulations required are all simple and the techniques utilized can be learned without difficulty. The essential requirements for the successful application of spot test procedures include: (1) a knowledge of the chemical basis of all of the details of the tests used so that every step of the procedures can be understood and executed intelligently; (2) strict observance of trustworthy experimental conditions; (3) scrupulous cleanliness of the laboratory and equipment; (4) the use of the purest reagents available. Whenever possible, tests should be repeated to insure reproducibility. It should always be a rule, also, to run both blanks and controls.

The most essential manipulation in spot test work is the actual "spotting" of reactants. It is frequently necessary, however, to undertake certain preliminary operations to provide the most advantageous reaction conditions. Particularly in cases where complex samples are to be investigated, preliminary separations, either physical or chemical, may be required. Various operations such as drying, evaporation, ignition, oxidation or reduction, and adjustment of pH, are often employed; in cases where organic substances are to be identified, it is often necessary to undertake syntheses on a small scale, and preparative operations are common.

On the basis of the preceding remarks, the following discussion is presented to summarize the laboratory and equipment needs for spot test analysis, together with some of the essential operational techniques.

2. Laboratory and Equipment Requirements

Laboratory needs for the application of spot tests vary greatly, depending on the number and type of analyses anticipated. Under any circumstances, the space devoted to spot test work should be so designed and arranged that it can be kept scrupulously clean and, if possible, free of laboratory fumes. Proper lighting is also of the utmost importance in spot test operations. Diffused daylight is ideal for observation of colors and precipitates encountered in the analyses, but good daylight-type fluorescent lighting is also very satisfactory and has the advantage that it is constant. The arrangement of the work sections of the spot test laboratory should be given careful consideration. The actual performance of the tests themselves can be restricted to a very small area because the amount and type of equipment needed is at a minimum. Most important is to have an ample supply of reagents immediately available to the analyst. Sections devoted to sample preparation, cleaning of glassware, preparation of solutions, and general macrochemical operations should be as remote as possible from the spot test workbench.

References p. 56

The requirements for a spot test laboratory suggested above can be met in a good many ways, depending on the amount of space available and the volume of work anticipated. For industrial laboratories or large research laboratories, a special room designed for spot test investigations is desirable. A separate laboratory is also of advantage in universities where instruction in spot test methods is introduced. In this connection, it is quite logical that the spot test laboratory can also be used in teaching other microchemical techniques. It is practical to include selected spot tests in the general qualitative analysis course, and in such cases a few sets of reagents and a spot test workbench can be incorporated in the general qualitative analysis laboratory.

In many cases where space is at a premium, it is necessary to restrict the area devoted to spot test work and make this section a part of some general laboratory. From the standpoint of the space involved, this is easily done because these procedures demand a minimum of equipment and the working area required is exceedingly small as compared to that needed for most analytical techniques. It should be kept in mind that most spot tests are extremely sensitive; consequently, it is necessary to have the test section so located that the amount of fumes and dusts can be kept at a minimum.

A very satisfactory arrangement for occasional spot test work is to set aside a portion of a hood. This special section of the hood should have access to gas, electricity, and water and should have excellent lighting. Because the number of working tools required is so small, it is entirely practical to keep most of the ordinary equipment on hand in the hood space. Where certain tests are anticipated as being routine, the necessary reagents can be kept also in the hood, while chemicals and special reagents for occasional use can be kept in a near-by section of the laboratory, together with such auxiliary apparatus as balances, ultraviolet lamps, centrifuges, dryers, ovens and furnaces.

The laboratory devoted to spot test analysis must be provided with a wide selection of chemicals. While the amount of chemicals required is quite small, it remains important to have a variety of reagents available. Certainly the various organic and inorganic reagents commonly used must be stocked and a complete supply of general acids, bases, solvents, and oxidizing and reducing agents should be on hand. It is highly desirable to have a set of solutions of various organic substances available so that controls can be run on tests. It is also highly desirable to have respresentatives of diverse classes of organic compounds for use in running controls. A collection of plant products and standard samples of chemicals, minerals, and various technical products is also of great value in the preliminary examination of

References p. 56

unknown materials. In this connection, it is well to have simplified procedures outlined on cards mounted near the required reagents.

Portable laboratories and portable kits can be designed for use in spot test work, whereby the necessary chemicals and equipment for a variety of tests can be included in a minimum amount of space. For greatest flexibility, however, the well-equipped spot test laboratory will include an assortment of general laboratory apparatus as well as supplementary reagents and chemicals.* The following list is suggested as a guide for the stocking of a versatile laboratory.

Glass- and porcelain-ware

Assorted sizes of beakers, volumetric flasks, Erlenmeyer flasks, suction flasks, round-bottom flasks, distillation heads, Conway cells [1], crystallizing dishes, evaporating dishes, filter sticks, separatory funnels, extraction pipets, fritted glass crucibles, graduated cylinders, pipets, burets, weighing bottles, storage bottles, vials, test tubes (macro-, semimicro- and micro-), centrifuge tubes in various sizes, microscope slides, cover glasses, and spot plates should be available.

In addition to these glass items, there should be some counterpart items made of porcelain and quartz. Especially in the case of porcelain items, there should be both white and black crucibles, dishes, and the well-known spot plates in assorted sizes. An important recent innovation is the introduction of polyethylene bottles, which are of great value in the storage of stock solutions. There is available, also, low actinic glassware, which is particularly useful in the storage of certain organic reagents that tend to decompose when exposed to low wavelength light.

It is absolutely necessary to keep a permanent stock of glass rods and pipettes.

Metal utensils

Platinum ware is standard for many spot test operations; and there should be available platinum dishes, foils, crucibles, and boats. Nickel crucibles and stainless steel beakers and dishes are also of value. Aluminum pans and dishes are very useful since because of their low cost they can be discarded

* The various laboratory supply houses stock microchemical apparatus along with standard laboratory items. In the United States, the Microchemical Specialties Company, Berkeley, California, and the Arthur H. Thomas Company, Philadelphia, Pennsylvania, have particularly good assortments of microchemical equipment. The British Drug Houses Ltd. (London) and the Eastman Kodak Company (Rochester) supply most of the reagents of importance for spot test analyses.

The publications of H. K. Alber and R. Belcher on standardization of microchemical apparatus should be noted. The Subcommission on the Standardization of Microchemical Apparatus, of the Commission of Pure and Applied Chemistry, is making cooperative studies on the standardization of laboratory equipment.

References p. 56

after use. Absolutely indispensable are forceps and spatulas of nickel plated steel.

Iron ware

There should be a good assortment of microburners as well as Tirrill and blast burners. Other essential metal ware items are sand baths, water baths, tripods, ring stands, clamps, rings, and buret holders.

Special equipment and apparatus

The following equipment and apparatus can be considered as essential in the spot test laboratory: trip scale, torsion balance, analytical balance, pH meter, ultraviolet lamp, infrared lamp, centrifuge, ovens, dryers (a common hair dryer is very useful in spot test operations), furnaces, hot plates, steam baths (an electric baby bottle warmer makes an excellent steam bath for microchemical use), microdistillation assembly, electrographic apparatus, and cooling blocks.

In addition to the above items, it is desirable to have a spectrophotometer (such as the Beckman model DU or model B), a stereoscopic microscope, chemical microscope (or, preferably, a polarized light microscope), electric timer, and a melting point apparatus.

3. Working Methods

Consideration of the methods used in the laboratory must start with a discussion of the chemicals and solutions used. In general, a very good assortment of reagent quality inorganic chemicals should be available. Quarter-pound bottles are more than adequate in most cases. Organic chemicals are widely used in spot test laboratories and a good selection of organic reagents should be on hand. In most cases a few grams of organic reagents is sufficient; and these materials may be stored in the original bottles, preferably in a darkened cabinet. In some cases the organic reagents should be kept under refrigeration.

The wide use of organic reagents requires an appreciation of the characteristics of such materials. Many such reagents tend to decompose upon standing, particularly when they have been made up in solution. It is often possible in such cases to retard deterioration through the use of low actinic glassware.

Polyethylene containers can be used to store almost any organic or inorganic compound. Such containers are so inert chemically that they offer great advantages in the storage of the more reactive reagents. Although polyethylene bottles cost somewhat more than standard glass bottles, they should soon pay for themselves in saving of chemicals, breakage and time that

References p. 56

would normally be required in the replacement of deteriorated solutions.

Chemicals and reagents actively used at the spot test workbench should be kept in containers most convenient for use. For dry chemicals it is usually sufficient to keep them in plastic-stoppered glass vials. Because the actual spot reactions usually require not more than a drop or two of liquid reagents, it is well to keep such materials in small dropping bottles. Bottles of the type shown in Fig. 1 are very satisfactory where there is no danger of the stopper becoming frozen. Bottles equipped with pipets are very convenient for most work, and the type shown in Fig. 2 or the Barnes dropping bottle shown in Fig. 3 are very satisfactory. All such bottles should be of resistant

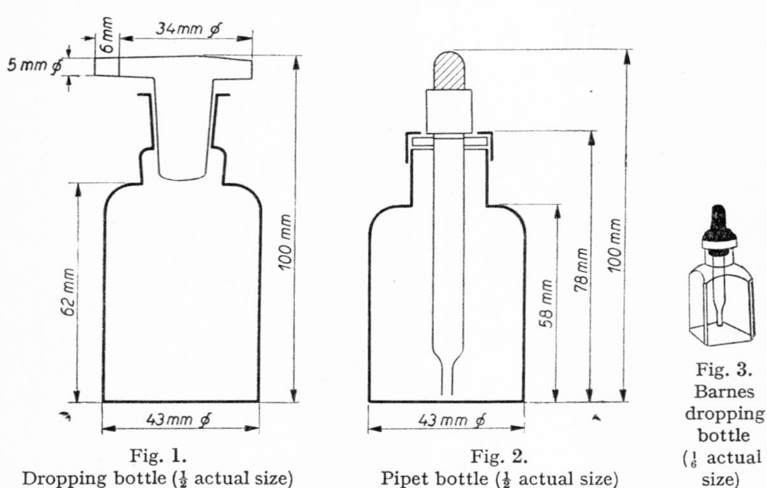

Fig. 1.
Dropping bottle ($\frac{1}{2}$ actual size)

Fig. 2.
Pipet bottle ($\frac{1}{2}$ actual size)

Fig. 3.
Barnes dropping bottle ($\frac{1}{6}$ actual size)

glass. Before being filled the bottles should be cleaned thoroughly and rinsed. Alkaline solutions that may attack glass may be stored in bottles coated inside with paraffin, although in most cases it is now preferred to store such solutions in polyethylene bottles. It is possible to obtain such bottles with nozzles so that reagents can be sprayed from the bottle by merely squeezing the sides; or droplets of reagent can be added by tipping the bottle and allowing reagent actually to run into the spray nozzle, after which the number of drops taken can be regulated by the amount of pressure applied to the sides of the bottle.

Sampling

In any analytical work one of the most important of all operations is the taking of samples. It is a total waste of time and effort to make careful analyses of samples that are not representative. For spot test operations it

References p. 56

must be kept in mind that such investigations can be utilized in the detection of inhomogeneities in the sample, or the tests can be utilized for ascertaining the average or general composition of a gross sample. Where selected portions of a sample are to be examined, it is often advisable to use low-power magnification and to sort the particles with the aid of a low-power wide-angle microscope. Sorting can be accomplished in such cases through use of fine glass fibers that have been moistened with glycerol or some other viscous and inert agent which can serve to collect the desired particles. Random impurities in many solid samples can best be studied by first dissolving the sample and then isolating the impurities through use of gathering agents, chromatographic techniques, or selective extractions. In some cases no such isolation methods need be applied; instead, highly sensitive spot tests are used directly for the detection or rough estimation of the impurity.

Materials in the gaseous state are sometimes subjected to spot test investigation. Materials of this type present peculiar problems because they are usually invisible substances. In handling gas samples, the desired constituent may be scrubbed from a given volume of the sample and subsequently examined, or a "catch" sample of the gas can be collected and the gross sample subjected to analytical investigation.

Samples of liquid unknowns present few problems for the selecting of representative samples, especially if the total sample can be stirred thoroughly to insure good mixing. In those cases where individual liquid samples must be taken and combined for later study, it is important that representative samples be secured by mixing like portions secured at the different locations.

The problem of proper sampling is so important that when there is a doubt concerning the reliability of samples for investigation, special treatises should be consulted for information on the products and sample methods used.

Grinding and mixing of solids

Solid samples are usually subjected to chemical treatment in the course of the analytical examination. These processes include: dissolution in water, acids, or alkalies; fusion with disintegrating agents such as sodium carbonates sodium peroxide, sodium pyrosulfate; volatilization of certain component-(simple distillation, sublimation, or the fuming of the sample with hydro, fluoric acid, hydrochloric acid, sulfuric acid, etc.), and extraction with organic solvents. Whatever treatment is employed, it is the general rule that the maximum reactive surface of the solid should be provided so that complete and rapid reactions may occur. This requirement is met in the case of slightly soluble compounds only when they are freshly precipitated and

References p. 56

gently dried. Ignited materials, especially natural or technical products, must usually be pulverized beforehand because a direct testing on compact surfaces is possible only in exceptional cases.

Agate mortars are preferred in the pulverizing of solid materials. Mortars and pestles of glass or of good quality porcelain are satisfactory for many purposes although they lack the resistance of agate. Very hard specimens should be broken up in a steel diamond mortar before final pulverizing in agate. Fig. 4 shows a micro mortar.

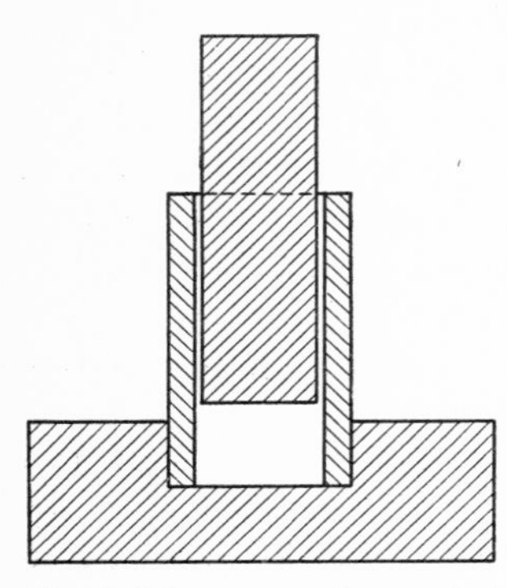

Fig. 4. Micro mortar (actual size)

To insure a uniform particle size, it is necessary to screen the powdered material through tightly stretched fine silk cloth. Such sifting is accomplished by adding the sample to the micro sieve and gently tapping the cloth so that the finest particles of the powder fall through the meshes and the coarser fragments are retained. The latter are then reground and the sifting and grinding continued until the total sample has been reduced to such a fine state of subdivision that it has passed through the cloth. Only when a perfectly homogenous material is being examined may the coarser residue be rejected and the first portion taken for chemical examination.

A very satisfactory arrangement for the sieving of samples is shown in Fig. 5. It consists of an open glass tube or micro beaker over which is stretched a circle of very fine silk. The cloth can be cemented in place or it can be fastened by means of a rubber band or piece of string.

Fig. 5.
Micro sieve (actual size)

Pulverized samples are mixed with powdered reagents only if the materials are thoroughly dry. The mixing can be done in crucibles, on watch glasses, or on glazed paper. Dry powders that are not hygroscopic can be weighed satisfactorily in tared aluminum dishes or on glazed paper. The mixing of powders can be performed with platinum wires or thin glass rods. Micro spatulas are very useful, and wooden toothpicks are of general value in the handling of powdered materials. Rather uniform specimens of some solid samples can be obtained by use of streak plates of unglazed porcelain. The sample is rubbed against the porcelain plate and the sample is collected in the

References p. 56

form of a streak which can be subsequently observed for its color and then taken up in appropriate reagents and analyzed by spot test procedures.

Evaporating, drying, igniting, fusing

The concentrating of solutions, removal of solvent or volatile constituents of solutions, and evaporation of solutions to dryness may be carried out, in most cases, in glass, porcelain, or platinum micro crucibles. The apparatus shown in Fig. 6 is very convenient for all of these operations. It consists of an aluminum block *Al* fitted with thermometer wells *T*; cavities for two or three micro crucibles are also provided in the block, and a small glass bell carrying a stopcock is fitted tightly to the top of the block. With this apparatus it is possible to remove water and volatile compounds under reduced pressure at temperatures below the normal boiling points. If the bell is placed on a well-fitted ground-glass plate the assembly can be used as a micro desiccator. (See also ref. [2].)

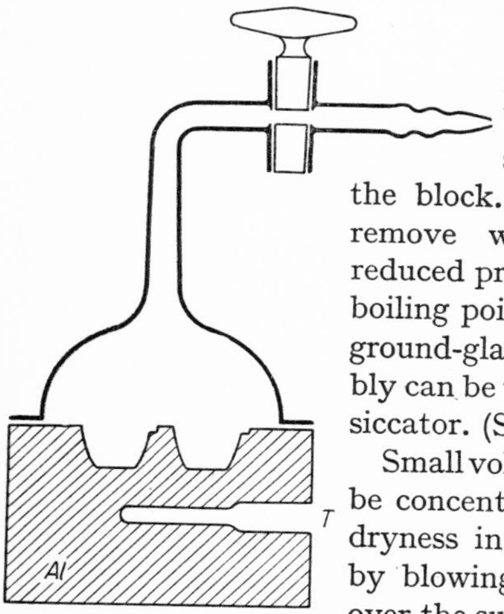

Fig. 6.
Aluminum block for concentrating solutions, etc. (½ actual size)

Small volumes of liquid can be concentrated or taken to dryness in centrifuge tubes by blowing dry, filtered air over the surface of the liquid while the centrifuge cone is immersed in a water bath. Fig. 7 shows the details of the apparatus for such an operation.

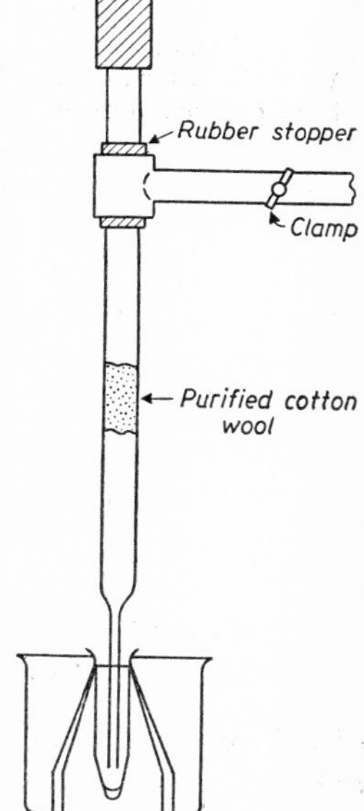

Fig. 7.
Set-up for concentrating a solution in a centrifuge tube (½ actual size)

If the temperature of a water bath is not sufficient to remove volatile compounds, an infrared lamp can be used advantageously,* and air or sand baths can be employed so as to obtain higher temperatures. A simple air bath is shown in Fig. 10. It consists of a nickel crucible with a copper wire triangle suspended in it through lateral slits. The triangle supports a micro crucible or micro beaker. The crucible can be heated directly by a burner or hot plate and the temperature checked by suspending a thermometer in the crucible. The thermometer is best protected

* A special radiator for analytical work has been described. [3]

by a metal shield. If the bath is filled with fine sand, the vessel containing the liquid to be evaporated can be placed on or in the sand.

When evaporation or fuming operations result in the evolution of clouds

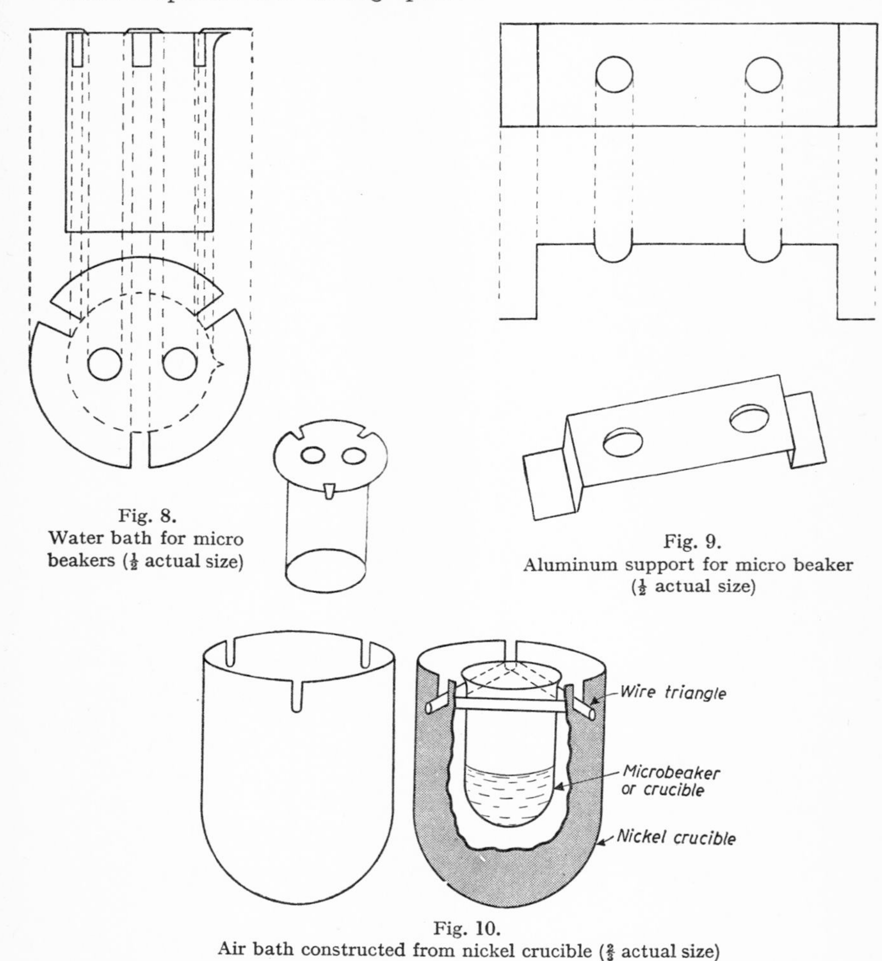

Fig. 8.
Water bath for micro beakers (½ actual size)

Fig. 9.
Aluminum support for micro beaker (½ actual size)

Fig. 10.
Air bath constructed from nickel crucible (⅔ actual size)

of acid vapors or noxious gases, the apparatus should be set up in a hood, or a hood arrangement should be constructed through use of an inverted glass funnel connected to an aspirator pump as shown in Fig. 11. *

Platinum spoons, or silica casseroles and watch glasses are very useful

* Appropriate apparatus are described by Gorbach. [3a]

when evaporating small volumes of liquid samples. Clear silica utensils are particularly useful in this case because, after the evaporation or ignition, the residue can be inspected directly over black glazed paper or other suitable backgrounds.

Fusing of powdered materials can be accomplished very conveniently, in most cases, through use of platinum spoons (Fig. 12) or platinum loops. Fluxes may be collected on platinum loops and then touched to powdered samples and the melt subjected to oxidizing or reducing flames, as desired. Platinum micro crucibles may also be used for fusions, particularly when muffle furnaces are available.

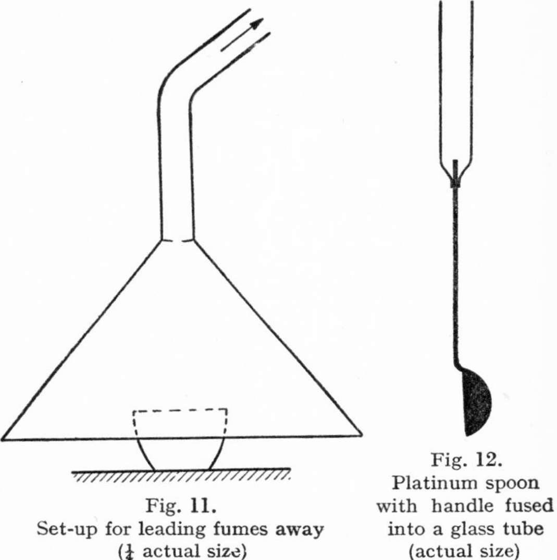

Fig. 11. Set-up for leading fumes away (¼ actual size)

Fig. 12. Platinum spoon with handle fused into a glass tube (actual size)

Addition and control of drops

The adding of a drop of test solution or reagent is basically a very simple operation and can be done in a variety of ways, depending on the circumstances. For the application of drops of reagent, it is generally unnecessary to maintain careful control of drop size. On the other hand, the measurement of the sample drop should, in many cases, be controlled carefully so that the same size drop is taken each time. This is particularly important when semi-quantitative estimates of materials are desired.

The simplest method of taking drops is to use a reagent bottle, such as the Barnes type (Fig. 3), which is stoppered with a dropper pipet. Control of the delivery of liquids can be maintained by means of the rubber cap. When drops of liquid are to be placed on paper or spot plates by means of such pipets, the latter should be held at right angles to the horizontal receiving surface. The pipet tip should be not more than one or two centimeters above the place where the drop is to be delivered. If drops of the test solution and the various reagents are to be placed on filter paper in succession, care must be taken to assure that successive drops all fall as near as possible to the center of the first drop. In a few exceptional cases it is desirable to use adjacent spots so that the reactants diffuse together and the course of the

References p. 56

reaction observed as the mixing occurs. Under any circumstance, it is important that the pipet tip remain free from any possible contamination. Therefore, all drops should be allowed to fall freely and the tip of the pipet must never be touched against the receiving surface.

Dropping bottles equipped with turn caps are also very convenient for the adding of reagents. It is somewhat more difficult to control the amount of reagent added with this type of bottle but, with a little practice, adequate control can be maintained.

Glass rods are very useful in transferring drops of solution or reagent where careful control of drop size is not required. A glass rod 3 millimeters in diameter delivers drops of about 0.05 ml volume, while smaller drops can be delivered by using rods of smaller diameter. The use of glass rods, however, is permissible only for exploratory work because close regulation of drop size is too difficult. If the rod is not wetted sufficiently there is danger that the drops will flow off too slowly and the operator will be tempted to touch the filter paper or other substance with the rod. This results in liquid being sucked off the rod with resultant loss of control of drop size. If rods are used they should be used only once and then placed in a receiving beaker containing a rinse solution.

Platinum loops are very convenient for the transfer of drops. A platinum loop can be fashioned permanently in a fine platinum wire; if desired, the size of the loop can be adjusted to deliver any predetermined size drop. Loops can be calibrated and the volume of sample delivered can be indicated by pasting a label on the handle of the loop holder.

Drops are often delivered by means of pipets of various descriptions. Calibrated capillary pipets are very useful where exact measurement of solution volumes is desired. Transfer pipets are also utilized widely and can be prepared readily in the laboratory. Dropper pipets with rubber bulbs are particularly useful and can be prepared in the laboratory or can be obtained from supply houses or drug stores (Fig. 13). Dropper pipets for accurate delivery of drops can be made by using large capillary tubing and blowing out storage bulbs if necessary. By retaining the flat surface of the tube end, more careful control of drop size is maintained. It is also helpful to coat the outside walls of the pipets in this case with silicone grease so that only the end of the pipet can be wetted.

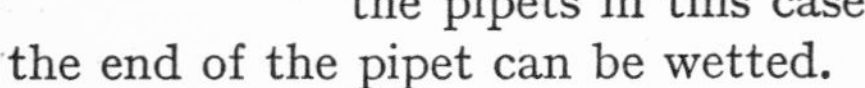

Fig. 13. Dropper pipet ($\frac{1}{2}$ actual size)

An extremely useful device for accurately delivering known amounts of

References p. 56

solution is the micro pipet-buret * described by Gilmont.[4] These pipets can be made to deliver from 0.0001 ml to 1 ml of solution. They have a glass reservoir and delivery tip and the volume of solution delivered is controlled by means of a synthetic ruby plunger, which is passed through a teflon gasket. The control of the volume of solution delivered is maintained by means of a micrometer screw activating the synthetic ruby plunger. A micrometer gauge reads directly in terms of volume delivered. The pipet can be mounted on a ring stand and a solution of the sample to be analyzed can be stored in the pipet and accurate volumes of sample delivered as desired. This device is also of importance in research work where new spot tests are being studied and if it is desired to maintain accurate control of volume of all reagents used in the procedure.

Figure 14 shows a small apparatus for delivering uniform drops of mercury; it consists of a storage vessel and is closed by a glass cock whose plug has a depression in place of the usual bore. The size of the drop is determined by the size of the depression and, consequently, different sized drops can be delivered by varying the depression. One advantage of this "mercury dropper", which can also be used to deliver drops of other liquids, is the tight closing of the storage vessel and the unvarying size of the drops discharged. The vessel is filled by taking out the cock and inserting a shortstem funnel in the delivery tube.

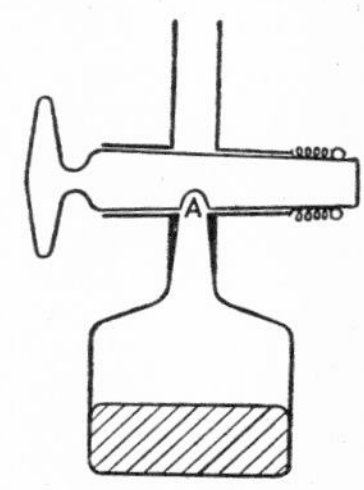

Fig. 14. Apparatus for delivering uniform drops of mercury ($\frac{1}{3}$ actual size)

Separation methods (operations)

While it is true in spot test work that the tests are unitized as much as possible in order to eliminate the tedium of separation procedures, it still remains necessary in many cases to isolate desired materials or to eliminate possible interfering substances through application of separation techniques. The most elegant method of separation is that of masking. In such operations, interfering materials are usually sequestered as soluble complexes or pseudo salts. Such methods have the obvious advantage that no subsequent separation of phases is required and so the separation can become an inherent step in a simplified spot test procedure. No special discussion of technique is required here because the only operation involved is the addition of the necessary conditioning agent.

Precipitation methods remain standard for analytical separations. The more complex the unknown, the greater the likelihood that separations of some sort are desirable or necessary. Particularly in the case of unknown

* These pipet-burets are available from the Emil Greiner Company, New York City.

inorganic mixtures and in the investigation of commercial products is it necessary to use some classification system of separation which will serve to isolate certain groups of substances for subsequent spot test investigation.

When spot reactions are made on paper, either by bringing together two drops or by spotting a reagent paper with a drop of the test solution, any insoluble compounds formed are precipitated directly in the paper, and the unchanged constituents of the solutions undergo capillary diffusion. The latter will then be present throughout the whole spotted area, particularly in the circular zone surrounding the precipitate. Precipitation and filtration are thus accomplished in the surface of the paper. Additional spot tests can be made then on the spotted area, either on the product that has precipitated there, or on the circular zone surrounding the precipitate. Spot reactions on paper not only accomplish a direct precipitation and filtration but also make it possible to purify precipitates by washing. This can be done by placing drops of water, or of a suitable wash liquid, on the center of the spot; the concentric ring around the precipitate is thus extended by capillary diffusion. If the filtrate is of no importance for additional tests, it is better to bathe the spotted paper in an appropriate wash liquid, which can be renewed if necessary. When it is desired to wash a precipitate by repeated treatment with drops of water, each drop should be completely absorbed before the next drop is added.

It is generally better to dry the spots before washing them. This fixes the precipitate more firmly in the capillaries of the paper and there is less likelihood of its being washed away. Spots are dried best and most quickly by a blast of warm air. This localizes the material that has remained in solution and undergone capillary diffusion; the accumulation will be greatest on the side of the paper toward the blast. This is an advantage if further tests by later spot reactions are to be made on the filtrate that has been separated by capillary action.

The testing for dissolved materials that have diffused out of a patch of precipitate should be made by spotting laterally. It is best to place a drop of the appropriate reagent on the dry paper beyond the primary spot. The reagent will spread uniformly from the point of application, and characteristic reaction pictures will be produced at the junction of the two spots. If colored reagents are used in this manner, even slight changes in color are quite apparent.

Solids can often be tested directly to determine their solubility in dilute acids, alkalis, and the like, by spot reactions on paper. A small quantity of the pulverized sample is heaped on a strip of filter paper and spotted with 1 or 2 drops of the solvent. The action is hastened by warming in a current of heated air. Complete solution is obviously established if the sample disappears. Partial solution can be detected by applying suitable reagents near the site of the original reaction. Direct spotting of solids on paper is not limited to the

References p. 56

determination of solubility, but can be used also if soluble colored reaction products are formed by the action of reagents and have then diffused away through the capillaries of the paper. Frequently, characteristic spot reactions can be made directly on white paper through this type of filtration.

The precipitation and filtration following spot reactions on paper may not be applied to all cases, because strongly acidic or alkaline solutions cannot be used, nor is it feasible to subject reaction mixtures to prolonged and intensive heating. Neither is it possible, as a rule, to detect and isolate small quantities of colorless reaction products on paper. Consequently, other means must be employed to separate solid and liquid phases. The choice of the method is determined by the particular needs of the moment.

When considerable quantities of liquid are involved, and if the solid or precipitate is of no further interest, a portion of the liquid can be withdrawn by a pipet for examination. The fine constricted end of the pipet is closed with a wad of cotton drawn out to a point. If the suspension is sucked into the pipet, the liquid which arrives in the tube will be free of precipitate. The pipet will deliver a perfectly clear liquid if the tip is carefully washed after removing the cotton.

A useful filter pipet [5] is shown in Fig. 15. It is constructed of glass tubing (6 mm diameter). A rubber bulb is attached to the short arm *A*; arm *B* is ground flat; arm *C* is drawn out to a fine capillary. A short piece of rubber tubing *D* is fitted over the top of *B*. A disk of filter paper of the same diameter as the outside diameter of the tube is cut from a sheet of filter paper by a sharp cork borer or hand punch and is placed on the flat ground surface of *B*. Tube *F* is placed on the paper, which is held in position by sliding the rubber tubing over it just far enough to hold it when *F* is removed. The filter pipet can be used either by placing a drop of the solution on the filter disk or by immersing *B* into the crucible, test tube, or other container holding the liquid to be filtered. The bulb is squeezed between the thumb and middle finger and the dropper point is closed with the index finger; the solution is thus allowed to pass through the paper when the bulb is released. The pipet is inverted over the spot plate, etc., in an inclined

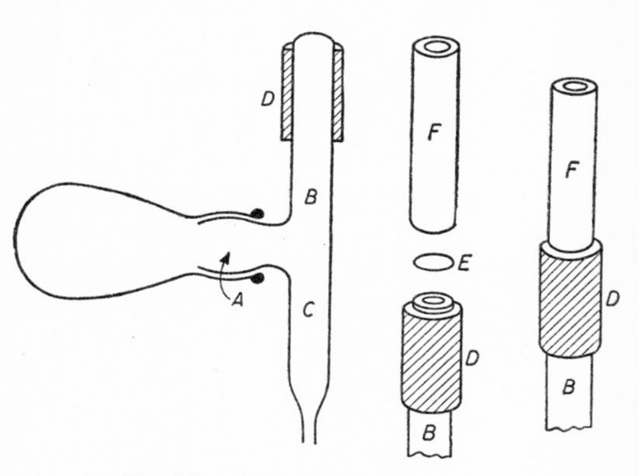

Fig. 15. Filtering pipet (actual size)

References p. 56

position with the bulb uppermost when it is desired to discharge drops of the filtered liquid. The liquid in the tip is forced onto the spot plate, etc., by manipulating the bulb. The precipitate on the paper can be removed for any further treatment by simply sliding the rubber tubing *D* over the arm *B*.

Another method of filtration employs an Emich filter stick, fitted into a heavy wall suction tube by means of a rubber stopper. The suction tube contains a micro test tube to receive the filtrate (Fig. 16). The filter stick contains a small asbestos pad.

Often filtration is not the best method of separating solid and liquid phases. Sedimentation of insoluble materials by centrifuging is sometimes preferable. In addition to greater speed, this procedure has the following advantages: no retention of the mother liquor by the filtering medium; the precipitate, freed from most of its moisture, is compressed into a small volume; the structure of the solid phase (crystalline or amorphous) has no effect on the sharp separation of the phases. The receptacles for centrifuging (centrifuge tubes) can be so chosen for size that the isolation of minute quantities of precipitate or of small volumes of filtrate can be effectively accomplished.

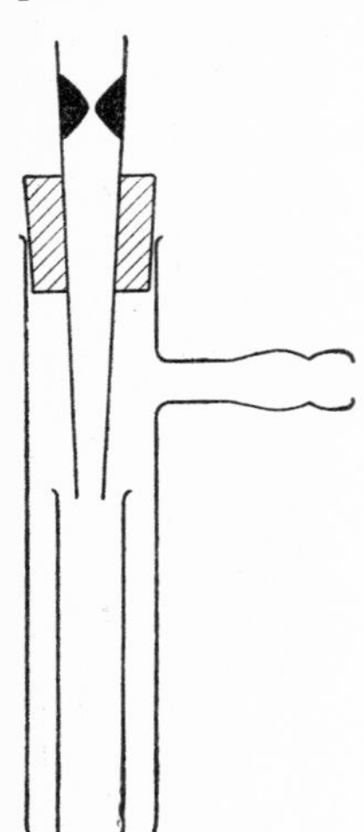

Fig. 16.
Set-up for microfiltration, using filter stick and suction (actual size)

A micro centrifuge tube is shown in Fig. 17, together with a glass support. This arrangement is useful for heating or evaporating on the water bath. A variety of centrifuge tubes with capacities of 0.5 to 3 milliliters should be available.

Centrifuge tubes are conveniently supported on a rack consisting of a wooden block provided with 9 to 12 holes, evenly spaced and 5/8 of an inch in diameter, 1/2 inch deep. Wide selections of micro centrifuges are now on the market. Those which are driven electrically (1500 to 3000 r.p.m.) are preferable to hand-operated centrifuges. The centrifuge should be provided with a metal shield and cover to protect the operator. Dangerous vibration of the instrument is avoided by always loading the carrier equally. This is done by counterbalancing the tube containing the sample by an opposing tube

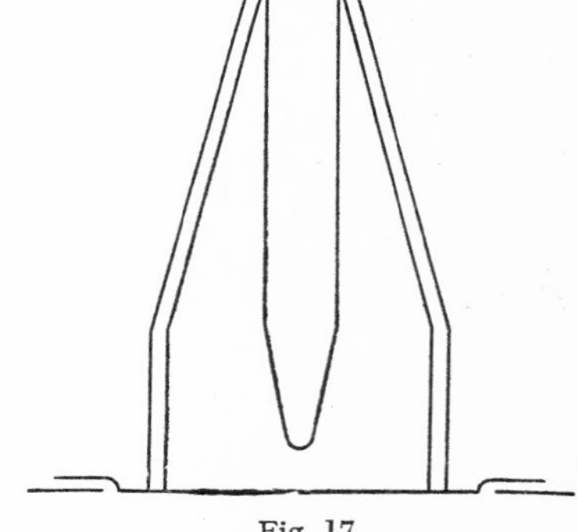

Fig. 17.
Micro centrifuge tube and support ($\frac{2}{3}$ actual size)

References p. 56

containing an equal weight of water or an approximately equal volume of the liquid being centrifuged. The cover of the centrifuge must not be lifted until the rotor has come to rest.

Precipitations are usually made in conical micro centrifuge tubes. The precipitate collects at the bottom of the tube when the suspension is centrifuged. A dropper pipet is usually used to remove the supernatant liquid because the liquid cannot conveniently be poured off directly. A dropper pipet suitable for this operation can be made easily from glass tubing; suggested dimensions are given in Fig. 13. A transfer capillary is convenient for removing the mother liquid or centrifugate, particularly from smaller tubes (0.5 to 2 ml capacity). The pipet is made of glass tubing (internal diameter about 2 mm) which can be drawn from wider

Fig. 18.
Removing supernatant liquid from a centrifuge tube by suction
(½ actual size)

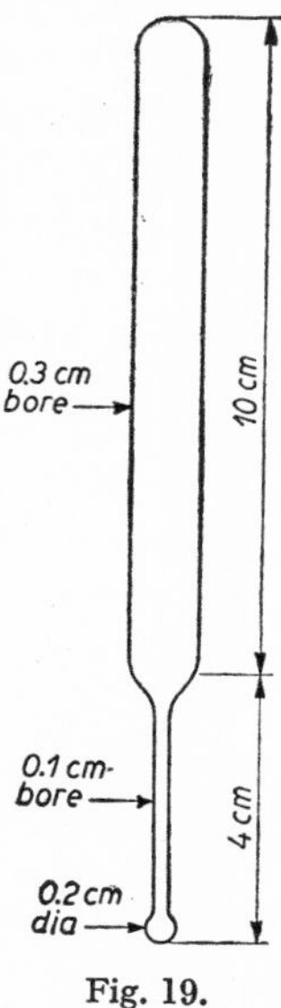

Fig. 19.
Stirrer (glass)
(½ actual size)

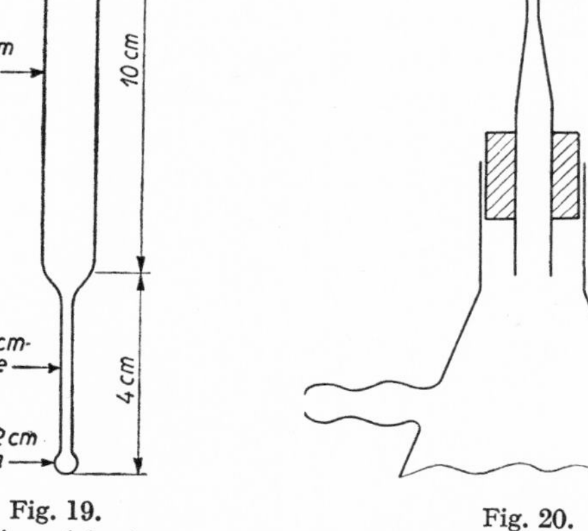

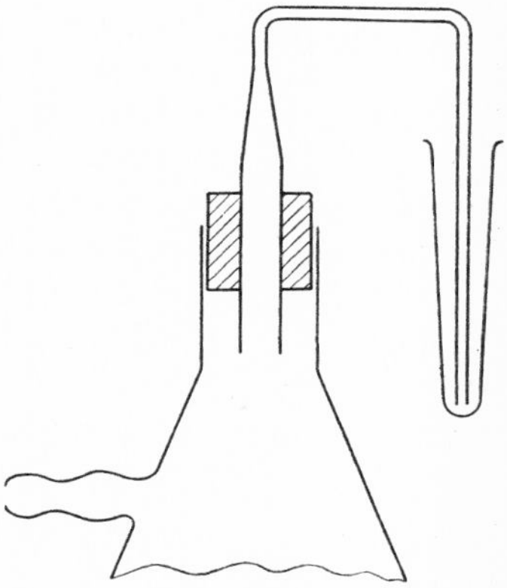

Fig. 20.
Device for withdrawing liquid from a centrifuge tube by suction
(½ actual size)

tubing. The length is 20 to 25 centimeters. One end is drawn to a tip with a fine opening by heating in a micro flame. The correct method of transferring the liquid to the capillary pipet is made evident by Fig. 18. The centrifuge tube is held in the left hand, and the pipet slowly pushed toward the precipitate so that the point of the capillary always remains just below the surface of the liquid. This is continued until almost the entire solution is in the pipet and the tip is about 1 millimeter above the precipitate.

References p. 56

The liquid is drained from the pipet into a clean, dry centrifuge tube.

Precipitates are washed by adding the wash solution directly to the precipitate in the centrifuge tube and stirring thoroughly either with a platinum wire or by means of a stirrer (Fig. 19). This is readily constructed from a glass rod. The suspension is then centrifuged and the clarified liquid removed with the aid of a pipet as just described. This operation may have to be repeated two or three times to insure complete washing.

Centrifuge tubes are cleaned with a feather or a small test tube brush. The tubes are filled then with distilled water and emptied by suction using the device shown in Fig. 20. After the suction has been started and the liquid drawn out, the tube is filled several times with distilled water without removing the suction device between emptyings. Dropper pipets are cleaned by repeated fillings with water; the bulb and tube are finally separated and both rinsed with distilled water from a wash bottle. Transfer capillary pipets are cleaned by blowing a stream of water from a wash bottle through them.

Small quantities of a precipitate can be collected by centrifuging in a micro centrifuge tube, and thus made more visible and accessible to further treatment. This method of separating solid and liquid phases can, therefore, be substituted for filtration in many instances. If the problem is merely the detection of formation of minimal quantities of precipitate that can produce not more than a slight opalescence if the precipitate is colorless, it is frequently necessary to centrifuge for considerable periods to accomplish the separation of the finely dispersed solid phase.

In some instances, a separation can be made quickly by means of flotation. This can be accomplished by shaking the suspension with an organic liquid that is not miscible with water. The surface tension is altered and the fine particles of the solid aggregate and collect as a thin film in the water-organic interface. This method is recommended particularly when it is necessary to detect the formation of a precipitate in a considerable volume of solution after a reagent has been added. The aggregation and localization by flotation or shaking-out succeeds best in neutral and acidic solutions. This treatment with an organic solvent is conveniently done in macro or micro test tubes provided with glass stoppers.

The flotation technique described above finds a number of applications in spot test work. More important is the use of liquid-liquid extractions (and liquid-solid extractions), which promise to become a very important means of analytical separation. Extraction procedures are of particular interest because they offer a means of separation comparable in efficiency to precipitation but still require much less time to perform. In the past the greatest number of applications of extraction techniques was found in the separation of organic compounds. While organic compounds are still very

References p. 56

well adaptable to separation in this way, it is important to note that many inorganic substances are now being separated by liquid-liquid extractions. In some cases chelate salts are formed and are then isolated by extraction; while, in many other instances, inorganic salts themselves are selectively extracted by suitable organic liquids. A number of devices are available for making extractions. Liquid-solid extractions can be accomplished in micro Soxhlet extractors, and liquid-liquid extractions can be carried out in small extraction funnels or in extraction pipets. A very convenient extraction pipet * has been described by Carlton.[6] The pipet consists of a capillary tip 6 cm long, 7 mm outside diameter, and 1.8 mm inside diameter; a bulb blown just above the capillary tip 4 cm long, 1.15 cm outside diameter, and about 2 to 3 ml capacity. An upper stem, 5 cm long, 7 mm outside diameter, and 5 mm inside diameter, is attached to the upper part of the bulb and a rubber bulb of 10 ml capacity is attached to the upper stem. None of the dimensions is critical. When low boiling liquids such as ether, chloroform, carbon tetrachloride, or carbon disulfide are used as extractants, a capillary of about 0.7 or 0.8 mm bore is recommended. Mixing is accomplished by drawing the liquids into the pipet and then expelling them, repeating the procedure several times. By using a rubber bulb of considerably greater capacity than the pipet, a large quantity of air is drawn into the pipet after the liquids have been drawn up; and the bubbling of this air through the two liquid layers provides a very efficient means of mixing. When thorough mixing has been accomplished, the two layers are permitted to separate and then the rubber bulb is squeezed until the lower phase has been removed from the pipet. By means of this device, efficient separations of many substances can be carried out in a matter of 10 to 20 seconds.

Three kinds of manipulation with gases or vapors are used in spot test analysis. Gases (vapors) are employed as auxiliary reagents for precipitation, alkalization, or oxidation of solutions. On the other hand, the liberation of small quantities of gases (vapors) as characteristic products which can be identified by subsequent reactions is the basis of certain tests. Finally, distillation of organic liquids may be an essential step in preparing organic samples for testing. The apparatus required for handling gases (vapors) is determined by the purpose at hand.

Spot reactions on paper involving the action of gases or vapors (H_2S, NH_3, halogens, steam) can be conducted by leading the gas directly from the generator, or by placing a strip of filter paper over the neck of an open flask filled with hydrogen sulfide water, ammonium hydroxide, etc. The steamer (Fig. 21) can be used as a gas generator, if the flask is filled with

* These extraction pipets are commercially available from the E. H. Sargent Company, Chicago, Illinois.

hydrogen sulfide water, bromine water, or ammonia water and the material to be gassed placed on the side arm. Heat is then applied to the flask.

The separation of certain groups of metals by treating the acidic or ammoniacal solution with gaseous hydrogen sulfide is a common step in chemical analyses. In spot test analysis, this precipitation can be accomplished by saturating a small volume of the solution with hydrogen sulfide in a micro centrifuge tube. The hydrogen sulfide is admitted through a fine capillary to prevent loss by spattering. The delivery tube is made by drawing out 6 mm glass tubing to form a capillary of 1 to 2 mm bore and 10 to 20 cm long. A plug of bleached cotton wool is inserted in the wide part of the tubing; then the capillary end is heated in a micro burner and drawn down to a finer tube of 0.3 to 0.5 mm bore and about 10 cm long. Fig. 22 shows the complete arrangement. The fine capillary delivers a stream of tiny bubbles; consequently, the solution does not spatter out of the micro centrifuge tube. The gas must be started through the tube before plunging the end of the capillary into the solution. Otherwise, the solution will rise in the capillary; and when the hydrogen sulfide is admitted, a precipitate will form in the capillary and clog it. The end of the sulfide precipitation can be easily detected through an increase in the size of the rising bubbles. At room temperature, this point is usually reached in about 3 minutes.

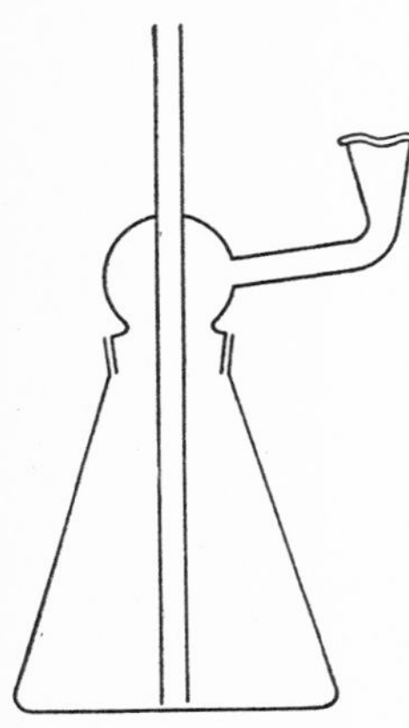

Fig. 21. Apparatus for treating paper with gases or vapors (actual size)

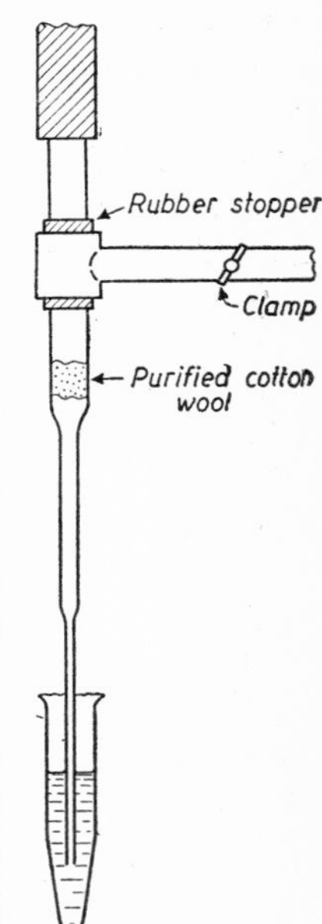

Fig. 22. Set-up for precipitating sulfides by leading in hydrogen sulfide (½ actual size)

An adequate supply of various types of special apparatus of small capacity must be kept on hand. These are required for the liberation of volatile compounds after decomposing small quantities of solid materials or solutions with acids or alkalis. An apparatus[7] designed for the detection of carbonate, sulfide, etc., is shown in Fig. 23. It consists of a micro test tube of about 1 milliliter capacity and can be closed with a small ground-glass stopper fused to a glass knob. The gas is evolved in the tube, aided if necessary by gentle warming, and is absorbed by the reagent. Since the apparatus is closed, no gas escapes; and if enough time is allowed, it is absorbed quantitatively. A drop of water may replace the reagent on the knob.

References p. 56

In this case the gas is dissolved, and the drop may then be washed onto a spot plate or into a micro crucible and treated there with the reagent. The apparatus shown in Fig. 24 is sometimes preferable, particularly when minute quantities of gas are involved. The tube is closed by a rubber stopper and the glass tube, blown into a small bulb at the lower end, may be raised or lowered at will. A change of color, or the presence of reaction products, may be made more distinct by filling the bulb with powdered gypsum or magnesia.* In some cases, it may be desirable to suspend a small strip of reagent paper from a glass hook fused to the stopper (Fig. 25). The apparatus shown in Fig. 26

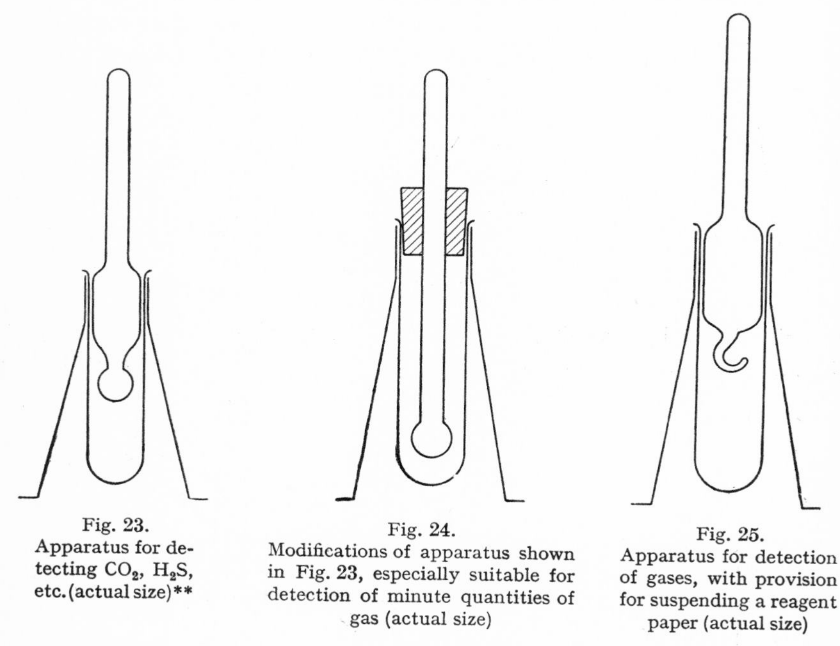

Fig. 23.
Apparatus for detecting CO_2, H_2S, etc. (actual size)**

Fig. 24.
Modifications of apparatus shown in Fig. 23, especially suitable for detection of minute quantities of gas (actual size)

Fig. 25.
Apparatus for detection of gases, with provision for suspending a reagent paper (actual size)

is used when a particular gas is to be identified in the presence of other gases. In this arrangement, the stopper of the micro test tube is a small glass funnel, and the impregnated filter paper is laid across it to absorb the gas. The impregnated paper permits the passage of the indifferent gases and retains only the gas to be detected. The latter forms a nonvolatile compound which can be identified by a subsequent spot test. Another useful apparatus (Fig. 27) consists of a micro test tube containing a loosely fitting glass tube narrowed at both ends. The lower end is filled to a height

* According to a suggestion by H. Kappelmacher (Vienna).

** An original improvement was recently described by Reckendorfer.[8]

of about 1 millimeter with an appropriate reagent solution. If the gas evolved forms a colored product with the reagent, it can be seen easily in the capillary.

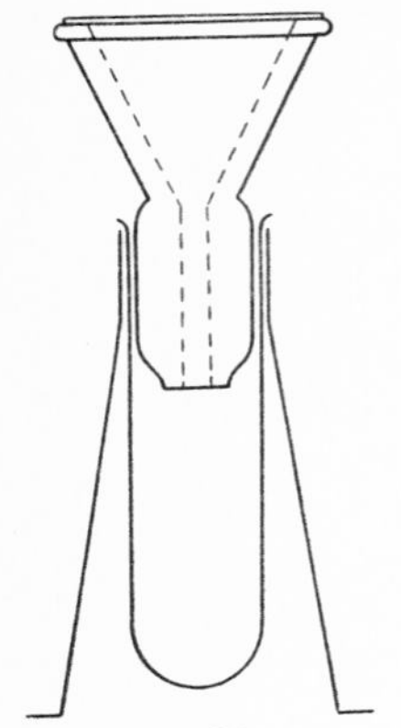

Fig. 26.
Apparatus for detecting a gas in the presence of indifferent gases (actual size)

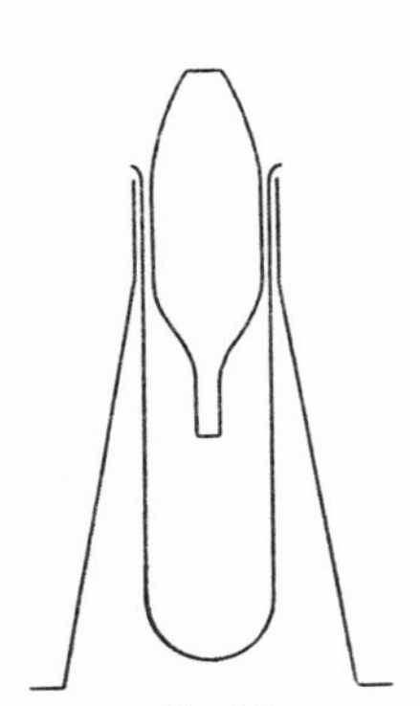

Fig. 27.
Apparatus for detecting a gas that forms a colored product with the reagent solution (actual size)

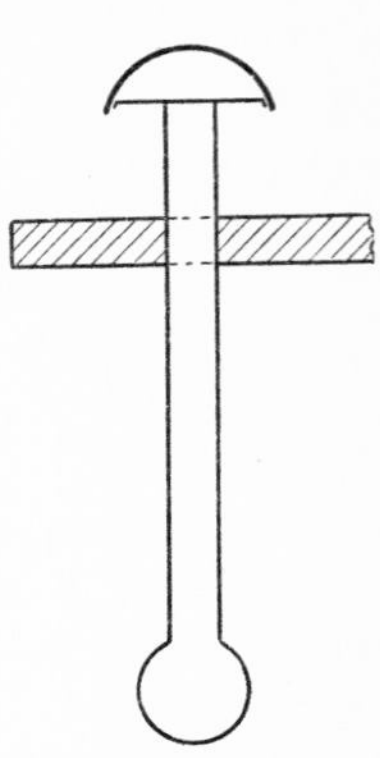

Fig. 28.
Apparatus for detecting a gas whose release requires high temperatures (actual size)

A simple hard glass tube, supported in a circular hole in an asbestos plate (Fig. 28), can be used if high temperature or ignition is required to free the

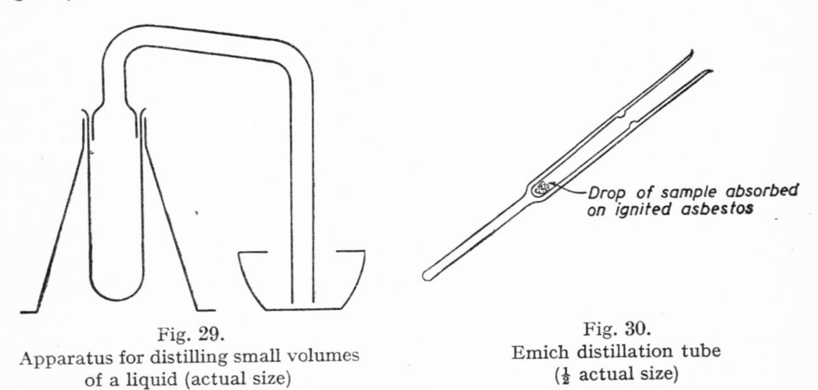

Fig. 29.
Apparatus for distilling small volumes of a liquid (actual size)

Fig. 30.
Emich distillation tube ($\frac{1}{2}$ actual size)

gas. The open end of the tube is covered with a small piece of reagent paper kept in place by a glass cap.

Micro distillation is sometimes required; the chromyl chloride test for chloride is an example. Very small quantities of material can be distilled in the apparatus shown in Fig. 29. A micro crucible or micro centrifuge tube

References p. 56

can be used as the receiver. The apparatus shown in Fig. 30 is satisfactory for most applications calling for distillation. Fractionation can be accomplished by heating the liquid and noting the rise of the vapors in the reflux tube. As the vapors rise they condense as rings of droplets in the tube. Various fractions can be collected as separate samples by adjusting the rate of rise of the condensate rings, using a thumb to close off the upper outlet.

One of the most useful methods of analytical separation is that of chromatography. This technique is applied when a solution contains several substances which can be adsorbed on a given adsorbent. The separation is accomplished by passing the solution through a column of the adsorbent so that the adsorbates will be found in the zones or bands on the adsorption column. The substance which is most strongly adsorbed forms a band at the top of the column. Succeeding bands are situated further down the column their position being determined by the respective strength of their adsorption. If the various constituents separated are colored substances, the respective bands will be colored. It is possible, however, to detect colorless materials by streaking the column after the chromatogram has been developed and the column extruded from the chromatographic tube. Of particular importance is the fact that chromatographic techniques can be applied to the separation of both organic and inorganic substances. Also, small amounts of material can be separated in this manner; and in many cases, separations can be effected that could not be accomplished with other techniques.

Procedures for chromatographic separations cannot be given here. It is generally necessary to evolve specific procedures for given separations; hence the choice of adsorbents, solvents, eluants, etc. will have to be made on the basis of the substances being separated. The general techniques and principles of chromatography have been discussed in a number of authoritative books.[9] In addition to the books on the subject, there are some excellent reviews that should be consulted. D. L. Clegg has reviewed the field of paper chromatography.[10] The annual reviews of analytical chemistry which are published in *Analytical Chemistry* each year, also contain excellent reviews of current developments in chromatography.

Electrochemical methods of separation are often used in spot test analysis. Differential diffusion is sometimes used, particularly in paper chromatography, in which an electric field is utilized to bring about desired separations. More important in spot test work itself is the use of electrographic methods which are utilized in the examination of metals, alloys, and ores. The principle of this separation is based on the application of anodic dissolution of metals. In practice the test substance is used as the anode with aluminum foil serving as the cathode. Filter paper, moistened with the proper reagent, is placed between these two electrodes and the migration of metals

References p. 56

when the proper voltage is impressed across the poles brings about the development of a print locating the exact position of the metals transferred from the surface of the test sample. A general discussion of these methods has been presented by Hermance and Wadlow.[11]

Various types of apparatus are commercially available for electrographic tests. A simple apparatus for electrographic work is shown in Fig. 31. It consists of an aluminum plate as the negative pole on which is laid first a layer of filter paper moistened with potassium chloride solution and then the reagent paper moistened with water or acid. A copper plate with a copper rod soldered to it serves as the positive pole. The current is furnished by a 1.5 volt dry cell and the potential drop across the electrodes can be controlled by a simple rheostat.

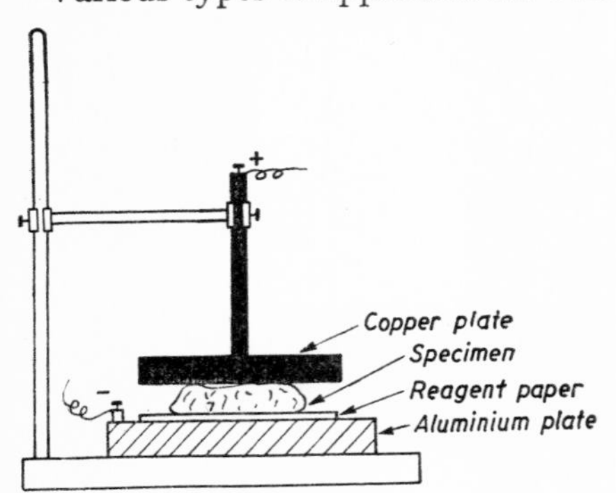

Fig. 31. Electrographic apparatus

4. Special Techniques

The majority of spot test reactions result in the formation of distinctive colors. Quite logically, many of these reactions lend themselves to colorimetric analyses; and it is possible to adapt such color reactions to special techniques having the high sensitivity and simplicity common to spot test procedures in general. Such methods are finding wide usage in industry where an estimate of quantities is desirable and some sacrifices in accuracies can be made in favor of speed, simplicity, and portability. The term "spot colorimetry" has been suggested for this type of technique by Tananaeff.

Spot colorimetry can be performed on spot plates, or the tests can be run on filter paper. The use of confined spots as introduced by H. Yagoda [12] has proved particularly attractive. Yagoda described the process for impregnating suitable spot test paper with paraffin rings so that the area of the spots can be confined to definite limits. Such papers are now commercially available. Spot colorimetry on confined spot test paper has the advantage that very often standard spots can be preserved for use as permanent standards. Confined spot testing can be run with reactions producing colored solutions and, in many cases, it is also applicable for use with colored precipitates.

In addition to the use of spot colorimetry, the tests which produce insoluble reaction products that are colorless or only slightly colored can be

References p. 56

utilized in "spot nephelometry." The nephelometric procedures are performed on black spot plates.

Spot reactions on paper do not always involve the union of a drop of the test solution and one of the reagent. Sometimes filter paper is impregnated with the proper reagent and the dry reagent paper is spotted with a drop of the test solution. This procedure, which assumes, of course, the availability of stable reagents, has the advantage that there is no mutual dilution of the reagent and test solution. A better localization and visibility of the reaction products at the place where the spot has formed is achieved, as compared with the result of bringing two drops together. A still better effect is obtained by impregnating filter paper with reagents which are so slightly soluble in water that no bleeding occurs when a drop of the test solution strikes the paper. Organic reagents which are only slightly soluble in water, but which dissolve readily in alcohol or other organic solvents, have this advantageous characteristic. Slightly soluble compounds, that can be precipitated on paper and in its capillaries by certain chemical reactions, can also be used in this way.

Spotting on reagent paper impregnated with an insoluble compound involves a reaction of dissolved materials with an insoluble reagent. This procedure cannot be used in macro analysis because compact materials, in general, react too slowly. If, however, these same solids are finely divided by precipitation in the capillaries of paper and are thus endowed with an extensive reactive surface, they will undergo chemical changes almost as rapidly as soluble reagents.[13] The localization of characteristic reaction products, with consequent better visibility and increase in the sensitivity of the test, is not the sole advantage of using reagent papers impregnated with insoluble compounds. In many cases, a highly desirable homogenizing and stabilizing can be accomplished by impregnating filter paper with insoluble compounds which then behave like soluble materials. For instance, it is not possible to prepare a good stable alkali sulfide paper; it oxidizes to sulfate too rapidly and furthermore the highly soluble alkali sulfide is washed away when the paper is spotted with an aqueous solution. On the other hand, it is easy to impregnate filter papers with slightly soluble sulfides (ZnS, CdS, Sb_2S_3, etc.). Such papers are stable; each has its maximum sulfide ion concentration (controlled by its solubility product) and hence it precipitates only those metallic sulfides whose solubility products are sufficiently low. Antimony sulfide paper precipitates only silver, copper, or mercury in the presence of lead, cadmium, tin, iron, nickel, cobalt, and zinc. Another striking example is the detection of iron by spotting on paper impregnated with the difficultly soluble white zinc ferrocyanide. In this form the test is far more sensitive than when it is made by uniting drops of a ferric solution and an

References p. 56

alkali ferrocyanide, or by spotting on potassium ferrocyanide paper. The latter also is less stable than zinc ferrocyanide paper. Consequently, if possible, it is always better to impregnate filter paper with "insoluble" reagents than with soluble ones.

It is easy to impregnate filter paper with reagents that are soluble in water or in organic solvents. The proper solutions are prepared in beakers or dishes and the strips of filter paper are bathed in them. Care must be taken that the strips do not cling to the sides of the container, that they do not touch each other or stick together, because this may prevent a uniform impregnation. The immersion should last for twenty to thirty minutes; the solution should be stirred quite frequently, or the vessel swirled, to produce uniformity. The strips are taken from the bath, allowed to drain, pinned to a cord (stretched horizontally) and allowed to dry in the air.

Instead of soaking the strips in the solution, reagents can be sprayed onto filter paper. The atomizing tip shown in Fig. 32 is excellent for this purpose. The impregnating solution is placed in a wide test tube which is then closed with the atomizing head. The paper is held horizontally and the spray expelled by blowing into the apparatus. The paper is sprayed first on one side and then on the other.

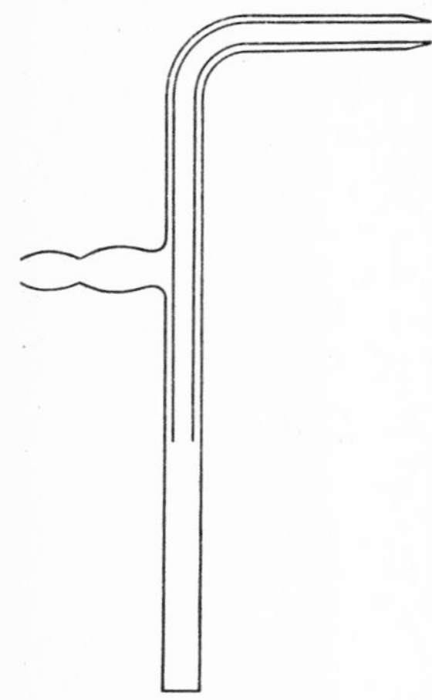

Fig. 32. Atomizer head for spraying reagents ($\frac{1}{2}$ actual size)

Filter paper is impregnated by soaking it in the appropriate solutions or by spraying when it is desired to prepare a stock of reagent papers. The following procedure [14] is recommended for single experiments or when, for special reasons, the spot produced by a drop of a solution must be dried before adding a drop of the other reactant. V-shaped strips of filter paper are spotted on each side, taking care that the spots stay in the center of the strips as nearly as possible. The strips then remain so stiff that they can be stood on the table and allowed to dry. The impregnated strips are cut at the crease before they are used.

Homogeneous impregnation can only be accomplished through gradual, uniform drying on all sides. If paper is soaked with a salt solution and then dried by exposing it to a stream of heated air from the drying apparatus, the rapid evaporation and the subsequent capillary diffusion always leads to an accumulation on the side of the paper turned toward the blast. This effect can be detected immediately in the case of colored reagents, because the color is far less intense on the side of the paper turned away from the blast. The localization of a reagent on one side of the paper is an advantage,

References p. 56

particularly for water-soluble reagents, because it is desirable to have the largest possible quantity of reagent available at the place where the spot is made in order that there will be a rapid and complete reaction with the materials in the test drop.

If strips of filter paper are dried in a blast of hot air, they should either be held on both ends with forceps or laid on a ribbed porcelain plate at such distance from the hot air apparatus that the current of warm air presses the paper against the porcelain plate. The completion of the drying is determined easily because the paper, which while moist adhered to the plate, now rises and flutters in the current of air.

There are no general procedures for impregnating filter paper with compounds produced by precipitation on and in the paper. As a rule, the strips are soaked with the solution of one of the reactants, dried, and then immersed in the solution of the appropriate precipitant. The excess reagents are then removed by washing and the paper dried. If this method is used, the order in which the solutions are applied, as well as their concentrations, makes a difference. The best conditions must always be determined by trial. It is a general rule that filter paper should be plunged quickly and uniformly into the particular reagent solution to avoid the production of zones (layers) of different concentrations. When highly impregnated reagent papers are being prepared, the precipitation of a difficultly soluble reaction product must never be attempted by a single treatment with concentrated solutions. The moistening and precipitation must be carried out separately with dilute solutions, and the reagent paper should be dried between the successive individual precipitations. If this procedure is not followed, a film rather than a homogeneous impregnation will result, and the reagent will come loose from the paper after it is either washed or dried. The excess liquid is removed after each phase of the impregnation by passing the paper through a small wringer at a uniform rate. The paper is best washed by spreading it on an inclined glass or porcelain plate. The spray of distilled water used for washing is distributed by means of a glass tube provided with a number of openings. In case the product precipitated in the paper is not particularly insoluble, it is best to wash slightly with water and then with dilute alcohol. Overheating must be avoided when drying the paper; as a rule, 60° to 80° C suffices. Sometimes it is preferable to use a reagent, if possible, in the gaseous form (hydrogen sulfide for precipitating sulfides or ammonia for precipitating oxides) rather than in solution; there is then no danger of washing away the precipitate. For the same reason, it is often advantageous first to form an adherent compound which cannot be washed off the paper and then carry out the reaction producing the desired reagent. For instance, a good lead sulfide paper is produced—not by soaking the paper in a lead salt solution

References p. 56

followed by treatment with hydrogen sulfide water or gas—but rather by forming zinc sulfide on the paper and converting this into lead sulfide by bathing in a solution of a lead salt. Steigmann [15] has developed an excellent method for impregnating filter paper with water-soluble acid organic reagents.

TABLE 1

RECOMMENDED FILTER PAPERS FOR SPOT TEST USE *

Manufacturer	*Designation*	*Remarks*
Schleicher and Schüll.	595	Fast absorption rate
Schleicher and Schüll.	601 (German)	Special spot test paper Medium absorption rate
Munktell	OK	Medium absorption rate
Whatman.	50	Slow absorption rate
Whatman.	120	Special spot test paper Fast absorption rate

* See Ph. W. West and W. C. Hamilton, *Mikrochem. ver. Mikrochim. Acta*, 38 (1951) 100.

Dannenberg [16] has recommended that inert, powdered, water-soluble or insoluble carriers (silica, starch, sugar, salts, etc.) be moistened with solutions of reagents and then dried. In this way, stable powder mixtures are obtained, which can be used in spot test analysis as dry reagents in place of reagent solutions. They often are better than the latter with respect to stability and economy.

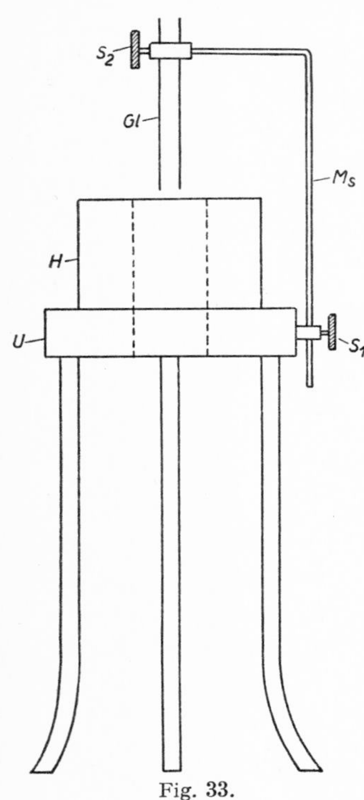

Fig. 33.

Ring oven method

The ring oven method of Weisz [17] can be employed for the separation of ions in one drop. The ring oven (Fig. 33) is an aluminum cylindrical heating block, provided with a central vertical perforation (22 mm inside diameter). One or more test drops are placed in the middle of a round quantitative filter paper by means of the capillary pipette (Fig. 34). One of the groups of materials is precipitated by means of an appropriate reagent and thus fixed in the fleck. The circle of filter paper is then placed on the ring oven in such manner that the fleck is precisely in the center of the hole in the heating block. The excess non-precipitated portions of the sample are then washed away by means of a suitable solvent (water, acid, ammonium hydroxide, alcohol, etc.) applied by a capillary pipette. The narrow glass tube *Gl* shown in Fig. 33 serves as a guide for the washing pipette. Because of the capillary action of the paper, the materials that have gone into solution in the wash liquid migrate outward and form a concentric front until they reach the edge of the opening of the heating block which should be at about 105°. The solvent evaporates there and the solutes are deposited forming a ring zone, 22 mm in diam., with sharply defined edges. A zone that is free of all portions of the sample lies between the initial fleck and the "ring". Consequently, the inner fleck can be easily cut out by means

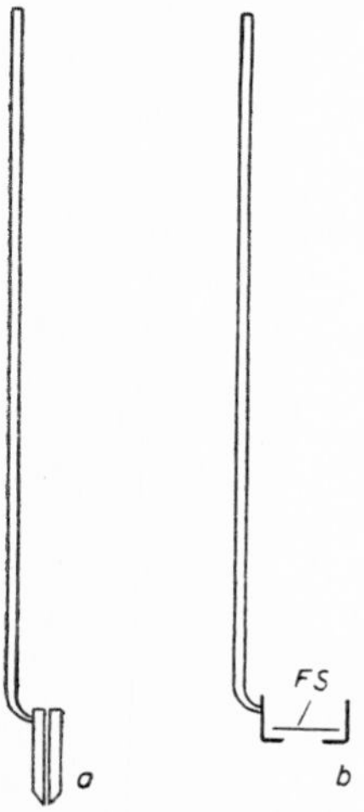

Fig. 34.

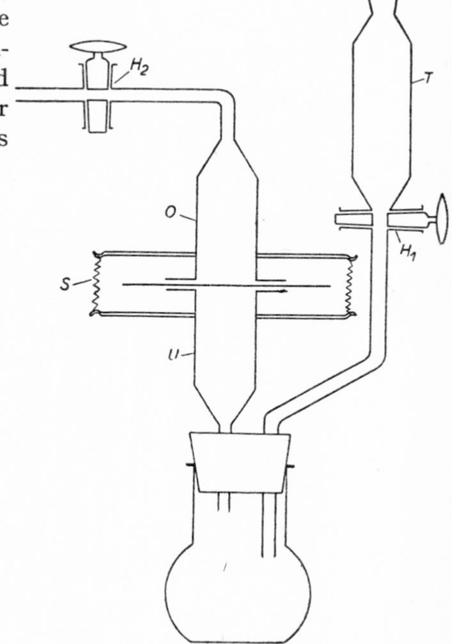

Fig. 35.

of a punch of the proper size. The precipitate on the little disk of filter paper can then be subjected to a proper preliminary treatment, such as oxidation, and then separated further by another precipitation. This is accomplished by placing the disk in a central position on a fresh circle of filter paper and extracting with an appropriate solvent on the ring oven, just as though the disk were simply an ordinary fleck. The glass holder shown in Figure 34b makes it easy to handle the small disk.

The resulting rings are about 0.1 to 0.3 mm in diameter. In other words, their surfaces are smaller than that of the initial fleck and consequently the concentration of the materials which have been washed out are actually

References p. 56

higher in the ring than in the original fleck. The filter paper is cut into several sectors. The individual sectors are then sprayed with suitable reagent solutions and the presence of the various members of the corresponding group is revealed by the resulting sharp circular arches.

The apparatus shown in Fig. 35 lends itself well to precipitations by gaseous reagents.

With the aid of this method there was worked out a separation procedure [18] which includes 14 ionic species, and which can be accomplished with a single drop (1.5 μl) which may contain as little as a few micrograms of solid test material.

The ring oven method is also suited for spot colorimetric analyses. [19, 20] It may also be employed, in combination with electrographic sampling, for the qualitative and quantitative analysis of steels and alloys.[21, 22]

The use of the ring oven technique has also been described for the examination of tiny amounts of radioactive materials with the aid of autoradiography.[23]

REFERENCES

1. Comp. E. J. Conway, *Microdiffusion Analysis and Volumetric Error*, 2nd ed., London, 1947, Part I.
2. T. S. Ma and R. T. E. Schenck, *Mikrochem. ver. Mikrochim. Acta*, 49 (1953) 245.
3. E. Abrahamczik, *Z. anal. Chem.*, 133 (1951) 144.
3a. G. Gorbach, *Microchemisches Praktikum*, Heidelberg, 1956.
4. R. Gilmont, *Anal. Chem.*, 20 (1948) 1109.
5. E. R. Caley, *Ind. Eng. Chem., Anal. Ed.*, 2 (1930) 77.
6. J. K. Carlton, *Anal. Chem.*, 22 (1950) 1072.
7. F. Feigl and P. Krumholz, *Mikrochemie*, 7 (1929) 83.
8. P. Reckendorfer, *Mikrochim. Acta*, (1955) 1008.
9. L. Zechmeister and K. Cholnoky, *Principles and Practice of Chromatography*, New York, 1943; H. H. Strain, *Chromatographic Adsorption Analysis*, New York, 1942; E. Lederer and M. Lederer, *Chromatography, A Review of Principles and Applications*, 2nd ed., Amsterdam, 1956; R. C. Brimley and F. C. Barrett, *Practical Chromatography*, New York, 1953; F. H. Pollard and J. F. W. McOmie, *Chromatographic Methods of Inorganic Analysis*, New York, 1953.
10. D. L. Clegg, *Anal. Chem.*, 22 (1950) 48. Consult also R. J. Block, R. Le Strange and G. Zweig, *Paper Chromatography*, New York, 1952; J. N. Balston and B. E. Talbot, *A Guide to Filter Paper and Cellulose Powder Chromatography*, London, 1952; F. Cramer, *Papierchromatographie*, 2nd ed., Berlin, 1953.
11. H. W. Hermance and H. V. Wadlow, *Electrography and Electro Spot Testing, Physical Methods in Chemical Analysis*, Vol. 2, New York, 1951, pp. 156-228.
12. H. Yagoda, *Ind. Eng. Chem., Anal. Ed.*, 9 (1937) 79.
13. F. Feigl, *Manual of Spot Tests*, Chapter 3, New York, 1943.
14. F. L. Hahn, *Mikrochemie*, 9 (1931) 34.
15. A. Steigmann, *J. Soc. Chem. Ind.*, 64 (1945) 88.
16. E. Dannenberg, *Anal. Chim. Acta*, 8 (1953) 310.
17. H. Weisz, *Mikrochim. Acta* (1954) 140.
18. H. Weisz, *ibid.* (1954) 376.
19. H. Weisz, *ibid.* (1954) 460.
20. H. Weisz, *ibid.* (1954) 785.
21. W. I. Stephen, *ibid.* (1956) 1531.
22. W. R. Nall and R. Scholey, *Metallurgia* (1956) 97.
23. H. Weisz and F. Scott, *Mikrochim. Acta* (1956) 1856.

Chapter 3

Tests for Metals

CATIONS, AND ANIONS OF METALLO ACIDS

In spot test analysis, as in qualitative macro- and semimicroanalysis, the metals are almost always detected by characteristic reactions of their mobile hydrated ions. Consequently, it is assumed that solutions of the material being investigated are available, or that they can be prepared. Since spot tests uniformly employ more sensitive and less disputable reactions, it is often possible to carry out decisive tests with as little as one drop of the dilute test solution, even though considerable quantities of other materials are present. This possibility permits the microchemical accomplishment of many so-called "ultimate" analyses, which demand only a decision as to the presence or absence of one particular substance. The detection of the constituents of a mixture of common metal ions can be successfully achieved, without a preliminary separation, by an intelligent application of specific and selective spot tests. In many cases, however, it is better to make a preliminary separation into definite groups and then to examine each group in turn. Such group separations can be carried out on a semimicro- or microchemical scale. It is then possible to identify the individual members of the groups by spot tests (see Chapter 6).

The sample should invariably be subjected to preliminary tests before beginning any systematic investigation. These tests will give information about the presence or absence of certain materials, and thus make possible a considerable simplification of the systematic analysis. Spot reactions can also be used for these preliminary tests, which will then require only small amounts of the sample (see Chapter 6).

In choosing spot reactions, it must be remembered that tests of varied sensitivities are available for most metals. Therefore, suitable micro-, semimicro- or macrochemical tests can be selected, depending on the problem at hand.

If the operator has had no previous experience with a particular spot test, it is essential that he try out the reaction with pure solutions containing a series of concentrations of the material to be detected. An acquaintance with the pertinent reaction picture is of great value in assessing the reliability of a test and in roughly estimating the quantity of material present. It is necessary to adhere strictly to the conditions prescribed in the procedure, to use reagents of proper concentration, to add them in correct order, etc.

In the following discussion, the cations to be tested for are arranged in groups. The latter are distinguished by the fact that the metals in any particular group are precipitated by a special group reagent. The data relative to the sensitivity of the various tests are valid, unless otherwise stated, for pure salt solutions tested under the conditions given in each case.

A. HYDROGEN SULFIDE GROUP

A 1. Basic Sulfide Group

1. Silver

(1) Test with manganese nitrate and alkali [1]

On mixing manganese and silver solutions with alkalis, a black precipitate [2] is formed; it consists of manganese dioxide and metallic silver:

$$Mn^{+2} + 2\ Ag^{+} + 4\ OH^{-} = MnO_2 + 2\ Ag^{\circ} + 2\ H_2O$$

The reaction takes place in very low silver ion concentration, so that even silver chloride, i.e., the slight amount of silver chloride which is dissolved by water, can be detected by the dark reaction product. The latter is probably not a mixture of metallic silver and manganese dioxide, but rather an adsorption complex. This is evidenced by the fact that its color is much more intense than that of a mere mixture of these materials (see p. 203).

Mercurous salts should not be present, since they also give a dark coloration with alkalis, due to finely divided elementary mercury. Salts of noble metals, stannous salts and mercuric salts react analogously with alkali and manganese salts. Therefore, when such interfering metal ions are present, the silver should be isolated as silver chloride beforehand. This step is also advisable when such metals as give colored hydroxides are present, since the latter reduce the sensitivity of the test.

Procedure. A drop of 0.1 *N* hydrochloric acid is placed on filter paper followed by a drop of the test solution in the middle of the moist fleck, and then a further drop of hydrochloric acid. The fleck is blackened on the addition of a drop of 0.1 *N* manganese nitrate and a drop of 0.1 *N* sodium hydroxide.

Limit of Identification: 2 γ silver
Limit of Dilution: 1 : 25,000

The sensitivity of the test may be increased as follows: One drop of hydrochloric acid is placed on a filter paper, and then a capillary containing about 0.025–0.03 ml of the test solution is touched to the moist fleck in such manner that only a portion of its contents escape to the paper. A second drop of hydrochloric acid is placed on the paper, and the capillary touched to it again. This alternate addition of the test solution and hydrochloric acid is repeated 3 to 5 times. Finally the manganese nitrate and alkali are added. In this way 0.3 γ of silver can be detected in a micro drop (0.025 ml).

(2) *Test with* p-*dimethylaminobenzylidenerhodanine* [3]

In acid solution, rhodanine (I) and similar compounds with an intact NH group form insoluble silver salts in which the silver atom replaces the hydrogen of the cyclic NH (imine) group. [4] The silver salt of *p*-dimethylaminobenzylidenerhodanine (II):

```
HN——CO           AgN——CO
|    |            |    |
SC   CH2          SC   C=CH—⟨C6H4⟩—N(CH3)2
  \S/               \S/
  (I)                      (II)
```

is red violet and is formed, when slightly acidified silver salt solutions are treated with a solution of the reagent in acetone. This selective action of the *p*-dimethylaminobenzylidenerhodanine occurs only in *acid* solution. In *alkaline* solution, owing to a tautomeric change of the radical component:

```
HN——CO           N==COH            N——CO
|    |      to   |    |       or   ‖    |
SC   CH2         SC   CH2         HSC   CH2
  \S/              \S/               \S/
```

it forms OH or SH groups in such quantity that a precipitation reaction with almost all heavy metals is possible. In part, this leads to a decomposition of the rhodanine molecule with formation of a metal sulfide.

In acid solution *p*-dimethylaminobenzylidenerhodanine also reacts with mercury, gold, platinum, palladium, as well as with cuprous salts. The latter, as usual, exhibit an ability to form insoluble compounds similar to the corresponding silver salts. However, under suitable conditions that are easily secured (see page 60), the detection of silver can be made decisively, even in the presence of mercury, gold, platinum and palladium salts.

Procedure. Filter paper (Whatman 120, double thickness) is impregnated with a saturated acetone solution of the reagent,[5] and when dry a drop of the slightly acid test solution is added. According to the silver content, a red-violet precipitate or stain is formed. Even with minute amounts of silver ion a positive reaction is easily distinguished against the yellow-brown color of the reagent, if the unused reagent is removed by bathing the reagent paper in acetone.

Limit of Identification: 0.02 γ silver (in 0.2 N HNO_3)
Limit of Dilution: 1 : 2,500,000

The test may also be carried out in a micro test tube; the unused reagent should be extracted with ether. As little as 0.2 γ Ag (violet flocks of the silver compound) may then be seen under the yellow ethereal layer.*

On a macro scale, this procedure [7] permits the detection of as little as 1 γ silver in 5 ml solution (1 : 5,000,000).

Test in the presence of mercury. In the presence of mercury salts, which also give colored precipitates with *p*-dimethylaminobenzylidenerhodanine, silver can be detected as follows:

Procedure. (*a*) A drop of the test solution is mixed on a spot plate with a drop of a 5 % potassium cyanide solution, whereby soluble $Hg(CN)_2$ and $K[Ag(CN)_2]$ are formed. If a drop of the reagent solution is added and then a drop of 2 N nitric acid, the colored silver *p*-dimethylaminobenzylidenerhodanine appears. The undissociated mercuric cyanide does not react with the reagent.

(*b*) A drop of the test solution is precipitated with the reagent, either on a spot plate or on filter paper, and then a few drops of dilute hydrochloric acid of ammonium chloride are added. The organic mercury compound dissolves to form sparingly dissociated mercuric chloride, while the silver precipitate remains, since the silver rhodanine is less soluble than silver chloride. Method (*b*) is recommended when *copper* as well as mercury is present, since the cuprous cyanide is sufficiently dissociated (when potassium cyanide is used) to react with the reagent to give a red insoluble cuprous salt, which resembles the silver compound.

To detect *silver in a mixture of lead, mercurous, and silver chlorides* as produced in the systematic group analysis, the mixture is treated with a 5 % solution of potassium cyanide, whereby both mercuric cyanide and metallic mercury are formed as well as $K[Ag(CN)_2]$. The mixture is filtered and the clear filtrate is treated on a spot plate with a drop of the acetone solution of the reagent and 2 drops of 2 N nitric acid.

In the presence of silver, a red coloration is formed, which should be compared with a blank, if small amounts of silver are suspected.

Limit of Identification: 0.63 γ silver } in the presence of 1000 times the
Limit of Dilution: 1 : 80,000 } amount of $PbCl_2$ and $HgCl_2$

Detection in the presence of gold, platinum, palladium.[8] Gold, platinum, and palladium salts also give colored precipitates in acid solution with *p*-dimethylaminobenzylidenerhodanine. Nevertheless, (in the absence of copper salts) silver can be detected in the presence of these ions, if the

* This kind of "analytical flotation" renders excellent service in improving the visibility of slight amounts of colored or colorless precipitates. [6]

reaction takes place in solutions containing cyanide ions. A solution of $K[Ag(CN)_2]$, after acidifying with dilute nitric acid, reacts with the reagent; gold, platinum, and palladium salts do not react under these conditions.

Procedure. A drop of the test solution, as slightly acid as possible, is stirred on a spot plate with a drop of 10 % potassium cyanide solution. A drop of the reagent solution is added and 1 : 4 nitric acid then stirred in until the mixture is acid. A pink coloration shows the presence of silver.

Limit of Identification: 1 γ silver
Limit of Dilution: 1 : 50,000
} in the presence of 4000 times the amount of gold

Limit of Identification: 2.5 γ silver
Limit of Dilution: 1 : 20,000
} in the presence of 300 times the amount of palladium

Reagent: 0.03 % solution of *p*-dimethylaminobenzylidenerhodanine in alcohol.

(3) *Test by physical development* [9]

After exposure to actinic light, an image is formed on a photographic plate (or film) on treatment with a developer. The latter is always a reducing substance which reacts with the silver bromide, that has been subjected to the influence of the light (latent image) to give a precipitation of silver. A physical developer,[10] which contains some silver salt as well as the reducing substance, may also be used. Such physical developers can precipitate [11, 12] appreciable amounts of silver itself on traces of silver bromide, which has been exposed to light. Physical development can be used as a spot test for silver in the following way:

Procedure. A circle (about 15 mm diameter) is drawn on a piece of filter paper (S & S 589) to designate the spot. A drop of the very dilute test solution is put in this circle. After the drop has soaked in, the paper is placed in a dish of 0.02 *N* potassium bromide for about half a minute. The bromide solution is then poured off and replaced by distilled water. The dish is kept in motion and the wash water changed every half minute (6 to 8 times) to remove all traces of soluble materials from the dish and paper. It is not necessary to expose the paper expressly, because the silver bromide retained in the fibers of the paper is attacked by the physical developer, even though the halide has not been photolyzed. After the last wash water is poured away, the developer is poured into the dish. In a few minutes, or as long as half an hour, according to the silver content of the test solution, a grey fleck of metallic silver, which turns blacker in time, develops within the circle. The developer, which is clear and colorless at first, becomes violet in time with separation of colloidal silver, and is finally quite cloudy. Usually, the action of the developer is complete before it becomes cloudy. However, in very great dilutions it may happen that the developer becomes cloudy before the fleck is distinct enough. In this case, the paper is removed, thoroughly washed, and placed in a fresh developer. The greatest cleanliness is essential for the success of this test.

If the filter paper is dirty or contains foreign bodies, black flecks may appear in the developer, due to the formation of silver on these impurities. However, as these flecks do not have the outline of a drop, and appear elsewhere than in the circle, they do not interfere with the correct interpretation of the test.

It is important that all the potassium bromide be washed out of the paper before development. Otherwise, any bromide that remains will combine with silver from the developer, and the resulting silver bromide, in its turn, will cause the deposition of silver. Consequently, the entire surface of the paper will be blackened.

Limit of Identification: 0.005 γ silver
Limit of Dilution: 1 : 10,000,000
Reagent: Developer: 500 ml water, 10 g metol (monomethyl-*p*-amidophenol), 50 g citric acid; immediately before use, 2 ml of a 0.1 *M* silver nitrate solution is poured into 50 ml of the metol solution.*

Silver sulfide [13] can also be used as a reduction nucleus, provided other metals precipitable by hydrogen sulfide and nitric acid are absent.

Procedure. One drop of the test solution is placed on filter paper and exposed to a hydrogen sulfide atmosphere for about 30 seconds. The spot is thoroughly washed and then developed.

Limit of Identification: 0.005 γ silver
Limit of Dilution: 1 : 10,000,000
Reagent: The stock solution contains 2 g pyrogallol and 2 g citric acid in 500 ml water. Sixty ml is treated with 2 ml 0.1 *N* silver nitrate shortly before using.

(4) *Test by catalytic reduction of Mn^{III} and Ce^{IV} salts* [14]

Solutions of manganese ^{III} and manganese ^{IV} salts are quite stable towards 2.5 *N* hydrochloric acid at room temperature, and liberate little or no chlorine. On the addition of silver nitrate, these brown solutions rapidly lose their color with evolution of chlorine.[15] The silver chloride, formed by the hydrochloric acid, acts as a catalyst in the reaction between the higher manganese oxides and hydrochloric acid. Ceric salts in hydrochloric acid solutions behave similarly; when alone, their yellow color fades very slowly with development of chlorine, but on the addition of silver nitrate the reaction is instantaneous. The fact that silver chloride has a catalytic effect

* The developer prepared as described will keep for a limited time only, owing to the formation of traces of nuclei of silver. The stability can be appreciably increased by the addition of 1 part per 1000 of gum arabic. (*Personal communication* from Dr. G. Schwarz, Antwerp.)

References pp. 247–257

only in the presence of chloride ions is due to an activation of hydrochloric acid by silver chloride. The compound $H[AgCl_2]$, which may be formed under these conditions, probably takes part in the reaction.[16] This catalyzed liberation of chlorine is effected by such small amounts of silver that the reaction may be used as a test for silver. It is carried out most conveniently as a spot test.

Procedure. In two adjoining depressions of a spot plate are placed 3 drops of the reagent solution (Mn^{III} or Ce^{IV} solution) and 2 drops of dilute hydrochloric acid. In one depression is placed a drop of the test solution and in the other a drop of water. The reagent is decolorized more or less quickly according to the amount of silver present.

Limit of Identification: 0.4 γ silver
Limit of Dilution: 1 : 120,000 } using Mn^{III} salts

Limit of Identification: 0.05 γ silver
Limit of Dilution: 1 : 1,000,000 } using Ce^{IV} salts

Reagents: Manganic Solution. 0.6 g manganese sulfate is dissolved in 60 ml water and 20 ml concentrated hydrochloric acid. Ten ml of a 0.1 *N* $KMnO_4$ solution is added, and the mixture well shaken. 15 ml of this solution is diluted with 50 ml of 1 : 2 hydrochloric acid.

Ceric Solution. 0.25 g cerium ammonium nitrate is mixed with 10 ml dilute nitric acid and the solution diluted to 100 ml with water.

The specific and sensitive catalytic test for silver can also be used to detect the latter in the mixture of precipitates formed by hydrochloric acid in the course of the systematic qualitative scheme. The precipitate, possibly containing chlorides of Ag, Hg, Pb, Tl, is washed with hot water, and a small portion heated in a micro crucible to weak glowing, to remove the Hg_2Cl_2 and TlCl. The residue in the crucible is treated, after cooling, with 1 to 2 drops of the Mn^{III} or the Ce^{IV} solution.[17]

Quantities of silver that are not too small can be detected in the chloride precipitate of the systematic scheme of analysis. The precipitate is stirred with ammonia water and 20 % KOH solution, and then formalin is added. Free silver is precipitated.[18]

(5) *Other tests for silver*

(*a*) On treating an acetic acid solution of a silver salt with K_2CrO_4, the silver may be recognized by the red Ag_2CrO_4 formed (*Idn. Limit:* 2 γ Ag). It is advisable to use the filtrate after precipitation with $(NH_4)_2CO_3$.

(*b*) By the reduction of ammoniacal silver solutions with $SnCl_2$, 1 γ Ag can be detected.[19]

(*c*) With a solution of dithizone in carbon tetrachloride [20] a violet precipitate, insoluble in CCl_4, is formed from neutral solutions of silver salts

(*Idn. Limit:* 0.05 γ Ag). Dithizone reacts with other metals, so special precautions must be taken when any of these are present.

(*d*) A black stain is formed on treating silver salts with stannous chloride and chromotropic acid.[21] With chromotropic acid alone, silver salts give a white precipitate, that rapidly turns brown (*Idn. Limit:* 0.1 mg Ag). [22] Free acids and Hg salts interfere with this reaction.

(*e*) Acid solutions of silver salts, spotted on a paste of CuCNS, form a smoke brown color.[23]

(*f*) When a drop of a silver solution is treated with a drop of a solution of phenothiazine, there results a grey coloration which changes toward blue-green (*Idn. Limit:* 0.54 γ Ag). [24]

(*g*) The acetone solution of *p*-dimethylaminobenzylidene thiobarbituric acid reacts with silver salts in a manner analogous to that of *p*-dimethylaminobenzylidenerhodanine (Test 2). The sensitivity as a spot reaction [25] is the same (*Idn. Limit:* 0.2 γ Ag). This test is subject to the same interference by Ag, Hg, and Pd salts.

(*h*) A solution containing complexone (ethylenediaminetetraacetic acid), $FeSO_4$, and sodium acetate reduces silver salts to elementary silver in the form of a black or grey precipitate (*Idn. Limit:* 0.05 γ Ag).[26]

(*i*) A red color results when a drop of the ammoniacal test solution is heated with a 1 % solution of $K_4Fe(CN)_6$ containing *α,α'*-dipyridyl (*Idn. Limit:* 2 γ Ag). [27] Only palladium and mercury react analogously (compare ferrocyanide test page 288).

(*j*) A red fleck is produced if a drop of the neutral test solution is placed on filter paper moistened with $K_2[Ni(CN)_4]$ solution containing dimethylglyoxime (*Idn. Limit:* 0.5 γ Ag). [28]

2. Mercury *

(*1*) *Test with diphenylcarbazone* [29,30]

Mercury salts, in acid solution, give violet to blue insoluble inner complex salts with diphenylcarbohydrazide (I) and its oxidation product diphenylcarbazone (II). The salt of the carbazone has the structure (III).[31]

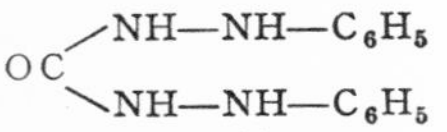

$$OC\begin{matrix}\diagup NH{-}NH{-}C_6H_5\\ \diagdown NH{-}NH{-}C_6H_5\end{matrix}\quad (I) \qquad OC\begin{matrix}\diagup N{=}N{-}C_6H_5\\ \diagdown NH{-}NH{-}C_6H_5\end{matrix}\quad (II) \qquad OC\begin{matrix}\diagup N{=}N(C_6H_5)\\ \diagdown NH{-}N(C_6H_5)\end{matrix}{>}Hg/2\quad (III)$$

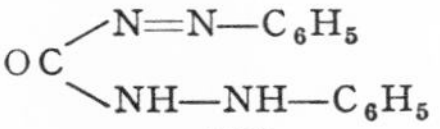

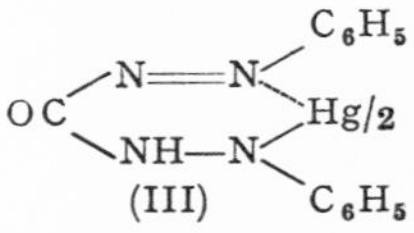

The sensitivity of the mercury test with diphenylcarbazone depends on the acidity of the test solution, and decreases with increasing acidity. In neutral or very weakly acid solutions, other heavy metals (Cu, Fe, Co and

* See page 280 concerning the detection of mercuric cyanide.

others) also give colored compounds with diphenylcarbazone [32] *. In 0.2 N nitric acid solution, the diphenylcarbazone reaction is specific for mercury, in the absence of chromates and molybdates (which give colored compounds under the same conditions, see page 167). The interference of chromates is prevented by reducing them, by means of sulfurous acid or hydrogen peroxide, to nonreacting chromic salts. Molybdates may be masked with oxalic acid, since the complex molybdenum-oxalic acid does not react with diphenylcarbazone.

The presence of chlorides affects the test through the formation of slightly dissociated $HgCl_2$ or $[HgCl_4]^{-2}$ ions. For the same reason the diphenylcarbazone test is less sensitive when applied to a $HgCl_2$ solution.

Procedure. A drop of the test solution is placed on a filter paper impregnated with a freshly prepared 1 % alcoholic solution of diphenylcarbazone. [33] According to the concentration of mercury, a violet or blue fleck appears.

Limit of Identification:	0.1 γ mercury	Without added nitric acid $Hg(NO_3)_2$ solution
	0.2 γ mercury	0.1 N nitric acid $Hg(NO_3)_2$ solution
	1.0 γ mercury	0.2 N nitric acid $Hg(NO_3)_2$ solution
Limit of Dilution:	1 : 500,000	
	1 : 250,000	
	1 : 50,000	

(2) *Test with cuprous iodide* [34]

Suspensions of white cuprous iodide, or filter paper impregnated with this compound, develop a deep red to orange color on contact with acidified solutions of mercury salts, the color depending on the quantity of mercury present. The reaction:

$$2\,Cu_2I_2 + Hg^{+2} \rightarrow Cu_2[HgI_4] + Cu_2^{+2}$$

is the basis of a quite selective test for bivalent copper, since cosolutes, which interfere by reacting with Cu_2I_2, can be made harmless by simple procedures.

A double decomposition occurs only with palladium salts:

$$Cu_2I_2 + Pd^{+2} \rightarrow PdI_2 + Cu_2^{+2}$$

The resulting palladous iodide blackens the Cu_2I_2. If desired, it is possible to remove the palladium, by precipitation with dimethylglyoxime in acid solution. Small amounts of palladium do not interfere with the detection of mercury in quantities above 0.1 γ, when a drop of the test solution is

* R. Alexejeff, *Chem. Abstracts*, 32 (1938) 451, states that many interfering metal ions are masked by adding alkali pyrophosphate to the solution at pH 7 to 7.2.

placed on filter paper impregnated with Cu_2I_2. Fractional precipitation then produces a black circle (PdI_2) surrounded by a red ring ($Cu_2[HgI_4]$).

Since cuprous iodide is a reductant, consideration must be given to the presence of oxidants. Salts of silver, mercury^I, gold, and platinum react with Cu_2I_2 and deposit the particular metal as a black coating. It is easy to remove Hg_2^{+2} and Ag^+ ions by precipitation with hydrochloric acid and filtration. Gold and platinum salts are inactivated by adding sodium bisulfite, which precipitates metallic gold and masks (decolorizes) platinum through the formation of $[Pt(SO_3)_3]^{-2}$ ions. Molybdates and tungstates react with cuprous iodide to produce lower oxides (molybdenum- and tungsten blue). This interference can be averted by adding sodium fluoride (formation of $[MoO_3F_2]^{-2}$ or $[WO_3F_2]^{-2}$ ions. If tungstate is present, its preliminary (incomplete) precipitation with hydrochloric acid is recommended. Oxidation of cuprous iodide by ferric, niobic, and ceric salts can be prevented by masking with phosphoric acid or sodium fluoride, or by reducing the ceric ions with sodium bisulfite. These procedures for removing interfering substances should precede the test with cuprous iodide. The consequent decrease in sensitivity is about 10 per cent.

Procedure [35]. The solution to be tested should be 1 *N* with respect to hydrochloric or nitric acid. One drop (0.05 ml) of potassium iodide-sodium sulfite solution is placed on a spot plate or a filter paper (S & S 598), followed by a drop of copper sulfate solution and finally, by means of a capillary, a drop of the test solution is added. Depending on the quantity of mercury present, a red or orange color appears.

Limit of Identification: 0.003 γ mercury (0.03 ml)

Reagents: *1)* Potassium iodide-sodium sulfite solution: 5 g KI and 20 g $Na_2SO_3 \cdot 7H_2O$ in 100 ml water

2) Copper sulfate solution: 5 g $CuSO_4 \cdot 5H_2O$ in 100 ml 1 *N* hydrochloric acid

(*3*) *Test with stannous chloride and aniline* [36]

Mercuric salts are reduced to metallic mercury by stannous chloride in the presence of aniline (compare page 111). A grey or black fleck is formed on paper. The specificity of the mercury test is affected by large amounts of silver salts, which are reduced by stannous chloride and aniline to give metallic silver. Bismuth salts, however, are without effect; they are reduced only by alkaline stannite solutions.

Procedure. A drop of the test solution is placed on test paper followed by a drop of freshly prepared stannous chloride solution and a drop of aniline. A black to brown color indicates mercury.

Limit of Identification: 1 γ mercury

Limit of Dilution: 1 : 50,000

(4) *Test with* p-*dimethylaminobenzylidenerhodanine* [37]

Mercury salts give a red-violet precipitate in weakly acid solutions with an alcohol or acetone solution of *p*-dimethylaminobenzylidenerhodanine (see page 59). Large amounts of chlorides and also free hydrochloric acid hinder the formation of the mercury rhodanine compound, because they reduce the normal slight dissociation of mercury chloride, and also lead to the formation of complex $[HgCl_4]^{-2}$ ions. However, in acetate-buffered solutions, the mercury gives a red color with the reagent even in the presence of excess chlorides. In hydrochloric acid solutions, buffered with acetate, the intensity of the red coloration bears no relation to the concentration of mercury and a relatively weak color develops even in fairly concentrated solutions. Obviously, only a portion of the mercury reacts with the rhodanine in the presence of chlorides.

Test in the absence of chlorides and large amounts of free acids

Procedure. A drop of the weak nitric or sulfuric acid test solution (the acidity should not be greater than 0.1 *N*) is mixed on a spot plate with a drop of reagent solution. According to the amount of mercury, a violet precipitate or pink color is formed. The reagent itself is colorless or faintly yellow in acid solution.

Limit of Identification: 0.33 γ mercury
Dilution Limit: 1 : 150,000

Test in the presence of chlorides and any amount of free acids

Procedure. A drop of the acid test solution is mixed on a spot plate with a drop of reagent and a few drops of saturated sodium acetate solution. In the presence of mercury there is a pink coloration. Since the reagent itself turns yellow to orange on the addition of acetate, a parallel test should be carried out with dilute hydrochloric or nitric acid.

Limit of Identification: 0.33 γ mercury
Limit of Dilution: 1 : 150,000
Reagent: Saturated alcoholic solution of *p*-dimethylaminobenzylidenerhodanine

Silver salts give a red-violet precipitate with the reagent under the same conditions (see page 59) and must be removed by precipitation with hydrochloric acid before the test. Copper salts give a dirty red coloration only in solutions that are neutral or strongly buffered with acetate. Under suitable conditions, mercury can be detected in the presence of even considerable concentrations of copper.

Test in the presence of copper

When an acid copper solution is treated with an excess of sodium phosphate, insoluble light green copper phosphate is formed, which reacts but slowly with rhodanine. The addition of sodium phosphate also increases the pH so much that mercury gives a red color with the rhodanine, which is clearly visible even in the presence of the green copper phosphate.

Procedure. A drop of the acid test solution (not more than 1 *N* acidity) is mixed on a spot plate with 5 drops of 10 % tertiary sodium phosphate solution and 1 drop of rhodanine solution. In the presence of mercury a violet to pink color is formed. When very small amounts of mercury are present, a blank test should be carried out, and the result observed at once.

Limit of Identification: 1 γ mercury } in the presence of 450 times the
Limit of Dilution: 1 : 50,000 } amount of copper

Reagent: Saturated alcoholic solution of *p*-dimethylaminobenzylidene-rhodanine

(5) Test by catalytic acceleration of the reduction of tin^{IV} salts [38]

Tin^{IV} salts are reduced to tin^{II} salts by hypophosphites:

$$2\ Sn^{+4} + H_2PO_2^- + 2\ H_2O = 2\ Sn^{+2} + PO_4^{-3} + 6\ H^+ \quad (1)$$

This reduction is slow and incomplete, but it can be remarkably accelerated[39] by mercury^{II} salts, which also react with hypophosphorous acid. Reaction (2) induces the reaction (1):

$$2\ Hg^{+2} + H_2PO_2^- + 2\ H_2O = 2\ Hg^{\circ} + PO_4^{-3} + 6\ H^+ \quad (2)$$

This induction effect, in which the production of mercurous ions probably plays an important role, occurs even with very small amounts of mercuric salts, so that the mercury can be detected by the formation of tin^{II} salts, which are then identified by the color reaction with cacotheline (see page 109). The test succeeds even with a solution of mercury cyanide, or of complex alkali mercury iodide, which yield exceedingly low concentrations of mercuric ions.

Salts of noble metals must be absent since they are reduced by hypophosphite to the finely divided, intensely colored metal and may hide the Sn^{+2}-cacotheline reaction. Silver salts and, to a greater extent, osmium salts, reduce the catalytic action of mercury.[40] Cupric salts are reduced by hypophosphite to cuprous salts and in this state react with cacotheline in the same way as Sn^{II} salts.

Procedure. A drop of the test solution is mixed with a drop of the reagent solution in a micro test tube, which is kept for about 15 seconds in a boiling water bath. According to the amount of mercury present, the yellow solution turns red or light pink. A blank should be carried out when very small amounts of mercury are suspected.

Limit of Identification: 0.1 γ mercury

Limit of Dilution: 1 : 500,000

Reagent: 10 g ammonium chlorostannate is dissolved in 80 ml 2 *N* hydrochloric acid and filtered after 24 hours. This solution is mixed with 50 ml of 5 % sodium hypophosphite solution and 8 ml of 1 % aqueous cacotheline solution [41] then added.

(6) *Test by activation of aluminum*

In contrast to the usual non-corrodibility of this metal, aluminum-amalgam (which is easily formed by placing mercury or mercury salts in contact with aluminum) is quickly converted into alumina in moist air.[42] The mercury thus liberated can amalgamate fresh aluminum, and the process then repeats itself, so that small amounts of mercury, e.g. mercury electrolytically precipitated on aluminum, can be detected by the formation of alumina.[43] Mercury from even one drop of solution can easily be electrically precipitated, on to aluminum foil. The small amounts of alumina formed can then be identified by the formation of a lake with alizarin (see page 185). Minimum amounts of mercury may be detected in this way.[44]

To use the alizarin test, the electrolytic precipitation of mercury must take place from neutral solution, since acid or alkaline solutions attack metallic aluminum sufficiently to give a strong reaction with alizarin. When the test solution is neutral, the mercury can be precipitated directly on an aluminum cathode, and the alizarin test carried out *in situ*. If, however, the test solution is acid or alkaline, the mercury must be deposited on a platinum wire cathode which, after washing with water, is connected as anode against a cathode of aluminum foil. On passing the current, the mercury dissolves from the anode and is redeposited on the aluminum and forms an amalgam.

Procedure. When mercury is to be detected in neutral solution, a drop is placed in a stamped depression in a piece of aluminum foil that has been roughened by sand blasting. A microdrop of dilute sodium sulfate solution is added as conductor and a drop of quinone solution as depolarizer. Then the tip of the platinum wire, which is used as anode, is immersed (the aluminum foil being the cathode), and the current from a flashlight battery is passed through for 15 minutes (see Fig. 36). The solution is then rinsed off the aluminum foil with water and the moistened portion dried with a clean cloth. After 5 minutes the alumina formed is moistened with a drop of an acetic acid solution of alizarin, allowed to stand for 3 minutes, rinsed with water, and again dried with the clean cloth; another drop of alizarin solution is placed on the spot, left for three minutes and then rinsed ten times in distilled water. If mercury is present, red dots of the aluminum lake of alizarin stick to the aluminum foil. For small amounts of mercury, a blank test must be carried out with distilled water. A red color is formed with alizarin, but it disappears on washing or drying.

Limit of Identification: 0.001 γ mercury

Limit of Dilution: 1 : 50,000,000

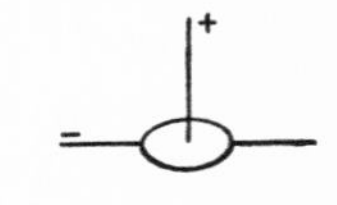

Fig. 36. Aluminum foil for electrolytic detection of mercury ($\frac{2}{3}$ actual size)

If an acid solution is to be examined, a drop should be placed in the apparatus shown in Figure 37 and electrolyzed for 15 minutes, with a platinum wire as cathode. The mercury collects on the wire. The latter is then washed and used as anode against the aluminum foil as cathode. On passing the current, the mercury goes anodically into solution and is deposited on the aluminum cathode, and can be detected there as described.

Limit of Identification: 0.007 γ mercury

Reagents: *1)* Sodium alizarin sulfonate: 0.2 g dissolved in 100 ml water and 0.5 ml acetic acid added

2) Quinone solution: cold saturated aqueous solution

3) Sodium sulfate: 0.02 g dissolved in 100 ml water

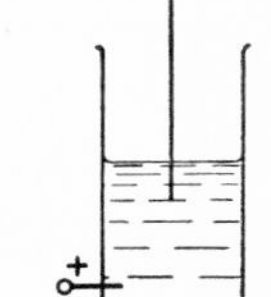

Fig. 37. Set-up for electrolytic deposition of mercury ($\frac{2}{3}$ actual size)

The preceding test can be recommended when minimal quantities of mercury are to be detected and, consequently, the tedious preparation and testing of suitable aluminum foil will seem to be justified. A shorter but less sensitive test is conducted as follows [45]: Aluminum foil is roughened with emery paper, washed with water, and etched with concentrated alkali until a uniform evolution of hydrogen occurs. After rinsing with water and rubbing with a cloth moistened with dilute alkali, the foil is dried. A drop of the test solution is placed on the prepared aluminum, allowed to react for 3 minutes, and the liquid is then taken up with filter paper. If mercury is present, white growths appear on the foil after about 5 minutes (*Idn. Limit:* 0.1 γ Hg). Considerable quantities of acids, or of metals that form a precipitate with alkalis, interfere with the test.

(7) *Test by catalytic decomposition of ferrocyanide* [46]

The action of Hg^{+2} ions on ferrocyanide ions in neutral or slightly ammoniacal solution causes the demasking of bivalent iron by removal of CN^- ions. A test for ferrocyanide is based on this effect (see page 288). The removal of CN^- ions occurs gradually. The first partial reaction leads to pentacyanoaquoferroate ions according to:

$$[(FeCN)_6]^{-4} + Hg^{+2} + H_2O \rightarrow [(FeCN)_5H_2O]^{-3} + CNHg^+ \qquad (1)$$

This is proved by the appearance of a violet color when organic nitroso compounds are present. This color reaction is known from qualitative organic analysis, where alkali pentacyanoaquoferroates are used as reagents for the identification of nitroso compounds (comp. Vol. II, Chap. 4):

$$[(FeCN)_5H_2O]^{-3} + RNO \rightarrow [(FeCN)_5RNO]^{-3} + H_2O \quad (2)$$

It is obvious, that (2) stops further removal of cyanide ions from the complex pentacyanoaquoferroate by mercury ions. It is assumed [47], that the $CNHg^+$ ions formed according to (1) are decomposed in acid solution:

$$CNHg^+ + H^+ \rightarrow Hg^{2+} + HCN \quad (3)$$

The mercury ions, thus liberated, enter newly into reaction (1) forming pentacyanoaquoferroate ions, followed by reaction (2). Mercury ions act, therefore, as catalysts, since the addition of (1), (2) and (3) leads to reaction (4):

$$[(FeCN)_6]^{-4} + RNO + H^+ \rightarrow [(FeCN)_5RNO]^{-3} + HCN \quad (4)$$

where the catalyst, namely the mercury ions, does not appear.

Procedure. A drop of a 9 : 1 mixture of saturated nitrosobenzene-water solution and 1 *M* acetate buffer (pH 4.1) is placed on a spot plate or in a micro test tube, then a drop of the test solution is added, followed by one drop of 0.2 % $K_4Fe(CN)_6$ solution. At higher concentrations of mercury, the violet color appears very rapidly even at room temperature. When lower mercury concentrations are dealt with, heating accelerates the reaction. When carrying out the test in a micro test tube, and heating it, the experiment takes a little over 2 minutes. On a porcelain spot plate the test must be warmed for about 10 minutes; because nitrosobenzene is a volatile substance, a drop of its saturated water solution must be added every two minutes. For very small amounts of mercury a blank test is recommended.

Limit of Identification: 0.002 γ Hg
Limit of Dilution: 1 : 25,000,000

It seems that nitrosobenzene can be substituted by other nitroso compounds, i.e. the water soluble nitroso R salt.

In view of their strong affinity to cyanide ions, gold, silver and palladium ions also behave analogously to mercury ions.

Fe^{+3}, Cu^{+2} and UO_2^{+2} ions interfere due to the formation of colored precipitates with ferrocyanide. Among the anions, only iodide ions interfere.

(*8*) *Other tests for mercury*

(*a*) Diphenylthiocarbazone (dithizone) forms an orange mercury salt, soluble in chloroform (*Idn. Limit:* 0.25 γ Hg). Since numerous other metals likewise form complex salts with dithizone, certain precautions are needed to make this mercury test indisputable.[48] A good method of isolating Hg^I and Hg^{II} is to treat the test solution with trichloroacetic acid and oxalic acid and then to extract the system with *n*-butyl acetate. The extract containing mercury is transferred to a spot plate and a drop of 0.002 % solution of dithizone in chloroform is added (*Idn. Limit:* 1 γ Hg in 1 ml).[49]

(*b*) Mercuric and mercurous salts may be detected by the formation of a brown or yellow precipitate with chromotropic acid.[50] The test is carried out on filter paper (*Idn. Limit:* 100 γ Hg). Acids and silver salts interfere with this reaction.

(*c*) A red color results if a drop of the test solution is heated with a drop of a 1 % ammoniacal solution of potassium ferrocyanide containing α,α'-dipyridyl (*Idn. Limit:* 0.5 γ Hg).[51] Silver and palladium react analogously (compare ferrocyanide test page 288) because they also demask ferrocyanide to yield ferrous ions.

3. Lead

(*1*) *Test with benzidine* [52]

Lead peroxide, in common with a number of other oxidizing agents, can oxidize benzidine (I) to "benzidine blue" (II). The latter is a meriquinoid oxidation product [53] consisting of one molecule of the amine, one molecule of the imine, and two equivalents of a monobasic acid:

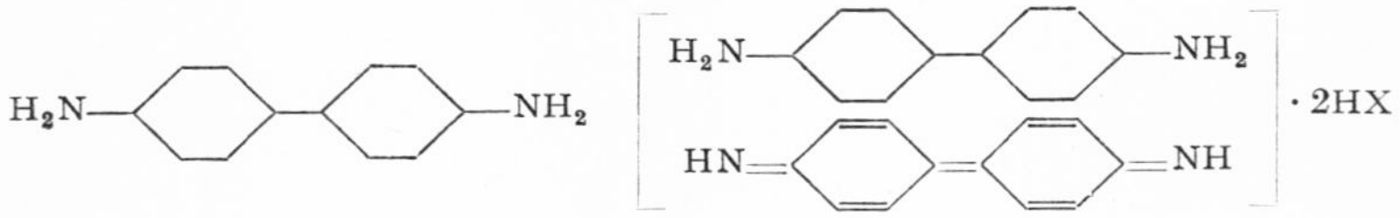

H_2N—⟨⟩—⟨⟩—NH_2

(I)

[H_2N—⟨⟩—⟨⟩—NH_2 / HN=⟨⟩=⟨⟩=NH] · 2HX

(II)

Alkali and bromine water (alkali hypobromite) are recommended for the conversion of the lead salt to PbO_2, because the excess hypobromite can be instantaneously destroyed by means of ammonia:

$$Pb^{+2} + 2\,OH^- + BrO^- \rightarrow PbO_2 + Br^- + H_2O$$
$$3\,BrO^- + 2\,NH_3 \rightarrow 3\,Br^- + N_2 + 3\,H_2O$$

This test for lead [54] is interfered with by cerium, manganese, bismuth, cobalt, nickel, silver and thallium salts. Under the same conditions, these metal ions also give higher oxides capable of oxidizing benzidine to benzidine blue. If, however, the lead test is carried out in an alkaline extract (plumbite solution), only thallium offers interference. All the other metal ions are precipitated as hydrous metal oxides and so do not enter the alkaline extract. When only bismuth salts are present, it is sufficient, before the addition of bromine water, simply to heat with the alkali. BiO(OH) is formed and it is not converted into the higher bismuth oxide by hypobromite.

Procedure. A drop of the test solution is placed on filter paper and treated successively with 2 *N* alkali and saturated bromine water. After the last drop has soaked into the paper, 2 drops of a 1 : 1 solution of ammonia are added, allowed to spread, and the excess ammonia removed by waving the paper over a small flame. The solution of benzidine in acetic acid is then added and, according to the amount of lead present, a light to deep blue fleck is formed. The blue color fades if only very small amounts of lead are present.

Limit of Identification: 1 γ lead
Limit of Dilution: 1 : 50,000

Lead can be detected in even greater dilutions by the following *procedure:* The test solution is treated with 3 ml of 3 *N* alkali and 2 ml of bromine water, boiled a few times, and filtered through quantitative filter paper. After washing with hot ammonia (1 : 1) followed by hot water, a drop of an acetic acid solution of benzidine is placed on the paper. In the presence of 10γ of lead in 10 ml test solution, which implies a dilution of 1 : 1,000,000, there is an appreciable blue coloration.

(2) *Test with sodium rhodizonate* [55]

The yellow aqueous solution of sodium rhodizonate (I) produces colored precipitates of basic lead rhodizonates from neutral or slightly acid lead solutions. Violet $Pb(C_6O_6)\cdot Pb(OH)_2\cdot H_2O$ is thrown down in neutral solution; scarlet-red 2 $Pb(C_6O_6)\cdot Pb(OH)_2\cdot H_2O$ is produced from weakly acid solutions. The deep color of these lead salts probably arises from the fact that the anionic component contains five double bonds. This hypothesis regarding the color intensity is supported by the fact that solid anhydrous alkali rhodizonates are deep brown.

(I) 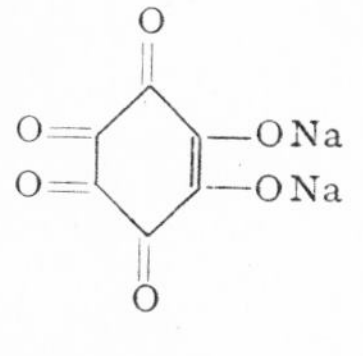 (II)

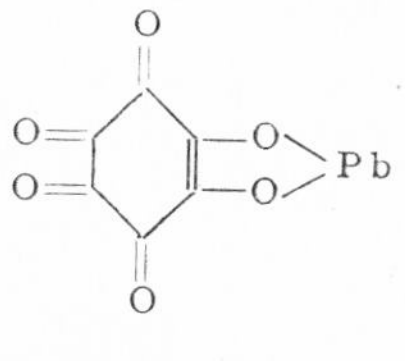

Both reactions are so sensitive that positive results are given immediately by even such slightly soluble materials as PbS, $PbSO_4$, $PbCrO_4$, etc.

Procedure. A drop of the test solution is placed on filter paper. After the liquid has been soaked up, the spot is touched with a drop of freshly prepared 0.2 % solution of sodium rhodizonate. A blue fleck or ring is formed if lead is present. When an intense reaction occurs, the blue fleck can be transformed to scarlet by spotting with a drop of buffer solution (pH = 2.8) containing 1.9 g sodium bitartrate and 1.5 g tartaric acid per 100 ml.

Limit of Identification: 0.1 γ lead
Limit of Dilution: 1 : 500,000

References pp. 247–257

Detection of lead in the presence of other metals

Sodium rhodizonate, at *p*H ~ 3, also forms colored compounds with Tl^{+}, Ag^{+}, Cd^{+2}, Ba^{+2} and Sn^{+2}. The sensitivity of some of these reactions is less than that with Pb^{+2}.

Lead can be identified easily in the ordinary qualitative scheme, where its chloride may be present along with the chlorides of silver, univalent mercury, and thallium. The mixed precipitate produced by hydrochloric acid is not washed, but is transferred directly to a crucible, dried by gentle warming, and then carefully heated to redness. Thallous and mercurous chloride volatilize. The cold residue is digested with 4 drops of strong ammonia water to dissolve any silver chloride. The contents of the crucible are then evaporated to dryness. Three drops of the buffer solution and one drop of sodium rhodizonate are added. If the original precipitate contained lead, a red precipitate or coloration will appear. It is necessary to dissolve the silver chloride because it melts and encloses lead chloride, which may thus be shielded from the action of the sodium rhodizonate.

This procedure, which can be used to detect lead in the presence of all of the other metals, is far more sensitive than the customary method of extracting the lead chloride in hot water and then adding a suitable reagent. The limiting proportion was determined in about 5 mg of the mixed chlorides; the ratio of Pb : Ag = 1 : 5000.

(3) Test with dithizone (diphenylthiocarbazone) [56]

Dithizone (I) precipitates the red inner complex lead salt (II) from neutral solution.

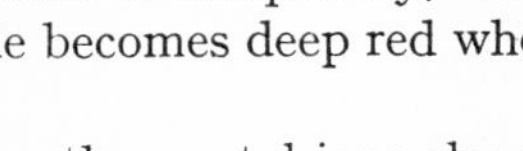

$$S{=}C\begin{cases}NH{-}NH{-}C_6H_5\\ N{=}N{-}C_6H_5\end{cases}$$

(I)

$$S{=}C\begin{cases}NH{-}N(C_6H_5){\searrow}\\ N{=}N(C_6H_5)\cdots{\nearrow}\end{cases}Pb/2$$

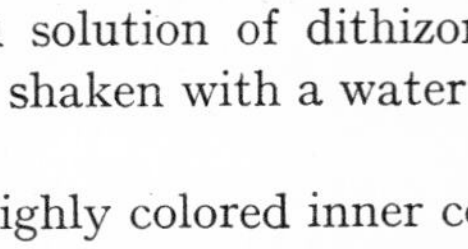

(II)

The precipitation also occurs from solutions containing alkali tartrate or cyanide. The lead salt is soluble in carbon tetrachloride and many other organic liquids. Consequently, the green solution of dithizone in carbon tetrachloride becomes deep red when it is shaken with a water solution of a lead salt.

Numerous other metal ions also form highly colored inner complex salts, which are soluble in organic liquids. However, if the conditions given here are observed, the test for lead is specific, although the sensitivity is less.

Procedure. A drop of the test solution is vigorously shaken in a small test tube with a drop of a carbon tetrachloride solution of dithizone. The green reagent assumes a brick-red color that is easily seen against a white background.

Limit of Identification: 0.04 γ lead (in neutral solution)
Limit of Dilution: 1 : 1,250,000
Reagent: A solution of 1 to 2 mg dithizone in 100 ml carbon tetrachloride

Sometimes the solution does not keep because the glass container furnishes traces of lead.

The fact that the reaction between lead salts and dithizone takes place despite the presence of considerable amounts of potassium cyanide and sodium potassium tartrate renders the test for lead specific among the heavy metals which are "masked" by the addition of these salts. Table 2 gives the limits of sensitivity and concentration limits to be achieved under the varying conditions.

TABLE 2

Foreign metal salt used	Limit of identification, γ in 0.05 ml	Foreign metal, γ in 0.05 ml	Potassium cyanide, γ in 0.05 ml	Other salts added, in 0.05 ml	Proportion limit, Pb : foreign metal
$AgNO_3$	0.2	11,600	10	—	1:58,000
$CuSO_4$·aq	0.1	1,800	9	—	1:18,000
$NiSO_4$·aq	0.1	1,600	10	—	1:16,000
$ZnSO_4$·aq	0.1	1,175	6	2 mg NH_4Cl	1:11,750
$Cd(C_2H_3O_2)$·aq . . .	0.1	1,600	10	2 mg NH_4Cl	1:16,000
K antimonyl tartrate.	0.1	930	2.5	5 mg Rochelle salt	1: 9,300

(*4*) *Other tests for lead*

(*a*) If a drop of dilute sulfuric acid and a drop of the test solution are placed successively on filter paper, the resulting fleck of $PbSO_4$ can be washed with dilute sulfuric acid and water. If it is then spotted with a solution of $Cd[SnI_4]$ containing iodide, the $PbSO_4$ is converted into orange-red 2 $PbI_2 \cdot SnI_2$ (*Idn. Limit:* 10 γ lead). [57, 58]

(*b*) Lead in neutral solutions, or as the hydroxide, can be detected by the formation of a violet lake color on the addition of a 1 % solution of the blue oxazine dyestuff, gallocyanine (*Idn. Limit:* 0.3 γ Pb).[59] When other metals are present, $PbSO_4$ should be precipitated on the paper before the treatment with gallocyanine. In this way 2–3 γ Pb may be detected in the presence of 10 to 15 times the amount of the foreign metal (Ag, Bi, Cu, Cd).

(*c*) Carminic acid (hydroxyanthraquinone derivative of complex structure) also forms a violet lake color with neutral solutions of lead salts, or with $Pb(OH)_2$, in the presence of ammonia (*Idn. Limit:* 1 γ Pb).[59] Under certain conditions this test for lead can also be used in the presence of small amounts of metals of the H_2S group.

4. Bismuth

(1) Test with cinchonine and potassium iodide [60]

Numerous, mostly univalent, organic bases, among them cinchonine, form insoluble (and some colored) double iodides of the general formula $BiI_3 \cdot B \cdot HI$ (B = base molecule) with weakly acid solutions of bismuth salts and potassium iodide. These double iodides can be regarded as salts of the particular bases with iodobismuthic acid $H[BiI_4]$. The cinchonine double iodide is orange-red.

Procedure.[61] Filter paper is impregnated with the cinchonine solution, and a drop of the slightly acid test solution is placed on it. An orange-red fleck indicates the presence of bismuth.

Limit of Identification: 0.14 γ bismuth

Limit of Dilution: 1 : 350,000

Reagent: Cinchonine solution: 1 g cinchonine is dissolved by warming in 100 ml water containing a little nitric acid. After cooling, 2 g potassium iodide is added.

Test for bismuth in the presence of other metals

Owing to the iodide content of the reagent solution, the bismuth test is affected by the presence of cupric, lead and mercury salts. Nevertheless, within certain limits, the test for bismuth may be carried out in the presence of these interfering metal ions, since they diffuse at different rates through capillaries of the paper, and are fixed in different zones. However, a more potent influence than the diffusion velocities is exerted by the fact that the products of the reaction of Hg^{+2}, Pb^{+2}, and Cu^{+2} with the I^- ions contained in the reagent are formed at different rates.

When a drop of test solution containing bismuth, copper, mercury and lead is placed on filter paper impregnated with the reagent, four different zones can be observed:

1) a white central ring, which contains the mercury,
2) an orange-colored ring of the bismuth compound,
3) a yellow ring of lead iodide,
4) a brown ring of iodine, liberated by cupric ions.

These rings or zones are wider or narrower according to the concentration of the metals.

The sensitivities of the bismuth test on filter paper, using capillary separation from the other metals, are:

a) Bismuth in the presence of lead

Limit of Identification: 12 γ bismuth

Proportion Limit: Bi : Pb = 1 : 112

b) Bismuth in the presence of copper
Limit of Identification: 12 γ bismuth
Proportion Limit: Bi : Cu = 1 : 168

c) Bismuth in the presence of mercury
Limit of Identification: 15 γ bismuth
Proportion Limit: Bi : Hg = 1 : 100

d) Bismuth in the presence of cadmium
Cadmium does not affect the test.

Bismuth in the presence of copper, lead, and mercury

Limit of Identification: 10 γ bismuth
Proportion Limit: Bi : Cu : Pb : Hg = 1 : 84 : 53 : 30

(2) *Test with alkaline stannite solution with the addition of lead salts* [62,63]

Bismuth salts and bismuth hydroxide are reduced to metallic bismuth by alkaline solutions of sodium or potassium stannite [64]:

$$2\ Bi(OH)_3 + 3\ Na_2SnO_2 = 2\ Bi^\circ + 3\ Na_2SnO_3 + 3\ H_2O \qquad (1)$$

This redox reaction (1) proceeds so rapidly that if a drop of 0.01 % bismuth solution is mixed on a spot plate with a drop of alkaline stannite solution, there is an immediate precipitation of metallic bismuth in the form of black flocks. When the same experiment is made with 1 % lead acetate solution, after 3–10 minutes standing there is a slight reduction to lead, which is revealed by a light brown coloration. This proves that the redox reaction:

$$Pb(OH)_2 + Na_2SnO_2 \rightarrow Pb^\circ + Na_2SnO_3 + H_2O \qquad (2)$$

proceeds only slowly and incompletely. However, reaction (2) is enormously accelerated by (1), and all the lead is reduced if the conditions are proper.* A visible effect can be produced by amounts of bismuth that are too small to be detected by the usual stannite reduction.

The induced reduction of lead through the reduction of bismuth is due probably to the fact that the reduction $Bi^{III} \longrightarrow Bi^\circ$ does not proceed directly. Lower oxides, perhaps Bi_2O, are formed as intermediates: $Bi^{III} \longrightarrow Bi^{I} \longrightarrow Bi^\circ$. The univalent bismuth may react with the lead $Pb^{II} + Bi^{I} \longrightarrow Bi^{III} + Pb^\circ$. Thus Bi^{III} is regenerated, is again rapidly reduced by stannite to Bi^{I}, which again can act on Pb^{II}, and so on. This hypothesis is supported by the fact that the bismuth reduction is capable of inducing other reductions accomplished with stannite.[65] The catalytic

* If for instance 1 ml of a 1 % $Bi(NO_3)_3$ solution is poured into 25 ml of 1.0 *N* $Pb(C_2H_3O_2)_2$ solution, and then an excess of alkaline stannite added, lead separates immediately. After a few minutes the filtrate gives no reaction for lead on acidifying with nitric acid and testing with sulfuric acid. A filtrate containing large amounts of lead is obtained if the reaction is carried out with no addition of bismuth nitrate.

References pp. 247–257

acceleration of the lead-stannite reaction by bismuth thus makes possible its sensitive detection in the absence of noble metals, copper, and mercury. The interference of the last two may be prevented as described below.

Procedure. A drop of the hydrochloric acid test solution, a drop of a saturated solution of lead chloride, and 2 drops of stannite solution are stirred on a spot plate. In the presence of large amounts of bismuth, a precipitate of lead appears at once. Smaller amounts require 1 to 3 minutes before a definite brown appears, which gradually intensifies until the lead is completely precipitated. Since lead alone is also reduced, although slowly, a blank test with a drop of hydrochloric acid, lead chloride solution, and 2 drops of stannite solution should be carried out, if small amounts of bismuth are suspected.

Limit of Identification: 0.01 γ bismuth
Limit of Dilution: 1 : 5,000,000
Reagent: *a*) 25 % sodium hydroxide solution
b) A solution of 5 g stannous chloride in 5 ml conc. hydrochloric acid, diluted to 100 ml with water.
The stannite solution is made up shortly before use by mixing equal volumes of *a*) and *b*).*

Test for bismuth in the presence of copper

Copper salts are slowly reduced by stannite solutions with deposition of red cuprous oxide, and thus interfere with the detection of small amounts of bismuth. The addition of potassium cyanide converts the copper ions into $K_4[Cu_2(CN)_6]$, which is quite stable to stannite. Consequently, even very small amounts of bismuth may then be detected in the presence of much copper.

Procedure. A drop of the test solution is stirred on a spot plate with a drop of a saturated solution of lead chloride, a drop of 2 *N* sodium hydroxide, a drop of 5 % potassium cyanide, and finally 2 drops of stannite solution. According to the bismuth content, a brown to black coloration appears either immediately or after a few minutes.

As little as 0.1 γ bismuth can be clearly recognized in one drop in the presence of 1 mg copper, i.e., in the ratio of 1 : 10,000, if a comparison test is run with a drop of 1 % copper solution.

Test for bismuth in the presence of mercury

Since stannite solution reduces mercury salts, and the free mercury makes it impossible to see the reduction of lead induced by bismuth, the mercury must be removed prior to the stannite test. A drop of the test solution is evaporated to dryness in a porcelain microcrucible, and then carefully ignited to volatilize the mercury. The residue is taken up in a drop of *N*

* A freshly prepared alkaline stannite solution is essential. On long standing, tin is often precipitated from alkaline stannite solutions, the stannite being partly oxidized, partly reduced: $2\ Na_2SnO_2 + H_2O = Na_2SnO_3 + 2\ NaOH + Sn°$.

hydrochloric acid and tested for bismuth as just directed. In this way, 0.05 γ bismuth can be detected in the presence of 0.5 mg mercury, which represents a ratio of 1 : 10,000.

(3) Test by luminescence of the hydrogen flame [66]

Bismuth gives a cornflower-blue coloration to the hydrogen flame. Extremely small amounts of bismuth can be detected when the sample, mixed with alkaline earth carbonate, is placed in the flame. Among the common metals, only antimony and manganese compounds give analogous flame reactions, the former green-blue, the latter yellow (see page 107). On the other hand, all the rare earth oxides or carbonates when mixed with calcium carbonate, or precipitated together with $CaCO_3$ (as collector) from very dilute solution, likewise exhibit a luminescence.[67] The colors shown are: Y, pale blue-violet; La, brick red; Ce, yellow-green; Pr, red; Nd, orange-red; Sm, yellowish green; Dy, pale green; Tm, yellowish green.

Procedure. The hydrogen is passed through two wash bottles containing water. After testing for freedom from the explosive admixture of oxygen, the gas is lit on the tip of a platinum wire or porcelain spatula. The flame should not be more than 0.5 cm long. Pure calcium carbonate and water are mixed to a thin paste, and a small amount is taken either with a loop of platinum wire, a Wedekind magnesia rod, or a narrow strip of mica.

The calcium carbonate is gently ignited in the flame, then the not more than slightly acid test solution added by means of a second platinum loop, and the ignition repeated. After cooling, the preparation is repeatedly placed in the lower part of the hydrogen flame for a short time. In the presence of bismuth, a cornflower-blue color is perceptible at the moment of lighting up of the flame. As soon as the lime begins to glow, the luminescence disappears, because it is masked by the yellow glow.

Limit of Identification: 0.004 γ bismuth

Limit of Dilution: 1 : 12,500,000

Reagent: Pure calcium carbonate: A solution of calcium nitrate is treated with ferric nitrate, precipitated with ammonia water, and filtered. Extremely pure $CaCO_3$ is thrown down from the filtrate on addition of ammonium carbonate. The traces of rare earths, which are present in many calcium salts, are removed by the precipitation of $Fe(OH)_3$.

(4) Test with potassium chromithiocyanate [68]

Bismuth salts, in solutions of mineral acids, react with potassium chromithiocyanate $K_3[Cr(CNS)_6]$:

$$Bi^{+3} + [Cr(CNS)_6]^{-3} = Bi[Cr(CNS)_6]$$

to give insoluble brick-red bismuth chromithiocyanate.

Only mercury, silver, thallium, and lead salts also give precipitates under these conditions, but the products are pink or yellow.

Procedure. A tiny drop (0.012 ml) of the test solution is placed on filter paper and dried over a warm asbestos plate, and the residue treated with a drop of an alcohol solution of the reagent. After drying, a drop of 1 : 2 sulfuric acid is added. According to the bismuth content, a brick-red fleck or ring is formed.

Limit of Identification: 0.4 γ bismuth
Limit of Dilution: 1 : 31,000
Reagent: 3 % alcoholic solution of potassium chromithiocyanate [69]

(*5*) *Other tests for bismuth*

(*a*) The stannite reduction of bismuth solutions containing lead (see page 77) can also be carried out in lead-free solutions, however with reduced sensitivity (*Idn. Limit:* 1 γ Bi).

(*b*) A solution of $K_4[Mn(CN)_6]$ may be used as reducing agent instead of stannite.[70] Black BiO and $K_3[Mn(CN)_6]$ are formed (*Idn. Limit:* 10 γ Bi in 0.01 ml). This test is considerably less sensitive than the tests 1 to 4 just described, but it may, however, be carried out in the presence of Hg, Ag, and Cu salts.

(*c*) Bismuth may be detected by forming black BiI_3 and converting this product, by means of hot water, to red BiOI (*Idn. Limit:* 25 γ Bi).[71]

(*d*) The orange-red double iodide of bismuth and the organic base is precipitated from nitric acid solutions of bismuth salts by an alcoholic solution of quinoline and potassium iodide (*Idn. Limit:* 1 γ Bi). [72] Pb^{+2}, Sb^{+3}, Hg^{+} and Ag^{+} ions give black precipitates. Interference of Cu^{+2} and Fe^{+3}, due to liberation of iodine, may be prevented by placing a drop of a solution of $NaHSO_3$ on the paper impregnated with the reagent.

(*e*) A yellow-green color develops if paper impregnated with thio-acetamide is spotted with the acidified bismuth solution (*Idn. Limit:* 7 γ Bi).[73]

(*f*) If bismuth iodide is extracted from an acid solution in the presence of alkali iodide by isobutyl ketone, the organic solvent turns yellow (*Idn. Limit:* 1 γ Bi). [74] Platinum and palladium salts interfere. Other interferences can be averted by adding hypophosphorous acid and oxalic acid.

5. Copper

(*1*) *Test by catalytic acceleration of the ferric-thiosulfate reaction* [75]

When a solution of a ferric salt reacts with a solution of an alkali thiosulfate a transient deep violet color results. The completed reaction can be expressed:

$$2\,Fe^{+3} + 2\,S_2O_3^{-2} \rightarrow 2\,Fe^{+2} + S_4O_6^{-2}$$

The intermediate color indicates that this equation is merely the net representation of the stoichiometric consumption of the reactants, whereas the real course of the reaction must be quite different. Actually, the change occurs in two stages.[76, 77] In the first (1) the violet complex anion $Fe(S_2O_3)_2^-$

is produced instantaneously, and then reacts at a measurably slow rate with the Fe^{+3} ions (2). The summation of these partial reactions gives the net equation of this redox reaction:

$$Fe^{+3} + 2\ S_2O_3^{-2} \rightarrow Fe(S_2O_3)_2^- \text{ (rapid)} \quad (1)$$
$$Fe\ (S_2O_3)_2^- + Fe^{+3} \rightarrow 2\ Fe^{+2} + S_4O_6^{-2} \text{ (slow)} \quad (2)$$
$$\overline{2\ Fe^{+3} + 2\ S_2O_3^{-2} \rightarrow 2\ Fe^{+2} + S_4O_6^{-2} \text{ (slow)}} \quad (3)$$

Copper^{II} salts accelerate the second partial reaction and therefore speed up the total reaction[78]. Consequently, the intermediate violet color due to the $Fe(S_2O_3)_2^-$ ions disappears far more rapidly when copper ions are present. This effect is probably due to an intermediate production of cuprous salt because the complex ion does not react with ferric ions as shown in (2), but rather with cupric ions as given in (3). The resulting Cu^I can then react very rapidly with Fe^{III} as shown in (5) regenerating Cu^{II}. The summation of (4) and (5) gives the partial reaction (2), in which the catalyst Cu^{II} does not appear as a reactant:

$$Fe(S_2O_3)_2^- + Cu^{+2} \rightarrow Cu^+ + Fe^{+2} + S_4O_6^{-2} \text{ (rapid)} \quad (4)$$
$$Cu^+ + Fe^{+3} \rightarrow Cu^{+2} + Fe^{+2} \text{ (rapid)} \quad (5)$$
$$\overline{Fe(S_2O_3)_2^- + Fe^{+3} \rightarrow 2\ Fe^{+2} + S_4O_6^{-2} \text{ (rapid)}} \quad (2)$$

The reaction between Fe^{+3} and $S_2O_3{}^-{}_2$ can be carried out in the presence of an alkali thiocyanate. Under these conditions, the CNS^- acts not only as indicator for the ferric salt, by producing the soluble red $Fe(CNS)_3$, but it also reduces the concentration of the Fe^{+3} ions, since $Fe(CNS)_3$ is only slightly dissociated. Consequently, the rate of the reaction between Fe^{+3} and $S_2O_3^{-2}$ ions, and therefore the velocity of all the partial reactions in which Fe^{+3} ions participate, is decreased. Accordingly, the catalyzed acceleration due to copper salts is particularly evident under these conditions. Therefore, extremely small amounts of copper are definitely revealed by the more rapid fading of the color that results from the action of thiosulfate on ferric thiocyanate.

The velocity of the $Fe(CNS)_3$–$Na_2S_2O_3$ reaction depends on variations in the concentration ratios of iron : thiocyanate : thiosulfate, on the acid content, on the temperature. Aluminum, zinc, nickel, and especially arsenic retard the reaction.

It must be noted that water and many reagents often contain traces of copper which introduce significant errors in comparative tests. Such traces can be removed from water by shaking it with calcium fluoride or talc and then separating the adsorbent by centrifuging.[79]

Procedure. A drop of the test solution and a drop of distilled water are placed in adjacent depressions of a spot plate. A drop of ferric thiocyanate solution and 3 drops of 0.1 *N* sodium thiosulfate solution are added to each, and stirred with a glass rod. The time of decolorization of a copper-free solution is 1½ to 2 minutes. The decolorization of the test solution containing as little as 1 γ of copper is almost instantaneous. The difference in time as compared with the blank test is appreciable with even very small amounts of copper.

Limit of Identification: 0.02 γ copper
Limit of Dilution: 1 : 2,500,000
Reagent: Ferric thiocyanate solution: 1.5 g ferric chloride and 2 g potassium thiocyanate in 100 ml water

(2) *Test with benzoinoxime* [80]

Benzoinoxime (I) behaves toward Cu^{+2} ions as a dibasic acid in that it forms, in neutral or ammoniacal solutions, a green amorphous precipitate of a copper salt with the presumable structure (II).

C_6H_5—CHOH—C(NOH)—C_6H_5

(I)

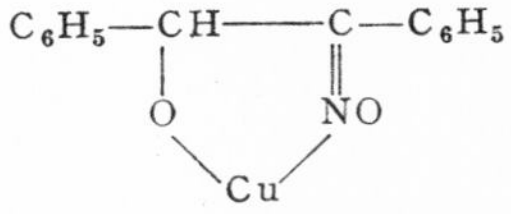

C_6H_5—CH——C—C_6H_5
O (single bond to CH) ‖ NO
Cu (bonded to O and NO)

(II)

Because of its resistance to ammonia, (II) can be viewed as an inner complex copper salt in which C_6H_5-groups or the CH-groups occupy coordination positions with respect to the Cu-atom and thus prevent the addition of NH_3 molecules to produce $[Cu(NH_3)_4]^{+2}$ ions. This seemingly obvious assumption is opposed by the insolubility of the organic copper salt in organic solvents of benzoinoxime. Therefore it is very probable that the salt has a polymer structure, in which the Cu-atoms are so situated that they are not subject to the action of ammonia molecules or to only a minor degree.[81] The following structural presentation exhibits this shielding by the hydrophobic phenyl groups ($R = C_6H_5$) contained in the benzoinoxime molecule:

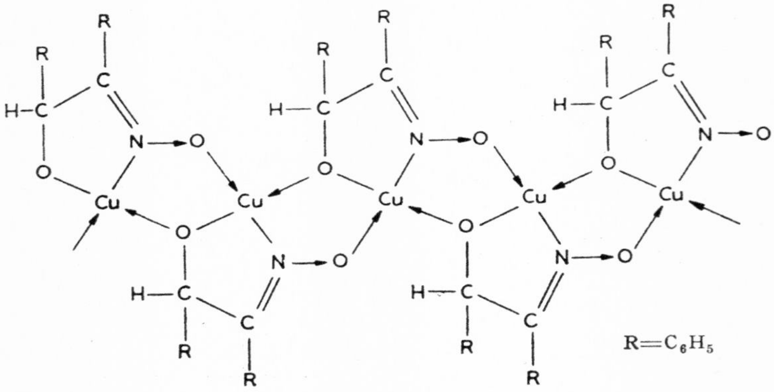

$R{=}C_6H_5$

References pp. 247–257

If other metallic salts which form precipitates with ammonia are present in such amount that they interfere with the recognition of the benzoin-oxime test, their precipitation can be prevented by the addition of sodium potassium tartrate. The sensitivity of the copper test is, however, somewhat reduced by considerable amounts of tartrate. Large amounts of ammonium salts also prevent the precipitation of the benzoin oximate. Consequently, if the test solutions are strongly acid or contain ammonium salts, a drop should be evaporated and ignited. The benzoinoxime reaction is carried out after the residue has been taken up in a drop of dilute hydrochloric acid.

Procedure. A drop of the weakly acid test solution is treated on filter paper with a drop of a 5 % alcohol solution of benzoinoxime,[82] and held over ammonia. A green coloration indicates copper. In the presence of large amounts of other ions precipitated by ammonia, a drop of a 10 % solution of Rochelle salt (K Na tartrate) is placed on the filter paper before the addition of the oxime.

The test may also be carried out on quantitative filter paper (S&S 589) impregnated with a saturated alcoholic solution of benzoinoxime.

Limit of Identification: 0.1 γ copper

Limit of Dilution: 1 : 500,000

If the test solution is strongly acid, a drop of ammonia is first placed on the filter paper, then a drop of the test solution, and the paper left for several minutes in a heated oven. In this way, as little as 0.5 γ copper can be detected in the presence of 1000 times the amount of hydrochloric acid.

The following procedure, which is based on adsorptive localization of traces of copper on filter paper, is highly recommended.[83] A quantitative filter paper is impregnated with 5 % alcoholic solution of benzoinoxime and, after drying, is cut into strips 1 mm × 10 mm. One of these strips is dipped into a warm drop of the test solution. After a few minutes, a distinct green zone develops and the color becomes stronger on treatment with ammonia.

Limit of Identification: 0.05 γ copper

Since this combination of adsorptive localization and the benzoinoxime reaction is a chromatographic test, it is generally not interfered with by any other metallic salts. Considerable amounts of neutral salts reduce the sensitivity somewhat.

(3) Test with salicylaldoxime [84]

Salicylaldoxime (I) gives a yellow greenish precipitate of the inner complex copper salicylaldoxime (II) with acetic acid solutions of copper^{II} salts.

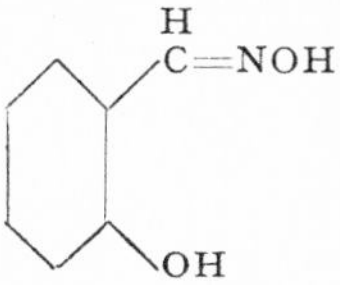

H
C=NOH
OH
(I)

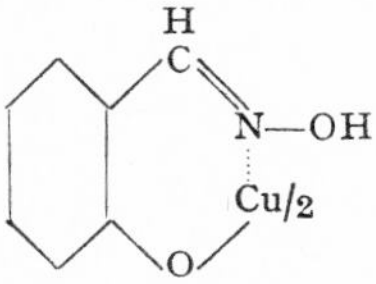

H
C
N—OH
Cu/2
O
(II)

Most other metal ions give precipitates with salicylaldoxime only in neutral or weakly alkaline solutions. Palladium and gold salts in acetic acid solutions also give precipitates of $Pd(C_7H_6O_2N_2)_2$ and metallic gold, respectively.[85]

Procedure. A drop of the test solution, previously neutralized and then slightly acidified with acetic acid, is mixed in a micro test tube with a drop of the reagent. According to the amount of copper present, a yellow greenish white precipitate or opalescence appears.

Limit of Identification: 0.5 γ copper

Limit of Dilution: 1 : 100,000

Reagent: 1 g salicylaldoxime [86] is dissolved in 5 ml alcohol and the solution added drop by drop to 95 ml water at 30° C. The oily suspension will almost disappear on gentle shaking. The reagent is then filtered.

(*4*) *Test with* o-*tolidine and ammonium thiocyanate* [87]

Benzidine blue is produced by oxidation when even very dilute solutions of copper^{II} salts are united with benzidine and alkali cyanide (see page 276) or alkali bromide, iodide or thiocyanate (see page 93). Since cupric salts, of themselves, do not oxidize benzidine, this effect obviously involves a raising of the oxidation potential of Cu^{II} salts by binding the Cu^{I} ions, formed during the oxidation, into complex or insoluble halides. Furthermore, it is likely that complex Cu-benzidine ions are formed.

If the homologous base *o*-tolidine is used, instead of benzidine, a more sensitive test for copper results. *o*-Tolidine has the advantages that its solutions are more stable and more easily oxidized by cupric salts.

Salts of silver, mercury ^{I}, iron, thallium^{III}, cerium^{IV} and gold, also chloroplatinic acid, interfere. Silver salts may be removed by precipitation as chloride, and the effect of iron salts can be prevented by adding alkali fluoride. The other strongly oxidizing metal salts, which interfere with the test through their oxidation of benzidine, may be reduced with bismuth amalgam and rendered harmless. Manganese salts interfere only in high concentrations, and give a slight blue color with the reagent after a considerable time. When this occurs it is advisable to carry out a blank with a saturated solution of manganese sulfate.

Procedure. A drop of the reagent solution is placed on filter paper, followed by a drop of the neutral or slightly acid test solution. A light or dark blue stain is formed according to the amount of copper present.

Limit of Identification: 0.003 γ copper (in 0.015 ml)

Limit of Dilution: 1 : 5,000,000

Reagent: A solution of 0.1 g *o*-tolidine and 0.5 g ammonium thiocyanate in 5 ml acetone

(*5*) *Test with phosphomolybdic acid* [88]

Molybdenum in the coordinated molybdenum trioxide molecules of phosphomolybdic acid $H_3PO_4 \cdot 12MO_3 \cdot aq$ is far more reactive than in free molybdic acid or in normal molybdates.[89] This increase in activity is due to the form of attachment of the molybdenum trioxide molecules, and is demonstrated by the fact that this reagent is reduced by certain organic and inorganic compounds which have little or no effect on molybdic acid and normal molybdates. Potassium cuprocyanide, formed from copper salts and potassium cyanide,

$$2\ CuSO_4 + 6\ KCN = K_2[Cu_2(CN)_4] + (CN)_2 + 2\ K_2SO_4$$

is among the compounds which can reduce the molybdenum in phosphomolybdates to "molybdenum blue", which is a colloidally dispersed mixture of lower molybdenum oxides.* This behavior may be utilized as a spot test for copper.

Procedure. A drop of the test solution is placed on filter paper, followed by a drop each of 1 % solution of potassium cyanide and 1 % solution of phosphomolybdic acid,[91] and finally a drop of dilute hydrochloric acid. According to the amount of copper present, a more or less deep blue color appears.

Limit of Identification: 1.3 γ copper
Limit of Dilution: 1 : 37,000

Large amounts of nitric acid may not be present. The other metals of the second group of the qualitative scheme affect the test for copper not at all or only in slight degree. Copper can be detected in the presence of other metals in the following limiting proportions:

2.5 γ copper in presence of 700 times the amount of mercury
2.5 γ copper in presence of 900 times the amount of lead
2.5 γ copper in presence of 400 times the amount of bismuth
4.0 γ copper in presence of 300 times the amount of mercury plus
300 times the amount of lead plus
120 times the amount of bismuth plus
280 times the amount of cadmium

(6) *Test with 1,2-diaminoanthraquinone-3-sulfonic acid* [92, 93]

The red-violet aqueous alkaline solution of 1,2-diaminoanthraquinone-3-sulfonic acid (I) turns deep blue on the addition of small amounts of a copper salt. Larger quantities produce a blue precipitate. In view of the constitution of the reagent, it is to be expected that this color change may be due to the formation of an addition compound of copper hydroxide in which the metal is coordinated as shown in (II) and (III).

* According to O. Glemser and G. Lutz [90], the chief constituent of molybdenum blue seems to be $2\ MoO_3 \cdot Mo_2O_5$.

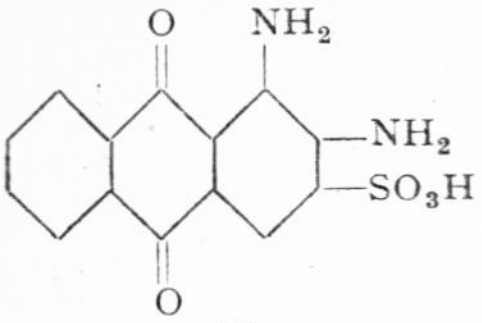

(I) 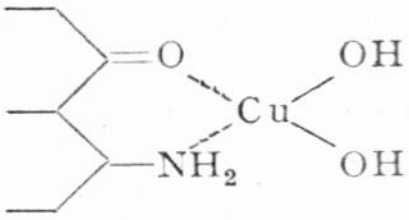 (II) 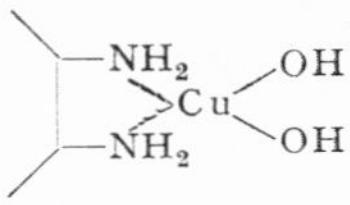 (III)

Analysis of the blue reaction product has shown that its composition does not correspond to the formulas just given.[94] Most of the colored material consists of hydrous copper oxide. In the light of studies [95] of metal salts reactions with alkaline solutions of dyestuffs, it is probable that this test involves a chemical adsorption of molecules of the 1,2-diaminoanthraquinone-3-sulfonic acid on gel particles of $Cu(OH)_2$, whereby the chelate bonds shown in (II) and (III) come into play. Such adsorption compounds, produced by surface reactions of dyestuffs, belong among the so-called color lakes. It is characteristic of all these lakes that no metal dyestuff compounds arise as a separate phase; the metal atoms do not leave their original phase association (compare aluminum-alizarin test, page 185).

In dilute solutions, cobalt and nickel salts give green-blue and dark blue colorations, respectively; flocculent precipitates with these same colors separate from more concentrated solutions. Other metal ions do not produce colored products, but it should be noted that certain accompanying materials may give precipitates because of the alkaline reaction medium. If the precipitates are colorless, the copper reaction can indubitably be recognized because when the reaction is positive any accompanying white precipitates assume a uniform blue color. The blue of the complex copper salt can be seen when colored precipitates, such as $Fe(OH)_3$, are present, if steps are taken to coagulate the precipitate (warming, stirring, centrifuging). It seems that greater quantities of mercury, magnesium and ammonium ions interfere.[95a] A further procedure for detection by isolation of cuprous thiocyanate will be discussed later.

Cyanides interfere with the reaction because they form stable complex alkali cuprocyanide. Under these circumstances, the procedure must include fuming with sulfuric acid, or ignition, to remove or destroy the cyanide. Tartrates, citrates, oxalates, and phosphates do not interfere since they are far weaker complexing agents for $copper^{II}$.

Procedure. A spot plate or micro crucible is used. A drop of the neutral or acid test solution is treated with a drop of the reagent and then made alkaline with 1 *N* sodium hydroxide. If the color changes from red-violet to cornflower-blue, copper is present. When very small quantities of copper are involved, a mixed color appears. In this case, it is best to use not more than a microdrop of the reagent and to compare the color with that obtained from a solution known to contain no copper.

Filter paper impregnated with the reagent solution may also be used. A drop of the neutral or extremely weak acid test solution is allowed to soak into the paper, and then is spotted with a drop of 1 *N* sodium hydroxide. A blue fleck or ring, depending on the copper content, appears on the red-violet reagent paper. The reagent paper is stable.

Limit of Identification: 0.02 γ copper

Limit of Dilution: 1 : 2,500,000

Reagent: A solution of 0.05 g of 1,2-diaminoanthraquinone-3-sulfonic acid [96] in 100 ml water. The solution is stable.

The following procedure, based on the interaction between cuprous thiocyanate and alkaline 1,2-diaminoanthraquinone-3-sulfonic acid, is recommended when considerable quantities of accompanying materials, particularly cobalt and nickel salts, are present. A drop of the test solution is placed on filter paper and dried (oven). The spot is then treated with a drop of potassium thiocyanate solution containing sulfur dioxide, and dried again. This treatment produces insoluble white cuprous thiocyanate, which remains fixed in the capillaries of the paper. Manganese, cobalt, nickel, or other heavy metal salts can be removed by bathing the paper repeatedly in potassium thiocyanate solution and finally in distilled water. The dried paper is then spotted with the alkaline reagent solution.

Reagents: 1) Potassium thiocyanate solution containing sulfurous acid: A freshly prepared mixture of 1 *N* KCNS and H_2SO_3 in the proportion 2 : 1

2) Alkaline dyestuff solution: 0.05 g 1,2-diaminoanthraquinone-3-sulfonic acid in 50 ml distilled water plus 50 ml 1 *N* NaOH

(7) *Test with rubeanic acid (dithiooxamide)* [97]

The yellow alcoholic solution of rubeanic acid (I) gives a dark green nearly black precipitate of copper rubeanate (III) from ammoniacal or weakly acid solutions of copper salts. The rubeanic acid (dithiooxamide) reacts in its tautomeric di-imido form (II) (compare test for cobalt, page 146):

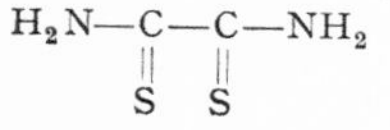

$$\underset{\text{(I)}}{H_2N{-}\overset{\displaystyle S}{\overset{\|}{C}}{-}\overset{\displaystyle S}{\overset{\|}{C}}{-}NH_2} \qquad \underset{\text{(II)}}{HN{=}C(SH){-}C(SH){=}NH} \qquad \underset{\text{(III)}}{HN{=}C(S{-})\text{—}C(S{-}){=}NH,\ S{-}Cu{-}S}$$

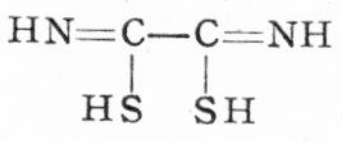

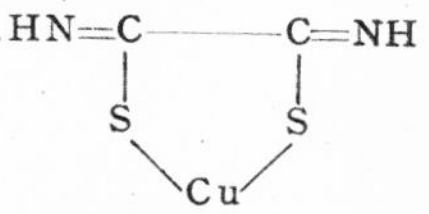

Because of the intense color and the insolubility in ammonia, it has been assumed that copper rubeanate (and likewise the cobalt and nickel salts)

are inner complex salts. However, this conjecture cannot be maintained if the distinguishing characteristic of inner complex salts is taken to be their solubility in organic liquids, because the rubeanates do not dissolve in such liquids. The resistance to ammonia is therefore probably to be ascribed to these salts being polymeric rubeanates, in which the metal atom is so located that it is shielded against the addition of ammonia molecules and hence against the formation of ammines. Perhaps the copper salt (and likewise the nickel and cobalt salts) can properly be given the structure of a linear polymer:

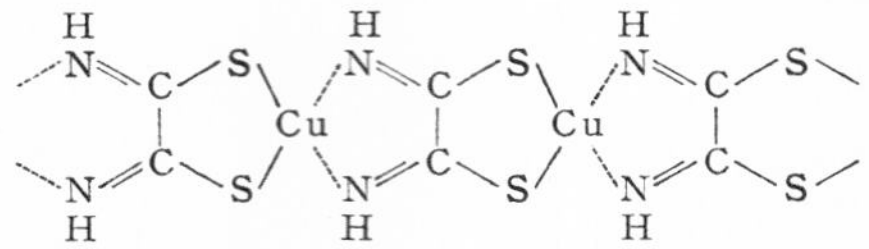

```
   H              H              H
..N=C-S\       /N=C-S\       /N=C-S\
    |    Cu        |    Cu        |
..N=C-S/       \N=C-S/       \N=C-S/
   H              H              H
```

This precipitation of copper with rubeanic acid also occurs in ammoniacal, tartrate-containing solutions, but not in alkaline cyanide solutions. Only cobalt and nickel ions react under the same conditions as copper ions; they give brown and blue-red precipitates respectively. The test is extremely sensitive, and under suitable conditions (see below) can also be carried out in the presence of cobalt and nickel salts. If the masking actions of malonic acid and ethylenediamine are employed (see page 90), copper can be detected with rubeanic acid with no interference whatsoever from other metal ions. However, the sensitivity is much less.

Procedure. A drop of the solution, as near neutral as possible, is placed on filter paper, held over ammonia, and treated with a drop of 1 % alcoholic solution of rubeanic acid.[98] A black or olive green fleck or circle indicates copper.

Even the traces of copper sometimes present in distilled water give a positive reaction. Therefore, in testing for small amounts of copper, a blank must be carried out with copper-free water (see page 81).

The following figures are for the limit of identification using a microdrop (0.015 ml).

The sensitivity of the reaction is reduced in the presence of large amounts of ammonium salts.

As little as 0.48 γ of copper can be detected in the presence of 400 times the amount of iron or cadmium.

Limit of Identification: 0.006 γ copper
Limit of Dilution: 1 : 2,500,000

Test for copper in the presence of nickel and cobalt by capillary separation[99]

Cobalt and nickel also produce colored rubeanates from neutral, ammoniacal or tartrate-containing solutions. The copper salt, however, has the lowest solubility product. This value is attained with even very small amounts of copper reacting with the slight, but adequate concentrations of the di-imide form of rubeanic acid found in alcohol solutions of this reagent. Nickel and cobalt are precipitated by rubeanic acid either slightly, or not at all, if the test solution contains free acetic acid. The determining factor is the relative concentration of acid and metal salt. Copper, on the other hand, is precipitated quantitatively.

Since, in practice, the nickel or cobalt content of the test solution is unknown, and therefore the correct proportion of acetic acid cannot be added exactly, this sensitive test for copper cannot be carried out in a test tube if cobalt or nickel are present, since they also may be precipitated. When, however, the test is carried out on filter paper, traces of copper can easily be detected, even in the presence of relatively large amounts of nickel or cobalt. Capillary separation or, more correctly, fractional precipitation is employed.

Procedure. A drop of the test solution, acidified with acetic acid, is placed on filter paper impregnated with rubeanic acid (spot test paper is best). Two zones are formed of differing acetic acid content. In the central zone the acid concentration is the greatest, and here only the copper is precipitated in the form of an olive green or black circle. The nickel diffuses farther and forms a blue to violet ring around the central zone. This ring develops more towards the middle as the acetic acid evaporates.

In the presence of cobalt, the same process occurs, except that the central zone is surrounded by a yellow-brown ring of cobalt rubeanate. When both cobalt and nickel are present, the copper can still be detected through capillary separation and fractional precipitation. The test solution should not contain more than 2 % cobalt and/or nickel.

Limit of Identification: 0.05 γ copper
Limit of Dilution: 1 : 1,000,000
} in the presence of 20,000 times the amount of nickel

Limit of Identification: 0.25 γ copper
Limit of Dilution: 1 : 200,000
} in the presence of 2000 times the amount of cobalt (in acetic acid solution)

When the test is concerned with the detection of copper in the presence of only cobalt, a drop of the neutral test solution may be placed on paper impregnated with rubeanic acid. The copper rubeanate formed in the central ring is surrounded by a concentric brown-yellow ring of cobalt rubeanate.

Limit of Identification: 0.05 γ copper
Limit of Dilution: 1 : 1,000,000
} in the presence or 20,000 times the amount of cobalt

References pp. 247–257

Test for copper through masking of interfering ions [100]

When spot tests are carried out with neutral solutions on filter paper, reactions with rubeanic acid are not obtained solely with copper, cobalt and nickel ions. The following ions also yield colored stains: ferric, faint orange; silver, yellow turning to greenish black; bismuth, tan; mercurous, brown turning to black; palladium, brown; platinum, rose. All these ions form soluble complex compounds when malonic acid and ethylenediamine are added. Among them, only the copper compound reacts with rubeanic acid to a sufficient extent so that a specific test can result.

Procedure. A drop of 20 % aqueous solution of malonic acid is placed on a spot test paper followed by a drop of the test solution (pH below 7). A drop of 10 % aqueous solution ethylenediamine is added and then a drop of 1 % solution of rubeanic acid in 95 % ethyl alcohol. If copper is present, a green stain appears.

Limit of Identification: 0.3 γ copper (0.03 ml drop)

Limit of Dilution: 1 : 100,000

(*8*) *Test with dithizone* [101]

Diphenylthiocarbazone (dithizone) reacts with many of the heavy metals to form insoluble colored inner complex salts, that can easily be extracted with organic solvents (see page 74). The copper salt is yellow-brown. It is formed in solutions that are neutral, ammoniacal, or that contain ammonium salts. When a dilute solution of the reagent in carbon tetrachloride is used, and a small amount of this reagent solution is shaken with the test solution, copper can be detected in the presence of large amounts of other heavy metals (with the exception of the noble metals and mercury). The greater tendency for copper dithizone to be formed as compared with the other heavy metal dithizonates has the result that the copper salt forms preferentially when only a limited amount of dithizone is present.

Procedure. A drop of the neutral or weakly ammoniacal test solution is shaken with a drop of a solution of dithizone in carbon tetrachloride in a stoppered micro test tube. The green color of the reagent changes immediately to yellow-brown.

Limit of Identification: 0.03 γ copper

Limit of Dilution: 1 : 1,660,000

Reagent: Solution of 1 to 2 mg dithizone in 100 ml carbon tetrachloride

Table 3 shows the sensitivity of the test in neutral copper solutions in the presence of other heavy metals which react with dithizone.

TABLE 3

Other metallic salt present	Limit of identification, γ in 0.05 ml	Other metals present, γ in 0.05 ml	Proportion limit, Cu : foreign metal
$Pb(C_2H_3O_2)_2 \cdot aq$	0.05	13,700	1 : 274,000
$ZnSO_4 \cdot aq$	0.07	2,000	1 : 28,500
$Cd(C_2H_3O_2)_2 \cdot aq$	0.05	7,200	1 : 144,000
$NiSO_4 \cdot aq$	0.1	4,000	1 : 40,000

The identification of copper by means of dithizone can also be carried out on filter paper or on a spot plate; a stronger reagent solution (about 10 mg of dithizone in 100 ml of carbon tetrachloride) should be used. The limit of identification is then 0.2 γ copper.

(9) Test with 2,2'-diquinolyl (cuproin) [102]

The alcohol-soluble colorless base, 2,2'-diquinolyl (I), forms a purple-red cuprous salt. The complex intercalation compound has the structure (II)

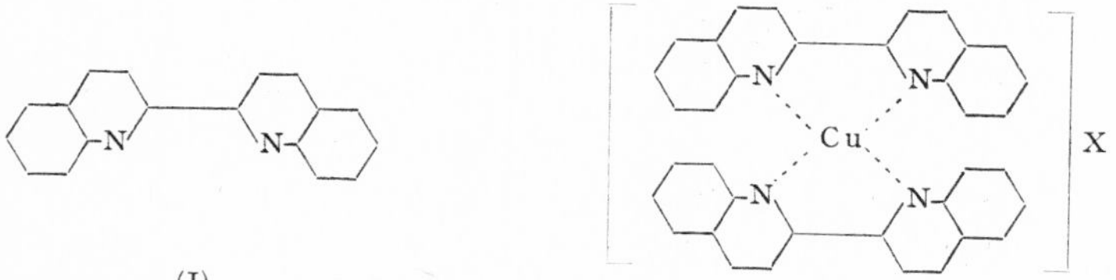

(I) (II)

X = univalent acid radical

The cuprous compound is formed in acid solution. It is not soluble in water, but dissolves to give a red color in certain organic liquids, some miscible, some immiscible with water. Only Cu^{+} ions react, even in dilute solution, with 2,2'-diquinolyl, which accordingly is one of the few absolutely specific reagents, and justly bears the name cuproin. Since the reagent responds only to cuprous ions, it is necessary to reduce Cu^{+2} to Cu^{+}, which can be accomplished by adding hydroxylamine hydrochloride before or along with the reagent.

When the following procedure is used the sensitivity of the test for copper is not decreased by foreign ions in the proportion 1 : 5000. Ions of the noble metals are reduced to the metal by the hydroxylamine hydrochloride, and thallium, silver, lead and mercurous ions are precipitated as chlorides. Such precipitates should be removed prior to the addition of the reagent, particularly if the quantity of copper to be detected is small. Tervalent iron, if present in high concentration, must be masked with tartaric acid and the

References pp. 247–257

pH then adjusted with ammonia. Colored ions in larger amounts impair the discernment of small quantities of (II). In such cases it is better to take advantage of the extractability of the copper salt in organic solvents.

Procedure. A drop of the test solution (pH above 3) is treated on a spot plate with several crystals of hydroxylamine hydrochloride and 3 drops of a saturated ethanol solution of cuproin [103]. A purple to pink color develops, according to the amount of copper present.

Limit of Identification: 0.05 γ copper
Limit of Dilution: 1 : 1,000,000

When copper is to be detected in the presence of large amounts of colored ions, it is advisable to place several drops of the test solution in an Emich tube and add several crystals of hydroxylamine hydrochloride and then 1 or 2 drops of a saturated solution of cuproin in isoamyl alcohol. The mixture is shaken and the layers allowed to separate. Depending on the quantity of copper present, the alcohol layer takes on a purple to pink color. This procedure will reveal copper in a solution at a 1 : 5,000,000 dilution and containing as much as 10,000 the quantity of foreign ions.

(*10*) *Test with alizarin blue* [104]

In view of its constitution, alizarin blue (I) might be expected to display the characteristics of alizarin (page 185) and 8-hydroxyquinoline (page 237). This is not the case, as shown by the facts that it is neither soluble in alkali as is alizarin nor in dilute acids as is oxine. Probably alizarin blue is a chelate compound as represented in (Ia). Solutions can be prepared in dioxane, pyridine, acetic anhydride, concentrated sulfuric acid. These solutions react with strongly acid solutions of copper^{II} salts to produce a cornflower-blue crystalline precipitate, whose constitution as an inner complex compound is shown in (II).

O OH —OH N O

(I)

H O O —O H N O

(Ia)

H O O —O Cu/2 N O

(II)

The precipitation reaction of alizarin blue with copper^{II} salts is strictly specific. Cobalt and nickel yield a blue color; the following test is not directly applicable in their presence. To detect small amounts of copper when the sample also contains nickel and/or cobalt, it is best to deposit the copper electrolytically on a platinum wire and then dissolve it in dilute nitric acid. These operations can be conducted on the micro scale.

Procedure. A drop of the violet reagent solution is placed in a depression of a spot plate and treated with a drop of the test solution, whose content of mineral acid is optional. For comparison, a drop of water is added to a drop of the reagent solution in an adjacent depression. An intense blue develops in both cases. On the addition of 1 or 2 drops of acetic anhydride, the blank turns honey-yellow, whereas a blue-violet precipitate remains in the copper-bearing solution. It is possible to detect 0.05 γ Cu (1 : 1,000,000) in this way.

A higher sensitivity is attained if the drop of test solution is taken to dryness in the depression of the spot plate, and the residue then treated with a drop of reagent solution and a drop of glacial acetic acid, added in succession. Tiny quantities of the copper salt of alizarin blue can then be detected by the blue-violet coating adhering to the porcelain.

Limit of Identification: 0.004 γ copper
Limit of Dilution: 1 : 12,500,000
Reagent: Saturated solution of alizarin blue [105] in pyridine

(*11*) *Other tests for copper*

(*a*) Filter paper is spotted successively with a drop of the test solution, a drop of 10 % solution of benzidine in acetic ester, and a drop of saturated KBr solution. A blue fleck appears (*Idn. Limit:* 0.6 γ Cu).[106] Oxidizing anions, and likewise Au^{+3} and Fe^{+3} ions interfere.[107]

(*b*) The addition of $(NH_4)_2Hg(CNS)_4$ to a solution containing both Cu^{+2} and Zn^{+2} does not produce a mixed precipitate of white $ZnHg(CNS)_4$ and yellow-green $CuHg(CNS)_4$, but, instead, a violet precipitate.[108] If one drop of the test solution, one of 1 % zinc acetate, and one of a solution containing 8 g $HgCl_2$ and 9 g NH_4CNS in 100 ml water are placed on filter paper or a spot plate, a violet precipitate forms (*Idn. Limit:* 0.1 γ Cu).[109] Nickel and cobalt interfere; iron must be masked by adding fluoride.

(*c*) Dimethylaminobenzylidenerhodanine (see page 59) reacts with cuprous salts (which behave in many ways similarly to Ag salts) to give a red precipitate (*Idn. Limit:* 0.6 γ Cu).[110] Cupric salts do not react, and must be reduced with H_2SO_3 before testing.

(*d*) A red stain is formed when a drop of an acetic acid solution of 8-hydroxyquinoline followed by a drop of the test solution and a drop of a 25 % potassium cyanide solution are placed on filter paper (*Idn. Limit:* 0.4 γ Cu).[111] The test depends on the formation of the complex alkali cuprocyanide and cyanogen. The latter reacts with the hydroxyquinoline, giving a red color. Neither oxidizing agents nor metallic ions giving colored oxine compounds should be present. The test is specific if conducted with ammoniacal cupric solutions.

(*e*) A violet to dark brown color appears if a drop of the test solution is treated with a drop of 25 % hydrobromic acid (*Idn. Limit:* 0.15 γ Cu).[112] Ag^+, Hg^+, Pb^{+2}, Cd^{+2}, Fe^{+3}, and Cl^- interfere.

(*f*) At *p*H 7 to 8, an orange coloration appears if a copper^{II} solution is treated with a 1 % water solution of 2-nitroso-1-naphthol-4-sulfonic acid (*Idn. Limit:* 0.01 γ Cu).[113] Co^{+2}, Ni^{+2} and Fe^{+2} give similar reactions. This test involves the production of soluble compounds in which the particular metal atoms are part of inner complex anions.

(*g*) Traces of copper accelerate the oxidation of phenetidine chloride by hydrogen peroxide (violet color).[114] The test may be carried out as a drop reaction in a micro crucible.

(*h*) Diphenylcarbazone and its substitution products, in neutral or slightly acid solution, may be used as sensitive but nonspecific reagents for copper [115] (*Idn. Limit:* 0.002 γ or 0.05 γ Cu in neutral or 0.1 *N* HCl solution, respectively, with di-*m*-nitrophenylcarbazone).

(*i*) A violet color is obtained when a drop of a solution of ammonium bromide in concentrated phosphoric acid reacts with a cupric salt (*Idn. Limit:* 0.1 γ Cu).[116] Many metal ions interfere.

(*j*) An indigo blue precipitate results on bringing together a drop of the neutral test solution, a drop of alkali tartrate in H_2O_2 solution, and finally a drop of an alcohol solution of *o*-hydroxyphenylfluorone (*Idn. Limit:* 0.4 γ Cu). [117]

(*k*) A brown-yellow fleck of the cuprous salt is obtained by spotting paper impregnated with white zinc diethyldithiocarbamate (*Idn. Limit:* 0.002 Cu). [118]

(*l*) A violet precipitate is given by the union on a spot plate of a drop of the test solution with one drop of acid potassium fluoride (2.5 g KHF_2 + 0.4 ml 6 *N* HCl + 5 ml water) and one drop 0.1 % alcoholic solution of *o*-hydroxyphenylfluorone (*Idn. Limit:* 1.7 γ Cu). [119] The color reaction with molybdate (compare page 118) and the precipitation reaction with cobalt can be prevented by prior addition of hydrogen peroxide or potassium tartrate.

(*m*) Traces of copper accelerate the autoxidation of alkaline resorcinol solutions.[120] A brown spot results on suitable filter paper (*Idn. Limit:* 0.005 γ Cu).[121] Silver gives a similar reaction.

6. Cadmium

(*1*) *Test with ferrous dipyridyl iodide* [122]

The base α,α'-dipyridyl (α,α'-dip) produces an intense red color in acid or neutral solutions of ferrous salts. The color is due to the formation of the stable complex cations $[Fe(\alpha,\alpha'\text{-dip})_3]^{+2}$ (see page 161). These products can combine with anions that have a large atomic volume to give slightly soluble, red crystalline compounds. The precipitation occurs, in part, from even very dilute solution. Accordingly, complex anions, such as $[HgI_4]^{-2}$, $[CdI_4]^{-2}$, $[Ni(CN)_4]^{-2}$, etc. in particular, and likewise considerable quantities of iodides, serve as precipitants for $[Fe(\alpha,\alpha'\text{-dip})_3]^{+2}$ ions.

The formation of red $[Fe(\alpha,\alpha'\text{-dip})_3]\cdot[CdI_4]$ can be used for the detection of cadmium if a saturated water solution of $[Fe(\alpha,\alpha'\text{-dip})_3]I_2$ containing an excess of I^- ions is used as reagent:

$$[Fe(\alpha,\alpha'\text{-dip})_3]^{+2} + 4\ I^- + Cd^{+2} \rightarrow [Fe(\alpha,\alpha'\text{-dip})_3]\cdot[CdI_4]$$

A red-violet precipitate is obtained. It is characterized by the fact that both its cationic and anionic constituents are complexes. This unusual combination accounts for the remarkable sensitivity and selectivity of this test for cadmium.

Procedure. The reaction is carried out on thick filter paper (S & S 601). A drop of the weakly acid, neutral, or ammoniacal test solution is placed on the paper and before the drop has been absorbed by the paper it is treated with a drop of the reagent. Reaction occurs at once. After the liquid has soaked in, the cadmium precipitate is left as a red fleck or ring, and because of its intense color it stands out against the reagent solution that has spread through the paper.

The test solution and the reagent can also be brought together in a centrifuge tube. In this case, it is best to place a drop of the reagent in the tube and then introduce a drop of the test solution, and to centrifuge without previously mixing the two. A red precipitate is produced in the constricted end of the tube. It is rather striking that the sensitivity of the test is considerably less if the drops are mixed before centrifuging.

Limit of Identification: 0.05 γ cadmium
Limit of Dilution: 1 : 1,000,000

Reagent: 0.25 g α,α'-dipyridyl and 0.146 g $FeSO_4\cdot 7H_2O$ are dissolved in 50 ml water, 10 g KI added, and after shaking vigorously for 30 minutes the deep red $[Fe(\alpha,\alpha'\text{-dip})_3]I_2$ is filtered off. The resulting saturated solution of this salt contains excess iodide ions to accomplish the formation of the $[CdI_4]^{-2}$ ions essential to the test. The solution is stable; in case it becomes turbid on long standing, it should be filtered before use.

Test for cadmium in the presence of other metals

The direct use of $[Fe(\alpha,\alpha'\text{-dip})_3]I_2$ solution containing an excess of I^- ions as precipitant for cadmium is interfered with by those metal ions which form slightly soluble or complex iodides. Solutions of Ag, Tl, or Pb salts may be precipitated by the iodide in the reagent and the precipitates are made red by adsorption of $[Fe(\alpha,\alpha'\text{-dip})_3]^{+2}$ ions. Cu^{+2} ions in neutral or acid solution

References pp. 247–257

liberate iodine forming Cu_2I_2, which likewise combines with α,α'-dipyridyl to form colored addition products.[123] Hg, Sn, Sb, and Bi salts form soluble complex iodides which also yield red precipitates with $[Fe(\alpha,\alpha'\text{-dip})_3]^{+2}$ ions, although the sensitivity is less in these cases.

The cadmium test can be made almost specific if the test solution is treated with ammonia, filtered, and the ammoniacal filtrate then tested with the reagent. The $Cd(NH_3)_4^{+2}$ ions in the filtrate react promptly, whereas any $Cu(NH_3)_4^{+2}$ and $Zn(NH_3)_4^{+2}$ ions are inactive. Only Tl^+ and $Ag(NH_3)_2^+$ ions, which likewise enter the ammoniacal filtrate, can interfere, since they form AgI and TlI by combination with the iodide in the reagent. Consequently, if cadmium is to be detected in the presence of silver and other ions that form slightly soluble iodides, the following procedure should be used. The test solution is treated first with dilute hydrochloric acid. Any precipitate (AgCl, TlCl, Hg_2Cl_2, $PbCl_2$) is filtered off, the filtrate made ammoniacal, refiltered if necessary, and the ammoniacal filtrate tested with the reagent. All these operations can be accomplished with one or two drops of the test solution in a microcentrifuge tube in which any precipitates can easily be separated by centrifuging.

The concentration limits of mixtures of cadmium and copper, and cadmium and zinc, which are of practical importance, were determined by preparing ammoniacal solutions of the respective mixtures. The following quantities of cadmium could be detected in one drop:

0.08 γ cadmium in the presence of 5000 times this quantity of copper;

0.01 γ cadmium in the presence of 5000 times this quantity of zinc.

If the detection of about 0.2 γ Cd is adequate, still better limiting concentrations can be obtained.

(2) *Test with di-*p*-nitrophenylcarbazide* [124, 125]

Numerous metal hydroxides, including cadmium hydroxide, are colored by di-*p*-nitrophenylcarbazide (I) and its oxidation product di-*p*-nitrophenylcarbazone (II).

$$O{=}C\begin{cases} NH{-}NH{-}C_6H_4(NO_2) \\ NH{-}NH{-}C_6H_4(NO_2) \end{cases} \quad \text{(I)} \qquad\qquad O{=}C\begin{cases} NH{-}NH{-}C_6H_4(NO_2) \\ N{=}N{-}C_6H_4(NO_2) \end{cases} \quad \text{(II)}$$

The color is probably due to adsorption complexes (lakes).[126] Cadmium hydroxide precipitated in the presence of (I) is brown. On standing for some time, the color changes to green-blue, because the carbazide is oxidized to the carbazone (II). The color change is accelerated by formaldehyde; the reason is as yet not known. This fact can be used to test for cadmium, even when copper is present. It is necessary to provide conditions which will

References pp. 247–257

permit the precipitation of cadmium hydroxide, but not of copper hydroxide. The diverse behavior of the complex cyanides toward formaldehyde can be utilized. Alkali cyanides react with formaldehyde to produce the alkali salt of glycolic nitrile, whereby cyanide ions are consumed:

$$KCN + CH_2O \rightarrow CH_2(OK)CN \quad \text{or} \quad CN^- + CH_2O \rightarrow CH_2\begin{matrix} \diagup O^- \\ \diagdown CN \end{matrix}$$

The complex cyanides of cadmium and univalent copper are slightly dissociated (omitting the intermediate stages):

$$[Cd(CN)_4]^{-2} \rightleftarrows Cd^{+2} + 4\ CN^- \quad (1)$$

$$[Cu_2(CN)_4]^{-2} \rightleftarrows Cu_2^{+2} + 4\ CN^- \quad (2)$$

The concentration of the cyanide ion in equilibrium (1) is always sufficient to react with formaldehyde, thus continually disturbing the equilibrium of the system, with the result that sufficient cadmium ions are formed to exceed the solubility product of cadmium hydroxide. The cadmium hydroxide formed under these conditions seems to be especially suited for the formation of the absorption color with the carbazide. Unlike the cadmium compound, the complex potassium cuprocyanide is so stable that the cyanide ion concentration in equilibrium (2) is not adequate to react visibly with formaldehyde. The demasking of $[Cd(CN)_4]^{-2}$ ions permits the detection of cadmium in the presence of copper.

Procedure. A drop of the acid, neutral, or ammoniacal test solution is mixed on a spot plate with a drop of 10 % sodium hydroxide and one of 10 % potassium cyanide. Then a drop of the reagent and two drops of 40 % formaldehyde are added, with stirring. In the presence of cadmium, a blue-green precipitate or coloration is formed. The reagent itself is red in alkaline solution and is colored violet by formaldehyde. Hence, with small amounts of cadmium, the color should be compared with that of a blank test.

Limit of Identification: 0.8 γ cadmium
Limit of Dilution: 1 : 62,000
Reagent: 0.1 % alcoholic solution of di-*p*-nitrodiphenylcarbazide [127]

Test for cadmium in the presence of other metals

To prevent the reaction between copper and dinitrodiphenylcarbazide, a sufficient excess of potassium cyanide must be used.

Procedure. A drop of the test solution, which should contain not more than 0.04 g copper in 1 ml, is stirred on a spot plate with a drop of 10 % sodium hydroxide and 3 drops of 10 % potassium cyanide until the copper hydroxide

References pp. 247–257

or cyanide is completely dissolved. Then 2 drops of the reagent and 3 drops of 40 % formaldehyde are mixed with the solution. By carrying out a parallel test with a 15 % solution of copper sulfate, the following values are attained:

Limit of Identification: 4 γ cadmium } in the presence of 400 times the
Limit of Dilution: 1 : 12,500 } amount of copper

If cadmium is to be detected in the presence of ions of the hydrogen sulfide group, it is best to add ammonia to the test solution, centrifuge, and carry out the test for cadmium on the supernatant liquid (containing the copper and cadmium).

Large amounts of ammonium salts reduce the sensitivity. Nevertheless, if a blank test is carried out, 4 γ cadmium can be readily detected in the presence of 250 times the amount of copper, and 2 γ cadmium in the presence of 120 times the amount of copper.

(3) *Test with di-β-naphthylcarbazone* [124, 125]

Di-β-naphthylcarbazone, like di-*p*-nitrophenylcarbazone, produces a violet adsorption compound with cadmium hydroxide. Consequently, it is convenient to use the following test, in the same way as described in (2), for the detection of cadmium in the presence of copper. It should be noted that other metal hydroxides likewise produce colored products with the reagent.

Procedure. Two drops of the neutral or slightly acid test solution are mixed on a spot plate with 2 drops of the reagent and a drop of 2 *N* sodium hydroxide. If cadmium is present the red of the reagent turns to violet.

Limit of Identification: 0.2 γ cadmium
Limit of Dilution: 1 : 250,000
Reagent: 0.02 % solution of di-β-naphthylcarbazone [128] in methanol

Test for cadmium in the presence of copper

Copper interferes with the test. This complication can be obviated by the method described in (2), if cadmium hydroxide is liberated from the cyanide complex by adding formaldehyde. The presence of copper actually increases the sensitivity of the cadmium test. The excess reagent is decolorized by oxidation in the presence of copper salts, which probably act as catalyst. The lake of cadmium hydroxide with naphthylcarbazone retains its violet color.

Procedure. Two drops of the neutral or slightly acid test solution, which may contain a maximum of 0.5 % copper, are mixed with a drop each of 2 *N* alkali and 10 % potassium cyanide solution. After all the copper hydroxide has dissolved, 2 drops of the reagent and 2 drops of 40 % formaldehyde are added. On stirring, the color disappears almost entirely within several minutes when

References pp. 247–257

cadmium is absent. If cadmium is present, a more or less intense violet precipitate appears, and becomes distinctly visible as the color of the excess reagent disappears.

Limit of Identification: 0.1 γ cadmium } in the presence of 5000 times the
Limit of Dilution: 1 : 500,000 } amount of copper

(*4*) *Other tests for cadmium*

(*a*) Alcoholic solutions of diphenylcarbazide can be used for spot reactions.[129] Violet precipitates or colorations are formed with neutral, ammoniacal or acetate-buffered cadmium solutions (*Idn. Limit:* 4 γ Cd). Substituted diphenylcarbazides are better than the parent substance with respect to greater selectivity and higher sensitivity (see page 96).

(*b*) *p*-Nitrodiazoaminoazobenzene forms a color lake with cadmium hydroxide.[130] The reagent consists of a 0.02 % solution in alcohol, which is also 0.02 *N* with respect to KOH. One drop of the reagent is placed on thick filter paper and treated with a drop of the test solution that has been acidified with acetic acid. One drop of 2 *N* KOH is then added. If cadmium is present, a bright pink circle surrounded by a violet-blue ring appears (*Idn. Limit:* 0.025 γ Cd). The interference by Cu, Ni, Fe, Cr, Co and Mg is prevented by adding tartrate. Only Ag, Hg, and NH_4 will then interfere. Silver is removed as iodide, mercury as sulfide.

(*c*) The blue ammoniacal solution containing $[Cu(NH_3)_4]^{+2}$ and $[Cd(NH_3)_4]^{+2}$ ions is just decolorized by careful addition of cyanide solution. A drop is then placed on filter paper along with a drop of 10 % sodium sulfide solution. A yellow stain or ring of CdS appears (*Idn. Limit:* 0.1 γ Cd).[131]

A 2. Acid Sulfide Group

7. Arsenic

(*1*) *Test with stannous chloride* [132]

Compounds of both ter- and quinquevalent arsenic are reduced to elementary arsenic by solutions of stannous chloride strongly acidified with hydrochloric acid. The arsenic separates as a brownish black precipitate.*

$$2\,As^{+5} + 5\,Sn^{+2} \rightarrow 2\,As^{\circ} + 5\,Sn^{+4}$$
$$2\,As^{+3} + 3\,Sn^{+2} \rightarrow 2\,As^{\circ} + 3\,Sn^{+4}$$

The reduction is accelerated by heating. It takes place only in strongly acid solution, since only arsenic cations or undissociated arsenic halides

* The precipitate always contains tin; see the detailed studies by K. Zwicknagel, *Z. anorg. allgem. Chem.*, 151 (1926) 41.

react with stannous chloride. In weakly acidified solutions, arsenic anions are present but no cations, since they undergo hydrolysis:

$$As^{+3} + 3\ H_2O \rightleftarrows AsO_3^{-3} + 6\ H^+$$

Quinquevalent arsenic cations behave analogously. The same is true of quadrivalent and sexivalent selenium and tellurium ions.

Among the other metallic salts, only those of the noble metals and mercury (not, however, antimony salts) are reduced to the metallic state under these conditions, so that this sensitive "Bettendorf" test for arsenic with stannous chloride is fairly specific. To remove the interfering mercury by simply igniting the sample is not feasible, since some arsenic compounds (such as arsenic trioxide or arsenic halides) are also volatilized. When, however, the arsenic is converted beforehand into the heat-resistant magnesium pyroarsenate, as described below, there is no danger of losing arsenic in amounts that are significant, unless extremely small quantities are being sought.

The sensitivity of the test is decreased by the presence of colored salts in the solution. Examples are chromium, cobalt and nickel in considerable amounts. Copper salts are reduced by stannous chloride to the colorless cuprous chloride and do not interfere.

If arsenic is to be detected in the presence of large amounts of these interfering materials, it is best to carry out the test on the mixture of sulfides precipitated in acid solution. It must be noted, however, that when arsenic is present in the quinquevalent form, it is slowly and incompletely precipitated by hydrogen sulfide in the cold. The precipitation can be remarkably accelerated by the addition of small amounts of potassium iodide.*

Procedure. A drop of the test solution is mixed in a micro crucible with 1 or 2 drops of ammonia, 1 of 10 % hydrogen peroxide, and 1 of 10 % magnesium chloride solution. The mixture is evaporated slowly and then strongly ignited. The arsenic, whether present as sulfide, or in alkaline sulfide solution, is thus converted into heat-resistant $Mg_2As_2O_7$; the mercury salts are volatilized. The residue is mixed with 1 or 2 drops of a concentrated stannous chloride solution in 35 % hydrochloric acid and warmed gently. The formation of a brown-black precipitate or coloration indicates arsenic.

Limit of Identification: 1 γ arsenic **
Limit of Dilution: 1 : 50,000

* Potassium iodide, through its iodide ions, acts as a catalyst in that it reduces the arsenic acid to arsenious acid: $2\ AsO_4^{-3} + 4\ I^- + 4\ H^+ = 2\ AsO_3^{-3} + 2H_2O + 2\ I_2$; the iodine liberated is again changed to the ionic state by the hydrogen sulfide: $I_2 + H_2S \rightarrow 2\ H^+ + 2\ I^- + S$.

** The test is distinctly more sensitive if the reduction with $SnCl_2$ is carried out in a micro test tube, and then extraction done with ether or amyl alcohol. Very small amounts of arsenic are accumulated and form a perceptible black layer in the water-organic liquid interface.

References pp. 247–257

(2) *Tests by reduction to arsenic hydride and decomposition with silver nitrate* [133, 134] *or gold chloride* [135]

(*a*) *With silver nitrate (Gutzeit test)*

When arsenic hydride gas is directed against filter paper moistened with silver nitrate solution, the wetted areas are colored. If a solution of one part silver nitrate and one part water is used, arsenic hydride forms a lemon-yellow fleck of $AsAg_3 \cdot AgNO_3$ which, on treatment with water, blackens with the deposition of metallic silver. Antimony hydride (SbH_3) colors the periphery of the moistened fleck dark red to black. A capillary separation between the reaction products of the two hydrides is thus effected with silver nitrate. Filter paper that has been moistened with more dilute silver nitrate solutions (e.g., 25 %), or with ammoniacal silver nitrate solutions, is blackened immediately by both gases. Hydrogen sulfide and hydrogen phosphide give similar colorations (by sulfide or phosphide formation) so that these hydrides must not be present.* Mercury salts interfere, because metallic mercury is deposited on the surface of the zinc in the generator and slows up or prevents the formation of nascent hydrogen and consequently of AsH_3.[137]

Procedure. The glass gas apparatus, described on page 48, Fig. 26, is used. A drop of the test solution is mixed with a few grains of zinc and dilute sulfuric acid. The apparatus is closed with the funnel with a flat rim, on which is placed the filter paper moistened with 20% silver nitrate solution. In the presence of arsenic, a gray fleck is formed.

Limit of Identification: 1 γ arsenic
Limit of Dilution: 1 : 50,000

When antimony and mercury salts are present, it is best to warm the caustic alkali solutions with metallic aluminum.[138] Arsenic hydride is liberated, whereas antimony salts do not give stibine under these conditions. Mercury salts should be reduced beforehand through digestion with powdered antimony. If arsenates are present, they should be converted to arsenite or arsenious acid by treatment with sodium bisulfite in acid solution.

(*b*) *With gold chloride*

The decomposition of arsenic hydride by means of gold chloride is more sensitive than by silver nitrate. The metallic gold formed by the reaction:

$$AsH_3 + 2\ AuCl_3 + 3\ H_2O = 2\ Au^\circ + 6\ HCl + H_3AsO_3$$

is clearly visible, even in minute amounts, because it is very finely divided.

* It has been suggested [136] to fix H_2S, PH_3 and SbH_3 by means of cotton wool impregnated with Cu_2Cl_2, and then identifying the AsH_3 on mercuric chloride paper by the formation of the yellow $As(HgCl)_3$. This procedure can carried out in the spot test technique.

Procedure. The apparatus is charged with a piece of arsenic-free zinc (size of a small pea) and 3 drops of 2 *N* sulfuric acid added. The acid should be of highest purity. The funnel is covered with filter paper impregnated with 1 % solution of chlorauric acid. After 10 to 15 minutes in the dark, the paper is examined for any stain that might appear (due to arsenic in the reagents). If the paper remains bright yellow, it is moistened with a further drop of gold chloride solution, and 4 more drops of 2 *N* sulfuric acid are placed in the bottom of the apparatus followed by a drop of the test solution. The stopper is replaced at once. In the presence of arsenic, a blue to blue-red fleck appears on the paper after 10 to 15 minutes.

Limit of Identification: 0.5 γ arsenic
Limit of Dilution: 1 : 100,000

(*3*) *Test with silver nitrate*

Alkali and alkaline earth arsenates form red-brown silver arsenate, insoluble in acetic acid:

$$AsO_4^{-3} + 3\ Ag^+ \rightarrow Ag_3AsO_4$$

If the arsenic is present in the form of arsenious acid, arsenite, sulfide, alkali sulfoarsenite, or sulfoarsenate, it should be converted into arsenate by treatment with ammonia and hydrogen peroxide. The test is only specific in the absence of chromates and ferricyanides; they also give colored silver salts, insoluble in acetic acid.

Procedure. A drop of the test solution, or a few grains (or more) of the solid sulfide, according to the expected arsenic content, is placed in a micro crucible and warmed with a few drops of ammonia and 10 % hydrogen peroxide. Sulfides must be completely dissolved, or made colorless. The mixture is then acidified with dilute acetic acid, and 1 or 2 drops of 1 % silver nitrate solution added. A red-brown precipitate or coloration appears if arsenic is present.

Limit of Identification: 6 γ arsenious acid
Limit of Dilution: 1 : 8,000

(*4*) *Test for alkali arsenites in the presence of arsenates*

Arsenites can be detected in the presence of arsenates by method (*a*) or (*b*):
(*a*) *Reduction of iodine to iodide.* The reaction of arsenious acid with iodine, and of arsenic acid with iodides, is reversible:

$$As_2O_3 + 2\ I_2 + 2\ H_2O \rightleftarrows As_2O_5 + 4\ I^- + 4\ H^+$$

In alkaline bicarbonate or acetic acid solution the reaction proceeds from left to right, consuming iodine. Thus, in the absence of other iodine-consuming substances, a decolorization of iodine is characteristic of arsenious acid.

References pp. 247–257

Procedure. A drop of the test solution is mixed on a spot plate with a drop of dilute sulfuric acid and a few mg of sodium bicarbonate, or else the solution is acidified with dilute acetic acid. A drop of a starch-iodine solution is then added. If the blue color disappears on stirring with a glass rod, arsenious acid is present.

Limit of Identification: 5 γ arsenic
Limit of Dilution: 1 : 10,000
} in the presence of any amount of As_2O_5

(*b*) *Reduction of arsenious acid to arsine.* In acid solutions, arsine is formed by the action of nascent hydrogen on either arsenic pentoxide or trioxide. The gas can be detected by the "Gutzeit" test (see page 101). However, in alkaline media, such as sodium hydroxide solution, only arsenites are reduced to arsine by metallic aluminum, while arsenates remain unchanged.

Therefore the test for arsenic trioxide in the presence of arsenic pentoxide can be carried out as described on page 101, except that the reduction in the reaction vessel is effected with aluminum foil and sodium hydroxide.

(*5*) *Other test for arsenic*

As^{III} can be detected in a drop of test solution on filter paper by the addition of a drop of concentrated hydrochloric acid, a drop of 0.5 % kairine (*n*-ethyl-*o*-hydroxytetrahydroquinoline) and a drop of 1 % ferric chloride solution. The paper is gently warmed; a red-brown stain is formed (*Idn. Limit:* 0.005 γ As).[139] As^{V} compounds must be reduced beforehand with hydroxylamine sulfate. Hg, Pb, and Cu salts interfere.

8. Antimony

(*1*) *Test by reduction to antimony*[140]

Soluble and insoluble antimony compounds are reduced to elementary antimony by nascent hydrogen. If the reduction is carried out on platinum foil or on the lid of a platinum crucible, with zinc and hydrochloric acid (or tin and hydrochloric acid)[141] the antimony appears as a black precipitate or forms a brown stain on the platinum. The antimony is always deposited first on any rough portion of the platinum surface.

Arsenic compounds are mainly reduced to the gaseous hydride, although small amounts of free arsenic are deposited on the zinc or appear as black flecks in the liquid. Tin compounds are reduced to metallic tin, which forms a gray platinum-colored deposit on the platinum. The presence of tin does not interfere with the test for antimony, but rather increases the sensitivity by providing nuclei for the deposition of small amounts of the antimony.

Procedure. A drop of the test solution, which must be free from nitric acid, is mixed with a drop of concentrated hydrochloric acid. A small piece of zinc is added, and the mixture is allowed to stand on platinum foil for 5 to 30 minutes, according to the amount of antimony present. The test can also be carried out on a watch glass. [142] A short piece of platinum wire is placed in contact with the zinc, and a drop of the strongly acid test solution is added. The zinc and platinum must remain in contact. When antimony is present, the platinum begins to darken, and then the antimony separates, in the form of glittering plates, on the surface of the platinum. Any tin adheres, in spongy form, to the zinc.

Limit of Identification: 20 γ antimony
Limit of Dilution: 1 : 2,500

The reduction on platinum foil is recommended for the detection of antimony oxide or antimonic acid in solid metastannic acid. The sample is mixed with dilute hydrochloric acid on platinum foil, gently warmed for a few minutes, and then a small piece of zinc is added. A black or brown fleck is visible on the platinum, after the surface has been washed with water, if the sample contains antimony.

(2) *Test with phosphomolybdic acid* [143]

As discussed on page 85, molybdenum has an increased reactivity in the complex phosphomolybdic acid. It can be smoothly reduced to molybdenum-blue by certain substances which have little or no effect on normal molybdates. Antimony^{III} salts also have this property when warmed but they react only with free phosphomolybdic acid or its soluble salts, and not with insoluble phosphomolybdates (difference from stannous chloride, see page 108).

In the absence of stannous salts, the following test is specific for antimony. It is very useful, in the ordinary systematic analysis, for determining whether Sb_2S_3 is present in the sulfides precipitated by mineral acids from alkaline sulfo solutions. This test is carried out in the solution obtained by warming the precipitated sulfides with 1 : 1 hydrochloric acid. In this solution, the antimony is always present as trichloride, and the tin as tetrachloride; the latter does not interfere with the test.

It is simpler to warm a little of the alkaline sulfo solution with concentrated sulfuric acid until a clear solution is obtained. The solution then contains Sb^{III} sulfate, Sn^{IV} sulfate and arsenic acid.

Procedure. A drop of the test solution is placed on filter paper impregnated with a 5 % aqueous solution of phosphomolybdic acid,[144] and held over steam. In a few minutes a blue coloration appears; it is more or less intense according to the amount of antimony present.

Limit of Identification: 0.2 γ antimony
Limit of Dilution: 1 : 250,000

(*3*) *Test with 9-methyl-2,3,7-trihydroxy-6-fluorone* [145]

Antimony^{III} salts, in acid solution, react with aromatic *o*-dihydroxy compounds (e.g., pyrocatechol, pyrogallol) to produce slightly soluble, white, or light yellow crystalline precipitates. These can be formulated as cyclic esters of antimonous acid.[146] The yellow dyestuff 9-methyl-2,3,7-trihydroxy-6-fluorone (I), which can be viewed as a derivative of pyrocatechol, forms a red antimonous compound (II).

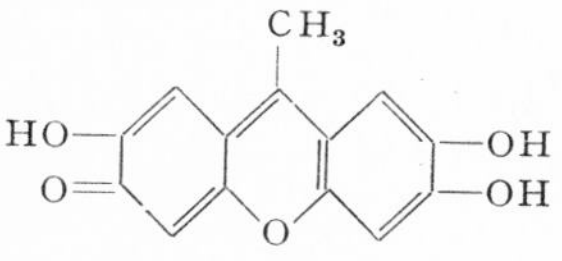

CH_3 / HO— / O= / —OH / —OH / O

(I)

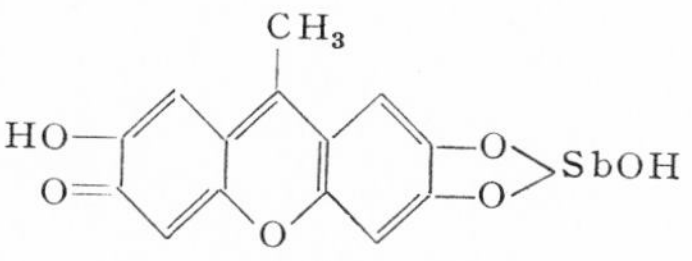

CH_3 / HO— / O= / —O / —O / >SbOH / O

(II)

In neutral or weakly acid solution, the reagent also forms orange-red or orange-brown precipitates or colorations with a number of other heavy metal ions. In 1 *N* hydrochloric acid solution, in addition to antimony^{III}, molybdenum (orange-red), germanium (orange), tin^{IV} (bright orange) also react. If the reaction is carried out as a spot test on paper with subsequent treatment with hydrochloric acid containing hydrogen peroxide, the molybdenum compound is decomposed with production of light yellow permolybdic acid. When antimony^{V} is present, preliminary reduction to antimony^{III} is necessary. Metallic magnesium is used as reductant.

Procedure.[147] A drop of the alcoholic reagent solution is placed on filter paper and dried in the air. A yellow fleck remains. (Paper prepared in this way is stable, whereas the alcoholic reagent solution decomposes on standing.) A drop of the test solution, which is approximately 1 *N* with respect to hydrochloric acid, is brought on the reagent paper and then treated with two or three drops of hydrochloric acid containing peroxide. A red fleck indicates the presence of antimony ^{III}.

Limit of Identification: 0.2 γ antimony
Limit of Dilution: 1 : 250,000
Reagents: *1*) Freshly prepared solution of 0.017 g of 9-methyl-2,3,7-trihydroxy-6-fluorone [148] in a mixture of 5 ml of 2 *N* hydrochloric acid and 5 ml of 95 % ethyl alcohol
2) Hydrochloric acid—hydrogen peroxide. (6 % H_2O_2 in 1 *N* HCl)

(*4*) *Test with rhodamine B* [149]

If solutions of antimony^{V} salts, in strong hydrochloric acid, are brought together with aqueous solutions of xantho dyestuffs, the red color changes to violet or blue, and a finely divided precipitate of the same color separates at once or after standing. Rhodamine B (tetraethylrhodamine), whose chloride has the structure (I), is a suitable reagent for antimony. The presence of strong hydrochloric acid is essential to the occurrence of the reaction or, if sulfuric acid solutions are presented for testing, it is necessary to add much alkali chloride. Antimony is present in such solutions as $H[SbCl_6]$. This fact, as well as an analogous precipitating action of molybdic acid and tungstic acid toward rhodamine B, argues for assigning structure (II) to the insoluble antimony^{V}—rhodamine B compound.[150]

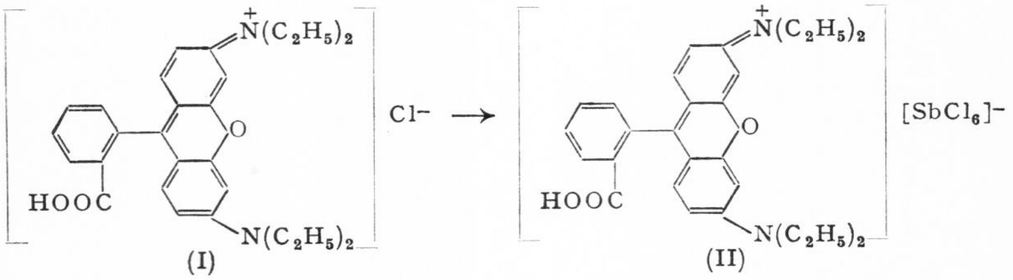

$\overset{+}{N}(C_2H_5)_2$; O ; HOOC ; $N(C_2H_5)_2$ (I) $Cl^- \rightarrow$ $\overset{+}{N}(C_2H_5)_2$; O ; HOOC ; $N(C_2H_5)_2$ (II) $[SbCl_6]^-$

Small amounts of antimony may be detected in the presence of much tin by means of rhodamine B. In the course of systematic analysis, antimony^{III} chloride is always formed, together with stannic chloride, when the acid sulfides precipitated from alkaline sulfo solutions are dissolved in 1 : 1 hydrochloric acid. Hence oxidation to antimony^{V} is necessary. This is easily accomplished by sodium nitrite.

Small amounts of iron do not interfere. Mercury, gold and thallium chlorides, and any basic bismuth chloride, precipitated by hydrolysis, likewise molybdates and tungstates in acid solution, give color reactions with rhodamine B similar to those given by antimony salts.

Procedure. One ml of the dyestuff solution is treated on a spot plate with a drop of the test solution, which should be made strongly acid with hydrochloric acid, and if necessary oxidized beforehand, by sodium nitrite. In the presence of antimony, the bright red (fluorescent) dye solution changes to violet.

Limit of Identification: 0.5 γ antimony } in the presence of 12,500 times the
Limit of Dilution: 1 : 100,000 } amount of tin
Reagent: Rhodamine B solution (0.01 g dyestuff in 100 ml water)

Antimony^{III} iodide, which exists in aqueous solution as $H[SbI_4]$, can be extracted with benzene. When the colorless benzene solution comes in contact with an aqueous solution of rhodamine B, a violet-red color appears.[151] This is probably due to an addition compound of SbI_3 and rhodamine B. It is claimed that this reaction is specific for the detection of small amounts of antimony.

Procedure. A drop of the solution to be tested is treated in a test tube with 5 drops of 1 : 3 sulfuric acid followed by a drop of 10 % potassium iodide solution. The solution is shaken vigorously with 1 ml of benzene. The benzene layer is removed with a pipette and placed in a depression of a spot plate. A drop of 0.2 % solution of rhodamine B is placed in the center of the depression. The presence of antimony is indicated by the violet antimony-rhodamine B complex diffusing into the benzene. If it is not certain that nitrites are absent, a few milligrams of solid urea should be added to the test solution prior to adding the potassium iodide. If oxidizing agents are present, solid sodium sulfite should be added, just prior to the extraction, until the color due to free iodine is discharged. A blank conducted on 1 to 3 milliliters of sulfuric acid shows only a very faint pinkish hue in the final benzene layer.

Limit of Identification: 0.2 γ antimony

Limit of Dilution: 1 : 250,000

The same identification limits (at greater dilution) are obtained if larger volumes of the test solution are used.

If a freshly prepared benzene solution of the dyestuff is used rather than an aqueous solution of rhodamine B, it is possible to detect as little as 0.05 γ antimony in drops (1 : 1,000,000). [152] Under these conditions, however, red benzene solutions are given by bismuth, gold, and mercury salts, and also by molybdates and tungstates, though only to a slight extent.

(*5*) *Test by luminescence in the hydrogen flame* [153]

Antimony salts impart a green-blue luminescence to the hydrogen flame. Extremely small amounts of antimony can be detected by this effect. The sample is mixed with alkaline earth carbonate, ignited gently, and touched with the hydrogen flame under suitable conditions. For the method of carrying out luminescence tests see page 79.

9. Tin

(*1*) *Test with dithiol* [154]

Dithiol is the trivial name of 4-methyl-1,2-dimercaptobenzene (I). Its solution in alcohol or dilute alkali reacts with weakly acid solutions of sulfide-forming metal ions to yield yellow to black precipitates. The magenta stannous salt (II or IIa) is formed from strong hydrochloric acid solutions of either Sn^{II} or Sn^{IV} salts. Sn^{IV} salts are reduced by the dimercaptan to give the aromatic disulfide. Since this disulfide is not soluble in water, it is better to use thioglycolic acid as reductant in all cases; it is oxidized to water-soluble dithioglycolic acid.

H_3C—C_6H_3(—SH)(—SH) (I) H_3C—C_6H_3(—S—)(—S—)Sn (II) H_3C—C_6H_3(—S—)(—S—)Sn(—S—)(—S—)C_6H_3—CH_3 (IIa)

Procedure. A micro test tube is used. One drop of the hydrochloric acid test solution is treated with 1 or 2 drops of the reagent solution and warmed. Depending on the quantity of tin present, a red precipitate or color is produced. When testing for tin^{IV} compounds, a little thioglycolic acid should be added prior to the introduction of the reagent solution.

Limit of Identification: 0.05 γ tin
Limit of Dilution: 1 : 1,000,000
Reagent: 0.2 g dithiol dissolved in 100 ml of 1 % sodium hydroxide.

The test just described is impaired especially by molybdate (even in small amounts) and also by considerable quantities of other metal salts. To attain complete specificity, the following procedure is recommended. The test solution is treated with an excess of yellow ammonium sulfide and then filtered or centrifuged. The clear filtrate (centrifugate) contains the sulfo salts of As, Sb, Sn, Te, Mo, W, V. If the solution is warmed with excess hydrogen peroxide, oxidation ensues with production of ammonium salts of the respective metallo acids and precipitation of $Sn(OH)_4$ and $Sb(OH)_5$. After centrifuging, the precipitate is dissolved in dilute hydrochloric acid and the test with dithiol carried out. If As, Sb, Mo, W, and V are absent, it is sufficient to acidify the sulfoalkaline solution prior to adding the organic reagent. The whole procedure can be translated to the technique of spot test analysis.

(2) *Test with ammonium phosphomolybdate*[155]

Stannous chloride reduces not only molybdates to the colored lower oxides,[156] but also reacts with phosphomolybdic acid and its salts. "Molybdenum blue" results. It is an important fact, especially for the detection of tin^{II} in the presence of antimony^{III} salts, that stannous chloride will reduce not only the soluble phosphomolybdic acid as does antimony trichloride (compare page 104), but also the insoluble phosphomolybdates (e.g., the potassium or ammonium salt). Concerning molybdenum blue see copper test *5*, page 84.

Procedure. Filter paper, impregnated with an aqueous 5 % solution of phosphomolybdic acid, is held for a short time over ammonia to form yellow, insoluble ammonium phosphomolybdate, and then dried. This paper will keep in the dark, in well stoppered bottles.

When a drop of the solution to be tested for stannous salt is placed on this paper, a more or less deep blue color forms, according to the amount of tin present.

Limit of Identification: 0.03 γ tin
Limit of Dilution: 1 : 670,000

Test for tin in the presence of antimony

If the tin occurs originally in the bivalent condition, the test may be carried out as described, and antimony salts do not interfere. If however, quadrivalent tin is present, it must be reduced to the stannous condition. This can be done by placing 1 or 2 drops of the test solution in a small porcelain crucible, and warming with either a small strip of magnesium ribbon or a tiny piece of zinc, plus a few drops of concentrated hydrochloric acid, until the metal dissolves. A drop of this solution is placed on ammonium phosphomolybdate paper. Any metallic antimony, which is precipitated as a black powder, does not interfere with the test for tin.

(3) Test with cacotheline[157]

The composition of cacotheline, which is a nitro compound of brucine, is not known.* When treated with acid tin^{II} solutions the light yellow solution turns violet due to the formation of a reduction product. Other reducing compounds such as hydrogen sulfide, sulfites, thiosulfates and antimony^{III} salts also react with cacotheline.[158] Iron^{II} salts (in the absence of fluorides and phosphates) do not react with this reagent; consequently, a tin solution, reduced with iron wire, can be used for the test. Colored metallic salts interfere, if present in considerable quantity. Ti^{III}, U^{IV}, Rh^{III}, Nb^{III} salts, and lower oxides of Mo and W [159] also interfere. A violet color is given by compounds of As, Sb and Te in the presence of metallic zinc and acids.

When carried out as a spot test on filter paper, the test is more sensitive than in a test tube (test tube *Idn. Limit:* 30 γ Sn in 3 ml).

Procedure. Filter paper is impregnated with a saturated aqueous solution of cacotheline[160]. A drop of the test solution is added before the paper is quite dry. According to the amount of tin present, a red circle or ring surrounded by a colorless zone is formed on the yellow paper.

Limit of Identification: 0.2 γ tin
Limit of Dilution: 1 : 250,000

(4) Test by flame color [161]

When hydrogen is evolved from zinc and hydrochloric acid in the presence of soluble tin compounds, the gas gives a characteristic blue color in the nonluminous flame of a Bunsen burner. The reason for this phenomenon is not known; it is not, however, due to the formation of tin hydride, as was first assumed.[162]

A portion of the solid or liquid is mixed with a considerable amount of concentrated hydrochloric acid in a porcelain dish, and a few granules of pure zinc stick added. The mixture is stirred with a test tube filled with cold water. If the portion of the test tube, which has been dipped in the mixture, is held in a nonluminous Bunsen flame, a characteristic blue flame

* According to N. Moufang and J. Tafel, *Ann.*, 304 (1898) 47, cacotheline is the nitrate of bisdimethylmononitrobrucine hydrate. The empirical formula is $C_{21}H_{21}(OH)_2NO_2 \cdot N_2O_3 \cdot HNO_3$. See also H. Leuchs, *Ber.*, 55 (1922) 564.

mantle forms around the test tube, revealing the presence of even very small quantities of tin. This flame color is specific for tin. The test may be carried out on a single drop of the solution, in which case the reducing power of the gases of the flame is sufficient.[163] The presence of arsenic, if more than equivalent to the amount of tin, causes this very characteristic reaction to fail. Hydrated metastannic acid, prepared in the wet way and ignited, responds decisively to this test, but native tinstone (cassiterite) does not. The mineral must be decomposed with sodium potassium carbonate (e.g., by fusion in a loop of platinum wire) prior to the test.

Procedure. A drop of the test solution is placed on a magnesia rod and evaporated by holding the support about 0.5 cm from a very small Bunsen flame. The residue is moistened with a drop of concentrated hydrochloric acid, and held in the reducing portion of a microflame. In the presence of tin, a blue flame mantle forms around the rod. If the amount of tin present exceeds 0.25 γ, the flame color will reappear several times, if the residue is moistened with concentrated hydrochloric acid between heatings.

Limit of Identification: 0.03 γ tin
Limit of Dilution: 1 : 1,660,000

(5) *Test through reaction with morin* [164]

Many water-insoluble colorless metal hydroxides and hydrated oxides are brought to vigorous fluorescence on contact with an alcohol or acetone solution of morin, i.e., 3,5,7,2′,4′-pentahydroxyflavone (for structure see page 183). The fluorescence, which is especially discernible in ultraviolet light, is due to an inner complex binding of the morin in the form of insoluble metal morinates or as adsorption compounds with the metal hydroxide (hydrated oxide). The reaction between morin and stannous or stannic hydroxide (alkali stannite and stannate also react) is extremely sensitive. For example, the addition of ammonia to 10 ml of a solution of $SnCl_2$ or $SnCl_4$ diluted to 1 : 500,000 produces no visible turbidity. If, however, a drop of this dilute tin solution is placed on filter paper, developed over ammonia, and then spotted with an acetone solution of morin, a fleck appears whose intense light green fluorescence is stable against dilute acetic acid. Since other metal hydroxides also produce fluorescences with morin, this procedure is not applicable for the direct detection of tin. Nevertheless, the morin test is specific if one proceeds from a sulfoalkaline solution, where SnS_3^{-2} ions are present. The solution of alkali- or ammonium sulfostannate is readily decomposed by hydrogen peroxide, the total sulfide- and polysulfide sulfur being oxidized to sulfate, and the tin is converted to stannic hydroxide or alkali stannate. These tin compounds respond to spotting with morin even though hydrogen peroxide is present. Any excess morin, which has a yellow self-fluorescence, can be removed by bathing the paper in an alcohol-acetic acid mixture.

Procedure. The test solution is treated with an excess of yellow ammonium sulfide and filtered. A drop of the filtrate is treated in a micro test tube with a drop of 5 % hydrogen peroxide; decolorization ensues. After 1–2 minutes, a drop of the solution is placed on filter paper (Schleicher & Schüll, No. 589 Red Ribbon), the fleck held over ammonia and then spotted with a drop of 0.02 % morin solution in acetone. The paper is bathed in a 20 : 1 alcohol-glacial acetic acid mixture for about four minutes and then examined in ultraviolet light. A yellow-green fluorescence indicates the presence of tin. A comparison test with a drop of dilute ammonia is necessary only when minimal amounts of tin are suspected.

Limit of Identification: 0.05 γ tin
Limit of Dilution: 1 : 1,000,000

(6) Other tests for tin

(*a*) Sn^{II} salts can be detected in acid solution by their reducing action on Fe^{III} salts. The red color produced with ammoniacal dimethylglyoxime serves to detect the Fe^{II} salt formed (see page 164) (*Idn. Limit:* 0.04 γ Sn).[165] α,α'-Dipyridyl can be used in place of dimethylglyoxime to reveal the Fe^{II} produced [166] (compare page 161).

(*b*) The blue azo dyestuff, diazine green S(K), is reduced by Sn^{II} salts, giving a violet to red color (*Idn. Limit:* 2 γ Sn). [167]

(*c*) Filter paper impregnated with the green photolysis product of 2-benzylpyridine is turned red on treatment with a hydrochloric acid solution of $SnCl_2$ (*Idn. Limit:* 1.3 γ Sn in 0.01 ml).[168] As^{III} and Sb^{III} salts do not interfere, but the same color change is effected by sulfites.

(*d*) Tin may be detected by placing a drop of the solution, which has been reduced, if necessary, with metallic magnesium, on filter paper impregnated with mercuric chloride.[169] The spot is then alkalinized with aniline. Only tin^{II}, but not $antimony^{III}$, is thus reduced to metal (brown fleck). (*Idn. Limit:* 0.6 γ Sn.)

(*e*) One drop of the hydrochloric acid solution containing Sn^{IV} is treated on a spot plate with a drop of 0.1 % solution of 1,2,7-trihydroxyanthraquinone in methyl alcohol. The mixture is first made alkaline with ammonia and then acidified with 33 % acetic acid. The orange coloration is due to the formation of an adsorption compound (lake) between stannic acid and the dyestuff (*Idn. Limit:* 0.2 γ Sn). [170] Ti, Zr, Al, Fe and Cr salts produce colorations, Bi^{+3} and MoO_4^{-2} give colored precipitates.

(*f*) The red solution of anthraquinone-1-azo-4-dimethylaniline hydrochloride gives a blue-violet precipitate, $(C_{22}H_{17}O_2N_3)_2 \cdot H_2SnCl_6$, with Sn^{IV} salts in hydrochloric acid solution.[171] Filter paper impregnated with the reagent may be used (*Idn. Limit:* 0.01 γ Sn). Zn, Cd, $[PbCl_6]^{-2}$, and Hg ions react similarly; Sb^{III}, U, Mo, Ir, Tl, and Ga ions, in considerable quantities, produce violet colorations.

(*g*) Strong hydrochloric acid solutions of $SnCl_2$ reduce the red-brown oxidation product of *o*-aminophenols to a green compound. When conducted as a spot reaction on paper, a central red fleck is surrounded by a green to blue-green ring (*Idn. Limit:* 0.06 γ Sn).[172]

10. Germanium

(*1*) *Test by the formation of an acid complex with mannite* [173]

Germanic acid, like boric acid, forms complex compounds with the polyvalent alcohols, such as glycerol, mannite, glucose.[174] The resulting compounds are strong monobasic acids.[175] Therefore, on mixing a germanate solution or Ge^{IV} oxide with alkali and mannite, a decrease in pH indicates the presence of germanium. In the absence of boron (compare page 343) this reaction is specific for germanium.

Procedure. A drop of the slightly acid germanate solution is mixed with a drop of phenolphthalein solution, and 0.01 *N* sodium hydroxide is added until the red color appears. On the addition of mannite (solid) the red color is partly or completely discharged.

Limit of Identification: 2.5 γ germanium
Limit of Dilution: 1 : 20,000

(2) *Test with ammonium molybdate and benzidine* [176]

In solutions containing mineral acids, germanates react with excess molybdates to form the water-soluble germanimolybdic acid $H_8Ge(Mo_2O_7)_6$. In such heteropoly acids, the complexly bound MoO_3-molecules have a greater tendency to enter into redox reactions than the Mo^{VI} in normal molybdates or molybdic acid. The enhanced reactivity is due to the coordinative binding of the molybdenum; this is proved by the ready reducibility of the molybdenum in phosphomolybdic, arsenimolybdic and silicomolybdic acid (compare page 335). When benzidine is treated with an acid solution of germanimolybdic acid and buffered with sodium acetate or ammonia, lower molybdenum oxides (molybdenum blue, see page 104) and a blue oxidation product of benzidine (benzidine blue, see page 72) result. Under carefully controlled conditions, this color reaction affords a sensitive and specific test for germanium.

Procedure. A drop of the alkaline or slightly acid test solution is treated, on a spot plate or filter paper, with a drop of the molybdate solution, and then with a drop of 0.1 % acetic solution of benzidine in order to form the benzidine salt of germanimolybdic acid. The mixture is then buffered with a few drops of a saturated solution of sodium acetate, or, when the test is carried out on filter paper, held over ammonia. A light or dark blue stain is formed, according to the amount of germanium present.

Limit of Identification: 0.25 γ germanium (in 0.025 ml)

Limit of Dilution: 1 : 100,000

Reagent: Ammonium molybdate solution: 1 g $(NH_4)_2MoO_4$ dissolved in 10 ml water and 10 ml conc. HNO_3 added

This method for the detection of germanium is only decisive in the absence of certain other materials. Apart from compounds which reduce molybdates directly—e.g., Sn^{II}, Fe^{II}, As^{III}, and Se^{IV}—arsenic acid, phosphoric acid and silicic acid should not be present, as they also form heteropoly molybdic acids, which enter into the same redox reaction with benzidine. The germanium can however be distilled out of hydrochloric acid solution (3.5–4.0 *N*) as Ge^{IV} chloride. The molybdate–benzidine test is then carried out with the distillate.

Procedure. One or two drops of the slightly alkaline test solution is evaporated in the small glass beaker (4 ml capacity) of the apparatus shown in Figure 38. After cooling, 2 or 3 drops of 4 *N* hydrochloric acid are added. The beaker is covered with the small bulb filled with water, and gently heated over a wire gauze. The germanic chloride vapor condenses on the glass knob of the condenser. When the drop is large enough, the heating is discontinued, and the drop is carefully transferred to a spot plate. The test is then carried out using 5 or 6 drops of saturated sodium acetate solution to inactivate the excess of hydrochloric acid.

Limit of Identification: 2 γ germanium

The distillation from hydrochloric acid solution does not effect a complete separation from all interfering substances. Tervalent arsenic and quadrivalent selenium are also volatilized, and these considerably reduce the sensitivity of the germanium test. Furthermore, As^V may be reduced to As^{III} in the process of distillation from hydrochloric acid solution, and so arrive in the distillate as $AsCl_3$. The tests for germanium in the presence of selenium and arsenic can only be successfully carried out if the interfering elements are converted to the higher oxides beforehand by evaporation with 30 % hydrogen peroxide and 1 or 2 drops of ammonia. The dry residue is then treated with a small crystal of potassium permanganate and 2 drops of 4 *N* hydrochloric acid. The germanium chloride can then be distilled, and the test

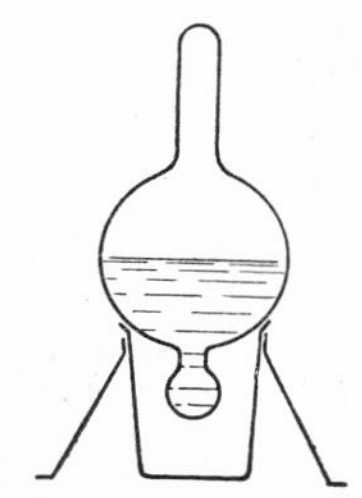

Fig. 38. Microbeaker fitted with water-filled bulb condenser ($\frac{3}{4}$ actual size)

carried out on a drop of the distillate. Any free chlorine is destroyed by means of a drop of a 5% sodium sulfite solution, after the addition of the molybdate reagent. In this way, 10 γ germanium may be detected in the presence of 1000 times the amount of selenium (as SeO_2), or 800 times the amount of arsenic.

(3) Test with 9-phenyl-2,3,7-trihydroxy-6-fluorone [177]

A pink precipitate or coloration is produced by the action of acid solutions of alkali germanates with the yellow alcohol solution of 9-phenyl-2,3,7-trihydroxy-6-fluorone (I). The composition of the reaction product is not known. In view of the phenolic nature of the dyestuff serving here as reagent, the possibility of ester formation as indicated by (II) is not excluded. Consideration must also be given to the production of an adsorption compound (color lake) [178] between GeO_2·aq and (I), with participation of the =C(OH)—CO— or the =C(OH)-group of the dyestuff molecule as bonding site.

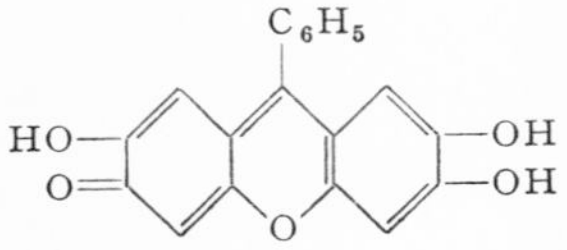

C_6H_5 HO— O= —OH —OH O

(I)

C_6H_5 HO— O= —O —O >GeO O

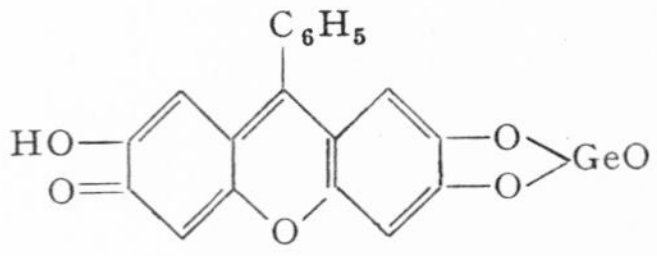

(II)

The acid-resistance of the reaction product is most noteworthy, and this characteristic, together with the color quality, results in a specific test for germanium.

Procedure. A drop of the alcoholic solution of the dyestuff is placed on filter paper (S & S 589 g). The yellow fleck, which remains after the alcohol has evaporated, is spotted with the acid test solution (between 3–6 *N* with respect to hydrochloric acid). Subsequent spotting with 2 or 3 drops of nitric acid produces a pink fleck.

Limit of Identification: 0.13 γ germanium
Limit of Dilution: 1 : 300,000
Reagent: 0.05 % solution of 9-phenyl-2,3,7-trihydroxy-6-fluorone [179] in 95 % alcohol, acidified with one drop of 6 *N* hydrochloric acid. The solution is stable.

(4) Other tests for germanium

(*a*) Addition of a drop of a saturated alcoholic solution of diphenylcarbazone to an alkaline germanium-molybdate mixture produces a purple color. The color changes to deep blue on the addition of concentrated hydrochloric or sulfuric acid. [180] The chemistry of the reaction may be analogous to that of Test *2* (*Idn. Limit:* 0.01 γ Ge).

(*b*) The similarity between boric acid and germanic acid (see page 112) is also shown in the behavior toward hydroxyanthraquinones, which give a color change in sulfuric acid solution with boric acid (see page 340). Germanium dioxide or germanic acid also react. In this way, 5 γ Ge (1 : 10,000) may be detected.[181] The dyestuff *p*-nitrobenzeneazochromotropic acid behaves similarly, but the reaction is less sensitive. These color reactions of germanic acid, like those of boric acid, probably are due to ester-formation with the phenolic OH-groups of the particular dyestuff.

(*c*) If one drop of a neutral germanium solution is warmed with a drop of freshly prepared solution of ammonium molybdate and hydroxylamine hydrochloride or alkali sulfite, a blue-green color appears. It fades on cooling and reappears when warmed again (*Idn. Limit:* 0.4 γ Ge).[182] Oxidizing and reducing agents interfere, and likewise considerable quantities of PO_4^{-3} or AsO_4^{-3} ions.

11. Molybdenum

(*1*) *Test with potassium thiocyanate and stannous chloride*[183]

Solutions of molybdates are colored yellow by the addition of potassium thiocyanate and hydrochloric acid. If a little zinc or other reducing agent is added, tervalent molybdenum is produced and combines with CNS^- ions to produce red, water-soluble $H_3[Mo(CNS)_6]$.

If phosphates, organic acids (oxalic and tartaric, etc.) as well as phosphoric acid, are present in the original solution, they also give stable complex compounds with molybdates, and the molybdate reaction may be completely inhibited or its sensitivity greatly reduced. Regard should also be paid to the presence of mercury salts and nitrites, which, under the conditions of the experiment, consume thiocyanate—by formation of the slightly dissociated mercury thiocyanate and soluble red nitrosyl thiocyanate (NOCNS), respectively. Tungstates form insoluble blue lower tungsten oxides, which decrease the sensitivity and specificity of the test. This interference may be lowered by using capillary separation.

Titanium, vanadium and uranium do not interfere with the reaction; only tungsten, as stated, gives a blue fleck of the lower tungsten oxides. This interference is prevented if a drop of hydrochloric acid is placed on the test paper; the insoluble tungstic acid is held back in the paper, while the molybdate diffuses through the capillaries to the edge of the drop, and can be detected there. The presence of formic or phosphoric acids reduces the sensitivity of the test.[184]

The test can be carried out on filter paper, using either stannous chloride or sodium thiosulfate as the reducing agent.[185]

Procedure. A drop of the test solution and a drop of a 10 % potassium thiocyanate solution are placed on filter paper previously moistened with 1 : 1

hydrochloric acid. In the presence of ferric iron, a red fleck (Fe^{III} thiocyanate) appears, which disappears on adding stannous chloride. A brick-red fleck of the Mo^{III} thiocyanate complex appears, if molybdate is present.

Limit of Identification: 0.1 γ molybdenum
Limit of Dilution: 1 : 500,000
Reagent: 5 % stannous chloride in 3 *N* hydrochloric acid

(2) *Test with potassium xanthate* [186]

In solutions containing mineral acids, a deep red-blue coloration is formed by the combination of molybdates with potassium ethyl xanthate $SC(SK)OC_2H_5$; with large amounts of molybdenum, there is a separation of almost black, oily drops. The product is the complex compound, $MoO_3 \cdot 2[SC(SH)(OC_2H_5)]$ [187] which is soluble in organic liquids, such as benzene, chloroform, carbon disulfide. To an extent, it is an organic analogue of $MoO_3 \cdot 2HCl$,* which likewise dissolves without decomposition in organic solvents. This very sensitive color reaction is quite specific for molybdates when starting from alkaline solutions in the absence of anions forming stable molybdenum complexes (fluorides, oxalates, tartrates etc.). Attention must be given to the fact that arsenious acid is xanthate-consuming since it forms $As[SCS(OC_2H_5)]_3$, which is soluble in chloroform.[188] Selenious acid seems to behave similarly.

Procedure. A drop of the nearly neutral or slightly acid test solution is mixed on a spot plate with a grain of solid potassium xanthate and treated with 2 drops of 2 *N* hydrochloric acid. A pink to violet color develops, depending on the quantity of molybdate present.

Limit of Identification: 0.04 γ molybdenum
Limit of Dilution: 1 : 250,000

The use of filter paper impregnated with water-insoluble zinc or cadmium xanthate can be recommended highly.[189] A red fleck or ring appears when a drop of the acid test solution is applied (*Idn. Limit:* 0.01 γ Mo). The reagent paper is prepared by immersing filter paper in $ZnSO_4$ or $CdSO_4$ solution, drying, and bathing in potassium xanthate solution. After washing with water, and drying, the paper is stable.

* This compound was first prepared by C. Debray, *Compt. rend.*, **46** (1858) **1101**. F. Ephraim, *Inorganic Chemistry*, 5th English ed., Londen and Edinburgh, 1948, p. 510, assigns structure (I) toit. The xanthate compound probably has the analogous structure (II). This assumption is supported by the fact that, as stated by Malowan (*loc. cit.*), the xanthic acid ethyl ester $SC(SC_2H_5)(OC_2H_5)$ does not react with molybdic acid. R. Montequi [*Chem. Abstracts*, **24** (1930) **3722**] assumes that the xanthate reaction involves a quinquevalent molybdenum compound.

$$\left[MoO\begin{matrix}(OH)_2\\Cl_2\end{matrix}\right] \qquad \left[MoO\begin{matrix}(OH)_2\\ [SCS(OC_2H_5)]_2\end{matrix}\right]$$

(I) (II)

(3) Test with phenylhydrazine [190]

Molybdates in acid solution give a blood-red color or red precipitate with phenylhydrazine. At high dilutions, a pink color appears. The chemistry of the reaction has not been fully explained; probably the phenylhydrazine is oxidized by molybdic acid to the diazonium salt. The latter couples with the excess phenylhydrazine and also involves a binding of molybdenum.

Procedure.[191] *a*) A drop of the test solution and of the reagent solution are mixed on a spot plate. A more or less deep red color appears according to the amount of molybdenum present. A comparison blank should be carried out for small amounts.

b) A drop of the reagent solution is placed on spot paper and, before it has soaked in, a drop of the test solution is added. A red ring forms around the spot after a few minutes, if molybdenum is present.

Limit of Identification: 0.32 γ molybdenum (on the spot plate); 0.13 γ molybdenum (on spot paper)

Limit of Dilution: 1 : 150,000 (on the spot plate); 1 : 318,000 (on spot paper)

Reagent: Mixture of phenylhydrazine and glacial acetic acid (1 : 2)

(4) Test with methylene blue and hydrazine [192]

Solutions of methylene blue with the cation (I) are not reduced by concentrated hydrazine sulfate solution even at the boiling temperature. If, however, tiny amounts of molybdate are added, the blue color disappears because of a rapid reduction to the leuco compound of the dyestuff, i.e. methylene white containing the cation (II). Accordingly, Mo^{VI} catalytically hastens the immeasurably slow redox reaction between methylene blue and

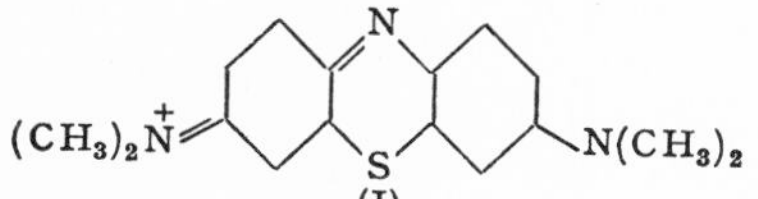

$(CH_3)_2\overset{+}{N}$ … N … S … $N(CH_3)_2$ (I) $(CH_3)_2\overset{+}{N}$ … H N … S … $N(CH_3)_2$ (II)

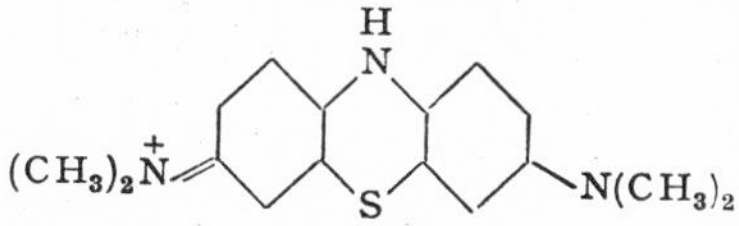

hydrazine. The basis of the catalytic action of the molybdenum is that Mo^{VI} is rapidly reduced to Mo^{V} by hydrazine in acid solution and on warming, and the Mo^{V} in turn reduces methylene blue to methylene white, with reoxidation to Mo^{VI}, which again is reduced by hydrazine, and so on. Therefore, this is an intermediate reaction catalysis made up of partial reactions (1) and (2) which involve Mo^{VI} and Mo^{V}. Addition of (1) and (2) yields (3), the net equation of the catalyzed redox reaction between methylene blue and hydrazine, in which molybdenum does not appear:

$$4\,MoO_3 + N_2H_4 \rightarrow 2\,Mo_2O_5 + N_2 + 2\,H_2O \qquad (1)$$

$$2\,Mo_2O_5 + 2\text{ methylene blue} + 2\,H_2O \rightarrow 4\,MoO_3 + 2\text{ methylene white} \qquad (2)$$

$$2\text{ methylene blue} + N_2H_4 \rightarrow 2\text{ methylene white} + N_2 \qquad (3)$$

References pp. 247–257

The catalytic action of molybdenum in the methylene blue-hydrazine system is so great that as little as 0.1 γ molybdenum can be detected in 10 ml of solution, which corresponds to a dilution of 1 : 10,000,000.

No analogous reaction is given by salts of U^{VI}, W^{VI}, Cr^{III}, Cr^{VI}, V^{IV}, V^{V}, Fe^{II}, nor by copper, cobalt and nickel salts. The presence of colored ions makes it difficult to see the decolorization of dilute methylene blue solutions. When tungsten is present, enough alkali fluoride must be added (production of $[WO_3F_2]^{-2}$ ions) to guarantee that no tungstic acid will be precipitated, since it would be partially reduced by hydrazine to blue insoluble W_2O_5, which likewise reduces methylene blue, though somewhat less rapidly than soluble Mo_2O_5.

When testing for molybdenum in mixtures containing metal salts, it is best to boil the test material with sodium carbonate or to fuse it with potassium sodium carbonate, to form water-soluble alkali molybdate. If a mixture of acid-insoluble sulfides is being tested, the sulfur can be roasted off and the alkali molybdate then dissolved out by digesting with alkali hydroxide.

Nitrates may not be present in the catalytic test for molybdenum, since in strongly acid solutions they react with hydrazine and also with methylene blue. Solid and dissolved nitrate can be destroyed completely by fuming with concentrated formic acid and ignition of the evaporation residue.[193]

Procedure.[194] A drop of the weakly acid test solution, in a micro test tube, is mixed with four drops of 0.0012 % aqueous methylene blue solution and about 20–30 mg of hydrazine sulfate. A parallel test is prepared with the same mixture and a drop of water. Both test tubes are placed in boiling water. With quantities exceeding 0.5 γ molybdenum, the color of the methylene blue is discharged within 3 minutes, whereas the blank remains unchanged. When smaller amounts of molybdenum are present, the fading of the blue is distinctly evident after 10 minutes heating and on comparison with the blank.

Limit of Identification: 0.012 γ molybdenum
Limit of Dilution: 1 : 4,160,000

(5) *Test with o-hydroxyphenylfluorone* [195]

A red or yellow color is produced by *o*-hydroxyphenylfluorone (9-phenyl-2,3,7-trihydroxy-6-fluorone) (compare detection of germanium, p.114) and its derivatives which contain one or two OH-groups in the benzene ring, when they are brought into reaction in 1 *N* hydrochloric acid solutions of antimony, tin, germanium, molybdenum, vanadium, iron, titanium and zirconium salts. The color reaction with *o*-hydroxyphenylfluorone is particularly intense and under the proper conditions it provides a specific test for molybdenum. The chemistry of the color reaction is not known. Corresponding to the phenol character of the reagent its action here may involve the production of a colored phenol ester of molybdic acid. It is remarkable that the color reaction is not impaired by alkali fluoride, a fact indicating that fluomolybdic acid or its ions behave in the same fashion as molybdate ions.

Procedure. Spot test paper (Mackeray, Nagel & Co. No. 640 W) is used. One drop of 0.1% alcoholic solution of *o*-hydroxyphenylfluorone is applied and dried at room temperature. The orange fleck is spotted with one drop of the 2 *N* hydrochloric acid test solution and then with a drop of 20% potassium fluoride solution followed by 2 or 3 drops of 0.5 *N* sulfuric acid. If molybdate is present a carmine fleck results. When dealing with slight amounts of molybdate it is best to conduct a blank.

Limit of Identification: 1.7 γ molybdenum
Limit of Dilution: 1 : 30,000

(*6*) *Other tests for molybdenum*

(*a*) On mixing molybdate and a tincture of cochineal, at pH 5.7 to 6.2, a flame-red fluorescence results. The test is carried out in capillary tubes (*Idn. Limit:* 0.02 γ MoO_3).[196] Boric acid gives a yellow fluorescence. With reference to interference of other ions, see the original paper.

(*b*) Molybdates on treatment with α,α'-dipyridyl and $SnCl_2$ give a violet color [197] due to a soluble addition compound of $MoCl_2$ and dipyridyl (*Idn. Limit:* 0.4 γ Mo). Interference due to tungstates can be prevented by complex salt formation with tartaric acid.

(*c*) Molybdate solutions containing mineral acid give a violet color with diphenylcarbazide [198] (*Idn. Limit:* 2.5 γ Mo). The effect is due to the diphenylcarbazone formed by oxidation through molybdenum^{VI}; this product reacts with molybdenum in some unexplained manner. It is likely that the reaction is analogous to that in the chromate-diphenylcarbazide reaction (see page 167). The test is more sensitive if di-β-naphthylcarbazone is used (*Idn. Limit:* 0.3 γ Mo). [199] Chromates interfere because of an analogous color reaction with carbazone (see page 167).

12. Tungsten

(*1*) *Test with 8-hydroxyquinoline (oxine)* [200]

If a solution containing an alkali tungstate is treated with a strong hydrochloric acid solution of oxine, colorless $WO_3 \cdot aq$ alone is precipitated, provided the concentration of WO_4^{-2} ions is adequate. In neutral or acetic acid solution, orange-yellow oxine tungstate (I) is precipitated on addition of a weakly acid or alcoholic solution of oxine.[201] This precipitate is not decomposed by concentrated hydrochloric acid but becomes red-brown.

Probably this is the result of the formation of an oxine ester of pyrotungstic acid (II). The conversion of (I) into (II) is the basis of a test for small quantities of tungsten.

(I) (II)

Solutions of alkali molybdate or vanadate containing 2000 γ Mo or 500 γ V per drop do not react in the same manner as tungstate; however even smaller amounts of these metallo acids prevent or impair the tungsten-oxine reaction. Nevertheless, the following procedure will reveal 100 γ tungsten in the presence of 500 γ molybdenum.*

Procedure. A spot plate is used. One drop of the neutral or weakly basic test solution is treated with one drop of a 5 % alcoholic solution of 8-hydroxyquinoline and then a drop or two of concentrated hydrochloric acid introduced. Depending on the amount of tungsten present, a brown precipitate or color appears.

Limit of Identification: 15 γ tungsten
Limit of Dilution: 1 : 33,333

(*2*) *Test with stannous chloride*

Tungstates are reduced by stannous chloride in strongly acid solutions to the blue lower tungsten oxides. This product, in contrast to "molybdenum blue" which forms under similar conditions, is sparingly soluble, and is stable in the presence of excess stannous chloride.

Procedure. One or two drops of the test solution are mixed on the spot plate with 3 to 5 drops of stannous chloride. In the presence of tungsten, a blue precipitate or coloration appears at once, or after a short time.

Limit of Identification: 5 γ tungsten
Limit of Dilution: 1 : 10,000
Reagent: 25 % solution of stannous chloride in conc. hydrochloric acid

Test in the presence of molybdenum

The test for tungsten in the presence of molybdenum can be carried out as a spot test.[203]

* It is possible that the interference by molybdate ions can by averted through their reaction with thiocarbazide. At pH = 2 there results a brown precipitate which can be extracted by a mixture (1 : 2) of acetone and *n*-butanol. Tungstate ions are not altered by this treatment. [202]

References pp. 247–257

Procedure. A drop of hydrochloric acid is placed on filter paper, and a drop of the test solution placed in the middle of the moist fleck. A bright yellow fleck of tungsten trioxide appears. When treated with potassium thiocyanate and stannous chloride (see page 115) the yellow fleck changes to blue, due to formation of lower tungsten oxides. Under these conditions, molybdenum gives a red fleck of $K_3[Mo(CNS)_6]$, which disappears on the addition of concentrated hydrochloric acid and hence does not interfere with the tungsten reaction.

Limit of Identification: 4 γ tungsten
Limit of Dilution: 1 : 12,500
Reagents: The same as in the molybdenum test, page 116

Ignited tungsten trioxide can be brought into solution by means of sodium peroxide, and detected as described. Similarly, tungsten in minerals can be detected after fusion with five times the bulk of sodium potassium carbonate.

Phosphoric acid and organic hydroxy acids (tartaric, citric, etc.), which form complex compounds with tungstic acid, hinder or prevent the precipitation of tungsten trioxide with mineral acids; certain other organic acids also impair the reaction to some extent.

(3) Test by catalysis of the Ti^{III}-malachite green reaction [204]

Many dyes, including malachite green (mal. gr.), are reduced rapidly to the corresponding colorless leuco compounds by titanium trichloride in warm neutral solution. The reduction is usually slow at room temperature and in acid solution. The slow reaction:

$$Ti^{III} + \text{mal. gr.} \rightarrow Ti^{IV} + \text{leuco mal. gr.}$$

is hastened by tungstate. This acceleration serves as the basis of a sensitive test for tungsten.

The catalysis rests on the fact that W^{VI} is reduced rapidly by Ti^{III} to W^V, which quickly reduces malachite green to the colorless leuco base. Consequently, this intermediate reaction catalysis involves two rapid partial reactions in which the catalyst (tungsten) participates. The summation of the partial reactions (1) and (2) gives the net catalyzed reaction (3) in which the catalyst does not appear.

$$Ti^{III} + W^{VI} \rightarrow Ti^{IV} + W^V \quad \text{(rapid)} \qquad (1)$$
$$W^V + \text{mal. gr.} \rightarrow W^{VI} + \text{leuco mal. gr.} \quad \text{(rapid)} \qquad (2)$$
$$Ti^{III} + \text{mal. gr.} \rightarrow Ti^{IV} + \text{leuco mal. gr.} \quad \text{(rapid)} \qquad (3)$$

The catalytic action of W^{VI} is nearly specific; only molybdenum reacts analogously, but the effect is less pronounced. Other metal ions that can be reduced by Ti^{III} have no such effect on the malachite green reduction.

Despite the specific catalytic action, consideration must be given to the presence of accompanying materials. First of all, metal ions interfere if they are reduced to lower valence stages or to the free metal, or if they form insoluble chlorides. $TiCl_3$ is consumed by these reductions, or the visibility of the reaction is impaired by the precipitated metal or chloride.

Alkali chlorides and chlorides of metals, which are not reduced by Ti^{III}, do not interfere seriously, but the time of a blank reaction is, in general, decreased. Alkali sulfates and magnesium sulfate markedly increase the speed of reaction between malachite green and Ti^{III} in the absence of tungsten, and consequently decrease the catalyzing effect. Nitrates and fluorides must be absent. The former are reduced by Ti^{III}; fluorides form stable complex salts with W^{VI} and thus inhibit the catalysis.

Although molybdenum catalyzes the reaction, its effect is much less pronounced than that of tungsten; in small amounts, it hardly interferes. Moreover, the presence of molybdenum is revealed by the appearance of a yellow-brown color, which fades rapidly when Ti^{III} is added to 0.01 *N* HCl solution containing molybdenum. When molybdenum and phosphate are both present, a permanent dark red-brown appears on adding Ti^{III} to the acid solution. However, the reduction of malachite green hardly proceeds more rapidly than in the absence of phosphate.

Procedure. One drop of the test solution, which may be neutral or 0.1 *N* with respect to hydrochloric acid, and which must be free from interfering substances, is treated on a spot plate with a microdrop of 1% titanous chloride solution and a microdrop of a 0.005% water solution of malachite green. Depending on the concentration of tungsten, the solution becomes colorless (very pale violet) more or less rapidly. With small quantities of tungsten, it is well to make a blank test.

Limit of Identification: 0.1 γ tungsten
Limit of Dilution: 1 : 500,000

(4) Test with diphenyline [205]

It is well known that benzidine (I), as its chloride, quantitatively precipitates white or light yellow amorphous "benzidine tungstate" from solutions of neutral alkali tungstates.[206] This reaction is unsuitable for the detection of tungsten, especially in the presence of much molybdenum, since molybdates react similarly. Furthermore, sulfate is precipitated as white crystalline benzidine sulfate. If, however, diphenyline (II), an isomer of

$H_2N-C_6H_4-C_6H_4-NH_2$ (I) $H_2N-C_6H_4-C_6H_4(NH_2)$ (II)

benzidine, is used in the form of its chloride, sulfates are not precipitated, and molybdates are precipitated only from solutions more concentrated than 10 %. Tungsten, however, is precipitated quantitatively as a white amorphous product.

The precipitation of tungstates by chlorides of polyatomic organic bases does not lead to formula-pure tungstates of the particular bases but to adsorption compounds of WO_3-gel and the bases. This is especially true of the precipitation of small amounts of tungsten from acid solution.[207]

Procedure. A drop of the test solution is mixed with a drop of diphenyline hydrochloride in a micro test tube. A precipitate, or cloudiness, indicates the presence of tungstate. For very small amounts, a blank test should be carried out and compared with the test, after both have stood for 15 minutes.

Limit of Identification: 6 γ tungsten

Limit of Dilution: 1 : 8500

Reagent: 1 % solution of diphenyline chloride [208] in 2 *N* hydrochloric acid

13. Vanadium

(*1*) *Test with hydrogen peroxide* [209]

A solution of vanadium containing sulfuric acid turns red-brown to blood-red, or, in very dilute solution, pale brown-pink, on the addition of hydrogen peroxide. Excess hydrogen peroxide causes partial decoloration. The reaction involves the formation of the colored peroxovanadium salt (I) which, in the presence of excess hydrogen peroxide, reacts:

$$\underset{\text{(I)}}{\left(V\!\begin{smallmatrix}\diagup O\\ |\\ \diagdown O\end{smallmatrix}\right)_2(SO_4)_3} + 6\,H_2O \underset{H_2SO_4}{\overset{H_2O_2}{\rightleftharpoons}} 2\underset{\text{(II)}}{\left(V\!\begin{smallmatrix}\diagup O\\ |\\ \diagdown O\end{smallmatrix}\right)(OH)_3} + 3\,H_2SO_4$$

and is converted into yellow orthoperoxyvanadic acid (II).[210] An excess of hydrogen peroxide and sulfuric acid is consequently to be avoided. The interference due to the color of iron^{III} salts can be prevented by the addition of phosphoric acid or fluoride (formation of colorless $[Fe(PO_4)_2]^{-3}$ or $[FeF_6]^{-3}$ ions). Titanium salts, which give a yellow per compound with hydrogen peroxide (see page 195), can be converted into complex non-reactive $[TiF_6]^{-2}$ ions by the addition of alkali fluorides.

Cerium salts, molybdates, chromates, iodides and bromides, as well as large amounts of colored metallic salts, reduce the sensitivity of this test for vanadium.

Procedure. A drop of the test solution is mixed with a drop of 20 % sulfuric acid, on a spot plate or in a porcelain micro crucible. After a few minutes, a drop of 1 % hydrogen peroxide solution is added. A further drop of peroxide may be necessary.

According to the amount of vanadium present, a red to pink coloration appears.

Limit of Identification: 2.5 γ vanadium
Limit of Dilution: 1 : 20,000

(2) Test through reduction of iron^{III} by vanadium^{IV}

Alkali vanadate is quantitatively reduced to vanadium^{IV} chloride when heated with concentrated hydrochloric acid; the same is true of free vanadic acid and its anhydride:

$$V_2O_5 + 10\ HCl \rightarrow 2\ VCl_4 + 5\ H_2O + Cl_2^{\circ} \quad (1)$$

Vanadium^{IV} salts are strong reductants and are oxidized in acid solution at room temperature by iron^{III} salts:

$$V^{IV} + Fe^{III} \rightarrow Fe^{II} + V^{V} \quad (2)$$

so that Fe^{II} equivalent to the V^{IV} is formed. Accordingly, vanadium can be detected indirectly by carrying out (1) and then testing for the iron^{II} produced in (2). Available sensitive tests for iron^{II} are the red color produced in acid solution with dipyridyl (page 161) or the red color in ammoniacal solution with dimethylglyoxime (page 164). In the following procedures, the interference by excess Fe^{III} salt with these color reactions is averted by the addition of alkali phosphate. In ammoniacal solution, the latter yields insoluble, light yellow $FePO_4$; in acid solution, colorless $[Fe(PO_4)_2]^{-3}$ ions result.

Procedure I.[211] A drop of the neutral test solution is heated to boiling along with a drop of concentrated hydrochloric acid. After evaporation to one-half of the original volume, the solution is cooled and a drop of 1 % ferric chloride solution is added and thoroughly mixed. Then a drop of saturated disodium phosphate solution and a drop of a 2 % solution of a,a'-dipyridyl in 0.1 N hydrochloric acid are introduced. A red or pink color indicates the presence of vanadium.

Limit of Identification: 0.1 γ vanadium
Limit of Dilution: 1 : 500,000

Procedure II.[212] The reduction is carried out as in Procedure I. The solution is then treated with one drop of 1 % ferric chloride solution followed by two drops of a 1 % solution of dimethylglyoxime in ethyl alcohol and one drop of saturated disodium phosphate solution. After thorough mixing, the solution is made alkaline with one or two drops of concentrated ammonia. A red or pink color indicates the presence of vanadium.

Limit of Identification: 1 γ vanadium

(3) Test with a-benzoinoxime [213]

In strongly acid solution, vanadates give a yellow precipitate with α-benzoinoxime (I). Only molybdates and tungstates react analogously under this condition; they yield white precipitates,[214] whose composition is unknown. Probably there is anhydride formation between OH-groups of the metallo acids and the OH- or NOH-group of the benzoinoxime molecule. Formulation (II) or (IIa) could correspond to a mixed anhydride of this kind in the case of metavanadic acid:

$$\begin{array}{ccc} C_6H_5{-}C{=}NOH & C_6H_5{-}C{=}NO{-}VO_2 & C_6H_5{-}C{=}NOH \\ | & | & | \\ C_6H_5{-}CH{-}OH & C_6H_5{-}CH{-}OH & C_6H_5{-}CH{-}O{-}VO_2 \\ (I) & (II) & (IIa) \end{array}$$

The formation of the yellow, acid-insoluble vanadium benzoinoxime compound is specific for vanadium and serves as a test. Colored ions interfere with the detection of small amounts if they tend to conceal the yellow precipitate. The effect of the color of the Fe^{+3} ion on the benzoinoxime test can be eliminated by adding phosphate. If the test is made in a micro centrifuge tube, the yellow vanadium compound of benzoinoxime can, if necessary, be collected along with the white compounds of benzoinoxime with molybdic and tungstic acid and, after digestion with water, the yellow solution can be poured off.

Procedure. Two drops of benzoinoxime solution are added to a drop of the test solution and followed by one drop of 3 *N* sulfuric acid. A yellow precipitate or color indicates the presence of vanadium.

Limit of Identification: 1 γ vanadium
Limit of Dilution: 1 : 50,000
Reagent: Saturated solution of benzoinoxime[215] in ethyl alcohol

(4) *Test with 8-hydroxyquinoline*[216]

8-Hydroxyquinoline (oxine) reacts with numerous metal ions, in neutral, ammoniacal, or acetic acid solution, to produce inner complex compounds of the coordination structure (II). Anions of metallo acids (molybdic, tungstic, vanadic) react in acetic acid solution to form insoluble compounds of a different structure: probably the oxine esters of the particular metallo acid are formed.[217] The vanadium compound is represented by (III).[218]

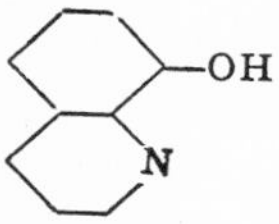

[Structures: (I) 8-hydroxyquinoline with —OH and N; (II) metal chelate with —O, N·····Me; (III) two quinoline rings linked —O—V(=O)(OH)—O—, each with N]

(I) 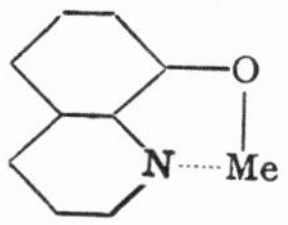 (II) 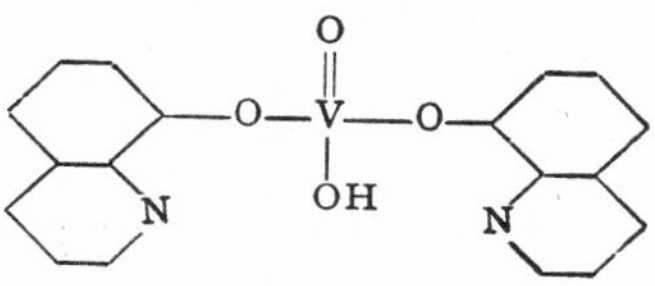 (III)

References pp. 247–257

With the exception of the iron^{III}, uranyl, and vanadium^{V} compounds of oxine, all metal oxinates are yellow to orange. The ferric, uranyl, and vanadium compounds are black-green, red-brown, and black-brown, respectively. Oxinates of the types (II) and (III) can be formed not only by the wet method but also by melting metal salts and oxides with oxine (m.p. 75° C). The solubility of metal oxinates in molten oxine appears to play a role in the realization of the fusion reactions. The production of dark oxine melts is indicative of the presence of iron or vanadium. Accordingly, if vanadium is to be indisputably detected by an oxine fusion, iron and uranium must be absent. They can be removed by igniting a small portion of the test material and extracting the ignition residue with ammonia. Any ammonium vanadate dissolves out.

Procedure. A drop of the ammonium vanadate solution is taken to dryness in a micro crucible; the residue is strongly ignited. After cooling, about 0.3 g oxine is introduced and the temperature raised to 250° C. The formation of a brown-black melt indicates the presence of vanadium.

Limit of Identification: 0.5 γ vanadium
Limit of Dilution: 1 : 100,000

If vanadium is to be detected by a fusion reaction with oxine in the presence of MoO_3 or WO_3, the melt must be heated not higher than 120° C. Otherwise, the oxine may reduce MoO_3 or WO_3 to colored lower oxides. It was possible to detect:

2 γ vanadium in the presence of 75,000 γ MoO_3 or 14,000 γ WO_3

(5) *Test with 3,3'-dimethylnaphthidine* [219]

Derivatives of benzidine retain its oxidizability (see page 72).[220] 3,3'-Dimethylnaphthidine (I) is particularly susceptible to the action of oxidants in weakly acid solution, and yields a blue-violet coloration. The production of a *p*-quinoidal di-imine (II) is probably essential to this color reaction.

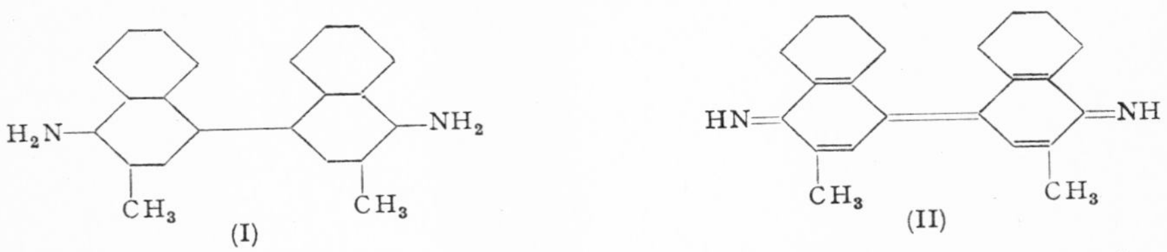

H_2N — NH_2 ; CH_3 CH_3
(I)

HN= =NH ; CH_3 CH_3
(II)

Because of the change $V^{V} \rightarrow V^{IV}$, acidic vanadate solutions are as active toward 3,3'-dimethylnaphthidine as acidified permanganate and chromate solutions. However, the oxidation products do not seem to be identical, because the color given by (I) plus $Cr_2O_7^{-2}$ or MnO_4^- fades in about 30 minutes, whereas the color produced with VO_3^- does not weaken even after

24 hours standing. Possibly the stability of the color is due to the production of a vanadium compound of (II).

Procedure. A drop of the acidified vanadate solution (0.1 *N* with respect to sulfuric acid) is transferred to a spot plate and mixed with a drop of reagent solution. In the absence of other oxidants, a red-violet color indicates vanadate.

Limit of Identification: 0.1 γ vanadium
Limit of Dilution: 1 : 500,000
Reagent: 1 % solution of 3,3′-dimethylnaphthidine [221] in glacial acetic acid

6) Other tests for vanadium

(*a*) The well-known catalytic oxidation of aniline to aniline black [222] by vanadium salts may be applied as a test for vanadium (*Idn. Limit:* 3 γ V). [223] This test, carried out on filter paper, in acid solution, requires the absence of other oxidizing agents.

(*b*) The oxine test may be carried out in a micro crucible by adding a drop of an acetic acid solution of oxine to the hot test solution, which has previously been treated with sodium tartrate (*Idn. Limit:* 0.27 γ V in 0.04 ml). [224] The blue-black vanadium oxine compound may be extracted with chloroform in a micro test tube (*Idn. Limit:* 0.1 γ V).[225]

(*c*) Vanadium^{IV}, a strong reducing agent, may be detected with phosphomolybdic acid by the formation of molybdenum blue [226] (*Idn. Limit:* 0.4 γ in 0.04 ml, in the absence of Ce^{III}, Co, Sn^{II}, Fe^{II} and chromates). Vanadium^{V} salts must be reduced with oxalic acid beforehand.

(*d*) Addition of alkali tungstate to phosphoric acid solutions of alkali vanadate produces a yellow to orange color (*Idn. Limit:* 0.3 γ V).[227]

(*e*) Acidified vanadate solutions give a violet coloration with diaminobenzidine hydrochloride. Oxidizing and sulfate ions should be avoided (*Idn. Limit:* 0.25 γ V).[228]

(*f*) Approximately neutral vanadate solutions give a green precipitate with a saturated solution of diphenylbenzidine in glacial acetic acid. Slight amounts of vanadium yield a yellow color (*Idn. Limit:* 0.05 γ V). [229] Oxidizing ions should be absent.

(*g*) Addition of 0.5 % solution of quercetin yields a green water-soluble product which is adsorbed on the reagent precipitated by dilution (*Idn. Limit:* 0.25 γ V).[230]

14. Gold

(1) Test with p-*dimethylaminobenzylidenerhodanine* [231]

Water-soluble gold^{III} salts give a red-violet precipitate with *p*-dimethylaminobenzylidenerhodanine (structure see page 59) in neutral and weakly acid solutions. It is likely that this reaction is not due to the formation of a gold^{III} salt of *p*-dimethylaminobenzylidenerhodanine. The latter probably reduces Au^{III} to Au^{I}, and a gold^{I} salt of the reagent is produced.

Silver, mercury and palladium salts must be absent as they also react with *p*-dimethylaminobenzylidenerhodanine to give colored insoluble products (see pages 59, 67 and 141).

The interference of silver in this test for gold can be avoided by precipitation and removal of silver chloride. The concentration of silver ions in a saturated solution of AgCl in 0.1 *N* hydrochloric acid is so low that in one milliliter the reaction with the reagent is no stronger than that of 0.1 γ gold.[232] In the presence of chloride, the reactivity of mercury is also decreased and even completely inhibited for small amounts. Palladium salts can be removed by precipitation in acid solution with dimethylglyoxime; the yellow crystalline inner complex Pd-dimethylglyoxime is formed.

Procedure. A drop of the neutral or weakly acid test solution is placed on rhodanine paper. In the presence of gold, a violet fleck or ring is formed.

Limit of Identification: 0.1 γ gold

Limit of Dilution: 1 : 500,000

Reagent: Spot paper (S & S No. 601) is impregnated with a saturated alcoholic solution of *p*-dimethylaminobenzylidenerhodanine, and dried.

(*2*) *Test with benzidine* [232a]

Gold^{III} salts oxidize acetic acid solutions of benzidine to benzidine blue which is a quinoneimine type of dyestuff (compare page 72). Colored gold hydrosols are produced at the same time. Considerable quantities of heavy metals or alkali salts flocculate the product and so prevent the formation of the sols and consequently impair the test. Platinum salts, oxidants, and autoxidizable substances, which act on benzidine, must be absent.

Procedure.[233] A drop of the test solution and one of benzidine solution are placed on filter paper. A blue color indicates gold.

Limit of Identification: 0.02 γ gold in 0.001 ml

Limit of Dilution: 1 : 50,000

Reagent: 0.05 % solution of benzidine in 10 % acetic acid

(*3*) *Test by thermal decomposition to metallic gold* [234]

Gold salts, like those of other noble metals, are reduced to the free metal by evaporating their solutions and heating the residue. These processes may be carried out in a glass capillary where, if the glass has been sufficiently heated, the reduced gold is enclosed by the molten glass or forms a colloidal dispersion in it. After cooling, the glass acts as a lens, and this magnification makes it possible to detect the small flakes of gold. For amounts of gold less than 1 γ in the drop, no particles of gold are formed but, instead, a red colloidal dispersion that is easy to see against a dark background. The only other metal which reacts similarly is copper, but the two metals may be distinguished by the behavior on diluting the test solution.

Procedure. A drop of the solution of the gold salt, which may either be acid or alkaline, is sucked into the fine capillary end of a glass tube fitted with a rubber cap. A small space is left between the drop and the end of the capillary. The drop is carefully evaporated by heating over a microburner; the rubber cap is then removed. The glass is then heated until the particles of gold are enveloped in a ball of melted glass, which later serves as a lens.

Limit of Identification: 1 γ gold

Limit of Dilution: 1 : 50,000

Gold solutions containing less than 1 γ gold in a drop give a reddish thread in the glass. This is easy to see against a dark background. As little as 0.05 γ gold may be detected by this means.

The procedure is applicable in the qualitative examination of gold alloys, colloidal gold solutions, toning and fixing baths, etc.

(4) Test with rhodamine B [235]

Hydrochloric acid solutions of gold^{III} chloride contain $[AuCl_4]^-$ ions, which yield a red-violet precipitate with an aqueous solution of the xanthone dyestuff rhodamine B. [236] As shown in (I), the precipitate is a salt of the basic dyestuff with chlorauric acid. An analogous compound is produced with $[AuBr_4]^-$ ions. These gold halogen compounds belong to the same salt type as the water-insoluble rhodamine compounds with $[SbCl_6]^-$, $[TlCl_4]^-$, and $[TlBr_4]^-$ ions, whose formation constitutes the basis of the sensitive tests for antimony (page 105) and thallium (page 158). All of these compounds, and likewise those rhodamine compounds precipitated with large quantities of $[HgBr_4]^{-2}$ and CNS^- ions, are soluble in benzene where they give a red solution. These benzene solutions, with the exception of that of the $[SbCl_6]$ compound, display an intense orange-yellow fluorescence in ultraviolet light. It is probable, that the fluorescence is due to the fact that in the benzene solution of the rhodamine salt of the type (I) there is a bond change resulting in the formation of isomeric salts. For the rhodamine salt of chlorauric acid, it is possible that (I) goes over into (II):

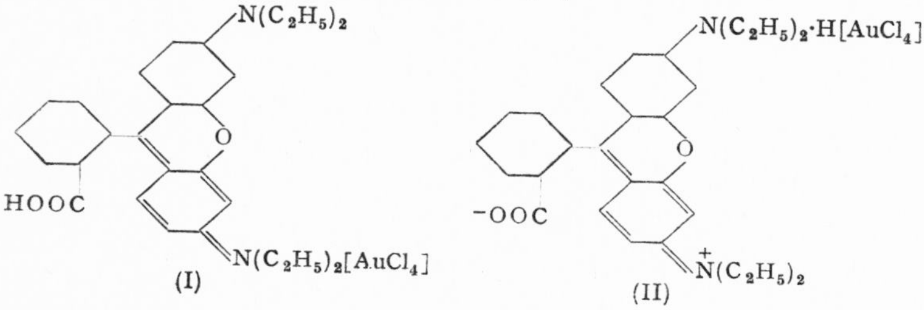

$N(C_2H_5)_2$ | O | HOOC | $N(C_2H_5)_2[AuCl_4]$

(I)

$N(C_2H_5)_2 \cdot H[AuCl_4]$ | O | ^-OOC | $\overset{+}{N}(C_2H_5)_2$

(II)

Analogous transformations may be assumed for the $[AuBr_4]$, $[TlCl_4]$, $[TlBr_4]$, etc. compounds. It should be pointed out that the transformation does not occur instantaneously in the benzene solutions of the gold compounds. Immediately after its preparation, the benzene solution shows no fluorescence, but the latter appears only after considerable exposure to daylight. In ultraviolet light, the solution begins to fluoresce in about 1 minute and the intensity of the fluorescence then steadily increases. The red benzene solution remains unchanged for hours in the dark, and there is no fluorescence.

The red benzene solutions, which fluoresce orange-red in ultraviolet light, can be obtained with gold solutions that are so dilute that they show no change on the addition of rhodamine (without extraction with benzene). Under the conditions prescribed here, in which a dilute solution of the dye functions as reagent, the test for gold is impaired by the presence of thallium^{III} salts, and also by considerable amounts of $HgBr_2$, since their rhodamine compounds also fluoresce orange-red in benzene solution. Larger quantities of antimony^{V} compounds consume the reagent, so that on extraction with benzene there is production only of benzene-soluble non-fluorescing antimony compounds, but no formation of fluorescent gold compounds. Nevertheless it is possible to detect gold specifically with rhodamine B, since antimony, mercury, and thallium compounds are easily removed (see below).

Procedure. A drop of the test solution is placed in a micro test tube and mixed with one drop of hydrochloric acid and one drop of rhodamine B solution. The mixture is shaken with 6–8 drops of benzene. If gold is present, the benzene layer turns red-violet to pink, according to the quantity present, and after about one minute under the quartz lamp displays an orange fluorescence.

Limit of Identification: 0.1 γ gold
Limit of Dilution: 1 : 500,000

Detection of gold in the presence of antimony, mercury and thallium

Even small quantities of gold may be precipitated in the metallic state from dilute solutions by the addition of stannous chloride and mercuric chloride. The gold is precipitated along with metallic mercury and is thus separated from antimony and thallium. Mercury acts as collector for the slight amount of metallic gold, [237] probably through the production of an amalgam. If the amalgam is separated from the solution and then heated, the mercury volatilizes, and the gold in the residue can be dissolved in hydrochloric acid containing bromine to form $[AuBr_4]^-$ ions. It can then be detected by means of rhodamine B.

References pp. 247–257

Procedure. One drop of the test solution is treated in a micro test tube with one drop of 1 % solution of mercuric chloride and one drop of 10 % stannous chloride solution (in 1 : 1 hydrochloric acid). After 5 minutes, the suspension is centrifuged, the clear liquid poured off, and replaced several times with dilute hydrochloric acid. After decanting, or better pipetting off the last wash liquid, the test tube is kept in a drying oven for about 30 minutes and then heated more strongly (muffle) to drive off the mercury. After cooling, 2 drops of bromine-hydrochloric acid (equal volumes of bromine water and concentrated hydrochloric acid) are run down the sides of the test tube by means of a fine pipette. The remainder of the procedure is as given above, but the excess bromine should be removed by means of a drop of a 10 % water solution of sulfosalicylic acid (see detection of chromium, page 168).

This procedure revealed 0.5 γ gold in the presence of 1000 γ antimony$^{III, V}$, mercury, or thallium.

(5) Other tests for gold

(*a*) A reagent paper is prepared by impregnating filter paper with a reducing substance (such as $SnCl_2$, Hg_2Cl_2, benzidine, pyrogallol, etc.) and dried at not more than 40° C for 2 hours. On adding a drop of a solution containing gold, a colored stain of metallic gold forms.[238] By comparing this with the fleck obtained from a solution of known gold content, the test may be made semiquantitative.

(*b*) One or two drops of the acidic test solution is saturated with potassium chloride and shaken with one or two ml of *n*-butyl alcohol. One drop of the alcohol solution is treated on the spot plate with a drop of a 1 % solution of α-naphthylamine in *n*-butyl alcohol. A violet color develops after 1–3 minutes[239] (*Idn. Limit:* 1 γ Au). Mercurous salts, chromates, and other oxidants interfere, and likewise alkali cyanide.

(*c*) One drop of the test solution is treated on filter paper with 1 or 2 drops of a formic acid solution of ascorbic acid-Complexone III. The reduction to metallic gold yields a violet color (*Idn. Limit:* 1 γ Au).[240] The sensitivity is enhanced by drying the fleck. To detect gold in the presence of palladium and platinum, the fleck should be spotted beforehand with one drop of 0.1 *N* Complexone solution. Silver and mercury salts, as well as large quantities of colored ions, interfere.

15–17. Platinum Metals

(1) Catalytic tests for platinum metals

(a) By condensation and activation of hydrogen on the finely divided metal [241]

Iridium, palladium, platinum, and rhodium condense gases, particularly hydrogen, by absorption and adsorption. In so far as these gases react with oxygen of the air (in the case of hydrogen, producing water) the heat liberated is taken up by the metal and can ignite inflammable gases. Finely divided platinum metals are especially reactive owing to their extensive free surface. When a drop of a solution of a salt of a platinum metal is placed on a strip of asbestos paper, and the moistened portion strongly ignited, finely divided metal is deposited. When cold, the strip is held in a stream of gas (ordinary city gas, hydrogen). The spotted portion begins to glow, and if sufficient metal is present the gas catches fire. The activation of molecular hydrogen by finely divided platinum metals must be brought into relationship with the equilibrium: $H_2^{o} \rightleftarrows 2\,H^{o}$. In this equilibrium, which lies far to the left, only the atomic hydrogen functions as reductant. Without contact with platinum metals, the quantity of atomic hydrogen is too slight and when consumed the reestablishment of the equilibrium is too slow to effectuate energetic reductions. On contact with platinum metals, absorption and adsorption result in such rapid reestablishment of the equilibrium that the constant supplying of atomic hydrogen permits the accomplishment of strong reducing actions.[242]

Procedure.[243] A strip of asbestos paper (2 × 6 cm) about 0.5 mm thick is well moistened at one end and a depression made in the paper with a rounded glass rod. This is best done on a cork, provided with a depression of the required depth. A point is made in the middle of the depression of the paper with a drawn-out end of a glass rod. If the asbestos is slightly pierced no harm results.

The asbestos is then strongly ignited in a reducing Bunsen flame and a drop of the test solution is placed precisely on the asbestos tip, which is ignited again. Before the asbestos is quite cold, it is held with the concave side of the depression over the mouth of a tube delivering a stream of hydrogen. By moving the strip slightly up and down, it is easy to discover the position at which the glowing is strongest. This glow indicates the presence of platinum metals.

The hydrogen is generated from purest zinc and hydrochloric acid, and passed through fine soda lime. It is then led upward through a capillary tube which is drawn so that the flame is not more than 5–10 mm high. The proper size of the opening must be determined by trial.

The intensity of the glowing depends mainly on the concentration of the solution, and only slightly on the drop size. A large volume of liquid is spread over a large surface, all of which glows after heating and treating with hydrogen. A small volume of the same solution, however, placed exactly on the tip of the point, gives an almost point-sized glowing area which is very easy to see, through contrast with the surrounding area.

When using a microdrop (1 mm^3) of a solution of platinic chloride, the following are seen:

References pp. 247–257

Concentration of the solution of platinum/(mm^3)	Result
$1 \cdot 10^{-3}$	the hydrogen ignites
$1 \cdot 10^{-4}$	the fleck glows very brightly
$8 \cdot 10^{-5}$	the fleck glows very brightly
$6 \cdot 10^{-5}$	the fleck glows clearly
$4 \cdot 10^{-5}$	the fleck glows weakly
$2 \cdot 10^{-5}$	the fleck glows only at times
$1 \cdot 10^{-5}$	the fleck does not glow at all

Accordingly, the following sensitivity data hold for platinum:
Limit of Identification: 0.04 γ platinum
Limit of Dilution: 1 : 250,000

Palladium, iridium and rhodium can also be detected by this catalytic reaction (*Idn. Limits:* 0.18 γ Ir; 0.02 γ Rh; 0.01 γ Pd).[244]

This test for the platinum metals is remarkably indifferent to the presence of foreign elements. Iron, uranium, copper, molybdenum, nickel, and cobalt salts do not interfere with the detection of platinum, even when present 1000 times in excess. Although arsenic poisons the catalysis, 0.07 γ of platinum can still be detected in the presence of 50 times the amount of arsenic (as arsenic trioxide).

(*b*) *By induced reduction of nickel salts* [245]

Sodium hypophosphite acts but slowly and incompletely on aqueous solutions of nickel salts to give elementary nickel (with varying nickel phosphide content):

$$2\,Ni^{+2} + H_2PO_2^- + 2\,H_2O \rightarrow 2\,Ni^\circ + PO_4^{-3} + 6\,H^+$$

This reaction is remarkably accelerated by small amounts of palladium (or other platinum metals).[246] Since palladium solutions are rapidly reduced by hypophosphite to elementary palladium, the action may be due to colloidal palladium, the hypophosphite being decomposed on the extensive surface with liberation of hydrogen. Apparently, the reducing effect on the nickel salts is due to the hydrogen in the nascent state or in an adsorbed and activated form of molecular hydrogen (see page 132).[247]

Procedure. Two test tubes, which have been rinsed with alcohol and ether to remove traces of fat, are charged with 10 ml of 1 % nickel acetate solution, and 1 ml of saturated sodium hypophosphite solution. One test tube contains also 1 ml of the neutral or weakly acid test solution, in the other 1 ml of water. The test tubes are then placed in a beaker containing boiling water. In a few minutes, hydrogen begins to appear in generous amounts, and in 2 to 30 minutes, according to the content of palladium, nickel is deposited, partly as

References pp. 247–257

a black powder and partly as a metallic mirror. The blank remains green, and nickel is deposited only after much longer warming.

Limit of Identification: 0.0025 γ palladium (in 1 ml)
Limit of Dilution: 1 : 400,000,000

The addition of potassium iodide, which forms insoluble PdI_2, completely inhibits the catalytic acceleration of the nickel reduction. A saturated water solution of PdI_2 has no action, in contrast to a saturated solution of palladium dimethylglyoxime, where one drop has a distinct effect.

Salts of platinum, osmium and ruthenium behave similarly to palladium, but with less sensitivity. A slight content of palladium may be responsible for the action in these cases.

Limit of Identification: 1.5 γ platinum, 0.5 γ osmium, 0.5 γ ruthenium (in 10 ml)

Limit of Dilution: 1 : 6,600,000 platinum; 1 : 20,000,000 osmium; 1 : 20,000,000 ruthenium

15. Platinum

(1) *Test with stannous chloride, in presence of other noble metals* [248]

Platinum salts, like those of gold, palladium, osmium, in acid solution, are reduced to the elementary state by stannous chloride. The platinum can be detected, even in the presence of these other noble metals, by a spot reaction. It is fixed on filter paper by conversion into insoluble ammonia-resistant $Tl_2[PtCl_6]$. The fleck is then spotted with an acid solution of stannous chloride.

Gold and palladium salts react with thallium nitrate to produce brown flecks of $Tl[AuCl_4]$ and $Tl_2[PdCl_4]$. In contrast to the analogous platinum compound, these products are soluble in ammonia.

Procedure. A drop of saturated thallium nitrate solution is placed on filter paper, followed by a drop of the test solution, and another drop of $TlNO_3$ solution. The paper is washed with ammonia. Any double salts of gold or palladium go into solution, and are washed away, while $Tl_2[PtCl_6]$ remains behind. On treatment with a drop of stannous chloride solution in strong hydrochloric acid, a yellow to orange-red stain is formed, according to the amount of platinum present.

Limit of Identification: 0.025 γ platinum (in 0.002 ml)
Limit of Dilution: 1 : 250,000

(2) *Test by reaction with alkali iodide* [249]

In neutral or acid solution, platinum^{IV} salts react with excess alkali iodide and produce a brown color due to complex platinum hexaiodo-ions: $Pt^{+4} + 6\,I^- \rightarrow [PtI_6]^{-2}$. The brown color disappears when alkali sulfite or sulfurous acid is added, because $[PtI_6]^{-2}$ ions are transformed into colorless complex $[Pt(SO_3)_3]^{-2}$ ions.

This reaction is suitable for the detection of platinum, provided compounds which oxidize iodide ion to iodine in acid solution are not present; palladium and gold^{III} salts likewise must be absent. With excess iodide the former produce red-brown $[PdI_4]^{-2}$ ions (compare page 138); the latter yield iodine and gold^{I} iodide, which dissolves in alkali iodide to form $[AuI_2]^{-}$ ions. The interference by palladium and gold can be avoided by making the solution basic with ammonia and then adding oxalic acid. The ammonia produces $[Pd(NH_3)_2]^{+2}$ or $[PdNH_2]^{+1}$ ions, which are resistant to oxalic acid and do not react with alkali iodide. On gentle warming, the gold salts are quantitatively reduced to the metal.

Procedure. A drop of the test solution, as weakly acid as possible, is treated with a drop of 5 % potassium iodide solution. A more or less intense brown-red color appears, depending on the quantity of platinum present.

Limit of Identification: 0.5 γ platinum

Limit of Dilution: 1 : 100,000

16. Palladium

(1) *Test with stannous chloride in the presence of gold and platinum* [250]

Palladium salts are also readily reduced to the elementary state by stannous chloride in acid solution. To detect palladium in the presence of gold and platinum metals, use is made of the fact that palladium salts react, in acid solution, with mercuric cyanide to give insoluble palladium cyanide, whereas gold and platinum salts form soluble complex cyanides. The white palladium cyanide reacts with stannous chloride in the same manner as soluble palladium salts.

Procedure. One drop of a saturated mercuric cyanide solution, followed by a drop of the test solution, and a further drop of mercuric cyanide solution, are placed on filter paper. The palladium is precipitated as palladous cyanide in the middle of the fleck, whereas the gold and platinum complex cyanides diffuse outwards, and can be completely washed away with several drops of water. Stannous chloride is then placed on the middle of the fleck. A dark orange to yellow-gold color appears if palladium is present.

Limit of Identification: 0.04 γ palladium (in 0.002 ml)

Limit of Dilution: 1 : 50,000

(2) *Test by catalysis of the reduction of phosphomolybdates by carbon monoxide*[251]

Acidified molybdate solutions remain unaltered even after long continued passage of carbon monoxide. Nevertheless, they rapidly turn blue (because of formation of Mo_2O_5) if the reducing gas also comes in contact with metallic platinum or palladium.[252] The carbon monoxide is adsorbed and activated to bring about the reaction:

$$2MoO_3 + CO \rightarrow Mo_2O_5 + CO_2 \qquad (1)$$

When carbon monoxide is allowed to act on an acid solution containing palladium chloride and molybdate, the first product is metallic palladium:

$$Pd^{+2} + CO + H_2O \rightarrow Pd^{o} + CO_2 + 2\,H^{+} \qquad (2)$$

The finely divided metal is particularly adapted to activate the gas for participation in reaction (1). It is still better to use phosphomolybdic acid, in which the complexly bound Mo^{VI} has a far greater tendency to enter into redox reactions than the Mo^{VI} in normal molybdates (compare page 84). The following test for palladium is based on a combination of the activation of carbon monoxide by adsorption on finely divided palladium and the activation of Mo^{VI} in phosphomolybdic acid. Through these dual effects, palladium salts bring about a catalytic acceleration of the redox reaction (1) between carbon monoxide and Mo^{VI} in phosphomolybdic acid.

Procedure. Two drops of the slightly acid test solution are mixed with a drop of phosphomolybdic acid in a microcentrifuge tube. The tube is gently warmed and a rapid stream of carbon monoxide is passed into the mixture by means of a capillary tube. According to the palladium content, a more or less intense blue color appears, at once or in a minute or two. For very small amounts of palladium, a blank on pure phosphomolybdic acid should be carried out to compare the colors.

Limit of Identification: 0.025 γ palladium

Limit of Dilution: 1 : 4,000,000

Reagents: *1*) 10 % phosphomolybdic acid (for preparation, see page 104)

2) The carbon monoxide is best prepared by dropping strong formic acid into concentrated sulfuric acid at about 100° C

Because of their intense self-color, large amounts of ruthenium and osmium interfere with the test for small amounts of palladium. This interference is overcome, if, after the carbon monoxide has been passed through the reaction mixture, the solution is shaken with a few drops of amyl alcohol, which extracts the molybdenum blue. As little as 0.12 γ palladium can thus be detected in the presence of 1000 times the amount of osmium.

$Iron^{III}$, mercury and gold salts reduce the sensitivity of the test. Other metal salts do not interfere appreciably if their concentration does not exceed 0.25 to 1 %.

When platinum salts are to be tested for palladium, the test solution should contain not more than 0.1 % platinum; larger amounts of platinum reduce the sensitivity of the test for palladium.

Arsenites and arsenates affect the test when their concentration exceeds 0.05 %. Arsenites have a reducing action; arsenates accelerate the carbon monoxide reduction of phosphomolybdic acid, even in the absence of palladium.

(3) *Test with nickel dimethylglyoxime* [253]

If a suspension of red nickel dimethylglyoxime (see page 149) is treated with a neutral or acetic acid solution of a palladium salt, there is no visible change, such as conversion into the yellow palladium dimethylglyoxime. Not more than traces of nickel can be detected in the filtrate. When dilute mineral acid is then poured over the residue on the filter, the solid, in contrast with nickel dimethylglyoxime, does not dissolve.[254] This effect is probably due to an adsorption of Pd^{+2} ions or $PdCl_2$ on the surface of the Ni-dimethylglyoxime, or to an exchange of palladium for nickel on the surface of the nickel dimethylglyoxime. In any event, on treatment with acid, a layer of acid-resistant Pd-dimethylglyoxime is formed immediately and protects the underlying Ni-dimethylglyoxime against dissolution in the acid. A sensitive and reliable spot test for palladium can be based on this protective layer effect. Paper impregnated with Ni-dimethylglyoxime is used.

Procedure. A drop of neutral or acetic acid test solution is placed on nickel dimethylgloxime paper and almost dried by gentle warming (waving over a flame or in a blast of heated air). The red paper is then bathed in dilute hydrochloric acid until the surface surrounding the fleck becomes white. The paper is then washed in cold water. A pink to red spot remains, depending on the quantity of palladium present.

Limit of Identification: 0.05 γ palladium
Limit of Dilution: 1 : 1,000,000
Reagent: Nickel dimethylglyoxime paper. Filter paper (S & S 589 g) is bathed in cold saturated alcohol solution of dimethylglyoxime. After drying, the paper is placed in 2 *N* $Ni(NO_3)_2$ that has been made barely ammoniacal; nickel dimethylglyoxime precipitates. After thorough washing with water, the paper is bathed briefly in alcohol and dried.

(4) *Test with mercuric iodide* [255]

When alkali iodides are added to solutions of palladium salts, the initial brown-black precipitate of palladium iodide dissolves in excess iodide to yield red-brown complex tetraiodopalladium ions:

$$Pd^{+2} + 2\,I^- \rightarrow PdI_2 \qquad (1)$$

$$PdI_2 + 2\,I^- \rightarrow [PdI_4]^{-2} \qquad (2)$$

Because of the low solubility product of palladium iodide, water suspensions of difficultly soluble metal iodides, such as Cu_2I_2 (white), AgI (yellow) and HgI_2 (red), also react with Pd^{+2} ions. Because of the low iodide concentration available, these reactions proceed only to the formation of insoluble PdI_2 as shown in (1); there is no production of soluble $[PdI_4]^{-2}$ ions as shown in (2). It is possible that actual topochemical reactions occur with these solid iodides,[256] i.e., their free surfaces may function as active reaction sites. Mercuric iodide can be obtained in a particularly surface-rich and therefore more reactive form by dissolving it in dioxane and then diluting the solution with water. (The resulting white precipitate is probably an addition compound of mercuric iodide and dioxane.) When a dioxane solution of mercuric iodide comes in contact with a Pd^{+2} solution, the formation of black palladium iodide through reaction with the white product is far more visible than when a suspension of red mercuric iodide is used.

Dioxane solutions of mercuric iodide react also with Ag^+ and Au^{+3} ions. If the quantity of palladium is not too small, these interferences can be avoided by throwing down silver chloride, or by precipitating metallic gold with oxalic acid, before making the test for palladium. Platinum salts do not react with mercuric iodide, in contrast to their behavior toward alkali iodides (compare page 134).

It should be noted that solutions of palladium salts, which have been alkalized with ammonia and then reacidified, do not react with mercuric iodide. The reason is that $[Pd(NH_3)_2]^{+2}$ and probably also $[PdNH_2]^+$ ions are formed resulting in an extensive decrease of the concentration of Pd^{+2} ions.

Procedure. A drop of the weakly acidified test solution is placed in a depression of a spot plate and a drop of a saturated dioxane mercuric iodide solution is added. Depending on the quantity of palladium present, a black precipitate or a grey suspension appears.

Limit of Identification: 0.08 γ palladium
Limit of Dilution: 1 : 630,000

(5) *Test with* p-*nitrosodiphenylamine* [257]

When the light yellow alcoholic solution of *p*-nitrosodiphenylamine (I) is added to neutral or slightly acid solutions of palladium salts, a dark purplish brown precipitate appears. It consists of two molecules of *p*-nitrosodiphenylamine and one molecule of palladium chloride (nitrate, etc.). This may involve the production of an addition compound, in which the nitrogen atom of the imido group or the oxygen or nitrogen atom of the nitroso group of the nitrosoamine may function as the coordination position.[258] Consequently, the structures (II), (IIa), and (IIb) may be involved in the palladium chloride addition compound:

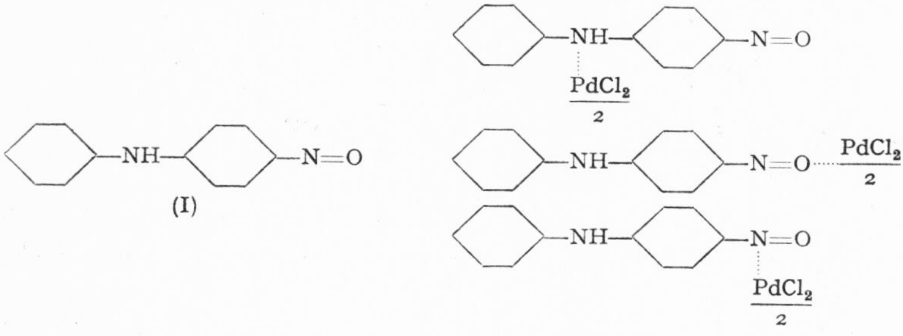

C_6H_5—NH—C_6H_4—N=O (I)

C_6H_5—NH($\frac{PdCl_2}{2}$)—C_6H_4—N=O (II)

C_6H_5—NH—C_6H_4—N=O···$\frac{PdCl_2}{2}$ (IIa)

C_6H_5—NH—C_6H_4—N(=O)($\frac{PdCl_2}{2}$) (IIb)

The behavior of *p*-nitrosodiphenylamine is rather selective with respect to palladium. Colored precipitates are produced only with gold, platinum, and silver salts, but the sensitivity is much less than with palladium ions.

Procedure. One drop of the weakly acid test solution is placed in a depression of a spot plate and a drop of reagent solution is added. Depending on the quantity of palladium, a purplish brown precipitate or a red color appears.

Limit of Identification: 0.005 γ palladium

Limit of Dilution: 1 : 10,000,000

Reagent: 5 mg *p*-nitrosodiphenylamine is dissolved in 50 ml of 95% alcohol and diluted to 100 ml. The solution will keep for several months.

When hydrochloric acid solutions of noble metal salts are being tested, the presence of silver salts need not be considered.* Gold chloride can be removed by shaking the strong hydrochloric acid solution with ether [259] or ethyl acetate.[260] The gold-free aqueous solution should be taken to dryness to remove the excess hydrochloric acid before the test with *p*-nitrosodiphenylamine is tried. There is as yet no method for the removal of platinum which is suitable for the spot test technique. Since the identification limit of the platinum reaction with *p*-nitrosodiphenylamine is 10 γ, the palladium test just given is not impaired by small amounts of platinum.

(6) *Test with phenoxithine* [261]

The addition of an acetone solution of phenoxithine (I) to a weakly acidic solution of palladium chloride gives a red-brown precipitate or a yellow color depending on the quantity of palladium present. The product is an addition compound (II)

* It should be pointed out that Yoe (*loc. cit.*) found that considerable amounts of palladium are adsorbed by the silver chloride when the latter is precipitated from solutions of palladium nitrate.

References pp. 247–257

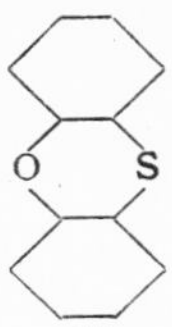

(I) 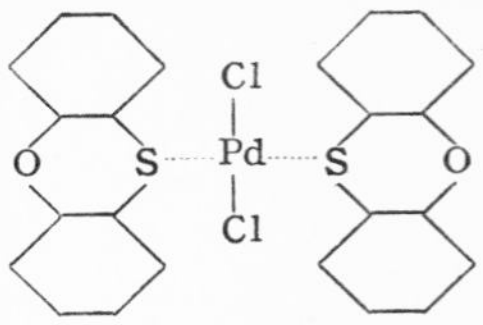 (II)

The formation of (I)) can be seen on a spot plate or filter paper, and conducted as a spot test the reaction provides a quite selective test for palladium.

Gold must be removed prior to the test because it will cause the development of a deep ruby red in the spot plate test and a diffused violet spot on the paper, apparently due to the reduction of the gold ions to the colloidal metal. Interference may also arise from OsO_4^{-2}, Os^{+3}, Ru^{+3}, and $RuCl_6^{-2}$ ions because they have distinct self-colors. Mercurous ion causes partial interference by the reduction of part of the palladium to the elementary state, but a positive response can still be seen. It is possible to detect 1 part of palladium in the presence of 200 parts of platinum or 100 parts of rhodium. Less favorable ratios should be avoided because of the color of these salts. No interference is caused by mercuric and iridic chloride, but free ammonia, ammonium ions, stannous, cyanide, thiocyanide, fluoride, oxalate, and tetraborate ions do interfere. Dimethylglyoxime, lead, silver, ferrous, ferric, stannic, cobaltous, nickel, cupric, nitrite, sulfate, chloride, and bromide ions do not interfere.

Procedure. Filter paper is bathed in a 10 % acetone solution of phenoxthine [262] and dried. One drop of the test solution, which should be approximately 0.1 *N* with respect to hydrochloric acid, is brought on the reagent paper. If palladium is present, a yellow to red-brown sharply delineated fleck appears.

Limit of Identification: 0.1 γ palladium
Limit of Dilution: 1 : 100,000

(7) *Other tests for palladium*

(*a*) 3-Hydroxy-1-*p*-sulfonatophenyl-3-phenyltriazine [263] is specific for palladium in dilute acid medium (*Idn. Limit:* 0.05 γ Pd).

(*b*) An orange-yellow fleck is produced by spotting the test solution on filter paper with an alcohol solution of naphthalene-4-sulfonic acid-1-azo-5-ortho-8-hydroxyquinoline (*Idn. Limit:* 2 γ Pd).[264]

References pp. 247–257

(*c*) A red-violet fleck results if filter paper impregnated with *p*-dimethylaminobenzylidenerhodanine is spotted with the acidified test solution (*Idn. Limit:* 0.004 γ Pd).[265] Special measures are required when Au, Pt, Ag and Hg salts are present; they give analogous reactions.

(*d*) Palladium may be detected by a neutral or slightly acidified solution of mercuric cyanide containing diphenylcarbazide.[266] The colorless solution turns violet because insoluble white $Pd(CN)_2$ and Hg^{+2} ions are formed; the latter then react with the diphenylcarbazide (see page 64). The test is carried out on a spot plate (*Idn. Limit:* 0.5 γ Pd). Considerable quantities of chloride impede the reaction.

(*e*) Among the colored water-insoluble salts of mercaptobenzeneindazole, the orange palladium salt is the only one that is resistant to dilute sodium hydroxide. The precipitation of the metallo-organic compound and the treatment with the basic solution can be carried out on filter paper impregnated with the reagent (*Idn. Limit:* 0.25 γ Pd).[267]

(*f*) Palladium cyanide, produced by spot reactions on filter paper impregnated with mercuric cyanide (compare Test *1*), is washed free of acid with water and then spotted with an alcohol solution of methyl yellow. The indicator changes to red (*Idn. Limit:* 0.4 γ Pd).[268]

(*g*) A faded brown color or a brown precipitate results when drops of $PdCl_2$ solution and 0.01 % solution of *p*-fuchsin are brought together (*Idn. Limit:* 0.01 γ Pd).[269] Gold, mercury, and large amounts of platinum salts react analogously.

(*h*) In approximately neutral solutions, a dark brown precipitate, which can be extracted to give a violet solution in chloroform, is produced by α-nitroso-β-naphthol in glacial acetic acid.[270] The test is specific within the noble metal group (*Idn. Limit:* 0.5 γ Pd).

(*i*) A red color results when a drop of an ammoniacal palladium solution is heated with one drop of a 1 % ammoniacal solution of $K_4Fe(CN)_6$ containing α,α'-dipyridyl (*Idn. Limit:* 2 γ Pd).[271] Silver and mercury react analogously (compare ferrocyanide test, p. 288).

17. Osmium and Ruthenium

(*1*) *Test by activation of chlorate solutions*[272]

Neutral or weakly acid solutions of alkali chlorates have but slight oxidizing action. On the other hand, they are strong oxidants in highly acid solutions. Alkali chlorate solutions are so activated by the addition of osmium tetroxide or ruthenium hydroxide that they give up their oxygen rapidly to organic and inorganic compounds with reduction to chloride.[273] For instance, iodine can be liberated from potassium iodide in weakly acid solutions, when merely a trace of osmium tetroxide (osmic acid) is present.* The action of the osmium tetroxide is due to the formation of a complex compound with the chlorate ion, in which the latter has an enhanced redox reactivity. The direct action of osmium tetroxide on potassium iodide according to:

$$OsO_4 + 4\ KI + 2\ H_2SO_4 = OsO_2 + 2\ K_2SO_4 + 2\ H_2O + 2\ I_2$$

is only of importance when larger amounts of osmium tetroxide are present. In any case, it is best to dilute the osmium test solution so that the reaction with potassium iodide does not proceed.

This test for osmium and ruthenium is decisive in the absence of other oxidizing compounds and colored ions.

Procedure. A drop of potassium chlorate-potassium iodide solution, acidified with a drop of dilute (1 : 1,000) sulfuric acid, is placed on a spot plate. A drop of 1 % starch solution and a drop of the neutral test solution are added. According to the osmium and ruthenium content, blue starch-iodine is formed, either at once or in a short time. A blank test should be carried out for small amounts of osmium tetroxide.

Limit of Identification: 0.005 γ osmium tetroxide
Limit of Dilution: 1 : 10,000,000

If the test and blank are allowed to stand for several hours and then compared, OsO_4 can be detected in a dilution of 1 : 50,000,000.

Reagent: Solution of 1 g potassium chlorate and 1 g potassium iodide in 100 ml water

This test for osmium may be made specific by utilizing the volatility of OsO_4. The acidified test solution is gently warmed in the apparatus shown in Figure 23 (page 47) and the volatile OsO_4 is taken up in a drop of water on the glass knob. The reaction with potassium chlorate and potassium iodide using this drop, is carried out, on a spot plate.

Limit of Identification: 0.01 γ osmium
Limit of Dilution: 1: 5,000,000

(2) *Test for ruthenium with rubeanic acid* [274]

Rubeanic acid (I) reacts in ammoniacal solution in its aci-form (II) with copper, cobalt and nickel salts giving insoluble colored compounds (see page 153). In solutions containing strong mineral acid, rubeanic acid reacts only with certain elements of the platinum group. With palladium and

*) For strongly acid alkali chlorate solutions various catalysts are known, e.g., vanadic acid [R. Luther, *Z. Elektrochem.*, 13 (1907) 437] and vanadyl salts [R. Luther and T. F. Ruttner, *Z. anal. Chem.*, 46 (1907) 521]. The reduction of $HClO_3$ by KI, which does not occur in dilute solutions, is, however, catalyzed by both ferrous and ferric salts. [See C. F. Schönbein, *J. prakt. Chem.*, [I], 75 (1858) 109; W. F. Green, *J. Phys. Chem.*, 12 (1908) 389].

platinum salts, red crystalline precipitates are obtained; they are inner complex salts (III) of a semi-aci-form of rubeanic acid. Ruthenium salts give a soluble blue compound, whose structure is not known.

$$\begin{array}{c}S{=}C{-}NH_2\\ |\\ S{=}C{-}NH_2\\ \text{(I)}\end{array}\qquad\begin{array}{c}HS{-}C{=}NH\\ |\\ HS{-}C{=}NH\\ \text{(II)}\end{array}\qquad\begin{array}{c}S{=}C{-}NH_2\\ |\quad\quad >Pd/2\\ HN{=}C{-}S\\ \text{(III)}\end{array}$$

It is remarkable that osmium, which in its other reactions is so similar to ruthenium, does not react with rubeanic acid; consequently, this reagent can be used for the detection of ruthenium in the presence of osmium. The reaction can also be used to detect ruthenium in the presence of palladium and platinum. The insoluble compound is filtered or centrifuged off, and the blue color of the clarified solution observed.

Procedure. A drop of the hydrochloric acid test solution is mixed in a micro crucible with 1 or 2 drops of a 0.2 % solution of rubeanic acid [275] in glacial acetic acid, and the mixture gently warmed over a micro burner. According to the amount of ruthenium present, a more or less deep blue color appears.

Limit of Identification: 0.2 γ ruthenium
Limit of Dilution: 1 : 250,000

(3) *Other test for osmium*

A blue or green stain is formed on paper by the reaction between osmium and a saturated solution of benzidine acetate or $K_4Fe(CN)_6$ (*Idn. Limit:* 0.01 γ Os in 0.001 ml). [276] For the detection of still smaller amounts, OsO_4 is volatilized in a glass tube drawn out to a capillary, and the vapor directed against paper impregnated with the reagent (*Idn. Limit:* 0.008 γ Os in 0.001 ml).

B. Ammonium Sulfide Group

18. Cobalt

(1) *Test with sodium thiosulfate* [277]

If a neutral solution of a cobalt salt is added to a concentrated solution of sodium thiosulfate a blue color results. This effect is probably due to the formation of complex $[Co(S_2O_3)_2]^{-2}$ ions. This test for cobalt is specific within the group of ammine-forming metal ions (Zn, Cd, Cu, Ni, Pd). Since iron^{III} salts give a transient violet color with thiosulfate (compare detection of copper, page 80) the test described here is well suited for application in the presence of ions of the ammonium sulfide group, provided high sensitivity is not essential.

Procedure. Several cg of solid sodium thiosulfate are treated on a spot plate with one drop of the neutral test solution. According to the quantity of cobalt present, a more or less intense blue color appears at once or within a few minutes. The color reaction is speeded up by the addition of a drop of alcohol.

Limit of Identification: 8 γ cobalt
Limit of Dilution: 1 : 6,250

(2) *Test with nitrosonaphthol* [278]

α-Nitroso-β-naphthol (I) in its tautomeric oxime form (II), gives colored precipitates with cobalt and certain other metal ions. In acetic acid, neutral, or ammoniacal solution, cobalt salts give red-brown precipitates, which, once formed, are not soluble even in mineral acids. The product is an inner complex Co^{III} salt (III).

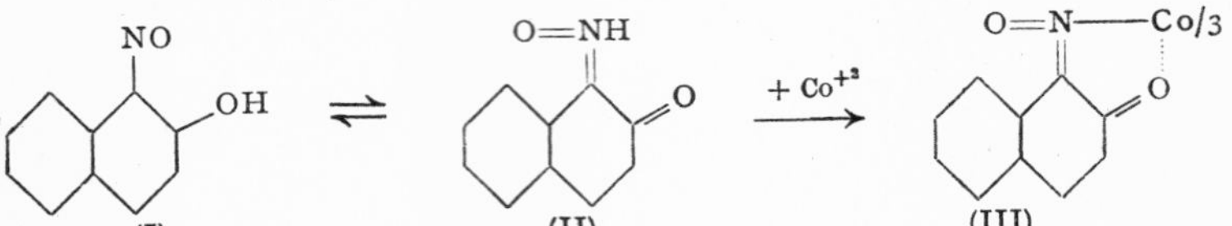

NO, OH (I) $\rightleftharpoons$ O=NH, O (II) $\xrightarrow{+ Co^{+3}}$ O=N—Co/3, O (III)

The conversion of Co^{II} into Co^{III} is due to the reagent itself which acts as an oxidizing agent since it is an orthoquinoid compound and a derivative of nitrous acid. In addition, oxidation by air probably occurs.

Ferric, uranyl, copper and palladium salts also give insoluble colored compounds with the nitrosonaphthol. Under suitable conditions, however, the test for cobalt can be conducted successfully even in the presence of considerable iron, uranium, and copper.

Procedure. A drop of the neutral or weakly acid test solution and a drop of the reagent solution are placed in succession on filter paper. A brown stain indicates the presence of cobalt. If the test solution is strongly acid, the spot should be exposed to ammonia before adding the reagent. The fleck should then be treated with a drop of 2 *N* sulfuric acid.

Limit of Identification: 0.05 γ cobalt
Limit of Dilution: 1 : 1,000,000
Reagent: 1 g α-nitroso-β-naphthol is dissolved in 50 ml of glacial acetic acid and the solution diluted to 100 ml with water.

As little as 0.006 γ cobalt can be detected in one drop (0.01 ml)[279] if an alkaline reagent solution is used.[280] The latter is prepared: 0.1 g of the reagent is dissolved by warming with 20 ml of water containing 1 ml of dilute alkali. The suspension is filtered and the filtrate made up to 200 ml. If the isomeric β-nitroso-α-naphthol is used, even smaller, about one tenth quantities of cobalt can be detected.[281]

Test for cobalt in the presence of iron

Ferric salts form brown compounds with nitrosonaphthol and thus interfere with the test for cobalt. Although iron^{III} nitrosonaphthol is soluble in acids, the precipitate dissolves quite slowly when it has been deposited in the pores of filter paper. In the presence of iron, it is best to precipitate the phosphates on filter paper [$Co_3(PO_4)_2$ and $FePO_4$] and then to add the reagent. The yellow-white ferric phosphate does not react with the nitrosonaphthol, whereas cobalt phosphate reacts instantaneously.

Limit of Identification: 0.21 γ cobalt
Limit of Dilution: 1 : 240,000
} in the presence of 1850 times the amount of iron

Limit of Identification: 0.84 γ cobalt
Limit of Dilution: 1 : 59,500
} in the presence of 2100 times the amount of iron

When the test solution is acid the following procedure should be used.

Procedure. A drop of the acid test solution is mixed on a spot plate with a drop of reagent solution and a few drops of a 10 % solution of trisodium phosphate. In the presence of cobalt, a brown coloration appears, which should be compared with a blank when testing for small amounts.

Limit of Identification: 0.5 γ cobalt
Limit of Dilution: 1 : 100,000
} in the presence of 1000 times the amount of iron

Reagent: 1 % solution of α-nitroso-β-naphthol in acetone

Test for cobalt in the presence of uranium

Uranyl salts react with α-nitroso-β-naphthol to form a yellow precipitate. It is therefore necessary to take precautions in the presence of uranium, especially when testing for small amounts of cobalt. Uranium may be converted into the nonreacting phosphate by treatment with ammonium phosphate. The test is then carried out as in the presence of iron salts.

Limit of Identification: 0.25 γ cobalt
Limit of Dilution: 1 : 200,000
} in the presence of 1000 times the amount of uranium

Test for cobalt in the presence of copper

Copper^{II} salts likewise react with α-nitroso-β-naphthol to yield a brown precipitate and so they interfere with a decisive test for cobalt. Copper is converted into cuprous iodide, which does not react with nitrosonaphthol, and cobalt may then be detected even in the presence of large amounts of copper. The free iodine resulting from the addition of potassium iodide: $2\ Cu^{+2} + 4\ I^- = Cu_2I_2 + I_2$ can be removed by sodium sulfite. The white insoluble cuprous iodide does not interfere with the recognition of cobalt^{III} nitrosonaphthol.

References pp. 247–257

Procedure. A drop of the acid test solution, a drop of 2 *N* hydrochloric acid, and one of 10 % potassium iodide, together with a little solid sodium sulfite, are mixed on a spot plate. Then a drop of a 1 % solution of nitrosonaphthol in acetone and a few drops of saturated sodium acetate solution are added. In the presence of cobalt, a more or less deep brown color appears.

Limit of Identification: 0.2 γ cobalt
Limit of Dilution: 1 : 250,000
} in the presence of 2500 times the amount of copper

(*3*) *Test with rubeanic acid* [282]

Cobalt is quantitatively precipitated by an alcohol solution of rubeanic acid, in the presence of ammonia or sodium acetate. The yellow-brown amorphous product is a cobalt salt of the di-imido form of rubeanic acid. Once formed, the salt (or rather its polymer) is no longer dissolved by dilute mineral acids. See page 153 regarding the mechanism of the formation of the salt.

Nickel and copper ions react under these same conditions, yielding respectively blue and black precipitates.

Procedure. A drop of the test solution is placed on filter paper, held over ammonia, and then spotted with a drop of a 1 % solution of rubeanic acid.[283] In the presence of cobalt, a brown fleck or ring is formed.

A large excess of ammonium salts reduces the sensitivity.

Limit of Identification: 0.03 γ cobalt
Limit of Dilution: 1 : 1,660,000

(*4*) *Test with ammonium thiocyanate and acetone* [284]

Acid solutions of cobalt salts, which contain sufficient ethyl alcohol or acetone, give an intense blue color on the addition of alkali thiocyanates.[285] As in the case of the Fe^{III}—CNS^{-1} system (compare page 164), the result is not due to the formation of a single product but depending on the concentration ratios there is produced a series of ions, ranging from $Co(CNS)^{+1}$ through $Co(CNS)_2$, $Co(CNS)_3^{-1}$ to $Co(CNS)_4^{-2}$, with equilibria being established between the various members of this series of ions and compounds.[286] Ether or ether-amyl alcohol mixture can extract the deep blue $Co(CNS)_2$ or $Co[Co(CNS)_4]$ from the equilibrium system.* Nickel salts interfere only if present in large amounts; they produce a light blue color. Ferric salts give the familiar deep red solution of [$Fe(CNS)_3$ or $Fe(CNS)_2^+$, etc.] and since the product is soluble in the same organic liquids as the cobalt salt, the blue due to small amounts of cobalt may be hidden. However, cobalt can be detected in the presence of large amounts of iron by taking proper measures (see below).

* A higher sensitivity is obtained when 3-*n*-butylamine is added to the ether-amyl alcohol extraction mixture.[287]

Copper^{II} salts, which give a red-brown coloration or precipitate of $Cu(CNS)_2$, should be removed as white acid-insoluble cuprous thiocyanate by the addition of alkali sulfite.

Procedure. A drop of the test solution is mixed with 5 drops of a saturated acetone solution of ammonium thiocyanate on a spot plate. According to the cobalt content, a green to blue color appears.

Limit of Identification: 0.5 γ cobalt
Limit of Dilution: 1 : 100,000

When nickel salts are to be tested for cobalt, the concentration of the solution should not exceed 0.2 % nickel. A solution of this strength, when it contains nickel alone, gives only a yellowish green color with an acetone solution of thiocyanate. If as little as 0.5 γ cobalt is present in a drop (corresponding to a proportion Co: Ni = 1 : 200), a definite blue-green color develops on addition of thiocyanate.

Test for cobalt in the presence of iron [288]

Ferric salts are converted by alkali fluorides into fairly insoluble colorless complex iron fluorides of the general formula $M_3[FeF_6]$. The stability of these complexes is so great that in aqueous solution the Fe^{+3} ion concentration is insufficient to give the red ferric-thiocyanate product. Since the cobalt reaction is not hindered by the presence of fluorides, cobalt can then be detected in the presence of much iron.

Procedure. One or two drops of the slightly acid test solution is mixed with a few milligrams of ammonium fluoride on a spot plate; then 5 drops of 10 % ammonium thiocyanate solution in acetone are added. In this way, 1 γ cobalt in 3 drops of a cobalt solution (1 : 50,000) can be detected by the blue color in the presence of 1000 times the amount of iron.

(5) Test with p-*nitrophenylhydrazone of diacetylmonoxime* [289]

The light yellow alcoholic solution of *p*-nitrophenylhydrazone of diacetylmonoxime is turned intensely violet by alkali hydroxide or carbonate. This color change, which can be reversed by adding acid or much ammonium salt, is due to the formation of *p*-quinoidal anions of water-soluble alkali salts of the aci-form of the nitro compound:

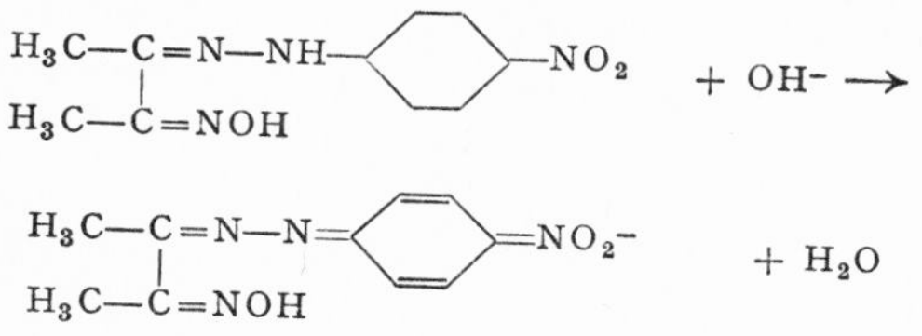

$$\begin{matrix} H_3C-C=N-NH-C_6H_4-NO_2 \\ | \\ H_3C-C=NOH \end{matrix} + OH^- \rightarrow$$

$$\begin{matrix} H_3C-C=N-N=C_6H_4=NO_2^- \\ | \\ H_3C-C=NOH \end{matrix} + H_2O \qquad (1)$$

This rearrangement (1) proceeds only to a slight extent with ammonia because the latter furnishes too few OH^- ions and consequently only an orange color results instead of the violet color. However, if Co^{+2} ions are added to an alcoholic-ammoniacal solution of the hydrazone, a violet color also ensues, the intensity depending on the cobalt content. The colored product has not been isolated but is likely that the following reaction (2) is involved with production of the violet inner complex cobalt-bearing anions:

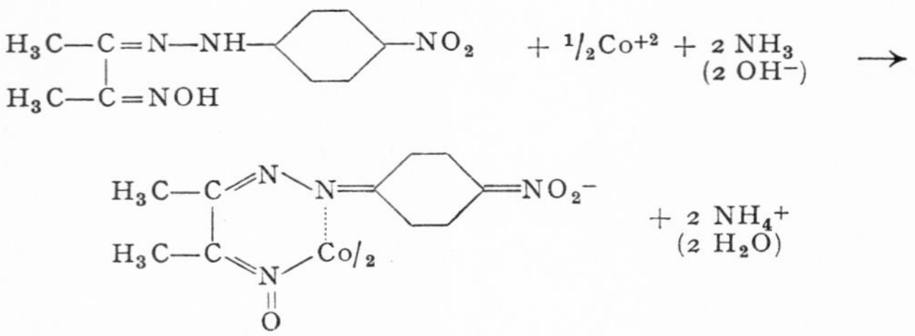

$$\begin{array}{l} H_3C{-}C{=}N{-}NH{-}C_6H_4{-}NO_2 \\ \quad\;\;| \\ H_3C{-}C{=}NOH \end{array} + {}^1/_2Co^{+2} + \underset{(2\,OH^-)}{2\,NH_3} \rightarrow$$

$$\begin{array}{l} H_3C{-}C{=}N{-}N{=}C_6H_4{=}NO_2^- \\ \quad\;\;| \qquad\quad \vdots \\ H_3C{-}C{=}N{-}Co/_2 \\ \qquad\quad\;\; \| \\ \qquad\quad\;\; O \end{array} + \underset{(2\,H_2O)}{2\,NH_4^+} \qquad (2)$$

The assumption that (2) is initiated by the relatively slight ionic concentration of the ammonia is supported by the analogous action of other OH^- donors such as magnesium oxide.

It should be noted that once the colored cobalt compound has been formed it is resistant to alkali cyanide as well as ammonium salts.

Color reaction (2) is specific for cobalt because other ammine-forming metal ions (Cu, Cd, Ni, Pd) do not react with the reagent to yield highly colored products.

Procedure. The test is made in a micro test tube. One drop of the test solution is treated with a drop of concentrated ammonia water and two drops of the reagent solution. A violet or pink color appears, the shade depending on the quantity of cobalt present. When slight amounts are involved, it is best to extract the excess reagent with ether.

Limit of Identification: 0.1 γ cobalt

Limit of Dilution: 1 : 500,000

Reagent: 0.1 % solution of *p*-nitrophenylhydrazone of diacetylmonoxime in alcohol. The reagent can be prepared by dissolving 1 g of diacetylmonoxime in 10 ml of water and adding a solution of 1.5 g of *p*-nitrophenylhydrazine in 100 ml of hot water. The mixture is kept in a water-bath for about 30 minutes. A light yellow solid separates on cooling; it is recrystallized from ethyl alcohol. (Yield: about 40 %).

The procedure is particularly recommended for testing ammoniacal solutions, if need be after filtering off any considerable amounts of insoluble metal hydroxides. If much copper or nickel is present, it is difficult or impossible to discern slight amounts of cobalt because of the blue color. In such cases, after addition of the reagent, drops of 1 % potassium cyanide solution are added until the blue $[Cu(NH_3)_4]^{+2}$, or $[Ni(NH_3)_4]^{+2}$ ions are converted into colorless $[Cu_2(CN)_4]^{-2}$ or light yellow $[Ni(CN)_4]^{-2}$ ions. Any slight pink color is then easily seen.

(6) *Other tests for cobalt*

(*a*) Cobalt can be detected (in the presence of 1000 times as much nickel) by means of a 1 % water solution of 2-nitroso-1-naphthol-4-sulfonic acid at pH 7–8.[290] Red coloration (*Idn. Limit:* 0.01 γ Co). Copper and ferrous salts also give color reactions (pages 94 and 167).

(*b*) The brown-yellow solution of chromotropic acid dioxime (2,7-dinitroso-1,8-dihydroxynaphthalene-3,6-disulfonic acid) gives a blue color in ammoniacal cobalt solutions (*Idn. Limit:* 0.05 γ—0.1 γ Co).[291]

(*c*) A greenish-blue fleck of the cobalt salt is formed by spotting filter paper impregnated with sodium pentacyanopiperidine ferroate (*Idn. Limit:* 0.05 γ Co).[292]

(*d*) In neutral solutions, a green color is produced by adding 3 % hydrogen peroxide and a saturated solution of sodium bicarbonate (*Idn. Limit:* 5 γ Co).[293] The sensitivity is lowered by colored ions or by those yielding precipitates with bicarbonate.

(*e*) If drops of test solution, 10 % ammonium acetate, manganous sulfate, and freshly distilled butyraldehyde are brought together, a brown color ensues (*Idn. Limit:* 1 γ Co).[294] The test is based on the catalytic acceleration by Co^{+2} ions of the oxidation of Mn^{+2} ions by butyraldehyde to give MnO_2. The reaction is impaired by NO_2^-, CNS^-, I^-, $S_2O_3^{-2}$, PO_4^{-3} ions. Considerable amounts of Au, Bi, Hg, Ce, Zr, Th, Sb ions decrease the sensitivity.

(*f*) A drop of the test solution is placed on filter paper and spotted with a drop of a saturated water solution of sodium azide. The fleck is exposed to the vapors of a saturated aqueous solution of sulfurous acid. A yellow color appears which changes to blue on treatment with a drop of a 2 % acetic acid solution of *o*-tolidine (*Idn. Limit:* 0.5 γ Co).[295] The test is based on the fact that the oxidation of complex Co^{II}-azide to complex cobalt^{III}-azide is catalyzed by the autoxidation of sulfurous acid. The color reaction with *o*-tolidine is due to the action of the tervalent cobalt formed. Copper and iron ions interfere and should be previously removed or masked. The test can be carried out in the presence of as much as 200 times the amount of nickel.

19. Nickel

(1) *Test with dimethylglyoxime*[296]

Dimethylglyoxime (diacetyldioxime), as well as many other α-dioximes of the general formula R—C(NOH)—C(NOH)—R, gives a bright red insoluble salt with nickel salts in neutral, acetic acid or ammoniacal solutions. In the case of dimethylglyoxime the inner complex compound has the structure:

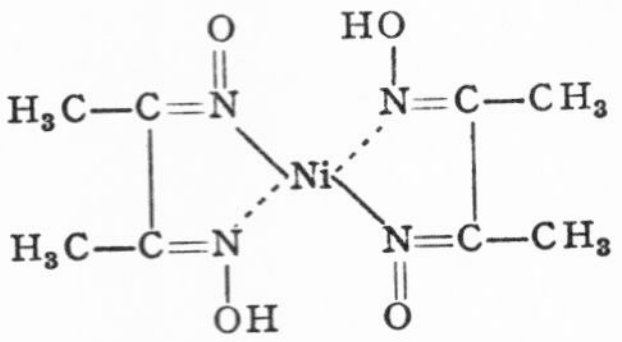

```
             O        HO
             ||        |
   H3C—C=N         N=C—CH3
        |     \     .·  |
        |      Ni<      |
        |    .·    \    |
   H3C—C=N         N=C—CH3
             |        ||
            OH        O
```

The precipitation of nickel from dilute solutions may be prevented by the presence of large amounts of oxidizing substances (hydrogen peroxide, halogens, nitrates, etc.). A red color develops (see page 152). Palladium gives a yellow, acid-insoluble, ammonia-soluble inner complex salt, whose structure is analogous to that of the nickel salt. Ferrous salts, in ammoniacal solutions containing tartrate, produce a red color (see page 164). Cobalt salts react with dimethylglyoxime to give soluble complex compounds of bi- and tervalent cobalt. The dimethylglyoxime is thus consumed, and so the sensitivity of the nickel test is very much reduced in the presence of considerable amounts of cobalt. A special procedure must be used for the detection of traces of nickel in cobalt salts (see page 446). Copper salts, in ammoniacal solution, give violet, soluble complex compounds with dimethylglyoxime. Gold salts, in acid solution, are reduced to the metal.

When the test for nickel is to be carried out in ammoniacal solution, and other metals, precipitable by ammonia, are present, the hydrous oxide precipitation must be prevented by the addition of citrates or tartrates. Care must be taken that ferric and cobalt salts are not both present. Although either of these alone is not precipitated by dimethylglyoxime from ammoniacal tartrate solutions, when they are present together a red-brown precipitate ($FeCoC_{12}H_{19}N_6O_6$) is formed. [297] Special attention must be given to this possibility (see page 152).

Procedure. A drop of the test solution and one of 1 % alcoholic dimethylglyoxime solution are placed on filter paper and held over ammonia, or they are mixed on a spot plate and a small drop of dilute ammonia added. The formation of a red fleck or circle on paper, or of a precipitate or color on the spot plate, indicates the presence of nickel.

Limit of Identification: 0.16 γ nickel

Limit of Dilution: 1 : 300,000

The test is considerably more sensitive when a drop of the test solution is placed on dried filter paper that has been impregnated with the reagent. The impregnation is best carried out using a warm saturated solution of the reagent in acetone. In this way as little as 0.015 γ nickel can be detected.[298]

Test in the presence of iron

Nickel can be detected by dimethylglyoxime in the presence of considerable amounts of iron^{III} salts after masking the iron.

Procedure. A drop of the test solution, a drop of a saturated sodium tartrate solution, 2 drops of a saturated sodium carbonate solution, and finally, a drop of alcoholic dimethylglyoxime solution are successively placed on a spot plate. Owing to the low surface tension of the alcohol, the red nickel dimethylgloxime precipitate is formed at the edge and surface of the liquid.

Limit of Identification: 0.5 γ nickel
Limit of Dilution: 1 : 100,000
} in the presence of 1000 times the amount of iron in 2 *N* hydrochloric acid solution

Test in the presence of cobalt, copper, and manganese

(*a*) *On spot plate*

When as much as 50 times the amount of cobalt is present in ammoniacal solution, the detection of 2 γ of nickel becomes difficult. If still larger amounts of cobalt are present, the test cannot be used because dimethylglyoxime is consumed by the cobalt. Furthermore, small quantities of nickel dimethylglyoxime are soluble in solutions of the cobalt dioxime compound.

When, however, cobalt is converted to the tervalent form before addition of the dimethylglyoxime, nickel can be detected in the presence of 200 times the amount of cobalt.

Procedure. A drop of the acid test solution, a drop of 3 % hydrogen peroxide, and a drop of a saturated solution of sodium carbonate are placed, in turn, on a spot plate. When large amounts of cobalt are present a green precipitate is formed. (It slowly darkens since it finally forms cobaltic oxide.) Smaller amounts give a green coloration. If a drop of the 1 % alcoholic dimethylglyoxime solution is then added, red nickel dimethylglyoxime is produced and slowly rises to the surface of the liquid, because of the lower surface tension of the alcohol.

Limit of Identification: 1.25 γ nickel
Limit of Dilution: 1 : 40,000
} in the presence of 200 times the amount of cobalt

When both cobalt and manganese are present, the same procedure may be used, although the sensitivity is reduced, owing to the formation of manganese dioxide. Nevertheless, 2.5 γ nickel may readily be detected in the presence of 80 times the amount of cobalt and manganese.

(b) On impregnated paper

The procedure described on page 150, employing paper impregnated with dimethylglyoxime, can be used for the detection of small amounts of nickel in the presence of much cobalt, copper or manganese.

Procedure. A drop of the test solution is placed on paper impregnated with dimethylglyoxime. The paper is then immersed in a bath of dilute ammonia, and kept in gentle motion. The colored dimethylglyoxime compounds of cobalt and copper are dissolved; the red fleck due to nickel remains on the white paper.

Limit of Identification: 0.8 γ nickel *Limit of Dilution:* 1 : 62,500	in the presence of 1250 times the amount of cobalt
Limit of Identification: 1.7 γ nickel *Limit of Dilution:* 1 : 29,400	in the presence of 590 times the amount of copper

The procedure is not applicable in the presence of manganese, owing to the formation of manganese dioxide. If, however, ammonium carbonate is used as the bath liquid, manganese dioxide is not formed. Since the copper and cobalt dimethylglyoxime compounds are not completely dissolved out, this procedure is only advisable when testing for nickel in the presence of manganese alone.

Limit of Identification: 0.1 γ nickel *Limi tof Dilution:* 1 : 500,000	in the presence of 10,000 times the amount of manganese

Test for small amounts of nickel in the presence of cobalt and iron salts [299]

Neither cobalt nor iron^{III} salts alone give insoluble compounds with dimethylglyoxime, but together they give a red-brown precipitate. This is a complex compound, whose empirical formula is $FeCoC_{12}H_{19}N_6O_6$.[297] Its formation can interfere with the detection of small amounts of nickel. This interference can be prevented by carrying out the test under special conditions (for mechanism, see page 446).

Procedure. One or two drops of the test solution are placed in an Emich centrifuge tube and warmed with a little saturated potassium cyanide solution until the initial precipitate is dissolved. A few milligrams of solid dimethylglyoxime and a few drops of formaldehyde are added; the mixture is stirred with a fine glass rod and then centrifuged. The deposit of dimethylglyoxime is colored red with a coating of nickel dimethylglyoxime if nickel is present.

(2) Test with dimethylglyoxime in the presence of oxidizing agents [300]

In the presence of large amounts of nitrates and other oxidizing agents, small amounts of nickel are not precipitated by dimethylglyoxime. Red to orange solutions are formed instead. If sufficient oxidizing agent is present, the precipitation of even large amounts of nickel is prevented.

When nickel dimethylglyoxime is suspended in alkali hydroxide solution and heated with a suitable oxidant, a pink solution results. Two nickel compounds appear to be formed, which contain either two or four dimethylglyoxime molecules per atom of nickel.[301] Because of the difficulty of isolation, it is not yet clear whether a dimethylglyoxime compound of quadrivalent nickel is involved [302] or a compound of divalent nickel with an oxidation product of dimethylglyoxime.[303] In any event, the water-solubility and the color characteristics of the reaction product indicate compounds in which the nickel is a constituent of an inner complex anion.

The formation of the red soluble nickel-dimethylglyoxime compound makes it possible to raise the sensitivity of the dimethylglyoxime test. The discernibility of the colored compound requires a lower ionization product and accordingly a smaller concentration of nickel than is required for the precipitation of the normal insoluble nickel^{II} dimethylglyoxime. In agreement with this finding is the fact that the colored water-soluble nickel compound is stable towards potassium cyanide, whereas nickel dimethylglyoxime dissolves readily. The following test for nickel can, therefore, also be carried out in the presence of cyanides.

Procedure. A drop of the test solution is mixed on a spot plate with 1 or 2 drops of saturated bromine water,* and after 1–2 minutes rendered alkaline (disappearance of the bromine color) with excess ammonia. Then a drop of 1 % alcoholic dimethylglyoxime solution is added. A red or orange color remains if nickel is present.

Limit of Identification: 0.12 γ nickel
Limit of Dilution: 1 : 400,000

(3) Test with rubeanic acid [304]

Rubeanic acid, the diamide of dithio-oxalic acid, presents an equilibrium mixture with its tautomeric (aci-) di-imido form:

$$\begin{matrix} SC-NH_2 \\ | \\ SC-NH_2 \end{matrix} \rightleftharpoons \begin{matrix} HSC{=}NH \\ | \\ HSC{=}NH \end{matrix} \rightleftharpoons 2\,H^+ + \left[\begin{matrix} SC{=}NH \\ | \\ SC{=}NH \end{matrix}\right]^{-2}$$

The aci-form combines with nickel, cobalt, and copper ions to give colored water-insoluble inner complex salts of the constitution

$$\begin{matrix} HN{=}C & - & C{=}NH \\ | & & | \\ S & & S \\ & \searrow\swarrow & \\ & Me & \end{matrix}$$

where Me = copper^{II}, cobalt, nickel. Polymers of these salts (see page 88) are formed when the concentration of the aci-form of rubeanic acid is large enough to exceed the solubility product of the rubeanate in question. This

* Instead of bromine water, persulfate (with the addition of a small amount of silver nitrate) may be used as the oxidizing agent (Communication from F. L. Hahn).

condition is attained by decreasing the hydrogen ion concentration in the equilibrium by the addition of alkali acetate, ammonia, or alkali. The copper, cobalt, and nickel compounds are quantitatively precipitated from strongly ammoniacal solutions but, once precipitated, they are distinctly resistant to dilute mineral acids. Cobalt, copper, and nickel rubeanates are soluble in potassium cyanide and consequently are not precipitated from alkaline cyanide solution.

Procedure. A drop of the test solution is placed on filter paper, held over ammonia, and then treated with a drop of the reagent solution. According to the nickel content, a blue to blue-violet stain forms.

Limit of Identification: 0.012 γ nickel (in 0.015 ml)

Limit of Dilution: 1 : 1,250,000

Reagent: 1 % alcoholic solution of rubeanic acid (see page 146)

It must be noted that the presence of much ammonium salt considerably reduces the sensitivity of the test.

Test for nickel in the presence of cobalt, iron, or copper by capillary separation of the ammine salts [305]

As just stated, rubeanic acid reacts to give colored complex compounds, not only with nickel, but also with cobalt and copper salts. Nevertheless, within certain concentration limits, nickel can be detected in the presence of these metals. Use is made of the different diffusion velocities of the metal ammine salts in thin filter paper, the diffusion velocity of nickel being the greatest. When a drop of an ammoniacal solution of copper, cobalt and nickel salts is placed on paper, or a drop of a neutral solution on paper is held over ammonia, the nickel accumulates in the outer zone of the spot. If a drop of the alcoholic reagent solution is then placed at the side and the drops coalesce, a blue ring of nickel rubeanate forms around the brown to green or brown circle due to the cobalt and copper compounds.

A drop (0.015 ml) of a solution containing 1 % cobalt and 0.0021 % nickel still shows a recognizable blue ring round the yellow-brown precipitation zone of the cobalt compound. This corresponds to a *limit of identification* of 0.32 γ nickel in a *limiting proportion* of Ni : Co = 1 : 480.

In the same way, 0.032 γ nickel can be detected in the presence of 4800 times the amount of iron. The iron is fixed in the center of the fleck as Fe^{III} hydrous oxide, while the nickel ammine salt diffuses outward and reacts with rubeanic acid in a separate zone.

20. Thallium

(1) Test with potassium iodide

Potassium iodide precipitates bright yellow thallous iodide from acid solutions. The test is affected by the presence of other metals giving insoluble iodides, namely mercury, silver and lead. In the presence of mercury, thallium can be detected by using a large excess of potassium iodide, whereby the mercuric iodide goes into solution as complex $K_2[HgI_4]$. At the same time, through mass action, the solubility of the thallous iodide is reduced. In the presence of silver and lead,[306] it is best to digest the iodide precipitate with sodium thiosulfate, or to carry out the precipitation in the presence of a large amount of thiosulfate. Silver and lead iodides are dissolved forming the complex anions $[Ag_2(S_2O_3)_2]^{-2}$ and $[Pb(S_2O_3)_2]^{-2}$, whereas thallous iodide remains unchanged.

Procedure. A drop of the weakly acid test solution is mixed with a drop of 10 % potassium iodide on a watch glass placed on a black background. When a precipitate appears, a drop or two of 2 % sodium thiosulfate is added. A yellow precipitate, insoluble in sodium thiosulfate, indicates thallium.

Limit of Identification: 0.6 γ thallium

Limit of Dilution: 1 : 80,000

(2) *Test with benzidine* [307]

When thallium is present in the Tl^{III} state, it can be detected, after conversion into TlO(OH), by the oxidation of benzidine to "benzidine blue" (see page 72). Thallous salts must be oxidized beforehand. This oxidation is conveniently accomplished by means of potassium ferricyanide in alkaline solution [308]:

$$Tl^+ + 2\ Fe(CN)_6^{-3} + 3\ OH^- \rightarrow TlO(OH) + H_2O + 2\ Fe(CN)_6^{-4}$$

If this reaction is carried out at room temperature as a spot reaction in a porcelain crucible or on a white spot plate, most or all of the TlO(OH) adheres to the porcelain, where it forms a brown mirror. The adhesion is so firm that the coating cannot be removed by washing with water. If the benzidine reaction is used to detect TlO(OH), quantities of thallium can be revealed that are too slight to produce a visible stain on porcelain.

Under these same conditions, lead, manganese and cobalt salts yield higher oxides which react with benzidine. Therefore, the following procedure is recommended when alkaline solutions are to be examined for the presence of thallium.

Procedure.[309] A drop of the alkaline test solution is mixed on a spot plate with a drop of 1 % potassium ferricyanide solution. According to the quantity of thallium, a brown turbidity appears at once or after several minutes. No change is visible with quantities below 10 γ. After about 30 minutes, the spot plate is carefully rinsed with water and spotted with a drop of benzidine solution. A blue color indicates thallium.

Limit of Identification: 0.5 γ thallium

Reagent: Benzidine solution: benzidine is warmed with dilute acetic acid; the cooled solution is filtered.

To detect thallium in the presence of lead, a drop of the acid or neutral test solution is treated with a drop of ammonia. After gentle warming, the mixture is centrifuged or filtered. After the $Pb(OH)_2$ precipitate has been washed with a drop of water, a drop of the filtrate or centrifugate is placed on a spot plate. Two drops of 1 N alkali and one drop of 1 % potassium ferricyanide solution are added. After 30 minutes, the solution is poured off and the invisible slight deposits adhering to the spot plate are tested with benzidine solution. This procedure will reveal 4.5 γ Tl in the presence of 5000 γ Pb.

(3) Test with phosphomolybdic and hydrobromic acid [310]

A yellow precipitate of thallium^{I} phosphomolybdate is formed by the interaction of thallous salts and phosphomolybdic acid. When this precipitate is treated with concentrated hydrobromic acid, molybdenum blue and thallium^{III} bromide are formed.

The phosphomolybdic acid is apparently activated by its combination with thallium since it normally reacts very slowly with hydrobromic acid. Possibly the capacity of the bromide ion to carry the positive charge from Mo^{VI} to Tl^{I} is connected with the proximity in space of these metals in the Tl^{I} phosphomolybdate molecule.* This very singular chemical reaction affords a means of testing for thallium^{I} salts, but care should be taken that no reducing substances are present, such as copper^{I}, tin^{II}, antimony^{III}, iron^{II} and Hg^{I}, since they are able to reduce phosphomolybdate to molybdenum blue.

Procedure. A drop of a saturated solution of phosphomolybdic acid (see page 104) is placed on filter paper, and the test solution is run from a capillary into the middle of the spot. A drop of 50 % hydrobromic acid is then added. A dark or light blue stain is formed, according to the amount of thallium present.

Limit of Identification: 0.13 γ thallium (in 0.025 ml)

Limit of Dilution: 1 : 200,000

(4) Test with dipicrylamine [311]

Dipicrylamine (I), or preferably the yellow aqueous solution of its sodium salt, reacts with thallous salts, in the absence of free acids, to produce a red crystalline precipitate of thallium dipicrylamine (II). Once this salt has been precipitated, it is rather resistant towards dilute mineral acids. The mechanism of the salt-formation is discussed on p. 231 in connection with the precipitation of potassium by means of this reagent.

* This hypothesis is supported by the observation that the formation of molybdenum blue is much slower when a mixture of phosphomolybdic and hydrobromic acid is treated on filter paper with a drop of a thallium salt. Another explanation involving the enhancement of the reactivity of Mo^{VI} in phosphomolybdic acid is given in Feigl, Ref. 4, p. 117.

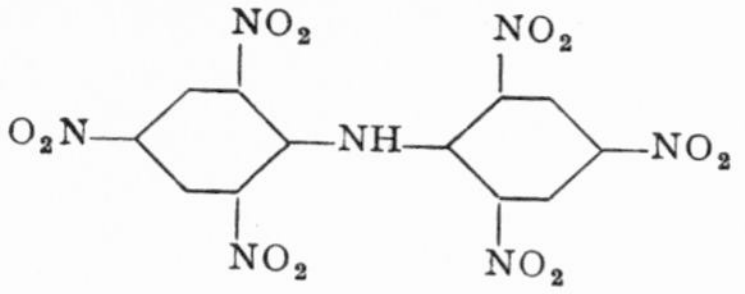

NO_2 NO_2
O_2N—NH—NO_2
NO_2 NO_2
(I)

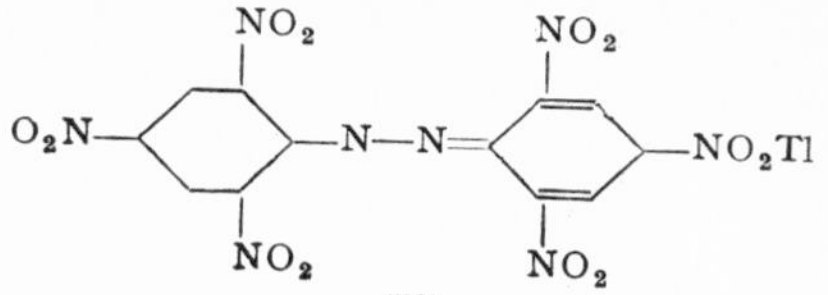

NO_2 NO_2
O_2N—N—N=—NO_2Tl
NO_2 NO_2
(II)

Thallium^{I} can be detected by placing a drop of the neutral test solution on filter paper impregnated with sodium dipicrylamine (preparation, page 232). Thallium dipicrylamine is formed. If the spotted paper is bathed in 0.1 *N* nitric acid, the unchanged orange-red sodium dipicrylamine is decomposed, forming bright yellow dipicrylamine, whereas the red thallium salt remains unaltered. In this way, 20 γ thallium can be detected.

Since the thallium dipicrylamine reaction exhibits a relatively low sensitivity for a spot test, and since, furthermore, potassium and ammonium salts as well as considerable quantities of magnesium and sodium salts interfere, it is better to use the dipicrylamine merely to identify thallium within the chloride group of the qualitative scheme after oxidation of Hg^{I} to Hg^{II}. In other words, the element should be isolated as thallous chloride, which is only slightly soluble in water. Any AgCl or $PbCl_2$ that may have been precipitated with the TlCl will not react with sodium dipicrylamine, whereas TlCl produces red dipicrylamine compound when spotted with the reagent.

Procedure. One or two milliliters of test solution are treated with dilute hydrochloric acid at room temperature. The precipitate is collected and washed free of acid with alcohol. The paper is then spread on a glass or porcelain plate, spotted with a drop of sodium dipicrylamine solution, and dried in a blast of heated air. One or two drops of 0.1 *N* nitric acid are applied. If a red color develops, TlCl is present.

Reagent: Solution of sodium dipicrylamine: 0.2 g dipicrylamine is dissolved in 2 ml 2 *N* sodium carbonate, then diluted with 75 ml water, and filtered

(5) *Test by oxidation to thallium*^{III} *in acid solution* [312]

Bromine (chlorine) water converts thallous to thallic ions in acid solution. Since excess bromine (chlorine) is readily removed by adding sulfosalicylic acid [formation of bromo (chloro) sulfosalicylic acid]. Tl^{+3} ions remain* and can liberate iodine from iodide ions. The reactions are:

$$Tl^{+} + 2\,Br^{0} \rightarrow Tl^{+3} + 2\,Br^{-} \quad (1)$$

$$C_6H_3(OH)(COOH)SO_3H + 2\,Br^{0} \rightarrow C_6H_2Br(OH)(COOH)SO_3H + HBr \quad (2)$$

$$Tl^{+3} + 3\,I^{-} \rightarrow TlI + 2\,I^{0} \quad (3)$$

* The complex ions $[TlCl_4]^-$ or $[TlBr_4]^-$ are produced in solutions which contain Cl^- or Br^- ions. These products behave like Tl^{III} ions in redox reactions.

Accordingly, the release of iodine through (1)–(3) constitutes a reliable test for thallium, in the absence of materials that oxidize iodide to iodine. Such oxidants include Fe^{III}, Cu^{II}, As^{V}, Sb^{V}, Ce^{IV} salts, also permanganate, chromate, molybdate, ferricyanide, nitrite, alkali halogenates.

Treatment of the solution with alkali carbonate will separate Fe^{III}, Cu^{II}, Sb^{V} and Ce^{IV} from Tl^{I}, since its carbonate is water-soluble. If Mn^{VI} and Cr^{VI} are first reduced to Mn^{II} and Cr^{III} by sulfurous acid, they can likewise be removed by carbonate precipitation. Molybdates can be quantitatively precipitated, in strong sulfuric acid solution, by means of benzoinoxime.[313] Alkali halogenates must be reduced by sulfurous acid. Nitrites are easily decomposed by adding urea:

$$2\ HNO_2 + CO(NH_2)_2 \rightarrow CO_2 + 2\ N_2 + 3\ H_2O$$

before carrying out reaction (3), which constitutes the test for thallium.

Procedure. A drop of the acid test solution is treated with a drop of bromine water in a depression of a spot plate. If the color is discharged, the addition is continued until a permanent yellow is obtained. The excess bromine is removed by adding a little solid sulfosalicylic acid. A drop of starch-cadmium iodide solution * is added to the decolorized solution. Depending on the amount of thallium present, a deep or a light blue appears immediately.

In case nitrite or much nitrate is present, the solution, after decolorization with sulfosalicylic acid, should be stirred with a little solid urea. The iodide solution is added several minutes later.

Limit of Identification: 0.2 γ thallium
Limit of Dilution: 1 : 250,000
Reagent: 5 % solution of cadmium iodide in 1 % solution of cadmium sulfate. (Starch solution to be added before using.)

(*6*) *Test with rhodamine B* [314]

Acid solutions of the dyestuff rhodamine B, which contain the dyestuff cation (I), react with solutions of thallium^{III} chloride or bromide in hydrochloric or hydrobromic acid to give violet precipitates.[315] With slight quantities of thallium, the red of the dyestuff turns toward violet. Since alkali halide solutions of $TlCl_3$ and $TlBr_3$ contain $[TlCl_4]^-$ and $[TlBr_4]^-$ respectively, insoluble dyestuff salts of these complex ions may be formed. For example, the bromide could produce (II):

* The use of CdI_2 solution in place of KI solution has the advantage that the release of small amounts of iodine by dissolved oxygen and traces of nitric oxide is much slower, because of the slight dissociation of CdI_2. Compare T. E. Moore and J. Lambert, *J. Am. Chem. Soc.*, 71 (1949) 3260.

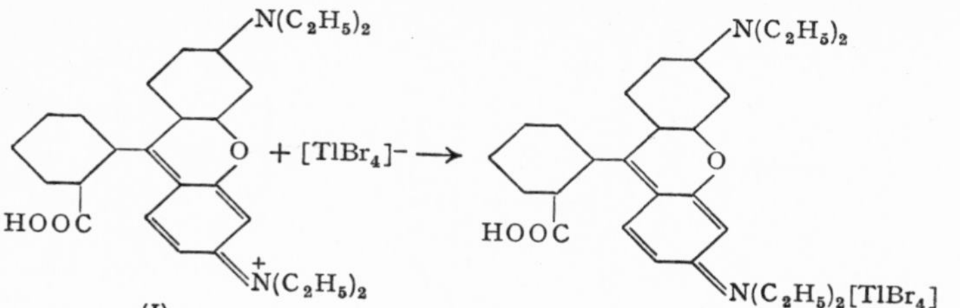

$N(C_2H_5)_2$ ... $N(C_2H_5)_2$

O + $[TlBr_4]^-$ → O

HOOC HOOC

$\overset{+}{N}(C_2H_5)_2$ $N(C_2H_5)_2[TlBr_4]$

(I) (II)

If a suspension of (II) (or the analogous $[TlCl_4]$ compound) in hydrochloric acid is shaken with benzene, the product dissolves in the benzene to form a red solution and the latter displays an intense orange-yellow fluorescence. In contrast, no fluorescent benzene layer is produced when hydrochloric acid solutions of the dyestuff are shaken with benzene. Furthermore, the solid water-insoluble thallium^{III}-rhodamine B salts show no fluorescence. In view of the fact that the molybdates and tungstates of (I), which are not soluble in dilute acids, are also not soluble in benzene, it is likely that the fluorescence of the benzene layer is due to (II). It seems a fair assumption that on extraction with benzene (ether acts analogously) (II) is transformed into the isomeric benzene-soluble salt (III):

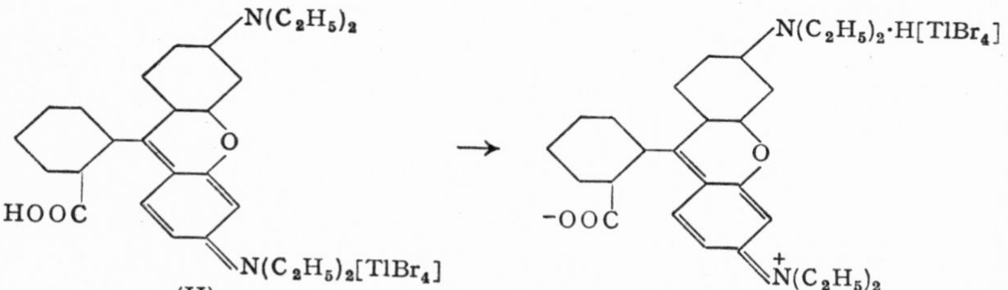

$N(C_2H_5)_2$ ... $N(C_2H_5)_2 \cdot H[TlBr_4]$

O → O

HOOC ^-OOC

$N(C_2H_5)_2[TlBr_4]$ $\overset{+}{N}(C_2H_5)_2$

(II) (III)

The formation of the fluorescent benzene solution of (III) will reveal quantities of thallium which are too slight to give either a visible precipitate in acid solution or to change the color of the reagent. If the conditions prescribed here are maintained, only salts of gold, mercury, and antimony interfere. The first two form rhodamine compounds which are soluble in benzene and give the solution an orange fluorescence (compare the test for gold, below). The antimony^V rhodamine compound dissolves in benzene with a red color [316] without fluorescence. The interference due to antimony, especially when large amounts are present, is due to the consumption of the reagent and not to the inhibition of the action of $[TlBr_4]^-$. If the conditions are properly controlled, it is possible nevertheless to achieve a specific detection of thallium with rhodamine B.

Procedure. The solution to be tested for thallium is oxidized with bromine water as described in test *5* and the excess bromine is removed by 10 % water solution of sulfosalicylic acid. These steps can be easily accomplished in a micro test tube with one drop of the test solution. One drop of concentrated hydrochloric acid and 1 or 2 drops of 0.05 % rhodamine B solution in concentrated hydrochloric acid are added and the mixture is then vigorously shaken with 6–8 drops of benzene. After separation of the layers (centrifuge if necessary) the benzene layer appears red-violet to pink, depending on the quantity of thallium present, and displays a yellow fluorescence in ultraviolet light. A comparison blank is advisable, because benzene often has a blue fluorescence.

Limit of Identification: 0.03 γ thallium

Limit of Dilution: 1 : 1,660,000

This procedure will reveal 0.04 γ thallium in the presence of 1000 γ iron, antimony, lead, bismuth, molybdate, etc.

Detection of thallium in the presence of antimony, gold, and mercury

When thallium is to be detected in the presence of antimony, gold, and mercury, the acid test solution should be treated with copper or brass wire. The gold, antimony, and mercury are reduced to the metallic state and adhere to the wire as a coating. The thallium^{I} salts are not affected; if thallium^{III} salts are present, they are reduced to thallium^{I}.

Procedure. A drop of the test solution is placed in a micro test tube, a drop of concentrated hydrochloric acid is added, and a thin copper or brass wire, whose end is twisted into a spiral, is hung into the solution. The test tube is kept in boiling water for about 5 minutes and the wire is then removed. The solution is filtered or centrifuged, if necessary, to remove the metal particles. The procedure for the detection of thallium is then used, after the solution has been treated with bromine water, sulfosalicylic acid, etc.

This procedure revealed 0.1 γ thallium in the presence of 500 γ antimony, gold, or mercury.

(*7*) *Other tests for thallium*

(*a*) Metal ions that form insoluble sulfides can, with a single exception, also be precipitated as carbonates or basic carbonates by adding an alkali carbonate. The exception is Tl^{+} ion, whose carbonate is quite soluble in water. Consequently, thallium can be separated from other metals by treating the test solution with an excess of sodium carbonate and filtering. If a drop of the filtrate is placed on thick filter paper (S & S 601) and the fleck held over ammonium sulfide, a black-brown spot or ring (Tl_2S) appears, according to the quantity of thallium present (*Idn. Limit:* 0.5 γ Tl). If considerable quantities of metal ions that are precipitated by sodium carbonate are present it must be remembered that small amounts of thallous carbonate may be entrained through adsorption. Furthermore, the carbonate precipitation of Fe, Pb and other heavy metal ions may be partially prevented if organic hydroxy compounds that form complexes are present. In this case, the filtrate will contain other metals that react with hydrogen sulfide.

(*b*) A solution of equimolar quantities of $AuCl_3$ and $PdCl_2$ in hydrochloric acid gives a cinnamon-brown precipitate with Tl^+ ions; the product turns black when moistened with caustic alkali (*Idn. Limit:* 0.4 γ Tl in 0.001 ml).[317]

(*c*) Tl^{III} hydroxide, produced by precipitation with alkaline ferricyanide solution, oxidizes the colorless acetic acid solution of leuco nitro diamond green to the blue-green dyestuff (*Idn. Limit:* 0.01 γ Tl).[318]

(*d*) The yellow-green fluorescence of neutral or weakly acid solutions of UO_2SO_4 is made to disappear by Tl^I salt solutions (*Idn. Limit:* 1 γ Tl).[319]

(*e*) The red fluorescence of acid or neutral solutions of rhodamine B is changed to violet by Tl^{III} (*Idn. Limit:* 0.5 γ Tl).[320]

(*f*) After oxidizing Tl^I by bromine and removing the excess by sulfosalicylic acid (compare test *5*) and buffering, Tl^{III} oxinate can be formed by shaking with a chloroform solution of oxine. The product dissolves in chloroform with a yellow color (*Idn. Limit:* 0.5 γ Tl).[321]

21. Iron

(*1*) *Test with potassium ferrocyanide*

Potassium ferrocyanide forms Prussian blue, $Fe_4[Fe(CN)_6]_3$, with ferric salts in acid solution. This compound, like Turnbull's blue $Fe_3[Fe(CN)_6]_2$, is reduced to white ferro-ferrocyanide [322] by strong reducing agents such as sodium hydrosulfite.

Procedure. The test can be carried out by placing one drop of the test solution and one of potassium ferrocyanide solution on filter paper or a spot plate.

Limit of Identification:	0.1 γ iron (on paper)
	0.05 γ iron (on a spot plate)
Limit of Dilution:	1 : 500,000 (on paper)
	1 : 1,000,000 (on a spot plate)

The Prussian blue reaction may be either masked or completely prevented in the presence of large proportions of other metal salts which give colored precipitates with potassium ferrocyanide or that form complexes with the Fe^{III} ion (fluorides, oxalates,[323] etc.). In such cases it is best to use test (*2*).

(*2*) *Test with α,α′-dipyridyl* [324]

$Iron^{II}$ salts in solutions of mineral acids react with the organic base *α,α′*-dipyridyl (I) to give a soluble deep red very stable complex cation; [325] it has the coordination formula (II).

(I) (II)

Iron^{III} salts do not react under these conditions; consequently very small amounts of iron^{II} salts can be detected in the presence of large proportions of iron^{III} as shown below. On the other hand, after reduction, tervalent iron can also be detected by means of α,α'-dipyridyl.

Other ammine-forming metallic ions in acid solution also react with α,α'-dipyridyl, but give weak colored or colorless soluble addition compounds, so that when sufficient reagent is used they do not interfere with the test for iron, except when traces are involved (compare Chap. 7, Section 28). Large amounts of halides, sulfates, etc. reduce the solubility of the ferro-dipyridyl compound in water, so that a red precipitate is sometimes formed. However, it does not interfere with the test.

The α,α'-dipyridyl test for iron is especially useful when compounds are present which interfere with the usual thiocyanate test (page 164). The cheaper α,α'-phenanthroline can be substituted for α,α'-dipyridyl.[326]

Procedure. A drop of the test solution is treated with a drop of the reagent solution on a spot plate, or a drop of the test solution is placed on filter paper (S & S 589) which has been impregnated with an alcoholic solution of the reagent, and dried. According to the amount of iron present, a red or pink circle is formed.

Limit of Identification: 0.03 γ iron
Limit of Dilution: 1 : 1,666,000
Reagent: 2 % solution of α,α'-dipyridyl [327] in alcohol

If iron^{III} salts are present, it is well to use a 2 % solution of α,α'-dipyridyl in thioglycolic acid as reagent.[328] The latter reduces iron^{III} almost instantaneously with formation of dithioglycolic acid:

$$2\,Fe^{+3} + 2\,HS{-}CH_2{-}COO^- \rightarrow 2\,Fe^{+2} + 2\,H^+ + \begin{array}{l} S{-}CH_2{-}COO^- \\ | \\ S{-}CH_2{-}COO^- \end{array}$$

Considerable quantities of cobalt and nickel ions may not be present because they produce soluble brown thioglycolates.

Test for iron^{II} in the presence of iron^{III} salts

Large amounts of iron^{II} salts can be detected in the presence of iron^{III} salts, by the normal procedure (as described) since the pink color of iron^{II} dipyridyl is easily visible, even in the presence of the yellow solution of iron^{III} salts. When, however, traces of iron^{II} are being sought, the slight pink coloration may easily be blanketed by the yellow of a concentrated ferric solution. In this case, it is best to convert the Fe^{+3} ions into colorless $[FeF_6]^{-3}$ ions by adding potassium fluoride. The pale pink due to the traces of ferrous salts can then easily be seen.

Procedure. A drop of the acid test solution is placed in a porcelain micro crucible coated with paraffine. The yellow solution is decolorized by adding a few crystals of potassium fluoride. Then a drop of the reagent solution is added. In the presence of iron^{II} salts, a red or pink coloration appears. In the absence of ferrous salts, the solution remains colorless, or a precipitate (complex potassium ferric fluoride) may separate.

(3) *Test with 8-hydroxyquinoline-7-iodo-5-sulfonic acid (ferron)* [329]

Acid solutions of iron^{III} salts react with water solutions of 8-hydroxyquinoline-7-iodo-5-sulfonic acid (I) to produce a green color due to the inner complex anion (II).

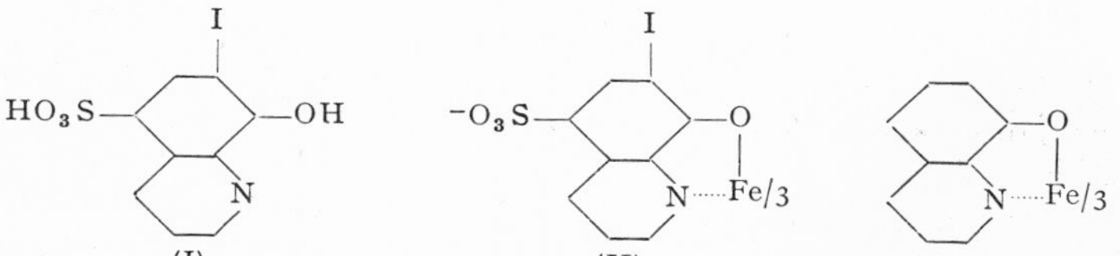

I HO$_3$S— —OH N (I)

I $^-$O$_3$S— —O N·····Fe/3 (II)

—O N·····Fe/3 (III)

The parent compound of the reagent, namely, 8-hydroxyquinoline, which is difficultly soluble in water, forms an insoluble green-black inner complex ferric salt (III). Consequently, the SO_3H group, which confers increased solubility in so many other cases, here brings about the formation of an inner complex anion, containing the iron atom.

The color test for iron^{III} with 8-hydroxyquinoline-7-iodo-5-sulfonic acid (ferron) is specific. The only interference arises from considerable quantities of colored ions or of strong oxidizing agents that partially destroy the reagent. It is important to maintain the acidity at pH 2 to 3.5. Consequently, if stronger or less acid solutions are at hand they must be buffered or acidified until a distinct acid reaction toward methyl orange paper (pH = 3.5) is obtained.

Procedure. One drop of the acid test solution (pH = 3.5) is treated on a spot plate with a drop of the orange-yellow water solution of the reagent. A green color indicates the presence of iron^{III}.

Limit of Identification: 0.5 γ iron

Limit of Dilution: 1 : 100,000

Reagent: 0.1 % water solution of 8-hydroxyquinoline-7-iodo-5-sulfonic acid.[330]

References pp. 247–257

(4) Test with potassium thiocyanate

Iron^{III} ions react with thiocyanate ions in acid solution to yield a red color. Several colored compounds result, which are in equilibrium with each other.[331] If the CNS^- concentration is low, $Fe(CNS)_2^{+1}$ ions are the chief product. However, when the CNS^- concentration reaches 0.1 *M* or higher, the more intensely colored $[Fe(CNS)_4]^{-1}$ and $[Fe(CNS)_6]^{-3}$-ions result. These facts indicate that $Fe(CNS)_3$ is also present in the red solution in equilibrium with anionic ferrithiocyanate ions. If the red solution is shaken with ether, all of the iron is extracted as $Fe(CNS)_3$ or $[Fe(CNS)_6]Fe$ to yield a red ether solution.

Phosphates, arsenates, oxalates, tartrates, and other organic hydroxyl compounds, fluorides, and many compounds which give stable complex salts with iron^{III}, can, according to the proportion present, so reduce the concentration of ferric ions that the ionic product necessary for the color reaction is not reached. Even considerable amounts of iron may then fail to be detected by the sensitive thiocyanate reaction. The same interference occurs when great amounts of mercury salts are present, because they form slightly dissociated mercuric thiocyanate or double thiocyanates, e.g., $K_2[Hg(CNS)_4]$, and thus consume the thiocyanate ions. In such cases test *2* should be used.

The presence of nitrites is to be avoided, since in acid solution they form nitrosyl thiocyanate NO·CNS. This compound gives a red color (which will disappear on heating) very similar to that of the iron thiocyanate.[332]

Procedure. A drop of the test solution is mixed on a spot plate with a drop of 1 % potassium thiocyanate solution. A more or less deep red color appears in the presence of iron^{III}.

Limit of Identification: 0.25 γ iron
Limit of Dilution: 1 : 200,000

Cobalt, nickel, chromium, copper and zinc salts reduce the sensitivity of the iron test because of the color of their ions, or of their reaction products with thiocyanate. However 1.25 γ iron can be detected in the presence of 320 times the amount of cobalt or chromium, and 0.025 γ iron in the presence of 640 times the amount of nickel or copper.[333]

Test for iron in the presence of fluorides, see page 449.

(5) Test with dimethylglyoxime [334]

Ferrous salts form a soluble red inner complex compound with dimethylglyoxime in ammoniacal solution or in the presence of organic bases. The complex compound was long regarded,[335] in analogy to nickel dimethylglyoximate, as Fe^{II} dimethylglyoximate, with two base molecules coordinated on the iron atom as shown in (I). However, the isomeric formulation (II) is more likely; it shows the compound as the salt of a Fe^{II}-dimethylglyoxime acid with an inner complex iron-bearing anion.[336]

```
       O         OH
       ||   B    |
H3C—C=N\    :  .N=C—CH3
    |    \Fe.     |
H3C—C=N'    :    \N=C—CH3
       |    B      ||
       OH          O
           (I)
```

```
 [          O        O          ]
 [          ||       |          ]
 [  H3C—C=N\       .N=C—CH3     ]
 [      |    >Fe<.    |         ]  H2·2B
 [  H3C—C=N'       \N=C—CH3     ]
 [          |         ||        ]
 [          O         O         ]
                (II)
```

(B = ammonia, pyridine or a primary aliphatic amine)

The dimethylglyoxime test for Fe^{II} is highly sensitive and selective. If $iron^{III}$ is also present, which is almost invariably the case, it is necessary to add citric or tartaric acid before making the solution ammoniacal to prevent the precipitation of ferric hydroxide. On the other hand, the test may be modified to serve as a general test for iron if Fe^{III} is previously reduced to Fe^{II} by the addition of hydrazine sulfate.

The presence of nickel interferes, since the precipitation of the red insoluble nickel dimethylglyoxime renders it difficult to see the red color of the solution. If, however, potassium cyanide is added, the nickel dimethylglyoxime dissolves, and only the red ferrous salt, which is stable in the presence of potassium cyanide, remains.

Cobalt and copper salts combine with dimethylglyoxime to give soluble brown complex salts, and thus interfere when present in large proportions.

Procedure. A drop of the test solution (previously reduced if necessary) is mixed with a crystal of tartaric acid, and then a drop of 1 % alcoholic dimethylglyoxime is added, followed by a little ammonia. According to the iron content, a more or less intense red coloration appears. The color fades on standing in the air because the ferrous complex oxidizes.

Limit of Identification: 0.4 γ iron
Limit of Dilution: 1 : 125,000

(6) Test with disodium 1,2-dihydroxybenzene-3,5-disulfonate[337]

Disodium 1,2-dihydroxybenzene-3,5-disulfonate (I) produces water-soluble, colored compounds with ferric salts. The iron atom is incorporated in an inner complex anion. In solutions whose pH is less than 5, the complex (II) is deep blue. The cautious addition of a base to the blue solution results in a sharp change to violet (III) at pH 5.7 to 6.5; at pH 7 the color changes to red (IV). The structural formulas (II) and (IV) have been established for the blue and red complex salts, respectively, through their analogy to

References pp. 247–257

the like-colored complex iron salts of pyrocatechol, which were isolated quite some time ago.[338] It is probable that (III) correctly represents the violet complex salt.

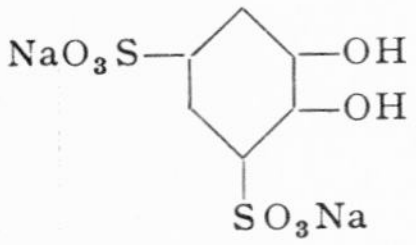

(I) Reagent

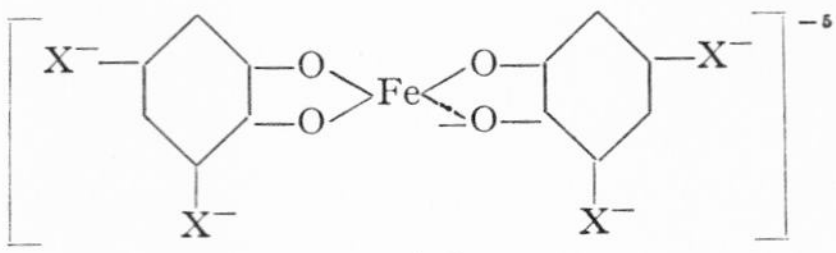

(II) Blue complex

$X^- = SO_3^-$

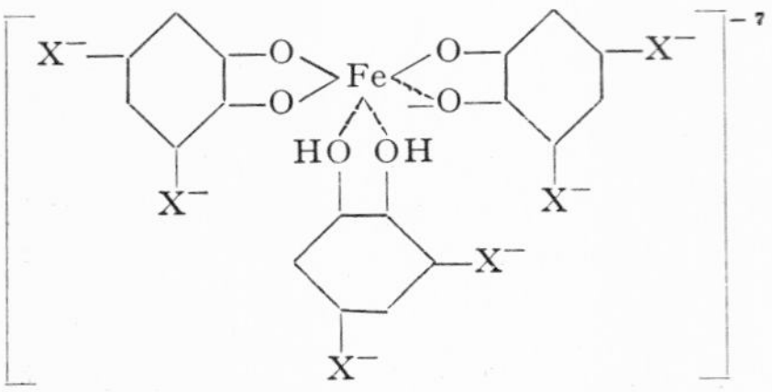

(III) Violet complex

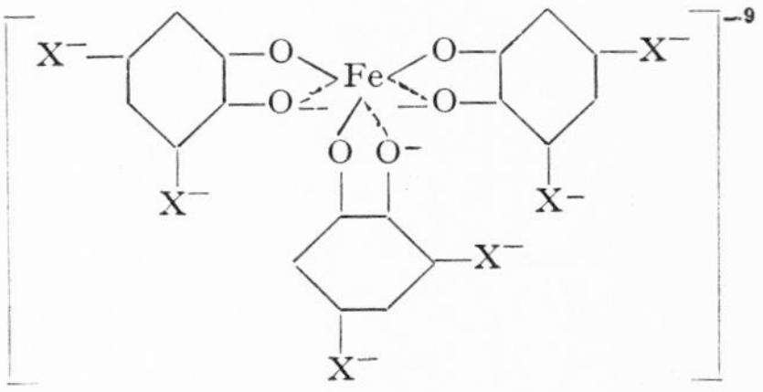

(IV) Red complex

The complex ferric salts (II)–(IV) are so stable and require for their formation such slight concentrations of Fe^{+3} ion that they are produced even in the presence of alkali fluorides, tartrates, citrates, etc., which prevent or impair many other tests for ferric iron. A further advantage is that the reagent can be applied as a colorless water solution.

It should be noted that even minute quantities of titanium salts give an orange color with this reagent; copper salts produce a green-yellow. The yellow color developed by considerable amounts of molybdenum impairs the test for iron.

Procedure. A drop of the test solution is treated in a depression of a spot plate with a drop of the reagent solution and a drop of buffer solution. A red to pink appears, depending on the amount of iron present.

Limit of Identification: 0.05 γ iron

Limit of Dilution: 1 : 1,000,000

Reagents: *1)* Solution of 0.0113 g disodium 1,2-dihydroxybenzene-3,5-disulfonate in 100 ml water

2) Buffer solution (pH = 9.8): 1 g $NaHCO_3$ and 0.5 g Na_2CO_3 in 100 ml water

(*7*) *Other tests for iron*

(*a*) If one drop of weakly acid iron solution is placed on filter paper impregnated with 5 % alcoholic oxine solution, a green-black fleck appears due to iron^{III} oxinate (*Idn. Limit:* 10 γ Fe).[339] If the test solution is strongly acid, the fleck should be exposed to ammonia. When other ions which react with oxine are present, the warm solution should be treated with zinc oxide paste, filtered, and the iron-bearing precipitate dissolved in the least volume of dilute hydrochloric acid.

(*b*) A purple color is produced in ammoniacal solution by the action of thioglycolic acid with iron^{II} and iron^{III} salts.[340] One drop each of the test solution, 6 *N* ammonia, and thioglycolic acid are mixed on a spot plate (*Idn. Limit:* 0.01 γ Fe). Sulfites interfere; zinc salts decrease the sensitivity.

(*c*) Ferric salts spotted on paper soaked with 0.2 % quercetin or quercitrin produce an olive-green ring (*Idn. Limit:* 3 γ Fe).[341] Uranium salts form rust-brown flecks (see page 209).

(*d*) Iron^{II} may be detected by spotting with 1 % aqueous solution of 2-nitroso-1-naphthol-4-sulfonic acid at pH = 5.[342] A green color is formed (*Idn. Limit:* 0.01 γ Fe). Copper and cobalt salts interfere.

(*e*) Isonitrosobenzoylmethane reacts with iron^{II} salts to yield a blue inner complex salt that is soluble in benzene. One drop of the test solution is placed on filter paper and spotted with a 1 % alcoholic solution of the reagent. When held over ammonia, a green fleck appears (*Idn. Limit:* 0.02 γ Fe).[343] The test may be conducted likewise on a spot plate or in a micro crucible. In the latter case, benzene extraction is possible. Copper, cobalt, nickel interfere; silver, lead, zinc, manganese, chromium lower the sensitivity.

22. Chromium

(*1*) *Test with diphenylcarbazide, after conversion into chromate* [344]

Strong acid solutions of chromates react with diphenylcarbazide (I) to give a violet coloration, even though only traces of chromium^{VI} are present. The color reaction involves an action that has not been observed in other cases [345]: the chromate, in acid solution, oxidizes diphenylcarbazide (I) initially to colorless diphenylcarbazone (II) and then to light yellow diphenylcarbadiazone (III):

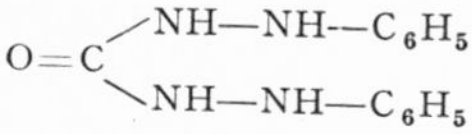

$$O{=}C\begin{matrix}\diagup NH{-}NH{-}C_6H_5\\ \diagdown NH{-}NH{-}C_6H_5\end{matrix} \qquad O{=}C\begin{matrix}\diagup NH{-}NH{-}C_6H_5\\ \diagdown N{=}N{-}C_6H_5\end{matrix} \qquad O{=}C\begin{matrix}\diagup N{=}N{-}C_6H_5\\ \diagdown N{=}N{-}C_6H_5\end{matrix}$$

(I) 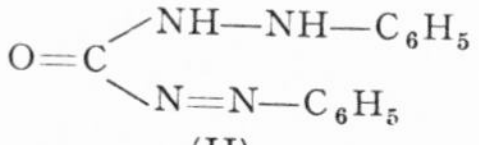 (II) 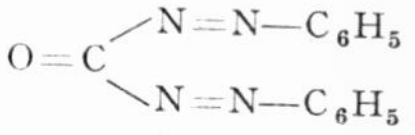 (III)

However, in distinction to all other known redox reactions of the chromate ion, there is no production of chromic ions but rather chromous ions are formed and the latter yield a red-violet inner complex salt (V) by action with the enol form (IV) of the carbazone.

$$\begin{array}{l} \quad N\text{—}NH\text{—}C_6H_5 \\ C\text{—}OH \\ \quad N{=}N\text{—}C_6H_5 \end{array} \qquad \begin{array}{l} \quad N\text{—}N\text{—}C_6H_5 \\ C\text{—}O\text{—}Cr\,(H_2O)_4 \\ \quad N{=}N\text{—}C_6H_5 \end{array}$$

(IV) (V)

That the reaction really involves the formation of an inner complex salt of chromium^{II} is shown by the fact that chromous salts react with carbazone only and not with carbazide or carbadiazone to yield a violet color. Accordingly, this is a unique instance in which the formation of a complex not only stabilizes an anomalous valence state but also brings about an anomalous redox reaction.

Since chromium^{III} salts can easily be converted into chromates by oxidation in acid or alkaline solution, or by oxidizing fusion, the diphenylcarbazide reaction can be utilized as a sensitive test for chromium. Molybdenum, mercury, and vanadium salts must be absent (compare pages 119 and 64).

a) Oxidation in alkaline solution with alkali hypobromite

Chromium^{III} ions react with excess alkali hydroxide to form green chromite ions, which are quantitatively oxidized, by excess alkali hypobromite, even at room temperature, to chromate ions:

$$2\ CrO_2^- + 3\ BrO^- + 2\ OH^- \rightarrow 2\ CrO_4^{-2} + 3\ Br^- + H_2O$$

Before testing for chromate with diphenylcarbazide, the solution should be acidified to decompose the unused hypobromite:

$$Br^- + BrO^- + 2\ H^+ \rightarrow H_2O + Br_2^\circ$$

The free bromine must be removed because it impairs the diphenylcarbazide test. A suitable means of removal is to add sulfosalicylic acid, which immediately reacts with bromine to form bromosulfosalicylic acid:

$$Br_2^\circ + C_6H_3(OH)(COOH)(SO_3H) \rightarrow C_6H_2Br(OH)(COOH)(SO_3H) + HBr$$

If the Cr^{III} solution is free of Co^{+2}, Ni^{+2} or Mn^{+2} ions, which produce water-insoluble black oxyhydrates of the respective metals on treatment with alkali hydroxide and alkali hypobromite, the reaction conditions for the diphenylcarbazide test can be secured by merely acidifying the solution with sulfuric acid and adding a few crystals of sulfosalicylic acid. Small amounts of the higher oxides of the metals go into solution, even though they are *per se* rather resistant toward dilute sulfuric acid. This is true particularly of Mn^{IV} oxide.* In this case, the reversible redox reaction:

$$MnO_2 + 4\ H^+ + 2\ Br^- \rightleftarrows Mn^{+2} + 2\ H_2O + Br_2^\circ$$

* F. Feigl (unpublished studies) showed that even natural pyrolusite is quantitatively dissolved by warming with dilute sulfuric acid and adding alkali bromide plus sulfosalicylic acid.

is made to go to completion by removal of the free bromine by sulfosalicylic acid, in other words the Mn^{IV} oxide is made to disappear.

In the following procedure, the detection of very small amounts of chromium by the diphenylcarbazide reaction is made difficult by the presence of large quantities of colored ions (Cu^{+2}, Co^{+2}, Ni^{+2}, Fe^{+3}). In such cases, a comparison should be made with the color of a drop of the acidified test solution, to establish with certainty any development of a slight violet color. The yellow due to Fe^{III} salts can be discharged by phosphoric acid (production of colorless $[Fe(PO_4)_2]^{-3}$ ions).

Procedure.[346] A drop of the neutral or weakly acid test solution is treated, on a spot plate, with one or two drops of sodium hypobromite solution. After about a minute, a drop of concentrated sulfuric acid and one or two drops of a 20 % aqueous solution of sulfosalicylic acid are added and then a drop of a saturated alcoholic diphenylcarbazide solution. A violet color appears; its intensity depends on the quantity of chromium present.

Limit of Identification: 0.25 γ chromium

Limit of Dilution: 1 : 200,000. This identification limit holds also in the presence of 1000 times the quantity of manganese, nickel or cobalt.

Reagent: Freshly prepared alkaline hypobromite solution (8 ml bromine water plus 2 ml 5 *N* sodium or potassium hydroxide, plus 2 g potassium bromide)

(*b*) *Oxidation by fusion with sodium peroxide*

The conversion of chromium salts into chromates for the diphenylcarbazide test can also be carried out as follows.[347]

Procedure. A drop of the test solution is taken up in a platinum loop and carefully evaporated to dryness over a very small flame. The loop is then heated to redness and dipped in a mixture of equal parts of sodium potassium carbonate and sodium peroxide, and a bead is made by inserting the loop repeatedly in the flame. The chromium is completely oxidized to chromate. The cold bead is dissolved in 1 or 2 drops of 1 : 1 sulfuric acid on a spot plate, and 1 or 2 drops of a 1 % alcoholic diphenylcarbazide solution added. A violet color indicates chromium.

Limit of Identification: 0.5 γ chromium

Limit of Dilution: 1 : 100,000

Evaporation on platinum wire, and subsequent ignition in the flame, always results in the loss of small amounts of material by spattering. Therefore, when very small amounts of chromium are to be detected, it is preferable to carry out the evaporation, fusion, solution in sulfuric acid, and finally the addition of the diphenylcarbazide, in a porcelain micro crucible. In this way, 0.02 γ chromium in a drop can easily be detected.

Limit of Dilution: 1 : 2,500,000

Filter paper impregnated with diphenylcarbazide and phthalic anhydride can be used in the spot test for chromate in neutral or acid solution.[348]

(*c*) *Oxidation in acid solution with alkali persulfate*

Chromium^{III} salts can be oxidized to chromates by heating with potassium persulfate in acid solution:

$$2\ Cr^{+3} + 3\ S_2O_8^{-2} + 8\ H_2O = 2\ CrO_4^{-2} + 6\ SO_4^{-2} + 16\ H^+$$

This reaction is greatly accelerated by the presence of soluble silver salts,[349] probably because of the intermediate formation of silver^{II} compounds (see page 173). Sufficient chromate ion is formed[350] at room temperature, in a few minutes, for the diphenylcarbazide reaction. When carrying out this procedure large amounts of halides should not be present; they stop the catalytic action of the silver ions by forming insoluble silver halides.

Manganese salts are oxidized by persulfate in the presence of silver ions to permanganate, whose violet color may prevent the discernment of the occurrence of the chromate-diphenylcarbazide reaction. However, sodium azide may be added; it completely reduces permanganates, but affects chromates only slightly (see page 345).

Procedure. A drop of the test solution is mixed on a spot plate with a drop of a saturated solution of potassium persulfate and a drop of a 2 % silver nitrate solution, and allowed to stand for 2 or 3 minutes. On adding a drop of 1 % alcoholic diphenylcarbazide[351] solution, a violet to red color is formed; this fades on long standing.

Limit of Identification: 0.8 γ chromium
Limit of Dilution: 1 : 62,500

Test for chromium in the presence of mercury and molybdenum[352]

Mercury salts, molybdates, and also vanadates give blue to violet compounds with diphenylcarbazide in acid solution, and therefore interfere with the chromium test. The interference can be prevented by the addition of suitable compounds which lower the ionic concentration of the interfering elements below that required for the diphenylcarbazide reaction. For mercury, it is sufficient to add an excess of hydrochloric acid or alkali chloride; the usual low dissociation of mercury chloride is thus even further reduced. (The formation of the complex $[HgCl_4]^{-2}$ ions also helps to lower the ionic concentration of mercury.) When chromium is to be detected in the presence of mercury, hydrochloric acid is used to acidify the alkaline chromate solution. In this way, 0.25 γ chromium is easily detected by a spot reaction in the presence of 2.5 mg mercury (1 : 10,000).

The simplest method to avoid the interference of molybdates is to form the complex molybdenum oxalic acid $H_2[MoO_3(C_2O_4)]$. The molybdate concentration is so reduced that neither the diphenylcarbazide, nor any other sensitive molybdate reaction takes place.[353] The oxalic acid (1 drop of saturated solution) must be added to the test solution before the addition of diphenylcarbazide. Otherwise, the molybdic acid is partly reduced by reaction with diphenylcarbazide, forming molybdenum blue which is stable toward oxalic acid. This procedure will reveal 0.5 γ CrO_3 in the presence of 20 mg MoO_3.

(2) *Test with benzidine or 2,7-diaminodiphenylene oxide after conversion into chromate* [354,355]

Chromates oxidize benzidine in acetic acid solution to the blue quinoidal "benzidine blue" (see page 72). Since all chromium^{III} salts are easily oxidized to chromates by sodium peroxide, and the benzidine-chromate reaction is very sensitive, a simple and definite test for chromium is available.

Procedure. A drop of a freshly prepared moderately concentrated sodium peroxide solution and a drop of the test solution are successively placed on filter paper. The resulting chromate diffuses through the fibers of the paper into the outer zone of the moist fleck. It can be detected there by spotting with a drop of an acetic acid-benzidine solution. A blue ring (benzidine blue) is formed.*

Limit of Identification: 0.25 γ chromium

Reagent: Acetic acid-benzidine solution (see page 175)

If a particularly high sensitivity of the chromate test is sought, it is better to replace the benzidine by its rather unavailable derivative, 2,7-diaminodiphenylene oxide (I). It too is oxidized to a quinoidal di-imine. The blue product has the structure (II):

H_2N—(2,7-diaminodiphenylene oxide)—NH_2 [HN=(quinoidal diphenylene oxide)=NH] · 2 HX

(I) (II) X = univalent acid radical

The reagent solution is prepared by warming 0.375 g of 2,7-diaminodiphenylene oxide [356] with a mixture of 5 ml glacial acetic acid and 45 ml water.

This test (as a spot reaction) will reveal 0.003 γ chromium^{VI}.

Limit of Dilution: 1 : 6,000,000

* According to L. Kul'berg, *Mikrochemie*, 20 (1936) 244, the sensitivity of this test for chromates with benzidine may be doubled by using a freshly prepared mixture of equal parts of a 1 % alcoholic solution of benzidine and 20 % H_2O_2. Production of perchromate is responsible for this enhancement, since perchromic acid would convert the reagent into benzidine blue more completely than chromic acid.

Test for chromium in the presence of other metals

The benzidine test is definite for chromium in the absence of vanadates which also oxidize benzidine. It can be carried out in the presence of lead, cobalt, silver and copper salts (which give the blue color with benzidine under the same conditions) by taking advantage of the capillary migration of the chromates. It is necessary to vary the conditions only in the presence of manganese, whose higher oxides react with benzidine (see page 175). The manganese dioxide is formed in a highly dispersed form, which partly diffuses to the outer zone and, although invisible there, reacts with the benzidine in the same way as chromates. This interference can be avoided by using a special method of precipitation, so as to cause complete aggregation of the manganese dioxide, before carrying out the spot test.

Procedure. A drop of the solution, which may contain cations of all the groups, is mixed on a watch glass with an excess of sodium peroxide, and stirred with a glass rod while being gently heated. A portion of the suspension is transferred to filter paper with the aid of a capillary tube. The precipitate forms a fleck on the paper. The liquid phase, which for small amounts of chromium is almost colorless, forms a colorless outer zone. If the paper is moistened from the outside of the wet spot with a drop of an acetic acid solution of benzidine, a blue ring is formed when it reaches the colorless zone and slowly extends to the center of the spot.

The following procedure may also be used: A short strip of filter paper (about 1 cm wide) is folded over twice, and put on a glass plate with the folded portions lying above each other. The precipitate and solution from the watch glass are then brought on to the filter paper. When the undermost layer becomes moist, most of the precipitate remains on the topmost strip. The second layer allows very little to pass through, the third section is often free, and the fourth always free from the precipitate. If the last (precipitate-free) layer of filter paper is treated with benzidine, the development of a blue color indicates the presence of chromate.

(3) Other tests for chromium

(*a*) The dyestuff "acid alizarin RC" forms an orange lake color with chromium^{III} salts [357] when the test is carried out on filter paper; the color is stable toward both acids and ammonia (*Idn. Limit:* 0.6 γ Cr). Although the test may be carried out in the presence of not too large amounts of the other elements of the ammonium sulfide group, nevertheless, it is advisable to carry out the simple conversion to chromate.

(*b*) In the absence of interfering ions, chromates can be detected by a spot reaction through the formation of red Ag_2CrO_4 or yellow $PbCrO_4$ (*Idn. Limit:* 6 γ Cr).

References pp. 247–257

(*c*) In phosphoric acid solution, chromates oxidize pyrrole to "pyrrole blue" (*Idn. Limit:* 5 γ Cr).[358]

(*d*) Chromates react with 1 % solution of strychnine in concentrated sulfuric acid to give a blue-violet to red color (*Idn. Limit:* 0.9 γ CrO_4).[359]

23. Manganese

(*1*) *Test by catalytic oxidation to permanganate in acid solution* [360]

Manganous salts in faintly acid solution do not react in the cold with alkali persulfates. On heating, part of the manganese separates as hydrated manganese dioxide. In the presence of small amounts of soluble silver salts, the oxidation proceeds to permanganate, so that a violet color forms. The oxidation can be represented:

$$2\ Mn^{+2} + 5\ S_2O_8^{-2} + 8\ H_2O = 2\ MnO_4^- + 10\ SO_4^{-2} + 16\ H^+$$

Silver ions catalyze this reaction, as well as other oxidation reactions of persulfates.[361]

This effect is related to the fact that the quantitative hydrolysis of alkali persulfate:

$$S_2O_8^{-2} + H_2O \rightarrow 2\ SO_4^{-2} + 2\ H^+ + O$$

which by itself occurs only after protracted heating, proceeds quickly and completely, even at room temperature, when silver salts are present. The catalysis probably involves the formation of labile and therefore very reactive Ag^{II} compounds.[362] They accomplish the oxidation and the production of nascent oxygen by the above hydrolysis. In the cycle, Ag^{I} is regenerated, is again converted to Ag^{II} by persulfate, and so on.

The formation of permanganate by activated persulfate in acid solution provides the basis of a sensitive test for manganese. It is best done when small amounts of manganese and silver are present from the beginning; otherwise, along with the red-violet coloration, there will be some separation of MnO_2 and Ag_2O_2. Chlorides should be absent, or present in not more than small amounts; otherwise the catalyzing silver ions are removed as silver chloride. The same holds for bromides, iodides, or other materials that precipitate silver compounds. It should be noted that Cr^{III} salts are oxidized under these same conditions, forming chromate. See page 344 regarding the detection of permanganate in the presence of chromate.

Procedure. A drop of the test solution is mixed with a drop of concentrated sulfuric acid in a micro crucible. A drop of 0.1 % silver nitrate solution is stirred in, followed by a few milligrams of ammonium persulfate; the mixture is then gently heated. A red-violet color indicates manganese.

Limit of Identification: 0.1 γ manganese
Limit of Dilution: 1 : 500,000

(2) Test by catalytic oxidation to permanganate in alkaline solution [363]

In the reaction between manganese salts and alkali hypobromites, a brown precipitate of manganese dioxide is formed:

$$Mn^{+2} + BrO^- + 2\ OH^- = MnO_2 + H_2O + Br^-$$

If the reaction is carried out by heating gently in the presence of small amounts of copper, the oxidation proceeds further, and a solution of permanganate is produced: *

$$2\ Mn^{+2} + 5\ BrO^- + 6\ OH^- = 2\ MnO_4^- + 5\ Br^- + 3\ H_2O$$

Cobalt and nickel salts, to a lesser degree, possess the same catalytic action. This is related to the fact that these salts, even at great dilution, bring about a decomposition of hypobromite into bromide with evolution of oxygen.[364]

The catalytic action of the copper is probably due to an intermediate formation of copper^{III} oxides, which oxidize MnO_2 to MnO_4^-, with regeneration of Cu^{II}. In support of this theory, is the fact that periodic and telluric acids, which form stable Cu^{III} salts, prevent this catalytic action of copper (see pages 302 and 350).

The formation of permanganate through the "activation" of hypobromite can be used as a test for manganese.[365] It has the advantage that it is not affected by the presence of colored metallic ions, since they are all precipitated as hydrous oxides or hydrates of higher oxides. Only chromium is an exception; it is oxidized to chromate, and renders the detection of small amounts of permanganate difficult. It is to be noted that when testing for manganese in the presence of much nickel or cobalt, an amount of copper exceeding the quantity of nickel (or cobalt) should be added to the test solution before heating with hypobromite.

Procedure. Two milliliters of 1 % copper sulfate solution are placed in a test tube, followed by a drop of the test solution and 8 to 10 ml of freshly prepared 0.1 *N* sodium hypobromite solution. After boiling for a short time, the supernatant liquid is red-violet if manganese is present.

Limit of Identification: 2.5 γ manganese

Limit of Dilution: 1 : 25,000

(3) Test with periodate and tetrabase [366]

Manganous salts are oxidized in acetic acid solution to permanganate on warming with an alkali periodate:

$$2\ Mn^{+2} + 5\ IO_4^- + 3\ H_2O = 2\ MnO_4^- + 5\ IO_3^- + 6\ H^+$$

* K. M. Filimanowitsch, *Z. anal. Chem.*, 86 (1931) 234, discovered this reaction and applied it to detect copper; 0.2 γ Cu may be revealed by its catalytic effect. In this case precautions must be taken in the presence of Co- and Ni-salts which behave like copper salts.

References pp. 247–257

Larger amounts of manganese can be detected directly by the permanganate color, while smaller amounts are revealed by adding a solution of tetrabase (tetramethyl-*p*-diaminodiphenylmethane) in chloroform. An intense blue color is formed, due to an oxidation product of the tetrabase.[367,368]

In this redox reaction with tetrabase, the permanganate is reduced to manganous salt, and the latter is again oxidized to permanganate by the periodate. Thus, the catalytic action of manganese is the basis of this test.

Chromium salts should not be present; they are oxidized by periodate to chromate, which also gives a blue color with tetrabase.

The presence of colored salts interferes when small quantities of manganese are to be detected.

Procedure. A drop of the neutral test solution is placed on a spot plate, followed by a drop of saturated potassium periodate solution, a drop of 2 *N* acetic acid, and two drops of 1% solution of tetrabase[369] in chloroform. According to the amount of manganese present, a more or less deep blue color is formed.

Limit of Identification: 0.001 γ manganese

Limit of Dilution: 1 : 50,000,000

(*4*) *Test with benzidine*[370]

Benzidine and its water-soluble salts can be converted by a number of oxidizing agents, as well as by autoxidation processes which occur in the presence of this base, into a blue oxidation product known as "benzidine blue" (see page 72). If precipitated manganous hydroxide is brought into contact with air, hydrated MnO_2 forms quickly through autoxidation:

$$2\ Mn(OH)_2 + O_2 = 2\ MnO_2 \cdot H_2O$$

This product reacts with benzidine to yield benzidine blue. The reaction sequence: precipitation – autoxidation – formation of benzidine blue, can be used in the form of a spot reaction on filter paper to detect quantities of MnO_2 that are too slight to be seen. It is interesting to note that the test is not sensitive if carried out on a spot plate or in a test tube. The paper functions somewhat as a participant in the reaction, probably through its capillary action and the resultant high dispersion of the MnO_2.

Procedure. A drop of the test solution is treated on filter paper with a drop of 0.05 *N* potassium or sodium hydroxide, and then with a drop of benzidine solution. A blue color results where the manganese dioxide was formed. The intensity depends on the manganese content. The color fades on drying, but reappears with fresh addition of benzidine solution.

Limit of Identification: 0.15 γ manganese

Limit of Dilution: 1 : 330,000

Reagent: Benzidine solution: 0.05 g benzidine base or hydrochloride dissolved in 10 ml acetic acid, diluted to 100 ml with water and filtered

Test for manganese in the presence of other metals [371]

The manganese–benzidine reaction, as just described, cannot be used in the presence of other oxidizing agents or autoxidizable substances which likewise oxidize benzidine (chromates, ferricyanides, cobalt, thallium^{III}, silver, cerium$^{III, IV}$ salts). In such cases, special procedures must be used.

It must also be noted that large amounts of indifferent metallic hydroxides may coat the manganese hydroxide and thereby reduce the sensitivity of the test. For instance, ferric hydroxide, although it has no effect on benzidine,* reduces the sensitivity of the manganese test so considerably that instead of 0.15 γ (as in pure manganese salts) not less than 2.5 γ can be detected in a drop when 250 γ iron is also present. If, however, the iron precipitation is prevented by the addition of Rochelle salt, manganese may be detected in the presence even of large amounts of iron. The entire procedure can be accomplished on filter paper.

Limit of Identification: 1 γ manganese
Limit of Dilution: 1 : 50,000
} in the presence of 1000 times the amount of iron

To detect *manganese in the presence of cobalt* (which under the normal conditions of the test gives the autoxidizable cobalt hydroxide), the cobalt should be converted into the very stable $K_3[Co(CN)_6]$ by adding potassium cyanide and warming. The excess potassium cyanide interferes with the manganese test by forming the complex $K_2[Mn(CN)_4]$, which does not react with alkali hydroxide to yield $Mn(OH)_2$ that is essential to the benzidine reaction. The cyanide should be removed by addition of a few drops of dilute hydrochloric acid (hood). The acid also destroys any manganese cyanide, whereas the cobalt complex ion is unchanged. This series of reactions can be carried out in a microcentrifuge tube with one drop of the test solution. It must be noted that the potassium cyanide solution of the complex cobalt salt will gradually develop a red-violet color with benzidine.

Limit of Identification: 0.5 γ manganese
Limit of Dilution: 1 : 100,000
} in the presence of 1200 times the amount of cobalt

To detect *manganese in the presence of cerium*, which under the normal conditions of the test yields autoxidizable $Ce(OH)_3$, the neutral or weakly acetic acid test solution must first be warmed for 2–3 minutes with a little freshly prepared calcium fluoride. The cerium is precipitated as the insoluble fluoride, and is entrained in the unchanged calcium fluoride. The manganese test can be carried out on the filtrate.

Limit of Identification: 5 γ manganese
Limit of Dilution: 1 : 10,000
} in the presence of 1000 times the amount of cerium

* It should be noted that ferric salts likewise oxidize benzidine, if free acids are absent.

Copper salts form copper hydroxide and basic copper acetate, which can hide the benzidine blue. When *small amounts of manganese* are to be detected *in the presence of large amounts of copper*, the test should be carried out as in the presence of cobalt. Cuprous cyanide is formed; it should be filtered off before proceeding with the test for manganese.

Limit of Identification: 1.6 γ manganese
Limit of Dilution: 1 : 30,000
} in the presence of 500 times the amount of copper

To detect *manganese in the presence of silver and thallium*[I], the latter are best precipitated as chlorides, the precipitate removed by filtering or centrifuging, and the test carried out on the clear liquid. Thallic salts can be reduced to thallous sulfate with sulfur dioxide, and then precipitated with sodium chloride. The sensitivity of the test remains the same as for pure manganese salts. The silver and thallium must be removed before carrying out the test since addition of alkali to a solution containing these ions gives the reaction:

$$2\ Ag^+ + Tl^+ + 3\ OH^- \rightarrow 2\ Ag^0 + Tl(OH)_3$$

Thallic hydroxide oxidizes benzidine, and silver oxide also has a distinct similar action on this reagent.

(5) *Test with silver ammine salts* [372]

The test for silver (see page 58) utilizing the reaction between silver, manganese and hydroxyl ions [373] can also be carried out in ammoniacal solution:

$$Mn^{+2} + 2\ [Ag(NH_3)_2]^+ + 4\ OH^- = MnO_2 + 2\ Ag^0 + 4\ NH_3 + 2\ H_2O$$

The formation of a dark precipitate containing manganese dioxide and finely divided silver can be used as a sensitive test for manganese. It can be applied in the presence of Al^{+3}, Cr^{+3}, Fe^{+3}, Co^{+2}, Ni^{+2}, Zn^{+2}, Hg^{+2}, Pb^{+2} and Bi^{+3} ions. It is especially useful for the detection of manganese in alloys and special steels. The sensitivity is always decreased when colored metallic hydroxides are formed at the same time.

Procedure. A drop of the ammoniacal silver nitrate solution is placed on spot paper, followed by a drop of the test solution. In the presence of manganese, a black fleck appears, and on warming it becomes more distinct.

It must be noted that ammoniacal silver solutions are slightly reduced to metallic silver by filter paper. Consequently a blank test is essential for small amounts of manganese.

Limit of Identification: 0.05 γ manganese
Limit of Dilution: 1 : 1,000,000
Reagent: Silver ammine salt solution: A saturated solution of silver nitrate is treated with concentrated ammonia until the initial precipitate is dissolved; then an equal volume of ammonia is added

24. Zinc

(*1*) *Test with ferricyanide and diethylaniline* [374]

Alkali ferricyanides oxidize acid solutions of diethylaniline [375] and other aromatic amines, as well as certain monoazo dyestuffs; a change of color results and ferrocyanide is formed. This reversible redox reaction, proceeds very slowly and incompletely. However, if the ferrocyanide ions are removed as insoluble white zinc ferrocyanide, the oxidation proceeds rapidly. This acceleration is a result of raising the oxidation potential of ferricyanide ions through removal of ferrocyanide ions. The white zinc ferrocyanide is deeply tinted by adsorption of the colored quinoidal oxidation products of the amines, and thus affords a sensitive test for zinc. The test is especially useful in the presence of chromium and aluminum. It can also be used in other instances provided no other cations are present, which form colored precipitates with potassium ferrocyanide (Co^{+2}, Ni^{+2}, Fe^{+2}, Mn^{+2}, Cu^{+2}). Traces of iron, which in practice are always to be reckoned with, do not interfere with the zinc test. Anions which oxidize the amines must be absent, e.g., permanganate, chromate, vanadate, persulfate, iodate.

Procedure. In the depression of a spot plate, mix 15 drops of potassium ferricyanide solution with 10 drops of a solution of diethylaniline a) or b). The sulfuric acid solution remains yellow and clear during the time of observation of the reaction (5 min). On addition of a solution containing zinc, the yellow changes to light brown to red-brown, and a precipitate or cloudiness appears. The phosphoric acid solution is dark yellow to red-yellow, and also remains clear during the observation time (5 min). Upon adding a drop of a solution containing zinc, the color becomes redder or dark red-brown, and a precipitate or cloudiness is observed.

Table 4 presents the results observed after 5 minutes.

Reagents: *1*) Potassium ferricyanide solution (2 g salt in 100 ml water)
2) Diethylaniline solution (free from monoethylaniline)
a) 0.25 g diethylaniline (mono-free) in 200 ml sulfuric acid (1:1)
b) 0.5 g diethylaniline in 200 ml 50 % phosphoric acid

(*2*) *Test with dithizone* [376]

Diphenylthiocarbazone (dithizone) (see page 71) forms insoluble colored inner complex salts with a number of heavy metals. These products are easily extracted by organic solvents. The purple-red zinc salt, formed in neutral, alkaline, and acetic acid solutions, is soluble in carbon tetrachloride without change of color.

Procedure. A drop of the test solution and a drop of a solution of dithizone in carbon tetrachloride are shaken together in a small stoppered test tube. In the presence of zinc, the green color of the reagent changes to purple-red.

TABLE 4

1 drop test solution contains (in mg):	Reaction picture after 5 minutes using:	
	Diethylaniline acidified with H_2SO_4	Diethylaniline acidified with H_3PO_4
1.0 Zn	Dense brownish-red precipitate	Dense dark brown-red precipitate
0.1 Zn	Pale brown-red precipitate and turbidity	Dense brown-red turbidity
0.1 Zn + 0.8 Al + + 0.3 Cr + 0.67 Na	Pale brown-red precipitate and turbidity	Dense brown-red turbidity
0.01 Zn	Red-yellow coloration and turbidity	Yellowish to brown-red color and turbidity
0.01 Zn + 0.8 Al + + 0.3 Cr + 0.67 Na	Slight brownish-yellow color and turbidity	Definite turbidity
0.001 Zn	Brownish yellow to red color and turbidity	Definite turbidity
0.001 Zn + 0.8 Al + + 0.3 Cr + 0.67 Na	Turbidity	Definite turbidity
0.0001 Zn	No reaction	Marked turbidity
0.0001 Zn + 0.8 Al + + 0.3 Cr + 0.67 Na	No reaction	Marked turbidity

Limits of Identification:
0.025 γ zinc in neutral zinc sulfate solution
0.06 γ zinc in 2 % ammonia
0.06 γ zinc in the presence of ammonia, ammonium salts, and tartrates
0.9 γ zinc in 2 % sodium hydroxide
0.9 γ zinc in 10 % acetic acid
0.05 γ zinc in 10 % acetic acid and 5 % ammonium acetate
Reagent: Solution containing 1 to 2 mg dithizone [377] in 100 ml carbon tetrachloride

When the zinc-dithizone reaction is carried out on filter paper or on a spot plate, 0.4 γ and 0.05 γ zinc, respectively, can be detected by a red fleck or coloration, if a solution of 10 mg dithizone in 100 ml carbon tetrachloride is used as reagent.

This test for zinc is especially useful in the presence of nickel or aluminum. In a drop of neutral solution, 0.2 γ zinc can be detected in the presence of 1600 times the amount of nickel, and 0.05 γ zinc in the presence of 2200 times the amount of aluminum, when 1 mg ammonium acetate is added.

The dithizone test for zinc is very sensitive in neutral or acetic acid solution, but is not specific, because many other metals cause a color change in the carbon tetrachloride layer. The use of doubly distilled water is necessary to avoid interference by traces of heavy metals. It is more satisfactory to carry out the test in alkaline solution even though it is then less sensitive. The solution turns red in the presence of zincates only, and the color is quite evident even in the presence of colored precipitates.[378] When copper or mercury salts are present, they should be removed by precipitation with hydrogen sulfide. This operation is carried out in an Emich microcentrifuge tube.

At *p*H 4 – 5.5, sodium thiosulfate considerably prevents the reaction of As, Hg, Au, Bi, Pb, and Cd, while permitting that of zinc to proceed.[379]

Procedure. A drop of the test solution is mixed on a watch glass with a drop of 2 *N* sodium hydroxide and a few drops of 0.01 % dithizone solution in carbon tetrachloride. The carbon tetrachloride is evaporated by blowing on the solution while stirring with a glass rod. A raspberry-red solution shows the presence of zinc. The color of any precipitate can be disregarded.

Limit of Identification: 5 γ zinc
Limit of Dilution: 1 : 10,000

In this way 50 γ zinc can be detected in the presence of 50 γ Ag, As, Fe^{III}, Mn, Pb, Sb, or Cd.

(3) Test by mutual precipitation of zinc and cobalt with mercury thiocyanate [380]

Zinc ions react with mercury thiocyanate ions:

$$Zn^{+2} + [Hg(CNS)_4]^{-2} = Zn[Hg(CNS)_4]$$

to yield a white crystalline precipitate. In the presence of small amounts of zinc, the precipitate appears quite slowly because of supersaturation. Cobalt salts react similarly, producing blue crystalline $Co[Hg(CNS)_4]$. If a very dilute cobalt solution, which gives a blue precipitate with the thiocyanate only after long standing, is mixed with a dilute solution of a zinc salt, a blue precipitate is formed quickly. Obviously the supersaturation with respect to the two double thiocyanates is overcome, probably because of the formation of mixed crystals.

Procedure. A drop of the test solution, a drop of the cobalt solution, and a drop of the mercury thiocyanate solution are mixed on a spot plate or in a micro crucible. The sides of the vessel are rubbed with a glass rod for about 15 seconds. According to the amount of zinc present, a blue precipitate is formed, either at once, or at the longest within 2 minutes. In the absence of zinc, the precipitation begins only after 2 to 3 minutes.

	Neutral	0.5 *N* HCl
Limit of Identification:	0.2 γ zinc	0.5 γ zinc
Limit of Dilution:	1 : 250,000	1 : 100,000

References pp. 247–257

When only very small amounts of zinc are present, a blank test should be carried out on a drop of water in place of the test solution.

Table 5 shows the identification and proportion limits obtained in the presence of various foreign materials.

TABLE 5

Accompanying compound	Identification limit, in γ zinc		Proportion limit	
	In neutral solution	In 0.5 N HCl	In neutral solution	In 0.5 N HCl
NH_4Cl	2	2	1 : 250	1 : 520
K_2SO_4	0.5	1	1 : 1000	1 : 500
$AlCl_3$	0.5	1	1 : 700	1 : 350
$Cr(NO_3)_3$	0.5	0.5	1 : 100	1 : 100
$MnSO_4$	2	2	1 : 100	1 : 100
$FeCl_3$	2	2	1 : 120	1 : 120
$CdCl_2$	2	2	1 : 200	1 : 200
$Pb(NO_3)_2$	1	2	1 : 500	1 : 250

In the presence of iron^{III} salts, red soluble ferric thiocyanate is formed and interferes with the discernment of the blue mutually precipitated mercurithiocyanates. In this case, the mixture should be treated with a few milligrams of alkali fluoride, one or two minutes after the addition of the reagents. The red color due to the ferric thiocyanate disappears, because of the production of colorless $[FeF_6]^{-3}$ ions.

Reagents: *1*) Cobalt sulfate solution: 0.02 % Co in 0.5 N HCl

2) Alkali mercury thiocyanate solution: 8 g $HgCl_2$ and 9 g NH_4CNS in 100 ml water. Allow to stand several days.

(*4*) *Test with ferricyanide and 3,3'-dimethylnaphthidine* [381]

The system: alkali ferricyanide–3,3'-dimethylnaphthidine–dilute acid shows no change after short standing. The oxidizing ability of the ferricyanide ions is not sufficient to convert the diamine into a colored quinoidal compound (compare page 126). However, the addition of zinc ions, which remove ferrocyanide ions from the redox equilibrium as acid-insoluble zinc ferrocyanide, raises the oxidation potential of ferricyanide to such extent that a red-violet oxidation product of the 3,3'-dimethylnaphthidine is produced. Even minimal amounts of zinc are sufficient to bring about this activation, and the effect may be used as a test for zinc. This test is based on the same principle as test *1* (page 178) and is subject to the same interferences.

Procedure. A drop of the reagent solution and a drop of the weakly acid test solution are brought together on a spot plate. If zinc is present, a red-violet color appears immediately.

Limit of Identification: 0.1 γ zinc

Limit of Dilution: 1 : 500,000

Reagent: *1*) 5 % solution of potassium ferricyanide in distilled water

2) Saturated water solution of 3,3'-dimethylnaphthidine [382] chloride

The reagent solution consists of a freshly prepared mixture of one part *1*) and two parts *2*).

(*5*) *Other tests for zinc*

(*a*) A blue color results if resorcinol is added to an ammoniacal zinc solution.[383] One drop of an alcohol solution of the reagent is mixed on a spot plate with 1 drop each of 6 *N* ammonia and the test solution. The reaction may also be carried out in a micro test tube with 1 ml of the reagent. The blue color, whose nature is not known, does not appear at once, but after several minutes (*Idn. Limit:* 2 γ Zn).

(*b*) If a drop of a zinc solution is placed on paper impregnated with $K_3[Co(CN)_6]$ and the paper then ashed, "Rinmann's green" (cobalt zincate) is formed.[384] In this way 0.6 γ Zn in 0.002 ml (*Dilution Limit:* 1 : 3300) may be detected.

(*c*) The increase in oxidation potential of ferricyanide toward *p*-phenetidine (compare page 178) resulting from the precipitation of zinc ferrocyanide is applied as a test for zinc.[385] A mixture of 6 drops of 2 % $K_3Fe(CN)_6$, 2 drops of *N* H_2SO_4, and 12 drops of 1 % *p*-phenetidine chloride is freshly prepared, and 0.1 ml of the brown-yellow solution is placed on a spot plate. A microdrop of the test solution is added without mixing. After 2 or 3 minutes, a blue precipitate or color appears at the boundary between the drops (*Idn. Limit:* 1 γ Zn). Using smaller drops, even less zinc may be detected (down to 0.05 γ Zn). The proportion limit for the test in the presence of Pb, Bi, Sn, alkali and alkaline earth metals may be 1 : 1000.

(*d*) Zinc can be detected by placing a drop of an acetic acid test solution on paper impregnated with uranyl ferrocyanide.[386] A white stain forms on the brown paper (*Idn. Limit:* 0.1 γ Zn). The presence of metal ions, which form insoluble ferrocyanides, interferes as do large amounts of neutral salts.

25. Aluminum

(*1*) *Test with morin* [387]

Morin, the coloring matter of fustic, is 3,5,7,2',4'-pentahydroxy flavanol (I). Its alcohol solutions react with aluminum salts in neutral or acetic acid solution to give an intense green fluorescence in daylight and ultraviolet light. The fluorescence is due to the formation of a colloidally dispersed inner complex aluminum salt of morin [388] with the probable structure (II), or to an adsorption compound of morin with alumina. Beryllium,[389] indium, gallium, thorium and scandium salts [390] also form fluorescent compounds with morin. The *p*H of the system has much influence in these reactions.[391] The only metal ion whose reaction with morin is independent of *p*H is Zr^{+4} or its hydrolysis product (compare page 201).

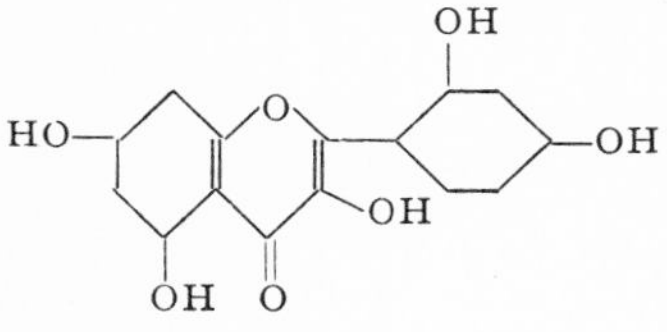

(I) 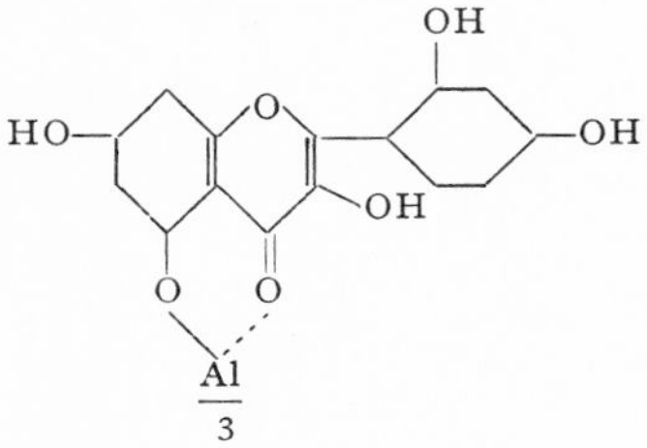 (II)

Procedure. To test for aluminum in the presence of other metals which give color or precipitation reactions with morin, the test solution is treated with excess 2 *N* potassium hydroxide and filtered. A drop of the filtrate is acidified on a black spot plate with 2 *N* acetic acid, and a drop of a saturated morin solution in methyl alcohol added. A green fluorescence appears in the presence of aluminum. For small amounts, comparison with a blank test is advisable.

Limit of Identification: 0.2 γ aluminum
Limit of Dilution: 1 : 250,000

The following procedure is more sensitive [392]. Filter paper is impregnated with freshly prepared morin solution, dried, and treated with a drop of the neutral or slightly acid test solution, and again dried. On spotting with 2 *N* hydrochloric acid, a fleck fluorescing light green can be observed under the quartz lamp.

Use is made here of the fact that aluminum morinate, once it has been formed, is stable against an acidity at which its formation does not occur to any significant extent.

Limit of Identification: 0.005 γ aluminum
Limit of Dilution: 1 : 10,000,000

If aluminum is to be detected in the presence of other members of the ammonium sulfide group, it is necessary to dissolve away the morin compounds of these elements by several more drops of dilute hydrochloric acid. Besides aluminum, only the zirconium compound resists this treatment.

(2) Test with alizarin sulfonic acid (alizarin S) [393]

Aluminum salts give a red precipitate with the violet solution of alizarin S (I) in ammonia. In accord with its constitution (especially because of the juxtaposition of an acid phenolic group to a CO-group), the reagent is a complex-former. The precipitate, nevertheless, does not have the composition and constitution of a definite inner complex aluminum salt as shown in (II). Rather, it is an adsorption compound of alumina gel and alizarin sulfonic acid. Adsorption systems of this kind containing metal hydroxide gel as principal and a dyestuff as collateral constituent are called lakes. The lake-formation is a result of chemical adsorption, i.e., of a chemical binding of dyestuff molecules to the surface of metal hydroxide gel particles. (Compare aluminum-alizarin lake, page 185).

The color lake, once it has been formed in ammoniacal solution, is stable to dilute acetic acid, whereas the violet color of the solution goes toward yellow (color of alizarin S) on acidification with acetic acid.

(I) (II)

Procedure. A drop of the caustic alkaline test solution containing aluminate is treated with a drop of aqueous 0.1 % alizarin S and 1 *N* acetic acid, until the violet color disappears, and then with a further drop of acetic acid. According to the amount of aluminum present, a red precipitate or color appears, which intensifies on long standing.

For the detection of small amounts of aluminum, a blank test on a drop of 1 *N* alkali alone should be carried out, and the color compared with that of the test solution.

Limit of Identification: 0.65 γ aluminum

Limit of Dilution: 1 : 77,000

Reagents: *1)* 0.1 % solution of sodium alizarin sulfonate in water

2) 1 *N* sodium hydroxide. This should be stored in a bottle made of resistant glass. A blank should give no color or only a faint pink

Test for aluminum in the presence of other ions

Iron, cobalt and copper salts give lakes with alizarin which are partially stable towards acid. These elements must therefore be removed before the test. This is best done by precipitation as hydroxides with sodium hydroxide; the aluminum remains in solution as aluminate.

Zinc salts do not interfere. Alkaline earth salts in concentrated solutions also give red alizarin compounds, a fact that must be remembered when carrying out this test for aluminum.

In the precipitation of the hydroxides, alkali of not more than 1 *N* strength should be used, and not in too great an excess, since alkali almost invariably contains traces of aluminum. Consequently, a blank should be carried out using exactly the same amount of alkali as in the test to avoid any error from this source.

Chromates interfere, because of their high color. However 2 γ aluminum may be clearly detected in the presence of 120 times the amount of chromium, if a comparative test is carried out on a 1 % potassium chromate.

(3) Test with alizarin[394]

Aluminum salts react with ammoniacal violet solutions of alizarin (I) to produce a red-violet precipitate, whose behavior toward dilute acetic acid is like that of the precipitate obtained with alizarin S (compare test *2*). Here again, it is not a matter of the formation of an inner complex aluminum alizarinate (II) or a mixture of this salt with aluminum hydroxide.

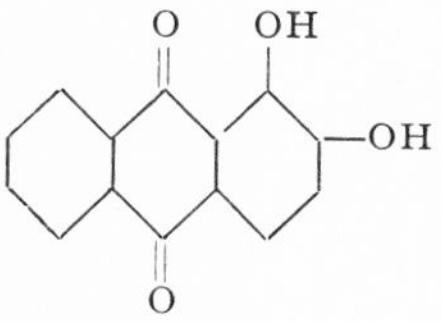

HAliz
(I)

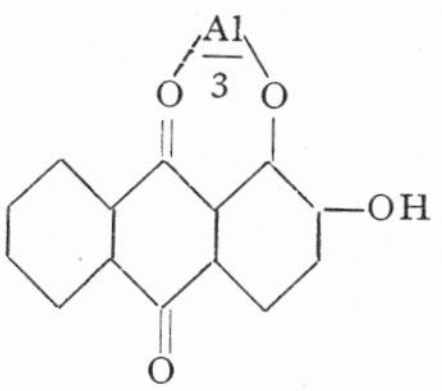

$^1/_3$ Al(Aliz)$_3$
(II)

There are good reasons [395] for believing that the precipitate is an adsorption system (lake) of $Al(OH)_3$ with alizarin. The production of this lake involves chemical adsorption, i.e., alizarin or its anion is bound on the surface of $Al(OH)_3$ gel particles:

$$[Al(OH)_3]_x + HAliz \longrightarrow [Al(OH)_3]_{x-1} \cdot Al\begin{matrix} \diagup Aliz \\ -OH \\ \diagdown OH \end{matrix} + H_2O$$

$$[Al(OH)_3]_x + Aliz^- \longrightarrow [Al\ (OH)_3]_{x-1} \cdot Al\begin{matrix} \diagup Aliz \\ -OH \\ \diagdown OH \end{matrix} + OH^-$$

alumina gel Al-alizarin lake

This scheme is intended to show that the reaction of alumina gel with alizarin does not result in the production of $Al(Aliz)_3$ as a separate phase, but instead to indicate that the Al-atoms in the surface of the gel take over the bonding of alizarin radicals without leaving their phase association. The places of attachment in the alizarin molecule are probably the same as shown in (II), i.e., there is chelate bonding, but with the difference that one Al-atom never holds three alizarin radicals. This latter condition would denote the production of $Al(Aliz)_3$ as a separate phase, whose color and chemical behavior, however, do not agree with those of the Al-alizarin lake, which is regarded as the product of a chemical adsorption.

Procedure. A drop of the test solution is placed on alizarin paper, and held over ammonia until the fleck turns violet. If large amounts of aluminum are present the red color is visible immediately. When small amounts of aluminum are suspected the paper is dried in an oven. The violet due to ammonium alizarinate disappears, since it is decomposed into ammonia and yellow alizarin; the red color of the aluminum lake then becomes clearly visible.

Limit of Identification: 0.15 γ aluminum

Limit of Dilution: 1 : 333,000

Reagent: Alizarin paper. Quantitative filter paper is soaked in a saturated alcohol solution of alizarin, dried, and kept in a stoppered container.

Test for aluminum in the presence of other metals of the ammonium sulfide group

Iron, chromium, uranium and manganese also give colored alizarin lakes. In some cases (compare page 187) a separation of the components of a solution occurs in the capillaries of the paper so that the various metal lakes are formed in different zones of the paper. Unfortunately, this procedure is successful only in the aluminum-uranium separation; it is applicable only within definite concentration differences for the other metals. However, in contrast to all the other ions of the ammonium sulfide group, aluminum gives no insoluble compound with potassium ferrocyanide. Thus, when the reaction is carried out as a spot test on filter paper impregnated with potassium ferrocyanide, the aluminum salt diffuses through the capillaries to the zone or water ring encircling the fleck of precipitate and can be detected there by the alizarin reaction.

Procedure. A small drop of the neutral or weakly acid test solution is placed on filter paper impregnated with a 5 % water solution of $K_4Fe(CN)_6$. The other ions are "fixed" in the paper as insoluble ferrocyanides. The aluminum travels with the water beyond the precipitate. Most of the aluminum can be washed into the outer zone by the addition of another drop of water to the middle of the fleck. A drop of alcoholic alizarin is added; the paper is held over ammonia and dried. A red ring of alcoholic alizarin-aluminum lake is formed around the ferrocyanide precipitate. To make the product visible, it is advisable, after the reaction is complete, to place the paper in hot water for about 2 minutes. Some of the metal ferrocyanides are washed away and the aluminum lake becomes more visible.

The *limits of identification* for aluminum in the presence of varying amounts of the other metals, are:

4.0 γ Al	in the presence of	800 γ Fe
3.0 γ Al	,, ,, ,, ,,	300 γ Fe + 750 γ Cr
1.4 γ Al	,, ,, ,, ,,	3912 γ Mn
0.6 γ Al	,, ,, ,, ,,	60 γ Fe + 372 γ Mn
0.6 γ Al	,, ,, ,, ,,	30 γ Fe + 180 γ Zn
0.1 γ Al	,, ,, ,, ,,	30 γ Zn + 20 γ Mn
0.5 γ Al	,, ,, ,, ,,	100 γ U
0.6 γ Al	,, ,, ,, ,,	30 γ Fe + 72 γ Co

The following should be noted with reference to the detection of aluminum in the presence of other metals: *When much iron is present,* it may happen that the potassium ferrocyanide in the paper in the area of the drop is insufficient to bind all the iron. When the fleck is then moistened with the wash water, unchanged ferric chloride diffuses outward, extends the Prussian blue fleck, and renders the detection of the aluminum difficult. It is best to use very small drops, so that the fleck is at most 5 mm in diameter, and also to carry out a comparative test on an iron solution under exactly the same conditions on the same paper. The difference is then quite easy to see.

When testing for *aluminum in the presence of uranium* there is danger of spreading the uranium ferrocyanide fleck, because of its slimy consistency (especially when very little uranium is present). This spreading can interfere with the detection of very small amounts of the aluminum lake. It is advisable to soak the filter paper in an ammonium carbonate solution after the test is completed. The uranium precipitate dissolves and forms $(NH_4)_4[UO_2(CO_3)_3]$. The aluminum lake can then be seen clearly.

In detecting a small amount of *aluminum in the presence of much iron and chromium,* a control test should be carried out on an aluminum-free iron-chromium solution. After addition of the alizarin, the paper should be placed in hot water for 1 minute and then dried. The potassium ferrocyanide is removed, and the paper becomes white. The lake color is thus rendered more visible. The control flecks show only a very slight color change, if any, because the chromium ferrocyanide is also washed away. The concentrations of iron and especially chromium in the control solution should approximate those in the test solution. However, it has been found that even though the concentrations of iron and chromium vary within wide limits, the solutions will still give useful comparison results.

Test for aluminum in the presence of uranium by capillary separation

Uranium salts give a blue lake with alizarin. When a drop of a solution containing both aluminum and uranium is placed on alizarin paper, the blue

References pp. 247–257

uranium lake forms in the center of the drop. The aluminum diffuses outward and, after being developed with ammonia, encircles the blue fleck with a red ring. The width of the ring depends on the aluminum content. In this way, 0.1 γ aluminum may be detected in the presence of 200 times the amount of uranium.

Test for aluminum in the presence of cobalt and nickel

By the lake method 0.1 γ aluminum may be directly detected in the presence of 0.014 mg of cobalt or nickel. Larger amounts of these metals lower the sensitivity, partly by lake formation, and partly because of the formation of colored ammine salts. Under such conditions, the following procedure should be used.

Procedure. A drop of the test solution is placed on alizarin paper, held over ammonia, dried, and then laid in hot 0.01 *N* hydrochloric acid. The coloration due to cobalt or nickel lakes or ammines disappears, and the aluminum lake can be redeveloped over ammonia. By this method 0.9 γ aluminum can be detected in the presence of 120 γ cobalt.

(4) Test with quinalizarin [396]

Quinalizarin (1,2,5,8-tetrahydroxyanthraquinone) (see page 192) which is especially useful for the detection of beryllium, is also a sensitive reagent for aluminum, with which it forms a lake stable to acetic acid. It is necessary to carry out the reaction in ammoniacal solution, as in the formation of the alizarin lake, and then acidify with acetic acid.

Procedure. A drop of the test solution is placed on quinalizarin paper; the latter is held briefly over concentrated ammonia, and then over glacial acetic acid, until the initial blue color (ammonium salt of quinalizarin) disappears and the unmoistened paper regains its brown color (free quinalizarin). According to the aluminum content, a red-violet or faint red fleck remains.

Limit of Identification: 0.005 γ aluminum (in 0.01 ml)
Limit of Dilution: 1 : 2,000,000
Reagent: Quinalizarin paper: 10 mg quinalizarin is dissolved in 2 ml pyridine, diluted with 20 ml acetone, and quantitative filter paper soaked in this solution

This test with quinalizarin is especially useful for the detection of aluminum in the presence of magnesium. The limiting ratio, as in many other spot tests, depends on the concentration of the test solution. For instance:

Concentration	Ratio, Al : Mg	Detection
1 γ per ml Al	1 : 10^{-4}	Fair
2 γ per ml Al	1 : 10^{-4}	Good
5 γ per ml Al	1 : 10^{-4}	Very good

(5) *Test with chrome fast pure blue B* [397]

When a mineral acid solution of an aluminum salt is treated with a water solution of the oxytriphenylmethane dye known as chrome fast pure blue B (I), the precipitate consists solely of the dye acid which is only slightly soluble in water. If, however, the system is neutralized before or after the introduction of the dye or made basic with ammonia, and then acidified with dilute sulfuric acid, and then shaken with ether, the ether layer becomes yellow and the water layer is red. The former contains the unused dye acid, the latter the inner complex aluminum-dyestuff compound (II), in which the aluminum is a constituent of the anion.

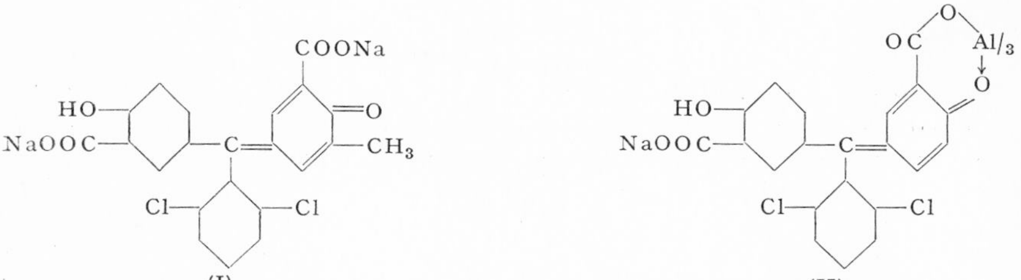

COONa HO— NaOOC— C= =O —CH_3 Cl— —Cl

(I)

OC O Al/3 O HO— NaOOC— C= Cl— —Cl

(II)

The fact that (II) is formed only in neutral or basic solution but not in mineral acid solution may be explained as follows: The dye acid underlying (I) contains the group *A* in the nonreactive hydrogen chelate form *A'*, which in its turn is in a pH-dependent equilibrium with the reactive anion *A"*. Accordingly, an adequate quantity of *A"* ions is obtained from this equilibrium (1) on alkalization:

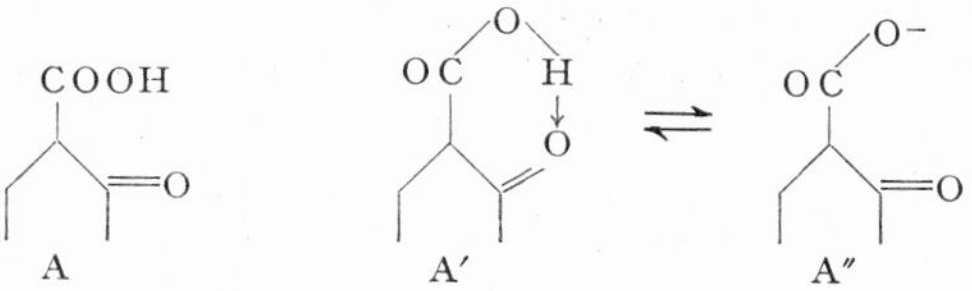

COOH =O (A) OC O H O (A') ⇌ OC O− =O (A") + H^+ (1)

Corresponding to this viewpoint, all hydroxytriphenylmethane dyes containing the group *A* react in the same fashion as chrome fast pure blue B when they come into contact with aluminum salts. To these color reactions the general formulation can be given:

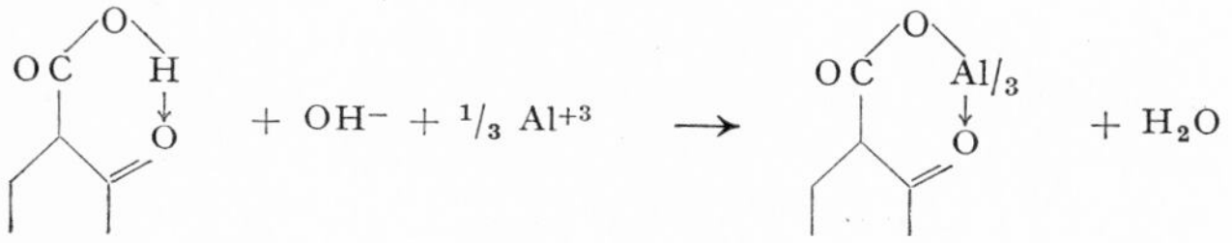

OC O H O + OH^- + $^1/_3$ Al^{+3} → OC O Al/3 O + H_2O (2)

The chelate compounds, produced as shown in (2), are not decomposed at once by even (1 : 1) sulfuric acid and therefore a sensitive test for aluminum has been developed on this basis. (Tervalent chromium salts exhibit an analogous behavior.) The color reactions are selective if the test solution is treated with a slight excess of magnesium carbonate before the reagent is introduced. It is essential that sulfuric acid be used for the acidification; if hydrochloric acid is employed the zirconium compounds formed at the same time remain unchanged. The sulfuric acid probably decomposes the colored zirconium compounds through formation of complex zirconium sulfuric acids. Chrome fast pure blue B is the best reagent because its dye acid can be completely extracted by ether, which is not true of some of the related dyes.

Procedure. The test is conducted in a micro test tube. Pulverized magnesium carbonate is added cautiously to one drop of the acidic test solution until the evolution of carbon dioxide stops. The excess of $MgCO_3$ should be as slight as possible. One drop of a 5 % water solution of chrome fast pure blue B is added and the solution then acidified with a drop of 1 : 1 sulfuric acid. If the quantity of aluminum is small, the excess dye acid will precipitate, whereas the dye may be completely consumed if large amounts of aluminum are present. The system is now shaken with 10–15 drops of ether. Any excess dye enters the ether and the lower water layer becomes magenta or pink according to the aluminum content.

Limit of Identification: 0.1 γ aluminium

Limit of Dilution: 1 : 500,000

This procedure revealed:

2 γ Al in the presence of 1000 γ Be, Zr, Cu, Ni, Co
0.2 γ Al ,, ,, ,, ,, 1000 γ Zn or Mn
5 γ Al ,, ,, ,, ,, 2500 γ Fe

If Fe^{III} ions are reduced by thioglycolic acid prior to adding the magnesium carbonate, 0.5 γ Al can be detected in the presence of 2500 γ Fe.

When detecting *aluminum in the presence of chromium,* one drop of the neutral or weakly acid test solution should be treated with a drop of 5 *N* sodium hydroxide and one drop of 30 % hydrogen peroxide and the mixture kept in a water-bath for about 10 minutes. The cold solution is treated with one drop of the 5 % dye solution and one drop of 1 : 1 sulfuric acid, shaken with 10–15 drops of ether, and the water layer examined for a red or orange-red color. This procedure will reveal 5 γ aluminum in the presence of 1000 γ chromium. A more favorable limiting ratio can be secured in the following way: After oxidation in a larger volume of liquid, the solution is made acid with a drop of concentrated hydrochloric acid, one drop of 2 % ferric chloride solution is introduced, and the mixture made basic with

References pp. 247–257

ammonia. The precipitate of $Fe(OH)_3 + Al(OH)_3$ (compare page 454) is separated from the dissolved chromate by centrifuging and washing. The mixed precipitate is dissolved in one drop of concentrated hydrochloric acid containing a drop of concentrated thioglycolic acid. One drop of the dyestuff solution is added to the chromate-free test solution, the mixture is made basic with ammonia, treated with sulfuric or hydrochloric acid, and the excess dye acid taken up in ether.

(6) *Other tests for aluminum*

(*a*) Pontachrome blue-black R (zinc salt of 4-sulfo-2,2′-dihydroxyazo-naphthalene) gives an orange-red fluorescence with aluminum salts in acetic acid solution (*Idn. Limit:* 2 γ Al in 1 ml).[398] This test appears to be specific for aluminum; Be, Zn, and Ga, which behave similarly to aluminum in the fluorescence reaction with morin, do not react with this dyestuff. It is probable that this reaction can be carried out as a spot test.

(*b*) Ashless filter paper is impregnated with a solution containing 0.1 % of the ammonium salt of aurintricarboxylic acid (aluminon)[399] and 1 % of ammonium acetate. One drop of an acid-aluminum solution is placed on the paper, developed over ammonia, and dried. A red lake is formed (*Idn. Limit:* 0.16 γ Al).[400] A number of other metal ions likewise form lakes with this reagent, but the products are decomposed by ammonia or ammonium carbonate.[401]

26. Beryllium

(1) *Test through reaction of beryllium hydroxide with morin*[402]

Beryllium, in alkaline solution, i.e. as alkali beryllate, reacts with morin (see page 183) to form a water-soluble compound (of unknown composition) which fluoresces yellow-green.[403] Probably the beryllium is bound in a fashion analogous to that of aluminum in aluminum morinate. However, the latter is decomposed by alkali hydroxide with production of aluminate, while beryllium morinate is stable and, because of its phenolic OH-groups, is soluble in alkali hydroxides. A bonding between beryllium and morin leading to fluorescence also occurs when beryllium hydroxide comes in contact with alcohol or acetone solutions of morin. The same is true of almost all water-insoluble metal hydroxides. However, a far-reaching selective precipitation of $Be(OH)_2$ can be secured with ammonia, from solutions containing an excess of the alkali salts of ethylenediaminetraacetic acid (I).[404] This acid or its alkali salts react with many 2-, 3-, and 4-valent metal ions to produce water-soluble chelate compounds with inner complex anions. These products are resistant to ammonia. The coordination structure of the inner complex anions with 2-valent metals (e.g. nickel) is shown in (II):

```
              H2C——CH2
HOOCH2C\      |      |   /CH2COOH
         N              N
HOOCH2C/                 \CH2COOH

                 (I)

                    H2C————CH2
                     |       |
     -OOCH2C—N           N—CH2COO-
              /  ↘     ↙  \
          H2C       Ni       CH2
             \    /    \    /
          O=C—O        O—C=O

                 (II)
```

Masking by ethylenediaminetetraacetate is conveniently accomplished by means of its disodium salt (Versene, Complexone III). This action does not extend to Be^{+2}, Ti^{+4}, Sn^{+4} and Cr^{+3} ions. Among their hydroxides, only $Be(OH)_2$ is brought to a green-yellow fluorescence, which is stable against ammonia. This exceptional behavior is the basis of a specific test for beryllium.

Procedure. A drop of the neutral or weakly acid test solution is placed in a depression of a spot plate and treated with three drops of Complexone III solution, one drop of a 0.02 % acetone solution of morin and one drop of concentrated ammonia. The mixture is transferred by means of a pipette to the filtering device shown on page 42, filtered, and the precipitate washed successively with two drops each of Complexone solution, water, and finally acetone. On examination in ultraviolet light, the filter paper, used in place of asbestos, shows a green-yellow fluorescing zone.

Limit of Identification: 0.07 γ beryllium

Limit of Dilution: 1 : 710,000

This test succeeds in the presence of 2000 times the quantity of aluminium, iron, magnesium, calcium.

Reagent: A cold saturated solution of disodium salt of ethylenediaminetetraacetate in ammonia (1 : 10)

(2) *Test with quinalizarin* [405]

When solutions of beryllium salts are brought together with red-violet solutions of quinalizarin [1,2,5,8-tetrahydroxyanthraquinone (I)] in ammoniacal or caustic alkali solution, a blue-violet precipitate or color appears. Although quinalizarin, as a derivative of alizarin, is a lake-forming dyestuff, which produces red to red-violet adsorption compounds with oxyhydrates of aluminum, zirconium, thorium, etc., its beryllium reaction product seems to be a stoichiometrically defined compound rather than an adsorption complex. In conformity with the fact that the blue product contains two atoms of beryllium combined with one molecule of quinalizarin, it seems proper to view the material as a basic beryllium salt with the structure (II) or (IIa):

References pp. 247–257

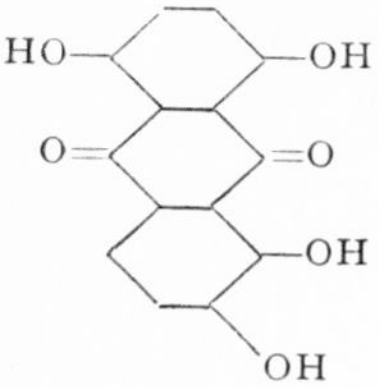

(I) 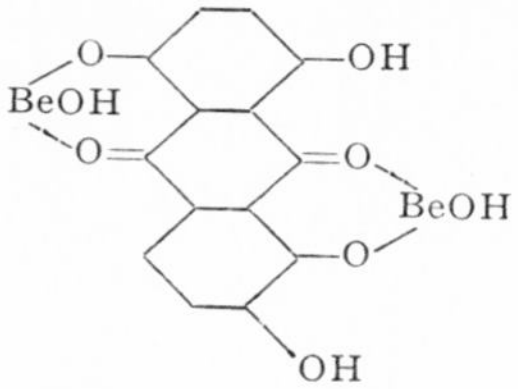 (II) 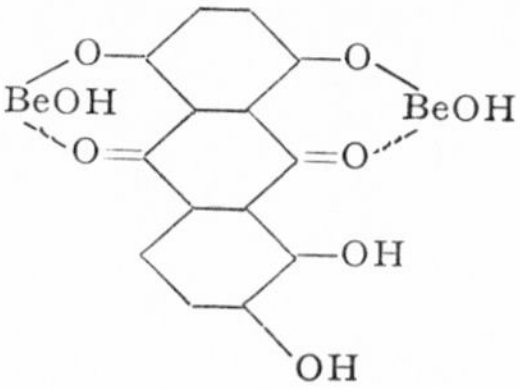 (IIa)

Structure (II) or (IIa) is in harmony with the fact that the product dissolves readily in alkali hydroxide. This solubility obviously is due to salification of the two free phenolic OH-groups with production of soluble alkali salts in which the beryllium is a constituent of a complex anion.

The blue beryllium-quinalizarin compound (II or IIa) is only slightly soluble in water. This is the basis of the high sensitivity of the quinalizarin-beryllium reaction and of its occurrence in solutions of complex alkali beryllium fluoride, in which $[BeF_4]^{-2}$ ions predominate. Another consequence of the very low solubility is that water suspensions of the beryllium compound are resistant to bromine water (and other oxidants), in contrast to ammoniacal solutions of the dyestuff, whose color is discharged at once on the addition of an oxidizing agent. The precipitability from solutions containing fluoride ion and the resistance against oxidation provide a possibility of enhancing the selectivity of the beryllium-quinalizarin reaction, if not too small quantities of beryllium are present.

This test is affected by large amounts of ammonium salts; they should be removed beforehand. Aluminum, in not too great excess, lead, zinc, and antimony salts do not interfere. Solutions of copper and nickel salts can be decolorized by adding potassium cyanide, whereas cobalt salts form the light yellow soluble potassium cobalticyanide $K_3[Co(CN)_6]$, which does not interfere with the recognition of the blue of the quinalizarin compound. $Iron^{III}$ salts may be masked by means of tartrates. However, tartrates interfere if aluminum is present because a red color is formed in the caustic alkaline solution of quinalizarin. If both iron and aluminum are present, the test solution should be treated with 5 ml of 1 *N* NaOH, cooled, diluted to 15 ml, and then filtered if necessary. The clear solution contains any beryllium as beryllate and can be tested as described below.

Magnesium also gives a blue compound with quinalizarin, which is useful in testing for this ion (see page 224). Beryllium can, nevertheless, be detected in the presence of magnesium by utilizing the divergent stabilities of their lakes toward oxidizing agents (see later). It must be noted that Nd, Pr, Ce, La, Zr, and Th salts also form lake colors with quinalizarin.[406]

Procedure. A drop of the test solution is mixed on the spot plate with a drop of freshly prepared solution of quinalizarin in sodium hydroxide. For very small amounts of beryllium, an alcoholic solution of the dyestuff may be used. A drop of dilute ammonia or alkali is then added. A blue color (compared with a blank) indicates the presence of beryllium; larger amounts yield a blue precipitate.

Limit of Identification: 0.14 γ beryllium
Limit of Dilution: 1 : 353,000
Reagent: Saturated alcoholic solution of quinalizarin, or freshly prepared solution of 0.05 g quinalizarin in 100 ml 0.1 N sodium hydroxide. The solution keeps.

Test for beryllium in the presence of magnesium [407]

Magnesium salts give a blue precipitate or color on the addition of an alkaline solution of quinalizarin. Probably an adsorption compound is formed between the dyestuff and magnesium hydroxide (see page 225). In sodium hydroxide solution, bromine destroys the colored beryllium compound, but the color of the magnesium lake persists longer. On the other hand, in ammoniacal solution, the magnesium color is completely destroyed or inhibited, but the beryllium lake retains its blue color. This procedure will reveal 10 γ beryllium in the presence of 1000 γ magnesium.

Procedure. A drop of the neutral test solution is treated on a spot plate with 2 drops of 0.05 % solution of quinalizarin in 2 N ammonia, and then 1 ml of saturated bromine water is added. The originally deep blue solution is made paler, but remains more or less permanently blue according to the beryllium content, whereas a comparison solution containing only magnesium is completely decolorized.

(*3*) *Test with* p-*nitrobenzeneazoorcinol* [408]

The yellow alkaline solution of the azo dye *p*-nitrobenzeneazoorcinol:

CH_3
O_2N—⟨ring⟩—N=N—⟨ring⟩—OH
OH

forms a reddish-brown lake with beryllium salts. This dye reacts analogously to quinalizarin (see page 192). It also reacts with magnesium salts (brown-yellow precipitate); but it is without action on the salts or hydroxides of the rare earths. Aluminum and the alkaline earths do not affect the beryllium test. Zinc hydroxide behaves similarly to beryllium hydroxide; however, the color disappears on addition of potassium cyanide, owing to the formation of the soluble $K_2[Zn(CN)_4]$. Similarly, the addition of cyanide prevents the precipitation of the hydroxides of nickel, cobalt, copper, cadmium, and silver, which interfere with the test.

Procedure. A drop of the alkaline solution of the dyestuff is placed on filter paper, and the tip of a capillary containing the test solution is touched to the middle of the yellow area so that the liquid runs slowly into the paper. The spot is then treated with a further drop of reagent. In the presence of beryllium, either the whole fleck, or only its center, is stained more or less intensively orange-red.

Limit of Identification: 0.2 γ beryllium (in 0.04 ml)
Limit of Dilution: 1 : 200,000
Reagent: 0.025 % solution of *p*-nitrobenzeneazoorcinol [409] in *N* sodium hydroxide

(*4*) *Other tests for beryllium*

(*a*) Chromeazurole, a fuchsin dye, gives a red-violet color lake with beryllium in weakly acid solution.[410] One drop of the test solution is treated on a spot plate with a drop of 2 *N* sodium acetate and one drop of the yellow alcoholic dye solution. Quantities of beryllium above 1 γ yield a deep violet color. Smaller amounts (down to 0.3 γ) give a pink color with a blue edge adhering to the spot plate. Iron, aluminum, zinc, and copper salts likewise give color lakes with chromeazurole. These interferences can be avoided by masking.

(*b*) Acidic solutions of beryllium salts (pH 4.5) give a color lake with 2-(*o*-hydroxyphenyl)-benzthiazole. The product fluoresces blue in ultraviolet light. A drop of the test solution is mixed with 3 drops of 0.001 *M* alcoholic reagent solution on a black spot plate (*Idn. Limit:* 0.03 γ Be). [411] Interferences by Fe^{III}, Cr, Al, Bi, Sb, Sn^{IV}, Zr, Ti can be overcome by sodium potassium tartrate.

27. Titanium

(*1*) *Test with hydrogen peroxide* [412]

Hydrogen peroxide causes a yellow color to develop in acid titanium solutions. In solutions containing sulfuric acid the color is due to the complex peroxidic anion furnished by the free peroxodisulfatotitanic acid [413] which is produced by the reaction:

$$Ti^{+4} + H_2O_2 + 2\,SO_4^{-2} \rightarrow \left[Ti\begin{matrix}\diagup O\\ |\\ \diagdown O\end{matrix}(SO_4)_2\right]^{-2} + 2\,H^+$$

Excess fluoride ions prevent the reaction through the formation of stable $[TiF_6]^{-2}$ ions. However, by demasking, it is possible to detect titanium in the presence of even large amounts of fluoride (see below). Large amounts of acetates, nitrates, chlorides, bromides, and colored ions reduce the sensitivity. Chromates, vanadates, and molybdates, should not be present because they too give color reactions with hydrogen peroxide. Iron salts yield a violet color with the hydrogen peroxide, but it can be discharged by adding syrupy phosphoric acid.

The detection of titanium by means of hydrogen peroxide in solutions containing fluoride is possible if beryllium chloride or sulfate is added.[414] The tendency of Be^{+2} ions to form $[BeF_4]^{-2}$ ions is so great that the reaction:

$$2\ [TiF_6]^{-2} + 3\ Be^{+2} \rightarrow 3\ [BeF_4]^{-2} + 2\ Ti^{+4}$$

occurs and the titanium in the $[TiF_6]^{-2}$ ion is transformed to Ti^{+4} ions, which then produce yellow peroxotitanate anions with hydrogen peroxide. If a drop of H_2O_2 solution that has been saturated with fluoride is added to a titaniferous solution and is then treated with beryllium sulfate, the resulting yellow color will reveal as little as 4 γ titanium in the presence of 6100 γ ammonium fluoride.

Procedure. A drop of the hydrochloric or sulfuric acid test solution is treated with a drop of 3 % hydrogen peroxide on a spot plate. According to the titanium content, a more or less intense yellow color results.

Limit of Identification: 2 γ titanium
Limit of Dilution: 1 : 25,000

(2) Test with pyrocatechol [415]

Pyrocatechol (I) gives a yellowish-red color with weakly acidified solutions of titanium salts. This color reaction is due to the formation of titanium pyrocatechol acid (II) whose inner complex anion provides the color.[416]

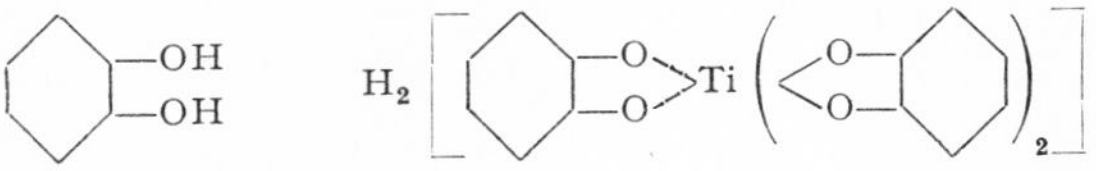

$$\text{C}_6\text{H}_4(\text{—OH})_2 \qquad H_2\left[\text{C}_6\text{H}_4\text{O}_2\text{>Ti}\left(\text{<O}_2\text{C}_6\text{H}_4\right)_2\right]$$

(I) (II)

Large concentrations of free mineral acids interfere. Colored ions, alkali hydroxides and carbonates reduce the sensitivity. $Iron^{III}$ salts should not be present, since they also give soluble colored compounds with pyrocatechol. A number of other elements interfere with the pyrocatechol reaction.[417] Consequently the test is especially suited for the identification of titanium after it has been separated from other metals.

Procedure. A drop of the sulfuric acid test solution is placed on filter paper impregnated with pyrocatechol. According to the quantity of titanium present, a more or less intense yellow-red fleck appears.

References pp. 247–257

Limit of Identification: 2.7 γ titanium
Limit of Dilution: 1 : 19,000
Reagent: 10 % aqueous pyrocatechol solution (freshly prepared)

(3) *Test with chromotropic acid* [418]

Color reactions of chromotropic acid (I) with titanium salts proceed in dilute acidic solution as well as in strong sulfuric acid solution. The chemistry of the reaction, which was long unknown or only unsatisfactorily explained, has now been clarified.[419] In weak acid solution (pH = 2.5 – 5) a wine red colored chelation of a titanium chromotropic acid (II) is obtained:

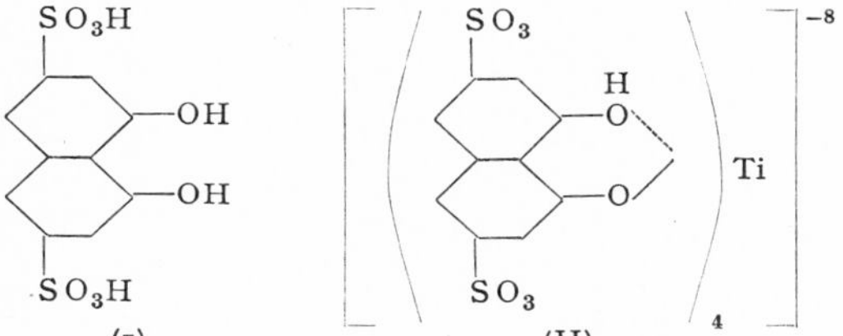

(I) (II)

In concentrated sulfuric acid solution, chromotropic acid is probably oxidized by Ti^{IV} to a violet *p*-quinoidal compound, which according to photometric studies is identical with the oxidation product obtained from chromotropic acid and alkali chromate in 2 *N* sulfuric acid. The reaction product yielded in concentrated sulfuric acid by chromotropic acid and titanium salts contains 2 Ti : 3 chromotropic acid. This ratio can be brought into harmony with (III), namely a binuclear compound in which the oxidation product of chromotropic acid assumes the coordinative binding of Ti^{III}.

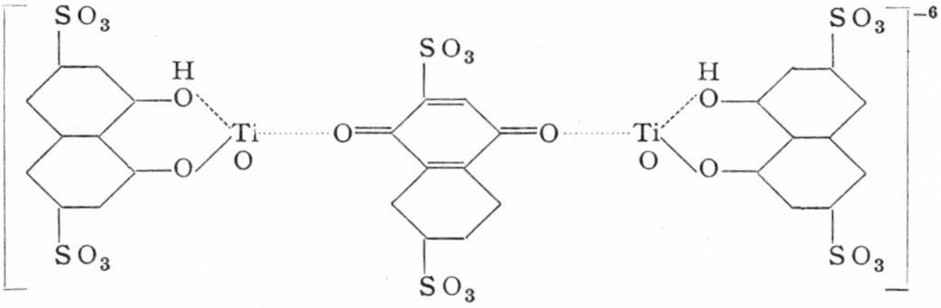

It is noteworthy that Ti^{III} salts do not react with chromotropic acid.[420]

Procedure. A drop each of the test solution and of a 5 % water solution of the sodium salt of chromotropic acid are mixed on filter paper or on a spot plate. A red-brown fleck or coloration of the solution results.

Limit of Identification: 3 γ titanium
Limit of Dilution: 1 : 16,700

References pp. 247–257

Test for titanium in the presence of other elements [421]

A special procedure must be used when titanium is not present alone since chromotropic acid reacts also with other metal salts. Ferric salts give a deep green color, uranyl salts a brown. These colors can be easily discharged by the addition of a hydrochloric acid solution of stannous chloride; the resulting ferrous and uranous salts do not react with chromotropic acid. Those mercury salts which dissociate give a yellow color and silver salts form a black stain on the paper with chromotropic acid. The presence, however, of these two products does not materially interfere, since the titanium color is still perceptible. The same is true of iron, because the titanium color usually appears as a brown-red fleck in the center of the green iron stain. When there is uncertainty as to the identity of the other elements present, the following procedure should be used.

Procedure. A large drop of the test solution is mixed on a watch glass with stannous chloride (a great excess is to be avoided). On gently heating, the precipitate balls together. A drop of chromotropic acid is placed on filter paper followed by a drop of the clear solution from the watch glass. In the presence of titanium, a red-brown fleck appears. The precipitate, as well as the clear solution, may be placed on the filter paper, because the precipitate remains fixed and the solution diffuses away. If the moist portion is then treated with chromotropic acid, a red-brown circle is formed.

An improvement on this procedure is the use of a solution of chromotropic acid in concentrated sulfuric acid.[422] A violet product appears and only large amounts of colored metallic salts or nitrates and other oxidizing agents interfere. The latter may be completely removed by fuming with concentrated sulfuric acid.

Procedure. A drop of the test solution containing sulfuric acid is well mixed on a spot plate or in a micro crucible with 5 drops of the reagent solution. A violet color indicates the presence of titanium. For very dilute solutions, comparison with a blank test is advisable.

Limit of Identification: 0.1 γ titanium
Limit of Dilution: 1 : 500,000
Reagent: 0.02 g chromotropic acid dissolved, with warming, in 20 ml concentrated sulfuric acid

(*4*) *Test with morin* [423]

An intense brown color is given by the light yellow alcohol solution of morin when it reacts in mineral acid solution with titanium salts. The composition of the product is unknown. Probably, a chelate compound is formed in which the titanium is included in a colored anion.

References pp. 247–257

Zirconium and thorium ions behave in an analogous manner. These ionic species can be removed by precipitation with phosphoric acid. $Iron^{III}$ ions must be reduced to $iron^{II}$ ions (by means of zinc and hydrochloric acid, or perhaps with thioglycolic acid). Up to 1000 times the quantity of colored ions do not interfere.

Procedure. One drop of the test solution, which is approximately 0.5 *N* with respect to hydrochloric acid, is treated on filter paper with one drop of a saturated methyl alcohol solution of morin. A yellow-brown fleck develops if titanium is present. Since morin leaves a lemon yellow stain on filter paper, a comparison test is advisable when slight amounts of titanium are suspected.

Limit of Identification: 0.01 γ titanium
Limit of Dilution: 1 : 5,000,000

(*5*) *Other tests for titanium*

(*a*) The tannin-antipyrine reaction produces a red precipitate when applied to titanium salts.[424] Filter paper soaked with 10 % tannin solution is treated successively with a drop of 20 % antipyrine and a drop of test solution (*Idn. Limit:* 0.2 γ Ti). The positive reaction of other metals (except molybdenum) is prevented by spotting at the close with 1 : 4 sulfuric acid. In this way, titanium can be detected in a solution whose dilution is 1 : 20,000 and containing as much as a thirtyfold excess of accompanying materials.

(*b*) One drop of the acidic test solution is treated with 3 or 4 drops of 0.025 % methylene blue solution and a tiny particle of metallic zinc.[425] The discharge of the dye is catalytically accelerated by the redox system: $Ti^{III} \rightleftharpoons Ti^{IV}$. Comparison with a blank makes this effect distinctly evident (*Idn. Limit:* 0.05 γ Ti). This test is suitable for the detection of titanium in the presence of iron, chromium, uranium, manganese. If instead of methylene blue, a drop of the light yellow saturated aqueous solution of alizarinsulfonic acid is used, the catalytic action of the redox system is evidenced by a change to green (formation of quinhydrone) (*Idn. Limit:* 0.5 γ Ti).[426] Mo^{VI}, W^{VI}, V^{V}, Cr^{III} interfere.

(*c*) A drop of test solution, buffered to pH = 4.5, gives a red-orange coloration with a drop of an alcohol solution of resoflavine (*Idn. Limit:* 0.1 γ Ti).[427]

28. Zirconium

(*1*) *Test with β-nitroso-α-naphthol* [428]

Colored precipitates of the empirical formula $ZrO(C_{10}H_6O_2N)_2$ are produced by treating acid solutions of zirconium salts with an alcoholic solution of either of the two isomers: α-nitroso-β-naphthol or β-nitroso-α-naphthol. The zirconium salt of the α-nitroso compound is green-yellow, that of the β-compound is red; both are regarded as inner complex salts. The salt formation is probably not due to the phenol groups of the nitrosonaphthols but to a NOH-group, because these nitroso compounds may react in their isomeric quinoidal oxime forms (see page 144). For instance the β-nitroso-α-naphthol functions:

$$C_{10}H_6(OH){-}NO \rightleftarrows C_{10}H_6(=O){=}NOH + {}^1/_2 ZrO^{+2} \longrightarrow C_{10}H_6(=O\cdots ZrO/2){=}NO + H^+$$

When an alcoholic solution of β-nitroso-α-naphthol is added to a hydrochloric acid solution of zirconium, a deep red color with no precipitation results; the product coagulates only on the addition of sodium acetate. The reaction is hindered in the presence of large amounts of sulfates or fluorides. They lower the concentration of Zr^{+4} or ZrO^{+2} by forming $[Zr(SO_4)_3]^{-2}$ or $[ZrF_6]^{-2}$ ions. Interference of sulfates can be prevented by the addition of barium chloride. The fluoride interference can be averted by adding beryllium salts.

Procedure. Filter paper is impregnated with a 2 % alcoholic solution of β-nitroso-α-naphthol, [429] and a drop of the hydrochloric acid test solution is added. According to the zirconium content, a more or less intense red fleck is formed on the yellow-brown paper. The fleck becomes more vivid on adding a little dilute acid.

Limit of Identification: 0.2 γ zirconium
Limit of Dilution: 1 : 250,000

(2) *Test with alizarin* [430]

Zirconium salts, in not too strong acid solution, give a red to dark violet precipitate on the addition of an alcoholic solution of alizarin or an aqueous solution of alizarin sulfonic acid. Other polyhydroxyanthraquinones behave similarly. The precipitation is accelerated by warming.

The precipitability by polyhydroxyanthraquinones from weakly acid solution is specific for zirconium (and hafnium). The precipitate is an adsorption compound (lake) of zirconium hydroxide and alizarin (compare Al-alizarin lake, page 185). The production of the lake involves the binding, through chemical adsorption, of alizarin on the surface of the $Zr(OH)_4$ sol particles, which are present in solutions of zirconium salts as a result of the hydrolysis: $Zr^{+4} + 4\,H_2O \rightarrow Zr(OH)_4 + 4\,H^+$. The hydrolysis equilibrium is constantly disturbed by the removal of $Zr(OH)_4$, so that, in a not too acid solution, there is extensive precipitation of zirconium in the form of the alizarin lake. The red-violet zirconium-alizarin lake is also produced by precipitating solutions of zirconium salts with ammoniacal solutions of alizarin. The lake is stable against dilute hydrochloric acid. In strong hydrochloric acid solutions of zirconium salts, alizarin produces a fairly stable hydrosol of the lake (compare the test for fluoride, page 269).

The test is hindered or prevented by the presence of fluorides, sulfates, phosphates, and organic hydroxyacids, also by molybdates and tungstates. These materials reduce the ionic concentration of zirconium, either through precipitation or complex ion formation. The interference due to sulfates can be prevented by the addition of barium chloride.

The following procedure [431] is for the detection of zirconium in the presence of beryllium, aluminum, titanium and thorium salts.*

Procedure. A drop of the test solution, which should be as nearly neutral as possible, is mixed in a micro crucible with a drop of an alcoholic alizarin solution, and boiled once. Zirconium and also the other metals give red to violet adsorption compounds. A drop of 1 *N* hydrochloric acid is then added. Only the zirconium compound remains unaffected. Much zirconium gives a deep red-violet color and precipitate; a little gives red flakes.

Limit of Identification: 0.5 γ zirconium

Limit of Dilution: 1 : 100,000

Reagent: Alizarin solution. An alcoholic solution of commercial alizarin is treated drop by drop with dilute hydrochloric acid until the pure yellow color develops; the solution is diluted with an equal volume of alcohol and filtered.

(3) *Test with morin* [432]

The constitution of morin as 3,5,7,2′,4′-pentahydroxyflavone (see page 182) reveals that it is clearly analogous to alizarin (and its derivatives) with respect to groups that are capable of chelation. In fact, it also reacts with solutions of all the metal salts which produce inner complex salts or adsorption compounds with alizarin. The compounds with alizarin exhibit characteristic colors, but this is not true of the compounds of the light yellow morin; in daylight the latter are uniformly pale yellow. However, in ultraviolet light, both morin and its compounds, the latter in increased measure, exhibit a strong yellow or yellow-green fluorescence. The analogy

* According to F. Pavelka, *Mikrochemie*, 4 (1926) 199, placing a drop of the solution of the pure salt on paper impregnated with alizarin, and holding over ammonia, gives the following *limits of identification:*

0.18 γ titanium (1 : 166,000) violet-red fleck
0.29 γ zirconium (1 : 103,000) raspberry fleck
0.24 γ thorium (1 : 125,000) violet fleck

between alizarin and morin appears plainly in their behavior toward zirconium. Just as the latter is the only metal which yields acid-stable red-violet lakes with alizarin (and its derivatives), it similarly forms an acid-resistant, bright green fluorescing lake with morin. The statements regarding the formation of the red zirconium-alizarin lake (test *2*) apply in this case also. The stability of the zirconium-morin lake toward hydrochloric acid is remarkable; it exceeds that of the alizarin lake. After it has once been produced as a hydrosol, the morin lake is stable toward even concentrated hydrochloric acid. The following fluorescence reaction is specific for zirconium. The test is impaired by ions which have a yellow or red color in 10 N hydrochloric acid; they quench or prevent the fluorescence. Morin is oxidatively decomposed in acid solution; consequently Cu^{+2}, Fe^{+3}, VO_3^{-1} and CrO_4^{-2} ions, and likewise ions of noble metals, interfere. Here again, quenching effects are involved, at least in part. Fluorides prevent the fluorescence reaction. On the other hand, even large amounts of sulfate do not impair the test.

Procedure. A drop of the test solution is mixed on a spot plate with a drop of 0.001 % morin solution in ethyl alcohol and five drops of concentrated hydrochloric acid. A green fluorescence under ultraviolet light indicates the presence of zirconium.

Limit of Identification: 0.1 γ zirconium
Limit of Dilution: 1 : 500,000

(*4*) *Test with* p-*dimethylaminoazophenylarsonic acid* [433]

Zirconium salts in acid solution give sparingly soluble white precipitates with arsenic acid and its organic derivatives [434] such as phenylarsonic acid $C_6H_5AsO(OH)_2$. The latter is the parent compound of different derivatives including some that give colored insoluble zirconium salts. The best are the azoarsonic acids, of which the *p*-dimethylaminoazophenylarsonic acid (I) is especially useful, because of its availability and solubility in acids. Its zirconium salt forms in acid solution as a brown precipitate (II) which, when produced by a drop reaction on filter paper, stays in the pores of the paper as a brown flock, while any excess of colored reagent is being washed out with dilute acid.

$(CH_3)_2N-C_6H_4-N{=}N-C_6H_4-AsO_3H_2$

(I)

$(CH_3)_2N-C_6H_4-N{=}N-C_6H_4-As(=O)(O-)_2\ Zr/2\ (ZrO)$ [435]

(II)

References pp. 247–257

In concentrations exceeding 1 N, sulfuric acid causes a marked reduction in the sensitivity of the test, because of the formation of complex zirconium-sulfuric acids. Phosphates, fluorides and organic acids, that give either precipitates or stable soluble complex compounds with zirconium salts, hinder the reaction. Antimony^{V} salts yield a brown coloration similar to that formed by zirconium salts, but it disappears after 2 or 3 minutes treatment with hydrochloric acid. Thus, 0.25 γ zirconium may be detected in the presence of 1000 times the amount of antimony. Antimony^{III} salts give a slight colored fleck that immediately disappears in the acid bath. Thorium salts give a similar reaction to that of zirconium, but the color disappears at once in the acid bath, so that 0.2 γ zirconium may be detected in the presence of 2500 times the amount of thorium. Gold salts, in concentrated solution, give a violet fleck that disappears at once when bathed in hydrochloric acid.

Procedure. A drop of the acid test solution is placed on filter paper (S & S, Black band No. 589) that has been impregnated with the reagent and dried. Large amounts of zirconium give a brown stain at once. On soaking the paper for a short time in a bath of 2 N hydrochloric acid at 50° to 60°, the reagent is quickly washed out of the paper. The brown zirconium fleck or ring (at higher acidities of the test solution) becomes clearly visible.

Limit of Identification: 0.1 γ zirconium } in 1 N hydrochloric acid solution
Limit of Dilution: 1 : 500,000 } in 1 N hydrochloric acid solution
Reagent: Solution of 0.1 g *p*-dimethylaminoazophenylarsonic acid [436] in 100 ml alcohol containing 5 ml concentrated hydrochloric acid

Test for zirconium in the presence of molybdenum, tungsten, and titanium

Molybdates, tungstates, and titanium salts give red-brown flecks when they are spotted on paper impregnated with *p*-dimethylaminoazophenylarsonic acid. The flecks are acid resistant. The addition of hydrogen peroxide converts any molybdate and tungstate into the corresponding per-acids; the titanium goes over into the peroxo-compound (see page 195). These products exhibit little if any reaction toward the arsonic acid.

Procedure. Three drops of the test solution, which should contain not more than 1 % molybdenum, 1.5 % titanium, or 0.5 % tungsten, are mixed on a spot plate with a drop of concentrated hydrochloric acid and a drop of 30 % hydrogen peroxide. A drop of this mixture is placed on the reagent paper. On bathing the paper in warm hydrochloric acid, a red fleck appears in the middle of the moist spot; this soon disappears. When zirconium is present, a brown ring of the zirconium salt shows around the fleck. At the same time, some of the stain due to tungsten (and to a slight extent also to molybdenum) remains, even after the paper has been in the acid bath for some time. The zirconium thus induces the coprecipitation of the tungstate and molybdate compounds. The simultaneous formation of an acid-stable brown ring with a red center makes it possible to detect very small amounts of zirconium in the presence of much tungsten and molybdenum. If the molybdenum or tungsten concentration is small, only the brown zirconium ring is formed.

Limit of Identification: 0.5 γ zirconium
Limit of Dilution: 1 : 100,000
} in the presence of 1000 times the amount of molybdenum or 500 times the amount of tungsten

Limit of Identification: 1 γ zirconium
Limit of Dilution: 1 : 50,000
} in the presence of 500 times the amount of titanium

Test for zirconium in the presence of tin^{IV}

Stannic salts give a brown precipitate with the azoarsonic acid. The precipitate, however, is not formed in high concentration of acid and low concentration of tin.

Procedure. Three drops of the test solution, which should contain at most 0.1 % tin, are mixed with a drop of concentrated hydrochloric acid on a spot plate. A drop of the mixture is placed on the reagent paper. After brief bathing in the warm hydrochloric acid, a slight red fleck remains, at most. In the presence of zirconium, a more or less intense brown ring or fleck is formed.

Limit of Identification: 0.2 γ zirconium
Limit of Dilution: 1 : 250,000
} when 250 times the amount of tin is present

(5) Other tests for zirconium

(*a*) Acid zirconium solutions give color reactions with carminic acid or gallocyanin.[437] The formation of the zirconium salt causes a change from yellow to red, or from pink to blue. The reaction with carminic acid may be carried out as a spot test (*Idn. Limit:* 0.5 γ Zr in 0.001 ml).

(*b*) An acetone-water solution of chlorobromoamine acid produces a red precipitate or color with acidified solutions of zirconium salts (*Idn. Limit:* 0.5 γ Zr).[438] Considerable quantities of tervalent metals, or of materials that form complexes with Zr^{+4} ions, interfere.

(*c*) Azo dyes derived from mandelic acid and its derivatives give pink flecks on filter paper with even small amounts of zirconium. On confined spot test paper, the method becomes semi-quantitative for small contents of zirconium.[439]

29. Uranium

(1) Test with 8-hydroxyquinoline (oxine) [440]

In neutral or masked alkaline solutions of uranyl salts, a quantitative precipitation of a red-brown product is obtained by adding 8-hydroxyquinoline (oxine). In contrast to other metal oxinates, which for the most

References pp. 247–257

part are inner complex phenolates [441], the uranium compound contains also a molecule of oxine as neutral part according to the formulation $UO_2(C_9H_6NO)_2 \cdot C_9H_6NOH$.[442] Probably this compound should not be viewed as uranyl oxinate but rather as the oxine ester of uranic acid, in other words as oxine uranate.

The precipitation through oxine also occurs from solutions of the complex alkali uranyl double carbonates, which yield $[UO_2(CO_3)_3]^{-4}$ ions. The latter are produced by adding an excess of alkali carbonate to solutions of uranyl salts. Since all metal ions forming oxinates are precipitated by alkali carbonate, it is thus possible to separate the uranium before conducting the actual procedure.

Procedure. The test solution is treated with an excess of ammonium carbonate solution. Any precipitate is filtered off or removed by centrifuging. One drop of the clear liquid is placed on a spot plate or filter paper and treated with a drop of 5 % alcohol solution of oxine. A red-brown precipitate or stain indicates uranium.

Limit of Identification: 10 γ uranium

Limit of Dilution: 1 : 5,000

Reagent: Ammonium carbonate solution: 2 g of the salt is dissolved in 10 ml of concentrated ammonia and diluted with 10 ml of water

(2) Test with potassium ferrocyanide

Neutral or acetic acid solutions of uranyl salts give a red-brown precipitate with potassium ferrocyanide. Very dilute solutions give only a coloration. Uranyl potassium ferrocyanide or uranyl ferrocyanide is formed.

The test is specific in the absence of ferric and copper salts, which also give colored ferrocyanides. However, under the correct conditions (see page 206), the test may also be applied in the presence of these metals.

Procedure.[443] A drop of the slightly acid test solution is placed on filter paper impregnated with 3 % potassium ferrocyanide, or a drop of the test solution and then the ferrocyanide are placed on filter paper. According to the concentration of uranium, a more or less intense brown stain is formed.

Limit of Identification: 0.92 γ uranium*

Limit of Dilution: 1 : 50,000

Test for uranium in the presence of iron

When the iron concentration is not too great, the uranium can be detected by treatment of the stain with a drop of water. Owing to its slimy consistency the uranium precipitate spreads more than the iron precipitate, and appears

* F. Hernegger and B. Karlik, *Chem. Abstracts*, 30 (1936) 408, state that as little as 0.05 γ uranium (1 : 100,000) can be detected if a microdrop (0.001 ml) is used.

outside the central stain of Prussian blue as a jagged brown fringe. When the proportions of the two elements are such that this test cannot be used, advantage may be taken of the fact that uranates precipitated by alkali are soluble in ammonium carbonate with the formation of complex carbonates, such as $(NH_4)_4[UO_2(CO_3)_3]$. This solution, after acidifying with hydrochloric acid, may be used for the uranium test (compare test *1*).

Procedure. A drop of concentrated ammonia is placed on filter paper and, before it is completely taken up, a drop of the test solution is added and rubbed with a glass rod. The paper is then dried, spotted with ammonium carbonate, and the site of the fleck rubbed with a glass rod to insure that all parts of the fleck come into contact with the ammonium carbonate. After redrying, dilute hydrochloric acid is spotted around the fleck, but without touching it (to avoid formation of Prussian blue when the ferrocyanide is added). The brown color is especially distinct at the edges of the hydrochloric acid ring adjoining the original fleck and extends to where the two flecks have been superimposed, producing a starlike picture.

Test for uranium in the presence of iron and copper[444]

The test for uranium with potassium ferrocyanide can also be carried out in the presence of ferric and cupric salts, if these metals are converted, before the addition of the ferrocyanide, into the nonreacting cuprous and ferrous forms. Reduction with iodide ions in acid solution serves this purpose:

$$2\ Cu^{+2} + 4\ I^- = Cu_2I_2 + I_2$$
$$2\ Fe^{+3} + 2\ I^- = 2\ Fe^{+2} + I_2$$

If the liberated iodine is decolorized (reduced) with thiosulfate, the uranium may then be detected with potassium ferrocyanide.

Procedure. A drop of concentrated potassium iodide solution is placed on filter paper and, after it has soaked in, a drop of the acid test solution is added; iodine is liberated. To complete the reduction, a further drop of potassium iodide is added, and then a drop of sodium thiosulfate to remove the elementary iodine. A drop of potassium ferrocyanide is placed on the decolorized fleck. A more or less deeply colored brown or yellowish circle is formed, according to the amount of uranium present.

It is better to carry out the reduction to ferrous and cuprous ions with sodium thiosulfate alone and to use a spot plate:

$$2\ Fe^{+3} + 2\ S_2O_3^{-2} = 2\ Fe^{+2} + S_4O_6^{-2}$$
$$2\ Cu^{+2} + 2\ S_2O_3^{-2} = Cu_2^{+2} + S_4O_6^{-2}$$

When copper is present it acts as a catalyst for the reduction of the iron by thiosulfate (see page 80).

(3) *Fluorescence test* [445]

Uranyl salts fluoresce best in the crystalline form, but only slightly in solution. If a dilute solution of uranium salts is allowed to evaporate slowly on a microscope slide, and the residue examined, single fluorescent crystals can be observed. Traces of impurities or too rapid evaporation of the solution interfere with the test because they prevent the formation of good crystals.

Borax beads containing uranium exhibit an appreciable green fluorescence. Fluoride beads of the alkalis and alkaline earth metals [446] fluoresce especially well. Sodium fluoride beads light up to a deep yellow color and are most striking. They can be used to detect uranium.

The shape of the bead is very important when testing for uranium by means of activated beads. Thin flat beads are better than the round type, because the ultraviolet light penetrates farther. Neither SiO_2, TiO_2 nor sulfates, etc., should be present, nor any other material that liberates hydrofluoric acid or forms complex compounds with fluorides. Iron should be avoided because it makes the bead yellow and so absorbs the ultraviolet light at the surface. Manganese salts, which color the beads blue, do not interfere so much as iron. Thorium salts also greatly reduce the luminosity, but it may still be perceptible provided sodium fluoride is present in excess. Only niobium [447] as well as greater quantities of beryllium (exceeding 1 mg/ml) give a similar fluorescence, but it is relatively so weak as to be of no importance.

Procedure. Sodium fluoride is fused in a loop of platinum wire (diam. 1 mm). When cold, the bead appears only slightly violet in ultraviolet light (reflected light). By means of a calibrated loop of platinum, 0.001 ml of the neutral test solution is placed on the bead and evaporated. After fusing for a short time, the bead is cooled and examined in ultraviolet light.

Limit of Identification: 0.001 γ uranium in 0.001 ml
Limit of Dilution: 1 : 1,000,000

(4) *Test through photolytic decomposition of oxalic acid* [448]

Uranium salts are light-sensitive in the presence of many organic materials (organic anions and also neutral compounds). When intensely illuminated, particularly in ultraviolet light, U^{VI} is reduced to U^{IV} and the organic material is oxidized with production of carbon dioxide.[449] Since U^{IV} can be reoxidized to U^{VI} by atmospheric oxygen, even small amounts of uranium can be photolytically active in appropriate systems.

The uranyl-sensitized decomposition of oxalic acid produces formic acid:

$$H_2C_2O_4 \rightarrow HCOOH + CO_2 \qquad (1)$$

In acetic acid solution, formic acid reduces mercuric chloride to water-insoluble mercurous chloride, which produces white mercuri-amido chloride and metallic mercury on contact with ammonia. Because of the high dispersion of the mercury, the reaction product is deep black:

$$2\ HgCl_2 + HCOOH \rightarrow Hg_2Cl_2 + 2\ HCl + CO_2 \quad (2)$$

$$Hg_2Cl_2 + 2\ NH_3 \rightarrow HgNH_2Cl + NH_4Cl + Hg^{\circ} \quad (3)$$

The realization of (2) and (3) makes possible a rather sensitive test for formic acid,[450] which is not impaired by large amounts of oxalic acid. A test for uranyl ions has been developed by combining (2) and (3) with (1) as effected by uranyl salts and illumination.

Procedure. A drop of the neutral or slightly acidified (nitric acid) test solution is treated with a drop of saturated sodium oxalate solution in a micro crucible and the mixture taken to dryness (water-bath, drying oven). The residue is placed under a quartz lamp for 30 minutes. A drop of 1 % mercuric chloride solution and a drop of sodium acetate solution are added and the contents of the crucible then evaporated to dryness at 100° in a drying oven. The cooled residue is moistened with a drop of 1 : 1 ammonia. A black or gray coloration indicates the presence of formic acid and hence of uranium.

Limit of Identification: 2.5 γ uranium

Limit of Dilution: 1 : 20,000

Reagent: Acetic acid-sodium acetate solution: 1 ml glacial acetic acid plus 1 g sodium acetate diluted to 100 ml with water

(5) *Test with rhodamine B* [451]

Rhodamine B, a red dye, is a sensitive reagent in acid solutions for monobasic complex metal halogeno acids (compare pages 105, 129, 158, 215). Its almost colorless solutions in benzene contain an equilibrium mixture of the lacto form (I) with minimal amounts of the red quinoid form (II):

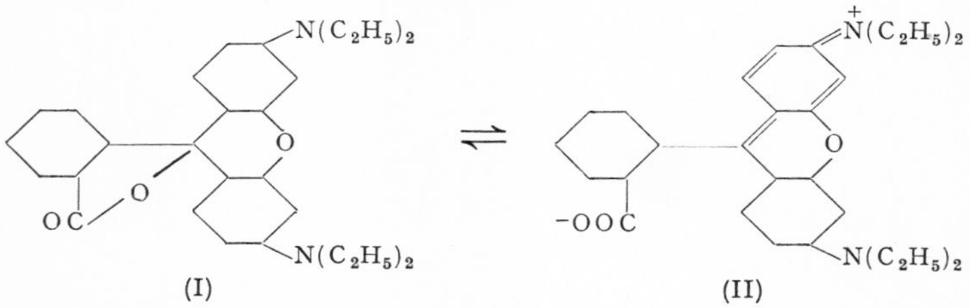

$N(C_2H_5)_2$ O OC O $N(C_2H_5)_2$ ⇌ $\overset{+}{N}(C_2H_5)_2$ O ^{-}OOC $N(C_2H_5)_2$ (1)

(I) (II)

When neutral solutions of uranyl-, ferric-, or bismuth nitrate (chloride) are shaken with a benzene solution of rhodamine B, the benzene layer turns red and exhibits an intense orange fluorescence in ultraviolet light. This effect is surprisingly heightened if a little benzoic acid or some other benzene-soluble carboxylic acid is added to the benzene solution. In the case of uranyl salts, the color (fluorescence) reaction is so marked that a sensitive test can be based on this finding. Although the benzene-soluble uranyl compound has not been isolated, the underlying chemistry is probably that the union of uranyl ions with benzoic acid yields a slight quantity of a complex uranyl benzoic acid (2) which then produces a red benzene-soluble salt (3):

$$UO_2^{+2} + 3\ C_6H_5COOH \longrightarrow H[UO_2(C_6H_5COO)_3] + 2\ H^+ \qquad (2)$$

$$H[UO_2(C_6H_5COO)_3] + (II) \longrightarrow \qquad (3)$$

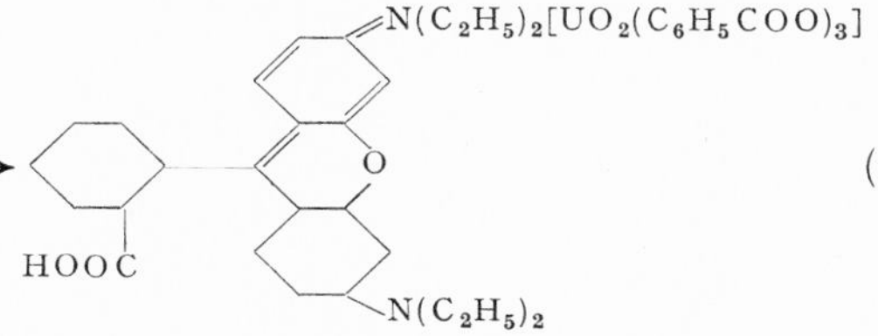

$N(C_2H_5)_2[UO_2(C_6H_5COO)_3]$

O

HOOC

$N(C_2H_5)_2$

On this basis, the formation of the benzene-soluble dye salt constantly disturbs equilibria (1) and (2) because the products contained in them are replenished after consumption and so suffice to accomplish the color reaction.

Procedure. The test is conducted in a micro test tube. A drop of the neutral test solution is treated with 5 drops of the reagent solution and shaken. A red of pink benzene layer results if uranium is present, the shade depending on the quantity of the latter.

Limit of Identification: 0.05 γ uranium

Limit of Dilution: 1 : 1,000,000

Reagent: A 0.5 % solution of benzoic acid in benzene is treated with an excess of rhodamine B, shaken, and then filtered. The solution keeps.

If iron and bismuth are also present, they must be removed because they show analogous behavior. The test solution is warmed with an excess of sodium carbonate and the precipitate removed. The filtrate which contains $[UO_2(CO_3)_2]^{-2}$ is taken to dryness with nitric acid. The residue contains uranyl nitrate, and can be tested by the procedure just described. As little as 0.5 γ uranium can be detected in the presence of 2500 γ iron, starting with one drop and operating within the bounds of spot test technique.

(*6*) *Other tests for uranium*

(*a*) A rust-brown fleck is produced by uranium when spotted on paper with 0.2 % solution of quercetin or quercitrin (*Idn. Limit:* 3 γ U). [452] See also page 167.

(*b*) A drop of the neutral test solution on treatment with a drop of 0.5 *N* Na_3PO_4 gives a precipitate of $(UO_2)_3(PO_4)_2$ which exhibits a strong yellow fluorescence (*Idn. Limit:* 2.5 γ U). [453]

(c) The red fluorescence of a drop of an alkaline solution of cochineal disappears on the addition of a drop of a solution of a uranyl salt (*Idn. Limit:* 2.5 γ U).[453]

(d) A red color appears if a drop of 0.12 % of fluorescein solution and 3 drops of 5 % ammonium chloride are added to a drop of a uranyl solution on a spot plate (*Idn. Limit:* 0.12 γ UO_2^{+2}).[454]

(e) Reduction of the weakly acid test solution yields U^{+4} ions which reduce Fe^{+3} to Fe^{+2}. Accordingly, a red color results if the reduced solution is treated with a $FeCl_3$ solution containing phenanthroline (*Idn. Limit:* 1 γ uranium).[455] If only slight amounts of uranium are suspected, it is advisable to add thorium nitrate after the test solution has been reduced and to precipitate ThF_4 and UF_4 jointly by means of ammonium fluoride. The test is then conducted with the precipitate.

30. Cerium

(1) *Test with hydrogen peroxide and ammonia* [456]

Cerium^{III} and^{IV} salts react with ammonia and hydrogen peroxide and precipitate yellow or red-brown cerium perhydroxide,[457] $Ce(OH)_2(OOH)$ and $Ce(OH)_3(OOH)$. This test is characteristic for cerium in a mixture of the rare earths, but may not be used directly in the presence of iron, because the color of cerium perhydroxide is similar to that of ferric hydroxide. The precipitation of iron must be prevented by the addition of tartrate; however, this masking reduces the sensitivity of the cerium test. The presence of large amounts of colored ions renders the detection of small amounts of cerium difficult.

Procedure. Single drops of the test solution, 3 % hydrogen peroxide, and dilute ammonia are mixed in a porcelain micro crucible and gently warmed. In the presence of cerium, a yellow precipitate or coloration appears.

Limit of Identification: 0.35 γ cerium
Limit of Dilution: 1 : 143,000

(2) *Test with ammoniacal silver nitrate* [458]

Ammoniacal silver nitrate reacts with solutions of cerous salts:

$$Ce^{+3} + Ag(NH_3)_2^+ + 4\ OH^- = Ce(OH)_4 + Ag^\circ + 2\ NH_3$$

The yellow cerium hydroxide precipitate is blackened by the finely divided elementary silver. It may be that the redox reaction results in an adsorption compound because the product is incomparably blacker than if the components are produced separately and then mixed.

References pp. 247–257

The test is characteristic for cerium in a mixture with other rare earths. Manganese^{II}, iron^{III} and cobalt^{II} react in the same way as cerium^{III} to give the higher metal oxides and elementary silver.

Procedure. A drop of the neutral test solution and a drop of an ammoniacal silver nitrate solution are mixed on a watch glass and gently warmed. According to the amount of cerium present, a black precipitate or brown coloration is formed.

Limit of Identification: 1 γ cerium

Limit of Dilution: 1 : 50,000

Reagent: Ammoniacal silver nitrate solution: 0.4 N silver nitrate solution is mixed with sufficient dilute ammonia to redissolve the initial precipitate

(3) *Test with benzidine* [459]

Benzidine is converted into "benzidine blue" by a number of oxidizing agents and autoxidizable substances (see page 72). Both Ce^{IV} hydroxide and the autoxidizable Ce^{III} hydroxide react in this way. The reaction, applied as a spot test, is characteristic for cerium among the rare earths.

The test is not directly applicable in the presence of manganese, cobalt, copper, silver, thallium, or chromates because they too convert benzidine to the blue compound. In these instances it is advisable to precipitate cerous fluoride from the neutral or slightly acid test solution with hydrofluoric acid, then to produce Ce^{III} hydroxide by spotting with alkali, and apply the benzidine test.

Procedure. A drop of the test solution is placed on filter paper, followed by a drop of dilute sodium hydroxide, and then by a drop of benzidine solution. According to the cerium content, a more or less deep blue coloration results.

Limit of Identification: 0.18 γ cerium

Limit of Dilution: 1 : 275,000

Reagent: Benzidine solution: 0.5 g benzidine base or hydrochloride dissolved in 10 ml acetic acid, diluted to 100 ml with water, and filtered

The following tests were carried out with regard to the detection of cerium in the presence of iron.

Sixty milliliters of a solution containing Ce^{III} and Fe^{III} (as sulfates) were treated with alkali until a permanent precipitate formed. Then a few drops of hydrofluoric acid were added, the mixture boiled, and filtered. The precipitate was washed once with water, covered with dilute alkali, and tested with an acetic acid solution of benzidine. In this way the blue color caused by 70 γ cerium can be discerned in the presence of 227 mg iron. This corresponds to a ratio Ce : Fe = 1 : 3230.

(4) Test with phosphomolybdic acid [460]

The enhanced oxidizing action of molybdic acid in certain heteropoly-acids, for instance, phosphomolybdic acid $H_3PO_4 \cdot 12\ MoO_3 \cdot aq.$ (compare page 85), is also exhibited in the behavior of this reagent toward Ce^{III} salts. In the presence of alkali, there is a redox reaction: $Ce^{III} + Mo^{VI} \rightarrow Ce^{IV} + Mo^{V}$, i.e., $CeO_2 \cdot aq.$ and molybdenum blue results. Since all the other rare earths, as well as zirconium and thorium salts, are indifferent toward phosphomolybdic acid, cerium may accordingly be identified in a mixture of these elements as obtained, for instance, in the systematic scheme of analysis.

Procedure. A drop of the test solution and of phosphomolybdic acid solution are mixed on a spot plate, and a drop of 40 % sodium hydroxide solution is added. According to the cerium content, a more or less intense blue color or precipitate appears.

Limit of Identification: 0.52 γ cerium
Limit of Dilution: 1 : 61,000
Reagent: Saturated aqueous solution of phosphomolybdic acid [461]

This method of detecting cerium in the presence of other rare earths (especially zirconium and thorium salts) gives better limiting proportions than the benzidine test (*3*). The comparative figures for 10 γ cerium are:

Benzidine reaction	Phosphomolybdic acid reaction
Ce : Pr = 1 : 13	Ce : Pr = 1 : 1100
Ce : Nd = 1 : 13	Ce : Nd = 1 : 90
Ce : La = 1 : 10	Ce : La = 1 : 2130
Ce : Th = 1 : 17	Ce : Th = 1 : 521
Ce : Zr = 1 : 1	Ce : Zr = 1 : 230

(5) Other tests for cerium

(*a*) Leuco malachite green is oxidized to malachite green by $Ce(OH)_4$, which is formed from $Ce(OH)_3$ by autoxidation. This reaction may be applied as a test for cerium.[462] Interference due to $Mn(OH)_2$, $Co(OH)_2$, $Tl(OH)_3$ and Ag_2O, which react in the same way, may be prevented by hydroxidic precipitation of the cerium by means of potassium cyanide solution (*Idn. Limit:* 0.03 γ Ce in 0.1 ml).

(*b*) Solutions of Ce^{IV} salts, containing much hydrochloric acid, form precipitates with ammonium naphthoate (black-blue), with ammonium anthranilate (dark red-brown), with ammonium salicylate (dark brown).[463] As spot reactions, the precipitates are formed from 10^{-3} to 10^{-4} N ceric solutions. Th, Pr, Nd, Zr salts do not form precipitates.

31. Indium and Gallium

(1) *Test for indium with alizarin or quinalizarin* [464]

Indium forms color lakes with polyhydroxyanthraquinones under appropriate conditions. These products are similar to the hydroxyanthraquinone lakes of aluminum, beryllium, and magnesium. The indium lake with alizarin is dark red, with quinalizarin a violet lake results. The lakes may be formed on paper impregnated with alizarin or quinalizarin.

This color reaction with hydroxyanthraquinones is not specific; numerous metals form similar lakes. However, under carefully controlled conditions, indium may be detected in the presence of Al, Zn, Ni, Co, Mn, Cs, Fe, provided the proportion of foreign metal is not too high (see page 214).

Procedure. Filter paper is impregnated with a saturated alcoholic solution of alizarin, or with a solution of quinalizarin in a mixture of pyridine and acetone. A drop of the test solution is placed on the dried paper. The test solution should be neutral or slightly acid with acetic acid. The moist fleck is held over ammonia and then immersed in a saturated aqueous solution of boric acid. The violet ammonium alizarinate or quinalizarinate decompose, and the red or violet indium lake is easy to see against the yellow or red paper.

Limit of Identification: 0.05 γ indium (in 0.025 ml)
Limit of Dilution: 1 : 500,000
Reagent: Saturated alcoholic solution of alizarin, or solution of 0.01 g quinalizarin in 2 ml pyridine and 20 ml acetone

Test for indium in the presence of elements of the ammonium sulfide group

Almost all the elements of the ammonium sulfide group give colored lakes with polyhydroxyanthraquinones and interfere with the direct test for indium. In a few cases the test may be carried out nevertheless. *In the presence of aluminum,* the interfering ion may be masked by formation of the complex $[AlF_6]^{-3}$ ion. A drop of the test solution is mixed with 3 or 4 drops of a saturated solution of sodium fluoride in a micro crucible. The test, as described, is then carried out using a drop of this mixture.

Limit of Identification: 1 γ indium } in the presence of 375 times the amount of aluminum
Limit of Dilution: 1 : 25,000 }

Probably Be, Zr, Th ions, which also interfere with the test, could be similarly masked, since they also form complex or insoluble fluorides.

To detect *indium in the presence of zinc, nickel, cobalt, manganese,* which also give colored lakes resistant to boric acid, these ions are converted to the corresponding unreactive complex cyanide ions. The paper impregnated with the dye is treated with a drop of 5 % potassium cyanide solution, followed by a drop of the test solution, and then another drop of the cyanide solution. After immersion in boric acid, the characteristic stain due to indium remains on the paper. The following sensitivities were obtained, using this procedure on a drop (0.025 ml):

Limit of Identification: 0.13 γ indium *Limit of Dilution:* 1 : 192,000	in the presence of 2200 times the amount of zinc
Limit of Identification: 0.06 γ indium *Limit of Dilution:* 1 : 400,000	in the presence of 5900 times the amount of nickel
Limit of Identification: 0.1 γ indium *Limit of Dilution:* 1 : 250,000	in the presence of 2400 times the amount of cobalt
Limit of Identification: 0.6 γ indium *Limit of Dilution:* 1 : 40,000	in the presence of 550 times the amount of manganese

To detect *indium in the presence of chromium,* 0.5 *N* sodium hydroxide is substituted for potassium cyanide in the foregoing procedure. The chromium is thus converted to chromite, which does not react.

Limit of Identification: 0.6 γ indium *Limit of Dilution:* 1 : 40,000	in the presence of 800 times the amount of chromium

To detect *indium in the presence of iron,* a drop of the test solution is placed in a micro crucible and treated drop by drop with a concentrated solution of sodium thiosulfate until no more violet color forms. A crystal of sodium sulfite and 6 to 8 drops of 5 % KCN solution are then added and the mixture warmed until the precipitate dissolves. The solution now containing the iron as potassium ferrocyanide may be used for the indium test by spotting the impregnated paper and subsequent treatment with boric acid.

Limit of Identification: 1 γ indium *Limit of Dilution:* 1 : 25,000	in the presence of 450 times the amount of iron

(2) Test for gallium with ferrocyanide and manganese^{II} salts [465]

Manganese^{II} ions form a brown insoluble ferricyanide. The same salt is precipitated when an acid manganese^{II} salt solution is mixed with an oxidizing agent (KNO_2, $KBrO_3$) followed by potassium ferrocyanide. The concentrations of the oxidizing agent and the manganese^{II} salt may be so selected that no brown precipitate of the manganese salt is formed. When a gallium salt is added to such a solution, a red-brown color or precipitate forms. The chemistry of this reaction has not been elucidated. Probably the oxidation potential of the ferricyanide is raised by removing the ferrocyanide from the ferricyanide-ferrocyanide equilibrium (see page 178) through formation of acid-insoluble gallium ferrocyanide.

Procedure. A drop of the test solution is mixed with 4 drops of a dilute solution of a manganous salt in hydrochloric acid. One drop of a mixture of potassium ferrocyanide and potassium bromate (10 drops of 0.1 *N* $KBrO_3$ to 20 ml of 2.5 % $K_4[Fe(CN)_6]$) is added. In the presence of gallium, a red-brown precipitate or turbidity forms.

Limit of Identification: 5 γ gallium (in 0.04 ml)

Limit of Dilution: 1 : 8000

Reagent: Acid manganous chloride solution: 0.5 % $MnCl_2 \cdot 4H_2O$ in 6 *N* HCl

The test is only specific in the absence of metals which react with ferrocyanide. If such are present, the gallium must be separated beforehand. For instance, extraction of $GaCl_3$ with ether from strongly acid solutions can be used. For details, the original paper should be consulted.

An excellent separation of aluminum and gallium is based on the finding that gallium alone is extracted by chloroform at pH = 2 from a mixture of the oxinates. This procedure which permits the separation of 1 part of gallium from 10,000 parts of aluminum,[466] may probably be useful in spot test analysis.

(3) Test with rhodamine B [467]

The tests described in this Chapter for the detection of antimony, gold, and thallium^{III} by means of rhodamine B are based on the use of this basic dye as a precipitant for the complex metal halogen acids: $HSbCl_6$, $HSbI_4$, $HAuCl_4$, $HTlCl_4$, $HTlBr_4$ or their anions, and the orange-red fluorescence of the benzene solutions of these rhodamine B salts. Gallium, which yields $[GaCl_4]^-$-ions in strong hydrochloric acid solution, can also be detected by means of rhodamine B.

Procedure. The test is conducted in a micro test tube. A drop of the hydrochloric acid test solution is treated with three drops of rhodamine B solution and the mixture is shaken with 3–5 drops of high purity benzene. If gallium is present, the benzene layer turns red or pink, and exhibits an orange-red fluorescence in ultraviolet light.

Limit of Identification: 0.5 γ gallium

Limit of Dilution: 1 : 100,000

Reagent: 0.2 % solution of rhodamine B in 1 : 1 hydrochloric acid

If metals, whose complex halogen acids react with rhodamine B are present (Sb, Au, Tl, platinum metals), use can be made of the fact that most of these metals are precipitated quantitatively from solutions of their salts by copper or brass wire, and consequently gallium can be detected without interference in the clear filtrate or centrifugate by means of rhodamine B. Thallium^{III} is reduced to Tl^{I} by copper (brass) and thallous salts do not react with rhodamine B.

C. Ammonium Carbonate Group

32. Barium

(1) Test with sodium rhodizonate [468]

The yellow aqueous solution of sodium rhodizonate (I) produces colored precipitates (II) with neutral solutions of the salts of bivalent heavy metals. [469]

$$\underset{\text{(I)}}{C_6O_4(ONa)_2} + MeCl_2 \longrightarrow \underset{\text{(II)}}{C_6O_4(O_2Me)} + 2\,NaCl$$

The precipitation capacity of sodium rhodizonate is particularly interesting within the group of the alkaline earth metals. Barium gives a red-brown precipitate of barium rhodizonate. Strontium salts react in the same way with rhodizonates, but calcium salts do not. Under suitable conditions, however, barium may be identified in the presence of strontium. The rhodizonate test is thus recommended only for the detection of barium within the ammonium carbonate (alkaline earth) group, or for its detection when bivalent heavy metals are known to be absent.

Procedure. A drop of the neutral or slightly acid test solution is placed on filter paper and then a drop of an aqueous 0.2 % solution of sodium rhodizonate. [470] According to the amount of barium present, a more or less intense red-brown stain is formed.

Limit of Identification: 0.25 γ barium
Limit of Dilution: 1 : 200,000

The paper may be impregnated with sodium rhodizonate. It must be dried *in vacuo* (over concentrated H_2SO_4) and stored in the dark. Otherwise, the finely divided rhodizonate is oxidatively decomposed. [471]

Test for barium in the presence of strontium

Strontium salts react similarly to barium salts with sodium rhodizonate. However, strontium rhodizonate is easily soluble in dilute hydrochloric acid in the cold, whereas the barium compound is converted into an insoluble bright red acid salt. It must be noted that small amounts of barium rhodizonate, prepared in a test tube in neutral solution, and then treated with hydrochloric acid, seem to dissolve, because the red barium salt is formed in a fine state of division and is hard to see. If, however, the reaction is carried out on paper, the brown-red stain of barium rhodizonate is turned to a vivid red by dilute hydrochloric acid (1 : 20). Probably the red barium salt is adsorbed on the paper so that even small amounts become plainly visible.

If no brown fleck is formed on treating a drop of the test solution with a drop of sodium rhodizonate on filter paper, the absence of both barium and strontium is indicated; a brown fleck may mean that either ion is present, or both. If the fleck disappears on treatment with dilute hydrochloric acid, only strontium is present; if it turns red, barium (or barium and strontium) is present.

Limit of Identification: 0.5 γ barium
Limit of Dilution: 1 : 100,000
} in the presence of 50 times the amount of strontium

The following is of interest with respect to the detection of barium in the presence of strontium [472]: Barium carbonate is more soluble than calcium or strontium carbonate. If it is thoroughly washed and then spot-tested directly with sodium rhodizonate, a brown-red precipitate appears. The other two carbonates do not react under these conditions. However, this differential behavior is only useful to distinguish between pure $BaCO_3$ and pure $SrCO_3(CaCO_3)$. It cannot be used indiscriminately to detect barium in a mixed carbonate precipitate, because barium rhodizonate is formed only if a very large excess of $BaCO_3$ is present. When a mixture of the carbonates is thrown down, the barium carbonate may be enveloped by $CaCO_3$ or $SrCO_3$, and thus shielded from the action of the sodium rhodizonate. In this case the following procedure can be used: The alkaline earth metals are precipitated as carbonates by the addition of ammonium carbonate to the solution obtained in the course of the usual qualitative scheme of analysis. To detect barium, or strontium, or both, in this precipitate, it suffices to place about 1 mg of the thoroughly washed precipitate in the depression of a spot plate, add 1 drop of 0.2 % solution of sodium rhodizonate and stir. If a positive reaction is obtained, barium is present. If no reaction is observed, a drop of 1 *N* acetic acid is added. A red precipitate or coloration will indicate the presence of barium, strontium, or both.

If neutral or weakly acid solutions are to be examined, the following procedure may be used for the detection of barium and strontium in a micro drop [473]: One drop of 0.5 % sodium rhodizonate solution is placed on Whatman filter paper No. 1. After the liquid has soaked in, the center of the yellow spot is treated with a drop of the test solution. A brownish or reddish-brown stain appears at once if barium or strontium is present. The fleck is then spotted with 1 or 2 drops of a mixture of equal parts of a

References pp. 247–257

saturated water solution of dimethylamine hydrochloride and 95 % ethanol. If barium is present, the color of the stain changes to bright red, whereas the color becomes gradually a distinctive violet-blue if the red-brown stain is due to strontium. If both barium and strontium are present, they may be detected side by side as follows: After treating the fleck with the dimethylamine hydrochloride-alcohol mixture, the resulting bright red spot of barium rhodizonate will be encircled by a violet-blue ring of the strontium compound, which migrates along with the reagent to the periphery of the fleck.

(2) *Test by induced precipitation of lead sulfate* [474]

Lead sulfate is readily soluble in an acetic acid solution of ammonium acetate; complex lead acetates are formed. If such a solution, which contains free sulfate ions, is mixed with a barium salt, then not only barium sulfate is precipitated, but also a considerable amount of lead sulfate. Mixed crystals or an addition compound of lead sulfate and barium sulfate probably results. This induced precipitation [475] may be utilized to increase the sensitivity of the barium sulfate reaction. Solutions that are so dilute with respect to barium as to give no visible precipitation with sulfuric acid, yield a definite cloudiness on addition of an acetate solution of lead sulfate.

Strontium and calcium sulfates behave similarly. Consequently, a positive reaction indicates merely the presence of alkaline earth cations.

Procedure. A drop of the neutral or slightly acid test solution is treated on a watch glass or in a micro test tube with a drop of the reagent solution. In the presence of barium, a more or less dense cloudiness or white precipitate is formed. Observation against a black background is advisable.

Limit of Identification: 0.4 γ barium

Limit of Dilution: 1 : 125,000

Reagent: 2 ml of 10 % lead acetate solution is mixed with 2 ml of 2 *N* sulfuric acid; the lead sulfate is dissolved by addition of solid ammonium acetate. The reagent keeps.

(3) *Test by precipitation of barium sulfate in the presence of permanganate* [476]

Barium sulfate is not affected by exposure to a solution of potassium permanganate. If, however, it is formed in the presence of this salt, the precipitate contains potassium permanganate, included in the lattice, and so acquires a distinct violet color.[477] The permanganate retained by barium sulfate is quite resistant to reducing agents that normally attack it at once (H_2O_2, Fe^{II} salts, SO_2, oxalic acid, etc.]. The retardation of the redox activity obviously is due to the absence of free, mobile, hydrated MnO_4^- ions. This resistance can be used under the conditions given here for the detection of barium (or sulfate, see page 314).

Lead sulfate behaves like barium sulfate in this respect. Strontium sulfate is only slightly colored by permanganate. Accordingly, it is possible to detect as little as 5 γ barium in the presence of 2500 γ strontium, if comparison tests with pure strontium salts are made. Calcium sulfate does not take up permanganate.

Procedure. Sulfate paper can be prepared by soaking strips of filter paper in sodium sulfate solution and drying. Three drops of the solution to be tested for barium are mixed with a drop of saturated permanganate solution on a spot plate or in a micro crucible. If preferred, a drop of the test solution may be treated with about one third its volume of permanganate solution. A drop of this violet mixture is placed on dry sodium sulfate paper and kept for 7 to 10 minutes at 70° to 80° C in a drying oven. The original violet color disappears. because the cellulose reduces the permanganate; the paper becomes uniformly brown by deposition of manganese dioxide. The dried paper is then soaked in a solution of sulfurous acid until the MnO_2 has disappeared and the parts of the paper which were not spotted become white. The complete reduction and solution of the brown oxide and the excess permanganate require about 1 or 2 minutes. The areas of the paper in which barium sulfate was formed with simultaneous adsorption of permanganate then show up on the perfectly white paper as violet flecks or rings, according to the quantity of barium present.

Limit of Identification: 5 γ barium
Limit of Dilution: 1 : 10,000

The test may also be made in a small centrifuge tube. A drop of the test solution is treated with 3 drops of cold, saturated permanganate solution, several drops of dilute sulfuric acid are added, and the color then discharged by dropwise addition of sulfurous acid. The violet tinted barium sulfate precipitate is brought into the tip of the tube by centrifuging. Its color can be distinctly seen against a white background, using a magnifying glass if necessary.

Limit of Identification: 2.5 γ barium
Limit of Dilution: 1 : 20,000
Reagent: Sodium sulfate paper, prepared with 0.5 *N* Na_2SO_4

(4) Other tests for barium

(*a*) Tetrahydroxyquinone can be used instead of rhodizonic acid, whose action it resembles.[478] A small quantity (tip of knife blade) of the solid reagent, together with potassium chloride, is placed in a depression of a spot plate. One drop of the test solution is added, followed by a drop of water. If barium is present a brown color appears, at first in the vicinity of the undissolved reagent. It changes to red on the addition of 0.1 *N* HCl (*Idn. Limit:* 5 γ Ba).

(*b*) Nitro-3-hydroxybenzoic acid is proposed as reagent for detecting barium.[479] To carry out the test, place a drop of a 1 % solution of the

References pp. 247–257

reagent in alcohol or acetone on a filter paper. After the solvent evaporates, add a drop of the solution to be tested and expose the paper to ammonia vapor. Heat the paper carefully over a hot plate to drive off the colored ammonium salt of the reagent. A bright red ring indicates the presence of barium (*Idn. Limit:* 1.5 γ Ba). Calcium, strontium, cobalt, zinc, manganese and some other ions sometimes form a yellow spot. Magnesium and lead increase the sensitivity.

33. Strontium

(*1*) *Test with sodium rhodizonate* [480]

Among the alkaline earth metals, only barium and strontium salts react in neutral solution with sodium rhodizonate (see page 216); they form brown-red precipitates. To detect strontium in the presence of barium, the latter must therefore be converted into some compound which does not react with sodium rhodizonate. Insoluble barium chromate is suitable, since strontium chromate is sufficiently soluble in water to react normally with sodium rhodizonate.

Procedure. A drop of the test solution is placed on filter paper impregnated with potassium chromate; the chromates are formed. After a minute, a drop of an aqueous 0.2 % solution of sodium rhodizonate is placed on the spot. A brown-red fleck or circle indicates the presence of strontium.

Limit of Identification: 3.9 γ strontium } in the presence of 80 times the
Limit of Dilution: 1 : 12,800 } amount of barium

Reagent: Potassium chromate paper. Filter paper impregnated with a saturated solution of potassium chromate, and dried.

If an aqueous solution of strontium is used (instead of the chromate), as little as 0.45 γ of strontium can be detected by this spot test, in the absence of other metals reacting with sodium rhodizonate.

34. Calcium

(*1*) *Test with ammonium ferrocyanide*

Calcium salts precipitate crystalline, white calcium ammonium ferrocyanide from ammoniacal, neutral or acetic acid solutions, when treated with ammonium or potassium ferrocyanide in the presence of ammonium salts *:

$$Ca^{+2} + 2\,NH_4^+ + Fe(CN)_6^{-4} = Ca(NH_4)_2Fe(CN)_6$$

* Potassium ferrocyanide and ammonium salts produce calcium potassium ammonium ferrocyanide of varying composition. Compare F. Flanders, *J. Am. Chem. Soc.*, 28 (1906) 1509.

Unlike this mixed ferrocyanide, the pure alkali or alkaline earth ferrocyanides are soluble.

Magnesium behaves analogously to calcium, but strontium is not precipitated from concentrated solutions. Thus the ferrocyanide reaction can be used to test for calcium in the alkaline earth group.

Procedure. A drop of the test solution is stirred on a watch glass with a few drops of a concentrated solution of ammonium ferrocyanide, then a drop of alcohol is added and stirred. A crystalline precipitate or cloudiness indicates the presence of calcium. The watch glass is best placed on glazed black paper, or a black spot plate may be used.

Limit of Identification: 25 γ calcium *

Limit of Dilution: 1 : 2000

(2) Test with the osazone of dihydroxytartaric acid [481]

Aqueous solutions of the sodium salt of the osazone of dihydroxytartaric acid (I) give bright yellow flocculent precipitates (II) with the alkaline earth metals:

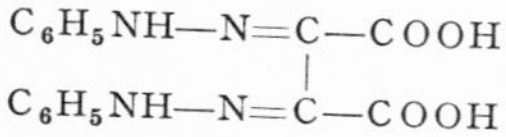

$$\begin{array}{l} C_6H_5NH{-}N{=}C{-}COOH \\ \qquad\qquad\quad | \\ C_6H_5NH{-}N{=}C{-}COOH \end{array} \quad \text{(I)} \qquad \begin{array}{l} C_6H_5NH{-}N{=}C{-}COO \\ \qquad\qquad\quad | \\ C_6H_5NH{-}N{=}C{-}COO \end{array}\!\!>Me \quad \text{(II)}$$

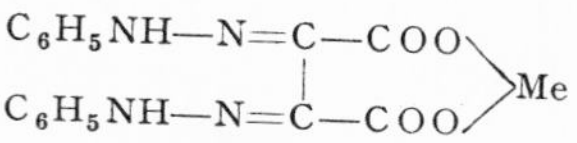

(Me = Ca, Ba, Sr)

Of these, the calcium salt is notable for its extremely slight solubility in water. This property is utilized to test for calcium in very low concentrations. The reagent is not specific and other metals, with the exception of alkali and ammonium salts, should not be present. Magnesium does not interfere provided its quantity is not more than 10 times the amount of calcium present. In higher concentrations, the magnesium salt is precipitated in the cold; on heating, it is relatively soluble. At the same time, for reasons that are not known, the precipitation of the normally insoluble calcium compound is completely prevented.

If magnesium salts are absent, the use of the *solid* reagent provides a very sensitive test for calcium. The reagent gives a perfectly clear solution in distilled water, but with traces of calcium a precipitate remains, which is not formed when the same amount of dissolved reagent is employed. The difference is due to the fact that while the solid reagent is dissolving, the calcium salt separates at the place where solution is occurring; the precipitate coats the solid reagent and protects it from solution. This protective layer

* If the ferrocyanide-calcium test is carried out in a micro test tube, the reaction succeeds down to dilutions of 1 : 1,000,000; see F. Feigl and F. Pavelka, *Mikrochemie*, 2 (1924) 85. Small traces of cloudiness are always difficult to see in single drops, because the layer observed is too thin.

effect [482] makes it possible to distinguish not only between tap and distilled water, but even after 1 part of tap water has been mixed with 30 parts of distilled water.

Procedure. A drop of the neutral test solution is treated with several grains of the reagent in the depression of a black spot plate or on a black watch glass. In the absence of calcium, the reagent dissolves completely. When calcium is present, a fine skin covers the drop of the test solution and, after a short while, a more or less dense precipitate is formed, according to the quantity of calcium present. A blank test with distilled water should be carried out when very small amounts of calcium are suspected.

Limit of Identification: 0.01 γ calcium
Limit of Dilution: 1 : 5,000,000
Reagent: Sodium salt of dihydroxytartaric acid osazone [483] (solid)

(3) *Test with sodium rhodizonate and alkali hydroxide* [484]

Neutral solutions of calcium salts, which do not react with sodium rhodizonate (in contrast to barium and strontium salts), immediately form a violet precipitate with alkaline solutions of this reagent. The reaction occurs also if a neutral solution is treated with the rhodizonate and then made basic with ammonia, caustic alkali, alkali carbonate or ammonium carbonate. In view of this mode of formation, the reaction may be due to the production of a basic calcium rhodizonate, whose structure may be

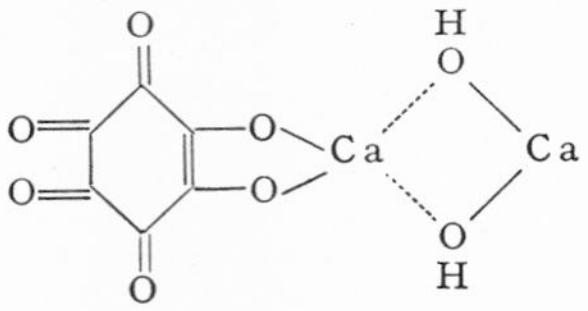

The solubility of calcium sulfate is high enough to allow reaction with an alkaline solution of sodium rhodizonate. In contrast, the far less soluble calcium carbonate, oxalate, phosphate, and fluoride do not react with this reagent.

The following procedure is recommended provided barium, strontium, and lead salts are absent.

Procedure. The test is conducted in a depression of a spot plate or on filter paper (S & S No. 601).* One drop of the neutral or weakly acid test solution is treated with a drop of freshly prepared 0.2 % sodium rhodizonate solution and one drop of 0.5 N sodium hydroxide. The solution is mixed by blowing briefly

* When this type of filter paper is used, the slow absorption of the liquid causes the reaction to occur mostly outside and not within the capillaries of the paper. After the liquid is soaked up, the precipitate remains on the surface and is quite visible against the white background.

through a pipette. A comparison test should be run on a drop of water. The presence of calcium is shown by the production of a violet precipitate.

Limit of Identification: 1 γ calcium

Limit of Dilution: 1 : 50,000

Test for calcium in the presence of barium and strontium

The following compilation shows the behavior of the alkaline earth metal sulfates toward neutral and basic solutions of sodium rhodizonate:

	Na-rhodizonate	Na-rhodizonate + NaOH
$BaSO_4$	—	—
$SrSO_4$	+	—
$CaSO_4$	—	+

These color reactions (violet) are applied in the test for strontium in the presence of barium and calcium, and also in the detection of calcium in the presence of barium and strontium.

If in the course of the qualitative scheme it is desired to detect calcium in the presence of barium and strontium, a slight quantity of the mixed carbonate precipitate can be converted into the corresponding sulfate mixture. Alkaline sodium rhodizonate can then be applied for the detection of $CaSO_4$ in the presence of $SrSO_4$ and $BaSO_4$.

Procedure. The test is conducted in a micro crucible. One drop of the test solution (prepared by dissolving the carbonate precipitate in dilute hydrochloric acid) is taken to dryness and about 0.5 g of solid ammonium sulfate is added to the residue. The contents of the crucible are heated gradually and finally brought to a temperature sufficient to drive off the ammonium sulfate completely (no more white fumes). The cold mass is treated with a drop of sodium rhodizonate solution and one drop of dilute alkali (see above). Depending on the quantity of calcium present, the white precipitate is tinted more or less violet.

This procedure will reveal 5 γ calcium in the presence of 20,000 γ barium.

Strontium sulfate, in mixture with barium sulfate and/or calcium sulfate, can be detected by this procedure if a neutral solution of sodium rhodizonate is employed. A violet color appears.

(*4*) *Test glyoxal-bis (2- hydroxyanil)*, see Addendum, p. 534.

D. Alkali Metals, Ammonia, and Derivatives of Ammonia

35. Magnesium

(*1*) *Test with alkali hypoiodite* [485]

Washed freshly precipitated magnesium hydroxide is tinted deep brown-red by a solution of iodine in potassium iodide. The color fades on digestion with potassium iodide, alcohol, potassium hydroxide or other solvents for iodine. It is also discharged by treatment with sulfite or thiosulfate. Likewise, when magnesium hydroxide is formed in the presence of free iodine, the precipitate is brown-red. It is obviously an adsorption compound of magnesium hydroxide and iodine. The most suitable conditions for the formation of this colored product (even in dilute solution) are attained when an iodine solution is decolorized with sodium or potassium hydroxide immediately before the addition of the magnesium salt. In the hypoiodite solution, the following equilibrium prevails:

$$I_2 + 2\,OH^- \rightleftharpoons I^- + IO^- + H_2O$$

The precipitation of magnesium as magnesium hydroxide removes the hydroxyl ions, and thus the iodine required for the adsorption is produced. The use of a freshly prepared hypoiodite solution is necessary because, on long standing, the reaction $3\,IO^- = IO_3^- + 2\,I^-$ occurs, and the concentration of the iodine-producing hypoiodite ions is thus decreased. Taking also into account the fact that an excess of hydroxyl ions is necessary for complete precipitation of magnesium hydroxide, the following procedure can be recommended for the detection of magnesium.

The hypoiodite reaction is quite selective for magnesium. Considerable quantities of reducing agents, ammonium and aluminum salts, and such ions which form colored hydroxides or higher oxides interfere with the test. Phosphate and oxalate ions interfere by yielding crystalline magnesium phosphate and oxalate. In contrast to the surface-rich hydroxide, these products are not capable of adsorbing free iodine.

Procedure. A drop of the neutral or acid test solution is placed on a spot plate along with a microdrop of 1 *N* potassium hydroxide or saturated lime water [486] followed by a microdrop of iodine solution; the mixture is stirred with a glass rod. The solution should be definitely brown; if necessary more iodine should be added. After a minute, the alkali content is increased until the solution becomes lemon yellow. In the presence of magnesium, the brown flecks of the adsorption compound made up of magnesium hydroxide and iodine are clearly visible in the yellow solution. A blank is necessary only when very small amounts of magnesium are suspected.

Limit of Identification: 0.3 γ magnesium
Limit of Dilution: 1 : 165,000
Reagent: 1 *N* iodine in 20 % potassium iodide solution

(2) *Test with quinalizarin* [487]

Magnesium salts give a blue precipitate or cornflower blue coloration, according to the concentration, with alkaline solutions of quinalizarin (1,2,5,8-tetrahydroxyanthraquinone) (see page 192). The colored product is a magnesium-quinalizarin lake, i.e., an adsorption compound of magnesium hydroxide with the dyestuff (compare Al-alizarin lake, page 185). The lake is present in the blue precipitate as a gel, in the blue solution as a hydrosol; on long standing, the latter coagulates.

The test for magnesium with quinalizarin is not interfered with by alkaline earth metals, or (in sufficient concentration of alkali) by aluminum. If large amounts of other metals which are precipitated by alkalis are present, it is best to carry out the magnesium test after the usual ammonium carbonate separation.

Large amounts of ammonium salts decrease the sensitivity of the test; likewise phosphate ions.

Lanthanum and beryllium salts react with quinalizarin in a similar manner, although the beryllium quinalizarin compound is differentiated from the magnesium compound by its stability towards oxidizing agents (see page 194). It must be noted that Nd, Pr, Ce, La, Zr and Th salts react with quinalizarin in the same way as magnesium salts.[488]

Procedure. A drop of the test solution and a drop of distilled water are placed in adjoining depressions of a spot plate and mixed with 2 drops of an alcoholic 0.01–0.02 % solution of quinalizarin. If the solution is acid, it will be colored yellow-red by the reagent. 2 *N* sodium hydroxide is added drop by drop until a change to violet occurs, and then an excess of about $\frac{1}{4}$ to $\frac{1}{2}$ of the volume then present. According to the amount of magnesium present, a blue precipitate or coloration appears; the blank remains blue-violet. The difference in shade is intensified on long standing, because the dyestuff is gradually decomposed in magnesium-free solutions (oxidation), whereas the colored magnesium compound is stable.

Limit of Identification: 0.25 γ magnesium *

Limit of Dilution: 1 : 200,000

(3) *Test with* p-*nitrobenzeneazo-α-naphthol* [489]

Magnesium hydroxide exhibits a remarkable adsorptive property for certain dyestuffs. This adsorption is to a certain degree the organic counterpart of the adsorption of elementary iodine on magnesium hydroxide (see test (2). For certain dyestuffs of the azo and anthraquinone series, there is a difference in color between the adsorption product (color lake) and the non-adsorbed dyestuff. Since the adsorption is instantaneous, it affords an excellent test for magnesium.

* By means of even more refined technique, F. L. Hahn, *Mikrochemie* (*Pregl Festschrift*), 1929, p. 133, has been able to identify as little as 0.001 γ in a microdrop.

Two compounds especially useful for this test are *p*-nitrobenzeneazoresorcinol (I) [490] and *p*-nitrobenzeneazo-α-naphthol (II):

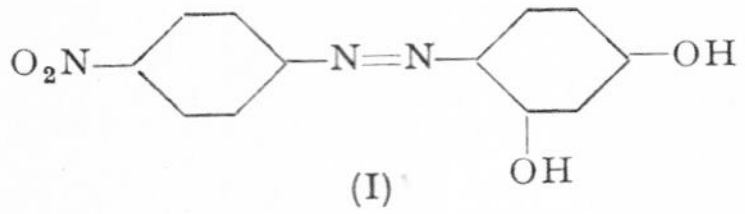

$$O_2N-C_6H_4-N{=}N-C_6H_3(OH)-OH \quad \text{(I)}$$

$$O_2N-C_6H_4-N{=}N-C_{10}H_6-OH \quad \text{(II)}$$

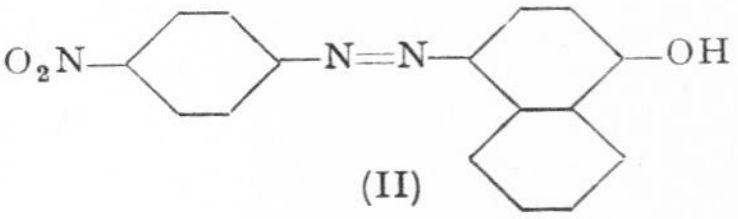

These azo dyes are red to red-violet in alkaline solution; when taken up by magnesium hydroxide deep blue products result. Doubtless this coloration of magnesium hydroxide by (I) and (II) is due to the adsorption of their alkali salts. The pure dyestuffs, dissolved in chloroform or alcohol, are adsorbed by magnesium hydroxide without change in color, whereas the color changes immediately when alkali is added. The sensitivity of the test obviously depends on the hydroxyl ion concentration, for if the dyestuff solution is made up with ammonia or other weak base instead of sodium hydroxide, it gives a much smaller effect.*

It is remarkable that the violet alkaline solution of *p*-nitrobenzeneazo-α-naphthol is taken up with a blue color not only by magnesium hydroxide, but also by pure filter paper quite free from magnesium.** This is additional evidence that the color change is due to adsorption and not to compound formation. Obviously this spot test should be carried out not on paper but on porcelain.

The magnesium test is best conducted in such manner so that the precipitation and tinting of the hydroxide are simultaneous, namely by adding an alkaline solution of the dyes to the test solution. A blue precipitate is immediately formed with as little as 0.25 mg magnesium. With smaller amounts only a blue coloration is seen, owing to the colloidal dispersion of the tinted magnesium hydroxide particles. The intensity of the blue color is dependent on the magnesium content.

The hydroxides of nickel, cobalt, and cadmium are also colored by *p*-nitrobenzeneazoresorcinol; the hydroxide and carbonate of calcium are colored by *p*-nitrobenzeneazo-α-naphthol. Special conditions are necessary to detect magnesium in the presence of these ions (see below).

Procedure. A drop of the test solution is treated on a spot plate with 1 or 2 drops of the alkaline dyestuff solution. According to the magnesium content, a blue precipitate is formed, or there is a color change from red-violet to blue.

* L. Kul'berg, *Chem. Abstracts*, 33 (1939) 3781, has advanced a different view as to the mechanism of the color change.

** Kolthoff (see p. 228) has observed the same effect with Titan yellow, and states that filter paper (S & S 589) impregnated with 0.01 % solution of the dyestuff containing 0.05 % $MgCl_2$, still gives a red ring on treating with 0.005 *N* NaOH.

It is always necessary to be sure that sufficient alkali is present; if the solution is too acid, the yellow of the free dyestuff appears, and a few drops of strong alkali should be added. If only traces of magnesium are to be detected, a comparative test on distilled water is necessary.

This reaction will reveal *magnesium in tap water (drop)*. The *magnesium in the ash of qualitative filter paper*, after treatment with dilute acid, also responds positively.

Sensitivities

	p-Nitrobenzene-azoresorcinol	*p*-Nitrobenzene-azo-α-naphthol
Limit of Identification:	0.5 γ magnesium *	0.19 γ magnesium
Limit of Dilution:	1 : 100,000	1 : 260,000

Reagents: The reagent solutions of both dyes are made up by dissolving 0.001 g dye [491] in 100 ml 2 *N* alkali

Test for magnesium in the presence of other metals

When carrying out the test for magnesium in the presence of other metal ions, interference can arise from the formation of hydroxides, which either consume the dyestuff with color formation, or are themselves colored. Aluminates, zincates, stannates, etc. can prevent the adsorption of the dyestuff on the magnesia, more or less, in favor of adsorption of the aluminate, zincate, etc. Finally, the sensitivity of the test is decreased when large amounts of other hydroxides are formed at the same time, even though they are colorless and take up no dyestuff. Their coprecipitation in the alkaline solution renders it difficult to see the color change. Alkaline earth hydroxides, bismuth hydroxide, etc. are cases in point.

The interference by those metal ions which form complex cyanide ions may be greatly decreased by the addition of a large excess of potassium cyanide to the test drop on the spot plate. Then 1.3 γ magnesium may be detected in the presence of:

1000 times the amount of *nickel, cadmium, copper*
200 ,, ,, ,, ,, *cobalt*
600 ,, ,, ,, ,, *zinc*

Lead, which is brought into solution as plumbite with sodium hydroxide, affects the test very little.

Iron^{III}, *chromium*^{III}, *aluminum*, and *tin* interfere seriously. The following procedure must be used if they are present:

* J. Stone, *Science*, 72 (1930) 322, detects 5 γ magnesium as follows: filter paper is soaked in an 0.01 % alcoholic solution of the azo dyestuff, dried, and a drop of the acid test solution is placed on the paper. After drying, the paper is dipped in 1 % sodium hydroxide solution. A blue stain of the magnesium complex shows up against a red background. The reagent paper will keep for months in dark bottles.

Procedure. Two drops of the test solution along with a few small crystals of sodium nitrite in a small test tube are heated in boiling water. Iron, aluminum, chromium, and tin are thus precipitated as hydroxides. The mixture is then centrifuged and the supernatant liquid transferred to a spot plate. The rest of the test is carried out as described.

In this way 13 γ *magnesium* may be detected *in the presence of 1000 times the amount of iron*^{III}, *chromium*^{III}, *tin*^{II}, *aluminum.*

In spite of their deep color, chromates affect the magnesium test only slightly.

To detect *magnesium in the presence of manganese,* 2 drops of the test solution are treated with colorless ammonium sulfide and, without heating, the suspension is centrifuged. The supernatant liquid is transferred to a spot plate by means of a capillary siphon and treated with the dyestuff solution. The latter is reduced within a few minutes by the excess ammonium sulfide to form a red aminoazo dye. It is therefore essential to carry out a blank test.

Proportion Limit: 13 γ magnesium in the presence of 500 times the amount of manganese

In testing for magnesium among the alkaline earth metals, it must be remembered that calcium hydroxide is colored blue by *p*-nitrobenzeneazo-α-naphthol in the same way as magnesium hydroxide. Because of the larger solubility and lesser coloration of the calcium hydroxide, 0.13 γ *magnesium* may still be identified *in the presence of 5000 times the amount of calcium* if the following procedure is used:

Procedure. A drop of the test solution, on a spot plate, is barely acidified and then treated with 1 or 2 drops of a slightly alkaline solution of the dyestuff. Dilute alkali is added drop by drop until a slight precipitate of calcium hydroxide is seen. If the solution appears blue compared with a blank test conducted under exactly similar conditions on a calcium solution, magnesium is present. The calcium hydroxide precipitate is tinted blue only if a large excess of alkali is present.

Magnesium can be detected in the presence of strontium and barium because their hydroxides are not colored by the dye. A large excess of alkali should be avoided since too much precipitate tends to mask the magnesium.

(4) Other tests for magnesium

(*a*) Alkaline yellow-brown solutions of Titan yellow are turned to bright flame red [492] by $Mg(OH)_2$ (*Idn. Limit:* 1.5 γ Mg).

(*b*) Volatile acids and ammonium salts are removed by evaporation of the test solution with sulfuric acid and heating. A drop (0.005 ml) of the

References pp. 247–257

aqueous solution of the residue and a drop of ammonia are placed on phenolphthalein paper and dried. The paper turns pink on moistening with water (*Idn. Limit:* 6 γ Mg).[493]

(*c*) Behaviors analogous to that described in test (*3*) are shown by ten other monoazo dyes. The color differences between the alkaline solutions of the dye and the hydrosol of $Mg(OH)_2$-dye lake are readily seen if the test is made on a spot plate (*Idn. Limit:* 0.1 γ Mg).[494] The reagent solutions plus 10 ml alcohol contain 10 mg dye dissolved in 10 ml 15 % sodium hydroxide and diluted to 100 ml.

36. Sodium

(*1*) *Test with zinc uranyl acetate*[495]

Sodium is precipitated from neutral or acetic acid solution by an acetic acid solution of zinc uranyl acetate. Yellow, crystalline sodium zinc uranyl acetate, $NaZn(UO_2)_3(CH_3COO)_9 \cdot 9\,H_2O$ precipitates. The reaction is specific for sodium. Salts of Cu, Hg, Cd, Co, Ni, Al, Mn, Zn, Ca, Sr, Ba, Mg, and NH_4 affect the sensitivity only in concentrations greater than 5 g per liter. Lithium salts are precipitated from solutions containing 1 g per liter and potassium salts from solutions exceeding 5 g per liter.

The sensitivity of the test is increased in dilute alcoholic solution; sodium may then be detected in potassium salts.

Procedure I. The test solution should be made as neutral as possible (if necessary by preliminary treatment with ammonia or zinc oxide). A drop of the clear solution is placed on a dark spot plate or black watch glass and stirred with 8 drops of the reagent solution. The formation of a yellow turbidity or precipitate indicates the presence of sodium.

Limit of Identification: 12.5 γ sodium
Limit of Dilution: 1 : 4000

Considerably smaller amounts may be detected by utilizing the fact that solid sodium zinc uranyl acetate gives a bright greenish-yellow fluorescence in ultraviolet light, whereas the reagent solution does not fluoresce.[496]

Procedure II. A drop of the test solution is placed, by means of a capillary pipet, on a strip of filter paper (S & S 601). (The paper should previously have been washed twice with dilute hydrochloric acid and 4 to 6 times with distilled water and dried. This treatment removes all sodium salts.) After the drop of test solution has soaked in, the paper is carefully dried over a microflame. One drop of reagent solution is added to the center and another from the side of the treated portion. The fleck is examined under the quartz lamp while still moist. In the presence of larger amounts of sodium, a green-yellow fluorescent fleck is seen at once; amounts smaller than 10 γ require 1 to 4 minutes to develop a similar fluorescing border.

References pp. 247–257

Limit of Identification: 2.5 γ sodium
Limit of Dilution: 1 : 20,000
Reagent: Solution of zinc uranyl acetate: (*a*) 10 g of uranyl acetate is dissolved by warming in 6 g of 30 % acetic acid, and diluted with water to 50 ml. (*b*) 30 g zinc acetate is stirred up with 3 g 30 % acetic acid, and diluted with water to 50 ml. Warm suspensions (*a*) and (*b*) are mixed and give a clear solution. A trace of sodium chloride is added, and after 24 hours the slight precipitate of sodium zinc uranyl acetate is filtered off and discarded. The reagent keeps.

37. Potassium

(*1*) *Test with sodium cobaltinitrite and silver nitrate* [497]

Neutral or slightly acid solutions of potassium salts give a yellow crystalline precipitate with sodium cobaltinitrite, $Na_3[Co(NO_2)_6]$:

$$3\ Na^+ + Co(NO_2)_6^{-3} + 2\ K^+ \rightarrow K_2Na[Co(NO_2)_6] + 2\ Na^+$$

The dipotassium salt is the chief product, though, according to the precipitation conditions, varying amounts of the mono- and tripotassium salts are also formed.[498]

The precipitation of potassium salts by means of sodium cobaltinitrite is appreciably more sensitive if silver salts are present; $K_2Ag[Co(NO_2)_6]$ is formed in this case. The increase in molecular weight by the substitution of silver for sodium makes the precipitate less soluble than the corresponding sodium compound. Obviously, the test solution must be halide-free.

Lithium, thallium, and ammonium salts must not be present; they also give crystalline precipitates with sodium cobaltinitrite. Ammonium salts may be removed by evaporation and ignition.

Procedure. A drop of the neutral or acetic acid test solution is placed on a black spot plate and treated with a drop of 0.05 % silver nitrate solution and a little solid sodium cobaltinitrite. A precipitate or turbidity indicates the presence of potassium.

Limit of Identification: 4 γ potassium (without addition of silver nitrate)
Limit of Dilution: 1 : 12,500
Limit of Identification: 1 γ potassium (with addition of silver nitrate)
Limit of Dilution: 1 : 50,000

(*2*) *Test with dipicrylamine* (*hexanitrodiphenylamine*) [499]

In contrast to its parent substance, (weakly basic colorless diphenylamine), hexanitrodiphenylamine (I) is a fairly strong acid. This is shown by its ability to form stable salts with strong bases. Accordingly, the bright yellow dipicrylamine dissolves in alkali hydroxide or carbonate to produce orange-red solutions. This deepening of color is due to a change in structure.

Hydroxyl ions convert the baso-form of dipicrylamine (I) into the aci-form (II), in which a continuous chain of conjugated double bonds effects the deepening of color. H^+ ions convert the orange-red aci-form to the bright yellow water-insoluble baso-form:

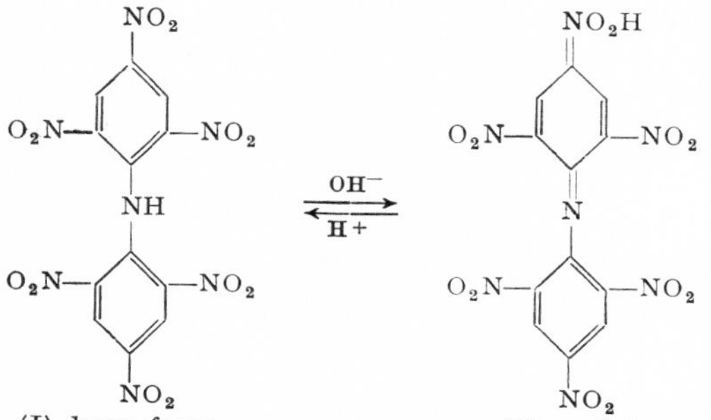

NO_2 / O_2N— —NO_2 / NH / O_2N— —NO_2 / NO_2 $\underset{H^+}{\overset{OH^-}{\rightleftarrows}}$ NO_2H / O_2N— —NO_2 / N / O_2N— —NO_2 / NO_2

(I) baso-form (II) aci-form

A solution of the sodium salt of dipicrylamine reacts with rubidium, cesium, and thallium^I, whose bases are strong. Red crystalline precipitates of the corresponding dipicrylaminates are formed. After they have once been precipitated, these products are distinctly resistant to dilute acids even though the same dipicrylaminates cannot be precipitated from acid solutions. This apparent anomaly is due likewise to the difference in structure of dipicrylamine in its two forms.

Ammonium dipicrylaminate is precipitated by considerable quantities of ammonium salts, but the latter, as well as all Tl^I halides can be removed by igniting the material to be tested. Other metal ions show no reaction, or they are precipitated as hydroxides or basic salts by the sodium dipicrylamine solution, which acts as an alkali because it is hydrolyzed in part. In such instances, dipicrylamine is precipitated in its water-insoluble, non-reactive baso-form (I). Consequently, considerable quantities of metal ions, which of themselves are unable to produce colored dipicrylaminates, can nevertheless impair the test for potassium by lowering the concentration of the aci-form (II), which is necessary for the precipitation of potassium. Therefore, if potassium is to be detected in the presence of considerable amounts of metal ions, which can be precipitated as hydroxides, it is better to warm the test solution beforehand with mercuric oxide to precipitate these ions as hydroxides. The filtrate from this preliminary step, or parts of it, is evaporated, ignited, and the ignition residue then tested for potassium.

The *identification limit* of this test is only 3 γ potassium, which consequently can be detected directly in the presence of sodium only if not more than 10 times the quantity of the latter is present. Otherwise, sodium dipicrylaminate is salted out. If potassium is to be detected in the presence of greater amounts of sodium, the latter should be removed beforehand as sodium zinc uranyl acetate (see page 229). The filtrate is evaporated to dryness, ignited, and the residue subjected to the test for potassium.

Procedure. Filter paper impregnated with sodium dipicrylaminate is spotted with a drop of the neutral test solution, or several grains of the ignition residue are placed on the paper. In the latter case, the material is then moistened with a drop of water. The paper is dried in a current of heated air, and placed in 0.1 *N* nitric acid. If potassium is absent, or present in quantities less than the identification limit, the original orange-red paper turns bright yellow (color of free dipicrylamine). If sufficient potassium is present to form its dipicrylaminate, a red fleck or ring remains at the site of the spot.

Limit of Identification: 3 γ potassium

Limit of Dilution: 1 : 16,000

Reagent: 0.2 g dipicrylamine [500] is dissolved in 2 ml of 2 *N* sodium carbonate and 15 ml water. The solution is filtered if necessary. Strips of filter paper (3 × 6 cm) are soaked in the solution, the excess liquid drained off, and the paper dried in a blast of heated air. If the paper moistened with the reagent solution is dried in the air, it is not so effective, though uniformly impregnated, as paper that has been spread on glass and then dried in a blast of heated air. The latter treatment results in an accumulation of the reagent on that side of the paper struck by the warm air. The reagent paper should be freshly prepared.

(3) *Test with sodium tetraphenyl boron* [501]

The water-soluble sodium tetraphenyl boron $Na[B(C_6H_5)_4]$ is a sensitive precipitant (Kalignost) for potassium in neutral or acetic acid solution. It is frequently recommended for the quantitative determination of this element.[502] Actually, $K[B(C_6H_5)_4]$ is less soluble than $KClO_4$. Within the alkali metal group, cesium, rubidium, and ammonium ions react in analogous fashion with this reagent. Ammonium salts can be removed by warming with dilute caustic soda. Heavy metal ions interfere with this test for potassium since they too are precipitable in many instances by this reagent. The procedure given here is particularly valuable if potassium is to be detected in the presence of much sodium.

Procedure. A black spot plate is used. A drop of the neutral or weak acetic acid test solution is treated with a drop of a 2 % water solution of sodium tetraphenyl boron. A white precipitate appears if potassium is present, or a turbidity is formed if the quantity is small. A blank test is advisable if slight quantities are suspected.

Limit of Identification: 1 γ potassium

Limit of Dilution: 1 : 50,000

38. Lithium

(1) *Test with alkali iron^{III} periodate* [503]

Periodates give precipitates with solutions of ferric salts and also with most of the tervalent and quadrivalent metals. These precipitates are soluble in excess potassium hydroxide and periodate. When iron^{III} periodate dissolves in alkali hydroxide, there probably is formed an alkali salt of iron^{III} para-periodic acid with the anion $[IO_6Fe]^{-2}$. Most of the reactions of iron appear to be masked in the solutions of this alkali salt.[504] The alkaline solution of the iron^{III} periodate complex is a selective reagent for lithium, which alone among the alkali metals forms a precipitate with the reagent, even from dilute solutions and in the cold. The composition of the yellow-white product which is difficultly soluble, depends on the experimental conditions; with excess of alkali periodate, the ratio of the components is 1 lithium : 2 iron : 2 periodic acid.

The test for lithium with the complex iron^{III} periodate is specific in the presence of even large amounts of sodium, potassium, rubidium, and cesium.

Saturated solutions of sodium chloride do not give a visible reaction at room temperatures. At 90° to 100° C, high concentrations of sodium chloride (above 20 γ in a drop) give a precipitate. If, however, small amounts of lithium are also present, a precipitate is formed in a few seconds even at 45° to 50° C. The precipitation of lithium induces that of sodium with iron^{III} periodate, and the sensitivity of the lithium test may be increased by this stratagem.

Ammonium salts interfere by reducing the alkalinity of the solution; they must be removed before the test by boiling with potassium hydroxide, or by igniting the solid sample. Bivalent metals also form precipitates with the complex iron^{III} periodate; they can, however, be removed by precipitation with 8-hydroxyquinoline in potassium hydroxide solution. The lithium test is then carried out on the filtrate.

Procedure. A drop of the neutral or alkaline test solution is mixed in a micro test tube with a drop of a saturated solution of sodium chloride, followed by 2 drops of the reagent. A blank test with water is started at the same time. Both tubes are dipped in water of 45° to 50° C for 15 to 20 seconds. In the presence of lithium, a yellow-white turbidity develops, while the blank remains clear.

Limit of Identification: 0.1 γ lithium
Limit of Dilution: 1 : 500,000

Reagent: Alkaline iron^{III} periodate solution: 2 g potassium periodate is dissolved in 10 ml of 2 *N* KOH (freshly prepared), diluted to 50 ml water and mixed with 3 ml of 10 % $FeCl_3$ solution and made up to 100 ml with 2 *N* KOH. The reagent is stable.

The following data are available with reference to the detection of lithium in the presence of other alkali chlorides.

0.05 γ Li may be detected in 1 drop of a saturated solution of NaCl.

Proportion Limit: 1 : 100,000.

0.2 γ Li may be detected in 1 drop of a saturated solution of KCl (after addition of NaCl).

Proportion Limit: 1 : 30,000.

In a drop of solution containing 0.35 γ Li in the presence of either 0.025 g RbCl or CsCl, an appreciable turbidity forms (compare with a pure Rb or Cs solution). This corresponds to ratios of:

Li : Rb = 1 : 50,000 Li : Cs = 1 : 60,000

(2) Test by etching of soda glass [505]

In fusion reactions, pulverized mixtures are heated and usually the reactant in greater proportion melts completely and the binary reaction occurs then in the interface between a solid and a melt. Such interfacial reactions also occur between an excess of the solid reactant and minimal amounts of the fusible partner to give rise to a so-called crypto fusion reaction.[506] An example is found in the behavior of lithium nitrate (m.p. circa 300° C) toward soda-lime glass. There is an exchange of sodium atoms of the sodium silicate for lithium atoms and the glass becomes dull. Since even slight quantities of lithium can bring about this effect, the latter may serve as a test for lithium.

The test may be conducted in the presence of salts of other elements but the latter may reduce the effectiveness of the lithium by forming mixed crystals and by etching or roughening the glass surface. Pyrex glass is not affected by hot lithium salts.

Procedure. A drop of the test solution (preferably nitrate) is placed on a microscope slide and taken to dryness. The residue is cautiously melted by careful passing of the slide through a flame. The cold slide is washed and dried. If lithium is present, the glass will show a white area which can be seen with the naked eye or with a magnifying lens in case no more than slight amounts of lithium are involved.

Limit of Identification: 1 γ lithium

Limit of Dilution: 1 : 50,000

39. Cesium

(1) Test with gold platinibromide [507]

Cesium salts give a deep black precipitate of $Cs_2Au_2PtBr_{12}$ on reaction with a mixture of gold and platinum bromides ($HAuBr_4$ plus H_2PtBr_6). The other alkali metals (as chlorides), with the exception of rubidium salts

References pp. 247–257

which behave similar to cesium in concentrations exceeding 2 %, do not react with the reagent.

Procedure. A drop of the reagent and then a drop of the neutral test solution are placed on filter paper. In the presence of cesium, a gray to black stain is formed.

Limit of Identification: 0.25 γ cesium (in 0.001 ml)
Limit of Dilution: 1 : 4000
Reagent: Solution of 0.36 g platinum bromide and 0.60 g gold bromide in 10 ml water

(*2*) *Test with potassium bismuth iodide* [508]

Cesium salts form a bright red precipitate of the double iodide Cs_2BiI_5 with the pale yellow solution of $KBiI_4$ or its acid. Among the other alkalis or univalent metals similar to the alkalis, only thallium reacts with this reagent; it gives a cinnamon brown precipitate. When thallium is present, or any metal that is precipitated by iodide, it is advisable to precipitate with potassium iodide beforehand and carry out the cesium test in the filtrate.

Procedure. A drop of potassium bismuth iodide is mixed with a drop of the test solution on filter paper. An orange or yellow stain, compared with a blank test, indicates the presence of cesium.

Limit of Identification: 0.7 γ cesium (in 0.001 ml)
Limit of Dilution: 1 : 1400
Reagent: Potassium bismuth iodide: 1 g Bi_2O_3 is dissolved, by boiling, in a saturated aqueous solution of 5 g potassium iodide, and then 25 ml acetic acid is added in small portions

(*3*) *Other test for cesium*

A black-brown fleck of $Cs_2AuPdCl_7$ is formed [508] on mixing a drop of a solution of a cesium salt with a drop of a 10 % equivalent solution of $AuCl_3$ and $PdCl_2$ (*Idn. Limit:* 1 γ Cs in 0.001 ml). Rubidium and thallium salts behave in the same way.

40. Ammonia (Ammonium Salts)

(*1*) *Test with* p-*nitrobenzenediazonium chloride* [509]

A good sensitive test for ammonia depends on the red coloration produced when *p*-nitrobenzenediazonium chloride (I) is shaken with a solution containing an ammonium salt and 10 % sodium hydroxide added drop by drop. The yellow-red ammonium salt of *p*-nitrophenylnitrosamine (II) is formed:

$$\underset{\text{(I)}}{O_2N\text{—}C_6H_4\text{—}\overset{+}{N}{\equiv}N\;Cl^-} + NH_3 + OH^- \longrightarrow \underset{\text{(II)}}{O_2N\text{—}C_6H_4\text{—}N{=}NONH_4} + Cl^-$$

This test may not be carried out on paper because the slight excess of alkali cannot be added conveniently, and a deep red may be formed even in the absence of ammonium salts. However, the test can be carried out on a spot plate if calcium oxide is used as alkalizer.

Procedure. [510] A drop of the slightly acid or neutral test solution is placed on a spot plate followed by a drop of the organic reagent. Then a particle of calcium oxide is added. In the presence of ammonium salts, a red zone forms at once around the calcium oxide. A blank test should be carried out on a drop of water when very small amounts are suspected.

Limit of Identification: 0.67 γ ammonia
Limit of Dilution: 1 : 75,000
Reagent: *p*-Nitrobenzenediazonium solution: One g *p*-nitraniline is dissolved by warming in 20 ml water and 2 ml dilute hydrochloric acid; the solution is diluted with 160 ml water while being vigorously shaken. On cooling, 20 ml of 2 to 5 % sodium nitrite solution is added; after a little shaking, everything dissolves. This solution becomes cloudy on standing, but may be used after filtering.

Diazotized *p*-nitraniline can be substituted by other diazotized primary aromatic amines.[511]

(2) *Tests after liberation of ammonia*

The release of ammonia by means of alkali from solid or dissolved ammonium salts (see below) serves as a characteristic test for ammonium ions even in the presence of other cations.* Cyanides should be absent or taken into account; they also give ammonia with warm alkalis: $CN^- + 2\ H_2O \rightarrow HCOO^- + NH_3$ and can thus be confused with ammonium salts.[512] The interference can be greatly reduced if mercury salts are added before the alkali. Undissociated mercuric cyanide, which is stable to alkali, is produced.

Three very sensitive tests will be given; they are based on the liberation of ammonia and can be carried out in the gas-generating apparatus described on pages 47 and 48. The familiar Nessler reaction and other tests for ammonia can also be made in this apparatus.

(*a*) *Test with red litmus paper*

Red litmus paper is preferable to the neutral paper for a spot test, because the color change is more distinct.

Procedure.[513] The apparatus is shown in Fig. 25 (p. 47). A drop of the test solution or a little of the solid is mixed with a drop of 2 *N* sodium hydroxide in the

* Volatile amines are likewise set free from their salts by alkali. I. B. Boussingault, *Ann. Phys.*, 29 (1850) 576, states that this interference can be avoided by alkalizing with magnesium oxide, which does not decompose salts of volatile bases. Mercury-amido compounds behave in the same way as NH_4 salts. $HgI_2 \cdot HgNH_2I$ is an exception and it is only slowly decomposed. This compound is formed in the familiar Nessler reaction. Compare L. M. Nichols and C. O. Willits, *J. Am. Chem. Soc.*, 56 (1934) 769.

apparatus. A strip of moist red litmus paper is hung on the hook, the apparatus is closed, and then warmed to about 40° C for 5 minutes. For small amounts of ammonia, the color of the paper should be compared with that of moist red litmus paper.

Limit of Identification: 0.01 γ ammonia
Limit of Dilution: 1 : 5,000,000

(*b*) *Test with manganese sulfate and silver nitrate* [513]

A black precipitate is formed on treating solutions of manganese and silver salts with alkalis (compare page 58):

$$Mn^{+2} + 2\ Ag^{+} + 4\ OH^{-} = MnO_2 + 2\ Ag^{0} + 2\ H_2O$$

This reaction can be used as a sensitive test for ammonia, since a neutral solution of manganese and silver salts (sulfate or nitrate) on being treated with a drop of ammonia, or exposure to ammonia gas, gives a black precipitate or dark color, according to the amount of ammonia involved. The sensitivity can be increased if the precipitate is treated with a drop of an acetic acid solution of benzidine (compare the manganese test described on page 175). The resulting benzidine blue will plainly reveal traces of a precipitate that was practically invisible.

Procedure. The apparatus shown in Fig. 23 or 24 (p. 47) is used. A drop of the test solution is mixed with a drop of 2 *N* alkali, a drop of the reagent solution placed on the knob of the stopper, or on the thickened end of the glass rod. The apparatus is closed and heated on an asbestos plate to about 40° C for 5 minutes. The drop of the reagent solution is then wiped on to a strip of quantitative filter paper. A black or gray fleck is left. Several drops of water are added and then a drop of an acetic acid solution of benzidine (see page 175). The fleck turns more or less blue. A blank is necessary when conducting the benzidine test for traces of MnO_2, because a blue color may appear in case the fleck has not been sufficiently washed.[514]

Limit of Identification: 0.05 γ ammonia
Limit of Dilution: 1 : 1,000,000
Reagent: Manganese nitrate-silver nitrate solution: 2.87 g $Mn(NO_3)_2$ is dissolved in 40 ml water and a solution of 3.35 g $AgNO_3$ in 40 ml water is added after dilution to 100 ml, dilute alkali is added, drop by drop, until a black precipitate is formed; this is filtered off. The reagent solution will keep if stored in a dark bottle.

(*c*) *Test through formation of zinc oxinate* [515]

When 8-hydroxyquinoline (oxine) (I) is added to an ammonical or caustic alkaline solution of a zinc salt, there is quantitative precipitation of the light yellow crystalline zinc oxinate (II), which displays a yellow-green fluorescence in ultraviolet light.[516] The precipitate is readily soluble in excess mineral acid. Consequently, the precipitation of zinc oxinate from neutral solutions of zinc salts is incomplete because of the rapid establishment of the equilibrium:

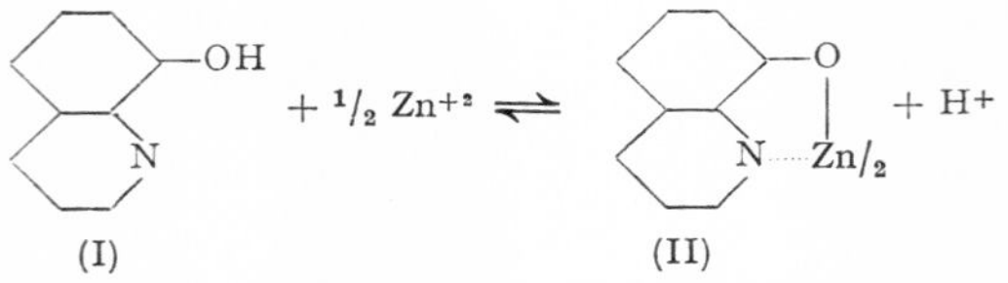

$$\text{(I)} + \tfrac{1}{2}\,Zn^{+2} \rightleftharpoons \text{(II)} + H^{+}$$

(I) (II)

Accordingly, if a solution of a zinc salt is treated with an excess of hydroxyquinoline and the precipitate is removed by filtration, the clear filtrate consists of the zinc oxinate equilibrium solution, from which fluorescent zinc oxinate can be precipitated by hydroxyl ions and consequently also by contact with ammonia vapors. Quantities of zinc oxinate which are too slight to give a visible turbidity nevertheless yield an easily discernible fluorescence.

Procedure.[517] The apparatus shown Fig. 23 or 24 (p. 47) is used. A drop of the test solution is mixed with a drop of 2 *N* alkali, and a drop of the reagent solution is placed on the knob of the stopper or on the thickened end of the glass rod. The apparatus is closed and placed in boiling water for five minutes. The hanging drop is then examined in ultraviolet light. If ammonia is present, the yellow-green fluorescence will be distinctly visible. The entire procedure can be conducted directly under the quartz lamp. A comparison blank test is advised when small quantities of ammonia are suspected.

Limit of Identification: 0.2 γ ammonia

Limit of Dilution: 1 : 250,000

Reagent: Zinc oxinate equilibrium solution: 1 % aqueous solution of zinc chloride is treated with excess 1 % alcoholic solution of 8-hydroxyquinoline and the precipitate is filtered off. The reagent solution keeps.

(3) Other tests for ammonia

(*a*) A mixture of silver nitrate and formaldehyde is instantly reduced by ammonia.[518] A silver nitrate-tannin mixture is similarly reduced (*Idn. Limits:* 0.05 γ NH_3 and 0.1 γ NH_3).[519] Both tests may be carried out in the apparatus described on page 47.

(*b*) The well known Nessler reaction may be carrried out as a spot test.[520] A drop of the test solution is mixed with a drop of concentrated alkaline hydroxide on a watch glass; a drop of the resulting suspension is placed on filter paper with a capillary tube and treated with a drop of Nessler solution. A yellow or orange-red stain or ring of $HgI_2 \cdot HgNH_2I$ is formed (*Idn. Limit:* 0.3 γ NH_3 in a drop of 0.002 ml). In the presence of Ag, Hg, Pb salts, the paper is treated with a drop of 2 *N* alkali, a drop of the test solution, a further drop of alkali, and finally the Nessler solution added crosswise (*Idn. Limit:* 2 γ NH_3 in a drop of 0.002 ml). It has been suggested [521] that it is advisable to liberate the ammonia in the gas apparatus (Fig. 23), and to place a drop of the Nessler solution on the knob of the apparatus. Very small amounts of the yellow precipitate can readily be identified, after wiping on to a strip of filter paper (*Idn. Limit:* 0.025 γ NH_3).

(*c*) Instead of Nessler solution, a drop of a neutral $MnSO_4$-H_2O_2 solution can be used; MnO_2 is formed. It can be smeared on filter paper and spotted with benzidine solution. A blue color develops (see page 237) (*Idn. Limit:* 0.03 γ NH_3).

41. Hydrazine

(*1*) *Test with salicylaldehyde (o-hydroxybenzaldehyde)* [522]

Benzaldehyde and other aromatic aldehydes condense with hydrazine or its salts to form water-insoluble colorless aldazines. Owing to its greater solubility in water, salicylaldehyde (I) is especially useful as specific reagent. It reacts with hydrazine:

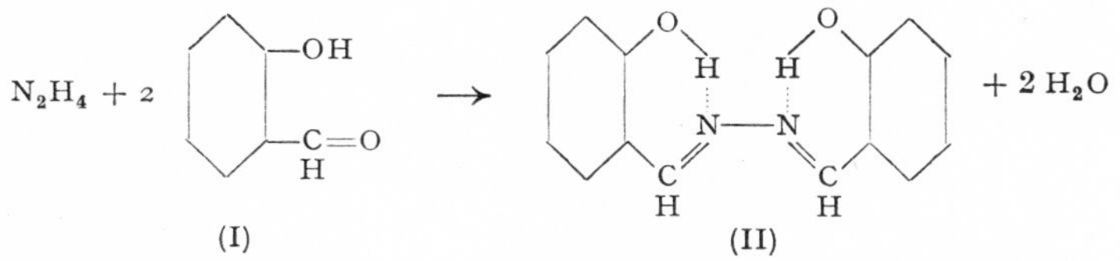

$$N_2H_4 + 2\ C_6H_4(OH)(CHO)\ \text{(I)} \rightarrow \text{(II)} + 2\ H_2O$$

The slight solubility of salicylaldazine and its immediate precipitation in acid solution (which is not the case with the condensation products of hydrazine and *m*- and *p*-hydroxybenzaldehyde) is probably connected with the fact that it may be formulated as a chelate compound as shown in (II).[523] The test for hydrazine by the aldazine reaction is not affected by the presence of hydroxylamine; the latter yields a soluble oxime on reaction with salicylaldehyde. However, it must be remembered that the reagent is consumed in the formation of the oxime and large quantities of hydroxylamine thus become harmful. Solid salicylaldazine displays an intense sulfur yellow fluorescence in ultraviolet light; its solutions in chloroform, ether, etc. are not fluorescent.

Procedure. A drop of the reagent is mixed with a drop of the neutral, or weakly acid or basic test solution in a micro test tube. Depending on the quantity of hydrazine present, a white precipitate or cloudiness appears immediately or after a short time. The fluorescence color of the aldazine may be used as a supplementary test (see below).

Limit of Identification: 0.1 γ hydrazine
Limit of Dilution: 1 : 500,000
Reagent: 5 g salicylaldehyde is boiled with 600 ml water and 20 ml 50 % acetic acid until the oil disappears. After cooling, the mixture is filtered from any undissolved salicylaldehyde. The solution will keep, but after considerable time it should be refiltered, because traces of salicylaldehyde frequently separate.

A very convenient test for hydrazine is based on the fluorescence of salicylaldazine and on the fact that salicylaldehyde (m.p. 2° C) has a distinct vapor pressure at room temperature. Accordingly, when salicylaldehyde vapor comes into contact with an acetate-buffered solution of a hydrazine salt, the fluorescing salicylaldazine is formed within a very short time.[524]

Procedure. About 0.5 ml salicylaldehyde is placed in a micro crucible which is then covered with a disk of filter paper, on which is brought a drop of the acid test solution, which has been buffered with solid sodium acetate. According to the quantity of hydrazine present, the fleck fluoresces yellow (quartz lamp) either at once or after several minutes.
Limit of Identification: 0.1 γ hydrazine
Limit of Dilution: 1 : 500,000

(2) *Test by reduction of ammoniacal cupric solutions* [525]

Hydrazine and its salts reduce ammoniacal solutions of cupric salts:

$$4\ Cu^{+2} + N_2H_4 \rightarrow 2\ Cu_2^{+2} + 4\ H^+ + N_2$$

Spot tests for hydrazine can be based on this reaction if carried out in the presence of anions which form cuprous salts that are insoluble in ammonia. Ferricyanide and ferrocyanide ions, whose cupric salts are easily soluble in ammonia, are suitable.

Procedure. A drop of the reagent solution is mixed with a drop of the test solution on a watch glass. A white precipitate or turbidity appears, depending on the quantity of hydrazine present. Slight turbidities quickly vanish because of oxidation (by atmospheric oxygen) to copper^{II} ferricyanide, which is soluble im ammonia.
Limit of Identification: 0.3 γ hydrazine
Limit of Dilution: 1 : 150,000
Reagent: Ammoniacal copper^{II} ferricyanide solution. A slight excess of freshly prepared potassium ferricyanide solution is added to 0.5 *N* copper sulfate solution. The light brown precipitate is dissolved in ammonia. The emerald green solution will keep for several days

(3) *Test with* p-*dimethylaminobenzaldehyde*

When neutral or acid solutions of hydrazine salts are treated with an alcohol solution of *p*-dimethylaminobenzaldehyde (I), the resulting orange to yellow color permits the detection of small amounts of hydrazine.[526] As little as 0.01 γ hydrazine is revealed if the reaction employing the aldehyde solution described below is conducted on a spot plate. The color reaction involves a condensation of hydrazine with the aldehyde, quite analogous to the reaction employed in test *1*, but with the difference that the aldazine (II) formed in reaction (1) is soluble in acids, as shown in (2), with formation of the quinoidal cation (III):

$$\begin{array}{c}NH_2\\|\\NH_2\end{array} + 2\,OCH{-}C_6H_4{-}N(CH_3)_2 \longrightarrow \begin{array}{l}N{=}CH{-}C_6H_4{-}N(CH_3)_2\\|\\N{=}CH{-}C_6H_4{-}N(CH_3)_2\end{array} \qquad (1)$$

(I) (II)

$$\begin{array}{l}N{=}CH{-}C_6H_4{-}N(CH_3)_2\\|\\N{=}CH{-}C_6H_4{-}N(CH_3)_2\end{array} + H^+ \longrightarrow \begin{array}{l}NH{-}CH{=}C_6H_4{=}\overset{+}{N}(CH_3)_2\\|\\N{=}CH{-}C_6H_4{-}N(CH_3)_2\end{array} \qquad (2)$$

(II) (III)

The production of (III) can be recognized also by the fact that a drop of the solution brought on filter paper leaves a red or pink stain, which fluoresces intensely in ultraviolet light.[527] This fluorescence will still reveal quantities which are too slight to be discerned through the colored ion. Since solutions of salts of *p*-dimethylaminobenzaldehyde do not fluoresce, and since only the *solid* red hydrochloride of (III) fluoresces blood red in ultraviolet light, it is probable that an adsorption occurs on filter paper and the fluorescence is due to an adsorbate. The acid constituent of the cellulose or the finely divided traces of silica, which the paper carries, may possibly be the adsorbent. A salt-like binding of the cation (III) with release of equivalent quantities of H^+ ion should be considered in this connection, or there may be a coordinative binding of the undissociated chloride of (III) through the nitrogen atom of the $—N(CH_3)_2$ group. The former would be an exchange adsorption, the second an addition adsorption.[528] If the paper is bathed in ammonia, the salmon fluorescing fleck becomes blue-green fluorescent. This change in the fluorescence color, which can be reversed by acids, obviously is due to conversion of adsorbed (III) into adsorbed (II) and vice versa. The aldazine (II) does not fluoresce either in the solid state nor when suspended in water, but if a drop of the water suspension is placed on filter paper, a blue-green fluorescence results.

The detection of hydrazine through the formation of soluble salts of *p*-dimethylaminobenzaldehyde and their fluorescing adsorbates on filter paper

References pp. 247–257

is specific. If hydroxylamine, which condenses with the aldehyde to form a soluble oxime, is present, a corresponding excess of the reagent must be used. Under these conditions, the *identification limit* given below remains valid even in the presence of 10,000 the quantity of hydroxylamine.

Procedure. A drop of the test solution is mixed with a drop of reagent solution, and the mixture is placed on filter paper. The fleck is bathed in 1 : 250 hydrochloric acid and then viewed in ultraviolet light. If hydrazine is present, the fleck fluoresces salmon. When bathed in dilute ammonia, the fleck exhibits a blue-green fluorescence, which is plainly visible when not too small quantities of hydrazine are at hand. If traces are being sought, it is advisable to make a blank test.

Limit of Identification: 0.001 γ hydrazine

Reagent: *p*-Dimethylaminobenzaldehyde solution: 0.4 g of the aldehyde is dissolved in 20 ml alcohol and acidified with 2 ml concentrated hydrochloric acid. If hydrazine is to be detected in very concentrated solutions of hydroxylamine salts, the reagent solution should be prepared with 4 g of the aldehyde in 20 ml alcohol.

(*4*) *Other tests for hydrazine* [529]

(*a*) Solutions of complex alkali copper pyrophosphate $Na_2CuP_2O_7$, which contain some dimethylaminobenzylidenerhodanine or alkali thiocyanate, on addition of alkaline hydrazine solutions give a precipitate of red cuprous dimethylaminobenzylidenerhodanine or white cuprous thiocyanate (*Idn. Limit:* 0.1 γ or 2 γ N_2H_4).

(*b*) Reduction accompanied by lightening or discharge of color occurs when a drop of ammoniacal hydrazine solution is placed on filter paper impregnated with lead^{IV} oxide or thallium^{III} oxide (*Idn. Limit:* 5 γ N_2H_4).

(*c*) Colorless alcoholic caustic alkali solutions of *o*(*p*)-dinitrobenzene turn blue-violet (orange) when warmed with hydrazine salts (*Idn. Limit:* 5 γ N_2H_4). Hydroxylamine and also sulfides react in the same manner.

42. Hydroxylamine

(*1*) *Test with diacetylmonoxime and nickel salt* [530]

Diacetylmonoxime (I) condenses with hydroxylamine to produce diacetyldioxime (dimethylglyoxime) the well known reagent for nickel. Consequently, if the condensation occurs in the presence of an ammoniacal nickel solution, the familiar red, slightly soluble nickel dimethylglyoxime (II) is formed directly from the components:

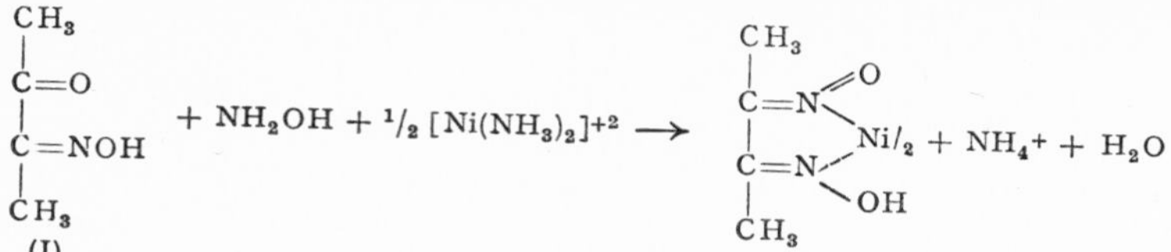

$$CH_3-C(=O)-C(=NOH)-CH_3 \text{ (I)} + NH_2OH + {}^1/_2\,[Ni(NH_3)_2]^{+2} \longrightarrow CH_3-C(=N{\to}O)-C(=N-OH)-CH_3\;Ni/_2 \text{ (II)} + NH_4^+ + H_2O$$

It is advisable to saturate the reagent (a mixture of the ketone and ammoniacal nickel solution) with nickel dimethylglyoxime, and to carry out the test as a spot reaction on filter paper.

This test should not be used in the presence of large amounts of hydrazine, which reacts with the reagent solution to give a red-brown precipitate. However, as little as 5 γ hydroxylamine can be detected in the presence of 300 γ hydrazine.

Procedure. Filter paper is impregnated with the reagent solution and dried. According to the amount of hydroxylamine present, a more or less intense red fleck or ring of nickel dimethylglyoxime is formed on the yellow paper when a drop of the test solution is placed on it.

Limit of Identification: 1 γ hydroxylamine

Limit of Dilution: 1 : 50,000

Reagent: 1.2 g diacetylmonoxime [531] is dissolved in 35 ml hot water, 0.95 g $NiCl_2 \cdot 6H_2O$ is added and, when cold, 2 ml concentrated ammonia. This solution is poured into a solution of 0.12 g hydroxylamine chloride in 200 ml water. After 24 hours, the precipitate of nickel dimethylglyoxime is removed. The red-brown solution is used as the reagent.

(2) Test with salicylaldehyde and copper salts [532]

Salicylaldehyde (I) with hydroxylamine and copper acetate gives the following reaction:

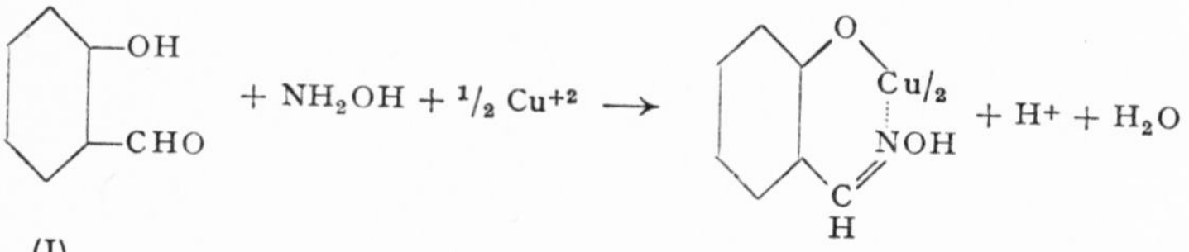

$$C_6H_4(OH)(CHO) \text{ (I)} + NH_2OH + {}^1/_2\,Cu^{+2} \longrightarrow C_6H_4(O\text{-}Cu/_2)(CH{=}NOH) \text{ (II)} + H^+ + H_2O$$

Pale yellow inner complex copper salicylaldoxime (II) is precipitated. The test for hydroxylamine based on this behavior, like (*1*), consists therefore in the formation of an inner complex salt from its components. A stock mixture of salicylaldehyde and copper acetate cannot be used as a reagent, because, depending upon the concentration, a green copper salt of salicylaldehyde separates after short or long standing. The reagents must be added separately to the solution in the prescribed order, and concentrations.

Procedure. A drop of the reagent solution (*1*) is mixed in a micro test tube with a drop of the test solution and 2 drops of the reagent solution (*2*). A bright yellow precipitate or cloudiness is formed according to the amount of hydroxylamine present.

Limit of Identification: 1 γ hydroxylamine

Limit of Dilution: 1 : 50,000

Reagents: *1*) Salicylaldehyde solution: 5 g salicylaldehyde is boiled in 600 ml water and 20 ml 50 % acetic acid until the oil globules disappear. After 24 hours the solution is filtered.

2) Copper acetate solution: 5 g salt is warmed with 150 ml water and 2 ml 50 % acetic acid, and filtered after cooling.

Detection of hydrazine and hydroxylamine when together

The formation and properties of salicylaldazine and copper salicylaldoxime, both using salicylaldehyde as reagent, may be applied as tests for hydroxylamine and hydrazine when together.

The relatively low solubility of hydrazine sulfate may be applied for the detection of *a little hydroxylamine in the presence of much hydrazine.* A saturated solution of hydrazine sulfate (at 32° C) contains 2.95 g hydrazine sulfate in 100 ml water, which is about 360 γ in a drop. This amount can be precipitated by salicylaldehyde as aldazine, without appreciably increasing the volume. Thus the main portion of the hydrazine can be removed as sulfate. All further manipulations for the detection of hydroxylamine in the presence of hydrazine are carried out with the filtrate.

Procedure. 1.5 ml of the test solution is treated with 4 : 10 sulfuric acid until a drop (tested separately) gives no further precipitate with sulfuric acid. The mixture is centrifuged after 10 minutes, and 0.5 ml of the centrifugate is rendered alkaline with strong ammonia and then acidified with 1 or 2 drops of 2 *N* sulfuric acid. Then 2.5 ml of salicylaldehyde solution are added, the salicylaldazine is centrifuged down after 30 minutes standing, and 3 drops of the supernatant liquid tested in a micro test tube with a drop of copper acetate solution. A yellow-green precipitate of copper salicylaldoxime is formed if hydroxylamine is present.

Limit of Identification: 10 γ hydroxylamine in the presence of 5000 γ hydrazine

The characteristic yellow fluorescence of salicylaldazine may be utilized for the detection of *a little hydrazine in the presence of much hydroxylamine.*

Procedure.[533] Filter paper is spotted with a drop of the acetate-buffered or weakly alkaline test solution and placed across the mouth of a small porcelain dish containing several drops of salicylaldehyde. With quantities of hydrazine exceeding 10 γ, the fleck exhibits a strong yellow fluorescence within 1–2 minutes. If small quantities of hydrazine are to be detected in the presence of much hydroxylamine, the fleck should be exposed to the salicylaldehyde vapors for about 15 minutes. The paper is then bathed in water or dilute acetic acid to

References pp. 247–257

remove the salicylaldoxime, which hides the yellow fluorescence. If hydrazine is present, a yellow fluorescing fleck will remain.

Limit of Identification: 0.1 γ hydrazine in the presence of 5000 γ hydroxylamine

For the preparation of reagents, see pages 240 and 244.

(3) Test with formaldehyde and iron^{III} salt [534]

Formaldehyde like other water-soluble aldehydes reacts with hydroxylamine in alkaline solution to yield an addition or condensation product, which is transformed into the corresponding hydroxamic acid by means of hydrogen peroxide: [535]

$$RCHO + NH_2OH \rightarrow RCHOH—NHOH \quad (1)$$

$$RCHOH—NHOH + H_2O_2 \rightarrow RCO—NHOH + 2\,H_2O \quad (2)$$

All hydroxamic acids react in weak acid solution with ferric ions to produce inner complex salts. In the case of formaldehyde, formhydroxamic acid (I) is produced by reactions (1) and (2) and on the addition of ferric ions its red inner complex salt (II) appears

```
HC=O          HC—O-.
|             |     Fe/3
HN—OH         HN—O/
 (I)           (II)
```

(II) is produced by bringing together formaldehyde, hydroxylamine, iron^{III} salts, and persulfate. If hydrazine is absent, the persulfate is not needed because the tervalent iron takes over the oxidation shown in reaction (2). Consequently, the test given here, like those described before, involves the formation of an inner complex salt from the components.

Procedure. A spot plate is used. A drop of the neutral test solution is treated with two drops of the iron-bearing formaldehyde solution and several mg of potassium persulfate is then added. On mixing, a red or pink color develops at once or after a little time, depending on the amount of hydroxylamine present.

Limit of Identification: 0.5 hydroxylamine

Limit of Dilution: 1 : 100,000

Reagent: 100 mg ferric alum dissolved in 10 ml of 20 % formaldehyde

This procedure revealed 10 γ hydroxylamine in one drop of saturated hydrazine solution.

(4) Test by conversion into nitrous acid [536]

Iodine, in acetic acid solution, oxidizes hydroxylamine rapidly and quantitatively to nitrous acid: [537]

$$NH_2OH + 2\,I_2 + H_2O \rightarrow HNO_2 + 4\,HI$$

If this oxidation is carried out in the presence of sulfanilic acid, the latter is diazotized and there is no redox reaction between the nitrous acid and the hydriodic acid (compare page 332). After removal of the excess iodine with thiosulfate, the diazonium salt may be coupled with α-naphthylamine to give a red azo dye. (See page 330 regarding this sensitive *Griess test* for nitrites.)

This test for hydroxylamine by conversion to nitrous acid and subsequent formation of an azo dye is decisive even in the presence of much hydrazine, provided a sufficient excess of iodine has been used in the oxidation. The test is, of course, inapplicable in the presence of nitrites.

Procedure.[538] A drop of the test solution is mixed either on a spot plate or in a micro crucible with a few milligrams of sodium acetate, 1 to 2 drops sulfanilic acid, and a drop of an acetic acid solution of iodine. The mixture is allowed to stand for 2 or 3 minutes. The excess iodine is then removed with 0.1 *N* thiosulfate, and a drop of α-naphthylamine solution is added. A pink to deep red color indicates hydroxylamine.

Limit of Identification: 0.01 γ hydroxylamine

Limit of Dilution: 1 : 5,000,000

Reagents: *1)* Iodine solution: 1.3 g iodine in 100 ml glacial acetic acid

2) Sulfanilic acid solution: 10 g acid in 750 ml water and 250 ml glacial acetic acid

3) α-Naphthylamine solution: 3 g of the base in 700 ml water plus 300 ml glacial acetic acid

(*5*) *Other tests for hydroxylamine*

(*a*) The test for ammonia involving the reaction: $2\ Fe(OH)_2 + NH_2OH + H_2O = 2\ Fe(OH)_3 + NH_3$ can be applied to the detection of hydroxylamine in the presence of hydrazine (*Idn. Limit:* 0.1 γ NH_2OH). [539]

Ammonium salts, nitrates, nitrites, azides must be absent. The latter are likewise reduced to ammonia by Fe^{II} hydroxide. The addition of ferrous hydroxide can be combined with the procedure described on page 236. It is sufficient to add a drop of ferrous sulfate solution before making the test solution alkaline.

(*b*) The addition of hydroxylamine or its salts to an ammoniacal solution of Cu^{II} ferricyanide produces a white precipitate of Cu^{I} ferricyanide (*Idn. Limit:* 0.2 γ NH_2OH).[540]

(*c*) A white or red precipitate of cuprous thiocyanate or cuprous *p*-dimethylaminobenzylidenerhodanine appears when ammoniacal hydroxylamine solution is added to a solution of $NaCu^{II}$ pyrophosphate containing alkali thiocyanate or *p*-dimethylaminobenzylidenerhodanine (*Idn. Limit:* 1 γ or 2 γ NH_2OH). [540]

(*d*) Colorless alcoholic caustic alkali solutions of *o*(*p*)-dinitrobenzene give a blue-violet (orange) color when warmed with hydroxylamine salts (*Idn. Limit:* 5 γ hydroxylamine). [541]

REFERENCES

1. N. A. Tananaeff and I. Tananaeff, *Z. anorg. allgem. Chem.*, 170 (1928) 120. Compare also *Z. anal. Chem.*, 106 (1936) 167.
2. Compare also D. Balarew, *Z. anal. Chem.*, 60 (1921) 392.
3. F. Feigl, *Z. anal. Chem.*, 74 (1928) 380.
4. F. Feigl, *Chemistry of SSS Reactions*, New York, 1949, p. 328.
5. For preparations see R. Andreasch and A. Zipser, *Monatsh.*, 26 (1905) 1203.
6. F. Feigl, Ref. 4, p. 442.
7. K. Heller and P. Krumholz, *Mikrochemie*, 7 (1929) 214.
8. F. Feigl, P. Krumholz, and E. Rajmann, *Mikrochemie*, 9 (1931) 295.
9. A. J. Velculescu, *Z. anal. Chem.*, 90 (1932) 111.
10. E. v. Angerer, *Wissenschaftliche Photographie*. Leipzig, 1931, p. 19.
11. The theory of this action is discussed by W. Ostwald, *Outlines of General Chemistry* (translated by W. W. Taylor), New York, 1912, p. 530.
12. See also F. Ephraim, *Inorganic Chemistry* (translated by P. Thorne and A. Ward), 3rd ed., New York, 1939, p. 252.
13. A. Velculescu, *Chem. Abstracts*, 34 (1940) 2278.
14. F. Feigl and E. Fränkel, *Ber.*, 65 (1932) 544.
15. R. Lang, *Z. anorg. allgem. Chem.*, 152 (1926) 201; *Ber.*, 60 (1927) 1389.
16. Compare F. Feigl, Ref. 4, p. 123.
17. R. Lang, *Z. anorg. allgem. Chem.*, 152 (1926) 201; *Ber.* 60 (1927) 1389.
18. A. v. Atta, *J. Chem. Education*, 16 (1939) 164.
19. N. A. Tananaeff, *Z. anorg. allgem. Chem.*, 140 (1924) 320.
20. H. Fischer, *Z. angew. Chem.*, 42 (1929) 1025; *Mikrochemie*, 8 (1930) 319.
21. G. Gutzeit, *Helv. Chim. Acta*, 12 (1929) 837.
22. E. A. Kocsis and G. Gelei, *Z. anorg. allgem. Chem.*, 232 (1937) 202.
23. V. I. Occleshaw, *J. Chem. Soc.*, 140 (1937) 1438.
24. R. Duval, *Anal. Chim. Acta*, 3 (1949) 21.
25. T. Pavolini and F. Gambarin, *Anal. Chim. Acta*, 3 (1949) 27.
26. R. Přibil, J. Doležal and V. Simon, *Z. anal. Chem.*, 140 (1953) 371.
27. F. Feigl and A. Caldas, *Anal. Chim. Acta*, 13 (1955) 526.
28. B. Sen and Ph. W. West, *Mikrochim. Acta*, (1955) 979.
29. P. Cazeneuve, *Compt. rend.*, 131 (1900) 346.
30. F. Feigl and F. Neuber, *Z. anal. Chem.*, 62 (1923) 370.
31. F. Feigl and F. Lederer, *Monatsh.*, 45 (1924) 63, 115.
32. Compare F. Welcher, *Organic Analytical Reagents*, Vol. III, New York, 1947, p. 43.
33. For preparation, see K. H. Slotta and K. R. Jacobi, *Z. anal. Chem*, 77 (1929) 346.
34. P. Artmann, *Z. anal. Chem.*, 60 (1921) 81.
35. E. van Dalen and B. van't Riet, *Anal. Chim. Acta*, 6 (1952) 101.
36. N. A. Tananaeff, *Z. anorg. allgem. Chem.*, 136 (1924) 373; 140 (1924) 321.
37. K. Heller and P. Krumholz, *Mikrochemie*, 7 (1929) 214.
38. F. Feigl and L. Badian, unpublished studies.
39. B. S. Evans, *Analyst*, 56 (1931) 177.
40. I. M. Kolthoff and R. S. Livingstone, *Ind. Eng. Chem., Anal. Ed.*, 7 (1935) 209.
41. For preparation, see N. Moufang and J. Tafel, *Ann.*, 304 (1898) 471.
42. Regarding this method of detecting aluminum, see F. Cucuel, *Mikrochemie*, 13 (1933) 323.
43. E. Schmidt and E. Tornow, *Chem.-Ztg.*, 56 (1932) 206.
44. F. Feigl and L. Badian, unpublished studies.
45. A. Langer, *Chem. Listy*, 32 (1938) 438; *Chem. Abstracts*, 33 (1939) 3717.
46. I. Kraljic and M. Mate, *Bull. Sci. Yougoslavie*, 3 (1956) 5.
47. S. Asperger and coworkers, *J. Chem. Soc.* (1953) 1041; (1955) 1449.
48. H. Wölbling and B. Steiger, *Z. angew. Chem.*, 46 (1933) 279.
49. Ph. W. West and McGee A. Duff, *Mikrochim. Acta*, (1955) 987.
50. E. A. Kocsis and G. Gelei, *Z. anorg. Chem.*, 232 (1937) 202.

51. F. Feigl and A. Caldas, *Anal. Chim. Acta,* 13 (1955) 526.
52. G. Denigès, *Précis de chimie analytique,* 4th ed., Paris, 1913, p. 72.
53. Compare W. Schlenk, *Ann.,* 363 (1908) 313.
54. F. Feigl and A. Singer, *Pharm. Presse,* 6 (1935) 37. See also F. Feigl and F. Neuber, *Z. anal. Chem.,* 62 (1923) 371.
55. F. Feigl and H. A. Suter, *Ind. Eng. Chem., Anal. Ed.,* 14 (1942) 840. See also I. M. Kolthoff, *Pharm. Weekblad,* 62 (1925) 1017.
56. H. Fischer, *Z. angew. Chem.,* 42 (1929) 1025; *Mikrochemie,* 8 (1930) 319.
57. N. A. Tananaeff, *Z. anorg. allgem. Chem.,* 167 (1927) 341.
58. A. S. Komarowsky and W. Owetschkin, *Z. anal. Chem.,* 71 (1927) 56.
59. F. Pavelka, *Mikrochemie,* 7 (1929) 303.
60. E. Léger, *Z. anal. Chem.,* 28 (1889) 374.
61. F. Feigl and F. Neuber, *Z. anal. Chem.,* 62 (1923) 373.
62. F. Feigl and P. Krumholz, *Ber.,* 62 (1929) 1138.
63. C. Porlezza, *Atti soc. sci, nat. toscana,* 37 (1928) 1.
64. L. Vanino and F. Treubert, *Ber.,* 31 (1898) 1113.
65. Compare Feigl, Ref. 4, p. 137.
66. J. Donau, *Monatsh.,* 34 (1913) 949.
67. A. Neunhofer, *Z. anal. Chem.,* 132 (1951) 91.
68. C. Mahr, *Z. anorg. allgem. Chem.,* 208 (1932) 313.
69. For preparation, see C. Mahr, *loc. cit.*
70. N. A. Tananaeff, *Z. anal. Chem.,* 105 (1936) 419.
71. G. Lochmann, *Z. anal. Chem.,* 85 (1931) 241.
72. M. V. Gaptschenko and O. G. Scheinziss, *Chem. Abstracts,* 32 (1938) 1212.
73. N. W. Wawilow, *Chem. Abstracts,* 32 (1938) 5725.
74. Ph. W. West, P Senise and J. K. Carlton, *Anal. Chim. Acta,* 6 (1952) 488.
75. F. L. Hahn and G. Leimbach, *Ber.,* 55 (1922) 3070.
76. J. Holluta and A. Martini, *Z. anorg. allgem. Chem.,* 140 (1924) 208.
77. H. Schmidt, *Z. physik. Chem.,* A 148 (1930) 321; *Z. Elektrochem.,* 36 (1930) 769.
78. Compare also A. C. Oudemans, *Z. anal. Chem.,* 6 (1867) 129.
79. F. Pavelka, *Mikrochemie,* 23 (1937) 202.
80. F. Feigl, *Ber.,* 56 (1923) 2032; *Mikrochemie,* 1 (1923) 76.
81. H. A. Suter and Ph. W. West, *Anal. Chim. Acta,* 13 (1955) 501.
82. For preparation, see A. Werner and P. Detscheff, *Ber.,* 38 (1905) 72.
83. W. Geilmann and K. Brünger, *Glastech. Ber.,* 87 (1929) 329.
84. F. Ephraim, *Ber.,* 63 (1930) 1928.
85. H. Holzer, *Z. anal. Chem.,* 95 (1933) 398. Compare also F. J. Welcher, *Organic Analytical Reagents,* Vol. III, New York, 1947, p. 259.
86. For preparation, see L. Claisen and O. Eisleb, *Ann.,* 401 (1913) 99.
87. L. Kul'berg, *Mikrochemie,* 20 (1936) 153.
88. F. Feigl and F. Neuber, *Z. anal. Chem.,* 62 (1923) 376.
89. Compare F. Feigl, Ref. 4, p. 116.
90. O. Glemser and G. Lutz, *Naturwissenschaften,* 34 (1947) 215.
91. For preparation see A. Linz, *Ind. Eng. Chem., Anal. Ed.,* 15 (1943) 459.
92. R. Uhlenhuth, *Chem. Ztg.,* 34 (1910) 887.
93. H. E. Ballaban, *Mikrochem. ver. Mikrochim. Acta,* 27 (1939) 62.
94. J. V. Dubský and V. Bencko, *Z. anal. Chem.,* 94 (1933) 19.
95. Compare Feigl, Ref. 4, p. 530.
95a. A. Okáč and J. Čelechovský, *Chem. Abstracts,* 45 (1951) 6530.
96. For preparation, see R. Uhlenhuth, *Chem. Zentr.,* 1911 I, 602; German Pat. 231,091; *Friedländer,* 10 (1910–12) 588.
97. Pr. Rây and R. M. Rây, *Quart. J. Indian Chem. Soc.,* 3 (1926) 118; Pr. Rây, *Z. anal. Chem.,* 79 (1929) 94.
98. For preparation, see R. Wollner, *J. prakt. Chem.,* 29 (1884) 129.
99. F. Feigl and H. J. Kapulitzas, *Mikrochemie,* 8 (1930) 239.
100. Ph. W. West, *Ind. Eng. Chem., Anal. Ed.,* 17 (1945) 740.

101. H. Fischer, *Mikrochemie*, 8 (1930) 319; *Z. angew. Chem.*, 42 (1929) 1025.
102. J. Hoste, *Anal. Chim. Acta*, 4 (1950) 23.
103. For preparation, see J. Wibaut and coworkers, *Rec. trav. chim.*, 54 (1935) 805.
104. F. Feigl and A. Caldas, *Anal. Chim. Acta*, 8 (1953) 117.
105. For preparation, see J. Auerbach, *J. Chem. Soc.*, 35 (1879) 799.
106. F. Feigl and K. Lettmayr, unpublished studies. See also F. Feigl and F. Neuber, *Z. anal. Chem.*, 62 (1923) 375.
107. A. Okáč and J. Čelechovský, *Chem. Abstracts*, 45 (1951) 6530.
108. R. Montequi, *Chem. Abstracts*, 21 (1927) 2858.
109. F. Feigl, *Mikrochemie*, 7 (1929) 6.
110. O. Funakoshi, *Chem. Abstracts*, 23 (1929) 4644.
111. A. S. Komarowsky and N. S. Poluektoff, *Z. anal. Chem.*, 96 (1934) 23.
112. S. Augusti, *Mikrochemie*, 22 (1937) 139.
113. L. A. Sarver, *Ind. Eng. Chem., Anal. Ed.*, 10 (1938) 378.
114. L. Szebellédy and M. Ajtai, *Chem. Abstracts*, 33 (1939) 4155.
115. P. Krumholz and F. Hönel, *Mikrochim. Acta*, 2 (1937) 180. See also Ref. 107.
116. F. Goldschmidt and B. R. Dishen, *Anal. Chem.*, 20 (1948) 373.
117. J. Gillis, A. Claeys and J. Hoste, *Anal. Chim. Acta*, 1 (1947) 421.
118. B. L. Clarke and H. W. Hermance, *Ind. Eng. Chem., Anal. Ed.*, 9 (1937) 292.
119. J. Gillis, A. Claeys and J. Hoste, *Anal. Chim. Acta*, 1 (1947) 421.
120. M. Ya. Shapiro, *Chem. Abstracts*, 44 (1950) 2887.
121. R. M. Rush and L. B. Rogers, *Mikrochim. Acta*, (1955) 821.
122. F. Feigl and L. I. Miranda, *Ind. Eng. Chem., Anal. Ed.*, 16 (1944) 141.
123. Compare G. Tartarini, *Gazz. chim. ital.*, 63 (1933) 597.
124. K. Heller and P. Krumholz, *Mikrochemie*, 7 (1929) 217.
125. P. Krumholz and F. Hönel, *Mikrochim. Acta*, 2 (1937) 181.
126. Compare F. Feigl, Ref. 4, p. 530.
127. For preparation, see P. Krumholz and F. Hönel, *Mikrochim. Acta*, 2 (1937) 177.
128. For preparation, see P. Krumholz and F. Hönel, *loc. cit.*
129. F. Feigl and F. Neuber, *Z. anal. Chem.*, 62 (1923) 378.
130. F. P. Dwyer, *Australian Chem. Inst. J. & Proc.*, 4 (1937) 26. Comp. also M. R. Verma and S. D. Paul, *Analyst*, 80 (1955) 399.
131. E. van Dalen and G. de Vries, *Anal. Chim. Acta*, 3 (1949) 567.
132. A. Bettendorf, *Z. anal. Chem.*, 9 (1870) 105.
133. M. Gutzeit, *Pharm. Ztg.*, 24 (1879) 263.
134. H. Beckurts, *Pharm. Zentralhalle*, 25 (1884) 197, 209, 223.
135. L. W. Winkler, *Z. angew. Chem.*, 30 (1917) 114.
136. E. Dowzard, *Proc. Chem. Soc.*, 17 (1901) 92; A. Gotthelf, *Z. anal. Chem.*, 24 (1905) 263.
137. J. Gangl and H. Dieterich, *Mikrochemie*, 19 (1936) 253.
138. D. Bezier, *Anal. Chim. Acta*, 1 (1947) 133.
139. W. Reppmann, *Z. anal. Chem.*, 99 (1934) 180.
140. C. R. Fresenius, *Z. anal. Chem.*, 1 (1862) 444.
141. G. Denigès, *Chem. Zentr.*, (1901) II, 1214.
142. E. Rupp, *Ber. deut. pharm. Ges.*, 32 (1923) 334.
143. F. Feigl and F. Neuber, *Z. anal. Chem.*, 62 (1923) 382.
144. For preparation, see A. Linz, *Ind. Eng. Chem., Anal. Ed.*, 15 (1943) 459.
145. P. E. Wenger, R. Duckert and C. P. Blancpain, *Helv. Chim. Acta*, 20 (1937) 437; *Mikrochim. Acta*, 3 (1938) 13.
146. Compare F. Feigl, Ref. 4, p. 177.
147. J. Gillis, J. Hoste and A. Claeys, *Anal. Chim. Acta*, 1 (1947) 291.
148. For preparation, see C. Liebermann and S. Lindenbaum, *Ber.*, 37 (1904) 1177, 2731.
149. E. Eegriwe, *Z. anal. Chem.*, 70 (1927) 400.
150. Compare V. J. Kuznetsov, *Chem. Abstracts*, 41 (1947) 1947.
151. Ph. W. West and W. C. Hamilton, *Anal. Chem.*, 24 (1952) 1025.
152. According to studies by V. Gentil.
153. J. Donau, *Monatsh.*, 34 (1913) 949.

154. R. E. D. Clark, *Analyst*, 61 (1936) 242.
155. F. Feigl and F. Neuber, *Z. anal. Chem.*, 62 (1923) 382.
156. Compare C. Zenghelis, *Z. physik. chem. Unterricht*, 24 (1911) 137.
157. G. Gutzeit, *Helv. Chim. Acta*, 12 (1929) 720.
158. Compare I. L. Newell and coworkers, *Ind. Eng. Chem., Anal. Ed.*, 7 (1935) 26 and H. Leuchs and coworkers, *Ber.*, 55 (1922) 564.
159. G. Beck, *Mikrochim. Acta*, 2 (1937) 287.
160. For preparation, see N. Moufang and J. Tafel, *Ann.*, 304 (1898) 47.
161. H. Meissner, *Z. anal. Chem.*, 80 (1930) 247. See also O. Schmatolla, *Chem. Ztg.*, 25 (1901) 468 and F. L. Hahn, *Z. anal. Chem.*, 82 (1930) 113.
162. Private communication from F. L. Hahn.
163. F. Feigl and H. J. Kapulitzas, unpublished studies.
164. F. Feigl and V. Gentil, unpublished studies.
165. F. Feigl, *Chem. Ztg.*, 43 (1919) 861.
166. V. A. Nazarenko, *Chem. Abstracts*, 36 (1942) 2489.
167. E. Eegriwe, *Z. anal. Chem.*, 74 (1928) 223.
168. H. Freytag, *Ber.*, 67 (1934) 1477.
169. N. A. Tananaeff, *Z. anorg. allgem. Chem.*, 133 (1924) 372.
170. E. Eegriwe, *Z. anal. Chem.*, 120 (1940) 81.
171. V. J. Kuznetsov, *Chem. Abstracts*, 35 (1941) 3190.
172. E. Ruzicka, *cf. Z. anal. Chem.*, 142 (1954) 385.
173. N. S. Poluektoff, *Mikrochemie*, 18 (1935) 48.
174. Compare F. Feigl, Ref. 4, p. 354.
175. A. Tschakirian, *Compt. rend.*, 187 (1928) 229.
176. A. S. Komarowsky and N. S. Poluektoff, *Mikrochemie*, 18 (1935) 66.
177. J. Gillis, J. Hoste and A. Claeys, *Anal. Chim. Acta*, 1 (1947) 302.
178. Comp. F. Feigl, *Chemistry of Specific, Selective and Sensitive Reactions*, New York, 1949, Chap. 10.
179. For preparation, see J. Gillis and coworkers, *loc. cit.*, Ref. 158.
180. G. S. Deshmukh, *Anal. Abstr.*, 2 (1955) 2684.
181. N. S. Poluektoff, *Mikrochemie*, 18 (1935) 49.
182. G. S. Deshmukh, *Naturwissenschaften*, 42 (1955) 70.
183. C. D. Braun, *Z. anal. Chem.*, 2 (1863) 36; F. C. Krauskopf and C. E. Swarte, *J. Am. Chem. Soc.*, 48 (1926) 3021; Y. Uzumasa and K. Doi, *Bull. Chem. Soc. Japan*, 14 (1939) 337; *Chem. Abstracts*, 34 (1940) 959.
184. J. Stěrba-Böhm and J. Vostřebal, *Z. anorg. allgem. Chem.*, 110 (1920) 82.
185. N. A. Tananaeff and G. A. Pantschenko, *Chem. Abstracts*, 24 (1930) 566.
186. S. Malowan, *Z. anorg. allgem. Chem.*, 108 (1914) 73; *Z. anal. Chem.*, 79 (1929) 202; J. Koppel, *Chem. Ztg.*, 43 (1919) 777.
187. M. Siewert, *Z. anal. Chem.*, 60 (1921) 464; *Z. ges. Naturw.*, 23 (1864) 5. See also J. Koppel, *loc. cit.*
188. N. Tarugi and F. Sorbini, *Chem. Abstracts*, 7 (1913) 1462.
189. B. L. Clarke and H. W. Hermance, *Ind. Eng. Chem., Anal. Ed.*, 9 (1937) 292.
190. L. Spiegel and Th. Maass, *Ber.*, 36 (1903) 512; E. Montignie, *Bull. soc. chim. France*, [4], 47 (1930) 128.
191. F. Feigl and H. J. Kapulitzas, unpublished studies.
192. R. Lang, *Z. anal. Chem.*, 128 (1948) 165.
193. F. Feigl and A. Schaeffer, *Anal. Chim. Acta*, 7 (1952) 507.
194. F. Feigl and H. Blohm, unpublished studies.
195. J. Gillis, A. Claeys and J. Hoste, *Anal. Chim. Acta*, 1 (1947) 421.
196. L. Szebellédy and J. Jónás, *Mikrochim. Acta*, 1 (1937) 46.
197. A. S. Komarowsky and N. S. Poluektoff, *Mikrochim. Acta*, 1 (1937) 264.
198. E. Lecocq, *Bull. soc. chim. Belg.*, 17 (1903) 412; *Chem. Zentr.*, (1904) I, 836.
199. P. Krumholz and F. Hönel, *Mikrochim. Acta*, 2 (1937) 177. Compare J. Gillis, *Mikrochem ver. Mikrochim. Acta*, 31 (1943) 58.
200. A. de Sousa, *Mikrochem. ver. Mikrochim. Acta*, 40 (1953) 104.

201. S. Halberstadt, *Z. anal. Chem.* 92 (1933) 86.
202. F. Feigl and V. Gentil, unpublished studies.
203. N. A. Tananaeff and G. A. Pantschenko, *Chem. Abstracts*, 24 (1930) 567.
204. E. B. Sandell, *Ind. Eng. Chem.. Anal. Ed.*, 10 (1938) 667.
205. F. Feigl, *Rec. trav. chim.*, 58 (1939) 471.
206. G. v. Knorre, *Z. anal. Chem.*, 47 (1908) 37.
207. Compare F. Feigl, Ref. 4, pp. 313 and 557.
208. For preparation, see H. Schmidt and G. Schultz, *Ann.*, 207 (1881) 330.
209. G. Werther, *J. prakt. Chem.*, 83 (1861) 195.
210. J. Meyer and A. Pawletta, *Z. anal. Chem.*, 69 (1926) 15.
211. Ph. W. West and L. J. Conrad, *Mikrochem. ver. Mikrochimica Acta*, 35 (1950) 443.
212. Ph. W. West and L. J. Conrad, *loc. cit.*, ref. 185; compare also F. Ephraim, *Helv. Chim. Acta*, 14 (1931) 1266.
213. Ph. W. West and L. J. Conrad, *loc. cit.*, Ref. 185.
214. H. Knowles, *U. S. Bur. Stand. J. Research*, 9 (1932) 1.
215. For preparation, see A. Werner and P. Delscheff, *Ber.*, 38 (1905) 72.
216. F. Feigl and L. Baumfeld, *Anal. Chim. Acta*, 3 (1949) 15.
217. Compare J. F. Welcher, *Organic Analytical Reagents*, Vol. I, N.Y., 1947, p. 297; compare also F. Feigl, Ref. 4, p. 194.
218. M. Borrel and R. Paris, *Anal. Chim. Acta*, 4 (1950) 267.
219. R. Belcher, A. J. Nutten and W. I. Stephen, *Analyst*, 76 (1951) 904.
220. R. Belcher and A. J. Nutten, *J. Chem. Soc.*, 154 (1951) 547.
221. For preparation, see K. Fries and W. Lehmann, *Ber.*, 54 (1921) 2922.
222. J. Lightfoot, *Jahresb. Fortsch. Chem.*, 25 (1872) 1076; A. Guyard, *Bull. soc. chim. France* [2], 25 (1876) 58.
223. N. A. Tananaeff and G. A. Pantschenko, *Chem. Abstracts*, 24 (1930) 567.
224. A. S. Komarowsky and N. S. Poluektoff, *Chem. Zentr.*, 1934, I, 3773.
225. R. Montequi and M. Gallego, *Chem. Abstracts*, 28 (1934) 3409.
226. A. S. Komarowsky and N. S. Poluektoff, *Chem. Zentr.*, 1934, I, 3773.
227. Ph. W. West and L. J. Conrad, *Mikrcchem. ver. Mikrochimica Acta*, 35 (1950) 446.
228. J. Hoste, *Anal. Chim. Acta*, 2 (1948) 403.
229. J.Hoste, *Anal. Chim. Acta*, 3 (1949) 36.
230. R. Přibil and I. Michael, comp. *Z. anal. Chem.*, 144 (1955) 140.
231. H. Holzer, *Mikrochemie*, 8 (1930) 275; see also F. Feigl, P. Krumholz, and E. Rajmann, *ibid.*, 165 (1931).
232. I. M. Kolthoff and E. B. Sandell, *J. Phys. Colloid Chem.*, 51 (1947) 210.
232a. G. Malatesta and E. di Nola, *Chem. Abstracts*, 8 (1914) 1397.
233. N. A. Tananaeff and K. A. Dolgow, *Chem. Abstracts*, 24 (1930) 1313.
234. C. Duval and P. Fauconnier, *Mikrochim. Acta*, 3 (1938) 30.
235. F. Feigl and V. Gentil, unpublished studies.
236. E. Eegriwe, *Z. Anal. Chem.*, 70 (1927) 400; C. C. Miller and A. J. Lowe, *J. Chem. Soc.*, 143 (1940) 1263.
237. E. B. Sandell, *Colorimetric Determination of Traces of Metals*, 2nd edition, New York, 1950, page 341.
238. R. N. Costeanu, *Z. anal. Chem.*, 104 (1936) 351.
239. Ph. W. West and Th. C. McCay, *Anal. Chem.*, 27 (1955) 1820.
240. J Doležal and P. Berman, comp. *Z. anal. Chem.*, 145 (1955) 44.
241. C. R. Fresenius, *Qualitative Analysis*, 17th ed. (translated by C. A. Mitchell), New York, 1921, p. 314. See also L. J. Curtmann and P. Rothberg, *J. Am. Chem. Soc.*, 33 (1911) 718.
242. Compare F. Feigl, Ref. 4, p. 139.
243. F. L. Hahn, *Mikrochemie*, 8 (1930) 77.
244. L. J. Curtmann and P. Rothberg, *J. Am. Chem. Soc.*, 33 (1911) 718.
245. F. Feigl and E. Fränkel, *Ber.*, 65 (1932) 540.
246. C. Paal and L. Friederici, *Ber.*, 64 (1931) 1766, 2561; see also R. Scholder and H. L. Haken, *ibid.*, 64 (1931) 2870.
247. C. Paal and L. Frederici, *Ber.*, 65 (1932) 540.

248. N. A. Tananaeff and G. T. Michaltschischin, *Z. anal. Chem.*, 94 (1933) 188.
249. F. Feigl and V. Gentil, unpublished studies.
250. N. A. Tananaeff and K. A. Dolgow, *Chem. Abstracts*, 24 (1930) 1313.
251. F. Feigl and P. Krumholz, *Ber.*, 63 (1930) 1917.
252. C. Zenghelis, *Z. anal. Chem.*, 49 (1910) 729.
253. F. Feigl, *Chemistry & Industry*, 16 (1938) 1161.
254. F. Feigl and C. M. Pinto, unpublished studies.
255. F. Feigl and V. Gentil, unpublished studies.
256. Compare F. Feigl, Ref. 4, Chapter 11.
257. J. H. Yoe and L. G. Overholser, *J. Am. Chem. Soc.*, 61 (1939) 2058; 63 (1941) 3224.
258. F. Feigl, Reference 4, page 326.
259. F. Mylius, *Z. anorg. Chem.*, 70 (1911) 203.
260. V. Lenher and C. H. Kao, *J. Phys. Chem.*, 230 (1926) 126.
261. O. König and W. R. Crowell, *Mikrochem. ver. Mikrochim. Acta*, 33 (1948) 298.
262. For preparation see *Organic Syntheses*, Vol. II, New York, 1944, p. 485.
263. N. C. Sogani and S. C. Bhattacharyya, *Anal. Chem.*, 29 (1957) 397.
264. G. Gutzeit and R. Monnier, *Helv. Chim. Acta*, 16 (1933) 233.
265. F. Feigl, P. Krumholz, and E. Rajmann, *Mikrochemie*, 9 (1931) 165.
266. F. Feigl, unpublished studies.
267. A. Steigmann, *J. Soc. Chem. Ind.*, 62 (1943) 42.
268. F. Feigl and G. B. Heisig, *J. Am. Chem. Soc.*, 73 (1951) 5634.
269. Ph. W. West and E. S. Amis, *Ind. Eng. Chem. Anal. Ed.*, 18 (1946) 400.
270. F. Feigl and J. E. R. Marins, unpublished studies.
271. F. Feigl and A. Caldas, *Anal. Chim. Acta*, 13 (1955) 526.
272. F. Feigl, studies with S. Pickholz.
273. K. A. Hofmann, *Ber.*, 45 (1912) 3329; K. A. Hofmann, O. Ehrhart and O. Schneider, *ibid.*, 46 (1913) 1657.
274. H. Wölbling and B. Steiger, *Mikrochemie*, 15 (1934) 295.
275. For preparation, see R. Wollner, *J. prakt. Chem.*, 29 (1884) 129.
276. N. A. Tananaeff and A. N. Romanjuk, *Z. anal. Chem.*, 108 (1937) 30.
277. D. Raquet, *Chem. Abstracts*, 42 (1948) 6698.
278. M. Ilinsky, *Ber.*, 17 (1884) 2592; see also F. Feigl and R. Stern, *Z. anal. Chem.*, 60 (1921) 31.
279. W. Böttger, *Mikrochemie* (*Emich Festschrift*), 1930, p. 28.
280. F. W. Atack, *J. Soc. Chem. Ind.*, 34 (1915) 641.
281. Compare B. N. Cacciapuoti and F. Ferla, *Ann. chim. applicata*, 29 (1939) 166.
282. Pr. Rây and R. M. Rây, *Quart. J. Indian Chem. Soc.*, 3 (1926) 118; Pr. Rây, *Z. anal. Chem.*, 79 (1929) 94.
283. For preparation, see A. Wollner, *J. prakt. Chem.*, 29 (1884) 129.
284. H. Ditz, *Chem.-Ztg.*, 25 (1901) 110; 46 (1922) 121; *Z. anorg. allgem. Chem.*, 219 (1934) 97; I. M. Kolthoff, *Mikrochemie*, 8 (1930) 176.
285. Compare H. W. Vogel, *Ber.*, 12 (1879) 2313; A. Rosenheim and coworkers, *ibid.*, 33 (1900) 1113; *Z. anorg. allgem. Chem.*, 27 (1901) 289; 49 (1906) 28.
286. M. Lehné, *Chem. Abstr.*, 45 (1951) 6117; comp. also A. K. Babko and O. P. Droho, *Chem Abstr.*, 47 (1953) 3175.
287. M. Ziegler and O. Glemser, *Z. anal. Chem.*, 152 (1956) 241.
288. I. M. Kolthoff, *Mikrochemie*, 8 (1930) 176.
289. F. Feigl and D. Goldstein, *Analyst*, 81 (1956) 709.
290. L. A. Sarver, *Ind. Eng. Chem., Anal. Ed.*, 10 (1938) 378.
291. A. Steigmann, *J. Soc. Chem. Ind.*, 62 (1943) 42.
292. F. Feigl and R. Uzel, *Mikrochemie*, 19 (1936) 132.
293. A. de Sousa, *Mikrochem. ver. Mikrochim. Acta*, 40 (1953) 252; comp. C. Duval, *Anal. Chim. Acta*, 1 (1947) 201.
294. Ph. W. West and L. A. Longacre, *Anal. Chim. Acta*, 6 (1953) 485.
295. P. Senise, *Mikrochim. Acta* (1957), in press.
296. L. Tschugaeff, *Ber.*, 38 (1905) 2520; O. Brunck, *Z. angew. Chem.*, 20 (1907) 1844; 27 (1914) 315. Compare F. J. Welcher, *Organic Analytical Reagents*, Vol. III, New York, 1907, p. 162.

297. I. G. Weeldenburg, *Rec. trav. chim.*, 43 (1924) 465; compare F. Feigl, Ref. 4, p. 277.
298. F. Feigl and H. J. Kapulitzas, *Mikrochemie*, 8 (1930) 244.
299. F. Feigl and H. J. Kapulitzas, *Mikrochemie* (*Emich Festschrift*), 1930, p. 128.
300. F. Feigl, *Ber.*, 57 (1924) 758; see also A. P. Rollet, *Compt. rend.*, 183 (1926) 212.
301. M. Hooreman, *Anal. Chim. Acta*, 3 (1949) 635.
302. F. Feigl, *Chemistry of Specific, Selective and Sensitive Reactions*, New York, 1949, p. 278.
303. A. Okáč and M. Polster, *Chem. Abstracts*, 43 (1949) 3740.
304. Pr. Rây and R. M. Rây, *Quart. J. Indian Chem. Soc.*, 3 (1926) 118; Pr. Rây, *Z. anal. Chem.*, 79 (1930) 95.
305 F. Feigl and H. J. Kapulitzas, *Mikrochemie*, 8 (1930) 239.
306. E. A. Werner, *Chem. News*, 53 (1886) 51.
307. F. Feigl, *Chem.-Ztg.*, 44 (1920) 689.
308. Ph. C. Browning and E. Palmer, *Am. J. Sci.*, 27 (1909) 379.
309. F. Feigl and D. Goldstein, unpublished studies.
310. N. S. Poluektoff, *Chem. Zentrbl.*, 1934, I, 3240.
311. F. Feigl and P. E. Barbosa, *Rev. soc. brasil quim.*, 10 (1941) 137.
312. F. Feigl and D. Goldstein, unpublished studies.
313. H. Knowles, *U. S. Bur. Stand. J. Research*, 9 (1932) 1.
314. F. Feigl, V. Gentil and D. Goldstein, *Anal. Chim. Acta*, 9 (1953) 393
315. E. Eegriwe, *Z. anal. Chem.*, 70 (1927) 400; C. C. Miller and A. J. Lowe, *J. Chem. Soc.* 143 (1940) 1258, 1263; C. C. Miller, *ibid.*, 144 (1941) 72.
316. S. H. Webster and L. T. Fairhall, *J. Ind. Tox.*, 27 (1945) 183.
317. N. A. Tananaeff, *Z. anal. Chem.*, 88 (1932) 343.
318. L. Kul'berg, *Mikrochemie*, 19 (1936) 183.
319. H. Gotô, *Chem. Zentrbl.*, (1941) I, 1068.
320. H. Gotô, *Chem. Abstracts*, 32 (1938) 5721; 35 (1941) 1720.
321. F. Feigl and L. Baumfeld, *Anal. Chim. Acta*, 3 (1949) 83.
322. M. Kohn, *Z. anorg. Chem.*, 49 (1906) 443.
323. M. Kohn, *Monatsh.*, 43 (1922) 373; 44 (1923) 97.
324. F. Feigl and H. Hamburg, *Z. anal. Chem.*, 86 (1931) 7.
325. F. Blau, *Monatsh.*, 19 (1898) 647.
326. P. E. Wenger and R. Duckert, *Helv. Chim. Acta*, 27 (1944) 757.
327. For preparation, see F. Hein and W. Retter, *Ber.*, 61 (1928) 1790; see also J. P. Wibaut and H. D. Tjeenk Willink Jr., *Rec. trav. chim.*, 54 (1935) 278.
328. F. Feigl and A. Caldas, *Anal. Chem.*, 29 (1957).
329. J. H. Yoe, *J. Am. Chem. Soc.*, 54 (1932) 4139.
330. For preparation, see A. Claus, *Friedländer*, 3 (1890–94) 964.
331. H. J. Schlesinger and H. B. van Valkenburgh, *J. Am. Chem. Soc.*, 53 (1931) 1212; S. Z. Lewin and R. Seiden, *J. Chem. Education*, 30 (1953) 445.
332. G. Nadler, *J. prakt. Chem.*, 99 (1866) 188.
333. K. Heller and P. Krumholz, *Mikrochemie*, 7 (1929) 219.
334. L. Tschugaeff and B. Orelkin, *Z. anorg. allgem. Chem.*, 89 (1914) 401; P. Slawik, *Chem.-Ztg.*, 36 (1912) 54; F. Feigl, *Chem. Ztg.*, 43 (1919) 861.
335. Compare F. J. Welcher, *Organic Analytical Reagents*, Vol. III, New York, 1947, p. 210.
336. Compare F. Feigl, Ref. 4, p. 277.
337. J. H. Yoe and A. L. Jones, *Ind. Eng. Chem., Anal. Ed.*, 16 (1944) 111.
338. H. Reihlen, *Z. anorg. Chem.*, 123 (1922) 173; R. Weinland and E. Walter, *Z. anorg. allgem. Chem.*, 126 (1923) 148; P. Karrer, *Organic Chemistry*, 4th English Edition, Amsterdam, 1950, p. 435; F. Feigl, Ref. 4, p. 363.
339. A. de Sousa, *Mikrochem. ver. Mikrochim. Acta*, 40 (1953) 265.
340. E. Lyons, *J. Am. Chem. Soc.*, 49 (1927) 1916.
341. E. A. Kocsis, *Mikrochemie*, 25 (1938) 13.
342. L. A. Sarver, *Ind. Eng. Chem., Anal. Ed.*, 10 (1938) 378.
343. F. Vlácil and V. Hovorka, *Chem. Abstracts*, 46 (1952) 3899.
344. P. Cazeneuve, *Compt. rend.*, 131 (1900) 346. *Bull. soc. chim.*, 23 (1900) 701; 25 (1901) 761.
345. M. Bose, *Anal. Chim. Acta*, 10(1954) 201, 209.

346. F. Feigl and W. A. Mannheimer, unpublished studies.
347. Studies with R. Nováček.
348. See J. F. Ege Jr., and L. Silverman, *Anal. Chem.*, 19 (1947) 693, regarding the preparation and use of this reagent paper.
349. F. Ibbotson and R. Howden, *Chem. News*, 90 (1904) 320.
350. F. Feigl, *Z. angew. Chem.*, 44 (1931) 741.
351. For preparation, see K. H. Slotta and K. R. Jacobi, *Z. anal. Chem.*, 77 (1929) 346.
352. F. Feigl and P. Krumholz, *Mikrochemie* (*Pregl Festschrift*), 1929, p. 80.
353. Compare F. Feigl, Ref. 4, p. 76.
354. N. A. Tananaeff, *Z. anorg. allgem. Chem.*, 140 (1924) 327
355. N. M. Cullinane and S. J. Chard, *Analyst*, 73 (1948) 95.
356. For preparation, see German Patent 48,709; F. J. Welcher, *Organic Analytical Reagents*, New York, 1947, Vol. II, p. 337.
357. F. Feigl and R. Stern, *Z. anal. Chem.*, 60 (1921) 28.
358. R. L. Dromljuk, *J. chim. appliqueé* (U.S.S.R.), 13 (1940) 157.
359. S. Augusti, *Mikrochemie*, 17 (1935) 17.
360. H. Marshall, *Z. anal. Chem.*, 43 (1904) 418.
361. Compare H. Marshall, *Z. physik. Chem.*, 37 (1901) 255 and W. Oelschlager, *Z. anal. Chem.*, 144 (1955) 27.
362. Compare Feigl, Ref. 4, p. 128.
363. F. Feigl and E. Fränkel, *Ber.*, 65 (1932) 541.
364. P. Fleury, *Compt. rend.*, 171 (1927) 957.
365. M. Duyk, *Ann. chim. applicata*, 12 (1907) 465, used KOCl and $CuSO_4$ in an analogous test for manganese.
366. H. H. Willard and L. H. Greathouse, *J. Am. Chem. Soc.*, 39 (1917) 2366.
367. J. Tillmans and H. Milden, *Chem. Abstracts*, 8 (1914) 3085.
368. W. Prodinger, *Mikrochem. ver. Mikrochim. Acta*, 36-37 (1951) 580.
369. For preparation, see J. Biehringer, *J. prakt. Chem.*, 54 (1896) 240.
370. F. Feigl, *Chem. Ztg.*, 44 (1920) 689; *Z. anal. Chem.*, 60 (1921) 24.
371. According to experiments carried out by S. Pickholz.
372. N. A. Tananaeff and I. Tananaeff, *Z. anorg. allgem. Chem.*, 170 (1928) 118.
373. Compare also D. Balarew, *Z. anal. Chem.*, 60 (1921) 392.
374. E. Eegriwe, *Z. anal. Chem.*, 74 (1928) 228.
375. W. H. Cone and L. C. Cady, *J. Am. Chem. Soc.*, 49 (1927) 2214.
376. H. Fischer, *Z. angew. Chem.*, 42 (1929) 1025; *Mikrochemie*, 8 (1930) 319.
377. For preparation, see E. Fischer, *Ann.*, 190 (1878) 16, 118; E. Fischer and E. Besthorn, *ibid.*, 212 (1881) 316.
378. G. Rienäcker and W. Schiff, *Z. anal. Chem.*, 94 (1933) 410.
379. Compare E. B. Sandell, *Colorimetric Determination of Traces of Metals*, 2nd ed., New York, 1950, p. 619.
380. P. Krumholz and J. V. Sanchez, *Mikrochemie*, 15 (1934) 114.
381. R. Belcher, A. J. Nutten and W. I. Stephen, *Analyst*, 76, (1951) 903.
382. For preparation, see K. Fries and W. Lohmann, *Ber.*, 54 (1921) 2922.
383. G. Sensi and R. Testori, *Ann. chim. applicata*, 19 (1929) 383; see also J. H. Yoe, *A Laboratory Manual of Qualitative Analysis*, New York, 1938, p. 176.
384. A. A. Benedetti-Pichler, *Ind. Eng. Chem., Anal. Ed.*, 4 (1932) 336.
385. L. Szebellédy and I. Tanay, *Z. anal. Chem.*, 106 (1936) 342.
386. F. Feigl, unpublished studies.
387. Fr. Goppelsroeder, *Z. anal. Chem.*, 7 (1868) 195; E. Eegriwe, *ibid.*, 76 (1929) 440.
388. E. Schantl, *Mikrochemie*, 2 (1924) 174.
389. H. L. Zermatten, *J. Proc. Acad. Sci. Amsterdam*, 36 (1933) 899.
390. G. Beck, *Mikrochim. Acta*, 2 (1937) 9, 287; *Mikrochemie*, 20 (1936) 194.
391. For a critical survey, see G. Charlot, *Anal. Chim. Acta*, 1 (1947) 231.
392. F. Feigl and H. E. Ballaban, unpublished studies.
393. F. W. Atack, *J. Soc. Chem. Ind.*, 34 (1935) 936; see also K. Heller and P. Krumholz, *Mikrochemie*, 7 (1929) 221.

394. F. Feigl and R. Stern, *Z. anal. Chem.*, 60 (1921) 9.
395. Compare Feigl, Ref. 4, p. 536.
396. F. L. Hahn, *Mikrochemie*, 11 (1932) 33.
397. F. Feigl and D. Goldstein, *Anal. Chem.*, 29 (1957) 456.
398. C. E. White and C. S. Lowe, *Ind. Eng. Chem., Anal. Ed.*, 9 (1937) 430.
399. L. Hammett and C. Sottery, *J. Am. Chem. Soc.*, 47 (1925) 142.
400. W. E. Thrun, *J. Chem. Education*, 14 (1937) 281.
401. Compare J. M. Yoe and W. L. Hill, *J. Am. Chem. Soc.*, 49 (1927) 2395; J. M. Yoe, *ibid.*, 54 (1932) 1022.
402. F. Feigl and D. Goldstein, unpublished studies.
403. H. L. Zermatten, *J. Proc. Acad. Sci. Amsterdam*, 36 (1933) 899.
404. Compare R. Přibil and K. Kucharský, *Chem. Abstracts*, 45 (1951) 3286.
405. H. Fischer, *Z. anal. Chem.*, 73 (1928) 54.
406. A. S. Komarowsky and I. M. Korenman, *Z. anal. Chem.*, 94 (1933) 247.
407. H. Fischer, *Z. anal. Chem.*, 73 (1928) 57.
408. A. S. Komarowsky and N. S. Poluektoff, *Mikrochemie*, 14 (1933) 315.
409. For preparation, see A. S. Komarowsky and N. S. Poluektoff, *Mikrochemie*, 14 (1933) 315.
410. M. Theiss, *Z. anal. Chem.*, 144 (1955) 192.
411. G. Holzbecker, comp. *Z. anal. Chem.*, 148 (1955) 201.
412. J. L. Schönn, *Z. anal. Chem.*, 9 (1870) 330.
413. R. Schwarz and W. Sexauer, *Ber.*, 60 (1927) 500; R. Schwarz, *Z. anorg. allgem. Chem.*, 210 (1933) 303.
414. F. Feigl and A. Schaeffer, *Anal. Chem.*, 23 (1951) 353.
415. J. Piccard, *Ber.*, 42 (1909) 4343.
416. A. Rosenheim, B. Raibmann, and G. Schendel, *Z. anorg. allgem. Chem.*, 196 (1931) 160.
417. Compare N. R. Pike and coworkers, *Ber.*, 68 (1935) 1023.
418. K. A. Hofmann, *Ber.*, 45 (1912) 2480.
419. A. Okáč and L. Sommer, *Z. anal. Chem.* 143 (1954) 52.
420. V. J. Kuznetsov, *Chem. Abstracts*, 39 (1945) 4562.
421. N. A. Tananaeff and G. A. Pantschenko, *Z. anorg. allgem. Chem.*, 150 (1926) 163.
422. F. Feigl and H. E. Ballaban, unpublished studies.
423. G. Almássy, comp. *Z. anal. Chem.*, 145 (1955) 62.
424. M. V. Gaptschenko and O. G. Scheinziss, *Chem. Abstracts*, 31 (1937) 8430.
425. O. K. Dobroljubski, *Chem. Abstracts*, 32 (1938) 4465.
426. P. Szarvas and G. Almássy, comp. *Anal. Abstr.*, 2 (1955) 3012.
427. G. Charlot, *Anal. Chim. Acta*, 1 (1947) 244.
428. J. Bellucci and G. Savoia, *Chem. Abstracts*, 18 (1924) 3333.
429. For preparation, see R. Henriques and M. Ilinsky, *Ber.*, 18 (1885) 704.
430. J. H. de Boer, *Chem. Weekblad*, 21 (1924) 404; *Chem. Abstracts*, 19 (1925) 793.
431. According to experiments carried out by E. Rajmann.
432. G. Charlot, *Anal. Chim. Acta*, 1 (1947) 234.
433. F. Feigl, P. Krumholz, and E. Rajmann, *Mikrochemie*, 9 (1931) 395.
434. Compare Feigl, Ref. 4, p. 291.
435. W. G. Hayes and E. H. Jones, *Ind. Eng. Chem., Anal. Ed.*, 13 (1941) 603.
436. For preparation, see F. Feigl, P. Krumholz and E. Rajmann, *Mikrochemie*, 9 (1931) 395.
437. F. Pavelka, *Mikrochemie*, 8 (1930) 345.
438. J. H. Yoe and L. G. Overholser, *Ind. Eng. Chem., Anal. Ed.*, 15 (1943) 373.
439. R. E. Oesper, R. Dunleavy and J. Klingenberg, *Anal. Chem.*, 24 (1952) 1492.
440. A. de Sousa, *Mikrochem. ver. Mikrochim. Acta*, 40 (1953) 319.
441. Compare F. J. Welcher, *Organic Analytical Reagents*, Vol. I, New York, 1947, p. 264ff.
442. F. Hecht and W. Reich-Rohrwig, *Monatsh.*, 53 54 (1926) 596; F. Hecht and H. Krafft-Ebing, *Z. anal. Chem.*, 106 (1936) 321.
443. F. Feigl and R. Stern, *Z. anal. Chem.*, 60 (1921) 39.
444. N. A. Tananaeff and G. A. Pantschenko, *Z. anorg. allgem. Chem.*, 150 (1926) 164.
445. F. Hernegger, *Anz. Akad. Wiss. Wien, Math.-naturw. Klasse*, 144 (1935) 217; F. Hernegger and B. Karlik, *Sitzber. Akad. Wiss. Wien, Math.-naturw. Klasse, Abt. IIa*, 144 (1935) 217;

Chem. Abstracts, 30 (1936) 408.
446. E. L. Nichols and M. K. Slattery, *J. Optical Soc. Am.*, 12 (1926) 449.
447. J. Papisch and L. E. Hoag, *Proc. Natl. Acad. Sci. U. S.*, 13 (1927) 726.
448. L. F. Carvalho (Rio de Janeiro), private communication.
449. Compare R. Fresenius and G. Jander, *Handbuch der analytischen Chemie*, Vol. IV, Berlin, 1948, p. 255.
450. F. Feigl and D. Goldstein, unpublished studies.
451. F. Feigl, V. Gentil and D. Goldstein, unpublished studies.
452. E. A. Kocsis, *Mikrochemie*, 25 (1938) 13.
453. H. Gotô, *Sci. Rep. Tohoku Univ.*, 29 (1940) 287.
454. M. Nageswara Rao and Ph. S. V. Raghawa Rao, *Z. anal. Chem.*, 142 (1954) 161.
455. F. Lucena Conde and L. Prat, *Mikrochim. Acta*, (1955) 799.
456. P. E. Lecoq de Boisbaudran, *Compt. rend.*, 100 (1885) 605.
457. L. Pissarjewsky, *Z. anorg. allgem. Chem.*, 31 (1902) 359; see also A. Lawson and E. W. Balson, *J. Chem. Soc.*, 138 (1935) 362.
458. W. Biltz and F. Zimmermann, *Ber.*, 40 (1907) 4979.
459. F. Feigl, *Oesterr. Chem.-Ztg.*, 22 (1919) 124.
460. A. S. Komarowsky and S. M. Korenmann, *Mikrochemie*, 12 (1932) 211.
461. For preparation, see A. Linz, *Ind. Eng. Chem., Anal. Ed.*, 15 (1943) 459.
462. L. Kul'berg, *Mikrochemie*, 21 (1936) 35.
463. F. A. Shemyakin and A. N. Belokon, *Chem. Abstracts*, 32 (1938) 4470.
464. A. S. Komarowsky and N. S. Poluektoff, *Mikrochemie*, 16 (1935) 227.
465. N. S. Poluektoff, *Mikrochemie*, 19 (1936) 248.
466. S. Lacroix, *Anal. Chim. Acta*, 1 (1947) 290.
467. H. Onishi, *Anal. Chem.*, 27 (1955) 872.
468. F. Feigl, *Mikrochemie*, 2 (1924) 188.
469. Compare Feigl, Ref. 4, p. 174.
470. For preparation, see G. Gutzeit, *Helv. Chim. Acta*, 12 (1929) 713.
471. Compare A. L. Godbert and R. Belcher, *Analyst*, 64 (1939) 346.
472. See F. Feigl and H. A. Suter, *Ind. Eng. Chem., Anal. Ed.*, 14 (1942) 840.
473. A. J. Llacer, *Mikrochim. Acta*, (1955) 921.
474. According to experiments carried out with J. V. Tamchyna and L. Weidenfeld. Compare F. Feigl, *Z. angew. Chem.*, 43 (1930) 550.
475. Compare Feigl, Ref. 4, p. 147.
476. F. Feigl and W. Aufrecht, *Rec. trav. chim.*, 58 (1939) 1127.
477. H. E. Wohlers, *Z. anorg. allgem. Chem.*, 59 (1908) 203; compare H. Yagoda, *J. Ind. Hyg. Tox.*, 26 (1944) 224.
478. J. H. Yoe, *A Laboratory Manual of Qualitative Analysis*, New York, 1938, p. 177.
479. L. M. Kul'berg and R. B. Liokumovich, *Chem. Abstracts*, 44 (1950) 2412.
480. F. Feigl, *Mikrochemie*, 2 (1924) 187.
481. Unpublished studies with S. Pickholz; see also F. Feigl, *Rec. trav. chim.*, 58 (1939) 472.
482. Compare Feigl, Ref. 4, p. 666.
483. For preparation, see J. H. Ziegler and M. Locher, *Ber.*, 20 (1887) 836.
484. F. Feigl and V. Gentil, *Mikrochim. Acta*, (1954) 435.
485. F. Schlagdenhauffen, *Z. oesterr. Apoth.-Ver.*, 16 (1878) 348; A. Hamy, *Compt. rend.*, 183 (1926) 129; W. J. Petraschenj, *Z. anal. Chem.*, 71 (1927) 291.
486. P. Remy-Gennete, *Bull. Soc. Ann. France*, 5 (1938) 666.
487. F. L. Hahn, H. Wolf, and G. Jaeger, *Ber.*, 57 (1924) 1394.
488. See A. S. Komarowsky and I. M. Korenman, *Z. anal. Chem.*, 94 (1933) 247.
489. According to experiments by K. Weisselberg, *Dissertation*, Vienna, 1930.
490. The reagent is also used by K. Suitsu and K. Okuma, *J. Soc. chem. Ind. Japan*, 29 (1926) 132, by W. L. Ruigh, *J. Am. Chem. Soc.*, 51 (1929) 1547, and by E. W. Engel, *J. Am. Chem. Soc.* 52 (1930) 1812. Other azo dyestuffs of similar behavior are described by E. Eegriwe, *Z. anal. Chem.*, 76 (1929) 354.
491. For preparation of *p*-nitrobenzeneazo-α-naphthol, see E. Bamberger and F. Meimberg, *Ber.*, 28 (1895) 848.

492. I. M. Kolthoff, *Chem. Weekblad*, 24 (1927) 254; *Mikrochemie* (*Emich Festschrift*), 1930, p. 180.
493. N. A. Tananaeff, *Z. anal. Chem.*, 88 (1932) 93.
494. E. Gagliardi and M. Theiss, *Z. anal. Chem.*, 144 (1954) 113, 264.
495. I. M. Kolthoff, *Z. anal. Chem.*, 70 (1927) 398.
496. F. Feigl, *Rec. trav. chim.*, 58 (1939) 473.
497. L. L. Burgess and O. Kamm, *J. Am. Chem. Soc.*, 34 (1912) 651.
498. L. de Koninck, *Z. anal. Chem.*, 49 (1910) 53.
499. N. S. Poluektoff, *Mikrochemie*, 14 (1933-34) 265; see also F. Feigl and P. E. Barbosa, *Rev. soc. brasil. quim.*, 10 (1941) 137.
500. For preparation, see K. H. Mertens, *Ber.*, 11 (1878) 845.
501. G. Wittig and coworkers, *Ann.* 563 (1949) 114, 118, 126.
502. Compare review by A. J. Nutten, *J. Ind. Chem.*, 30 (1954) 29, 57, 245.
503. O. Pročke and R. Uzel, *Mikrochim. Acta*, 3 (1938) 105.
504. Compare F. Feigl, Ref. 4, page 94.
505. J. Steward and I. W. Young, *J. Am. Chem. Soc.*, 57 (1935) 695.
506. Compare F. Feigl, *Proc. Int. Symposium on the Reactivity of Solids*, Gothenburg 1952, p. III.
507. E. S. Burkser and M. v. Kutschment, *Mikrochemie*, 18 (1935) 18.
508. N. A. Tananaeff, *Z. anal. Chem.*, 88 (1932) 343.
509. E. Riegler, *Chem.-Ztg.*, 21, Repertorium 307 (1897).
510. F. Feigl, *Mikrochemie*, 7 (1929) 12.
511. I. M. Korenman, *Z. anal. Chem.*, 90 (1932) 115.
512. Compare W. Böttger, *Qualitative Analyse anorganischer Verbindungen*, Leipzig, 1925, p. 478.
513. F. Feigl, *Mikrochemie*, 13 (1933) 132.
514. Compare H. Malissa and coworkers, *Mikrochem. ver. Mikrochim. Acta*, 38 (1951) 386.
515. L. Veluz and M. Pesez, *Ann. Pharm. France*, 4 (1946) 10.
516. R. Berg, *Z. anal. Chem.*, 71 (1927) 171.
517. F. Feigl, C. Costa Neto and J. E. R. Marins, unpublished studies.
518. C. Zenghelis, *Compt. rend.*, 173 (1921) 153.
519. G. K. Makris, *Z. anal. Chem.* , 81 (1932) 212.
520. N. A. Tananaeff and A. A. Budkewitsch, *Chem. Abstracts*, 30 (1936) 5905.
521. M. Ishidate, Tokyo, private communication.
522. F. Feigl, *Rec. trav. chim.*, 58 (1938) 474.
523. Compare F. Feigl, Ref. 4, p. 396.
524. F. Feigl and W. A. Mannheimer, *Mikrochem. ver. Mikrochim. Acta*, 40 (1952) 50.
525. F. Feigl and M. Steinhauser, *Mikrochem. ver. Mikrochim. Acta*, 35 (1950) 553.
526. M. Pesez and A. Petit, *Bull. soc. chim. France*, (1947) 122.
527. F. Feigl and W. A. Mannheimer, *Mikrochem. ver. Mikrochim. Acta*, 40 (1952) 355.
528. F. Feigl, Ref. 4, p. 530 ff.
529. F. Feigl and M. Steinhauser, *Mikrochem. ver. Mikrochim. Acta*, 35 (1950) 553, and unpublished studies.
530. W. N. Hirschel and J. A. Verhoeff, *Chem. Weekblad*, 20 (1923) 319; see also F. Feigl, *Rec. trav. chim.*, 58 (1939) 474.
531. For preparation, see O. Diels and H. Jost, *Ber.*, 35 (1902) 3292.
532. F. Feigl, *Rec. trav. chim.*, 58 (1939) 474.
533. F. Feigl and W. Mannheimer, *Mikrochem. ver. Mikrochim. Acta*, 40 (1952) 50.
534. O. A. Guagnini and E. E. Vonesch, *Mikrochim. Acta* (1954) 211.
535. G. Oddo and E. Deleo, *Ber.*, 69 (1936) 287.
536. J. Blom, *Ber.*, 59 (1926) 121; compare G. Endres, *Ann.*, 518 (1935) 109.
537. F. Raschig, *Schwefel- und Stickstoffstudien*, Verlag Chemie, Leipzig, 1924, p. 183.
538. F. Feigl and V. Demant, *Mikrochim. Acta*, 1 (1937) 132.
539. F. Feigl and R. Uzel, *Mikrochemie*, 19 (1936) 136.
540. F. Feigl and M. Steinhauser, *Mikrochem. ver. Mikrochim. Acta*, 35 (1950) 553.
541. F. Feigl, unpublished studies.

Chapter 4

Tests for acid radicals

ANIONS

The systematic search for acid radicals is usually made in solutions that contain cations only of the alkali metals and univalent thallium. Such solutions (sodium carbonate extracts) are obtained: *a*) by treating the test solution with sodium carbonate and removing off the resulting precipitate; or *b*) by fusing the solid sample with sodium carbonate (or $NaKCO_3$); the melt is extracted with water and filtered if need be. In both procedures, the metal cations are transformed into their slightly soluble carbonates or hydroxides, and thus separated from the water-soluble alkali salts of the acid radicals. Univalent thallium, whose carbonate is soluble, presents an exception. Likewise many complex anions formed by certain heavy metals with organic acids are not altered by treatment with sodium carbonate. Spot reactions to detect the various acid radicals (anions) can be successfully made on drops of the sodium carbonate extract.

When fusing mixtures of materials with sodium carbonate (or Na_2CO_3 + K_2CO_3), it should be noted that the high temperatures employed may bring about changes in the sample that do not occur when its water suspension is boiled with alkali carbonate. This applies particularly to thermo-decomposable materials such as ammonium salts, chlorates, iodates, perchlorates, persalts, etc., which give off ammonia or oxygen, and disappear completely even at temperatures below the fusion point of alkali carbonates. Formates and oxalates are converted into carbonates, the former with evolution of hydrogen and carbon monoxide, the latter with loss of carbon monoxide alone. Since these gases are liberated at high temperatures, they may immediately react with certain materials with which they are then in contact; e.g., nitrites and nitrates can be completely transformed into carbonate. Familiar instances of fusion reactions are the formations of chromate and permanganate in the systems: alkali carbonate + Cr^{III} (Mn^{II}) + alkali nitrate (or halogenate).

Because of the possibility of alterations during carbonate fusions, this procedure should never be used directly when testing for anions; it should be applied, if at all, only after the sample has been digested with sodium carbonate solution.

It is advisable to make preliminary tests before proceeding to the systematic search for acid radicals. Such tests should especially include a study of the behavior of the sample toward water and chemically active solvents, keeping in mind the cations whose presence has been established by a preliminary separate test. Furthermore, it must always be remembered that the proved presence of oxidizing anions excludes the presence of reducing anions and *vice versa*.

The discussion (page 57) should be referred to with regard to the choice of spot tests for acid radicals, giving due consideration to the objectives of the macro- or microchemical analysis, the necessity of adhering closely to the procedures as given, and the making of model tests when using unfamiliar methods of detection. What has been said there relative to tests for metals applies equally to acid radicals.

Since no really satisfactory spot test scheme is known for the separation of the acid radicals into definite groups,* the acid-forming nonmetal is used as the basis of arranging the acid radicals in this section. The sensitivity data given for the various tests are valid, unless otherwise stated, for solutions of pure alkali salts tested by the respective procedures.

1. Hydrochloric acid

(1) Test by precipitation of silver chloride in the presence of other halide ions [1]

In contrast to acetic acid or weak mineral acid solutions of chlorides, which remain practically unaffected on the addition of hydrogen peroxide, bromides and iodides are oxidized to elementary bromine and iodine, respectively. If this oxidation is conducted in the presence of 8-hydroxyquinoline (oxine), this phenolic compound is brominated or iodized by the free halogen. The acid solution of the halogenated oxine does not react with silver nitrate and consequently only the unchanged chloride ions will be precipitated as silver chloride. If thiocyanates are warmed with acetic acid solution of hydrogen peroxide, the addition of silver nitrate produces no precipitate. The probable reason is that hydrogen peroxide converts thiocyanic acid into sulfuric acid and hydrogen cyanide. The production of $(CNS)_2$ is also possible and the latter behaves like bromine or iodine toward oxine. Since cyanides are decomposed by acids, the following procedure is specific for chloride in the presence of other halide ions. At least, the chloride test succeeds in 2 % solutions of the alkali salts of other halogens.

* See, however, I. K. Taimni and M. Lal, *Anal. Chim. Acta*, 17 (1957) 367, 372.

Procedure. The test is made in a micro test tube. A drop of the test solution is warmed with a drop of oxine solution, a drop of hydrogen peroxide solution, and a microdrop of nitric acid for about 4 minutes. A drop of 1 % $AgNO_3$ solution is then added. If chloride is present, a colorless precipitate or turbidity will appear, the intensity of the response depending on the quantity of chloride involved.

Limit of Identification: 2 γ chlorine
Limit of Dilution: 1 : 25,000
Reagents: (*1*) 2 % solution of oxine in 1 : 4 acetic acid
(*2*) Hydrogen peroxide solution: 2 parts of 6 % H_2O_2 plus 1 part dilute acetic acid

(*2*) *Test by formation of chromyl chloride*

Chlorides which react with sulfuric acid to give hydrochloric acid, yield chromyl chloride on being mixed with potassium dichromate and treated with concentrated sulfuric acid:

$$6\ H_2SO_4 + K_2Cr_2O_7 + 4\ KCl \rightarrow 2\ CrO_2Cl_2 + 3\ H_2O + 6\ KHSO_4$$

This product boils at 116° C and can be distilled out of the reaction mixture. The sample must be in the solid form because water hydrolyzes chromyl chloride to dichromic acid and hydrochloric acid. Since it can be saponified to give chromate, which even in great dilution gives a deep violet color with diphenylcarbazide (see page 167), chlorides may be detected in a few milligrams of solid or the evaporation residue from a drop of liquid.[2] This test does not apply to insoluble chlorides such as AgCl or Hg_2Cl_2, but $PbCl_2$ and TlCl respond to this test because of their appreciable solubility.

Procedure. A few grains of the solid sample, or a drop of the test solution, is placed in the glass bulb of the apparatus shown in Figure 23 (see page 47) and evaporated to dryness. A little powdered potassium dichromate and a drop of concentrated sulfuric acid are introduced and the apparatus is then stoppered. The knob of the stopper carries a drop of dilute alkali. The apparatus is heated for a few minutes. After cooling, the knob is dipped into a drop of a 1 % alcoholic solution of diphenylcarbazide[3] that has been treated with a drop of dilute sulfuric acid (spot plate). When chlorides were present, the chromyl chloride causes the reagent to change to violet.

Limit of Identification: 0.3 γ chlorine
Limit of Dilution: 1 : 150,000

It must be noted that there is always also a liberation of chlorine:

$$K_2Cr_2O_7 + 6\ KCl + 7\ H_2SO_4 = 3\ Cl_2 + 4\ K_2SO_4 + Cr_2(SO_4)_3 + 7\ H_2O$$

This reaction increases with increase of bichromate relative to chloride, and obviously this is the reason why the chromyl chloride reaction is not very sensitive when chlorides are present.

If iodide is present and the molecular ratio iodide : chloride exceeds 1 : 15, the chromyl chloride formation is prevented completely in favor of the liberation of chlorine.[4] This result is due to the production of iodic acid from iodide and chromic acid. In the presence of concentrated sulfuric acid, and especially on warming, the iodic acid liberates chlorine from chlorides with regeneration of iodide.

Small amounts of bromide do not interfere although, under the conditions of the test, they are oxidized to elementary bromine. Larger amounts of bromine interfere because they oxidize the diphenyl-carbazide. The free bromine may be removed, without materially affecting the sensitivity of the chromyl chloride reaction, by adding phenol to the reagent solution (formation of tribromophenol).

Nitrates interfere because nitrosyl chloride may be formed instead of chromyl chloride. Other oxidizing materials, such as nitrites, chlorates, etc., liberate chlorine. Fluorides should be removed beforehand because hydrogen fluoride forms a volatile compound resembling chromyl chloride.

(3) *Test by volatilization of hydrochloric acid* [5]

When decomposable solid chlorides are heated with concentrated nitric acid, not only nitrosyl chloride but also hydrochloric acid is produced. The latter renders silver nitrate solution cloudy because of the formation of AgCl. Bromides and iodides behave similarly. The reaction is therefore especially suitable for a rapid sensitive preliminary test for halogen hydracids.*

Procedure. A few particles of the sample are placed in the bulb of the apparatus (Figure 23), or a drop of the test solution is allowed to evaporate there. Two drops of concentrated nitric acid are then added. A drop of dilute silver nitrate solution is placed on the glass knob of the stopper which is then fitted into position. The apparatus is warmed over an asbestos plate until the first bubbles rise. For very small amounts of hydrochloric acid, the turbidity due to silver chloride is rendered more visible by transferring the drop to a black porcelain plate.

Limit of Identification: 1 γ hydrochloric acid

Limit of Dilution: 1 : 50,000

* A preliminary test for halides, that may be carried out on a *neutral* solution of alkali halide in the absence of other anions precipitated by Ag^+, is to mix a little well-washed Ag_2CO_3 with a drop of the test solution and one of phenolphthalein solution on a spot plate. In the presence of halides, alkali carbonate is formed: $Ag_2CO_3 + 2\,Hal^- = 2\,AgHal + CO_3^{-2}$. The latter turns the phenolphthalein red [*Idn. Limit:* 0.5 γ KCl (KBr) and 1 γ KI]. F. Feigl and R. Seboth, unpublished studies.

References pp. 356–360

(4) Identification of silver chloride [6]

Silver chloride reacts with potassium ferrocyanide:

$$4\ AgCl + K_4[Fe(CN)_6] \rightarrow Ag_4[Fe(CN)_6] + 4\ KCl$$

This reaction cannot be detected by mere observation because both silver salts are white. If, however, a precipitate of silver chloride that has been treated with $K_4Fe(CN)_6$ is moistened with a drop of concentrated nitric acid, the silver ferrocyanide is converted into silver ferricyanide, and the color changes from white to red-brown:

$$3\ Ag_4[Fe(CN)_6] + 4\ HNO_3 = 3\ AgNO_3 + NO + 2\ H_2O + 3\ Ag_3[Fe(CN)_6]$$

Other silver halides react analogously with ferrocyanide and concentrated nitric acid.

Silver chloride may be isolated from a mixture of silver halides by warming with a mixture of 4 parts saturated ammonium carbonate solution and 1 part ammonia. The silver chloride is completely soluble in this mixture, and can be reprecipitated from the filtrate by acidifying with nitric acid. Only a trace of silver bromide dissolves in the NH_3—$(NH_4)_2CO_3$ mixture; silver iodide is insoluble.

2. Hydrobromic acid

(1) Fluorescein test [7]

Free bromine converts the yellow dye fluorescein (I)

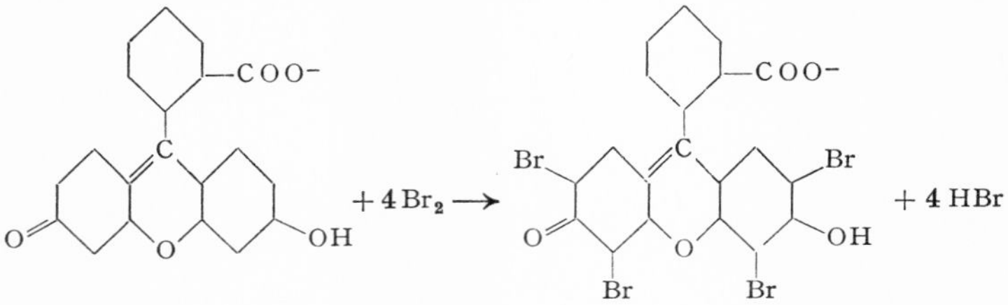

$$\text{(I)} + 4\ Br_2 \longrightarrow \text{(II)} + 4\ HBr$$

(I) (II)

into red tetrabromofluorescein (II) (eosin). Filter paper impregnated with an aqueous solution of fluorescein consequently turns red on coming into contact with bromine vapor.

To carry out this test for bromide, the latter must be oxidized to free bromine. This is best accomplished by heating with chromic acid, or with lead peroxide and acetic acid: *

$$2\ Br^- + PbO_2 + 4\ HAc \rightarrow Pb(Ac)_2 + 2\ Ac^- + 2\ H_2O + Br_2^0$$

* F. L. Hahn, *Mikrochemie*, 17 (1935) 228, recommended chloramine-T. It has the great advantage that the reaction takes place in a homogeneous medium, without going through the gas phase, and is applicable even in the presence of large amounts of chloride.

Chlorides do not affect the fluorescein test for bromides, since they remain practically unchanged after treatment with lead peroxide and acetic acid. Thus bromides may be detected in the presence of even large amounts (up to 1 : 10,000) of chlorides.[8] Iodides, however, are oxidized to free iodine under the conditions of the experiment, and can change the color owing to the formation of tetraiodofluorescein (erythrosine).* In this case, the bromide can be identified by procedure (*a*).

Procedure. (*a*) The gas apparatus shown on page 48 (Figure 26) is used. It is charged with a drop of the test solution and a few particles of lead peroxide and acetic acid, or a few drops of chromic acid. The apparatus is closed with the funnel stopper and filter paper moistened with fluorescein solution is laid over the funnel. The apparatus is gently warmed, and, according to the amount of bromine present, a circular red fleck is formed more or less rapidly on the yellow test paper. The apparatus of Figure 27 may alternatively be used.

Limit of Identification: 2 γ bromine
Limit of Dilution: 1 : 25,000
Reagent: Fluorescein solution: Saturated solution of the dyestuff in 1 : 1 alcohol. For the detection of traces of bromine, it is advisable to crystallize the dyestuff twice from 80 % alcohol.

The liberation of bromine from bromides and the formation of eosin also takes place on prolonged heating with an acetic acid-hydrogen peroxide mixture, in the presence of fluorescein. The following procedure [9] (*b*) is strongly recommended for the detection of bromine.

Procedure. (*b*) A spot plate or a micro crucible is used, a drop of the test solution is mixed with a microdrop of 1 % alcoholic fluorescein solution and a microdrop of a 10 : 1 mixture of glacial acetic acid and 30 % hydrogen peroxide, and evaporated to dryness over the water-bath. A pink to red color appears. A blank test on a drop of water is advisable when small amounts of bromine are suspected.

Limit of Identification: 0.3 γ bromine
Limit of Dilution: 1 : 165,000

If iodide is present, it is best to conduct the test on filter paper (procedure (*c*)) in which the iodide is oxidized to iodate by hydrogen peroxide and any elementary bromine liberated is absorbed by filter paper.[10]

Procedure. (*c*) One drop of the test solution is placed on filter paper, dried, and spotted with acetic acid-hydrogen peroxide solution (2 parts 6 % hydrogen peroxide + 1 part glacial acetic acid). It is important that the oxidizing mixture cover the fleck completely. The fleck is dried and if iodine is still visible the

* According to O. Frehden and C. H. Huang, *Mikrochem. ver. Mikrochim. Acta*, 26 (1939) 47, as little as 6 γ iodine produces a color change.

References pp. 356–360

treatment with the oxidant and drying must be repeated. The spot is then treated with a drop of 1 % alcoholic fluorescein solution. A red ring or stain develops on warming.

Limit of Identification: 0.3 γ bromine
Limit of Dilution: 1 : 160,000

(2) *Fuchsin test* [11]

The basic triphenylmethane dyestuff, fuchsin, containing the quinoidal cation

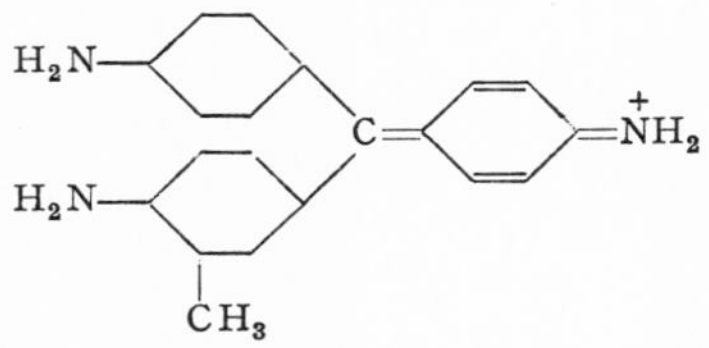

H_2N—
C= =$\overset{+}{N}H_2$
H_2N—
CH_3

is decolorized by bisulfite (see page 311). The colorless solution is turned blue by free bromine, due to the formation of a brominated dyestuff.[12] Neither free chlorine nor iodine tint fuchsin-bisulfite solution, and so the reaction is suitable for the detection of even small amounts of bromide in chlorides and/or iodides.

The liberation of bromine from bromides is best effected by a strong chromic acid solution; the bromine is allowed to act on filter paper impregnated with the reagent. In the absence of bromine, the paper becomes pink on warming; even small amounts of bromine give a distinct violet.

Procedure. A drop of the test solution, or a little of the solid substance, is placed in the bulb of the apparatus shown in Figure 27. Several drops of 25 % chromic acid solution are added and the apparatus is closed with the capillary fitting, which contains a drop of the reagent solution. The apparatus is then carefully warmed over a wire gauze (the liquid should not be allowed to boil). After a period, that varies according to the bromide content, the liquid in the capillary turns violet.

Limit of Identification: 3.2 γ bromine
Limit of Dilution: 1 : 15,600
Reagent: Fuchsin-bisulfite solution: 0.1 % fuchsin solution, decolorized with sodium bisulfite

(3) *Test with permolybdate and α-naphthoflavone* [13]

α-Naphthoflavone forms an insoluble blue-violet adsorption compound with iodine, and an orange-red compound with bromine.[14] Therefore, if bromine is liberated from bromides by means of a suitable oxidizing agent (e.g., permolybdic acid),[15] the colored bromonaphthoflavone compound is produced. The iodine compound is precipitated first if both halides are present. However, it is decomposed by further action of the oxidizing agent, and iodic acid results. Free chlorine does not react with naphthoflavone.

Procedure. A little solid molybdic acid is added to a drop of test solution, followed by a drop of a 0.1 % solution of α-naphthoflavone in glacial acetic acid, and finally 1 or 2 drops of a mixture of 3 parts 50 % sulfuric acid and 1 part 30 % hydrogen peroxide. An orange color develops; its intensity depends on the amount of bromide present.

Limit of Identification: 1 γ bromine
Limit of Dilution: 1 : 50,000

This procedure, including decomposition of any iodine compound by the addition of excess permolybdic acid, will reveal as little as 0.5 γ iodine and 0.5 γ bromine (as halide) in a drop of saturated solution of NaCl.

(4) Other tests for bromides

(*a*) A drop of the acid test solution is mixed with a few drops of a warm, saturated solution of $KMnO_4$ and a drop of $CuSO_4$ solution.[16] Bromine is liberated, and will blue iodide-starch paper placed over the crucible containing the mixture (*Idn. Limit:* 5 γ Br). This reaction is not applicable in the presence of iodides.

(*b*) A blue-violet ring is produced on filter paper if a drop of KBr—$CuBr_2$ solution is dried and then spotted with H_2SO_4 (1 : 1) (*Idn. Limit:* 0.02 γ Br).[17]

3. Hydriodic acid

(1) Test with palladous chloride

Palladous chloride reacts with iodide solutions to form a brown-black precipitate of palladous iodide, which is insoluble in mineral acids and soluble in ammonia. However, the product is distinctly soluble in concentrated solutions of certain salts, e.g., sodium chloride, magnesium chloride, etc. Palladous nitrate must not be used; it gives a red-brown precipitate of palladous bromide with bromides (especially from concentrated solutions). Even insoluble iodides, such as Cu_2I_2, HgI_2, AgI, produce palladous iodide on treatment with palladous chloride. The solid iodides are blackened.

Procedure. A drop of the test solution is spotted on filter paper with a drop of 1 % palladous chloride solution. A more or less intense black-brown fleck or ring forms, according to the quantity of iodide present.

Limit of Identification: 1 γ iodine
Limit of Dilution: 1 : 50,000

References pp. 356–360

(2) Test by oxidation to free iodine

Iodides react with a number of oxidizing agents in acid solution to yield free iodine, which gives the familiar test with starch. The latter effect involves the production of a deep blue adsorption compound of iodine and starch.[18] When only small amounts of iodine are present, the insoluble blue starch-iodine compound remains in colloidal suspension.

An acid solution of potassium nitrite (nitrous acid) is best for liberating the iodine from iodides:

$$2\ HI + 2\ HNO_2 \rightleftarrows I_2 + 2\ NO + 2\ H_2O$$

It should be noted that the starch-iodine reaction is more sensitive the lower the temperature. Blue starch-iodine solution loses its color when heated; the color reappears on cooling. The sensitivity is decreased by large amounts of alum, magnesium sulfate, alkali sulfate, and the like. Certain organic materials (proteins, resorcinol), when present in large amounts, completely prevent the starch-iodine reaction.[19] Cyanides, which impair the test by formation of cyanogen iodide,

$$HCN + I_2 \rightleftarrows CNI + HI$$

can be removed, before the test, by acidifying and heating.

The test for iodide by oxidation with nitrite can be carried out (*a*) on a spot plate or (*b*) on filter paper containing starch. The spot reaction [20] described under (*b*) is the more sensitive and is particularly recommended for neutral or alkaline test solutions.

Procedure. (*a*) A drop of the acid test solution is mixed on a spot plate with a drop of starch solution and a drop of 10 % potassium nitrite solution. A blue coloration indicates the presence of iodide.

Limit of Identification: 2.5 γ iodine
Limit of Dilution: 1 : 20,000
Reagent: Starch solution [21]: 1 g soluble starch and 5 mg mercuric iodide (preservative) are mixed with a little water and the paste poured into about 500 ml boiling water

Some varieties of filter paper suitable for spot reactions already contain sufficient starch. The test for iodine on paper is about 100 times as sensitive as on a spot plate. This heightened sensitivity is due to the fixation of the iodine on the surface of the paper and to the contrast with the white background.

Procedure. (*b*) Filter paper, tested beforehand for starch with positive results, is treated in succession with single drops of 2 *N* acetic acid, test solution, and 0.1 *N* potassium nitrite. A blue fleck or ring shows the presence of iodine.

Limit of Identification: 0.025 γ iodine
Limit of Dilution: 1 : 2,000,000

References pp. 356–360

(*3*) *Test by conversion to iodate* [22]

When very small amounts of iodide are to be detected (quantities below the *limit of identification* of test (*2*)), the iodide may be oxidized to iodate. The iodine can then be liberated by the addition of iodide and acid:

$$IO_3^- + 5\,I^- + 6\,H^+ = 3\,I_2 + 3\,H_2O \qquad (1)$$

Since one iodide ion gives one molecule of iodate, which in turn liberates six atoms of iodine, the sensitivity of the test is thus markedly multiplied. The oxidation of iodide to iodate may be carried out by means of bromine water in neutral solution:

$$I^- + 3\,Br_2 + 3\,H_2O = IO_3^- + 6\,H^+ + 6\,Br^- \qquad (2)$$

The bromine in excess must be removed before the detection of the iodate produced. The removal is accomplished by adding sulfosalicylic acid, which immediately consumes the bromine to form colorless bromosalicylic acid (see page 168).

The detection of iodide through conversion into iodate is not affected by chlorides and bromides. Halogenates do interfere because, in acid solution, they oxidize iodides to iodine.

Procedure. A drop of neutral or weakly acid test solution is mixed with a drop of bromine water in a depression of a spot plate. The excess bromine is removed by means of several small crystals of sulfosalicylic acid. Traces of bromine vapor remaining in the air above the test would liberate iodine from the iodide that must be introduced subsequently. Consequently, the cadmium iodide solution * (see p. 158) may not be added immediately after the solution has been decolorized with sulfosalicylic acid; instead it is necessary to wait 3–5 minutes, and during this period air should be blown at intervals through the test solution by means of a pipette to produce thorough mixing and aeration. A blank test with a drop of distilled water is advisable when small amounts of iodine are expected. A blue coloration that appears immediately after the addition of starch-cadmium iodide solution indicates the presence of iodide in the test solution.

Limit of Identification: 0.05 γ iodine
Limit of Dilution: 1 : 1,000,000
Reagent: 5 % solution of cadmium iodide in 1 % cadmium sulfate solution (Starch solution is added just before use.)

(*4*) *Test by catalytic acceleration of the reduction of cerium*^{IV} *salts* [23]

The reduction of yellow ceric solutions by arsenite in acid solution:

$$2\,Ce^{+4} + AsO_3^{-3} + H_2O = 2\,Ce^{+3} + AsO_4^{-3} + 2\,H^+ \qquad (1)$$

* The advantage of using CdI_2 solution in place of KI solution is that the liberation of small amounts of iodine by dissolved oxygen and traces of nitrous oxides is much slower. This effect is due to the slight dissociation of cadmium iodide. Compare J. Lambert, *J. Am. Chem. Soc.*, 71 (1949) 3260.

is very slow. The color is quickly discharged after the addition of a very small amount of alkali iodide.[24] The action of iodide ions can be explained: Ce^{+4} ions oxidize I^- ions very rapidly to I_2, and the latter enters into the familiar redox reaction with arsenite:

$$2\,Ce^{+4} + 2\,I^- \rightarrow 2\,Ce^{+3} + I_2 \qquad (2)$$

$$I_2 + AsO_3^{-3} + H_2O \rightleftarrows 2\,I^- + AsO_4^{-3} + 2\,H^+ \qquad (3)$$

Although, in mineral acid solution, equilibrium (3) lies almost completely to the left, oxidation of As^{III} occurs in the presence of Ce^{IV} ions because iodide ions are constantly removed by reaction (2). Accordingly, the acceleration by iodide ions of the reaction between Ce^{IV} and As^{III} is due to an intermediate catalysis reaction. Summation of (2) and (3) yields the stoichiometric equation of the catalytically accelerated redox reaction (1) in which iodide ions do not appear.

Alkali bromides, chlorides, sulfates, and nitrates interfere only when very small amounts of iodide are to be detected. Bromides have the greatest deleterious effect; however, when the amount of bromide is approximately known, small amounts of iodine may still be detected if a comparative test is carried out. Metal salts which give colored aqueous solutions interfere (Fe^{III}, UO_2, Ni, Cu, Co). Cyanides, mercuric, silver, and manganese salts impair the reaction, as do compounds which reduce Ce^{IV}. In such cases the difficulty may occasionally be averted by using more concentrated ceric solutions. Under the experimental conditions, barium and strontium salts are precipitated as sulfates, which are colored yellow by coprecipitation of ceric salt. Osmium salts behave similarly to iodides.

Procedure. A drop of the test solution, a drop of sodium arsenite, and a drop of Ce^{IV} solution are mixed on a spot plate. The yellow color disappears at a rate varying with the iodide content. For small amounts, a blank test should be carried out; the yellow color then usually disappears in about 30 minutes.

Limit of Identification: 0.05 γ potassium iodide

Limit of Dilution: 1 : 1,000,000

Reagents: *1)* 0.05 *N* ceric ammonium sulfate in 2 *N* sulfuric acid

2) 0.1 *N* sodium arsenite (neutral or slightly acid)

(*5*) *Other tests for iodides* (see also Addendum, p. 535)

(*a*) Filter paper is impregnated with a 0.1 *N* $AgNO_3$ solution and dried out of contact with air. A solution containing iodide forms a bright yellow fleck or ring of AgI (*Idn. Limit:* 5 γ iodine).[25] In mixtures which also contain chlorides and bromides, the silver halides are fractionated and form in concentric zones, so that 5 γ iodide may be identified in the presence of 5 mg Cl or Br.

(*b*) A drop of the test solution is placed on filter paper and exposed briefly to the fumes of concentrated hydrochloric acid. The fleck is then spotted with dilute thallous nitrate solution. Yellow thallous iodide is formed.[26]

(*c*) A drop of the test solution is placed on filter paper (Whatman No. 1) fumed for 1 minute over concentrated nitric acid and then spotted with starch solution. A blue color indicates iodide ions (*Idn. Limit:* 0.3 γ iodine).[27]

4. Hydrofluoric acid

(*1*) *Test with zirconium-alizarin solution* [28]

The addition of alizarin or alizarin sulfonate to dilute solutions of zirconium chloride containing hydrochloric acid results in a red-violet color due to the formation of hydrosols of a zirconium lake with these dyestuffs (see page 200). The dispersions turn yellow as soon as they are treated with excess fluoride. The zirconium combines with the latter to form colorless complex $[ZrF_6]^{-2}$ anions and is withdrawn from the alizarin lake. Consequently, only the color of the alizarin remains.

Other anions, which give either complex ions or a precipitate with zirconium, behave similarly to fluorides. Large amounts of sulfates, thiosulfates, phosphates, arsenates, and oxalates have this effect and must be taken into account when carrying out the spot test.[29] Fluorine in complex combination (silicofluoride, boron fluoride) behaves similarly to fluoride ion.[30]

Procedure. A strip of quantitative filter paper is impregnated with zirconium-alizarin solution. The dried paper is moistened with a drop of 50 % acetic acid and then a drop of the neutral test solution is placed on the moist fleck. A yellow spot appears in the presence of fluorides. When only small amounts of fluorine are present, it is advisable to hasten the reaction by warming the paper in steam.

Limit of Identification: 1 γ fluorine

Limit of Dilution: 1 : 50,000

Reagents: Zirconium-alizarin paper: Commercial zirconium oxide (ZrO_2) is digested with warm dilute hydrochloric acid and filtered. The solution should contain about 0.5 mg zirconium per ml. Several ml of the zirconium solution is treated with a slight excess of an alcoholic solution of alizarin. Excess of alizarin may be recognized by extracting a portion of the solution with ether, which becomes yellow. The zirconium-alizarin solution thus prepared is warmed for ten minutes in the water-bath. Filter paper is bathed in the warm solution and dried.

When testing for larger amounts of fluorine, the following preparation is advised.[31] Dry filter paper is soaked in a 5 % solution of zirconium nitrate in 5 % hydrochloric acid and, after draining, is placed in a 2 % aqueous solution of sodium alizarin sulfonate. The paper is colored red-violet by the zirconium lake. It is washed until the wash water is nearly colorless, and then dried.

Oxalates can be removed by igniting before beginning the test. In the presence of sulfates, benzidine hydrochloride should be added to the test solution, and then a drop of the benzidine sulfate suspension placed on the reagent paper. Any yellow coloration of the red reagent paper caused by fluorine can easily be seen on the under side of the paper.

Detection of fluorine in solid substances [32]

Direct detection of even insoluble fluorides is also possible by means of the zirconium-alizarin test. Use is made of the fact that in the presence of small amounts of acid, precipitated or native calcium fluoride is readily soluble in warm aqueous solutions of salts that form complex fluorides. Pure calcium fluoride gives a clear solution; impure products leave only the "gangue" behind.

The salts which may be used for "decomposing" calcium fluoride by this wet method, include borax, (when, no doubt, fluoboric acids are formed), and also beryllium,[33] aluminum, iron, and chromium salts which, as expected, have a solvent action, since the respective fluometallic acids are known, at least in the form of their alkali salts. The solvent action of these materials depends on their effect on the equilibrium:

$$\underset{\text{(solid)}}{CaF_2} \rightleftarrows Ca^{+2} + 2\,F^-$$

The equilibrium is not appreciably shifted by hydrogen ions, but removal of fluoride ions and their incorporation in complex anions (e.g., $[AlF_6]^{-3}$ or $[BeF_4]^{-2}$) disturbs the equilibrium until all the calcium fluoride is brought into solution.

A hydrochloric acid solution of the zirconium-alizarin compound also dissolves calcium fluoride with formation of $[ZrF_6]^{-2}$, Ca^{+2}, and free alizarin.* The test can be carried out on a microscope slide by stirring a very small amount of the powdered fluoride with a drop of the violet reagent solution. A yellow color is formed at once.

The test may be successfully carried out on a crystal face of native fluorspar by contact with the reagent solution for about 30–60 seconds.

* J. H. de Boer and J. Basart, *Z. anorg. allgem. Chem.*, 152 (1926) 213, were the first to observe that insoluble fluorides dissolve in zirconium chloride solutions, and that the fluoride ion can then be titrated, using alizarin as indicator.

Small amounts of powdered fluorapatite, $3\ Ca_3(PO_4)_2 \cdot CaF_2$ (ca. 3 % fluorine) also give a yellow color immediately.

Precipitated tertiary calcium phosphate, under the same conditions, reacts after a few minutes due to production of acid-insoluble zirconium phosphate. The decomposition of the zirconium-alizarin lake is not instantaneous as in the case of calcium fluoride probably because the color lake is adsorbed by zirconium phosphate, which is the initial product. This supposition is supported by the finding that zirconium phosphate is reddened immediately by contact with an acid zirconium-alizarin solution.

Calcium fluoride was easily detected in a 2 mg sample of a mixture with barium sulfate in the ratio 1 : 200, which represents an absolute quantity of about 10 γ CaF_2.

The fact that solid calcium fluoride reacts immediately with zirconium alizarinate renders it possible to detect fluoride in a mixture with phosphates and oxalates. These salts interfere with the detection of fluoride in aqueous solution, because they form either insoluble or complex zirconium phosphates (or oxalates) and thus destroy the red zirconium alizarin compound and also produce the yellow color. The fluoride in such mixtures may be isolated by treating the alkaline or neutral test solution with calcium chloride; the precipitate of calcium fluoride is ignited and digested with dilute acid. The residue then may easily be tested for fluoride by the zirconium-alizarin solution.[34]

Reagent: 0.05 g zirconium nitrate is dissolved in 50 ml water and 10 ml concentrated hydrochloric acid. The solution is mixed with a solution of 0.05 g sodium alizarin sulfonate in 50 ml water.

(2) *Test with zirconium azoarsonate*[35]

Brown insoluble zirconium azoarsonate, formed by the action of *p*-dimethylaminoazophenylarsonic acid on zirconium salts (see page 202), reacts with fluorides to produce colorless $[ZrF_6]^{-2}$ ions and red azophenylarsonic acid. If a drop of an acidic fluoride solution is placed on filter paper impregnated with zirconium azoarsonate, a change from brown to red indicates fluoride (in the absence of interfering compounds).

Procedure. Three drops of the neutral or alkaline test solution are acidified with a drop of 2 *N* hydrochloric acid on a spot plate or in a small porcelain crucible. A drop of this mixture is transferred to the light brown reagent paper. When fluoride is present, the middle of the moist fleck is colorless, and is surrounded by a red circle of the liberated azo dyestuff.

Limit of Identification: 0.25 γ fluorine

Limit of Dilution: 1 : 200,000

Reagent: Zirconium azoarsonate paper: 0.025 % solution of *p*-dimethylaminoazophenylarsonic acid[36] is made up in a mixture of 9 parts alcohol and 1 part concentrated hydrochloric acid. Filter paper (S & S 589g) is laid in this solution for a few minutes. After drying in the air, the bright red paper is placed in a 0.01 % zirconium oxychloride ($ZrOCl_2$) solution in 1 *N* hydrochloric acid. The paper turns brown at once. After 10 minutes, the paper is washed 5 minutes with cold, and then 5 minutes with warm (55° C) 2 *N* hydrochloric acid, then with water, finally rinsed with alcohol and ether, and then dried *in vacuo*.

With regard to interference with this test for fluoride, the same holds true as with the zirconium-alizarin test. Such anions interfere as give insoluble or—in the same way as the fluoride ion—stable soluble complex compounds. These interfering ions include phosphates, arsenates, sulfates, thiosulfates. The interference due to sulfates is appreciably less than in the zirconium-alizarin test. The test may be made completely specific, and also applicable to insoluble compounds, minerals, etc., by releasing the fluorine as hydrofluoric acid or H_2SiF_6 before the spot test is finally carried out. The apparatus described on page 47 (Fig. 23) is suitable. A few particles of the sample (or 1 or 2 drops of liquid which are evaporated there to dryness) are placed in the bulb of the dry apparatus. A drop of concentrated sulfuric acid is added, and the stopper is replaced, after placing a drop of dilute hydrochloric acid on the knob. The liberation of hydrofluoric acid is hastened by warming the apparatus over an asbestos plate. Since too lengthy heating causes volatilization of appreciable amounts of sulfuric acid, it is advisable to add a crystal of tartaric acid to the mixture and to stop the heating when charring begins. After 1 or 2 minutes heating, the drop on the knob of the apparatus is touched to the filter paper impregnated with zirconium azoarsenate. A red color appears if the test is positive.

In this way 0.4 γ CaF_2 was detected in the presence of 16,000 γ Al_2O_3, which represents a limiting proportion of 0.2 γ F in the presence of 80,000 times the amount of foreign substance.

(*3*) *Test by conversion into silicomolybdic acid*[37]

Both soluble and insoluble fluorides are converted into silicon tetrafluoride (b.p. –65° C) on warming with quartz sand (silica) and concentrated sulfuric acid:

$$MeF_2 + H_2SO_4 = MeSO_4 + 2\ HF$$
$$SiO_2 + 4\ HF = SiF_4 + 2\ H_2O$$

The gas may be absorbed in a drop of water; it reacts to produce silicic and fluosilicic acids:

$$3\ SiF_4 + 4\ H_2O = H_4SiO_4 + 2\ H_2SiF_6$$

Ammonium molybdate converts these products into silicomolybdic acid, $H_8[Si(Mo_2O_7)_6]$. The latter, unlike free molybdic acid, oxidizes benzidine in acetic acid solution to a blue dyestuff, with concurrent production of molybdenum blue (compare page 362).

The apparatus described on page 47 (Figure 23) is excellent for the hydrofluoric acid test. Silicon tetrafluoride, developed from the sample by sulfuric acid, is collected without loss in a drop of water hanging on the knob under the stopper, and is hydrolyzed there.

Procedure. The sample, mixed with very pure quartz sand or silica powder, is placed in the bulb of the apparatus and moistened thoroughly with 1 or 2 drops of concentrated sulfuric acid. The stopper, with its suspended water drop, is put in place and the apparatus gently heated for about 1 minute. After removing the flame, the apparatus is allowed to stand 3 to 5 minutes. The water drop is then washed into a micro crucible, 1 or 2 drops of ammonium molybdate added, and the mixture warmed until bubbles begin to rise. When completely cool, a drop of benzidine and a few drops of saturated sodium acetate solution are added. A blue color shows the presence of silicic acid and hence fluoride in the original sample.

Limit of Identification: 1 γ fluorine

Limit of Dilution: 1 : 50,000

Reagents: *1)* Quartz sand: commercial quartz sand is heated with concentrated sulfuric acid, washed with water, and dried

2) Ammonium molybdate solution: 15 g $(NH_4)_2MoO_4$ is dissolved in 300 ml water and poured into 100 ml nitric acid (sp. gr. = 1.2)

3) 1 % solution of benzidine in 10 % acetic acid

Samples containing carbonates, thiosulfate, sulfides and sulfites must be ignited before the fluoride test, to prevent any considerable evolution of carbon dioxide, and also to avoid interference with the silicomolybdate reaction by hydrogen sulfide and sulfur dioxide, which likewise produce molybdenum blue. Iodides and bromides interfere in so far as they evolve the halogen under the experimental conditions. The halogen is taken up in the drop of water and can oxidize the benzidine. It is advisable therefore, in the presence of these halides, to remove them by precipitation with silver nitrate, after "opening up" the solid sample. Large amounts of chlorides should also be removed to prevent frothing due to hydrogen chloride. Large amounts of nitrates also interfere; the nitric acid liberated by the concentrated sulfuric acid accumulates in the hanging drop of water and later oxidizes the benzidine to a soluble yellow product. Consequently, the solid sample or the evaporation residue must be ignited to convert the nitrates into oxides or carbonates. The fluorides remain unaltered. Adequate decomposition of the fusible alkali or alkaline earth nitrates requires strong and protracted ignition. When solutions are to be tested for fluoride, a small volume, after treatment with silver sulfate if necessary is evaporated on quartz sand and the dried mixture then tested as just described.

(*4*) *Test through chemical adsorption on glass* [38]

Vapors of hydrofluoric acid, liberated from fluorides by heating with concentrated sulfuric acid, have a corrosive effect on glass, as shown by a more or less intensive frosting. Studies [39] indicate that the anhydrous silica of the glass, through formation and saponification of silicon tetrafluoride, is converted into hydrated silicic acid. The decrease in transparency accompanying this change is the basis of the etching test for fluoride, which formerly was so widely used. Small amounts of hydrofluoric acid, which are insufficient to produce a visible etching, may nevertheless alter the surface of glass so that it is not wet by concentrated sulfuric acid. The latter no longer runs off smoothly, but collects in drops as water does on an oily hydrophobic surface. The decrease in wettability is a result of a surface reaction between silica of the glass and hydrofluoric acid, with participation of concentrated sulfuric acid as dehydrating agent: [40]

$$[SiO_2]_x + H_2F_2 \underset{H_2O}{\overset{H_2SO_4}{\rightleftarrows}} [SiO_2]_{x-1}\cdot OSiF_2 + H_2O \qquad (1)$$

This schematic presentation is intended to show that the surface reaction leads to a silicon-fluorine bonding, in which the participating Si-atoms preserve their original phase association. Such surface reactions constitute the essence of a chemical adsorption (compare the surface reaction: alumina-alizarin, page 185). In the familiar volatilization reaction of silica with concentrated hydrofluoric acid:

$$[SiO_2]_x + 2x\ H_2F_2 \rightarrow x\ SiF_4 + 2x\ H_2O \qquad (2)$$

the surface reaction (1) is the preliminary stage for the production of SiF_4 as an independent phase, an indispensable part also of the etching test.

The chemical adsorption of hydrofluoric acid on glass may be applied, under proper conditions, to the detection of fluoride, and also of silicofluorides, which behave in the same way. The presence of substances which form complexes with fluorine decreases the sensitivity. Silicic and boric acids are outstanding instances.

Procedure.[41] A few particles of powdered potassium dichromate are dissolved in 1 to 1.5 ml concentrated sulfuric acid in a narrow test tube. The walls should be thoroughly covered with the hot mixture to remove all traces of grease. When the walls are clean, the mixture runs evenly over the surface. A few grains of the solid sample or a drop of the test solution is added and the mixture warmed. On rotating the tube, an uneven flow of the sulfuric acid, along with unmoistened areas, indicates fluorine. Large amounts of fluoride give this result even without heating.

Limit of Identification: 0.5 γ fluorine
Limit of Dilution: 1 : 100,000

(5) *Test through prevention of the formation of fluorescing metal salts of 8-hydroxyquinoline* [42]

As pointed out on page 125 in connection with the test for vanadium by means of 8-hydroxyquinoline (oxine), many metal ions are precipitated as inner complex oxinates when this reagent is added to acetic acid or ammoniacal solutions of the metal ions. Most insoluble metal oxinates are yellow and exhibit a strong yellow-green fluorescence in ultraviolet light, whereas the reagent itself does not fluoresce as solid or in solution. Certain oxides and carbonates, notably calcium, magnesium, and aluminum, can be brought to fluorescence, if they are moistened with oxine dissolved in water or an organic liquid. Qualitative filter paper contains small amounts of calcium, magnesium, and aluminum oxide, along with silica; they are highly dispersed in the cellulose fibres. If a drop of a chloroform solution of oxine is placed on the paper and the solvent allowed to evaporate, the residual fleck is highly fluorescent because of the metal oxinates produced. Quantitative filter papers, from which such metal oxides have been removed by washing with dilute hydrofluoric acid, do not show this effect. If vapors of hydrofluoric acid are allowed to act on qualitative filter paper, fluorides of calcium, magnesium, etc. are formed at the exposed places and can no longer be transformed into the fluorescent oxinates when spotted with the chloroform oxine solution, whereas the rest of the paper fluoresces strongly. A reliable test for fluoride is based on this effect.

Procedure. Three drops of sulfuric acid and one drop of the test solution are placed in a micro crucible. The crucible is closed with a piece of qualitative filter paper, sufficiently greater in diameter than the crucible so that the area exposed to hydrogen fluoride can be compared easily with the unexposed portion of the paper. The charged crucible is kept at 50–60° C for five minutes. The paper is then spotted with a chloroform solution of oxine and examined in ultraviolet light. If the sample contained fluoride, the exposed area shows no or only a faint fluorescence compared with the surrounding portions of the paper.

Limit of Identification: 0.05 γ fluorine
Limit of Dilution: 1 : 1,000,000

Due to the formation of fluoboric acid, the sensitivity of this fluoride test is lower when boric acid is present. Nevertheless, 0.5 γ fluorine can be detected in the presence of 50 γ boric acid.

Reagent: Solution of 50 mg of 8-hydroxyquinoline in 100 ml chloroform

References pp. 356–360

5. Hydrocyanic acid

(1) *Test with copper acetate-benzidine acetate* [43]

Copper salts react with benzidine to give "benzidine blue" in the presence of cyanides (or other halides). This reaction (see page 84) takes place because the oxidation potential of copper^{II} salts is increased when the resulting copper^{I} ions are taken up through the formation of insoluble cuprous cyanide. The reaction may be applied as a test for copper or for cyanide. Consequently, by passing hydrocyanic acid into a solution of copper and benzidine acetates, or by allowing the gas to come into contact with filter paper impregnated with a solution of these materials, a blue color appears instantly.

This very sensitive color reaction is characteristic for prussic acid, provided oxidizing or reducing compounds are absent. The test can be used both for cyanides that are decomposed by acids,* and for those that are not. Either procedure (*a*) or (*b*) may be followed.

Procedure. (*a*) A drop of the test solution is mixed with a drop of dilute sulfuric acid in a porcelain micro crucible, which is then covered by a piece of filter paper moistened with a drop of the reagent. A small watch glass placed on top of the paper provides a sufficiently tight seal. A blue area is formed on the white paper, the shade depending upon the amount of hydrogen cyanide liberated.

Limit of Identification: 0.25 γ cyanogen
Limit of Dilution: 1 : 200,000

Procedure. (*b*) A drop of the test solution (or a few milligrams of the solid) is placed in the reaction bulb of the apparatus (page 48, Fig. 26). One or two thin pieces of zinc and 2 or 3 drops of dilute sulfuric acid are added. The apparatus is closed with the funnel attachment. Filter paper moistened with the reagent solution is laid across the funnel. The apparatus is warmed gently. The paper turns blue if hydrocyanic acid accompanies the hydrogen.

Limit of Identification: 1 γ cyanogen
Limit of Dilution: 1 : 50,000
Reagent: Copper acetate-benzidine acetate solution

The solutions of copper acetate and benzidine acetate are best stored separately in well-stoppered, dark bottles, and the reagent mixture prepared freshly each time it is needed. (The mixture of acetates will not keep longer than 2 weeks.) The reagent is prepared from equal volumes of (*a*) and (*b*).

* A. M. Schapowalenko (*Chem. Zentr.*, 1930, II, 588) recommends that the prussic acid to be identified with copper acetate-benzidine acetate be liberated from the test solution by heating with sodium bicarbonate, to prevent any interference with the test by volatile oxidizing or reducing compounds.

(*a*) 2.86 g copper acetate in a liter of water

(*b*) 675 ml of a solution of benzidine acetate, saturated at room temperature, and 525 ml water

(2) *Test with copper sulfide* [44]

Copper sulfide is readily soluble, in cyanide solution, even in the cold, giving cuprocyanide:

$$2\ CuS + 4\ KCN \rightarrow 2\ Cu(CN)_2 + 2\ K_2S$$
$$2\ Cu(CN)_2 \rightarrow Cu_2(CN)_2 + (CN)_2$$
$$Cu_2(CN)_2 + 2\ KCN \rightarrow K_2[Cu_2(CN)_4]$$

Consequently, a suspension of copper sulfide is completely clarified by cyanides. This fact may be used to detect cyanides even in the presence of ferrocyanides, ferricyanides, iodides, bromides, chlorides, and thiocyanates. The test may be made on paper impregnated with cupric sulfide; the sensitivity is greater.

Procedure. (*a*) A drop of a freshly prepared copper sulfide suspension is placed on filter paper or on a spot plate, and a drop of the test solution added. In the presence of cyanide, the brown fleck or color disappears at once.

Limit of Identification: 2.5 γ cyanide

Limit of Dilution: 1 : 20,000

Reagent: Copper sulfide suspension: 0.12 g crystallized copper sulfate is dissolved in 100 ml water. After adding a few drops of ammonia, the solution is brought to turbidity with a little hydrogen sulfide.

Procedure. (*b*) Filter paper is bathed in an ammoniacal solution of 1 g copper sulfate per liter and dried. Immediately before the test, hydrogen sulfide is blown onto the paper, turning it a uniform brown. A drop of the test solution placed on the reagent paper forms a white circle when cyanide is present.

Limit of Identification: 1.25 γ cyanide

Limit of Dilution: 1 : 40,000

(3) *Test by formation of thiocyanate*

A very simple test, applicable also in the presence of sulfide or sulfite, is the conversion of alkali cyanide into thiocyanate, which can be identified by the sensitive iron reaction. The thiocyanate is formed by warming with yellow ammonium sulfide:

$$CN^- + (NH_4)_2S_2 = CNS^- + (NH_4)_2S$$

Should thiocyanates be present from the beginning, it is necessary to isolate the cyanide beforehand, e.g., by precipitation as zinc cyanide. This method is applicable only if not too small quantities of cyanide are present, since the zinc cyanide precipitate must be thorougly washed with water before proceeding with the polysulfide treatment.

References pp. 356–360

Procedure. A drop of the test solution is stirred with a drop of yellow ammonium sulfide on a watch glass and warmed until a rim of sulfur is formed around the liquid. (Evaporation to dryness should be avoided.) After the addition of 1 or 2 drops of dilute hydrochloric acid, the mixture is allowed to cool, and 1 or 2 drops of a 1 % solution of ferric chloride are added. When zinc cyanide is to be treated with polysulfide, the zinc sulfide, which precipitates first, should be isolated by filtration or centrifugation. A red color indicates the presence of cyanide in the test solution.

Limit of Identification: 1 γ cyanide
Limit of Dilution: 1 : 50,000
} in the presence of 500 times the amount of sodium sulfite and sulfide

(*4*) *Test by demasking of alkali palladium dimethylglyoxime* [45]

Bright yellow inner complex palladium dimethylglyoxime (I) is not soluble in acids but it dissolves in alkali hydroxides. The yellow solution contains alkali palladium dimethylglyoxime (II), in which the palladium is a constituent of an inner complex anion.

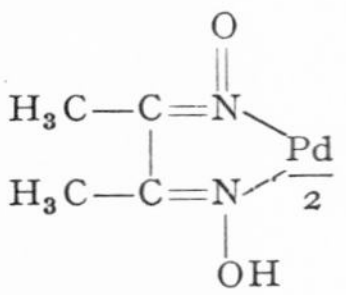

```
         O
         ||
H3C—C=N\
    |     Pd
H3C—C=N/   2
        |
        OH
```

(I) $Pd(DH)_2$
DH_2 = dimethylglyoxime
DH = univalent acid radical of DH_2
D = bivalent acid radical of DH_2

```
[         O          ]
[         ||         ]
[ H3C—C=N\           ]
[     |     Pd       ]  Me
[ H3C—C=N/   2       ]
[        |           ]
[        O           ]
```

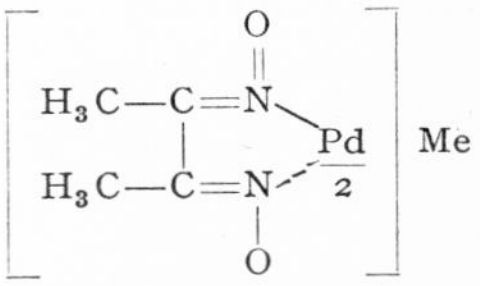

(II) $Me_2[PdD_2]$
Me = K, Na

The inner complexly bound dimethylglyoxime is masked in solutions of (II); i.e., no red Ni-dimethylglyoxime precipitate appears when nickel salts are added to the solution. Likewise, because of its inner complex binding, the palladium is masked against practically all precipitation reactions that are characteristic for Pd^{+2} ions. Cyanide ions present an exception. Added in excess to the yellow solutions of (II), they cause immediate discharge of the color, and addition of a nickel solution (containing ammonium chloride) brings down red Ni-dimethylglyoxime.*

The demasking of dimethylglyoxime is due to the reaction:

$$[PdD_2]^{-2} + 4\ CN^- \rightarrow [Pd(CN)_4]^{-2} + 2\ D^{-2}$$

Since the Ni-dimethylglyoxime reaction is highly sensitive (compare page 150), cyanide ions can be detected easily through the demasking of solutions containing $[PdD_2]^{-2}$ ions. An advantage of this test is that it can

* The addition of ammonium salts prevents the precipitation of green $Ni(OH)_2$.

be made in alkaline solution and, in contrast to the other methods, does not require the previous liberation and evolution of hydrogen cyanide.

It should be noted that larger quantities of ammonia gradually decolorize the yellow alkali palladium cyanide solution because of a demasking:

$$[PdD_2]^{-2} + 2\ NH_3 \rightarrow [Pd(NH_3)_2]^{+2} + 2\ D^{-2}$$

Therefore, the test described here cannot be recommended for the detection of very slight amounts of cyanide in solutions that are strongly ammoniacal or that contain much ammonium salts.

Procedure. A drop of the alkaline test solution and a drop of alkali palladium dimethylglyoxime are brought together on a spot plate; a drop of nickel-ammonium solution is then introduced. A red precipitate or pink appears, the shade depending on the amount of alkali cyanide present. A blank test is necessary only when small quantities of cyanide are suspected.

Limit of Identification: 0.25 γ cyanide

Limit of Dilution: 1 : 200,000

Reagents: *1)* Alkali palladium dimethylglyoxime: Purest Pd-dimethylglyoxime (prepared by precipitation of an acid solution of $PdCl_2$ with dimethylglyoxime and thoroughly washed) is shaken with 3 *N* potassium hydroxide. The undissolved portion is filtered off.

2) Nickel-ammonium solution: 0.5 *N* nickel chloride saturated with ammonium chloride

(5) *Test through reaction with alkali mercuri chloride* [46]

Aqueous solutions of mercuric chloride, a non-electrolyte, are but slightly acidic because of the hydrolysis of Hg^{+2} ions, which are present in very low concentration. This concentration can be decreased still more and the solution is brought to neutrality by adding alkali chloride. Soluble alkali mercuri chloride, e.g. $Na[HgCl_3]$, results, and the solution shows a neutral reaction to the usual acid-base indicators. Many precipitation and color reactions of Hg^{+2} ions are masked in solutions of these double chlorides since $[HgCl_3]^-$ ions are predominantly present. However, there is no masking against the action of hydrocyanic acid, which in accordance with:

$$Na[HgCl_3] + 2\ HCN \rightarrow Hg(CN)_2 + NaCl + 2\ HCl$$

produces soluble undissociated mercuric cyanide and a quantity of hydrochloric acid equivalent to the hydrocyanic acid. This reaction is the basis of a test for hydrocyanic acid, in the absence of other volatile acids.

References pp. 356–360

Procedure. A drop of the test solution is mixed with a drop of concentrated sulfuric acid in a micro crucible. The crucible is then covered with a sheet of indicator paper, which has previously been moistened with a solution of alkali mercuri chloride. The color change characteristic of the indicator occurs to an extent that depends on the quantity of hydrocyanic acid liberated. When Congo paper is used, the change red → blue is distinct with 5 γ cyanide. The test is more sensitive if "Accutint" indicator (wide range, pH 0-5) is employed; the color change is green → yellow.

Limit of Identification: 2.5 γ cyanide
Limit of Dilution: 1 : 20,000
Reagent: Solution of potassium mercuri chloride: 5 g mercuric chloride and 1.4 g potassium chloride in 100 ml water

(6) Other tests for cyanide

(*a*) Filter paper is bathed in a saturated solution of picric acid and then in sodium carbonate solution. One drop of the test solution is placed on the dried paper. A red color indicates the presence of cyanide.[47] Sulfides produce a similar result.

(*b*) Prussic acid combines with iodine ($HCN + I_2 \rightarrow HI + CNI$), to form colorless cyanogen iodide. Consequently, if the acid is liberated from cyanides and allowed to act on blue starch-iodine, the color is discharged. Warming with dilute sulfuric acid suffices for cyanides that decompose easily; insoluble or complex cyanides require the addition of metallic zinc. The evolution of HCN can be made from 1 drop of solution in the apparatus of Fig. 23, page 47, and Fig. 26, page 48. Potassium iodide starch paper spotted with acidified 0.1 % KIO_3 or 0.1 *N* iodine can be used for the decolorization test (*Idn. Limit:* 1 γ HCN). [48]

(*c*) Cyanides discharge the color of brown cupric ferricyanide, producing colorless $K_2[Cu_2(CN)_4]$[49]. This test can be carried out as a spot reaction under the conditions described on page 277.

(*d*) Alkali cyanides (also complex cyanides) catalyze the condensation of benzaldehyde to benzoin (benzoin rearrangement) in basic solution. The resulting benzoin is readily revealed by the violet color which appears on the addition of *o*-dinitrobenzene (*Idn. Limit:* 0.05 γ CN).[50]

6. Mercuric cyanide

(1) Test by precipitation with silver nitrate in alkaline solution

Mercuric cyanide is so little dissociated in aqueous solution that the addition of silver ions produces no precipitate, because the solubility product of silver cyanide is not attained. For the same reason, mercuric cyanide does not respond to the other cyanide reactions. For instance, on acidifying mercuric cyanide (or an alkali cyanide solution mixed with excess mercuric chloride) with mineral acids, no prussic acid is liberated. Likewise, the cyanide is not oxidizable, etc.[51] However, addition of alkalis to mercuric cyanide renders the cyanide reactive again. Silver cyanide is then precipitated by silver nitrate, and all the other characteristic cyanide reactions occur.[52] The reason for this increase in the reactivity of mercuric cyanide is that caustic alkali effects the formation of complex compounds, e.g. $K[Hg(CN)_2(H_2O)OH]$, whose complex anion dissociates:

$$[Hg(CN)_2(H_2O)OH]^- \rightleftarrows CN^- + H_2O + Hg(CN)OH$$

This anion is also formed:

$$Hg(CN)_2 + OH^- + H_2O = [Hg(CN)_2(H_2O)OH]^-$$

on the addition of sodium acetate or all alkali salts of weak acids whose solutions are basic because of hydrolysis. Cyanide may thus be detected in mercuric cyanide.

Procedure. A drop of the acid test solution is mixed on a watch glass with silver nitrate; the solution remains clear. On adding a drop of a concentrated solution of sodium acetate, silver cyanide is precipitated.

Limit of Identification: 5 γ mercuric cyanide
Limit of Dilution: 1 : 10,000

(2) Test by conversion into $[HgI_4]^{-2}$ and cyanide ions [53]

Due to the slight dissociation of mercuric cyanide in aqueous solution, the concentration of cyanide ions is so small that the liberation of prussic acid ($CN^- + H^+ \rightarrow HCN$) occurs to only a minute extent. The action of mineral acids on mercuric cyanide can be appreciably increased by excess potassium iodide:

$$Hg(CN)_2 + 4\ KI \rightleftarrows K_2[HgI_4] + 2\ KCN$$

The equilibrium lies so decidedly to the right, that hydrocyanic acid is formed in appreciable amounts when a mineral acid is added. The prussic acid can then be identified by the test given on page 276. A blank, from which the potassium iodide is omitted, should be run for comparison.

(3) Test by formation of cyanogen iodide [54]

Mercuric cyanide, a non-electrolyte, does not react in water solution with potassium iodide or silver nitrate to give mercuric iodide or silver cyanide, respectively. However, it reacts immediately with a solution of iodine in potassium iodide; cyanogen iodide is formed:

$$Hg(CN)_2 + 2\ KI + 2\ I_2 = K_2[HgI_4] + 2\ CNI$$

An iodine solution is thus decolorized by mercuric cyanide and, in the absence of other substances which react with iodine, this fact may be used as a decisive and sensitive test for mercuric cyanide.

Procedure. A drop of the test solution is mixed on a spot plate with a drop of 0.005 *N* iodine solution in potassium iodide. Decolorization will result if mercuric cyanide is present. The test may also be carried out by placing a drop of a dilute solution of iodine on potassium iodide-starch paper, and then adding a drop of the test solution to the moist fleck of starch-iodine. A white patch is formed on the blue background (formation of cyanogen iodide).

Limit of Identification: 0.25 γ mercuric cyanide

Limit of Dilution: 1 : 200,000

(*4*) *Other tests for mercuric cyanide and oxycyanide*

(*a*) A red color is produced by mercuric cyanide or oxycyanide in a potassium ferrocyanide solution containing α,α'-dipyridyl because of demasking the iron of the complex anion (compare the test for mercury, page 72, and for ferrocyanide, page 288). (*b*) Dry heating of these mercuric cyanides yields dicyanogen, which can be detected in the gas phase (page 371). (*c*) Mercuric cyanide and oxycyanide can be differentiated through the fact that the latter forms Nessler reagent with potassium iodide and can thus be detected through the red-brown precipitate of $HgI_2 \cdot HgNH_2I$ produced on the addition of ammonia.

7. Thiocyanic acid

(*1*) *Test by catalysis of the iodine-azide reaction* [55]

Both soluble and insoluble thiocyanates accelerate the reaction:

$$2\ NaN_3 + I_2 \rightarrow 2\ NaI + 3\ N_2$$

which, by itself, proceeds very slowly if at all. Consequently thiocyanates can be detected by the evolution of nitrogen that ensues when they are added to an iodine-azide solution. In the presence of sulfides and thiosulfates, which behave similarly, special precautions must be taken (see pages 284, 303, 318).

Procedure. A drop of the test solution and a drop of the iodine-azide solution are mixed on a spot plate or in the capillary tube shown on page 42, Figure 17. Gas bubbles (nitrogen) are formed in the presence of thiocyanates.

Limit of Identification: 1.5 γ potassium thiocyanate

Limit of Dilution: 1 : 33,000

This test is especially useful for the rapid detection of thiocyanate in the presence of oxalic, tartaric and other organic hydroxy acids, also phosphoric acid and ferrocyanic acid, and especially in the presence of iodides. It is well known that the thiocyanate reaction for iron^{III} does not occur in the presence of these organic acids or phosphoric acid, owing to formation of complex iron^{III} salts, while, with ferrocyanide, Prussian blue is formed. The test with ferric chloride cannot be used in the presence of iodides because the iodine liberated by the reaction:

$$2\ Fe^{+3} + 2\ I^- \rightarrow 2\ Fe^{+2} + I_2$$

renders it impossible to see the red iron-thiocyanate color reaction. The acceleration of the iodine-azide reaction by thiocyanates is unaffected by the presence of these anions (with the exception of phosphate and ferrocyanide). The sensitivity is somewhat decreased in the presence of much sodium phosphate or potassium ferrocyanide, probably because of the alkalinity of these salts.

Test for traces of potassium thiocyanate in potassium iodide

As stated, the ferric test for thiocyanates is not applicable in the presence of iodides; the iodine liberated makes it impossible to see the red ferric thiocyanate. However, the iodine-azide reaction may be carried out without modification if only slight amounts of iodide are present. For reasons that are not yet clear, large amounts of iodide considerably reduce the sensitivity of the test, and the reaction may be completely prevented. Since the interference is only due to the presence of too much iodide ions, it can be obviated by reducing the iodide ion concentration. Addition of mercuric chloride serves this purpose. The reaction: $Hg^{+2} + 4\ I^- = [HgI_4]^{-2}$ produces complex mercuri-iodide ions, which do not interfere with the catalysis of the iodine-azide reaction.

Procedure. A drop of the test solution is placed on a watch glass, and very small drops of a cold saturated solution of mercuric chloride are added (with a capillary tube) until a trace of a permanent precipitate of red mercuric iodide remains. A drop of the iodine-azide solution is added and the mixture observed for the evolution of nitrogen. The mercuric chloride may be added after the iodine-azide solution, if preferred.

Limit of Identification: 1.5 γ to 3 γ potassium thiocyanate
Limit of Dilution: 1 : 33,000 or 1 : 16,500
} in the presence of 3500 or 6000 times the amount of potassium iodide

A very simple test for thiocyanate in the presence of iodide is based on the fact that a solution of iron^{III} salt containing much iron^{II} salt reacts solely with CNS^- ions but not with I^- ions.[56] The lack of reaction is due to a lowering of the oxidation potential of Fe^{+3} ions by Fe^{+2} ions.

References pp. 356–360

Procedure. A drop of the weakly acetic acid test solution is spotted on filter paper with a drop of the reagent solution. The latter contains 1 g ferric chloride + 1 g $FeSO_4$ in 100 ml of dilute hydrochloric acid. A red-brown fleck or ring indicates thiocyanate.

Limit of Identification: 0.3 γ thiocyanate
Limit of Dilution: 1 : 160,000

Test for thiocyanates in the presence of sulfides and thiosulfates

Since sulfides and thiosulfates likewise catalyze the iodine-azide reaction (pages 303, 318) such compounds must be removed before thiocyanates can be detected by the catalytic action. The removal is best carried out by precipitation with mercuric chloride, which reacts with sulfides and thiosulfates:

$$S^{-2} + Hg^{+2} = HgS$$
$$S_2O_3^{-2} + Hg^{+2} + H_2O = HgS + 2\ H^+ + SO_4^{-2}$$

The test for thiocyanate can be carried out on the clear liquid obtained after filtering or centrifuging down the mercuric sulfide. (When excess mercuric chloride has been used, white $2HgS \cdot HgCl_2$ is formed.)

Limit of Identification: 10 γ potassium thiocyanate in the presence of 360 γ sodium thiosulfate

(2) *Test by conversion into prussic acid* [57]

Soluble thiocyanates may be converted into cyanide and sulfate by numerous oxidizing agents. The cyanide can then be identified by the benzidine reaction for prussic acid (see page 276). The conversion into cyanide by means of potassium permanganate is quantitative at room temperature:

$$5\ CNS^- + 6\ MnO_4^- + 8\ H^+ = 6\ Mn^{+2} + 5\ SO_4^{-2} + 5\ CN^- + 4\ H_2O$$

When testing for thiocyanate by the formation of cyanide, it is essential that no soluble cyanide should be present at the start. If this is the case, the prussic acid must be driven completely from the sample before the test by heating with dilute acid (hood).

Chlorides, bromides and iodides should not be present, because the corresponding halogens are liberated by potassium permanganate. They oxidize benzidine to benzidine blue. The production of any oxidizing or reducing gas lowers the specificity of the test.

Procedure. The apparatus of Fig. 26, page 48 is used. A drop of the test solution is mixed with 1 to 3 drops of an acid potassium permanganate solution. For small amounts of thiocyanates, the mixture is gently warmed. The hydrocyanic acid formed when thiocyanate is present is revealed by the blue color developed on copper acetate-benzidine acetate paper (see page 276) laid on the funnel of the apparatus or by any other prussic acid reaction.*

Limit of Identification: 1 γ thiocyanogen
Limit of Dilution: 1 : 50,000

(3) Test with cobalt salts [58]

Cobalt^{II} salts give a blue solution of $K_2[Co(CNS)_4]$ with thiocyanates in the presence of acetone. The reaction (see page 146) may be used as a sensitive test for thiocyanate.

Procedure. A drop of the acid test solution is mixed in a micro crucible with a small drop (about 0.02 ml) of 1 % cobalt sulfate solution, and evaporated to dryness. The residue is red-violet. The color fades on the addition of a few drops of acetone if no thiocyanate is present, whereas the acetone becomes blue-green to green in the presence of thiocyanates.

Large amounts of halides, especially iodides, give only a slight greenish color under these conditions. However, 6 γ thiocyanate may be detected in the presence of 1000 times the amount of iodide, bromide, chloride, thiosulfate and acetate. The latter two ions, on evaporation with cobalt salts, give a green color which, unlike that developed by thiocyanate, does not enter the acetone. Nitrites interfere because they react with thiocyanates to give red nitrosyl thiocyanate.

(4) Other test for thiocyanate

The evaporation residue from a drop of the neutral or weakly alkaline test solution is heated with solid ammonium chloride to 200–300° C. Hydrogen sulfide is evolved and can be detected by means of lead acetate paper (*Idn. Limit:* 10 γ potassium thiocyanate).[59]

8. Cyanic acid

Test by conversion into hydroxyurea [60]

Alkali salts of cyanic acid (cyanate ions) react with hydroxylamine hydrochloride (hydroxylammonium ions):

$$KCNO + NH_2OH.HCl \rightarrow NH_3OH \cdot CNO + KCl \quad (1)$$

$$[CNO^- + NH_3OH^+ \rightarrow NH_3OHCNO] \quad (1a)$$

The resulting hydroxylamine salt is in equilibrium with the isomeric hydroxyurea:

$$NH_3OHCNO \rightleftharpoons OC\begin{matrix}\diagup NH_2 \\ \diagdown NHOH\end{matrix} \quad (2)$$

* If the apparatus shown in Fig. 27, is used, as little as 0.12 γ thiocyanate can be detected.

which is analogous to the familar ammonium cyanate ⇄ urea equilibrium.

Hydroxyurea is an amidohydroxamic acid, which like other hydroxamic acids, gives a violet water-soluble inner complex ferric salt in weak acid solutions:

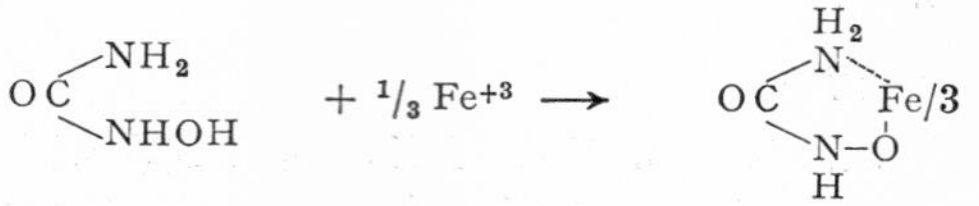

$$OC\begin{matrix}\nearrow NH_2 \\ \searrow NHOH\end{matrix} + {}^{1}/_{3}\,Fe^{+3} \longrightarrow OC\begin{matrix}\nearrow \overset{H_2}{N} \searrow \\ \searrow \underset{H}{N}-O \nearrow\end{matrix} Fe/3 \qquad (3)$$

A test for cyanate ions is based on the realization of (1), (1a), (2), (3). Thiocyanate ions must be absent since they yield red water-soluble ferric salts.

Procedure. The test is conducted on a spot plate. A drop of the neutral test solution is mixed with a drop of a 2 % water solution of hydroxylamine chloride and then one drop of a 2 % solution of ferric chloride is added. A transient violet color signifies the presence of cyanate ions. The instability of the color is due to the reducing action of the excess hydroxylamine.

Limit of Identification: 30 γ cyanic acid
Limit of Dilution: 1 : 1,600

9. Hydrazoic acid

(1) Test by formation of silver or ferric azide [61]

Hydrazoic acid reacts with Ag^+ ions to form white insoluble AgN_3. Red soluble $Fe(N_3)_3$ is produced by reaction with ferric salts. Since the acid, liberated from its soluble salts by the addition of dilute mineral acids, is completely expelled at 37° C, it can be easily and indisputably detected in its vapor state, even if other halogen hydracids which are not volatilized are present in the sample.

Sulfites, thiosulfates, and sulfides interfere because they are decomposed by acids; the resulting SO_2 or H_2S reduces ferric salts. Black silver sulfide is formed in the case of sulfides. These interferences can be avoided by preliminary oxidation of the neutral or alkaline test solution with several drops of hydrogen peroxide. If cyanides are present, prussic acid is liberated and the vapor will precipitate silver cyanide. Consequently, in such cases, the ferric chloride test for hydrazoic acid should be used.

Procedure. A drop of the test solution is placed in the bulb of the apparatus described on page 47 (Fig. 23) and treated with 1 or 2 drops of dilute hydrochloric acid. A drop of dilute silver nitrate solution, or of ferric chloride solution, is placed on the knob of the stopper. The apparatus is closed, and the bulb is heated gently. A turbidity, or red color, indicates hydrazoic acid.

Limit of Identification: 5 γ sodium azide
Limit of Dilution: 1 : 10,000

References pp. 356–360

Insoluble azides can be decomposed by means of zinc and acid. It is merely necessary to add a bit of the metal to the sample in the apparatus and to use somewhat more acid.

(*2*) *Test by reaction with nitrous acid* [62]

Hydrazoic acid reacts with nitrous acid:

$$HN_3 + HNO_2 \rightarrow H_2O + N_2O + N_2$$

In practice this reaction [63] is realized by acidifying a mixture of the alkali salts; it occurs so rapidly that it can serve for the complete removal of nitrites from nitrates (compare p. 328). Accordingly, if a dilute solution of alkali nitrite is treated with alkali azide and the mixture made acidic by means of acetic acid and then tested with Griess reagent for nitrite (compare p. 330), the color reaction does not occur, provided an excess of azide was present. Hydrazoic acid can be detected in this manner if no other compounds are present which react with nitrite in acid solution. Among these are: sulfamic acid, halogenates, hydrazine salts, hydrogen peroxide. If azide is to be detected in the presence of nitrite-consuming materials, it is necessary to liberate the hydrazoic acid by acidification and gentle warming and to allow its vapors to react with nitrite (see below).

Procedure. A spot plate is used. Two adjacent depressions are each charged with a drop of 0.001 % sodium nitrite solution; a drop of water is added to one, a drop of the test solution to the other. Both are then treated with a drop of the acetic acid Griess reagent (for preparation see p. 331). A pink color appears in the azide-free comparison mixture, whereas if azide was present in the other mixture, the color does not appear at all or its intensity is much less.

Limit of Identification: 1 γ sodium azide
Limit of Dilution: 1 : 50,000

If hydrazoic acid is to be detected in the gas phase, the procedure described in test *1* is used with the modification that a drop of 0.01 % sodium nitrite solution is placed on the knob of the gas absorption apparatus. After releasing the hydrazoic acid by warming to 50° C for 3–5 minutes, the suspended drop is transferred to a spot plate and stirred with a drop of Griess reagent. The color is compared with that of a drop of the sodium nitrite solution. The *identification limit* with this procedure is 5 γ sodium azide. The *dilution limit* is 1 : 10,000.

10. Ferrocyanic acid (Hydroferrocyanic acid)

(*1*) *Test with uranyl acetate* [64]

Soluble ferrocyanides give a brown precipitate of uranyl ferrocyanide $(UO_2)_2[Fe(CN)_6]$ in neutral or weakly acid solution. Ferricyanides react only in concentrated solutions and after long standing or on heating; they then give dirty yellow uranyl ferricyanide. Accordingly, uranyl salts may be used to detect ferrocyanides in the presence of ferricyanide and halide ions. When much ferricyanide is present, the precipitation of uranyl ferrocyanide induces that of the ferricyanide. Therefore, when the test is carried out on a spot plate, the sensitivity is increased by the presence of ferricyanides. The test may also be carried out on paper impregnated with uranyl acetate, but in this case ferricyanide should not be present. It is partially reduced in the paper to ferrocyanide, which reacts with the uranyl acetate and thus interferes with the test.

Procedure. A drop of the test solution is mixed on a spot plate with a drop of 1 *N* uranyl acetate solution, or placed on paper impregnated with uranyl acetate. According to the amount of ferrocyanide present, a brown precipitate or ring is formed.

Limit of Identification: 1 γ $K_4Fe(CN)_6$ } on the spot plate
Limit of Dilution: 1 : 50,000

Limit of Identification: 0.5 γ $K_4Fe(CN)_6$ } on uranyl acetate paper
Limit of Dilution: 1 : 100,000

(2) *Test by demasking of iron^II ions* [65]

The complex ferrocyanide ions are so stable that a direct test for Fe^{+2} or CN^- ions is not possible in either acid or alkaline solution. This signifies that the complex equilibrium

$$Fe(CN)_6^{-4} \rightleftharpoons Fe^{+2} + 6\ CN^-$$

lies almost completely to the left. Accordingly, no change is perceptible when either α,α'-dipyridyl or mercuric chloride is added to a solution of alkali ferrocyanide. However, if both reagents are added to an ammoniacal or neutral solution of potassium ferrocyanide, cyanide ions are withdrawn from the equilibrium because of the formation of undissociated mercuric cyanide and iron^{+2} ions are liberated. The latter give a red color with α,α'-dipyridyl (compare detection of iron, page 161). Therefore, the demasking reaction is:

$$Fe(CN)_6^{-4} + 3\ Hg^{+2} \rightarrow 3\ Hg(CN)_2 + Fe^{+2}$$

Ferricyanide ions are demasked in analogous fashion by mercuric chloride and iron^{+3} ions are set free. However, this does not detract from the detection of ferrocyanide because α,α'-dipyridyl reacts only with iron^{+2} ions.

The most favorable conditions for the realization of the demasking and the iron^{+2} dipyridyl reaction are obtained by warming the test solution with water-insoluble mercuri amidochloride. The net reaction of the demasking may be assumed to be*:

$$Fe(CN)_6^{-4} + 3\,HgNH_2Cl + 3\,H_2O \rightarrow Fe^{+2} + 3\,Hg(CN)_2 + 3\,NH_3 + 3\,Cl^- + 3\,OH^-$$

or

$$Fe(CN)_6^{-4} + 3\,HgNH_2^+ + 3\,H_2O \rightarrow Fe^{+2} + 3\,Hg(CN)_2 + 3\,NH_3 + 3\,OH^-$$

The use of mercuri amidochloride has the additional advantage that acid solutions can likewise be tested for ferrocyanide since neutralization is achieved through the reaction:

$$HgNH_2Cl + 2\,H^+ \rightarrow Hg^{+2} + NH_4^+ + Cl^-$$

Procedure. The test is made in a micro test tube. A drop of the neutral or sodium carbonate test solution is treated with one drop of a 1 % alcoholic solution of α,α'-dipyridyl (or o,o'-phenanthroline) and several mg of mercuri amidochloride. Depending on the amount of ferrocyanide present, a more or less intense red color appears on warming in the water-bath.

If acid solutions are presented for examination, the warming should be postponed for several minutes after the addition of the mercuri amidochloride.

Limit of Identification: 0.2 γ potassium ferrocyanide

Limit of Dilution: 1 : 250,000

This identification limit was obtained even in the presence of 1000 γ potassium ferricyanide.

See page 411 concerning the detection of acid insoluble ferrocyanides.

(3) *Test with iron*^{III} *chloride* [66]

Potassium ferrocyanide reacts with ferric chloride to give Prussian blue:

$$K_4Fe(CN)_6 + FeCl_3 \rightarrow 3\,KCl + \underset{\text{soluble Prussian blue}}{KFe^{III}[Fe^{II}(CN)_6]}$$

$$3\,KFe[Fe(CN)_6] + FeCl_3 \rightarrow 3\,KCl + \underset{\text{insoluble Prussian blue}}{Fe_4[Fe(CN)_6]_3}$$

These equations only summarize the course of the reaction. It is likely that the ferrocyanide acts as a reducer, so that Turnbull's blue (the ferrous salt of ferricyanic acid) $KFe^{II}[Fe^{III}(CN)_6]$, is also formed.[67]

By mixing a drop of potassium ferrocyanide and iron^{III} chloride on a spot plate or watch glass, 1.3 γ potassium ferrocyanide may be detected. The test is appreciably more sensitive when carried out on paper impregnated with ferric chloride since the capillary localization of the Prussian blue then comes into play. As little as 0.07 γ $K_4Fe(CN)_6$ may be detected.

* These equations are supported by the fact that a colorless solution of ferrocyanide containing phenolphthalein turns red on the addition of $HgNH_2Cl$ and subsequent warming.

When carried out on a spot plate, the test is not affected by the presence of ferricyanide, but this is not the case on filter paper. The reducing effect of the paper is such that some iron^{II} salt is always formed from the iron^{III} chloride, and the Turnbull's blue reaction occurs. In addition, when ferricyanide is placed on paper it is reduced in part to ferrocyanide, and thus gives a spurious positive reaction. It is only permissible to use paper when ferricyanides are absent.

Test with iron^{III} chloride in the presence of thiocyanates and/or iodides, using capillary separation *

Thiocyanates and/or iodides interfere with the sensitive Prussian blue test for ferrocyanides when carried out in a micro test tube or on a spot plate. There is simultaneous formation of red ferrithiocyanate or free iodine, which interfere with the decisive detection of small amounts of Prussian blue. If, however, the test is carried out on paper impregnated with ferric chloride, no difficulty is experienced because of thiocyanates or iodides. A capillary separation succeeds through the fact that ferric thiocyanate is readily decomposed by mercuric chloride, sodium fluoride, or sodium thiosulfate. If there is danger of interference by ferricyanide, the second procedure should be followed.

Procedure. (*a*) A drop of the acid test solution is placed on filter paper impregnated with iron^{III} chloride. The ferrocyanide reacts immediately with the ferric chloride paper producing a blue circle or ring. A concentric ring of red ferric thiocyanate appears some distance away from the Prussian blue.

Limit of Identification: 1.5 γ $[Fe(CN)_6]$ } in the presence of 2000 times the
Limit of Dilution: 1 : 33,000 } amount of thiocyanate

A far better sensitivity is attained for the detection of ferrocyanide in the presence of thiocyanate when use is made of the fact that ferric thiocyanate is decomposed by mercuric chloride (formation of $[Hg(CNS)_4]^{-2}$ ions), sodium fluoride (production of $[FeF_6]^{-3}$ ions), or sodium thiosulfate (reduction to Fe^{+2} ions).

Procedure. (*b*) When a drop of a solution containing considerable thiocyanate is placed on paper impregnated with ferric chloride, only a deep red fleck is seen. If treated with a solution of mercuric chloride, sodium fluoride, or sodium thiosulfate, the red color due to the ferric thiocyanate is discharged. The circle or ring of Prussian blue, due to ferrocyanide, is readily visible on the paper, which is now white.

Limit of Identification: 0.5 γ $[Fe(CN)_6]^{-4}$ } in the presence of 5000 times the
Limit of Dilution: 1 : 100,000 } amount of thiocyanate

* According to experimental work of R. Novacek. The capillary separation of $Fe(CN)_3^{-4}$ and CNS^- was first recommended by G. Gutzeit, *Helv. Chim. Acta*, 12 (1929) 829.

To detect *ferrocyanide in the presence of both thiocyanate and iodide*, a drop of the test solution is placed on ferric chloride paper. In the presence of much iodide, only a deep brown fleck of free iodine is formed. On addition of a drop of a concentrated sodium thiosulfate solution, the iodine, the ferric thiocyanate and also the yellow paper are completely decolorized. The fleck of Prussian blue, which was previously hidden by iodine and ferric thiocyanate, is now easily seen against the white background.

Limit of Identification: 0.5 γ $[Fe(CN)_6]^{-4}$
Limit of Dilution: 1 : 100,000
} in the presence of 3000 times the amount of thiocyanate and 2000 times the amount of iodide

11. Ferricyanic acid (Hydroferricyanic acid)

The usual test for ferricyanides (formation of Turnbull's blue with iron^{II} salts [68] is not very successful as a spot test. On filter paper, there may be reduction of the ferricyanide and also of any added iron^{III} salts by the cellulose of the paper, thus giving rise to a blue color. When the test is carried out on a spot plate, it must be remembered that ferricyanides often contain ferrocyanide which reacts with the iron^{III} salts which are almost always present in the iron^{II} salts, and thus form Prussian blue. In addition, white ferrous ferrocyanide turns blue on exposure to air. Furthermore, it is not feasible to precipitate a mixture of ferro- and ferricyanides with ferric chloride, and to form Prussian blue from the soluble ferric ferricyanide by adding a reducing agent to the filtrate. For reasons that are not clear, the ferricyanide is coprecipitated to a marked extent with the Prussian blue. Consequently, the only reliable test for ferricyanide must be based on its properties as an oxidizing agent.

(*1*) *Test with benzidine*[69]

Benzidine acetate is oxidized by soluble ferricyanides, with formation of insoluble blue meri-quinoid compounds (see page 72). This reaction is specific for ferricyanides in the absence of other oxidizing compounds (chromates, peroxo-compounds, etc.). The test can also be used in the presence of ferrocyanides.* However, it should be noted that the benzidine salt of ferrocyanic acid separates as a white precipitate, similar to benzidine sulfate. More reagent is consumed, and the detection of very small amounts of ferricyanide is rendered more difficult. To detect very small amounts of ferricyanide in the presence of large amounts of ferrocyanide, it is necessary to add sufficient lead salt to precipitate insoluble

* Ferrocyanide solutions, on standing and especially in direct light, are appreciably oxidized to ferricyanide.

lead ferrocyanide; ferricyanides remain in solution. Addition of benzidine, then causes the white $Pb_2[Fe(CN)_6]$ to turns blue, because of the adsorption of benzidine blue.

Procedure. A drop of the neutral test solution (if necessary, treated with a drop of 1 % lead nitrate solution) is mixed on a spot plate with benzidine acetate solution. A blue precipitate or coloration appears, depending on the ferricyanide content.

Limit of Identification: 1 γ potassium ferricyanide
Limit of Dilution: 1 : 50,000
Limit of Identification: 5 γ $K_3Fe(CN)_6$ } in the presence of 1000 times the
Limit of Dilution: 1 : 10,000 } amount of ferrocyanide
Reagent: Acetic acid-benzidine solution: 2 *N* acetic acid saturated in the cold with benzidine

(2) *Test with phenolphthalin* [70]

If a red alkaline solution of phenolphthalein (I) is heated with metallic zinc, reduction to colorless phenolphthalin (II) occurs. Oxidation regenerates the phenolphthalein:

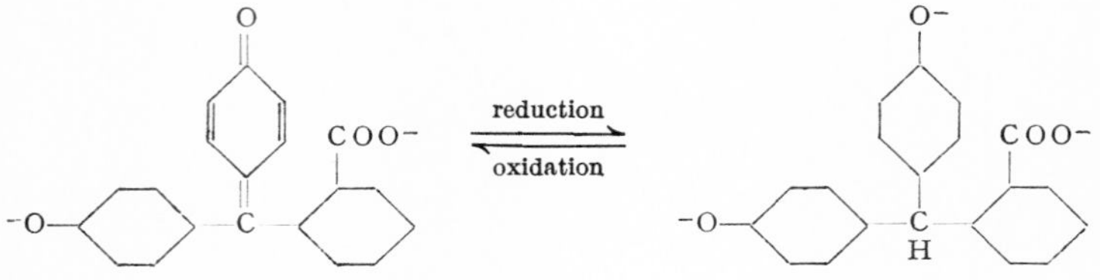

(I) $\underset{\text{oxidation}}{\overset{\text{reduction}}{\rightleftharpoons}}$ (II)

A colorless alkaline solution of phenolphthalin gradually turns pink in contact with air (oxidative regeneration of phenolphthalein). An intense red is produced immediately on addition of strong oxidizing agents. (In this connection, see page 355).

Alkali ferricyanides oxidize alkaline phenolphthalin solutions instantly. This same behavior is exhibited by insoluble metal ferricyanides that are decomposed by the free alkali in the phenolphthalin solution to produce alkali ferricyanide and insoluble hydrous metal oxide. A test for ferricyanide in the presence of other oxidizing materials is based on this behavior. The neutral or weakly acidified test solution is treated with zinc sulfate; insoluble zinc ferricyanide separates. The washed precipitate is tested with phenolphthalin solution.

Procedure. A drop of the neutral or slightly acidified test solution is placed on filter paper or a spot plate and treated with a drop of the reagent. A red or pink appears if ferricyanide is present.

Limit of Identification: 0.5 γ potassium ferricyanide
Limit of Dilution: 1 : 100,000
Reagent: Phenolphthalin solution: 2 g phenolphthalein, 10 g sodium hydroxide, 5 g zinc dust, and 20 ml water are warmed for about 2 hours under reflux. After cooling, the liquid is passed through a hardened filter paper and the colorless filtrate is made up to 50 ml with water. The solution must be stored in the dark; if necessary, it is decolorized by adding granules of zinc.

(3) *Test with zinc chloride and tetrabase* [71]

Tetrabase, which is tetramethyl-*p*-diaminodiphenylmethane with the cation (I), may be regarded as a stable leuco form of a diphenylmethane dye with the cation (II), whose intense blue color is due to the development of quinoidal linkings.[72] Strong oxidizing agents convert (I) into (II) in acid solution.

$$(CH_3)_2N-C_6H_4-CH_2-C_6H_4-\overset{+}{N}H(CH_3)_2 \qquad (I)$$

oxidation ↑↓ reduction

$$(CH_3)_2N-C_6H_4-CH{=}C_6H_4{=}\overset{+}{N}(CH_3)_2\cdot \qquad (II)$$

Ferricyanides or $[Fe(CN)_6]^{-3}$ ions by themselves have no action on tetrabase. However, immediate oxidation occurs when Hg^{+2} or Zn^{+2} ions are added. The reaction:

$$[Fe(CN)_6]^{-3} + \text{leuco dye} \rightleftarrows \text{dye} + [Fe(CN)_6]^{-4} + H^+$$

is reversible and the equilibrium of this redox reaction lies so far to the left that there is no discernible production of the quinoidal dyestuff from the leuco form. If, however, mercuric or zinc salts are introduced, the ferrocyanide ions are removed as insoluble mercuric or zinc ferrocyanide. The decrease in the concentration of $[Fe(CN)_6]^{-4}$ raises the oxidation potential of the ferricyanide sufficiently to enable the latter to oxidize the leuco dye in considerable amounts (see page 178).

This effect is the basis of a sensitive test for ferricyanide in weakly acid, neutral, or alkaline solution. Under the conditions prescribed here, other oxidizing agents such as nitrate, persulfate, iodate, etc., have no action on tetrabase. On the other hand, chromate, bichromate, and permanganate interfere by oxidizing tetrabase.

Procedure. One drop of an alcohol solution of zinc chloride and tetrabase is placed on filter paper. After the solvent has evaporated, the spot is treated with a drop of the weakly acidified test solution. A blue-violet fleck or ring appears if ferricyanide is present. The sensitivity of the test is very much lower in alkaline solution.

Limit of Identification: 0.1 γ potassium ferricyanide
Limit of Dilution: 1 : 500,000
Reagent: A 2% solution of zinc chloride in alcohol is saturated with an excess of tetrabase [73] and filtered

Capillary detection of ions that contain the cyanogen radical [74]

It is possible to detect the constituents in one drop of solution that contains $[Fe(CN)_6]^{-4}$, $[Fe(CN)_6]^{-3}$ and CNS^- ions. Spot reactions are used.

Procedure. Filter paper is soaked in a saturated solution of lead nitrate. A drop of the test solution is added. Insoluble $Pb_2[Fe(CN)_6]$ is fixed in the middle of the moist fleck. Ferricyanide and thiocyanate diffuse outward through the capillaries and, after the addition of a little water, remain in the outer zone. One side of the spot is then moistened with a dilute ferric solution (from a capillary) and the other side is similarly treated with a freshly prepared dilute ferrous solution. In the presence of the proper anions, a fleck of Prussian blue appears in the center, and Turnbull's blue and ferric thiocyanate at the border. A separate sample must be used in the test for cyanide ions.

(4) Other tests for ferrocyanide and ferricyanide

(*a*) Ferrocyanide can be detected, in slightly acid solution, with titanium tetrachloride. The reaction is carried out on a spot plate. A yellow to red-yellow precipitate, $Ti[Fe(CN)_6]$ is formed (*Idn. Limit:* 50 γ $K_4Fe(CN)_6$).[75]

(*b*) Filter paper is impregnated with a freshly prepared, saturated (70° C) solution of trithiourea (Thi_3) cuprous chloride, $[CuThi_3]Cl$. A gray to red-violet fleck of $[CuThi_3]_3[Fe(CN)_6]\cdot H_2O$ appears when the paper is spotted with a neutral solution of ferricyanide (*Idn. Limit:* 0.9 γ $K_3Fe(CN)_6$].[76]

(*c*) Soluble and insoluble ferricyanides yield brown $Tl(OH)_3$ when warmed with Tl_2SO_4 and NaOH. The product gives a blue color if spotted with an acetic acid solution of benzidine (*Idn. Limit:* 3.5 $Fe(CN)_6^{-3}$). [77] Oxidizing compounds must be absent.

12. Hypohalogenous acids

(1) Test with safranin [78]

The red alkaline solution of the phenazine dye safranin O yields a blue-violet precipitate or color on the addition of an alkali halogenite. The chemistry of the reaction has not been unravelled. Possibly an intermediate product of the oxidative decomposition of the dyestuff is involved. Evidence

References pp. 356–360

for this is the complete decolorization of safranin solutions by excess alkali hypohalogenite, and also the fact that ferricyanides react similarly to the hypohalogenites in strong alkaline solution.

Procedure. Filter paper is soaked in a saturated aqueous solution of safranin O and dried in the air. A drop of the test solution is placed on a piece of the reagent paper and the liquid allowed to soak in. The paper is then bathed in water. If hypohalogenite is present, a more or less intense violet stain or ring appears on the red paper. The bathing in water is necessary when dilute hypohalogenite solutions are involved, because strong alkali hydroxide produces a brown product on the paper, but this dye base is completely removed by the water.

Limit of Identification: 0.5 γ ClO^-, BrO^-, 2.5 γ IO^-

(*2*) *Test with thallous hydroxide* [78]

Colorless water-soluble thallous hydroxide yields brown insoluble thallic hydroxide with hypohalogenite:

$$Tl(OH) + NaHal\,O + H_2O \rightarrow Tl(OH)_3 + NaHal \quad (Hal = Cl, Br, I)$$

When dealing with colorless solutions (absence of alkali ferricyanide, chromate, permanganate) it should be remembered that only hydrogen peroxide and alkali peroxides react analogously to hypohalogenites. In the presence of these compounds, the hypohalogenite test cannot be considered because of the instantaneous reaction:

$$H_2O_2 + NaHal\,O \rightarrow H_2O + NaHal + O_2$$

Solutions of chloramine T have no effect on thallous hydroxide. The same holds for the tests described in (*1*) and (*3*).

Procedure. Filter paper is bathed in 2 % thallous sulfate solution and dried. A drop of the test solution is applied to a piece of the reagent paper. Depending on the quantity of hypohalogenite present, a more or less intense brown stain or ring appears on the colorless paper.

Limit of Identification: 0.5 γ alkali hypohalogenite
Limit of Dilution: 1 : 100,000

The test may also be conducted as a spot reaction on a black spot plate.

(*3*) *Test by reaction with ammonia* [78]

Ammonia is oxidized almost instantly by hypohalogenites:

$$2\,NH_3 + 3\,Hal\,O^- \rightarrow 3\,H_2O + 3\,Hal^- + N_2$$

Hypohalogenites are completely destroyed by excess ammonia. Use may be made of this fact when other oxidants, i.e. halogenates, chromates, etc., are to be detected in the presence of hypohalogenites by the well known redox reaction with HI (formation of elementary iodine).

In conformity with the above equation, when a solution of an ammonium salt is brought together with a hypohalogenite solution, ammonia is consumed. Since this consumption is readily revealed by a negative Nessler reaction (cf. page 238), a specific test for hypohalogenites is thus possible.

Procedure. A drop of 0.003 % ammonium chloride solution is placed in each of two adjacent depressions of a spot plate. To one is added a drop of the test solution, to the other a drop of an alkali hydroxide solution, that has been boiled. After mixing by blowing into the liquid through a pipette, a drop of Nessler solution is added to each mixture. The comparison test shows a distinct red color, whereas in the presence of hypohalogenite there is no or at best a much reduced color.

Limit of Identification: 0.1 alkali hypohalogenite
Limit of Dilution: 1 : 500,000

(4) Test for hypohalogenite in the presence of chromate, ferricyanide, permanganate

The test described in (3) for hypohalogenite is impaired by the presence of these colored anions and in general by dissolved or suspended colored compounds. This holds at least for the detection of slight amounts of hypohalogenite. In such cases, the test for the consumption of ammonia should be made in the gas phase.

Procedure. A drop of 0.003 % ammonium chloride solution and a drop of the test solution are mixed in the bulb of the gas absorption apparatus (Fig. 23, p. 47). A drop of Nessler solution is placed on the knob of the stopper. The closed apparatus is then placed in a boiling water-bath for 1–2 minutes. For comparison, a drop of the ammonium chloride solution is made alkaline and carried through the same procedure in a second absorption apparatus. If the two suspended drops are then wiped on filter paper, a yellow streak will appear only from the comparison test. Qualitative paper must be used because quantitative paper almost always contains traces of ammonium salts.[79]

This procedure revealed 0.8 γ alkali hypobromite in the presence of 3000 γ potassium chromate.

(5) General test and differentiation of hypohalogenites by liberation of halogen

Hypohalogenite solutions, according to their formation

$$2Hal^\circ + 2NaOH \rightarrow NaHalO + NaHal + H_2O$$

always contain the respective alkali halide and alkali hydroxide. If zinc chloride in excess is added to such solutions, OH^- ions disappear and free halogen is liberated according to:

$$2OH^- + Zn^{+2} \rightarrow Zn(OH)_2 \qquad (1)$$

$$HalO^- + Hal^- + H_2O + Zn^{+2} \rightarrow Zn(OH)_2 + 2Hal^\circ \qquad (2)$$

A general test for hypohalogenite, as well as for their differentiation, may be based on the realization of (1) and (2).

For the general test for hypohalogenites, (1) and (2) must be carried out in the presence of alkali iodide; iodine is then liberated and may be sensitively identified by the blue color obtained with starch or thiodene indicator. Under these conditions no other inorganic oxidants interfere with this test.

If only zinc chloride and starch (or thiodene indicator) are added to the test solution, a blue color indicates the presence of hypoiodite. Hypobromite is identified by the formation of red tetrabromofluorescein (eosin), specific for bromine, on the addition of zinc chloride and fluorescein (compare p. 262). Upon addition of alkali bromide (formation of bromine) hypochlorite shows the same behavior of hypobromite.

All tests may be carried out on a spot plate with one drop of the test solution. The reaction picture as well as the limits of identification obtained are shown on the following table:

Hypohalogenite	Zn^{+2}+I^-+ Thiodene	Zn^{+2} + Thiodene	Zn^{+2} + Fluorescein	Zn^{+2} + Br + Fluorescein
NaClO	blue color (2 γ)	–	–	red color (10 γ)
NaBrO	blue color (2.5 γ)	–	red color (7 γ)	red color (7 γ)
NaIO	blue color (2.5 γ)	blue color (2.5 γ)	–	–

Reagents: *1)* 20% solution of $ZnCl_2$ plus 0.1 g KI and 0.2 g Thiodene
2) as *1)*, without KI
3) Saturated alcoholic solution of fluorescein
4) as *3)*, saturated with KBr

When testing for hypohalogenites it must be kept in mind that, due to the interaction between the hypohalogenite ions, a mixture of them can never be expected, but only the presence of a single hypohalogenite. Thus, hypoiodite is not stable in the presence of hypochlorite or hypobromite, since it is oxidized by the latter to iodate. Equally, hypobromite is not stable in the presence of hypochlorite, which oxidizes it to bromate. Therefore, the data given in the above table are sufficient to identify with security the various hypohalogenites.

13. Chloric acid

Test with manganous sulfate and phosphoric acid [80]

In strong phosphoric acid solution, chlorates warmed with manganese sulfate form complex manganese^{III} phosphate ions:

$$ClO_3^- + 6\ Mn^{+2} + 12\ PO_4^{-3} + 6\ H^+ = 6[Mn(PO_4)_2]^{-3} + Cl^- + 3\ H_2O$$

The color of the complex manganese-containing anion is similar to that of the permanganate anion. Quantities of the $[Mn(PO_4)_2]^{-3}$ ion too small to be discerned by their color can be detected by adding diphenylcarbazide. A violet color results, similar to that produced by the chromate-diphenylcarbazide reaction (see page 167).

Persulfates and periodates react similarly (see page 301). Persulfates may be destroyed by evaporating the sulfuric acid solution with a little silver nitrate. Complete decomposition ensues:

$$2\ H_2S_2O_8 + 2\ H_2O = 4\ H_2SO_4 + O_2$$

(compare page 173). The test cannot be applied in the presence of periodates.

Procedure. A drop of the test solution is mixed with a drop of a 1 : 1 mixture of saturated manganous sulfate solution and sirupy phosphoric acid in a small porcelain dish. The mixture is briefly warmed over a micro burner, and allowed to cool. According to the chlorate content, a more or less deep violet coloration results. Very pale colors may be intensified by adding a drop of 1 % alcoholic solution of diphenylcarbazide.[81] A more distinct violet coloration results, due to the formation of an oxidation product of diphenylcarbazide.

Limit of Identification: 0.05 γ chlorate
Limit of Dilution: 1 : 1,000,000

14. Bromic acid

Test with manganous sulfate and sulfuric acid [82]

When treated with manganous sulfate and sulfuric acid, bromates give a red color initially, due to the intermediate formation of Mn^{III} sulfate. On boiling, or better on addition of alkali acetate, a precipitate of brown hydrated manganese^{III} oxide is formed:

$$HBrO_3 + 6\ MnSO_4 + 3\ H_2SO_4 \rightarrow 3\ Mn_2(SO_4)_3 + 3\ H_2O + HBr$$
$$Mn_2(SO_4)_3 + 3\ H_2O \rightarrow Mn_2O_3 + 3\ H_2SO_4$$

Chlorates and iodates, under the same conditions, give neither a red color or a precipitate. The test for bromate with manganese sulfate and sulfuric acid may be carried out as a spot reaction, when it is combined with the benzidine reaction for the detection of higher oxides of manganese (page 175).

Procedure.[83] A drop of the test solution is mixed in a microcentrifuge tube with a drop of 2 % manganous sulfate solution acidified with a little sulfuric acid, and warmed for 2 to 3 minutes on the water-bath. After cooling, a few drops of an acetic acid-benzidine solution (see page 175) and a few crystals of sodium acetate are added. A more or less deep blue color results, depending upon the quantity of bromate in the test solution.

Limit of Identification: 20 γ potassium bromate
Limit of Dilution: 1 : 2500

15. Iodic acid

(1) *Test with hypophosphorous acid* [84]

Even in the cold, iodates are reduced by hypophosphorous acid; iodides and phosphoric acid result. This redox reaction does not occur instantaneously and directly, but via partial reactions with different velocities. Phosphorous acid and iodide are formed initially (1); the latter then reacts with the iodate to produce free iodine (2), which then oxidizes the phosphorous acid produced in (1) to phosphoric acid:

$$IO_3^- + 3\ H_3PO_2 \rightarrow 3\ H_3PO_3 + I^- \quad (1) \quad \text{rapid}$$
$$5\ I^- + IO_3^- + 6\ H^+ \rightarrow 3\ H_2O + 3\ I_2 \quad (2) \quad \text{rapid}$$
$$H_3PO_3 + I_2 + H_2O \rightarrow H_3PO_4 + 2\ HI \quad (3) \quad \text{slow}$$

As shown by this series of equations, free iodine and phosphorous acid are formed as intermediate products, which react only slowly with each other. The intermediate production of iodine can be revealed by the starch-iodine test and thus made to serve as a test for iodate. Bromates and chlorates do not react under these conditions.

Procedure. A drop of the neutral test solution is mixed on a spot plate with a drop of starch solution and a drop of a dilute solution of hypophosphorous acid. A transitory blue appears in the presence of iodates.

Limit of Identification: 1 γ iodic acid
Limit of Dilution: 1 : 50,000

(2) *Test with potassium thiocyanate* [85]

Alkali thiocyanates react with iodates in acid solution:

$$5\ CNS^- + 6\ IO_3^- + 6\ H^+ + 2\ H_2O = 5\ HSO_4^- + 5\ HCN + 3\ I_2$$

The liberation of iodine thus indicates the presence of iodate.

Procedure. A drop of 5 % potassium thiocyanate solution is placed on starch paper and followed by a drop of the acid test solution. A more or less intense blue stain indicates the presence of iodate.

Limit of Identification: 4 γ sodium iodate
Limit of Dilution: 1 : 12,000

(3) Test with pyrogallol [86]

In acid solution, iodates oxidize pyrogallol to purpurogallin. Strong oxidizing agents, such as iodine and chromate, act in the same way. Persulfate, nitrate, and bromate give similar colors, which can be distinguished if parallel tests are made.

Procedure. Filter paper impregnated with acid pyrogallol is spotted with a drop of the test solution. A pink to red color appears according to the quantity of iodate present.

Limit of Identification: 0.25 γ sodium iodate
Limit of Dilution: 1 : 200,000
Reagent paper: Strips of filter paper are bathed in oxalic acid, dried, and then bathed in an acetone solution of pyrogallol

16. Perchloric acid

Test by fusion with cadmium chloride [87]

In analogy to alkali halogenates and periodates, alkali perchlorates lose oxygen on dry heating:

$$KClO_4 = KCl + 4\ O \tag{1}$$

If this thermal decomposition occurs in the presence of excess chloride, chlorine is produced by the action of nascent oxygen [88] at its point of origin. Accordingly, in a mixture of alkali perchlorate and cadmium chloride, which melts without decomposition at 568° C, there occurs in addition to (1) the redox reaction

$$CdCl_2 + O = CdO + Cl_2 \tag{2}$$

The resulting chlorine can be detected in the gas phase with thio-Michler's ketone, that is, by contact with reagent papers which contain 4,4'-bis-dimethylamino-thiobenzophenone, or fluorescein with alkali bromide. The yellow reagent papers turn blue or red, respectively. Concerning the chemistry of these two color reactions, see page 368.

Procedure. A micro test tube is used. A drop of the neutral or weakly alkaline solution is taken to dryness with a drop of 2 % $CdCl_2$ solution, and kept at 160° C for a short time to remove the water. The test tube is closed with a disc of reagent paper and heated over a micro burner until the $CdCl_2$ fuses. Depending on the quantity of perchlorate a more or less intense blue or red circular stain appears on the yellow reagent paper within 1 minute.

With thio-Michler's ketone:
Limit of Identification: 1 γ ClO_4^-
Limit of Dilution: 1 : 50,000

With fluorescein:

Limit of Identification: 5 γ ClO_4^-

Limit of Dilution: 1 : 10,000

Reagents: *1*) Thio-Michler's ketone paper: Quantitative filter paper is bathed in a 0.1 % benzene solution of 4,4 bis-dimethylamino-thio-benzophenone and dried in the air. The paper keeps well if stored in the dark.

2) Fluorescein-KBr paper: Strips of filter paper (S & S 598) are bathed in a weakly alkaline solution containing 0.1 g dye and 0.5–0.8 g KBr per 100 ml, and dried. The paper should be moistened before use.

The above procedure can not be used directly if halogenates or nitrates are present. The former likewise yield free chlorine when fused with $CdCl_2$. The latter yield $Cd(NO_3)_2$ initially, which then loses N_2O_4 on dry heating, and thus also oxidizes thio-Michler's ketone to a blue compound. These interferences can be eliminated by fuming the sample once or twice with concentrated hydrochloric acid. Halogenates are thus decomposed completely, and nitrates to a large extent. The evaporation can be conducted in the presence of $CdCl_2$ in a micro test tube. The evaporation residue can then be carried through the procedure given. It is thus possible to detect 1 γ ClO_4 in the presence of 1,000 γ $KClO_3$ or KNO_3.

17. Periodic acid

(*1*) *Test with manganous sulfate and phosphoric acid* [89]

Periodates react with manganous sulfate, on heating in strong phosphoric acid solution, to form complex manganese^{III} phosphate ions, which have extraordinary tinctorial power. The violet resembles that of permanganate ions. The reaction can be represented:

$$IO_4^- + 2\,Mn^{+2} + 4\,PO_4^{-3} + 2\,H^+ = 2\,[Mn(PO_4)_2]^{-3} + IO_3^- + H_2O$$

Persulfates and chlorates give a similar effect. Persulfates can be decomposed by evaporation of the sulfuric acid solution in the presence of a little silver nitrate. The test is not applicable when chlorates are present.

Procedure. A drop of the test solution is mixed with a drop of a 1 : 1 mixture of saturated manganous sulfate solution and syrupy phosphoric acid in a small porcelain dish. The mixture is warmed briefly over a micro burner and allowed to cool. A more or less deep violet is formed according to the periodate present. Very pale colors may be intensified by the addition of a drop of a 1 % alcoholic solution of diphenylcarbazide; a distinct violet appears.

Small amounts of the complex manganic ions can be revealed by adding diphenylcarbazide (see page 298).

Limit of Identification: 5 γ periodate

Limit of Dilution: 1 : 10,000

(2) Test with manganese salts and tetrabase [89]

Manganous salts are oxidized to permanganate on warming with alkali periodates in acid solution:

$$2\ Mn^{+2} + 5\ IO_4^- + 3\ H_2O = 2\ MnO_4^- + 5\ IO_3^- + 6\ H^+$$

This sensitive reaction, which is used for the detection of manganese (see page 174), may also be used to detect periodates. The identification of small amounts of permanganate is accomplished by the use of tetrabase (tetramethyl-*p*-diaminodiphenylmethane); a blue meri-quinoid oxidation product is formed (see page 293). Chlorates, bromates, and iodates do not interfere with the test. Persulfates react with manganese^{II} salts, giving higher manganese oxides, which also react with tetrabase. If necessary, persulfate may be removed by heating the sulfuric acid solution with a little silver nitrate. Colored metallic salts should not be present because they hide the light blue coloration.

Procedure. A drop of the test solution is mixed with a drop of the reagent solution. A blue coloration indicates periodate.

Limit of Identification: 0.5 γ periodate

Limit of Dilution: 1 : 100,000

Reagents: *1)* 2 *N* acetic acid is shaken in the cold with tetrabase [90] and filtered

2) 10 % solution of manganous chloride

Equal parts of solution *1)* and *2)* are mixed before using

(3) Test by the formation of complex copper^{III} *salts* [91]

A dark red color appears on warming alkaline solutions of copper salts with certain oxidizing agents (such as persulfate) in the presence of alkali periodate. The complex anion formed from tervalent copper and the periodate radical is responsible for the color.[92]

The formation of this colored ion may be used as a spot test for periodate [procedure (*a*)]. Alternatively, the interference by periodate, with the catalytic effect of copper^{III} salts on the reaction between Mn^{+2} ion and hypobromite may be utilized [procedure (*b*)]. This reaction normally proceeds with formation of manganese dioxide but, in the presence of even extremely small amounts of copper salts, permanganate is formed instead. The activation of the hypobromite is due to the intermediate formation of Cu^{III} oxide, which causes a further oxidation from MnO_2 to MnO_4^- ion, with regeneration of Cu^{II} oxide (see page 174). The formation of the stable complex ion of tervalent copper with periodate prevents the activation of the hypobromite.

The production of stable complexes of tervalent copper, under the given conditions, is limited to periodic acid and telluric acid (see page 350). The test is therefore extremely selective.

Procedure. (*a*) A drop of 1 : 50,000 copper sulfate solution is added to the alkaline test solution, followed by a drop of 2 *N* alkali hydroxide, and finally a little solid potassium or sodium persulfate. A micro crucible is used. The mixture is boiled. A yellow color indicates periodic acid.

Limit of Identification: 2 γ periodic acid
Limit of Dilution: 1 : 25,000

Procedure. (*b*) A drop of the test solution, a drop of the copper-manganese solution, and a drop of sodium hypobromite are mixed in a micro test tube. A blank test is carried out simultaneously on a drop of water plus the reagents. Both test tubes are placed in boiling water, or heated to boiling over a flame. In the presence of periodic acid, the solution remains colorless or is slightly yellow, while the blank clearly shows the permanganate color within a few minutes.

Limit of Identification: 0.5 γ periodic acid
Limit of Dilution: 1 : 100,000
Reagents: *1*) Copper-manganese solution: A drop of 4 % solution of $CuSO_4 \cdot 5\ H_2O$ is added to 100 ml of a 0.04 % solution of $MnCl_2 \cdot 4\ H_2O$
2) Sodium hypobromite: A freshly prepared solution of bromine in 2 *N* NaOH

18. Hydrogen sulfide

(*1*) *Test with sodium nitroprusside* [93]

The yellow alkaline solution of sodium nitroprusside, $Na_2[Fe(CN)_5NO]$, gives a red-violet color with soluble sulfides. The colored product with sodium sulfide has the empirical formula $Na_4[Fe(CN)_5NOS]$. It is not certain whether an addition compound is formed from the components, or whether a complex ion $[Fe^{II}(CN)_5N^{III}OS]^{-4}$ is produced, with reduction of the iron. The latter assumption is the more probable.[94] The color is discharged on the addition of acid.

Procedure. A drop of the alkaline test solution is mixed on a spot plate with a drop of 1 % sodium nitroprusside solution. A more or less intense violet appears, depending on the amount of sulfide present. The test may also be carried out on filter paper impregnated with an ammoniacal (5 %) solution of sodium nitroprusside.

Limit of Identification: 1 γ sodium sulfide
Limit of Dilution: 1 : 50,000

(*2*) *Test by catalysis of the iodine-azide reaction* [95]

The following test for sulfides (as well as the tests for thiocyanate and thiosulfate, pages 282 and 318) has been developed from the observation that solutions of sodium azide and iodine do not react to any visible extent, but on the addition of a crystal of sodium sulfide or sodium thiosulfate there is an immediate and vigorous evolution of nitrogen, with consumption of iodine.[96]

$$2\ NaN_3 + I_2 = 2\ NaI + 3\ N_2$$

Written in this manner, the equation implies that the sulfur-bearing material takes no part in the reaction, which may therefore be regarded as catalyzed. Even traces of sulfides start this reaction, whose occurrence is easy to see because of the evolution of nitrogen. The effect on this reaction thus provides a convenient test for sulfide (thiosulfate, thiocyanate).

The acceleration of the reaction between NaN_3 and I_2 can be explained as an intermediate reaction catalysis: the iodine is taken up by the thio compounds to form labile reactive compounds as shown in (1). These iodine-sulfur compounds react with sodium azide as shown in (2), whereby the pure sulfur compound, the catalyst, is regenerated unchanged. Summation of (1) and (2) gives the foregoing equation, in which the catalyst, namely, the sulfidic compound, does not appear. This series of reactions is repeated until either the iodine or the sodium azide is consumed.

$$MeS + I_2 \rightarrow IMe{-}SI \qquad \text{primary reaction (1)}$$
$$IMe{-}SI + 2\ NaN_3 \rightarrow MeS + 2\ NaI + 3\ N_2 \qquad \text{secondary reaction (2)}$$
$$2\ NaN_3 + I_2 \rightarrow 2\ NaI + 3\ N_2 \qquad \text{catalyzed reaction (1) + (2)}$$

(Me = bivalent metal)

The reactive intermediate compound in (1) and (2) is hypothetical. Although its postulation satisfactorily explains the mechanism of the catalysis, its existence has not been proved, i.e., the compound has not been isolated. This is not surprising, because the lability and reactivity of the intermediate compounds of a catalyst are the very characteristics responsible for the catalytic effect. The formulation IMe—SI corresponds to the addition of iodine to MeS to form $MeS \cdot I_2$ (compare page 306).

The catalyzed reaction makes it possible to detect extremely small quantities of sulfides. Furthermore, the effect is strictly specific for *sulfide sulfur*, in both inorganic and organic form (see Volume II). Sulfur combined in other ways (sulfate, sulfite) and free sulfur has as little effect on the reaction as selenides and tellurides, which in many other respects are very similar to metal sulfides. The fact that soluble as well as insoluble sulfides exert this catalytic action is of considerable analytical value.

Test for soluble sulfides

Procedure. A drop of the sample and a drop of iodine-azide solution are mixed on a watch glass. An immediate development of gas, rising in tiny bubbles through the liquid, indicates the presence of sulfide.

The test is only decisive in the absence of thiosulfate and thiocyanate; they likewise catalyze the iodine-azide reaction. When interfering sulfur compounds are present, the sulfide sulfur must be isolated beforehand. This operation is carried out by precipitation with zinc (or cadmium) carbonate in an Emich microcentrifuge tube. After centrifuging and thorough washing, the sulfide sulfur may be detected by applying the iodine-azide reaction to the precipitate.

Limit of Identification: 0.3 γ sodium sulfide

Limit of Dilution: 1 : 166,000

Reagent: Sodium azide-iodine solution: 3 g sodium azide is dissolved in 100 ml 0.1 N iodine. The solution is stable.

As little as 0.02 γ sulfide sulfur in one drop can be detected by using a solution of 1 g sodium azide, 1 g potassium iodide, plus a tiny crystal of iodine, in 3 ml water.

Limit of Dilution: 1 : 2,500,000

It is probable that all the tests with sodium azide-iodine solutions, as described in this text, would exhibit greater sensitivities if more concentrated reagent solutions were used.

Test for solid sulfides

All solid metal sulfides, natural or synthetic, and also the sulfo salts (e.g., $HgCl_2 \cdot 2HgS$ and similar molecular compounds of sulfides) react immediately with sodium azide solutions with evolution of nitrogen. They may thus be detected in the presence of sulfates, free sulfur, etc.

The test can therefore also be applied in the scheme of systematic qualitative analysis to determine the presence of sulfides in the "insoluble residue" (see page 407), and also to detect mercury in the residue (containing lead sulfate, mercury sulfide, or $Hg(NO_3)_2 \cdot 2HgS$ and sulfur) left after treatment of the basic sulfides with concentrated nitric acid. When this test is used to detect mercury sulfide, the solution in *aqua regia* or in hydrochloric acid containing hydrogen peroxide,[97] and the subsequent stannous chloride reaction, which are both otherwise essential, may be by-passed with considerable saving in time.

Procedure. Extremely small amounts of solid sulfides, e.g., in the dust from a powder, may be detected very simply by placing a drop of sodium azide-iodine solution either in a capillary or an Emich microcentrifuge tube. A particle of the test substance is introduced on the tip of a platinum wire bent at 45° C (Fig. 39). In the presence of sulfides, the gas bubbles rise in the capillary or in the constricted part of the conical tube. They can readily be seen with the naked eye or with the aid of a magnifying glass.

The detection of solid insoluble sulfides through catalysis of the iodine-azide reaction is also remarkable from the theoretical standpoint because it throws a characteristic light on the ability of solid materials in general to enter into reactions. If, in accord with the classical views of the ionic theory, it is assumed that chemical reactions in and with water solutions involve solely ionic and molecular species in solution, a calculation in the case of copper sulfide reveals the following if the solubility (in water) of CuS is $1.9 \cdot 10^{-17}$ grams per liter. One gram mole (95.6 g) of copper sulfide contains $6.06 \cdot 10^{23}$ molecules, so that about $12 \cdot 10^4$ molecules of CuS are present in one liter of saturated solution. If one drop (0.05 ml) of the reagent is placed on solid CuS, and the iodine-azide reaction is catalytically initiated, then under the (improbable) assumption that the solution is immediately saturated with CuS, it follows that no more than 6 molecules of CuS are reacting. Still smaller numbers are obtained if an analogous calculation is made for HgS or As_2S_3, which are actually less soluble than CuS in water. Even the most sensitive chemical reaction (including the iodine-azide catalysis by soluble sulfides) always involves many millions of molecules. So the assumption is justified that in the case of solid insoluble sulfides the catalysis is *not* due to dissolved molecules, whose number is inadequate, but rather that the effect is due to the metal sulfide molecules still held in the lattice of the solid phase. In the present instance, it is probable that the first stage consists of the addition of free iodine to the metal sulfide, forming $MeS \cdot I_2$. The latter is the tautomeric form of Me(SI)I which was assumed (page 304) to be the reactive intermediate compound in the iodine-azide catalysis. The idea that iodine is activated by addition to metal sulfide is supported by: (*a*) In conformity with the free surface of solids, large numbers of molecules must be involved when iodine is added; (*b*) Experiments have shown that heavy metal sulfides react in the solid state with iodine dissolved in organic liquids and produce the corresponding iodides. It seems much more likely that the course of these reactions is not direct, but that they rather involve the formation fo the intermediate $MeS \cdot I_2$. Accordingly,[98] $MeS + I_2 \rightarrow MeS \cdot I_2 \rightarrow MeI_2 + S$.

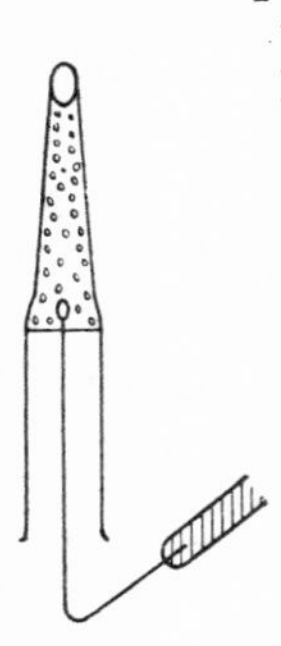

Fig. 39. Set-up for testing solids for sulfides (¾ actual size)

(3) Test for alkali hydrosulfides in the presence of alkali sulfides

Both sulfides and hydrosulfides are oxidized by iodine:

$$Na_2S + I_2 = 2\,NaI + S \quad \text{and} \quad NaHS + I_2 = NaI + S + HI$$

The solution of alkali sulfides which, owing to hydrolysis, originally had an alkaline reaction, becomes *neutral* after oxidation by iodine, if normal sulfides are present, and *acid* when hydrosulfides are present. Therefore, hydrosulfides may be detected in the presence of sulfides by the acid reaction toward appropriate color indicators after oxidation with iodine.

Procedure. A drop of the test solution on a spot plate is treated, drop by drop, with a 0.1 *N* iodine solution until the iodine color is permanent. A drop of carbon disulfide is added and the mixture stirred with a glass rod. The carbon disulfide dissolves both the sulfur formed in the reaction and the excess iodine. The acid reaction of the aqueous solution can then easily be tested by means of litmus paper.

(4) Test for free alkali, in the presence of alkali sulfides, with thallium nitrate [99]

To detect free alkali in the presence of alkali sulfides, which also give an alkaline reaction owing to hydrolysis, use is made of the fact that thallous sulfide is insoluble whereas thallous hydroxide is soluble in water and is almost as strong a base as potassium hydroxide. Free alkali may be detected by adding neutral thallous nitrate and then testing the solution for an alkaline reaction.

Procedure. A drop of the test solution is mixed on a spot plate with 1 or 2 drops of a saturated solution of thallium nitrate, and stirred. A strip of litmus paper immersed in the resulting suspension of thallous sulfide turns blue if free alkali is present.

(5) Other tests for sulfides

(a) Hydrogen sulfide can be detected through the formation of methylene blue.[100] This test can be carried out as a spot reaction in the apparatus described on page 47 (Fig. 23). The gas liberated from sulfides is absorbed in a drop of dilute NaOH, which is transferred to a spot plate and treated with a mixture containing 1 drop each of concentrated HCl, 0.1 *N* $FeCl_3$ and several granules of *p*-aminodimethylaniline. Methylene blue is formed (*Idn. Limit:* 1 γ H_2S).

(b) Sulfides may be detected with sodium plumbite solution (*Idn. Limit:* 1.8 γ Na_2S). [101]

19. Sulfurous acid

(1) Test with sodium nitroprusside [102]

(a) Detection in solutions

Sulfites and free sulfurous acid, even in concentrated solutions, give only a pale red color with sodium nitroprusside $Na_2[Fe(CN)_5NO]\cdot H_2O$. The color is appreciably deepened by the addition of a saturated solution of zinc sulfate or nitrate and a few drops of a potassium ferrocyanide solution; a red precipitate is formed. The chemistry of this change has not been worked out. Probably, as in the test for sulfides with sodium nitroprusside (see page 303), a complex compound is formed,* with reduction of the iron. On addition of acids, the red color disappears. Sulfates and thiosulfates do

* G. Scagliarini, *Chem. Abstracts*, 29 (1935) 2872, has prepared a cadmium compound with hexamethylenetetramine, $Cd[Fe(CN)_5NOSO_3]\cdot 12\,H_2O\cdot 2\,C_6H_{12}N_4$, and assumes that the complex ion $[Fe(CN)_5NOSO_3]^{-2}$ is present.

not interfere; sulfides in alkaline solution give a violet color. The reaction may be made somewhat more sensitive by changing the order of adding the reagents.

Procedure. A drop of a 1 *N* solution of potassium ferrocyanide is added to a drop of a cold saturated solution of zinc sulfate (or nitrate) and then a drop of 1 % sodium nitroprusside is added. White zinc ferrocyanide precipitates. A drop of the neutral test solution is introduced. The precipitate becomes red if sulfite is present.

Limit of Identification: 3.2 γ sodium sulfite
Limit of Dilution: 1 : 16,000

(b) Test by evolution of sulfur dioxide

When moist salmon-pink zinc nitroprusside $Zn[Fe(CN)_5NO]$ is brought into contact with sulfur dioxide it turns red. The test is more sensitive when the reaction product is held over ammonia, which decolorizes the unused zinc nitroprusside.[103] The composition of the product, which is decomposed both by alkalis and mineral acids, has not been determined. Possibly a loose addition compound is formed between SO_2 and $Zn[Fe(CN)_5NO]$. Other metal nitroprussides behave similarly.

The test for sulfites is carried out by liberating the sulfur dioxide with mineral acid, and then bringing the gas into contact with zinc nitroprusside. The apparatus described on page 47 (Fig. 23) is suitable.

Procedure. A grain of the solid sample or a drop of the test solution is placed in the reaction bulb. The sulfur dioxide is liberated by a drop of 2 *N* hydrochloric or sulfuric acid. The knob of the glass stopper is covered with a thin layer of the reagent paste and the stopper is inserted immediately after adding the acid. After the SO_2 has acted, the stopper is held for a short time over ammonia. A more or less deep red coloring of the reagent indicates the presence of sulfurous acid.

Limit of Identification: 3.5 γ sulfur dioxide
Limit of Dilution: 1 : 14,300
Reagent: Zinc nitroprusside paste is made by treating sodium nitroprusside with an excess of zinc chloride at the temperature of boiling water. The washed precipitate is kept in a dark bottle.

(c) Test for sulfites in the presence of sulfides and thiosulfates

Sulfites cannot be detected with certainty by the preceding test in the presence of sulfides and thiosulfates. The hydrogen sulfide liberated from sulfides by acid (apart from its reaction with alkaline nitroprusside itself) reacts with sulfur dioxide from the sulfite:

$$2\ H_2S + SO_2 = 2\ H_2O + 3\ S^\circ$$

References pp. 356–360

with precipitation of sulfur, and thus interferes with the nitroprusside test. Thiosulfates may be mistaken for sulfites since they are decomposed by mineral acids:

$$S_2O_3^{-2} + 2\,H^+ = SO_2 + S^\circ + H_2O$$

to give sulfur dioxide.

Thiosulfates and sulfides must therefore be removed before carrying out the test. This is conveniently done by means of mercuric chloride which reacts with both types of compounds:

$$S^{-2} + Hg^{+2} = HgS$$

$$S_2O_3^{-2} + Hg^{+2} + H_2O = HgS + SO_4^{-2} + 2\,H^+$$

with formation of acid-stable mercuric sulfide. The liberation of SO_2 from sulfites by the acid remains mostly unaffected when mercuric chloride is added.

Procedure. The apparatus shown in Figure 23, is used. A drop of the test solution is mixed with 2 drops of saturated mercuric chloride solution. After about a minute, the mixture is acidified with dilute hydrochloric or sulfuric acid. The test is carried out as described above.

In this way, 20 γ Na_2SO_3 may be detected in the presence of 900 γ $Na_2S_2O_3$ and 1500 γ Na_2S, which represents a ratio

$$Na_2SO_3 : Na_2S_2O_3 : Na_2S = 1 : 45 : 75$$

Still smaller amounts of sulfite can be detected, following removal of sulfide and thiosulfate, by test *2*.

(2) Test by induced oxidation of nickel^{II} hydroxide [104]

The autoxidation of sulfur dioxide induces the oxidation of green nickel^{II} to black nickel^{IV} hydroxide.[105] This effect is all the more remarkable since this change is otherwise only accomplished by strong oxidizing agents (free halogens, persulfates, etc.). Hydrogen peroxide is without action (see page 354).

A probable explanation of this effect is that the initial product, basic nickel sulfite, on contact with air is oxidized both cationically and anionically:

$$2\,Ni(OH)_2 + SO_2 \longrightarrow Ni_2(OH)_2SO_3 + H_2O$$

$$\begin{matrix} Ni{-}OH \\ \diagdown \\ \quad SO_3 \\ \diagup \\ Ni{-}OH \end{matrix} + O_2 + H_2O \longrightarrow Ni(OH)_4 + NiSO_4$$

The oxidation of $Ni(OH)_2$ by contact with SO_2 and atmospheric oxygen may be applied for the detection of sulfites; the change of color is very distinct when the experimental conditions are controlled. The formation of benzidine blue from benzidine acetate by means of higher oxides of metals [106] (see page 72) may be used for smaller quantities of sulfite, because the reaction succeeds even when no more than traces of $Ni(OH)_4$ are formed.

Procedure. A drop of the test solution (or a little of the solid) is placed in the bottom of the apparatus described on page 47 (Fig. 23). A little freshly precipitated nickel hydroxide, that has been washed free from alkali, is placed on the glass knob beneath the stopper. After adding 1 or 2 drops of 1 : 1 hydrochloric acid, the apparatus is closed, and the SO_2 liberated by warming gently. According to the amount of sulfite present, the green paste of $Ni(OH)_2$ turns gray to black. For very small amounts of sulfite, the paste is transferred (after the reaction) to quantitative filter paper, and treated there with a drop of an acetic acid solution of benzidine (see page 175). A blue color indicates sulfite.

Limit of Identification: 0.4 γ sulfur dioxide

Limit of Dilution: 1 : 125,000

Instead of the paste, paper impregnated with $Ni(OH)_2$ may be used for the detection of gaseous SO_2. For preparation see page 324.

The procedure cannot be employed directly if alkali sulfide is present, because the hydrogen sulfide set free on acidifying gives black nickel sulfide on contact with the nickel hydroxide. It should also be remembered that simultaneous liberation of sulfur dioxide and hydrogen sulfide will lead to loss of the former through the reaction: $SO_2 + 2\ H_2S \rightarrow 2\ H_2O + 3\ S^\circ$.

These interferences can be averted by adding bismuth nitrate which precipitates all the sulfide as bismuth sulfide. The test is applied to the clear liquid. If 1 : 20 hydrochloric acid is used, it is not necessary to filter because at this concentration the acid is without effect on bismuth sulfide.

The induced reaction may be used to show that sulfur dioxide is formed by the action of light on finely divided sulfur in contact with air (autoxidation). Powdered sulfur is placed on paper smeared with moist $Ni(OH)_2$. After exposure to sunlight for a few minutes, the sample is treated with benzidine acetate solution. A bright blue color develops.[107]

(3) *Test by induced oxidation of cobalt^{II}-azide* [108]

On adding an excess of sodium azide to neutral or slightly acid solutions of cobalt salts, a violet color appears which is due to complex cobalt^{II}-azide anions. In contact with air an oxidation to anionic azide-complexes of tervalent cobalt takes place, shown by a change of color from violet to yellow. This very slow autoxidation is enormously accelerated by sulfurous acid or sulfite ions. Probably the autoxidation of sulfite to sulfate induces the oxidation of the complex bounded cobalt. However, the color change violet → yellow is not sufficiently sensitive to serve as a test for this induction effect and therefore, for sulfite. It is better to identify the tervalent cobalt through a color reaction with an acetic acid solution of *o*-tolidine (formation of a blue chinoidal oxidation product of the base). Compare page 84.

Procedure. A drop of a slightly acid solution of a cobalt salt (containing about 0.5 mg Co) is placed on filter paper and spotted with a drop of a saturated water solution of sodium azide. A violet stain is formed, which is then spotted with a drop of the test solution (pH=5–6). A light yellow color appears which changes to blue on spotting with a drop of a 2 % acetic acid solution of *o*-tolidine.

Limit of Identification: 0.5 γ $NaHSO_3$

Limit of Dilution: 1 : 100,000

(4) *Test by decolorization of malachite green* [109]

Neutral sulfite solutions instantly decolorize aqueous solutions of triphenylmethane dyestuffs, e.g., malachite green or fuchsin. This discharge of color is due to a reaction between the dyestuff and sulfurous acid; the quinoidal structure is destroyed and consequently the color is lost. In the case of fuchsin (I), the colorless N-sulfinic acid of *p*-fuchsin-leuco-sulfonic acid (II) [110] results:

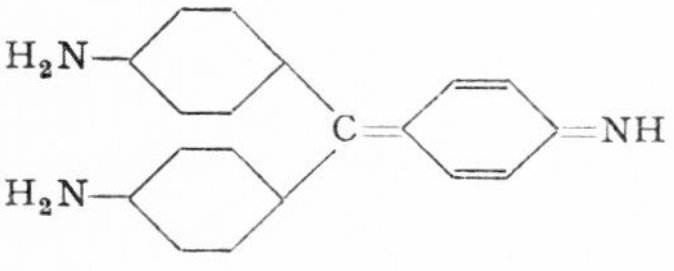

$(H_2N{-}C_6H_4)_2C{=}C_6H_4{=}NH$ (I)

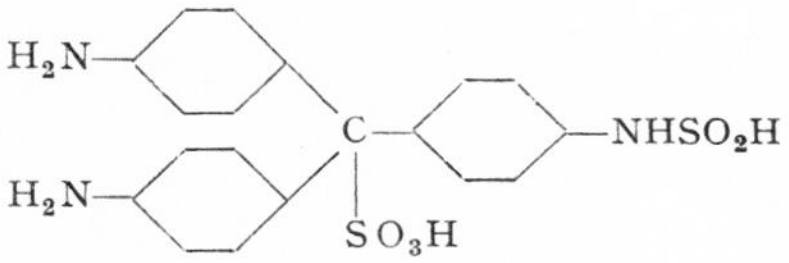

$(H_2N{-}C_6H_4)_2C(SO_3H){-}C_6H_4{-}NHSO_2H$ (II)

Addition of aldehydes, e.g., an aqueous solution of acetaldehyde, to the decolorized solution causes the color to reappear (concerning this effect see Volume II, Chapter 4). The alkali salts of the N-sulfinic acid of *p*-fuchsin-leuco-sulfonic acid are formed by the action of neutral sulfites.

When acid sulfites or free sulfurous acid are used, the decolorization of the solution of the dyestuff is incomplete. In such cases, previous neutralization with sodium bicarbonate is essential. However, free alkali also interferes with the decolorization reaction and must be removed beforehand with carbonic acid.

Thiosulfates, polythionates and hydrosulfides are without effect on solutions of triphenylmethane dyestuffs; mono- and polysulfides react in the same way as sulfites.

Procedure. A drop of the dyestuff solution is placed in a depression of a spot plate, and a drop of the neutral test solution added. The solution is decolorized in the presence of sulfite.

Limit of Identification: 1 γ sulfur dioxide
Limit of Dilution: 1 : 50,000
Reagent: Malachite green solution: 10 g malachite green in 400 ml water

Neutral salts (e.g., zinc, lead, cadmium salts) considerably reduce the sensitivity of this sulfite test. Consequently, it is not feasible [111] to prevent the interference due to sulfides (and hydroxides) by adding of $ZnSO_4$,[112] etc.

(5) Test for small amounts of sulfite in thiosulfate solutions

Alkali thiosulfates have no action on aldehydes, but neutral sulfites react with aldehydes to give soluble neutral aldehyde bisulfite compounds and simultaneously liberate an equivalent quantity of alkali:

$$RCHO + Na_2SO_3 + H_2O = RCH(OH)SO_3Na + NaOH$$

Thus sulfites can be detected in the presence of thiosulfates by the development of an alkaline reaction after the addition of an aldehyde. Formaldehyde acts most rapidly.

Procedure. A drop of a neutral 1 % solution of formaldehyde is added to 1 or 2 drops of the neutral thiosulfate solution (previously neutralized if necessary with $ZnCl_2$). The formation of alkali is detected with phenolphthalein. In this way 0.1 mg of Na_2SO_3 may be identified in 0.05 ml of 0.1 *N* $Na_2S_2O_3$.

Far smaller amounts of sulfite may be detected in the presence of any amount of thiosulfate, when free alkali is absent, if use is made of the fact that on oxidizing a sulfite-thiosulfate mixture with iodine:

$$SO_3^{-2} + I_2 + H_2O = SO_4^{-2} + 2\,I^- + 2\,H^+$$
$$2\,S_2O_3^{-2} + I_2 = S_4O_6^{-2} + 2\,I^-$$

hydrogen ions are formed *only* in the case of sulfite. The resulting acidity can be revealed by litmus paper.

Procedure. A drop of the neutral test solution is stirred with a 0.1 *N* iodine solution, on a spot plate, until the iodine color is permanent. A strip of litmus paper is placed in the brown solution, and then immediately dipped into a 0.1 *N* thiosulfate solution to remove the adsorbed iodine. If sulfites were present, the lower end of the litmus paper is reddened.

Limit of Identification: 5 γ sodium sulfite } in the presence of 1000 times the
Limit of Dilution: 1 : 10,000 } amount of sodium thiosulfate

References pp. 356–360

(6) Test for bisulfites in the presence of sulfites [113]

Because of hydrolysis, alkali sulfites have an alkaline reaction in aqueous solution; bisulfite solutions are neutral. The interaction of these compounds with barium chloride or hydrogen peroxide may be applied to detect bisulfite in the presence of sulfite. The reactions are:

$$SO_3^{-2} + Ba^{+2} = BaSO_3$$

$$HSO_3^{-} + Ba^{+2} = BaSO_3 + H^{+}$$

or

$$SO_3^{-2} + H_2O_2 = SO_4^{-2} + H_2O$$

$$HSO_3^{-} + H_2O_2 = SO_4^{-2} + H_2O + H^{+}$$

Procedure. A drop of the test solution is stirred with a drop of 1 % barium chloride solution. A second drop is stirred with neutral 3 % hydrogen peroxide. If the mixture remains neutral to litmus paper, only sulfite can be present, whereas an acid reaction indicates that bisulfite is present.

Reagent: 1 % solution of barium chloride, or 3 % hydrogen peroxide, shaken with $CaCO_3$, and filtered

(7) Other tests for sulfites

(*a*) Filter paper impregnated with the green product of the action of light on 2-benzylpyridine is turned red by sulfites (*Idn. Limit:* 7.2 γ $NaHSO_3$ in 0.1 ml). [114] Gaseous SO_2 has the same effect and also $SnCl_2$.

(*b*) Sulfurous acid can be detected through the reduction of $Fe[Fe(CN)_6]$ to Turnbull's blue by SO_2. The test can be made in a gas evolution apparatus (*Idn. Limit:* 5 γ SO_2). [115]

20. Sulfuric acid

(1) Test with barium rhodizonate [116]

Barium salts give a red-brown precipitate with rhodizonic acid or sodium rhodizonate (see page 216). The precipitate, produced on paper, is stable toward dilute hydrochloric acid, whereas sulfates and sulfuric acid cause immediate decolorization owing to the formation of insoluble barium sulfate. This fact can be applied to give a sensitive and specific test for sulfates.

Procedure. A drop of barium chloride is placed on filter paper followed by a drop of a freshly prepared 0.1 % solution of sodium rhodizonate. The red fleck is treated with a drop of the acid or alkaline test solution. In the presence of sulfates, the red fleck of barium rhodizonate disappears.

The test can also be made in a depression of a spot plate. A suspension of colored barium rhodizonate is prepared by bringing together one drop each of dilute barium chloride and sodium rhodizonate solution. The color is discharged when a drop of a solution containing sulfate is added. A blank test is necessary if small quantities of sulfate are being sought.

Limit of Identification: 5 γ sodium sulfate
Limit of Dilution: 1 : 10,000

A test for soluble sulfates, even better than the foregoing decomposition of barium rhodizonate, is provided by reducing barium sulfate to sulfide with metallic potassium. The sulfide is detected by the iodine-azide test (see *4*). The reduction to sulfide affords a means of detecting traces of sulfate in preparations of pure products, etc. (see page 494). If the sample is in acid solution, a drop should be evaporated to dryness (in a hard-glass tube) with a drop of barium chloride (sulfate-free) and the residue then treated as described on page 315.

(*2*) *Test with barium carbonate and phenolphthalein* [117]

Barium carbonate reacts rapidly with neutral alkali sulfate solutions:

$$BaCO_3 + Na_2SO_4 \rightarrow BaSO_4 + Na_2CO_3$$

The occurrence of the reaction on warming can be detected by means of phenolphthalein, which turns red owing to the alkaline reaction of the resulting sodium carbonate.

Among the anions, only silicofluorides interfere; they react similarly to sulfates. The test solution must necessarily be neutral. When alkaline solutions are present, as when testing a sodium carbonate extract, it is advisable to acidify a few drops of the test solution with hydrochloric acid, evaporate to dryness on the water-bath, and then to add barium carbonate. After this treatment, the limit of identification is about double that with neutral solutions.

Procedure. A drop of the neutral test solution is mixed with a drop of a pure barium carbonate suspension in a micro crucible and evaporated to dryness over the water-bath. A drop of 1 % solution of phenolphtalein in 1 : 1 ethyl alcohol is added. A pink or red color indicates sulfates. A blank comparison test is essential because water suspensions of barium carbonate almost invariably show a slight alkalinity.*

Limit of Identification: 5 γ sodium sulfate
Limit of Dilution: 1 : 10,000

(*3*) *Test by precipitation of barium sulfate in the presence of permanganate* [118]

If Ba^{+2} ions are precipitated by adding SO_4^{-2} ions to a solution also containing potassium permanganate, the barium sulfate includes perman-

* Private communication from T. Fukutomi.

ganate in its lattice.[119] Consequently, the precipitate is violet. The permanganate thus retained by barium sulfate is quite resistant to all reducing agents that ordinarily attack it at once.[120] Accordingly, the color of the excess permanganate in the liquid can be discharged and the violet barium sulfate then becomes plainly visible. The fixing of permanganate forms the basis of a test for barium (see page 218), as well as for the following method of detecting sulfate.

Procedure. Three drops of the solution to be tested for sulfate are mixed on a spot plate with a drop of saturated potassium permanganate solution. One drop of the mixture is placed on dry filter paper impregnated with barium chloride, and the paper kept at 70–80° C for 7 or 8 minutes in an oven. The excess barium chloride is removed by bathing the paper for about a minute in water, and rinsing. The paper is then placed in 1 *N* oxalic acid, which removes any unused $KMnO_4$ as well as the MnO_2 precipitated in the paper through the oxidation of the cellulose. A bright violet or pink fleck or ring remains if barium sulfate, i.e. sulfate, is present.

Limit of Identification: 2.5 γ sulfuric acid

Limit of Dilution: 1 : 20,000

Reagent: Barium chloride paper: Filter paper is impregnated with 0.5 *N* $BaCl_2$ and dried

(4) Test for insoluble sulfates [121]

Insoluble sulfates, e.g., alkaline earth sulfates or lead sulfate, can be reduced to sulfides and detected by the sensitive iodine-azide reaction described on page 303. The test is specific when insoluble sulfides or free sulfur are absent.

Insoluble sulfates (e.g., $PbSO_4$, $BaSO_4$) can be very simply reduced to sulfide by heating with metallic potassium in a glass capillary. The details of this reduction, and the subsequent azide reaction can be carried out as described in Volume II, where the detection of sulfur in organic compounds is discussed. As little as 0.5 γ $CaSO_4$ may thus be detected in a drop of a dilute solution of gypsum. The test is so sensitive that traces of lead sulfate dissolved in water (1 γ $PbSO_4$) can be detected. The resulting sulfide may alternatively be detected with sodium nitroprusside (see page 303), but since this test is less sensitive, larger amounts of the test substance must be taken for reduction with potassium.

Rather less sensitive, but adequate in most cases, is reduction with carbon (plus sodium carbonate as flux):

$$BaSO_4 + 4\,C = BaS + 4\,CO$$

Procedure. A sample (size of a pin head) is thoroughly ground in a small mortar with a few splinters of charcoal and four times the bulk of $NaKCO_3$

(sulfur-free). A portion of the mixture is fused on a Wedekind magnesia spoon with a blowpipe, using an alcohol flame. The sulfate is converted into sulfide. The melt (still warm) is transferred to a microcentrifuge tube, a drop of water is added, and the mixture stirred with a glass capillary. One or two drops of an iodine-azide solution (preparation, page 305) are added. If sulfide was formed by the reduction, the iodine-azide solution decomposes with evolution of nitrogen. The small bubbles can easily be seen as they rise through the liquid.

Limit of Identification: 36 γ barium sulfate *
Limit of Dilution: 1 : 1,400

Another general method of detecting insoluble sulfates is discussed on page 408.

(5) Test for free (unbound) sulfuric acid by charring of glucose [122]

Free sulfuric acid can be detected in the presence of volatile mineral and organic acids, and of mineral salts, by utilizing the dehydrating action of concentrated sulfuric acid. Cellulose (filter paper), starch, sugar, and the like, lose water and char when heated above 200° C.[123] Charring also occurs if solutions containing only a little sulfuric acid are evaporated to dryness in the presence of carbohydrates, and the temperature then raised to 120° C. Phosphoric acid, if present in considerable quantities, exerts an analogous dehydrating action. Free hydrochloric acid produces partial charring**, but because of its volatility it can be completely removed by evaporation with water, leaving behind unchanged any sulfuric acid.

Procedure. Several milligrams of glucose are placed in a micro crucible, a drop of the test solution added, and the crucible placed in an oven heated to 120° C. If considerable quantities of free sulfuric acid are present, distinct charring occurs very soon after the charge has gone to dryness. With small quantities of acid, a black or brown ring will appear after the sample has been kept at this temperature for about one hour. A blank test with glucose remains unchanged under this treatment.

* The limit of identification was determined as follows: a drop of a solution of Na_2SO_4 of known strength was placed on filter paper, treated with a drop of $BaCl_2$, and dried. A little Na_2CO_3 was added and the strip of filter paper was rolled, ashed on a magnesia spoon, and the fusion made as described.

** The charring of organic materials when heated at 120° C with hydrochloric acid seemingly presents, at first glance, a contradiction to the familiar fact that a dilute aqueous solution of this acid, on concentration, forms an azeotropic mixture containing 20.2 % HCl, which boils at 110° C. This constant-boiling mixture is obtained if the acid is heated in glass or porcelain. However, hydrochloric acid is adsorbed by paper and carbohydrates and, when retained in this fashion, gives off water at 120° C and then dehydrates the adsorbent to replace this evaporated water. This adsorption effect can be demonstrated by placing a drop of hydrochloric acid on filter paper and then keeping the paper in an oven at 120° C until partial charring has produced a grey fleck. If moistened Congo paper is then pressed against the fleck, the indicator turns blue[124], because of the hydrochloric acid adsorbed by the filter paper.

If hydrochloric acid is present, a drop of the test solution is evaporated in a micro crucible at 105° to 110° C. Then one or two drops of water are added and the evaporation is repeated. One drop of water and a pinch of glucose are then introduced and the test carried out as just described.

Limit of Identification: 8 γ sulfuric acid

Limit of Dilution: 1 : 6200

(6) *Test for free (unbound) sulfuric acid with methylenedisalicylic acid* [125]

If its action is prolonged sufficiently, and particularly when warmed, concentrated sulfuric acid not only dehydrates organic materials but also functions as a strong oxidant. This latter effect is exhibited in the familiar Kjeldahl digestion for the decomposition of non-volatile organic compounds. Consequently, it is very likely that the relatively low sensitivity of test *5* may be attributed to the fact that in addition to carbonization, there is also oxidation of the sugar with production of CO_2, SO_2, and water, whereby sulfuric acid is consumed. A possibility of utilizing the oxidative and dehydrating effects of concentrated sulfuric acid simultaneously is presented by its action toward methylenedisalicylic acid (I). The quinoidal formaurindicarboxylic acid (II) is formed*; it is an intense red compound: [126]

HO–⟨ ⟩–CH_2–⟨ ⟩–OH + H_2SO_4 →
HOOC COOH
(I)

HO–⟨ ⟩–CH=⟨ ⟩=O + $2 H_2O$ + SO_2
HOOC COOH
(II)

Methylenedisalicylic acid is not soluble in water; it melts at 238° C. Its oxidation is so rapid that small quantities of free sulfuric acid can be detected by the pink color which appears on warming to 150° C with (I). Concentrated phosphoric acid, which acts solely as a dehydrant, has no effect on methylenedisalicylic acid under the conditions of the test. When large quantities of free phosphoric acid are present, the test for free sulfuric acid is less sensitive because the formaurindicarboxylic acid is not obtained in solid form, but it remains dissolved in the concentrated phosphoric acid.

In place of methylenedisalicylic acid, *m*- or *p*-hydroxybenzaldehyde may be used as reagent for free sulfuric acid. When these low-melting aldehydes are warmed with concentrated sulfuric acid, a red color develops. Probably

* This acid was first prepared by oxidizing methylenedisalicylic acid with nitrosylsulfuric acid (conc. H_2SO_4 + KNO_2). [126]

the initial reaction is a condensation of three aldehyde molecules; this is followed by oxidation to quinoidal compounds. The test based on this effect is somewhat more sensitive than the test employing methylenedisalicylic acid, but these aldehydes are not so readily available as is methylenedisalicylic acid.

When very concentrated hydrochloric acid is evaporated, the sulfuric acid it contains in small amounts is volatilized. This is probably due to the formation of chlorosulfonic acid ($ClSO_3H$). Though such losses are not completely avoided, they are greatly lessened if water is added prior to the evaporation.

Procedure. The test solution is diluted with water if necessary and then one drop is mixed in a porcelain micro crucible with several milligrams of methylenedisalicylic acid.[127] The crucible is then kept in a drying oven at 150° C for three minutes. A dark- or light red residue remains, the color depending on the quantity of free sulfuric acid present. It is advisable to conduct a comparison blank test.

Limit of Identification: 2.5 γ sulfuric acid
Limit of Dilution: 1 : 20,000

21. Thiosulfuric acid

(*1*) *Test by catalysis of the iodine-azide reaction* [128]

The reaction:

$$2\ NaN_3 + I_2 = 2\ NaI + 3\ N_2$$

is catalyzed by soluble and insoluble sulfides, thiosulfates and thiocyanates (page 304). Extremely small amounts of the sulfur-containing compound may be detected by the evolution of nitrogen. The test is only applicable to the detection of thiosulfates in the absence of sulfides and thiocyanates.*

Procedure. On mixing a drop of the test solution and a drop of iodine-azide solution on a watch glass, a more or less vigorous development of bubbles ensues, according to the thiosulfate content.

Limit of Identification: 0.15 γ sodium thiosulfate
Limit of Dilution: 1 : 330,000

* L. Hamburg and L. Weidenfeld, showed that solutions of sulfides (even freshly prepared by mixing NaOH and H_2S) always contain some thiosulfate, though the amount may be very small. This may be demonstrated as follows: a drop of the sulfide solution is placed on a spot plate and a few small crystals of $Cd(NO_3)_2$ stirred in. A strip of filter paper (0.5 × 4 cm) is then dipped in this suspension of CdS. The clear liquid ascends the paper, leaving the CdS at the lower end. After 1 or 2 minutes, the paper is removed and the clear portion is cut off from that containing CdS, and placed on a watch glass. A drop of iodine-azide solution is placed on the moist paper; a perceptible development of bubbles of nitrogen results. Some of the bubbles adhere to the watch glass even after the paper is removed.

The thiosulfate in a micro drop (0.025 ml) of a 0.00001 *N* sodium thiosulfate solution may be detected as follows.[129] The drop is taken up in the loop of a platinum wire, evaporated by heating the wire adjacent to the loop, and then heated to glowing. The loop is then placed in a drop of iodine-azide solution, suspended in an inverted microcentrifuge tube. Tiny bubbles of gas are evolved and collect in the capillary end of the centrifuge tube (see page 305).

It is surprising that a solution of sodium thiosulfate which has been treated with an excess of iodine and thus "quantitatively" converted into tetrathionate ($2\ Na_2S_2O_3 + I_2 \rightarrow 2\ NaI + Na_2S_4O_6$), will still catalyze the sodium azide reaction with discharge of the color and evolution of nitrogen. In contrast, tetrathionate prepared by other methods and then recrystallized, shows no reaction with azide. Other sulfur compounds, which can initiate the iodine-azide catalysis and which of themselves can be oxidized by iodine, behave analogously. Traces of thiosulfate are probably responsible for this effect; they persist even after addition of excess iodine. This catalysis can be used as a test for thiosulfate, though it is admittedly less sensitive than other procedures. A drop of the thiosulfate solution is treated in a micro crucible with 0.005 *N* iodine-starch solution until the blue color persists. One drop of 2 % sodium azide solution is added and the mixture is watched for fading or discharge of the color.[130]

Limit of Identification: 0.025 γ sodium thiosulfate
Limit of Dilution: 1 : 2,000,000
Reagent: Sodium azide-iodine solution: Solution of 1 g sodium azide in 100 ml 0.1 *N* iodine

(2) *Test for thiosulfates in the presence of sulfides* [131]

Thiosulfates may be detected in the presence of small amounts of sulfide by applying the reactions of sulfides and thiosulfates with mercuric chloride; mercuric sulfide results:

$$S^{-2} + Hg^{+2} = HgS$$
$$S_2O_3^{-2} + Hg^{+2} + H_2O = HgS + SO_4^{-2} + 2\ H^+$$

Thus, free (sulfuric) acid is formed from thiosulfates; it can be detected by litmus paper. Aqueous solutions of mercuric chloride, because of hydrolysis, are weakly acid, and therefore this reaction with water must be repressed before testing for the acidity developed by the test reaction. Addition of potassium chloride serves this purpose; $K_2[HgCl_4]$ or $K[HgCl_3]$ is formed; its aqueous solutions are neutral to litmus.

Procedure.[132] A drop of the test solution and a drop of 2 % mercuric chloride solution are stirred together on a spot plate. If excess $HgCl_2$ is present, the white mercuric sulfo salt ($2\ HgS \cdot HgCl_2$) is formed. A little solid potassium chloride is mixed in. Litmus paper dipped into the liquid turns red if thiosulfate is present.

Limit of Identification: 8 γ sodium thiosulfate } in the presence of 50 times the
Limit of Dilution: 1 : 6200 } amount of sodium sulfide

22. Hyposulfurous acid (Dithionous acid) [133]

Hyposulfurous acid is known only as its alkali salts. These include sodium hyposulfite ($Na_2S_2O_4$) and sodium sulfoxylate-formaldehyde, which is an addition product ($CH_2O \cdot NaHSO_2 \cdot 2\ H_2O$) known as Rongalite. Sodium hyposulfite, which is better known as hydrosulfite, is stable only in the dry state; in solution it avidly absorbs oxygen from the air. In contrast, Rongalite is stable even in solution. Both compounds are powerful reducing agents because of conversion of the sulfur to higher valence states:

$$Na_2S_2O_4 + H_2O + 3\ O \longrightarrow 2\ NaHSO_4 \quad (1)$$

$$HO{-}CH_2{-}O{-}SONa + 2\ H_2O_2 \longrightarrow NaHSO_4 + CH_2O + 2\ H_2O \quad (2)$$

The marked reducing action is also demonstrated by the fact that they can function as hydrogen donors through withdrawal of oxygen from water:

$$Na_2S_2O_4 + H_2O \longrightarrow NaHSO_2 + NaHSO_3 \quad (3)$$

$$NaHSO_2 + 2\ H_2O \longrightarrow NaHSO_4 + 4\ H^\circ \quad (4)$$

$$HO{-}CH_2{-}O{-}SONa + 2\ H_2O \longrightarrow CH_2O + NaHSO_4 + 4\ H^\circ \quad (5)$$

According to (1) and (2), sodium hydrosulfite and Rongalite behave like other reductants. But as shown in (3)–(5), they occupy a special position with respect to their action as hydrogen donors. An appropriate indicator for compounds which act as reductants by furnishing hydrogen has been found, namely *o*-dinitrobenzene. The *para* isomer may also be used.

These dinitrobenzenes are soluble in alcohol; the solutions are colorless and turn slightly yellow when made alkaline. If the alkaline solutions are warmed with sodium hydrosulfite or Rongalite, colored quinoidal alkali salts of the aci-form of *o*(*p*)-nitroso-nitrobenzene results.[134] The anions are violet and red, respectively. The structure of the *ortho* quinoidal anion is shown in the following equation:

$$C_6H_4(NO_2)_2 + 4\ H^\circ + 2\ OH^- \longrightarrow C_6H_4({=}NO^-)({=}NO_2^-) + 3\ H_2O$$

The color reactions of salts of hyposulfurous acid with the isomeric dinitrobenzenes are quite selective. Among the inorganic reductants, only sulfides, stannites, hydroxylamine, and hydrazine act in the same manner. It is unlikely that the three latter will be encountered when searching for hydrosulfite and sulfoxylate. On the other hand, technical preparations may contain slight amounts of alkali sulfides. The interference due to sulfides

References pp. 356–360

when testing for Rongalite with the dinitrobenzenes can be removed if the test solution is shaken in the cold with lead carbonate (precipitation of lead sulfide) and the test made on the filtrate (centrifugate). A preliminary test for sulfide with nitroprusside is not feasible, since both sodium hydrosulfite and sodium sulfoxylate-formaldehyde yield a green to blue color with sodium nitroprusside.[135]

Procedure. The test is conducted in a micro test tube. A drop of a saturated alcoholic solution of *p*-dinitrobenzene is treated with a drop of concentrated ammonium hydroxide. A drop of the test solution is added and the mixture is heated for about 30 seconds over a bare flame. Depending on the quantity of Rongalite or sodium hyposulfite (freshly prepared solution) present, an orange or pink color is obtained.

Limit of Identification: 3 γ sodium sulfoxylate-formaldehyde
Limit of Dilution: 1 : 16,600

Differentiation of sodium hydrosulfite and Rongalite

A rapid differentiation of these materials is based on the fact that sodium hydrosulfite (dissolved or solid) reacts immediately at room temperature, and under cooling, with alcoholic ammoniacal solutions of *o*- or *p*-dinitrobenzene to give a violet or orange color, respectively, whereas the formaldehyde addition compound reacts slowly or not at all under these conditions.

Procedure. A micro test tube is used. A drop or two of a saturated alcoholic solution of *o*- or *p*-dinitrobenzene is treated with one drop of concentrated ammonium hydroxide and then several mg of the solid sample is introduced. Sodium hydrosulfite produces an immediate blue-violet or orange color, whereas Rongalite shows no reaction at first, followed in a few minutes by a pale violet or pink coloration, whose intensity slowly increases. If the salts are added to the reagent mixture that has been precooled with ice water, Rongalite gives no color reaction even after some hours, while sodium hydrosulfite produces the respective colors immediately. On standing, the colors fade and eventually disappear.

A reliable method for detecting Rongalite (and differentiating it from sodium hydrosulfite) is through a supplementary detection of its formaldehyde component. This is easily accomplished by the color reaction with chromotropic acid. For details see Volume II.

Detection of Rongalite by thermal decomposition [136]

Rongalite loses its water of crystallization at 120° C, and formaldehyde and hydrogen sulfide begin to come off at 125° C according to:[137]

$$2\,CH_2\begin{matrix}\diagup OH\\ \diagdown OSONa\end{matrix} \rightarrow 2\,CH_2O + H_2S + Na_2SO_4$$

Thermal decomposition of sodium hydrosulfite yields sulfur dioxide, sulfur, sodium sulfate, sodium thiosulfate, sodium sulfide or polysulfide. Therefore, Rongalite may be specifically detected through the hydrogen sulfide it yields on pyrolysis.

Procedure. The test is made in a micro test tube. A little of the solid (a few mg suffice) is heated over a micro flame. The mouth of the test tube is covered with filter paper moistened with lead acetate. A brown or black stain of lead sulfide appears if Rongalite is present.

The limit of identification could not be measured because, when solutions of Rongalite are taken to dryness, up to 80% of the material is oxidized to formaldehyde bisulfite and the latter yields no hydrogen sulfide when thermally decomposed.

If sodium hydrosulfite is evaporated with a micro drop of 40 % formaldehyde, the residue contains formaldehyde sulfoxylate and formaldehyde bisulfite:

$$Na_2S_2O_4 + 2\,CH_2O + H_2O \rightarrow CH_2(OH)SO_2Na + CH_2(OH)SO_3Na$$

Consequently, the evaporation residue then likewise yields hydrogen sulfide on thermal decomposition.

23. Persulfuric acid

(*1*) *Test with benzidine* [138]

Neutral or weakly acetic acid solutions of alkali persulfates turn blue on addition of benzidine acetate due to the formation of a quinoidal oxidation product of benzidine (see page 72). Alkali peroxides, perborates, and hydrogen peroxide do not react with benzidine under these conditions.

Chromates, permanganates, ferricyanides and hypohalogenites should not be present, because they also oxidize benzidine to benzidine blue. Chlorates, perchlorates, bromates, iodates and nitrates do not react under the conditions of the test, but they do reduce the sensitivity of the persulfate-benzidine reaction. Periodates give a brown coloration.

The benzidine derivatives: 2,7-diaminofluorene (I) [139] and especially 2,7-diaminodiphenylene oxide (II) [140] (see page 171) are even more sensitive than benzidine, but are not so readily available. They, like benzidine,

H_2N—(fluorene ring with CH_2 bridge)—NH_2 (I)

H_2N—(dibenzofuran ring with O bridge)—NH_2 (II)

form blue quinoid oxidation products and, in general, may be substituted for benzidine.

Procedure. A drop of the neutral or slightly acid (acetic) test solution and a drop of benzidine acetate are mixed on a spot plate or in a crucible. In the presence of persulfates, a more or less intense blue color appears.

Limit of Identification: 0.25 γ potassium persulfate } in neutral solution
Limit of Dilution: 1 : 200,000 }

Limit of Identification: 1 γ potassium persulfate } in acetic acid solution
Limit of Dilution: 1 : 50,000 }

In the presence of 3 mg potassium bromate, 2.5 γ potassium persulfate may be detected; in the presence of 3 mg potassium iodate, 0.5 γ persulfate.

Reagent: 2 % solution of benzidine in dilute acetic acid

(2) *Test by reaction with nickel hydroxide* [141, 142]

Light green insoluble Ni^{II} hydroxide reacts with an alkaline solution of persulfate to form black Ni^{IV} oxyhydrate:[143]

$$Ni(OH)_2 + S_2O_8^{-2} + 2\,OH^- \rightarrow NiO(OH)_2 + 2\,SO_4^{-2} + H_2O \quad (1)$$

$Ni(OH)_2$ can be converted into $NiO(OH)_2$ also by neutral and weakly acid solutions of persulfate with concomitant formation of nickel sulfate:

$$2\,Ni(OH)_2 + S_2O_8^{-2} \rightarrow NiO(OH)_2 + 2\,SO_4^{-2} + Ni^{+2} + H_2O \quad (2)$$

Redox reactions (1) and (2) are characteristic for persulfate, and their occurrence is easily noted because of the accompanying color changes. Hydrogen peroxide and other per-compounds (perborate, percarbonate, etc.) exert no oxidizing action on $nickel^{II}$ hydroxide, but are decomposed catalytically by contact with it.[144] Neutral and alkaline hydrogen peroxide reduce Ni^{IV} oxyhydrate to Ni^{II} hydroxide:

$$NiO(OH)_2 + H_2O_2 \rightarrow Ni(OH)_2 + H_2O + O_2$$

If persulfate is to be detected in a mixture with hydrogen peroxide, the latter must be decomposed if the objective is to produce $NiO(OH)_2$ by the action of persulfate (see below).

Solutions of halogenites oxidize nickel hydroxide:

$$Ni(OH)_2 + HalO^- \rightarrow NiO(OH)_2 + Hal^-$$

However, a mixture of halogenite and persulfate need not be taken into consideration; they are incompatible because of the reaction:

$$S_2O_8^{-2} + HalO^- + 2\,OH^- \rightarrow 2\,SO_4^{-2} + Hal^- + O_2 + H_2O$$

Halogenates have no effect on $nickel^{II}$ hydroxide.

In view of the oxidation of $Ni(OH)_2$ to $NiO(OH)_2$, filter paper impregnated with nickel hydroxide seemed to offer special promise with regard to developing a spot test for persulfate. Trials showed that an identification limit of 0.5 γ $K_2S_2O_8$ could be secured, but only with neutral solutions. The sensitivity is much less when the paper is spotted with alkaline persulfate solutions; at an alkalinity of 0.05 *N* the identification limit is 5 γ persulfate. In highly alkaline solution, small quantities of persulfate cannot be detected at all with $Ni(OH)_2$. This lowering of the sensitivity and also the fact that flecks of black $NiO(OH)_2$ produced on $Ni(OH)_2$ paper are not lasting but disappear on aging (and very quickly if warmed in an oven) are not due to a decomposition of $NiO(OH)_2$, which is entirely stable when dried. Instead, the cause resides in the reduction of this product by the cellulose of the paper, a reaction which proceeds rather rapidly in alkaline surroundings and when the temperature is raised. Because of this adverse effect on the sensitivity, $Ni(OH)_2$-paper can be recommended only when the persulfate test solution is neutral. The reagent paper gives good service under such conditions. The dark flecks disappear when gently warmed and the reagent paper is regenerated. The following procedure has general applicability.

Procedure. Single drops of test solution, 1 *N* sodium hydroxide and 1 % nickel sulfate (hexahydrate) solution are placed in succession in a depression of a spot plate. A black or grey precipitate appears, according to the quantity of persulfate present.

Limit of Identification: 2.5 γ potassium persulfate

Limit of Dilution: 1 : 20,000

Reagent: Nickel hydroxide paper. Preparation: Strips of filter paper are bathed in a solution of 10 g $NiSO_4 \cdot 6H_2O$ in 30 ml concentrated ammonia and then dried at 110° C. The strips are then kept for 5 minutes in a solution of 5 g sodium hydroxide and 2 g sodium carbonate in 100 ml water. After thorough washing with water, the homogeneously impregnated reagent papers are dried at 110° C. They are stable for months.

Detection of persulfate in the presence of hydrogen peroxide

The $Ni(OH)_2$ test for persulfate fails if hydrogen peroxide is present (page 354). The reason is not only the reduction by hydrogen peroxide of $NiO(OH)_2$, but, in addition, contact with $Ni(OH)_2$ seems to accelerate the reaction:

$$S_2O_8^{-2} + H_2O_2 \rightarrow 2\ SO_4^{-2} + 2\ H^+ + O_2$$

which consumes the persulfate. Hydrogen peroxide can be decomposed, with practically no loss of persulfate, by adding silver nitrate to the alkaline solution. The reaction:

$$2\ Ag^+ + H_2O_2 + 2\ OH^- \rightarrow 2\ Ag^o + 2\ H_2O + O_2$$

yields elementary silver which, in its turn, catalyzes the decomposition: $H_2O_2 \rightarrow H_2O + O$. Theoretically, an extremely small quantity of silver nitrate would suffice, but it is advisable to add larger amounts to obtain rapid destruction of the hydrogen peroxide. Small amounts of $NiO(OH)_2$ can be easily detected in the presence of precipitated silver and silver oxide if the following procedure is used.

Procedure. A drop of the test solution is placed in an Emich tube and followed in succession by a drop of *N* sodium hydroxide and a drop of 0.1 % silver nitrate solution. The mixture is stirred with a glass capillary until the evolution of oxygen has ceased. Several milligrams of barium sulfate are added and the suspension is centrifuged. A gray sediment forms. Several more milligrams of barium sulfate are added, and on centrifuging again, the sediment is covered with a white layer of barium sulfate. A drop of 1 % nickel sulfate solution is then added without stirring; after 30 seconds the centrifuging is repeated. Any $NiO(OH)_2$ can be easily discerned as a black or gray-green stratum over the white barium sulfate. A blank is recommended when small amounts of persulfate are suspected. This procedure will reveal 5 γ potassium persulfate in the presence of any quantity of hydrogen peroxide.

24. Aminosulfonic acid (Sulfamic acid)

Test by deamidation [145]

If organic amino compounds are warmed with excess nitrous acid, the NH_2-group is replaced by an OH-group. An analogous deamidation occurs readily in the case of aminosulfonic acid (better known as sulfamic acid). It reacts immediately at room temperature with nitrous acid:

$$NH_2SO_3H + HNO_2 \rightarrow H_2O + N_2 + H_2SO_4$$

Since all of the metal salts of this acid are water-soluble, this reaction is discernible through the precipitation of barium sulfate if barium ions are introduced. A qualitative test and a gravimetric method for nitrous acid (nitrite) are based on this precipitation.[146] The converse of the nitrite test permits the direct detection of sulfamic acid* by a spot test, provided sulfates and sulfides are absent.

Procedure. The test is conducted in a micro test tube. A drop of the test solution is treated with a drop of a 1 : 1 mixture of 1 % barium chloride and

* Aminosulfonic acid and its salts are not only analytical reagents but also find wide technical use.[147]

10 % sodium nitrite, and a drop of dilute hydrochloric acid. If aminosulfonic acid is present, a precipitate or turbidity of barium sulfate will appear. With slight quantities of sulfamic acid, the $BaSO_4$ can be brought down almost immediately by adding several drops of ether and swirling the test tube. This effect is due to the considerable solubility of ether in water which thus accelerates the normally sluggish separation of small amounts of barium sulfate.

Limit of Identification: 8 γ aminosulfonic acid
Limit of Dilution: 1 : 6,250

When sulfamic acid is to be detected in the presence of sulfite, sulfate, carbonate, etc., barium chloride solution should be added, the suspension clarified by filtration or centrifugation, and the clear liquid used for the test. Starting with one drop of the test solution, this procedure revealed 50 γ sulfamic acid in the presence of 4000 γ sodium sulfate.

25. Nitric acid

(1) Test with ferrous sulfate and sulfuric acid [148]

When a crystal of ferrous sulfate is brought into contact with a nitrate solution and concentrated sulfuric acid added very carefully,[149] a brown zone forms around the crystal. This color is due to an addition compound, $FeNO{\cdot}SO_4$ [150]. The reactions are:

$$2\ HNO_3 + 6\ FeSO_4 + 3\ H_2SO_4 = 3\ Fe_2(SO_4)_3 + 4\ H_2O + 2\ NO$$
$$FeSO_4 + NO = FeSO_4{\cdot}NO$$

Nitrites react similarly to nitrates. Iodides and bromides interfere (owing to liberation of the halogen) and must be removed by precipitation as the respective silver halides.* Large amounts of chloride reduce the sensitivity. Other interfering ions are: cyanide, ferrocyanide, thiocyanate, chromate, sulfite, thiosulfate, iodate. They react with ferrous sulfate and concentrated sulfuric acid to give a visible change, which makes it difficult to see the characteristic $FeNO{\cdot}SO_4$ ring. Tungstates and molybdates interfere, because they are reduced by iron^{II} to colored lower oxides.

Procedure. A crystal of ferrous sulfate (pin head) is placed on a spot plate together with a drop of the test solution. A drop of concentrated sulfuric acid is allowed to run in at the side. In the presence of nitrate, a brown ring is formed around the ferrous sulfate crystal.

Limit of Identification: 2.5 γ nitric acid
Limit of Dilution: 1 : 20,000

* A saturated solution of silver sulfate or acetate may be used. When large amounts of halides are present it is better to use a solution recommended by M. J. Murray and A. W. Avens, *Ind. Eng. Chem., Anal. Ed.*, 4 (1932) 58, consisting of 7.8 g Ag_2SO_4 in 25 ml 4 N ammonia made up to 100 ml with water.

References pp. 356–360

(2) *Test with diphenylamine or diphenylbenzidine*

The well known diphenylamine reaction of nitric acid [151] depends on the oxidation of diphenylamine (I) through colorless N,N′-diphenylbenzidine (II) to the blue quinoid imonium ion (III) [152]

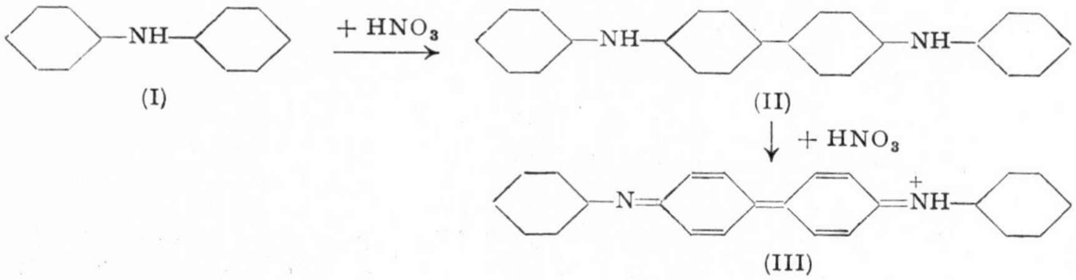

C_6H_5—NH—C_6H_5 (I) $\xrightarrow{+ HNO_3}$ C_6H_5—NH—C_6H_4—C_6H_4—NH—C_6H_5 (II)

$\downarrow + HNO_3$

C_6H_5—N=C_6H_4=C_6H_4=$\overset{+}{N}$H—C_6H_5 (III)

The oxidation to the blue dye takes place from either diphenylamine (I) or from diphenylbenzidine (II). The latter has the advantage that it better utilizes the oxidizing action of the nitric acid and prevents the oxidation from stopping at the diphenylbenzidine stage if an excess of diphenylamine is used.[153] The nitric acid test is therefore more sensitive when diphenylbenzidine is used in place of diphenylamine.[154]

It should be noted that the diphenylamine reaction proceeds especially well in the presence of chlorides. Probably free chlorine is produced. It, in turn, furnishes particularly active hypochlorous acid.[155]

$$Cl_2 + H_2O \rightarrow HCl + HOCl$$

This color reaction is not specific for nitric acid, because other oxidizing agents (e.g., nitrites, chlorates, bromates, iodates, chromates, permanganates, selenites, vanadates, molybdates, peroxides, higher metal oxides, antimony^{V}, iron^{III}) also transform diphenylamine or N,N′-diphenylbenzidine to the blue quinoidal compound. Interferences by alkali halogenates, perchlorates, periodates, permanganates, persulfates, peroxides and ferricyanides can be averted if a drop of the weakly alkaline test solution is taken to dryness and the residue heated to 400–500° C. The foregoing compounds are thermally decomposed, whereas alkali- and alkaline earth nitrates (nitrites) are unchanged. The test with diphenylamine or diphenylbenzidene can then be carried out directly on the ignition residue or its water extract. In this connection it should be noted that alkali iodides are oxidized to free iodine to a considerable degree by concentrated sulfuric acid and the brown color hides the blue of the quinoidal compound. Consequently, solutions containing more than 10 % iodine (as iodide) cannot be tested for nitrate by means of the diphenylamine (diphenylbenzidine) reaction.[156]

References pp. 356–360

Procedure. About 0.5 ml of the strongly acid (sulfuric) diphenylamine (or diphenylbenzidine) solution is placed on a spot plate, and a drop of the test solution is placed in the middle. A blue ring forms as the two liquids mix. The depth of color depends on the nitrate content.

(*a*) *With diphenylamine*

Limit of Identification: 0.5 γ nitric acid
Limit of Dilution: 1 : 100,000

(*b*) *With diphenylbenzidine*

Limit of Identification: 0.07 γ nitric acid
Limit of Dilution: 1 : 700,000
Reagent: Several crystals of diphenylamine or diphenylbenzidine [157] are covered with concentrated sulfuric acid. A little water is added; after solution is complete, more sulfuric acid is added. About 1 mg diphenylamine or diphenylbenzidine should be present in 10 ml of the finished reagent solution.

When testing for nitrate in the presence of oxidants (cations or anions) use can be made of their ready reduction by sulfurous acid. This procedure may not be applied directly for the removal of nitrite since characteristically nitrate is then also reduced.[156] Under such conditions it is best to decompose the nitrous acid beforehand by means of hydrazoic acid (compare page 287).

Procedure. A drop of the test solution is treated on a white spot plate with a few mg of sodium azide followed by two drops of 12 *N* sulfuric acid. After the evolution of gas has ceased, a few mg of sodium sulfite is added and two more drops of 12 *N* sulfuric acid. The mixture is stirred with a thin glass rod or a toothpick and then five to ten drops of diphenylamine reagent added. The formation of a blue color indicates the presence of nitrate.
Reagent: Diphenylamine solution: 250 mg of ammonium chloride is dissolved in 90 ml of water. A solution of 250 mg of diphenylamine in about 100 ml of concentrated sulfuric acid is cautiously added. After cooling, the volume is made up to 250 ml with concentrated sulfuric acid.

This procedure revealed 0.5 γ nitrate ion in the presence of 500 γ of the following oxidants: hydrogen peroxide, and ionic species: nitrite, chlorate, bromate, ferricyanide, chromate, permanganate, vanadate, molybdate, tungstate, ferric, ceric.

(*3*) *Test with brucine*[158]

Brucine is an extremely poisonous alkaloid obtained from seeds of the *Strychnos* family. Its constitution has not yet been completely unravelled. It gives a red color when its sulfuric solution is mixed with even minute amounts of nitrates. The constitution of the colored product has not been determined; it is probably an oxidation product.

Chlorates and nitrites give a similar color reaction. The removal of nitrites will be described later.

Procedure. A few drops of a solution of brucine in concentrated sulfuric acid are mixed on a spot plate with a few drops of the test solution. In the presence of nitrate, a red color appears. On standing it changes to yellowish-red.

Limit of Identification: 0.06 γ nitric acid

Limit of Dilution: 1 : 800,000

Reagents: 0.02 % brucine solution in sulfuric acid (prepared immediately before use). Concentrated sulfuric acid usually contains a trace of nitrate or nitrite, and hence gives a pink-red color with brucine. If such acid is diluted with water to specific gravity of 1.4 and kept boiling in a platinum dish for a long time, nitric or nitrous acids are removed completely. [159]

Test for nitrates in nitrites

The sensitive color reactions for nitrates with brucine or diphenylamine are not applicable in the presence of nitrites which give similar reactions. Hydrazoic acid can be used to destroy the nitrite beforehand: [160]

$$HN_3 + HNO_2 = N_2 + H_2O + N_2O$$

The removal of nitrite with sodium azide is complete in acid solution. To test nitrites for a nitrate content, sodium azide should be added to the test solution, which is then acidified and allowed to stand for a short time. The solution is then boiled to complete the reaction and to remove the excess hydrogen azide. Any of the usual procedures may be applied to test for nitrates in the cold test solution.

Another and even better method of destroying nitrites depends on the reaction with aminosulfonic acid (sulfamic acid): [161]

$$MeNO_2 + NH_2{\cdot}SO_3H = MeHSO_4 + H_2O + N_2$$

This reaction is very vigorous; considerable heat is developed even at moderate concentrations of the reacting substances. The reaction liquid froths because of the violent evolution of nitrogen. In concentrated solutions the reaction is complete in about 1 minute; in diluted solutions it is almost instantaneous. Nitrite can no longer be detected in a 0.00005 N potassium nitrite solution by even the most sensitive reagents immediately after mixing with 0.5 % aminosulfonic acid solution.

Because of the great velocity of the reaction, there is no need to fear a slight oxidation to nitric acid of the nitrous acid (formed as an intermediate), except in very concentrated solutions.

References pp. 356–360

All the operations for the removal of nitrite, by sodium azide or aminosulfonic acid,[162] and the nitrate test itself, may be carried out with a single drop or crystal of the test substance on a watch glass.

Even the purest nitrite preparations invariably show a slight nitrate content,[163] which must be considered when testing for nitrates in the presence of nitrites.

(*4*) *Test after reduction to nitrite* [164]

Metallic zinc reduces nitrates, in acetic acid solution, to nitrites or nitrous acid.* A sensitive test for nitrates results if the reduction is accomplished in the presence of the reagents required for the sensitive Griess test for nitrites (see test *1* below). Nitrites must be removed before carrying out the test. The removal is best done with sodium azide as just discussed.

Procedure. A drop of the neutral or acetic acid solution is mixed on a spot plate with a drop of sulfanilic acid solution and a drop of α-naphthylamine solution; a few milligrams of zinc dust are added. In the presence of nitrates, a red coloration gradually develops. It is due to the formation of the azo dyestuff *p*-benzenesulfonic acid - azo-α-naphthylamine.

Limit of Identification: 0.05 γ nitric acid

Limit of Dilution: 1 : 1,000,000

Reagent: Zinc dust (nitrite- and nitrate-free): Zinc dust is heated on the water-bath for an hour with dilute acetic acid; after cooling, the liquid is poured off, and the metal is again stirred up with dilute acetic acid, filtered with suction, washed with water, and dried.

(*5*) *Other test for nitrate*

On a spot plate, one drop of the test solution is mixed with one drop of a 0.05 % solution of the sodium salt of chromotropic acid in concentrated sulfuric acid, and, if necessary, a few drops of concentrated sulfuric acid are added. A yellow color due to a nitro derivative of chromotropic acid indicates a positive response (*Idn. Limit:* 0.2 γ NO_3^--ions).[165]

26. Nitrous acid

(*1*) *Test with sulfanilic acid and α-naphthylamine* (*Griess test*) [166]

Primary aromatic amines react in acid solution with nitrous acid to form diazonium cations:

$$[Ar—NH_3]^+ + HNO_2 \rightarrow [Ar—N{\equiv}N]^+ + 2\ H_2O$$

which may condense with the same (or with another) primary aromatic

* F. L. Hahn and G. Jaeger, *Ber.*, 58 (1925) 2335, discuss the use of metallic lead for this purpose.

References pp. 356–360

amine to produce colored cations of an amino azo compound with the general formula

$$Ar—N{=}N—Ar(Ar')—\overset{+}{N}H_3$$

The intense color of azo dyestuffs formed in this way from colorless components serves as a test for even very small amounts of nitrite.

The best amine to use is sulfanilic acid (I), with subsequent coupling of

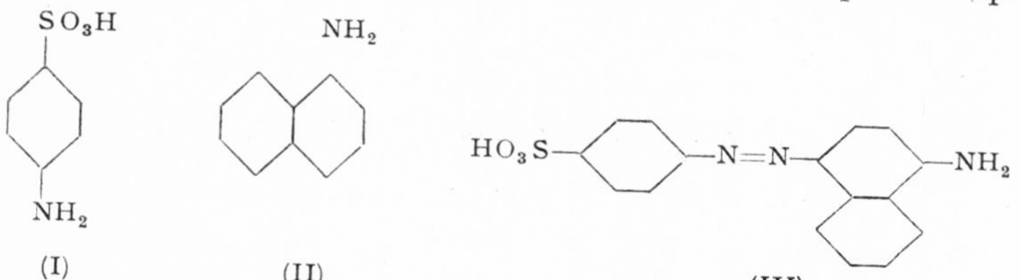

SO_3H / NH_2 (I) NH_2 (II) HO_3S—⬡—N=N—⬡—NH_2 (III)

the resulting diazonium compound with α-naphthylamine (II), to give the red *p*-benzene sulfonic acid-azo-α-naphthylamine (III), here shown in its baso-form.

Procedure. A drop of the neutral or acetic acid test solution is mixed on a spot plate with a drop each of sulfanilic acid and of α-naphthylamine solutions. According to the nitrite content, a red color is formed, either at once or after standing a short time.*

Limit of Identification: 0.01 γ nitrous acid

Limit of Dilution: 1 : 5,000,000

Reagents: *1*) Solution of sulfanilic acid in acetic acid: 1 g sulfanilic acid is dissolved by warming in 100 ml of 30 % acetic acid

2) Solution of α-naphthylamine-acetic acid: 0.03 g α-naphthylamine is boiled in 70 ml water; the colorless solution is decanted from the blue-violet residue and mixed with 30 ml glacial acetic acid.

(*2*) *Test with 1,8-naphthylenediamine* [167]

Nitrites react in neutral solutions with 1,8-naphthylenediamine (I):

$$C_{10}H_6(NH_2)_2 \text{ (I)} + HNO_2 \rightarrow C_{10}H_6<\begin{matrix}N\\NH\end{matrix}>N \text{ (II)} + 2\,H_2O$$

An orange-red precipitate of 1,8-aziminonaphthalene (II) is produced.[168]

* The diazo reaction for nitrite may be carried out as a spot reaction on filter paper impregnated with sulfanilic acid and α-naphthylamine oxalate. The *limit of identification* is 0.00005γ NO_2^- in a microdrop (0.5 mm³) if proper conditions are maintained. Compare F. L. Hahn, *Mikrochemie*, 9 (1931) 31.

References pp. 356–360

Only selenite yields a colored (brown) precipitate (see page 349); other acids, e.g., sulfuric acid, give colorless insoluble salts.

Procedure. A drop of the test solution is mixed with a drop of 1,8-naphthylenediamine in a micro crucible. An orange-red precipitate or coloration appears immediately or after warming, according to the nitrite content.

Limit of Identification: 0.1 γ nitrous acid

Limit of Dilution: 1 : 500,000

Reagent: 0.1 % solution of 1,8-naphthylenediamine[169] in 10 % acetic acid

(3) *Test by release of iodine on filter paper containing starch* [170]

Nitrites react with iodides in acid solution:

$$2\ NO_2^- + 2\ I^- + 4\ H^+ \rightarrow I_2 + 2\ H_2O + 2\ NO$$

If carried out on filter paper containing a little starch, even minimal quantities of the resulting blue starch-iodine are easily seen because of fixation on the surface of the paper, and through contrast with the white background. Most spot papers already contain sufficient starch. Oxidizing agents that liberate iodine from iodides in acid solution must be absent.

Procedure. The filter paper is tested beforehand for starch. One drop each of 2 *N* acetic acid, test solution, and 0.1 *N* potassium iodide, are placed successively on suitable filter paper. According to the quantity of nitrite present, a blue fleck or ring appears.

Limit of Identification: 0.005 γ nitrous acid

Limit of Dilution: 1 : 10,000,000

(4) *Other tests for nitrites*

(*a*) A freshly prepared alcoholic acid solution of chrysean (aminothioamide of thiazole) gives a red color and subsequently a dark brown precipitate with nitrites.[171] This reaction may be applied as a spot test (*Idn. Limit:* 0.25 γ nitrite).

(*b*) A microdrop (0.01 ml) of the test solution and one drop of an acetic acid solution of benzidine are brought successively on filter paper. The diazotization and coupling give a yellow or brown dye. (*Idn. Limit:* 0.7 γ $NaNO_2$).[172]

(*c*) The "ring test" (see page 326) due to the formation of $FeSO_4 \cdot NO$, can also be used as a test for nitrite if acetic acid is used in place of sulfuric acid.[173] Nitrate gives no reaction under these conditions (*Idn. Limit:* 2 γ HNO_2).

(*d*) The red acid solution of toluosafranine (Safranine T) turns blue on addition of alkali nitrite (*Idn. Limit:* 0.5 γ HNO_2).[174]

27. Phosphoric acid

(1) Test with ammonium molybdate and benzidine [175]

Phosphates react with molybdates to produce salts of the complex phosphomolybdic acid $H_7P(Mo_2O_7)_6$ in solutions containing mineral acids. The slightly soluble yellow ammonium salt is often used for the detection and determination of phosphoric acid. Germanic, arsenic and silicic acids behave analogously and react with molybdate in acid solution to form, respectively, germanimolybdic acid $H_8Ge(Mo_2O_7)_6$, arsenimolybdic acid $H_7As(Mo_2O_7)_6$, silicomolybdic acid $H_8Si(Mo_2O_7)_6$.

In these polyhetero acids the complexed molybdic acid has an enhanced oxidizing power toward many inorganic and organic compounds, which are oxidized slightly, if at all, by free molybdic acid and normal molybdates. For example, iodides are oxidized but slightly by molybdates in acid solution, even on warming, whereas phosphomolybdic acid liberates appreciable amounts of iodine, even in the cold.[176]

Because of the formation of phosphomolybdic acid, the addition of even slight quantities of PO_4^{-3} ions to acid molybdate solutions brings about an activation of the molybdenum with consequent entrance into redox reactions.

Benzidine (see page 72), which is unaffected by free molybdic acid and normal molybdates, is oxidized instantly in acetic acid solution by phosphomolybdic acid and even by the difficultly soluble ammonium salt. This oxidation is applied in the test for phosphates. The reaction is extremely sensitive, since two blue products are formed, namely benzidine blue (= oxidation product of benzidine) and molybdenum blue (= reduction product of molybdic acid). Solutions of phosphate, which are too dilute to yield a visible precipitate with ammonium molybdate, give a distinct blue with molybdate and benzidine.

Arseni- and silicomolybdic acids and their salts react similarly with benzidine. However, under suitable conditions, phosphate may be decisively detected even in the presence of large amounts of arsenic and silicic acids.

When the acid test solution contains much hydrogen peroxide, permolybdic acid is formed on the addition of ammonium molybdate. Most precipitation reactions of molybdates are masked in solutions of permolybdic acid, which contain MoO_5^{-2} ions in place of MoO_4^{-2} ions.[177] Consequently, no precipitation of $(NH_4)_3PO_4 \cdot 12\,MoO_3$ occurs. The hydrogen peroxide must be destroyed before carrying out the test. Similarly, oxalates and fluorides, which form stable complex compounds with molybdenum, interfere with the quantitative precipitation of phosphate as phosphomolybdate. The interference by fluorides can be avoided by adding beryllium salts.[178] This demasking of $[MoO_3F_2]^{-2}$ ions through formation of $[BeF_4]^{-2}$ ions is based on the reaction:

$$2[MoO_3F_2]^{-2} + Be^{+2} + 2H_2O \rightarrow 2MoO_4^{-2} + [BeF_4]^{-2} + 4H^+$$

Procedure. A drop of the acid test solution is placed on quantitative filter paper, followed by a drop of molybdate and a drop of benzidine solution. Then the paper is held over ammonia. When most of the free mineral acid has been neutralized, a blue stain appears, the intensity depending on the phosphate content.

Limit of Identification: 1.25 γ P_2O_5

Limit of Dilution: 1 : 40,000

Arsenates do not react under these conditions; the formation of arsenimolybdic acid proceeds very slowly in the cold. In this way, 1.5 γ phosphoric acid may be detected in the presence of 1.5 milligrams of arsenic acid (*Limiting proportion* 1 : 1000).*

An even greater sensitivity may be attained as follows: a drop of ammonium molybdate solution is placed on filter paper (S & S 589) and dried in an oven. A drop of the test solution is placed on the freshly prepared paper, followed by a drop of a benzidine solution and a drop of saturated sodium acetate solution. According to the amount of phosphate present, a blue fleck or ring is formed.

Limit of Identification: 0.05 γ P_2O_5

Limit of Dilution: 1 : 1,000,000

Reagents: *1)* Ammonium molybdate solution: 5 g salt is dissolved in 100 ml cold water and poured into 35 ml nitric acid (sp. gr. = 1.2)

2) Benzidine solution: 0.05 g base or its chloride is dissolved in 10 ml concentrated acetic acid and diluted with water to 100 ml

Test for phosphoric acid in the presence of silicic acid [179]

Soluble silicates also react with molybdic acid under the above conditions forming a soluble silicomolybdic acid. This product, in its turn, reacts with benzidine similarly to the phosphoric acid compound, thus impairing the decisive nature of the test for phosphate.

It is possible to detect phosphoric acid in the presence of silicic and arsenic acids by utilizing the fact that, under suitable conditions, the formation of silicomolybdic and arsenimolybdic acids can be prevented by tartaric acid. This masking depends on the formation of a stable complex compound of molybdic and tartaric acids, in which the molybdic acid does not react with arsenic and silicic acids, but is not masked toward phosphoric acid.

Procedure. A drop of the test solution is placed on filter paper followed by a drop of a tartaric acid-molybdate solution. The paper is held over a hot wire

* Interference due to As_2O_5 can be prevented by warming the test solution with H_2SO_3; the As_2O_3 formed by reduction does not react with molybdate. It is advisable to remove excess H_2SO_3 by addition of alkali nitrite to the acid solution. See N. A. Tananaeff and C. N. Potschinok, *Z. anal. Chem.*, 88 (1932) 273.

References pp. 356–360

gauze to accelerate the reaction. A drop of benzidine reagent is then added and the paper developed over ammonia.

Limit of Identification: 1.5 γ P_2O_5 } in the presence of 500 times the
Limit of Dilution: 1 : 33,000 } amount of silicic acid

Reagent: Tartrate-ammonium molybdate solution: 15 g crystalline tartaric acid is dissolved in 100 ml of the usual nitric acid-ammonium molybdate solution. Other reagents as in the previous test.

(*2*) *Test with o-dianisidine molybdate and hydrazine* [180]

Phosphomolybdic acid, or PO_4^{-3} plus MoO_4^{-2} ions, form insoluble phosphomolybdates in acid solution not only with ammonium salts but also with salts of organic bases. A brown precipitate results with salts of *o*-dianisidine (down to 10γ P_2O_5 per drop). If the precipitate is treated with excess hydrazine hydrate, the complexly bound Mo^{VI} is reduced to molybdenum blue and the precipitate disappears. The test for phosphate described here is based on this effect. It is superior to test *1* in that arsenate ions do not interfere. Only silicates react analogously to phosphates; they yield *o*-dianisidine-silicomolybdate. However, silicates can be rendered harmless by adding sodium fluoride to the test solution.

Sulfides interfere because they too reduce molybdate to molybdenum blue. Thiosulfates likewise give a blue coloration, but this fades completely on the addition of hydrazine hydrate.

Procedure. The test is conducted on a spot plate. A drop of the test solution is treated with one drop of the reagent solution; a brown precipitate or color appears. Then a drop or two of a 85 % solution of aqueous hydrazine hydrate solution is added. A positive response is indicated by the formation of a deep or light blue solution.

Limit of Identification: 0.05 γ phosphoric acid
Limit of Dilution: 1 : 1,000,000
Reagent solution: Dianisidine molybdate solution: A solution of 0.25 g dianisidine in 1 ml concentrated sulfuric acid is added drop by drop and with swirling to a solution of 2.5 g of sodium molybdate in 20 ml of 7.5 % sulfuric acid. The mixture is allowed to stand for 12 hours and then filtered if need be.

28. Silicic acid

(*1*) *Test with ammonium molybdate and benzidine* [181]

Soluble silicates form the complex silicomolybdic acid $H_4SiO_4 \cdot 12MoO_3 \cdot aq.$, when treated with molybdate in acid solution. Its ammonium salt differs from the ammonium salts of the analogous phosphoric and arsenic acid compounds in that it is soluble in water and acids and gives a yellow solution.* Silicomolybdic acid, like the corresponding phosphomolybdic acid (see page 333), oxidizes benzidine in acetic solution to a blue meriquinoid product, and simultaneously the molybdenum is reduced to "molybdenum blue". The silicomolybdic acid is formed rapidly and completely only at higher temperatures and consequently the silicate solution must be gently warmed, in a crucible or on filter paper, after the addition of the molybdate. Care must be taken that the liquid does not boil, as traces of silicic acid might dissolve out of the porcelain of the crucible. The result then would lead to false conclusions.

Phosphoric and arsenic acids form complex compounds similar to silicomolybdic acid and these products also give a color reaction with benzidine. These acids must therefore be removed before the test. Regarding other interferences, see page 333.

Procedure. (*a*) A drop of the test solution and one drop of the molybdate solution are placed on filter paper and gently warmed over a wire gauze. A drop of benzidine solution is added and the paper is developed over ammonia. A blue color indicates silicic acid.

Limit of Identification: 1 γ SiO_2

Limit of Dilution: 1 : 50,000

(*b*) A drop of the slightly acid test solution (the acidity should not exceed 0.5 *N*) is placed in a micro crucible, mixed with a drop of molybdate solution, and carefully warmed over a wire gauze until bubbles appear. After thorough cooling, a drop of benzidine solution is added and then a drop of sodium acetate. A blue color indicates silicic acid.

Limit of Identification: 0.1 γ SiO_2

Limit of Dilution: 1 : 500,000

The crucible must be tested for resistance to acids. A drop of water and a drop of molybdate solution are mixed in the crucible, warmed, and treated as described. Only those crucibles which give no trace of blue after addition of benzidine and sodium acetate are suitable.

Reagents: As for the phosphoric acid test, page 334

Test for silicic acid in the presence of phosphoric acid

When silicic acid is to be detected in the presence of much phosphoric acid, the test must be carried out on the filtrate obtained after precipitation with ammonium molybdate. The detection of silicic acid is simplified by the addition of oxalic acid, which decomposes small amounts of ammonium phosphomolybdate whereas silicomolybdic acid is stable toward oxalic acid.

* The detection of silicic acid in drinking water by the formation of the soluble, yellow silicomolybdate was first described by A. Jolles and F. Neurath, *Z. angew. Chem.*, 11 (1898) 315.

Procedure. A drop of the test solution (the P_2O_5 content should not exceed 0.1 mg) is mixed in a microcentrifuge tube with 2 drops of molybdate solution, and centrifuged. The supernatant liquid is transferred to a micro crucible by means of a capillary tube, and gently warmed. After cooling, 2 drops of 1 % oxalic acid solution are added to destroy any remaining traces of ammonium phosphomolybdate. A drop of benzidine and a few drops of saturated sodium acetate solution are then added. A blue color indicates silicic acid.

Limit of Identification: 6 γ SiO_2 } in the presence of 250 times the
Dilution Limit: 1 : 8300 } amount of P_2O_5
Reagents: See page 334.

29. Carbonic acid

(*1*) *Test with sodium carbonate and phenolphthalein* [182]

Solutions of alkali carbonates react basic toward phenolphthalein. A solution of sodium carbonate colored red by phenolphthalein is decolorized by free carbon dioxide, which reacts to form bicarbonate:

$$Na_2CO_3 + CO_2 + H_2O = 2\ NaHCO_3$$

In the presence of a small excess of free carbon dioxide (3 to 5 %), such solutions have an acidity which is above the change point of phenolphthalein.

A rather sensitive test for carbonates is based on this fact. The carbon dioxide is liberated by dilute acid and taken up in a drop of sodium carbonate solution containing phenolphthalein. The gas is detected by the discharge of the red color.

The concentration of the reagent solution must be so selected that it is not decolorized, under the conditions of the experiment, by the carbon dioxide in the air.

Procedure. The gas apparatus described on page 47 (Figure 23) is used. One or two drops of the test solution or a pinch of the solid is treated in the apparatus with 3 drops of 2 *N* sulfuric acid. A drop of a solution of sodium carbonate, reddened with phenolphthalein, is put on the knob of the stopper, which is then placed in position. The drop is decolorized immediately, or after a short time, according to the amount of carbon dioxide liberated. To avoid false conclusions due to the carbon dioxide in the air, a blank test should be carried out in a duplicate apparatus.

Limit of Identification: 4 γ carbon dioxide (in 2 drops of solution)
Limit of Dilution: 1 : 12,500
Reagent: 1 ml of a 0.1 *N* sodium carbonate mixed with 2 ml 0.5 % alcoholic phenolphthalein solution and 10 ml water

Sulfides, sulfites, thiosulfates, cyanides, azides, fluorides, formates, acetates and nitrites interfere because the respective volatile acids, liberated on acidifying their salts, also decolorize phenolphthalein-sodium carbonate solution. Under suitable conditions, however, the carbonate test may be carried out successfully even in the presence of large amounts of the materials listed below.

Test for carbonates in the presence of cyanides, azides, sulfides, sulfites, thiosulfates, fluorides and nitrites [183]

Cyanides or *azides* can be converted into insoluble silver cyanide or azide by addition of a few drops of a saturated solution of silver nitrate.

Limit of Identification: 5 γ CO_2 in the presence of 2500 times the amount of KCN or NaN_3

Sulfides, sulfites and thiosulfates can be quantitatively oxidized to sulfates by addition of a few drops of 10 % hydrogen peroxide.

Limit of Identification: 5 γ CO_2 in the presence of { 20,000 times the amount of $Na_2S_2O_3$, 10,000 times the amount of K_2S, 10,000 times the amount of Na_2SO_3 }

Fluoride interference may be prevented by formation of the complex $[ZrF_6]^{-2}$ ions; a drop of a concentrated solution of $ZrCl_4$ is added (see page 269).

Limit of Identification: 5 γ CO_2 in the presence of 1000 times the amount of K_2F_2

Nitrites are rendered harmless by treatment with aniline hydrochloride; the aniline is diazotized and the nitrous acid consumed.

Limit of Identification: 5 γ CO_2 in the presence of 1000 times the amount of KNO_2

Procedure. All the substances required to prevent interference may be added before the release of the carbon dioxide; if necessary, in the apparatus where the test is carried out. The rest of the procedure is as described.

Another method for removing interfering anions (including acetate and formate) is to precipitate the carbonate by adding calcium chloride. The precipitation, washing and isolation of the calcium carbonate can be accomplished in the bulb of the gas evolution apparatus, if the latter is placed in a centrifuge tube.

(2) *Test for bicarbonate in the presence of carbonate*

Calcium hydrogen carbonate, in contrast to calcium carbonate, is soluble in water and, on addition of ammonia, is converted into $CaCO_3$:

$$Ca(HCO_3)_2 + 2\,NH_3 = CaCO_3 + (NH_4)_2CO_3$$

Therefore, when testing for alkali bicarbonate in the presence of alkali carbonate, 1 or 2 drops of the solution (for very small amounts of bicarbonate, about 1 ml) is placed in an Emich microcentrifuge tube, or even a larger centrifuge tube, and treated with an excess of neutral calcium chloride solution. After centrifuging, a drop of dilute ammonia is added to the clear liquid. A turbidity of calcium carbonate appears, if bicarbonate is present.

Instead of treating the solution with ammonia, the clear bicarbonate solution, obtained as described, may be tested for carbon dioxide in the apparatus used in the previous procedure.

30. Boric acid

(*1*) *Test with tincture of curcuma* [184]

Free boric acid gives a deep red compound when evaporated with tincture of curcuma (turmeric). The boric acid changes the yellow curcumin (diferuloylmethane)

$$CH_2[CO—CH{=}CH—C_6H_3(OH)OCH_3]_2$$

into the isomeric red-brown rosocyanine.[185] A loose addition product of boric acid and curcumin is formed as an intermediate. The rosocyanine is turned blue to greenish-black by alkali; the original brown-red color is restored by acidification.

Ferric, molybdenum, titanium, niobium, tantalum, and zirconium compounds also turn curcumin red-brown. However, this color product, unlike that due to boric acid, does not change to blue or green on addition of alkali.

The curcuma reaction is hindered by oxidizing substances (hydrogen peroxide, chromates, nitrites, chlorates, etc.) and by iodides. They must be decomposed or removed before carrying out the reaction.

Procedure. (*a*) A drop of the test solution, acidified with hydrochloric acid, is placed on curcuma paper (3 × 7 mm) and dried at 100° C. A red-brown fleck, which turns blue to greenish-black on treatment with 1 % sodium hydroxide, indicates the presence of boric acid.

Limit of Identification: 0.02 γ boron

Limit of Dilution: 1 : 2,500,000

(*b*) A drop of the test solution is placed on a microscope slide and acidified with a drop of hydrochloric acid. A silk thread impregnated with curcuma [186] is placed in the solution, which is then evaporated by warming gently. In the presence of boric acid, the thread is colored pink-violet, and on treating with 1% sodium hydroxide, the color changes to that of Prussian blue. Molybdenum, titanium, zirconium, and tantalum salts do not interfere with the test carried out in this way. The sensitivity is decreased by free phosphoric and silicic acids.

References pp. 356–360

Reagents: 1) Tincture of curcuma: 20 g curcuma is boiled with 50 ml alcohol, filtered, and the extract diluted with 50 ml water

2) Curcuma paper: Quantitative filter paper is soaked in the tincture of curcuma and dried

3) Curcuma silk: Threads of viscose silk are soaked in the tincture after the addition of alkali (1 ml 10 % NaOH per 100 ml tincture). The wet threads are heated until the mass becomes syrupy. The threads are then immersed in 0.5 % alcohol, pressed between filter paper, then bathed in dilute sulfuric acid, washed with water, and dried.

(2) *Test with hydroxyanthraquinones* [187]

Many water-insoluble hydroxyanthraquinones dissolve in concentrated sulfuric acid and give deeply colored solutions. With boric acid these solutions give characteristic color changes [188] which may be applied also in the detection of anthraquinones. These changes of color are apparently due to the formation of inner complex boric acid esters involving the OH-groups in the *peri* position; the concentrated sulfuric acid functions as solvent and as dehydrating agent.[189] For example, in the case of alizarin (assuming the presence of metaboric acid) the following equation holds:

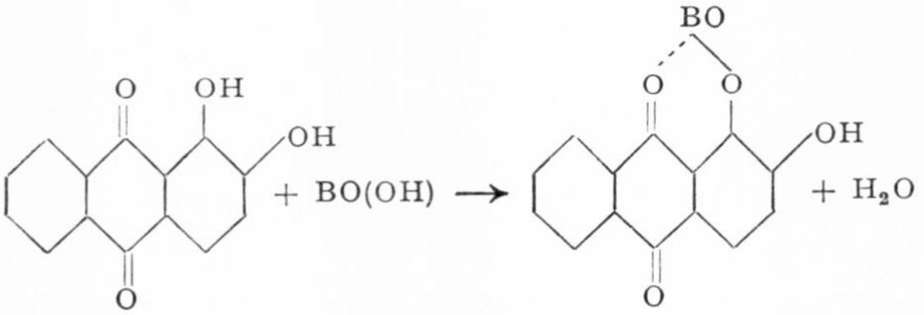

$$\text{(alizarin)} + BO(OH) \longrightarrow \text{(boric ester, BO bridging peri O atoms)} + H_2O$$

For the detection of boric acid, the best hydroxyanthraquinones are: 1,2-dihydroxyanthraquinone-3-sulfonic acid (alizarin S); 1,2,4-trihydroxyanthraquinone(purpurin); 1,2,5,8-tetrahydroxyanthraquinone(quinalizarin).

Procedure. A drop of the slightly basic test solution is evaporated to dryness in a micro crucible. The residue is treated with 2 or 3 drops of the reagent solution, and gently warmed. The evaporation must take place in the presence of alkali to avoid loss of boric acid in the steam. If boric acid is present, the following color changes occur:

Alizarin S: yellowish-red to red *Limit of Identification:* 1.0 γ boron
Purpurin: orange to wine red *Limit of Identification:* 0.6 γ boron
Quinalizarin: violet to blue *Limit of Identification:* 0.06 γ boron

The behavior of antimony^{III} salts toward sulfuric acid solutions of hydroxyanthraquinones is analogous to that of boric acid; they must be

References pp. 356–360

oxidized to antimony^{V} by means of chlorine water. Larger amounts of colored ions (Fe^{+3}, Cu^{+2}, Ni^{+2}, Cr^{+3} etc.) make it difficult to detect the color change of the reagent solutions. Be^{+2} ions lower the sensitivity of the test. Iodides interfere by releasing iodine; they can be removed by precipitation with Ag_2SO_4. Oxidizing acids impair the test (see test *3* regarding their removal).[190]

When alizarin S is used as reagent, the sensitivity may be considerably increased by viewing the reaction picture in ultraviolet light.[191] The solution of alizarin S in concentrated sulfuric acid is yellow in ultraviolet light, and is changed to brilliant red by boric acid (*Idn. Limit:* 0.02 γ boron). The interferences due to colored ions and Be^{+2} ions are obviated if the fluorescence colors are observed.

Reagents: *1*) 0.2 % solution of alizarin S
2) 0.5 % solution of purpurin
3) 0.01 % solution of quinalizarin
} in concentrated sulfuric acid

(*3*) *Test with* p-*nitrobenzeneazochromotropic acid* (*chromotrope 2 B*) [192]

The addition of borate to the blue solution of *p*-nitrobenzeneazochromotropic acid in concentrated sulfuric acid, changes the color to greenish-blue. The reason is probably analogous to that in the reaction of boric acid with hydroxyanthraquinones (test *2*), in which an inner complex phenol ester of boric acid is formed. The reagent has two OH-groups in *peri* position as shown in (I).

The reagent has an esterifiable phenolic OH-group in a coordination position with respect to a N-atom of the azo group. The chelate binding responsible for the color change could be (II) or (III), i.e., ester of ortho- or metaboric acid:

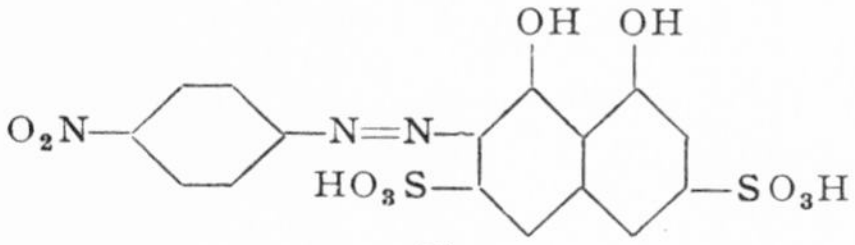

(I)

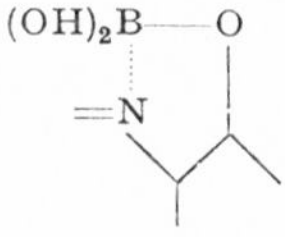

(II)

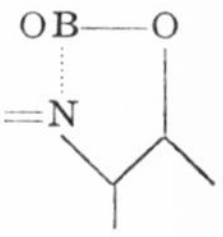

(III)

The detection of boron by this color reaction is adversely affected by oxidizing agents and fluorides. The former give a red or yellow color with the reagent; fluorides form fluoboric acid or boron fluorides and so prevent the production of the colored ester. Boron can be detected in the presence of these materials under certain conditions, but the sensitivity is then lower.

Procedure. A drop of the slightly alkaline test solution is evaporated to dryness in a porcelain dish. While the dish is still warm, the residue is stirred with 2 or 3 drops of the reagent. After cooling, the color is observed. In the presence of boric acid, the color changes from blue-violet to greenish-blue. A blank test is necessary only when very small amounts of boric acid are involved.

Limit of Identification: 0.08 γ boron (in 0.04 ml)

Limit of Dilution: 1 : 500,000

Reagent: 0.005 % solution of chromotrope 2B [193] in concentrated sulfuric acid

Test in the presence of oxidizing agents or fluorides

Nitrates and chlorates (and probably other oxidizing compounds) may be reduced and thus rendered harmless by evaporating with solid hydrazine sulfate. Fluorides can be removed by evaporation with silica and sulfuric acid (volatilization of SiF_4).

Procedure. One or two drops of the test solution is treated, in a porcelain crucible, with a little solid hydrazine sulfate or with a very small quantity of silica and 1 or 2 drops of concentrated sulfuric acid. The contents of the crucible are carefully heated until fumes of sulfur trioxide are evolved. The warm residue is treated with 3 or 4 drops of the reagent solution, and any color change noted after cooling.

Limit of Identification: 0.25 γ boron { in the presence of 13,000 times the amount of potassium nitrate; in the presence of 11,600 times the amount of potassium chlorate }

0.5 γ boron in the presence of 2700 times the amount of sodium fluoride

(*4*) *Test by decomposition of methyl borate with alkali fluoride* [194]

Boric acid reacts with alkali fluorides to produce alkali borofluoride and hydroxide:

$$B(OH)_3 + 4\ KF \rightarrow K[BF_4] + 3\ KOH$$

The free hydroxide can be detected through the formation of a black precipitate on treatment with silver nitrate-manganous nitrate solution (compare page 58). The presence of boric acid can be confirmed in this way.

Acid or alkaline solutions of borates are most commonly to be tested. Consequently the boric acid should be converted beforehand to its volatile methyl ester, b.p. 65° C, which is distilled into the fluoride-containing reagent solution. The ester is then saponified:

$$B(OCH_3)_3 + 3\ H_2O = B(OH)_3 + 3\ CH_3OH$$

regenerating free boric acid that attacks the fluoride, with production of OH^- ions.

Procedure. The distillation can be carried out in the apparatus described on page 48 (Fig. 29). A porcelain crucible, paraffined inside, serves as receiver and shoul dcontain about 1 ml of reagent solution. A drop of the test solution (plus a drop of alkali if the solution is acid) is placed in the distillation apparatus, and evaporated to dryness. Five drops of concentrated sulfuric acid and 5 drops of pure methyl alcohol are then added, and after stoppering the apparatus it is heated to about 80° C in a water-bath. The methyl borate distils into the receiver and undergoes saponification. The resulting alkali causes the formation of a black precipitate. For very small amounts, it is advisable to add a few drops of an acetic acid solution of benzidine to the micro crucible. Traces of manganese dioxide are revealed by the blue color (see page 175).

Limit of Identification: 0.01 γ boron

Limit of Dilution: 1 : 5,000,000

Reagents: *1*) Manganese-silver solution containing fluoride: 2.87 g $Mn(NO_3)_2$ and 1.69 g. $AgNO_3$ are dissolved in 100 ml water, a drop of dilute alkali added, and the black precipitate removed. A solution of 3.5 g potassium fluoride in 50 ml water is added to the filtrate; a white precipitate forms; on heating it turns gray. The mixture is again filtered, and the clear solution used as the reagent.

2) Acetic acid-benzidine solution (for preparation see page 175)

(*5*) *Other tests for boric acid*

(*a*) The increase in acidity with organic polyhydroxy compounds may be applied as a test for boric acid.[195] On mixing equialkaline solutions of boric acid and a polyhydroxy compound which can combine with boric acid, the pH of the mixture is lowered, owing to the production of an acid stronger than the boric acid.[196] This increase in acidity may be detected by the use of suitable indicators. It is advisable to use bromothymol blue as indicator, and to treat the test solution with acid or alkali, as may be required, until the indicator turns green; if mannite is then added, the color turns yellow in the presence of boric acid. For detecting extremely small amounts and low concentrations of boric acid, the mannite should be recrystallized beforehand from a solution neutralized to bromothymol with base, washed with pure acetone, and dried on the water-bath. The test solution, prepared as described, is divided into two portions. Mannite is added to one and then the colors of the two portions are compared. The color change is appreciable down to concentrations of 0.02 γ per ml. Two drops of solution on a spot plate are sufficient for the test (*Idn. Limit:* 0.001 γ B).

(*b*) A change from red to blue results on the addition of 0.05 % solution of carminic acid (in concentrated sulfuric acid) to solutions containing small amounts of B_2O_3 (*Idn. Limit:* 0.1 γ B in 0.03 ml).[197]

(*c*) A drop of the test solution, one drop concentrated H_2SO_4, and 5 drops

CH_3OH are mixed in a porcelain crucible. The mixture is warmed and the resulting methyl borate vapor set on fire. It burns with a green flame.[198]

31. Permanganic acid

Test for permanganate in the presence of chromate [199]

If no other colored ions are present, permanganate is easily recognized because the violet of MnO_4^- ions is quite visible even at high dilutions. The detection of permanganate in the presence of much chromate or bichromate is of importance since the intense color of the CrO_4^{-2} or $Cr_2O_7^{-2}$ ions may blanket the violet due to MnO_4^- ions, specially when dealing with small concentrations of the latter.

The method given here is based on the fact that oxidation of cellulose occurs when a drop of permanganate solution is placed on filter paper. Within a few minutes, finely divided manganese dioxide is formed and remains as a round fleck. It is dark to light brown depending on the extent of the reduction of the permanganate. Even very dilute permanganate solutions act on cellulose and the resulting slight precipitate is distinctly visible because of localization of the MnO_2 as well as through contrast with the white paper. Under similar conditions alkali chromate is practically without action on filter paper, even at high concentrations. Cellulose is oxidized by chromate very slowly; brown, difficultly soluble chromic chromate is produced. If a solution contains both chromate and permanganate, only the permanganate reacts at first. The manganese dioxide is retained as central fleck, while the chromate migrates through the capillaries and forms a concentric yellow ring around the fleck. The greater part of the chromate can be washed away by adding water drop by drop, or by soaking the paper in water. In this way even traces of permanganate can be detected by the production of MnO_2. It is necessary to wash away the chromate only when traces of permanganate are involved. Normally the fleck of MnO_2 is distinctly visible, even on filter paper colored yellow by chromate.

It should be noted that the fleck of MnO_2 formed from a very dilute solution of pure permanganate is less intense than a fleck produced from the same permanganate solution to which alkali chromate has been added. Moreover, when using solutions of pure permanganate, the limit of visibility of MnO_2 lies at much higher concentrations than in solutions which also contain chromate. This phenomenon probably arises from the fact that permanganate produces disintegration products of cellulose which in turn are then oxidized by the chromate. The resulting brown chromic chromate reinforces the color of the MnO_2 fleck. Consequently, this heightened sensitivity involves an oxidation by the chromate, a result induced by the preliminary redox reaction of the permanganate with filter paper cellulose.

References pp. 356–360

Procedure. A drop of the neutral or slightly acid test solution is allowed to soak into thick filter paper (S & S 601). If permanganate is present, a dark to light brown central circular stain appears, surrounded by a yellow zone. In the absence of permanganate, only a yellow circular spot forms. When very small amounts of permanganate are suspected, it is best to carry out a blank test with a solution of pure chromate. The paper is bathed in water to reveal the slight stain of MnO_2 left on the yellow spot, or on the washed paper. If a drop contains less than 5 γ $KMnO_4$, it is best to wait about 3 minutes (after spotting with the test solution) before making the comparison and washing out the chromate. All of the permanganate will have reacted by this time.

Identification Limit: 0.3 γ $KMnO_4$ in presence of 20,000 γ K_2CrO_4

32. Chromic acid *

(*1*) *Test for chromates in the presence of permanganates* [200]

The sensitive diphenylcarbazide reaction for chromates (page 167) is not directly applicable in the presence of permanganate because of its intense violet color. However, the different behavior of these ions toward hydrazoic acid may be applied to prevent interference of permanganate in the detection of small amounts of chromate. Hydrazoic acid has little appreciable effect on chromic acid, while permanganates are readily reduced, at room temperature, to manganese^{II} salts. (The reaction is not sufficiently understood to be expressed by an equation.[201] **) Manganese^{II} salts do not interfere with the diphenylcarbazide test for chromate, and a trace of chromate may be detected even in a saturated solution of potassium permanganate.

Procedure. A drop of the test solution is mixed on a spot plate with a drop of concentrated sulfuric acid. Small particles of sodium azide are added, while stirring with a glass rod, until the color of the permanganate has disappeared. A few milligrams of sodium azide suffice to reduce considerable amounts of permanganate immediately (with evolution of nitrogen). The permanganate-free solution is then treated with a drop of a 1 % alcoholic solution of diphenylcarbazide.[202] A more or less intense blue-violet to red color is formed according to the amount of chromate present.

Limit of Identification: 0.5 γ K_2CrO_4 }
Dilution Limit: 1 : 100,000 } in the presence of 6000 times the amount of $KMnO_4$

33. Selenious acid and Selenic acid

Heating or evaporation with concentrated hydrochloric acid reduces selenic acid and its salts quantitatively to selenious acid; chlorine is evolved. The resulting solution then displays the selenite reactions discussed below.

* See also Chromium (pages 167 *et seq.*).

** About one-seventh of the hydrazoic acid is oxidized to nitrogen; nitrate is also formed, and possibly N_2O.

(1) Test with hydriodic acid [203]

Selenious acid is reduced by iodides, in acid solution, to elementary selenium; free iodine also results:

$$SeO_3^{-2} + 4\ I^- + 6\ H^+ \rightarrow Se^o + 3\ H_2O + 2\ I_2$$

The color of the iodine is removed by adding thiosulfate, and the selenium is left as a red-brown powder. Under the same conditions, tellurous acid reacts with hydriodic acid, forming the complex red-brown $[TeI_6]^{-2}$ anion. However, this tellurium compound can be decomposed and decolorized by sodium thiosulfate. Selenium can thus be detected successfully, even in the presence of considerable quantities of tellurium.

Procedure. A drop of concentrated hydriodic acid (or concentrated potassium iodide solution plus a drop of concentrated hydrochloric acid) is placed on filter paper and a drop of the acid test solution is placed in the middle of the moist fleck. The black-brown stain that develops is completely decolorized by a drop of 5 % sodium thiosulfate if no selenium is present; otherwise a red-brown fleck of selenium is left.

Limit of Identification: 1 γ selenium (0.025 ml)
Limit of Dilution: 1 : 25,000

(2) Test with pyrrole [204]

Under suitable conditions, selenious acid oxidizes pyrrole to "pyrrole blue", a dyestuff of unknown constitution. This reaction can serve as a test for selenious acid, since selenic, tellurous, and telluric acids remain unaltered. Hence, the test can be used to distinguish between selenious and selenic acids.

The test with pyrrole is impaired by the presence of oxidizing materials which likewise produce pyrrole blue. They include VO_3^-, MoO_4^{-2}, MnO_4^-, CrO_4^{-2}, NO_3^-, BrO_3^-, IO_3^-, IO_4^-, $[PO_4 \cdot 12MoO_3]^{-3}$, Au^{+3}, Hg^{+2}, Sb^{+5}.

If the redox reaction between selenious acid and pyrrole is carried out in concentrated phosphoric acid, the sensitivity can be increased 25-fold by adding ferric salts. This singular effect cannot be ascribed to a participation of Fe^{III} in the oxidation of the pyrrole, because ferric salts are extremely weak oxidants in the presence of phosphoric acid, where they are mostly bound into complex $[Fe(PO_4)_2]^{-3}$ ions. It is probable that the equilibrium of the pyrrole blue reaction:

$$\text{pyrrole} + SeO_2 \rightleftharpoons Se^0 + \text{pyrrole blue}$$

is disturbed by removal of the "pyrrole blue" through capture by the complex ferriphosphoric acid, $H_3[Fe(PO_4)_2]$, with consequent rise of the oxidation potential of the selenious acid.

Procedure. A drop of a 5 % solution of ferric chloride and 7 drops of syrupy phosphoric acid are added to a drop of the test solution. The mixture is well stirred. A drop of the pyrrole reagent is then added and the mixture stirred again. A green-blue color indicates the presence of selenious acid. A blank test should be carried out when small amounts of selenious acid are suspected.

Limit of Identification: 0.5 γ selenium
Limit of Dilution: 1 : 100,000
Reagent: 1 % solution of pyrrole in aldehyde-free ethyl alcohol

(3) Test with asymmetric diphenylhydrazine [205]

Selenious acid is readily reduced to elementary selenium and is capable of oxidizing numerous organic compounds. In common with a number of other oxidants,[206] it reacts with asymmetric diphenylhydrazine (I) to produce violet quinoneanildiphenylhydrazone (II).

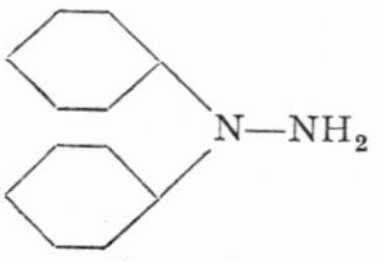

N—NH$_2$ (I)   —N= =N—N   (II)

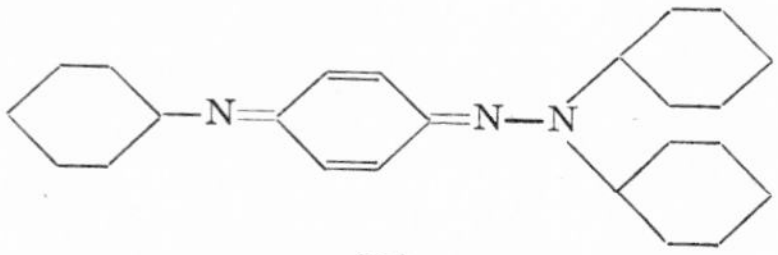

The action of other oxidizing agents may be prevented by suitable precautions; oxygen compounds of tellurium are without effect. This reaction may therefore be applied as a sensitive and specific test for selenium, either as element, selenide, selenite, or selenate, since they all are readily converted into selenious acid.

Procedure. Four drops of a solution of asym. diphenylhydrazine in glacial acetic acid are mixed with a drop of 2 *N* hydrochloric acid and a drop of the test solution on a spot plate. In the presence of selenious acid, a red color appears at once and soon changes to a bright red-violet. When extremely small amounts of selenious acid are present, the color appears after a few minutes, and a blank test is advisable.

Limit of Identification: 0.05 γ SeO_2
Limit of Dilution: 1 : 1,000,000
Reagent: 1 % solution of asym. diphenylhydrazine in glacial acetic acid. A freshly prepared solution is advisable when testing for small amounts of SeO_2.

Selenic acid also gives a red color under the same conditions (*limit of identification:* 100 γ). Since selenates are converted into selenious acid by heating with concentrated hydrochloric acid, it is preferable to heat a drop of the test solution with the acid in a micro crucible and test with asym. diphenylhydrazine after cooling. As little as 1 γ selenic acid may be detected in this way.

References pp. 356–360

Interference due to the presence of other oxidizing agents, which may also cause the formation of (II), may be prevented as follows: the solid test substance or a drop of the test solution is heated with a few drops of concentrated hydrochloric acid in a micro crucible to destroy oxidizing oxy-acids (e.g., HIO_3, $HBrO_3$, $HClO_3$, $HMnO_4$, etc.) and peroxides. Tungstates, molybdates, Fe^{III}, and Cu^{II} salts are not affected by this treatment. If any of these are present, oxalic acid should be added to the hydrochloric acid. Complex oxalates, which do not react with asym. diphenylhydrazine, are produced.

Commercial preparations of tellurous acid and alkali tellurites usually give a slight reaction, due to the presence of a trace of SeO_2. A content of 0.001 % SeO_2 in TeO_2 may be detected by means of the reaction just described. After strong ignition, TeO_2 containing SeO_2 no longer gives a positive reaction, because the SeO_2 is volatilized.

(*4*) *Test by catalytic hastening of the reduction of methylene blue by sulfide ions*[207]

Methylene blue can be reduced to the colorless leuco compound by suitable reducing agents acting in neutral, acid, or alkaline solution. (Regarding the formulas of methylene blue (MB) and its leuco compound (HMB) see page 117). The reduction of methylene blue by alkali sulfides is slow even when they are in large excess. However, if a slight quantity of selenium is dissolved in the alkali sulfide to produce $[S \ldots Se^{\circ}]^{-2}$ ions, the color of the methylene blue is quickly discharged. The pertinent redox reactions are:

$$2\,MB + S^{-2} + 2\,H_2O \rightarrow 2\,HMB + 2\,OH^- + S^{\circ} \quad (1)\ (slow)$$

$$2\,MB + [S\ldots Se^{\circ}]^{-2} + 2\,H_2O \rightarrow 2\,HMB + 2\,OH^- + S^{\circ} + Se^{\circ} \quad (2)\ (rapid)$$

Comparison of (1) and (2) reveals that the reducing action of S^{-2} ions is obviously enhanced as a result of the formation of $[S\ldots Se^{\circ}]^{-2}$ ions, produced by the addition of selenium to S^{-2} ions. The alkali sulfide, which is present in excess, immediately dissolves the free sulfur and also the free selenium produced in (2). The $[S \ldots Se^{\circ}]^{-2}$ ions are regenerated and again participate in the rapid reaction (2). Therefore, selenium acts as a catalyst.

To detect selenium by the method presented here, it must be transformed into active $[S\ldots Se^{\circ}]^{-2}$ ions. If present as selenite, the conversion is easily accomplished by adding excess alkali sulfide: The reaction:

$$SeO_3^{-2} + 2\,S^{-2} + 3\,H_2O \rightarrow Se^{\circ} + 2\,S^{\circ} + 6\,OH^-$$

is followed by the binding of the sulfur and selenium, through the agency of the excess alkali sulfide, into practically inactive $[S \ldots S^{\circ}]^{-2}$ ions and the catalytically active $[S \ldots Se^{\circ}]^{-2}$ ions.

Procedure. A drop of water and a drop of alkaline test solution are placed in adjacent depressions of a spot plate. A drop of 0.2 *M* sodium sulfide solution is added to each. A drop of methylene blue solution is added to the blank and another to the test solution. Any discharge of the blue color is noted. If the mixture contains considerable quantities of selenium, it loses its color almost at once; with smaller amounts the discharge of the blue color is still definitely faster than in the blank test.

Limit of Identification: 0.08 γ selenium (volume of drop 0.08 ml)

Limit of Dilution: 1 : 1,000,000

Reagent: 0.01 % solution of methylene blue, or 0.05 % solution of methylene blue for test tube procedure.

By test tube procedure, 0.3 γ selenium can still be detected in 2 ml of solution.

(5) *Other tests for selenious acid*

(*a*) Red elementary selenium is precipitated from selenite by solid thiourea (*Idn. Limit:* 2 γ Se). [208] Nitrites interfere as do large amounts of copper; Te, Hg, and Bi form yellow precipitates. 5 γ Se is detectable in the presence of 25 γ Te.

(*b*) A red-blue fluorescing precipitate is produced by mixing a drop of test solution with a drop of 10 % thiourea solution.[209] Heavy metals do not interfere (*Idn. Limit:* 5 γ Se).

(*c*) Selenites in solutions of mineral acids may be detected by reduction to Se with $FeSO_4$ (*Idn. Limit:* 10 γ Se) or with ascorbic acid (*Idn. Limit:* 1 γ Se).[210]

(*d*) A brown product is formed by the reaction between selenites in acetic acid solution and 1,8-naphthylenediamine (*Idn. Limit:* 1 γ SeO_2). [211]

(*e*) A strongly acid solution of 3,4,3',4'-tetraaminobenzidine can be employed in place of 1,8-naphthylenediamine. A yellow precipitate or color is produced with selenites. Selenium can be detected in 0.5 ml at a dilution of 1 : 1,000,000.[212]

(*f*) Colorless 4-dimethylamino-1,2-phenylenediamine and 4-methylthio-1,2-phenylenediamine react with selenious acid to give stable bright red and blue colors, respectively (*Idn. Limit:* 0.05 γ Se in a small drop). [212a]

34. Tellurous acid and Telluric acid

(1) *Test with hypophosphorous acid* [213]

Tellurites and tellurates are reduced to the element on evaporation with hypophosphorous acid:

$$TeO_3^{-2} + H_2PO_2^- = Te^0 + PO_4^{-3} + H_2O$$

or

$$2\ TeO_4^{-2} + 3\ H_2PO_2^- = 2\ Te^0 + 3\ PO_4^{-3} + 2\ H_2O + 2\ H^+$$

The test is decisive in the absence of Pt, Cu, Ag, Au salts, and selenious acid, which are also reduced by hypophosphorus acid to the metal or selenium, respectively. If necessary, selenious acid can be reduced to the elementary state by heating the concentrated sulfuric acid solution with sodium sulfite; this procedure leaves tellurium compounds unaffected. The tellurium can then be detected in the filtrate after removing the SO_2.

Procedure. A drop of the test solution containing mineral acid and a drop of 50 % hypophosphorous acid are mixed in a porcelain micro crucible and evaporated almost to dryness. According to the amount of tellurium present, black flakes or a gray color will appear. For very small amounts of tellurium, a blank test should be carried out.

Limit of Identification: 0.1 γ tellurous acid
0.5 γ telluric acid
Limit of Dilution: 1 : 500,000 or 1 : 100,000, respectively

(2) *Test with alkali stannite* [214]

Tellurites and tellurates are reduced to free tellurium by alkali stannites in alkaline solution:

$$TeO_3^{-2} + 2\ SnO_2^{-2} + H_2O = 2\ SnO_3^{-2} + Te^o + 2\ OH^-$$

$$TeO_4^{-2} + 3\ SnO_2^{-2} + H_2O = 3\ SnO_3^{-2} + Te^o + 2\ OH^-$$

The element separates as black flocks. Since alkali stannites are without effect on the analogous selenium compounds, tellurium may be detected in the presence of selenium. Silver, mercury, copper, bismuth, and antimony salts are also reduced to the metal by stannites; therefore it is advisable always to carry out the detection of tellurium in the sodium carbonate extract of the material to be tested.

Procedure. A drop of a solution of stannous chloride, a drop of 25 % sodium hydroxide, and a drop of the alkaline test solution are mixed on a spot plate. A black precipitate or gray color appears, according to the amount of tellurium present. The color develops only after 1 or 2 minutes with very small amounts. In such cases a blank test is necessary.

Limit of Identification: 0.6 γ tellurium (0.025 ml)
Limit of Dilution: 1 : 41,000
Limit of Identification in the presence of selenium: 0.8 γ tellurium in the presence of 100 times this amount of selenium
Reagent: Stannous chloride: 5 g crystalline $SnCl_2$ is dissolved in 5 ml concentrated HCl and made up to 100 ml with water

(3) *Test by formation of a complex copper*^{III} *salt* [215]

If certain oxidizing agents (e.g., persulfate) are added to copper hydroxide suspended in an alkaline solution of an alkali tellurate, a dark red-brown color develops on heating. This is due to a stable complex anion formed from tervalent copper and the tellurate radical.[216]

References pp. 356–360

The formation of the colored anion may be applied as a spot test for tellurates [procedure (*a*)]. Alternatively, the impeding effect by tellurate on the catalytic action of copper^{II} salts on the reaction between Mn^{+2} ions and hypobromite may be used [procedure (*b*)]. This latter reaction normally proceeds with the formation of manganese dioxide, but in the presence of even very small amounts of copper salts, permanganate is formed. The activation of hypobromite is due to the intermediate formation of copper^{III} oxide which causes the further oxidation of manganese dioxide to permanganate with regeneration of copper^{II} oxide (see page 174). The production of a stable complex ion containing tervalent copper and tellurate prevents the activation of the hypobromite.

The formation of a stable complex of tervalent copper under the given conditions is limited to telluric acid and periodic acid (see page 301), i.e., the test is extremely selective.

Using procedure (*a*), 1 γ Te may be detected in the presence of 500 times the amount of selenium (as selenic acid). Selenious acid interferes with the tellurium test, so that any selenite present must be oxidized to selenate (by bromine) before the reaction. Using procedure (*b*), 2.5 γ Te may be detected in the presence of 20,000 times the amount of Se (as selenic acid).

Procedure. (*a*) A drop of 1 : 50,000 copper sulfate and a drop of alkali hydroxide are added to a drop of the alkaline test solution and finally a little solid sodium or potassium persulfate. The mixture is boiled. A yellow color indicates telluric acid.

Limit of Identification: 0.5 γ tellurium (as H_6TeO_6)
Limit of Dilution: 1 : 100,000

Procedure. (*b*) A drop of the copper-manganese solution and a drop of sodium hypobromite solution are added to a drop of the test solution in a micro-test tube. A blank test on water plus the reagents is set up. Both tubes are heated for a few minutes in boiling water, or are brought to boiling over a flame. In the presence of telluric acid, the test solution becomes colorless or slightly yellow, while the blank test assumes the permanganate color.

Limit of Identification: 0.2 γ tellurium
Limit of Dilution: 1 : 250,000
Reagents: 1) Copper-manganese solution: A drop of 4 % solution of $CuSO_4 \cdot 5\,H_2O$ is added to 100 ml of 0.04 % solution of $MnCl_2 \cdot 4\,H_2O$
2) Sodium hypobromite: A freshly prepared solution of bromine in 2 *N* NaOH

(*4*) *Test with ferrous sulfate and phosphoric acid* [217]

Selenites and selenates are quantitatively reduced to elementary selenium when their acid solutions are warmed with ferrous sulfate. In contrast, tellurite and tellurate solutions remain unaltered because the redox potential of iron^{II} is not sufficient to accomplish the reaction: $Fe^{II} + Te^{IV(VI)} \rightleftarrows Fe^{III} + Te^{\circ}$. However, the reducing power of Fe^{+2} ions can be raised considerably if the Fe^{+3} ions are removed as soon as they are produced. Phosphoric acid is excellent for this purpose; it immediately converts the Fe^{+3} ions into $[Fe(PO_4)_2]^{-3}$ ions. Fluorides act analogously through the formation of $[FeF_6]^{-3}$ ions. Consequently, when tellurites or tellurates are warmed with a ferrous sulfate-phosphoric acid mixture there is complete reduction to free tellurium. The following test is based on this reaction.

Procedure. A drop of the test solution is mixed in a micro crucible with a drop of 10 % ferrous sulfate solution and a drop of syrupy phosphoric acid. The mixture is warmed. A black precipitate (free tellurium) indicates the presence of tellurite or tellurate.

Limit of Identification: 0.5 γ tellurium

Limit of Dilution: 1 : 100,000

If tellurium is to be detected in the presence of selenium, the acid solution should be reduced with ferrous sulfate alone. After the selenium has been completely precipitated, the suspension is filtered or centrifuged. A drop of phosphoric acid is added to a drop of the filtrate or centrifugate and the mixture reheated.

(*5*) *Test with ammonium polysulfide and sulfite* *

Alkali tellurites and tellurates are converted into alkali sulfotellurates by the action of alkali polysulfides,i.e., by sulfide ions and free sulfur:

$$TeO_3^{-2} + 3\ S^{-2} + S^{\circ} + 3\ H_2O \rightarrow TeS_4^{-2} + 6\ OH^-$$
$$TeO_4^{-2} + 4\ S^{-2} + 4\ H_2O \rightarrow TeS_4^{-2} + 8\ OH^-$$

When solutions of alkali sulfotellurate are warmed with an excess of alkali sulfite, free tellurium is precipitated:

$$TeS_4^{-2} + 3\ SO_3^{-2} \rightarrow 3\ S_2O_3^{-2} + S^{-2} + Te^{\circ}$$

This deposition of tellurium from sulfalkaline solution is strictly specific and permits the detection of tellurium in the presence of large quantities of sulfur and selenium, which are dissolved by alkali sulfide with formation of polysulfide or selenosulfide. The yellow solutions of polysulfide or the red-brown solutions of selenosulfide yield no precipitate with alkali sulfite;

* The test is based on a procedure described by A. Brukl and W. Maxymowicz, *Z. anal. Chem.*, 68 (1926) 14, for the gravimetric determination of tellurium and its separation from other elements.

References pp. 356–360

instead their color is discharged and alkali thiosulfate and selenosulfate result.

In the systematic scheme of qualitative analysis, tellurium is precipitated as free tellurium, along with the basic and acid sulfides, when the hydrochloric acid solution is treated with hydrogen sulfide. If the precipitate is digested with yellow ammonium sulfide, the tellurium, together with arsenic, antimony and tin sulfides, goes into solution as a sulfosalt. Since solutions of alkali sulfarsenate, sulfantimonate, and sulfostannate are not affected by the addition of alkali sulfite, tellurium can thus be detected in the presence also of arsenic, antimony, and tin.

Procedure.[218] A drop of the alkaline test solution is stirred with a drop of yellow ammonium polysulfide in a depression of a spot plate. A pinch (tip of knife blade) of solid sodium sulfite is then stirred in. The spot plate is placed on a heating block at 120° C. The yellow (or in the presence of selenium, red-brown) solution loses its color, and if tellurium is present, a black precipitate or gray coloration appears in 1–2 minutes. This procedure will reveal 2.5 γ tellurium. Greater sensitivity can be attained by evaporating to dryness and stirring the residue with a drop or two of water. Any red selenium, that may have come down, redissolves while the free tellurium remains unchanged.

Limit of Identification: 0.5 γ tellurium

This identification limit for tellurium is also reached in an ammonium polysulfide solution that is saturated with selenium (about 50 γ selenium per drop).

Limit of Dilution: 1 : 100,000.

35. Hydrogen peroxide

(*1*) *Test with lead sulfide* [219]

Lead sulfide (black) reacts with hydrogen peroxide to yield lead sulfate (white):

$$PbS + 4\,H_2O_2 = PbSO_4 + 4\,H_2O$$

When finely divided lead sulfide is used for the reaction, the decolorization is almost instantaneous. This is particularly true if paper impregnated with lead sulfide is spotted with hydrogen peroxide or solutions of alkali peroxides. Lead sulfide papers of varying strengths can be used, depending on the quantity of hydrogen peroxide expected. Only when much of the latter is present and when strongly impregnated lead sulfide paper is used, may the test solution contain small quantities of mineral acid. Consequently, the absence of free acid should be established because lead sulfide, when highly dispersed in paper, is much more reactive than in the compact form. The latter is completely resistant to dilute non-oxidizing acids.

References pp. 356–360

Procedure. A drop of the test solution, which should be almost neutral or not more than slightly acid, is placed on filter paper impregnated with lead sulfide. A white fleck or circle is formed on the light brown paper, depending upon the hydrogen peroxide content.

Limit of Identification: 0.04 γ hydrogen peroxide

Limit of Dilution: 1 : 1,250,000

Reagent: Filter paper (S & S 601) is soaked in a 0.5 % solution of lead acetate, H_2S blown on to the surface, and the paper then dried in a vacuum desiccator. The paper will keep in stoppered bottles.

The sensitivity of the test can be improved (*Idn. Limit:* 0.01 γ H_2O_2) by using S & S 602 filter paper and taking particular care in impregnating it with lead sulfide.[220]

(2) Test with ferric ferricyanide [221]

The brown acid solution of ferric chloride and potassium ferricyanide (ferric ferricyanide) reacts with reducing agents ($SnCl_2$, Na_2SO_3, $Na_2S_2O_3$, etc.) to form Prussian blue or a mixture of Prussian blue and Turnbull's blue. Hydrogen peroxide likewise reduces ferricyanide:

$$2\,[Fe(CN)_6]^{-3} + H_2O_2 = 2\,[Fe(CN)_6]^{-4} + 2\,H^+ + O_2$$

The resulting ferrocyanide reacts with the ferric chloride in the solution to yield Prussian blue.*

Procedure.[222] A drop of the reagent solution and a drop of water are placed on a spot plate; in an adjacent depression a drop of the reagent solution and a drop of the test liquid are used for comparison. A more or less intense blue color or precipitate is formed, depending on the hydrogen peroxide content.

Limit of Identification: 0.08 γ hydrogen peroxide

Limit of Dilution: 1 : 600,000

Reagent: A mixture of equal parts of 0.4 % ferric chloride and 0.8 % potassium ferricyanide

(3) Test by reduction of higher nickel oxides [222]

Hydrogen peroxide is catalytically decomposed by nickel hydroxide. Nickel^{II} oxide or hydroxide is regenerated by the action of hydrogen peroxide on higher nickel oxide: [223]

$$Ni_2O_3 + H_2O_2 = 2\,NiO + H_2O + O_2$$

The disappearance of the black color of the higher nickel oxides on treatment with hydrogen peroxide affords a means for the detection of the latter.

* M. Kohn, *Chem. Abstr.*, 37 (1943) 845 found that insoluble, yellow-green copper ferricyanide is also reduced by small amounts of H_2O_2 (in the presence of sodium acetate) forming brown Cu ferrocyanide (*Idn. Limit:* 1 γ H_2O_2).

References pp. 356–360

It is best to use a paste made from nickel^{III} hydroxide and an inert colorless substance, such as barium sulfate. It is not feasible to use filter paper impregnated with the oxyhydrate of ter- or quadrivalent nickel because such papers are not stable (compare detection of persulfate, page 323).

Procedure. Several small dabs of Ni_2O_3 paste are placed in two adjacent depressions of a spot plate. Water is added to one; a drop of the neutral or slightly acid test solution to the other. According to the amount of peroxide present, the paste turns paler or becomes colorless.

Limit of Identification: 0.01 γ hydrogen peroxide

Limit of Dilution: 1 : 5,000,000

Reagents: Nickelic hydroxide-paste: Baryta water is treated with bromine water, and the resulting barium hypobromite is mixed with nickel sulfate solution and warmed. The proportions of Ba and Ni should be so adjusted that a *gray* precipitate is formed. It is filtered and well washed. The wet precipitate is stored in a weighing bottle with a well-fitting stopper. The paste can be kept for a long time.

(4) *Test by catalyzed oxidation of phenolphthalin* [224]

If a red alkaline solution of phenolphthalein is warmed with metallic zinc, reduction to colorless phenolphthalin occurs.[225] The reaction is discussed on page 292. The colorless solution of phenolphthalin gradually turns red on standing in contact with the air. The reoxidation to phenolphthalein is more rapid if strong oxidizing agents are added. Hydrogen peroxide by itself acts quite slowly, but its action is accelerated when copper salts are present.[226] This effect is the basis of a test for peroxides, which is specific provided other weak oxidizing agents are absent.

Procedure. One drop each of the test solution, 0.01 *N* copper sulfate and phenolphthalin solution (for preparation, see page 293) are brought together on a spot plate. A pink to red coloration indicates the presence of peroxides.

Limit of Identification: 0.04 γ hydrogen peroxide in 0.04 ml

Limit of Dilution: 1 : 1,000,000

(5) *Other tests for hydrogen peroxide*

(*a*) H_2O_2 may be detected by the reduction of a neutral 0.01 % solution of $AuCl_3$, to free gold (*Idn. Limit:* 0.07 γ H_2O_2). [227]

(*b*) A yellow-red precipitate is formed with acid solutions of alkali thiocyanates (*Idn. Limit:* 0.7 γ H_2O_2). [227]

(*c*) Freshly prepared colorless solutions of potassium cerous carbonate [a solution of $Ce_2(SO_4)_3$ mixed with excess K_2CO_3] is converted by H_2O_2 into yellow or brown-yellow potassium perceric carbonate (*Idn. Limit:* 0.1 γ H_2O_2). [228]

References pp. 356–360

(*d*) When a mixture of equal parts of 1 % alkaline *o*-tolidine solution, sodium acetate buffer (pH = 4), and 1 % $FeSO_4$ is treated with a drop (0.025 ml) of a solution containing H_2O_2, a blue color appears.[229]

(*e*) An intense chemiluminescence is produced by oxidizing "luminol" (3-aminophthalhydrazide) with H_2O_2 in the presence of hemin.[230] Several drops of reagent solution (0.1 g luminol and 2 mg hemin in 100 ml 20 % sodium carbonate) are mixed with one drop of the test solution on a spot plate (*Idn. Limit:* 0.012 γ H_2O_2).

(*f*) Chromates are reduced by H_2O_2. Consequently, the color produced by the sensitive CrO_4^{-2}-diphenylcarbazide reaction (see page 167) is discharged by H_2O_2 and peroxides.[231] If very dilute K_2CrO_4 solutions are used, and a comparison test run, it is possible to detect 0.5 γ H_2O_2.

REFERENCES

1. H. Weisz, *Mikrochim. Acta* (1956) 1225.
2. F. Feigl, Studies with H. J. Kapulitzas and M. Ishidate.
3. For preparation, see K. H. Slotta and K. R. Jacobi, *Z. anal. Chem.*, 77 (1929) 344.
4. W. Krings, *Z. angew. Chem.*, 50 (1937) 414.
5. F. Halla and F. Ritter, *Mikrochim. Acta*, 1 (1937) 365.
6. N. A. Tananaeff and A. M. Schapowalenko, *Z. anal. Chem.*, 100 (1935) 344.
7. D. Ganassini, *Chem. Zentr.*, 1904, I, 1172.
8. R. G. Aickin, *Chem. Abstracts*, 31 (1937) 7358.
9. Communication from O. Heim, East Hampstead, England.
10. H. Weisz, *Mikrochim. Acta* (1956) 1225.
11. I. Guareschi, *Z. anal. Chem.*, 52 (1913) 451, 538, 607; 53 (1914) 490.
12. For more details about the mechanism of this color reaction reaction see H. Wieland and G. Scheuing, *Ber.*, 54 (1921) 2527.
13. F.Schulek, *Z. anal. Chem.*, 102(1935)111; R.Uzel, *Chem. Abstracts*, 29(1935)6523; 30(1936)1683.
14. F. L. Reith, *Pharm. Weekblad*, 66 (1925) 1097.
15. F. Feigl, *Chemistry of Specific, Selective and Sensitive Reactions*, New York, 1949, p. 300.
16. N. A. Tananaeff and A. M. Schapowalenko, *Z. anal. Chem.*, 100 (1935) 345.
17. F. Kirchhof, *Chem. Abstracts*, 33 (1939) 942.
18. Compare Feigl, Ref. 13, p. 561.
19. Fr. Goppelsroeder, (abstract) *Z. anal. Chem.*, 2 (1863) 398; J. Löwe, *J. prakt. Chem.*, [1], 74 (1858) 348.
20. C. L. Wilson, *Chemistry & Industry*, 59 (1940) 378.
21. Preparation according to M. Mutnianski, *Z. anal. Chem.*, 36 (1897) 220. Comp. also F. de Leo, R. Indovina and A. Bellino, *Ann. Chim. (Rome)*, 44 (1954) 859.
22. F. Feigl, Studies with E. Rajmann and D. Goldstein.
23. I. M. Kolthoff and E. B. Sandell, *J. Am. Chem. Soc.*, 56 (1934) 1426.
24. Compare H. H. Willard and Ph. Young, *J. Am. Chem. Soc.*, 50 (1928) 1372.
25. A. J. Velculescu and J. Cornea, *Z. anal. Chem.*, 94 (1933) 255.
26. G. Gutzeit, *Helv. Chim. Acta*, 12 (1929) 713.
27. H. Weisz, *Mikrochim. Acta* (1956) 1225.
28. J. H. de Boer, *Chem. Weekblad*, 21 (1924) 404.
29. F. Pavelka, *Mikrochemie*, 6 (1928) 149; compare also J. P. Alimarin, *Z. anal. Chem.*, 81 (1930) 8.
30. J. Stone, *J. Chem. Education*, 8 (1931) 347.
31. W. Biltz, *Ausführung qualitativer Analysen*, Leipzig, 1930, p. 44.
32. F. Feigl, *Mikrochemie*, 7 (1929) 13.

33. Compare F. Feigl and A. Schaeffer, *Anal. Chem.*, 23 (1951) 351.
34. Communication from W. Böttger, Leipzig.
35. F. Feigl and E. Rajmann, *Mikrochemie*, 12 (1932) 133.
36. For preparation, see F. Feigl, P. Krumholz, and E. Rajmann, *Mikrochemie*, 9 (1931) 395.
37. F. Feigl and P. Krumholz, *Mikrochemie* (*Pregl Festschrift*), 1929, p. 83.
38. B. Fetkenheuer, *Chem. Abstracts*, 17 (1923) 1398.
39. Compare H. A. Williams, *Analyst*, 75 (1950) 510.
40. Compare G. Canneri and A. Cozzi, *Anal. Chim. Acta*, 2 (1948), 321, also L. M. Dubnikov and J. F. Tikhomirov, *Chem. Abstr.*, 40 (1946) 7063.
41. S. K. Hagen, *Mikrochemie*, 15 (1934) 313.
42. F. Feigl and G. B. Heisig, *Anal. Chim. Acta*, 3 (1949) 563.
43. J. Moir, *Chem. News*, 102 (1910) 17; C. Pertusi and E. Gastaldi, *Chem.-Ztg.*, 37 (1913) 609; A. Sieverts and A. Hermsdorf, *Z. angew. Chem.*, 34 (1921) 3.
44. O. L. Barnebey, *J. Am. Chem. Soc.*, 36 (1914) 1092.
45. F. Feigl and H. E. Feigl, *Anal. Chim. Acta*, 3 (1949) 300.
46. F. Feigl and A. Krutman, unpublished studies.
47. G. Gutzeit, *Helv. Chim. Acta*, 12 (1929) 713.
48. F. Feigl, unpublished studies.
49. L. Rossi and collaborators, *Chem. Abstracts*, 33 (1939) 84.
50. F. Feigl and A. Caldas, *Mikrochim. Acta* (1955) 992.
51. K. A. Hofmann and H. Wagner, *Ber.*, 41 (1908) 317, 1628.
52. Compare K. A. Hofmann and H. Kirmreutter, *Ber.*, 41 (1908) 314, and also F. Feigl and J. V. Tamchyna, *ibid.*, 62 (1929) 1897.
53. I. M. Kolthoff, *Z. anal. Chem.*, 57 (1918) 14.
54. Studies with H. J. Kapulitzas.
55. Experiments by G. Hirsch; see also F. Feigl, *Mikrochemie*, 7 (1929) 12.
56. H. Weisz, *Mikrochim. Acta* (1956) 1225.
57. W. P. Malitzky and M. T. Koslowsky, *Mikrochemie*, 7 (1929) 98.
58. I. M. Kolthoff, *Mikrochemie*, 8 (1930) 176.
59. F. Feigl and H. E. Feigl, *Mikrochim. Acta* (1954) 85.
60. D. Davidson, *J. Chem. Education*, 17 (1940) 84.
61. F. Feigl, *Rec. trav. chim.*, 58 (1939) 476.
62. F. Feigl and D. Goldstein, unpublished studies; compare also H. Lees, *Biochem. J. Proc.*, 47 (1950) 44.
63. F. Sommer and H. Pincus, *Ber.*, 48 (1915) 1963.
64. F. Feigl, *Rec. trav. chim.*, 58 (1939) 477.
65. F. Feigl and A. Caldas, *Anal. Chim. Acta*, 13 (1955) 526.
66. F. Feigl, studies with H. Molterer.
67. Compare D. Davidson, *J. Chem. Education*, 14 (1937) 238; see also E. Müller, *J. prakt. Chem.*, [2] 90 (1914) 117, 123.
68. For the detection and distinction between Prussian blue and Turnbull's blue see M. Kohn, *Monatsh.*, 66 (1935) 394. For the formulae of the iron cyanide compounds see H. Reihlen, *Ann.*, 451 (1927) 75; 469 (1929) 30 and 475 (1929) 101.
69. F. Feigl, *Rec. trav. chim.*, 58 (1939) 478.
70. A. G. Kniga, *Chem. Abstracts*, 31 (1937) 7000; see also O. Baudisch, *Biochem. Z.*, 232 (1938) 34.
71. L. N. Lapin, *Chem. Abstracts*, 35 (1941) 4305.
72. Compare W. Prodinger, *Mikrochem. ver. Mikrochim. Acta*, 36-37 (1951) 580.
73. For preparation, see J. Biehringer, *J. prakt. Chem.*, 54 (1896) 240.
74. A. M. Schapowalenko, *Chem. Abstracts*, 24 (1930) 3458.
75. W. C. Oelke, *Ind. Eng. Chem., Anal. Ed.*, 12 (1940) 498.
76. E. Storfer, *Mikrochemie*, 17 (1935) 170.
77. L. Kul'berg, *Z. anal. Chem.*, 106 (1936) 30.
78. F. Feigl, D. Goldstein and R. A. Rosell, unpublished studies.
79. F. Feigl, *Z. anal. Chem.*, 152 (1956) 52.
80. F. Feigl, *Rec. trav. chim.*, 58 (1939) 479.
81. For preparation see K. H. Slotta and K. R. Jacobi, *Z. anal. Chem.*, 77 (1929) 346.

82. D. Vitali, *Bull. chim. pharm.*, 38 (1899) 201; *Chem. Zentr.*, 1899, I, 1083.
83. F. Feigl, experiments with S. Pickholz.
84. Compare A. Brukl and M. Behr, *Z. anal. Chem.*, 64 (1924) 23.
85. N. A. Tananaeff and A. M. Schapowalenko, *Z. anal. Chem.*, 100 (1935) 352.
86. A. L. Gottlieb, *Chem. Abstracts*, 32 (1938) 4469.
87. F. Feigl and D. Goldstein, *J. Microchem.*, in press.
88. F. A. Gooch and D. H. Kreider, *Am. J. Science*, 48 (1894) 38.
89. F. Feigl and H. E. Ballaban, unpublished studies.
90. For preparation, see J. Biehringer, *J. prakt. Chem.*, 54 (1896) 240.
91. F. Feigl and R. Uzel, *Mikrochemie*, 19 (1936) 133.
92. M. Vrtiš, *Rec. trav. chim.*, 44 (1925) 429; L. Malaprade, *Compt. rend.*, 204 (1937) 979.
93. H. Kràl, *Z. anal. Chem.*, 36 (1897) 696.
94. Compare K. A. Hofmann, *Ann.*, 312 (1900) 1; J. F. Virgili, *Z. anal. Chem.*, 45 (1901) 409; G. Scagliarini, *Chem. Abstracts*, 29 (1935) 2872.
95. F. Feigl, *Z. anal. Chem.*, 74 (1928) 369. Compare L. Metz, *ibid.*, 76 (1929) 347; also *Jahresbericht der Chemisch-Technischen Reichsanstalt*, Vol. V, Verlag Chemie, Leipzig, 1926, p. 143.
96. F. Raschig, *Ber.*, 48 (1915) 2088.
97. F. Feigl, studies with R. Schacherl.
98. F. Feigl, *J. Chem. Education*, 20 (1943) 177.
99. F. Feigl, unpublished studies.
100. E. Fischer, *Ber.*, 16 (1883) 2234.
101. N. A. Tananaeff and A. M. Schapowalenko, *Z. anal. Chem.*, 100 (1935). 346.
102. C. Bödecker, *Ann.*, 117 (1861) 193.
103. E. Eegriwe, *Z. anal. Chem.*, 65 (1924-25) 128.
104. F. Feigl and E. Fränkel, *Ber.*, 65 (1932) 545.
105. C. Wicke, *Z. Chem.*, 8 (1865) 89, 305; F. Haber and F. Bran, *Z. physik. Chem.*, 35 (1900) 84; W. Böttger and E. Thomä, *J. prakt. Chem.*, 147 (1936) 11.
106. F. Feigl, *Chem.-Ztg.*, 44 (1920) 689.
107. P. Reckendorfer, *Z. Pflanzenkrankh. Pflanzenschutz*, 45 (1935) 537.
108. P. Senise, *Mikrochim. Acta* (1957) 640.
109. E. Votoček, *Ber.*, 40 (1907) 414.
110. H. Wieland and G. Scheuing, *Ber.*, 54 (1921) 2527.
111. L. Rosenthaler, *Mikrochemie*, 8 (1927) 27.
112. F. Feigl, studies with H. J. Kapulitzas.
113. A. Villiers, *Bull. soc. chim.* [3], 47 (1887) 546.
114. H. Freytag, *Ber.*, 67 (1934) 1477.
115. G. B. Heisig and A. Lerner, *Ind. Eng. Chem., Anal. Ed.*, 13 (1941) 843.
116. G. Gutzeit, *Helv. Chim. Acta*, 12 (1929) 736.
117. F. Feigl, *Rec. trav. chim.*, 58 (1939) 479.
118. F. Feigl and W. Aufrecht, *Rec. trav. chim.*, 58 (1939) 1127.
119. H. Yagoda, *J. Ind. Hyg. Toxicol.*, 26 (1944) 224.
120. H. E. Wohlers, *Z. anorg. allgem. Chem.*, 59 (1908) 203.
121. F. Feigl, experiments carried out with L. Weidenfeld and L. Badian.
122. F. Feigl, studies with N. Braile.
123. Compare C. Duval, *Anal. Chim. Acta*, 2 (1948) 92.
124. F. Feigl, *J. Chem. Education*, 22 (1945) 344.
125. F. Feigl and Cl. Costa Neto, unpublished studies.
126. S. Kahl, *Ber.*, 31 (1898) 148.
127. For preparation, see E. Clemmsen and A. H. C. Heitman, *J. Am. Chem. Soc.*, 33 (1911) 737.
128. F. Feigl, *Z. anal. Chem.*, 74 (1928) 369.
129. F. Feigl, studies with R. Nováček.
130. Compare R. M. Fanstone, *Brit. J. Phot.*, 76 (1929) 714.
131. Compare A. Sander, *Chem. Ztg.*, 39 (1915) 945.
132. F. Feigl, studies with H. J. Kapulitzas.
133. F. Feigl and L. Hainberger, *Mikrochim. Acta* (1955) 105.
134. R. Kuhn and F. Weygand, *Ber.*, 69 (1936) 1969.

135. E. Commanducci, *Bull. chim. pharm.*, 57 (1918) 101.
136. F. Feigl and Cl. Costa Neto, *Chemist Analyst*, 44 (1955) 91.
137. G. Panizzon, *Melliand Textilber.*, 12 (1931) 119.
138. A. Monnier, *Chem. Abstracts*, 11 (1917) 426.
139. J. Schmidt and W. Hinderer, *Ber.*, 65 (1932) 87.
140. N. M. Cullinane and S. J. Chard, *Analyst*, 73 (1948) 95.
141. W. Lenz and E. Richter, *Z. anal. Chem.*, 50 (1911) 537.
142. F. Feigl and D. Goldstein, unpublished studies.
143. E. S. Tomula and coworkers, *Z. anal. Chem.*, 135 (1952) 265.
144. Compare F. Feigl, K. Klanfer and L. Weidenfeld, *Z. anal. Chem.*, 80 (1925) 5.
145. F. Feigl and C. Stark, unpublished studies.
146. P. Baumgarten and I. Marggraff, *Ber.*, 63 (1930) 1019.
147. Comp. W. E. Gordon and M. E. Cupery, *Ind. Eng. Chem.*, 31 (1939) 1237.
148. Desbassins de Richemont, *Chem. Zentr.* (1835) 782.
149. The basis of the "ring test" for nitrates is discussed by E. Schröder, *Z. anorg. allgem. Chem.*, 202 (1931) 382.
150. Compare W. Manchot, *Ber.*, 47 (1914) 1601.
151. Compare J. Tillmans and W. Sutthoff, *Z. anal. Chem.*, 50 (1911) 473; I. M. Kolthoff and G. E. Noponen, *J. Am. Chem. Soc.*, 55 (1933) 1448.
152. H. Wieland, *Ber.*, 46 (1913) 3296; 52 (1919) 886.
153. G. W. Monier-Williams, *Analyst*, 56 (1931) 397.
154. H. Riehm, *Z. anal. Chem.*, 81 (1930) 439.
155. Compare F. L. Hahn and G. Jäger, *Ber.*, 58 (1925) 2340; G. J. Barannikov, *Chem. Abstr.*, 33 (1939) 6196.
156. H. A. Suter and P. H. Suter, *Mikrochim. Acta* (1956) 1136.
157. For preparation, see M. Marqueyrol and H. Muraour, *Analyst*, 56 (1931) 397.
158. R. Kersting, *Ann.* 125 (1863) 254.
159. C. R. Fresenius, *Qualitative Analysis*, 17th edition (translated by C. A. Mitchell), New York, 1921, page 488.
160. F. Sommer and H. Pincus, *Ber.*, 48 (1915) 1963; J. Thiele, *ibid.*, 41 (1908) 2681.
161. P. Baumgarten and I. Marggraff, *Ber.*, 63 (1930) 1019.
162. For preparation, see P. Baumgarten, *Ber.*, 59 (1926) 1978.
163. F. L. Hahn and P. Baumgarten, *Ber.*, 65 (1930) 3028.
164. J. Blom, *Ber.*, 59 (1926) 121.
165. Ph. W. West, private communication.
166. P. Griess, *Ber.*, 12 (1879) 427. Compare F. L. Hahn and G. Jaeger, *ibid.*, 58 (1925) 2335 and J. Blom, *ibid.*, 59 (1926) 121.
167. F. Feigl, unpublished studies.
168. A. de Aguiar, *Ber.*, 7 (1874) 315.
169. For preparation, see R. Meyer and W. Müller, *Ber.*, 30 (1897) 775.
170. C. L. Wilson, *Chemistry & Industry*, 59 (1940) 378.
171. J. V. Dubský, J. Trtílek, and A. Okáč, *Mikrochemie*, 15 (1934) 99.
172. N. A. Tananaeff and A. M. Schapowalenko, *Z. anal. Chem.*, 100 (1935) 350.
173. C. J. Engelder, T. H. Dunkelberger, and W. Schiller, *Semi-micro Qualitative Analysis*, 2nd ed., New York, 1940, p. 183.
174. P. Carboni, *Chim. Ind. Ital.*, 36 (1954) 825.
175. F. Feigl, *Z. anal. Chem.*, 61 (1922) 454; 74 (1928) 386; 77 (1929) 299.
176. Further instances are given in Feigl, Ref. 13, p. 115.
177. F. Feigl and I. D. Raacke, *Anal. Chim. Acta*, 1 (1947) 317.
178. F. Feigl and A. Schaeffer, *Anal. Chem.*, 23 (1951) 352.
179. F. Feigl, *Z. anal. Chem.*, 77 (1929) 299.
180. J. W. Robinson and Ph. W. West, private communication.
181. F. Feigl and P. Krumholz, *Mikrochemie* (*Pregl Festschrift*), 1929, 82.
182. F. Feigl and P. Krumholz, *Mikrochemie*, 8 (1930) 131.
183. F. Feigl, unpublished studies.
184. F. W. Daube, *Ber.*, 3 (1870) 609.

185. M. E. Schlumberger, *Bull. soc. chim.*, [2], 5 (1866) 194; L. Clarke and C. L. Jackson, *Am. Chem. J.*, 39 (1908) 696.
186. E. M. Chamot and H. J. Cole, *Ind. Eng. Chem.* 10 (1918) 38.
187. F. Feigl and P. Krumholz, *Mikrochemie* (*Pregl Festschrift*), 1929, p. 77.
188. Compare J. Houben, *Das Anthrazen und die Anthrachinone*, Leipzig, 1928, p. 343.
189. Compare F. Feigl, Ref. 15, p. 355.
190. F. Feigl and A. Schäffer, *Anal. Chem.*, 23 (1951) 351.
191. L. Szebellédy and St. Tanay, *Z. anal. Chem.*, 107 (1936) 26.
192. A. S. Komarowsky and N. S. Poluektoff, *Mikrochemie*, 14 (1933-34) 317.
193. For preparation, see A. S. Komarowsky and N. S. Poluektoff, *loc. cit.*
194. F. Feigl and L. Badian, unpublished studies.
195. F. L. Hahn, *Compt. rend.*, 197 (1933) 762. Compare also: A. S. Dodd, *Analyst*, 54 (1919) 282; T. Burkhalter and D. Peabock, *Anal. Chem.*, 28 (1956) 1186.
196. Compare F. Feigl, Ref. 15, p. 356.
197. F. P. Sorkin, *Chem. Zentr.*, 1937, II, 632.
198. W. Stahl, *Z. anal. Chem.*, 101 (1935) 342.
199. F. Feigl and H. A. Suter, *Chemist-Analyst*, 32 (1943) 4.
200. F. Feigl, *Mikrochemie* (*Emich Festschrift*), 1930, p. 127.
201. Compare F. Raschig, *Schwefel- und Stickstoffstudien*, Leipzig, 1925, p. 201. Compare also H. H. Willard and Ph. Young, *J. Ind. Eng. Chem.*, 4 (1932) 187.
202. For preparation, see K. H. Slotta and K. R. Jacobi, *Z. anal. Chem.*, 77 (1929) 346.
203. N. S. Poluektoff, *Mikrochemie*, 15 (1934) 32.
204. R. Berg and M. Teitelbaum, *Mikrochemie* (*Emich Festschrift*), 1930, p. 23.
205. F. Feigl and V. Demant, *Mikrochim. Acta*, 1 (1937) 322.
206. H. Wieland, *Ber.*, 43 (1910) 3260. Compare also J. J. Postowsky, *ibid.*, 69 (1936) 1913.
207. F. Feigl and Ph. W. West, *Anal. Chem.*, 19 (1947) 351.
208. P. Falciola, *Chem. Abstracts*, 21 (1927) 3580. Compare N. S. Poluektoff, *Mikrochemie*, 15 (1934) 32.
209. M. H. Evans, *Chem. Abstracts*, 32 (1938) 8299.
210. V. E. Levine, *Chem. Abstracts*, 31 (1937) 8617; V. Gentil, unpublished studies.
211. F. Sachs, *Ann.*, 365 (1909) 150; O. Hinsberg, *Ber.*, 52 (1919) 21.
212. J. Hoste, *Anal. Chim. Acta*, 2 (1948) 406.
212a. E. Sawicki, *Anal. Chem.*, 29 (1957) 1376.
213. F. Feigl and L. Badian, unpublished studies.
214. N. S. Poluektoff, *Mikrochemie*, 15 (1934) 32.
215. F. Feigl and R. Uzel, *Mikrochemie*, 19 (1936) 133.
216. B. Brauner and B. Kužma, *Ber.*, 40 (1907) 3362.
217. F. Feigl and R. Seboth, unpublished studies.
218. F. Feigl, unpublished studies.
219. R. Kempf, *Z. anal. Chem.*, 89 (1933) 88.
220. H. Freytag, *Z. anal. Chem.*, 131 (1950) 77.
221. C. F. Schönbein, *J. prakt. Chem.*, [1], 79 (1859) 67. Compare also F. Feigl and E. Fränkel, *Mikrochemie*, 12 (1932-33) 309.
222. F. Feigl and E. Fränkel, *Mikrochemie*, 12 (1932–33) 304.
223. F. Feigl, K. Klanfer, and L. Weidenfeld, *Z. anal. Chem.*, 80 (1929) 5.
224. O. Schales, *Ber.*, 71 (1938) 447.
225. A. v. Baeyer, *Ann.*, 202 (1885) 80.
226. P. Thomas and G. Charpentier, *Compt. rend.*, 173 (1921) 1082.
227. F. Feigl and E. Fränkel, *Mikrochemie*, 12 (1932–33) 304.
228. E. Plank, *Z. anal. Chem.*, 99 (1934) 105.
229. L. Kul'berg and L. Matwejew, *Chem. Abstracts*, 30 (1936) 7492.
230. W. Langenbeck and U. Ruge, *Ber.*, 70 (1937) 367.
231. L. N. Lapin, *Chem. Abstracts*, 35 (1941) 1727.

Chapter 5

Tests for Free Elements

The great majority of all natural and artificial products consist of chemical compounds or mixtures of compounds and consequently the qualitative analysis of a sample seldom requires a direct search for uncombined metallic or nonmetallic elements. A sample can usually be sufficiently characterized by the statement that certain elements, or their ions, or certain stoichiometrically insoluble compounds are present. This information generally suffices to establish what sort of quantitative determinations are eventually conducted on the sample. Accordingly, no chemical analysis should be instituted until the sample has been treated with water or acids, and the resulting solution as well as any insoluble residue tested for the materials contained in them.

In certain instances, particularly when commercial materials are being examined, information regarding the presence or absence of free elements can be of considerable value, especially if tests can be conducted rapidly and with small amounts of the sample. The following sections contain instances illustrating the use of spot tests for such exploratory aims.

A. FREE METALS AND ALLOYS

1. General

Metallic appearance, luster, color, hardness, and sometimes magnetic properties, are characteristic of metals and alloys. However, certain metalloids or semimetals (e.g., As, Sb, Se, Te) look like metals, and some compounds of metals (FeS, Co_3O_4, Fe_3O_4, for instance) are magnetic. [1]

Neither the solubility of metals in dilute acids, nor the deposition of a metal from the solution of a salt by introducing a less noble metal, constitutes a decisive general test for free metals. For example, metals which precede hydrogen in the electromotive series, in other words, those that are less noble, are expected to be soluble in nonoxidizing acids, while the metals more noble than hydrogen ought not to dissolve. However, the facts are not always in accord with these expectations. Metals less noble than hydrogen are sometimes quite resistant to acids or go into solution very slowly.

References p. 382

Examples are Al, Fe, Zn (of high purity) and Pb. On the other hand, more noble metals (Cu, Co, Hg, etc.) are markedly attacked by dilute, nonoxidizing acids. The causes of these anomalies include: invisible layers of oxides, passivity, deposition of salts on the surface of the metal, exceptionally high purity of the metal, and sequestration of the metal ions produced, if they form complexes with the anions of the dissolving acid. The same factors also operate in cases of apparent departures from the usual rule of the oxidation of less noble metals by ions of more noble metals (cementation).

Two methods will be given for detecting free metals and alloys; the tests are based on redox reactions. They are decisive, provided they are applied in the absence of sulfides and insoluble compounds of metals at lower valences (cuprous halides, Hg_2Cl_2, etc.), which also act as reductants. Digestion of the sample with water will definitely and easily insure the absence of soluble reducing compounds.

(1) Test with phosphomolybdic acid[2]

A water solution of phosphomolybdic acid, $H_3PO_4 \cdot 12\,MoO_3 \cdot aq$, may dissolve metals and alloys first of all because of the H^+ ions in the solution. Furthermore, phosphomolybdic acid also acts as a strong oxidizing acid. The reason is that the Mo^{VI} or MoO_3 molecules of the complex acid are more readily reduced than normal molybdic acid.[3] Consequently, when metals or alloys are attacked by the yellow aqueous solution of phosphomolybdic acid, hydrosols of "molybdenum blue" are always formed. It is a mixture containing varying proportions * of the oxides of hexavalent and quinquevalent molybdenum. The solution of bivalent metals by phosphomolybdic acid, with production of Mo_2O_5 as the essential constituent of the "molybdenum blue," can be expressed:

$$Me^0 + 2\,H^+ \rightarrow Me^{+2} + 2\,H^0$$
$$2\,MoO_3 + 2\,H^0 \rightarrow H_2O + Mo_2O_5$$
$$Me^0 + 2\,MoO_3 \rightarrow MeO + Mo_2O_5$$
$$MeO + 2\,H^+ \rightarrow Me^{+2} + H_2O$$

The color quality of molybdenum blue is extraordinarily intense. Consequently, even a slight dissolution of metal is made clearly evident and the presence of free metal is established.

Procedure. A small quantity of the sample is added to 1 or 2 drops of a saturated water solution of phosphomolybdic acid in a depression of a spot plate.

* O. Glemser and G. Lutz, *Naturwissenschaften*, 34 (1947) 215, have shown that the chief constituent of molybdenum blue is Mo_4O_{11} ($= Mo_2O_5 \cdot 2\,MoO_3$).

References p. 382

If necessary, a drop of dilute sulfuric acid is then added. A blue or green color appears if free metal or an alloy is present. The reagent solution keeps if stored in a tightly stoppered brown bottle.

(2) Test with ferric ferricyanide

If a ferric solution is mixed with a solution of ferricyanide, a brown color appears. The resulting solution contains water-soluble ferric ferricyanide, $FeFe(CN)_6$. This compound reacts not only with soluble reducing agents,[4] but also with those that do not dissolve in water. A blue color or precipitate appears (Prussian blue plus Turnbull's blue). Ferric ferricyanide yields Fe^{+3} and $Fe(CN)_6^{-3}$ ions, and when it reacts with metals or alloys two different reactions may occur. Ferric ions may be reduced:

$$2\ Fe^{+3} + Me^{o} \rightarrow Me^{+2} + 2\ Fe^{+2} \quad (1)$$

or ferricyanide ions may be reduced:

$$2\ Fe(CN)_6^{-3} + Me^{o} \rightarrow Me^{+2} + 2\ Fe(CN)_6^{-4} \quad (2)$$

Both Fe^{+2} ions and $Fe(CN)_6^{-4}$ are produced and, in their turn, they react with the ions of ferric ferricyanide to produce ferrous ferricyanide (Turnbull's blue) and ferric ferrocyanide (Prussian blue). Accordingly, either type of reduction of ferric ferricyanide leads to a blue reaction product. Consequently, this test for uncombined metals is very sensitive. The test gives positive results even with metallic silver, and in a lesser degree, but in a still more visible form, even with metallic platinum.

Procedure. The sample is treated, in a depression of a spot plate, with one drop of the reagent solution. On stirring with a glass rod, a blue to green coloration appears if metals or alloys are present. In some cases (silver, for instance) the metal acquires a blue coating.

Reagent: Freshly prepared mixture of equal volumes of 0.4 % ferric chloride and 0.8 % potassium ferricyanide solutions.

2. Aluminum, Lead, Zinc, Tin

Detection through reduction of o-*dinitrobenzene*[5]

These metals are attacked by alkali hydroxide even at room temperature with evolution of hydrogen. For instance:

$$Pb + 2\ NaOH \rightarrow Pb(ONa)_2 + 2\ H^{o}$$

If a colorless alcoholic solution of *o*-dinitrobenzene is added to the caustic alkali solution, the nascent hydrogen reduces the nitro compound to the violet alkali salt of a quinoidal nitrol-nitro acid:[6]

References p. 382

$$C_6H_4(NO_2)_2 + 4\ H^o + 2\ NaOH \rightarrow C_6H_4(NONa)(NO_2Na) + 3\ H_2O$$

Accordingly, the test involves a redox reaction analogous to that in the tests for water-soluble hydrazine, hydroxylamine, sodium hydrosulfite, Rongalite (see Chapters 3 and 4), which function as hydrogen donors in alkaline solution. Accordingly, the absence of these reducing agents should be established beforehand and if present they must be removed by digestion of the sample with water.

Metallic magnesium, which is always coated with a thin layer of oxide if it has been in contact with air, does not react, whereas aluminum, zinc, lead, and tin react, even though they are covered with an oxide layer, because these respective oxides are soluble in strong alkali. Likewise, magnesium, which has been freed of its coating of oxide by immersion in acid, gives a positive response to the reagent by virtue of the reaction:

$$Mg^\circ + 2\ H_2O \rightarrow Mg(OH)_2 + 2\ H^\circ$$

No studies have been reported regarding the behavior of beryllium, but it is most likely that it too will react positively through the formation of alkali beryllate.

Procedure. The test is made in a micro test tube. A tiny portion of the sample is treated with a drop of a 1 % alcoholic solution of *o*-dinitrobenzene and a drop of 0.5 *N* caustic alkali. The test tube is then suspended in boiling water. A violet color indicates the presence of metals which are attacked by caustic alkali.

The test can be also carried out on metal foils or blocks by placing a drop of a mixture of equal volumes of the caustic alkali and reagent solution on the specimen.

3. Mercury (Mercury Vapor)

The detection of metallic mercury, particularly as vapor, has considerable practical significance because sometimes it is important to establish quickly the presence of amalgams and mercury compounds. Without exception, they are decomposed by heat with evolution of mercury vapor. The following tests are applicable for the detection of mercury vapor.

(1) Test by reduction of palladium chloride [7]

When mercury vapor comes into contact with filter paper impregnated with palladium chloride, the light brown paper turns deep black (deposition of finely divided palladium) if the period of exposure is long enough:

$$PdCl_2 + Hg^\circ \rightarrow HgCl_2 + Pd^\circ$$

References p. 382

It must be noted that mercury vapor reacts with greater sensitivity when the palladium chloride paper is dry. This is an advantage in the test described here.

Procedure. Several milligrams of the sample are placed in a small porcelain crucible, which is then covered with a small disk of the reagent paper. The heating should be gentle at first, the tip of the micro flame being directed against the bottom of the crucible. This starts the combustion. Later, the temperature is raised until the bottom of the crucible is red. If mercury is present, the exposed area of the paper will be gray or black. The stain can be made more visible by holding the paper briefly over concentrated ammonia water. The light brown paper loses its color [formation of $Pd(NH_3)_4Cl_2$] and the gray fleck of metallic palladium is then easily seen against the white background.

Reagent: Filter paper moistened with 1 % solution of palladium chloride and then dried

The test for mercury vapor through reduction of palladium chloride is so sensitive that it will reveal even the minute traces which enter the air when mercury stands at room temperature. This can be demonstrated by merely allowing a drop of mercury to stand in a micro crucible covered with a disk of the reagent paper. At 25° to 30° C, the deposition of palladium is quite distinct within ten minutes; on longer standing the light gray stain gradually blackens.

The presence of intermetallic mercury compounds (amalgams) can also be established by this test. Several milligrams of the sample are kept at about 100° C (drying oven or hot plate) for 30 to 60 minutes. The micro-crucible is covered with $PdCl_2$ paper. This treatment is sufficient to volatilize enough mercury to affect the paper, even though amalgams have a lower mercury tension than the free metal.

(2) *Test with selenium sulfide* [8]

Selenium sulfide, SeS_2, is likewise a sensitive reagent for mercury vapor. Black mercuric selenide and mercuric sulfide are formed rapidly even by contact with traces of mercury. Suitable reagent paper is prepared by bathing filter paper in a water solution of selenious acid, then exposing it to hydrogen sulfide, washing, and drying. It is still better to prepare particularly active SeS_2 by the following special procedure, and then rub the dry material on filter paper.

Finely divided free selenium reacts with mercury vapor to produce black mercury selenide.[9] Suitable reagent paper is prepared by soaking filter paper in a solution of potassium selenocyanate (KCNSe), draining, and exposing to an atmosphere of hydrogen chloride (hood!). The potassium selenocyanate

References p. 382

solution is readily prepared by vigorously shaking potassium cyanide solution with an excess of pure selenium powder in a stoppered flask:

$$KCN + Se \rightarrow KCNSe$$

The decomposition by hydrogen chloride:

$$HCl + KCNSe \rightarrow KCl + HCN + Se$$

requires only 15–30 seconds. Longer exposure markedly decreases the reactivity of the selenium deposited on the paper.

(3) Test with cuprous iodide [10]

If mercury vapor comes into contact with colorless cuprous iodide, red cupro tetraiodo mercuriate results:

$$2\ Cu_2I_2 + Hg^\circ \rightarrow Cu_2\,[HgI_4] + 2\ Cu^\circ$$

The reaction is analogous to that involved in the test for soluble mercury salts (page 65). Accordingly, the reaction is given also by contact of vapors of mercury salts with the reagent, which is not true of the tests *1* and *2*.

The procedure given in test *1* can be employed but the $PdCl_2$ paper is replaced by paper coated with cuprous iodide paste. (The test may also be conducted in the absorption apparatus shown in Fig. 23, page 47).

Reagent: Cuprous iodide paste: A solution of 5 g copper sulfate crystals in 75 ml water is mixed with a solution containing 5 g sodium sulfite and 11 g potassium iodide in 75 ml water. The resulting colorless precipitate is filtered or centrifuged, well washed with water, and stored moist. Just prior to the test, a little of the cuprous iodide is made into a thin slurry with water.

4. Arsenic

Detection through formation of arsenic hydride [11]

Sodium formate melts at 253° C and if heated above 300° C it decomposes:

$$2\ HCOONa \rightarrow (COONa)_2 + 2\ H^\circ \qquad (1)$$

In mixture with sodium hydroxide, heating above 205° C is sufficient to bring about the reaction: [12]

$$HCOONa + NaOH \rightarrow Na_2CO_3 + 2\ H^\circ \qquad (2)$$

If elementary arsenic is present during the thermal decomposition shown in (1) or (2), gaseous arsenic hydride is produced by the action of the nascent hydrogen. This product can be detected in the gas phase by means of

References p. 382

suitable reagents. Filter paper moistened with silver nitrate is recommended; a black stain of elementary silver is obtained (compare page 101).

No antimony hydride is formed if elementary antimony is subjected to the procedure just outlined. Accordingly, free arsenic can be detected in the presence of free antimony.

It should be noted that the nascent hydrogen produced in (1) or (2) will yield hydrogen sulfide with free sulfur and also with sulfides and alkaline earth sulfates, and consequently will blacken silver nitrate paper. (Lead sulfate is predominantly converted into lead sulfide.) In this procedure, the sample should be heated first of all with sodium formate and the vapors allowed to come into contact with lead acetate paper. If the latter is not darkened, a positive response with silver nitrate paper indicates the presence of elementary arsenic, provided no arsenate or arsenite is present since they likewise yield arsenic hydride by reaction with fused sodium formate.

Procedure. A small portion of the sample is heated in a micro test tube with several cg of the sodium formate-hydroxide mixture. The mouth of the test tube is covered with filter paper moistened with a 5 % solution of silver nitrate. If arsenic is present, the reagent paper will develop a grey or black stain.

Reagent: The evaporation residue of a solution containing 20 % sodium formate and 12 % sodium hydroxide.

5. Molybdenum, Tungsten, Vanadium

Detection through conversion into alkali salts of the metallo acids

These metals (as well as their oxides) produce the water-soluble alkali salts of the respective metallo acids when fused or sintered with alkali nitrite. (This was first observed with free tungsten.[13]) The heating may be conducted in a porcelain crucible. After the reaction mass has cooled, it is treated with water, and the clear filtrate (centrifugate) can be tested for tungsten, molybdenum, or vanadium by the methods outlined in Chapter 3.

If it is merely required to learn whether acid-forming metals which react with alkali nitrite are present, the activation of hydrogen peroxide by MoO_4^{-2}, WO_4^{-2}, or VO_3^- ions can be tried. The chemistry of this test is given on page 417. The procedure may not be used if metallic chromium is present because of the formation of yellow alkali chromate.

Procedure. A spot plate is used. A drop of the solution obtained after the nitrite fusion is treated with a drop of 3 % hydrogen peroxide and a drop of a 2 % solution of thallous nitrate in 2 *N* ammonia. If metallo acids are present, a brown precipitate or color appears.

B. FREE NON-METALS

6. Free halogens

(1) Test with fluorescein [14]

It is common practice to test for free halogens (Cl_2, Br_2, I_2) by means of starch-iodide paper which turns blue when exposed to a free halogen. Iodine forms the blue starch-iodine complex directly; chlorine and bromine react with the iodide in the paper, liberating iodine which then combines with starch. However, this test is not specific for free halogen, because many other oxidants (H_2O_2, per-salts, HNO_2, $KMnO_4$, etc.) likewise liberate iodine from the iodide of the reagent paper, and thus produce a blue starch-iodine complex.

Fluorescein provides a specific reaction for bromine and iodine; chlorine can also be detected if the other halogens are absent. The yellow dye reacts with free bromine to produce red eosin (tetrabromofluorescein), whereas iodine yields red-yellow erythrosine (tetraiodofluorescein) (see page 262). Chlorine is without direct action on fluorescein. If, however, a mixture of potassium bromide and fluorescein is used, the bromine set free by the chlorine immediately forms eosin. Thus fluorescein also provides a test (indirect) for chlorine. All these tests can be carried out as spot reactions on fluorescein and fluorescein-potassium bromide paper.

Procedure. A drop of the neutral test solution is placed on fluorescein paper. If a red spot appears, free bromine or iodine is present. It is not possible to differentiate between these halogens by this means.

If the response is negative, a drop of the test solution is placed on fluorescein-potassium bromide paper. If chlorine is present, a red spot results.

Limits of Identification: 1 γ chlorine; 2 γ bromine; 6 γ iodine

Reagents: *1)* Fluorescein paper: Strips of filter paper (S & S 598g) are bathed in a saturated water solution of fluorescein and dried

2) Fluorescein-potassium bromide paper: Strips of filter paper (S & S 598g) are bathed in a very weakly alkaline solution containing 0.1 g dye and 0.5–0.8 g KBr and dried

(2) Test with 4,4′*-bis-dimethylamino-thiobenzophenone* [15]

When dilute solutions of chlorine, bromine or iodine are shaken with the yellow benzene solution of 4,4′-bis-dimethylamino-thiobenzophenone (thio-Michler's ketone) a deep blue results. The same color reaction occurs when dissolved or gaseous halogens come into contact with filter paper impregnated with the thioketone.

The color reaction of the reagent (I) with free halogens is due to an oxidation of the zwitter ion (II), which is isomeric with (I), to yield a water-soluble compound. The latter, which is a disulfide, yields the *p*-quinoidal cation (III).

$$(CH_3)_2N\text{—}C_6H_4\text{—}\underset{\underset{S}{\|}}{C}\text{—}C_6H_4\text{—}N(CH_3)_2 \rightleftharpoons (CH_3)_2N\text{—}C_6H_4\text{—}\underset{\underset{S^-}{|}}{C}\text{=}C_6H_4\text{=}\overset{+}{N}(CH_3)_2$$

(I) (II)

$$2\ (CH_3)_2N\text{—}C_6H_4\text{—}\underset{\underset{S^-}{|}}{C}\text{=}C_6H_4\text{=}\overset{+}{N}(CH_3)_2 + 2\ Hal \longrightarrow$$

$$\begin{array}{l} (CH_3)_2N\text{—}C_6H_4\text{—}C\text{=}C_6H_4\text{=}\overset{+}{N}(CH_3)_2 \\ \qquad\qquad\qquad\ \ |\\ \qquad\qquad\qquad\ \ S \\ \qquad\qquad\qquad\ \ | \\ \qquad\qquad\qquad\ \ S \\ \qquad\qquad\qquad\ \ | \\ (CH_3)_2N\text{—}C_6H_4\text{—}C\text{=}C_6H_4\text{=}\overset{+}{N}(CH_3)_2 \end{array} + 2\ Hal^-$$

(III)

When this reagent is used for the detection of free halogen, it should be noted that analogous reactions are given by oxidizing cations [16], nitrous acid, and nitrogen tetroxide, but not by hydrogen peroxide or persulfates. Consequently, it is best to make the test for the halogens in the gas phase.

Reagent: Quantitative filter paper is soaked in a 0.1 % benzene solution of 4,4′-bis-dimethylaminothiobenzophenone and dried in the air. When the paper is stored away from light, it keeps very well.

If the reagent paper is spotted with aqueous solutions of the free halogens, it is possible to detect: 0.2 γ chlorine; 0.2 γ bromine; 0.5 γ iodine.

(3) *Detection of chlorine and bromine through formation of polymethine dyes* [17]

A sensitive test for pyridine is based on the fact that it reacts with bromcyanogen and primary aromatic amines to give colored Schiff bases of glutaconic aldehyde. These products are known as polymethine dyes (compare Volume II, Chapter 4). The net equation of this formation of the dyes is:

$$C_5H_5N\ (\text{pyridine}) + CNBr + 2\ NH_2R \longrightarrow \begin{array}{c} H \\ C \\ HC\qquad CH \\ |\qquad\ \ \| \\ RNCH\ \ HCNHR \end{array} + NH_2CN + HBr$$

Chlorcyanogen behaves in the same manner as bromcyanogen. Since these halogen cyanides are formed immediately when free halogens come into reaction with potassium cyanide, for instance:

$$Br_2 + KCN \rightarrow KBr + CNBr$$

the action of halogens on a mixture of pyridine, potassium cyanide, and a primary aromatic amine, such as benzidine, will produce the corresponding polymethine dye. This makes it possible to detect bromine and chlorine with adequate sensitivity. The procedure can be applied to aqueous solutions of these halogens and also to vapor phases.

Procedure. The reaction is conducted on a spot plate. A drop of the test solution is treated with a drop of 25 % potassium cyanide solution and a drop of a mixture of 25 % pyridine and 2 % benzidine hydrochloride solution. If elementary chlorine or bromine is present, a red precipitate or color results.

Limit of Identification: 2.5 γ chlorine or bromine

This test is very well suited for the detection of chlorinated water; about 1 ml is taken for the test. Halogens may also be driven out of the water by acidifying with sulfuric acid and boiling. The vapor should be brought into contact with filter paper moistened with the mixture of reagents prescribed above.

7. Dicyanogen

(*1*) *Detection through hydrolysis to hydrogen cyanide*

The detection of dicyanogen is of analytical interest because, with the exception of the heat-stable alkali and alkaline earth cyanides, all normal and complex heavy metal cyanides, as well as certain thiocyanates, split off dicyanogen when heated (compare page 371). This gaseous product exhibits the characteristics of a halogen, including the hydrolysis reaction with water:

$$(CN)_2 + H_2O \rightarrow HCN + HCNO$$

Accordingly, if filter paper moistened with a solution of copper- and benzidine acetate comes into contact with dicyanogen, the blue color characteristic of hydrogen cyanide appears (compare page 276). This test for dicyanogen is obviously reliable for dicyanogen only in the absence of hydrogen cyanide. Consequently, dry mixtures which contain cyanides and acidic compounds may not be tested in this manner for dicyanogen. The solution of this problem requires the prior removal of the cyanide by warming with dilute acid, or else the procedure given in test *2* must be applied.

(2) *Test with alkaline cyanide solution containing 8-hydroxyquinoline (oxine)* [18]

A raspberry color appears if dicyanogen is produced in a concentrated solution of potassium cyanide containing oxine.[19] This effect, which is the basis of a spot test for copper (compare page 93) makes possible a convenient test for gaseous dicyanogen split off when inorganic and organic compounds undergo thermal decomposition. The water-soluble red compound has not been isolated and its composition has not been established. However, certain facts are known regarding the manner of its formation. The color reaction occurs only when much alkali cyanide is present, a finding that leads to the assumption that—in analogy to polyhalides—polycyanide ions are produced from cyanide ions and dicyanogen:

$$CN^- + (CN)_2 \rightarrow [CN \ldots (CN)_2]^-$$

The saponification of the dicyanogen is prevented or at least diminished through the production of the polycyanide ions. It is not known with any certainty how the dicyanogen included in the polycyanide ions reacts with the oxine. Two explanations are given in Vol. II, Chapter 5.

The color reaction of dicyanogen with potassium cyanide and oxine is not as sensitive as the test described in test *1* but it has the advantage of being specific. The *limit of identification* is 1 γ dicyanogen. This value was obtained by heating a mixture of mercuric cyanide and mercuric chloride and assuming that the yield of dicyanogen is 50 %:

$$Hg(CN)_2 + HgCl_2 \rightarrow Hg_2Cl_2 + (CN)_2$$

as is true in the preparation of pure dicyanogen.[20]

Procedure. The test is conducted in a micro test tube. A granule of the dry sample is heated over a micro flame while the mouth of the test tube is covered with filter paper impregnated with oxine and moistened with 25 % potassium cyanide solution. If dicyanogen is released, a more or less intense red circular stain appears on the paper.

Reagent: Oxine paper: Filter paper is bathed in a 10 % ether solution of oxine and dried in the air. The impregnated paper keeps.

If normal cyanides are absent, it is well to take the dry sample to dryness in the micro test tube along with a drop of 0.5 % alcoholic solution of mercuric chloride and then to follow the procedure just given. Heating with mercuric chloride leads to quantitative release of the cyanide groups as dicyanogen. The treatment with mercuric chloride prior to the thermal decomposition is not permissible if cyanide groups are present since the latter form $Hg(CN)_2$, which then yields dicyanogen on dry heating.

The detection of dicyanogen by its action on concentrated solutions of

References p. 382

potassium cyanide containing oxine is reliable only in the absence of compounds which split off water when heated. Water vapor readily hydrolyzes dicyanogen to hydrogen cyanide and cyanic acid (compare test *1*).

8. Free sulfur

(*1*) *Test by formation of thallium polysulfide* [21]

When black thallous sulfide, finely dispersed in the capillaries of filter paper, comes into contact with a solution of sulfur, a red-brown polysulfide ($Tl_2S_3 \cdot 2Tl_2S$) is quickly formed. When considerable amounts (over 50 γ in drops) of sulfur are brought into reaction, the red-brown compound is clearly visible on the black paper, after the solvent has evaporated. The polysulfide differs from Tl_2S not only in color but also in its chemical behavior. At room temperature it is not attacked by dilute mineral acids and hydrogen peroxide; in contrast, thallous sulfide dissolves almost at once, because it forms a soluble thallous salt by direct action of the acid, or because of oxidation to soluble Tl_2SO_4. Accordingly, if black thallous sulfide paper is spotted with a drop of a solution of sulfur in colorless ammonium sulfide or an organic solvent, and then bathed in dilute acid or hydrogen peroxide, the paper will not be completely decolorized. A red-brown fleck will remain where the drop was applied. The stain will be brown if only small quantities of sulfur are present. Selenium behaves analogously (see page 375).

The union of free sulfur with thallous sulfide can be used as a rapid method of detecting sulfur. However, the test succeeds only on filter paper impregnated with Tl_2S. The reason is that two effects are utilized in the spot reaction; they are related to the capillary and surface action of the paper. One of these effects is the rapidity with which the Tl_2S, dispersed in the capillaries of the paper, reacts with sulfur or selenium. This enhancement of reaction rate is a consequence of the extensive free reaction surface. The second effect is the envelopment of the highly dispersed thallous sulfide by the polysulfide (or selenide) formed on its surface. This coating protects the underlying, still unaltered, Tl_2S against the solvent action of acids or hydrogen peroxide. The consequence of this protective layer effect is that the color of the polysulfide (or selenide) formed at the site of the spot is reinforced by the color of the protected Tl_2S. With very dilute solutions of sulfur or selenium, it is probable that the acid-stable fleck consists of practically nothing but unchanged thallous sulfide. The latter is shielded from the solvent action of acids or peroxide by a quantity of polysulfide that, of itself, is too slight to be visible.

The detection of sulfur, or selenium, by a spot reaction necessitates a preliminary extraction of the sample with an organic solvent. Since only a drop of the extract is needed for the test, minute amounts of the sample suffice (microextraction apparatus). As a rule, carbon disulfide is used; it dissolves all crystalline modifications of sulfur, but not the amorphous varieties.* Pyridine is a more reliable extractant; its solvent powers are lower, but it dissolves all varieties of sulfur.[22] Pyridine dissolves about 4 % of sulfur at room temperature, and about 25 % at the boiling point (114° C).

Procedure. Freshly prepared thallous sulfide paper is spotted with a drop of the solution to be tested for sulfur. The solvent (carbon disulfide, pyridine, etc.) is allowed to evaporate at room temperature or is removed by a blast of heated air. The spotted paper is then placed in 0.5 *N* nitric acid and swirled. If no sulfur (or selenium) is present, the whole paper turns white in about 30 seconds; a brown fleck is left if these elements are present. The intensity of the color gives an index of the quantity of sulfur. It is well to place the paper in water after the fleck has developed; this prevents any further action of the acid and a gradual fading of the spot.

Parallel tests with solutions of known sulfur content can be used to determine whether the unknown contains a trace or considerable amounts of the free element. The flecks can be preserved if the developed paper is thoroughly washed with water and then dried.

Limit of Identification: 3 γ sulfur
Limit of Dilution: 1 : 17,000

Thallous sulfide paper is prepared by bathing filter paper (S & S 589 or Whatman 42) in 0.5 % solution of thallous carbonate or acetate for several minutes. The liquid is allowed to drain off and the paper is dried in a blast of heated air. Thallous sulfide is deposited by placing the paper across a beaker containing ammonium sulfide solution warmed to 80° C. The conversion to thallous sulfide is rapid; the paper turns perfectly black on the side exposed to the fumes. Cut into strips, it is ready for immediate use. It is best always to use freshly prepared reagent paper when testing for sulfur. The paper deteriorates on standing because the sulfide is partially oxidized.

(2) *Test by conversion into thiocyanate* [23]

Free sulfur is converted into sulfide, polysulfide, and thiosulfate by boiling with alkali hydroxide. Evaporation of polysulfide and thiosulfate with a solution of potassium cyanide produces thiocyanate, which is easily detected by its reaction with ferric chloride. The underlying reactions are:

* R. Edge, *Ind. Eng. Chem., Anal. Ed.*, 4 (1930) 371, states that amorphous sulfur can be quantitatively converted into the crystalline form by keeping the sample for several hours at 105° C.

References p. 382

$$4\ S + 6\ NaOH = 2\ Na_2S + Na_2S_2O_3 + 3\ H_2O$$
$$S + Na_2S = Na_2S_2$$
$$Na_2S_2 + KCN = KCNS + Na_2S$$
$$Na_2S_2O_3 + KCN = KCNS + Na_2SO_3$$
$$Fe^{+3} + 3\ CNS^- = \text{red soluble complex}$$

Free sulfur, as well as sulfur loosely combined in sulfides, horn substance, egg yolk, etc. may be detected by this conversion into thiocyanate.

Procedure. A few milligrams of the test substance are placed in a porcelain microcrucible together with a few drops of dilute sodium hydroxide and cautiously evaporated to dryness over a very small flame. The evaporation is repeated after the addition of a few drops of a 0.1 % potassium cyanide solution. The residue is taken up with dilute sulfuric acid and ferric chloride is added. In the presence of sulfur, a distinct red color appears.

(3) Test by conversion into thiosulfate or mercury sulfide [24]

When crystalline free sulfur is to be detected in a mixture with other solids, it can be extracted with carbon disulfide (compare page 373). After evaporation of the solvent, the sulfur in the residue can be converted into sodium thiosulfate by warming with a solution of sodium sulfite. The thiosulfate is revealed by using the sensitive test based on the catalytic acceleration of the reaction: $2\ NaN_3 + I_2 \rightarrow 2\ NaI + 3\ N_2$.

The carbon disulfide used must be *absolutely* sulfur-free. Since even "pure" or "for analysis" carbon disulfide often contains sulfur, it is necessary to purify the solvent before the test by shaking with mercury and then distilling.

Slight amounts of sulfur dissolved in carbon disulfide can be detected by shaking with mercury; the resulting mercuric sulfide collects on the surface of the metal.[25]

Procedure. About 6–8 ml of the carbon disulfide is placed in a hard glass test tube or measuring cylinder. A drop of mercury is added and shaken vigorously. If even slight amounts of free sulfur are present, the surface of the metal is stained black or shows an iridescent film of mercuric sulfide, which can be further confirmed by the iodine-azide test (see page 303). The carbon disulfide is decanted and the contaminated mercury is placed on a watch glass and heated in a current of steam (see page 46) to remove the final traces of carbon disulfide. The mercury is then covered with the iodine-azide reagent and a foam of nitrogen bubbles appears if free sulfur was present in the carbon disulfide. The use of the mercury dropper (page 39) is recommended.

The iodine-azide test will reveal traces of mercuric sulfide which are too small to give a visible alteration of the drop of mercury. The complete removal of the carbon disulfide is essential because it too like other mercapto- and thiocompounds catalyzes the iodine-azide reaction.

References p. 382

(4) Detection through conversion into hydrogen sulfide

Benzoin (I) is a reducing agent; when heated with sulfur it produces hydrogen sulfide, and is transformed into benzil (II). This result has been employed for the detection of elementary sulfur by heating glycerol, containing iron, with sulfur and benzoin; black iron sulfide results.[26] It is better to use the following procedure in which the solid sample is allowed to react with molten benzoin (m.p. 137° C). Moderate heating of the melt brings about the reaction:

$$\underset{(I)}{C_6H_5CHOHCOC_6H_5} + S^\circ \rightarrow H_2S + \underset{(II)}{C_6H_5COCOC_6H_5}$$

The resulting hydrogen sulfide is conveniently and sensitively detected by lead acetate paper.

This test is specific for elementary sulfur. Polysulfides and thiosulfates do not react. Likewise, elementary selenium is not changed by the action of molten benzoin.

Procedure.[27] A micro test tube is used. A slight amount of the sample is mixed with several cg of benzoin and the mouth of the test tube is closed with moist lead acetate paper. The test tube is plunged into a glycerol bath previously heated to 130° C. The temperature is then raised to 150° C. According to the amount of free sulfur, the paper acquires rapidly a black or brown stain.

Limit of Identification: 0.5 γ sulfur.

This procedure will reveal elementary sulfur in the presence of sulfides or elementary selenium. A distinct test was obtained from a mixture containing 10 γ sulfur plus 2500 γ selenium.

9. Free selenium

(1) Test through addition to thallous sulfide[28]

Selenium, like sulfur, when dissolved in organic liquids or colorless ammonium sulfide, reacts with thallous sulfide paper (see page 372). An addition compound is formed. The first stage probably is the formation of thallous selenide:

$$Tl_2S + Se^\circ \rightarrow Tl_2Se + S^\circ$$

The resulting free sulfur then unites with the selenide to produce a polysulfide-like complex which is resistant to dilute acids and hydrogen peroxide. Since it is black, it cannot be seen on the black thallous sulfide paper, in contrast to the analogous red-brown compound formed by considerable quantities of sulfur. However, the unaltered thallous sulfide can be dissolved away by bathing the paper in dilute acid or hydrogen peroxide solution, and the spot at which the drop of selenium solution was applied then shows up as a dark brown to black fleck. The latter consists of the selenide-sulfur complex, or of thallous sulfide, "protected" from the action of the solvents by a coating of the complex.

This method of detecting free selenium is decisive only when free sulfur is absent. Selenium exists in several allotropic modifications. The usual red β-form, obtained by reducing a solution of selenious acid, is slightly soluble in carbon disulfide. The more stable dark α-form, obtained by heating the red modification, is almost insoluble in this solvent.

Procedure. Use the method given on page 373.
Limit of Identification: 1 γ selenium

(2) Test by formation of silver selenide [29]

Although selenium is only slightly soluble in carbon disulfide, the dilute solution reacts almost instantly with metallic silver to produce black silver selenide, Ag_2Se. Carbon disulfide solutions of sulfur also react with metallic silver, but the sulfide is formed very slowly, even though much dissolved sulfur may be available because of the greater solubility of sulfur in this solvent. Consequently, it is possible to detect selenium in the presence of considerable quantities of sulfur, particularly if a comparative test is made on a carbon disulfide solution containing an approximately equal amount of sulfur. It is essential to use carbon disulfide as the solvent if selenium is to be detected in the presence of sulfur as by the procedure given here. It has been found [30] that solutions of sulfur in pyridine, benzene, and toluene react quickly with silver foil, despite the fact that sulfur is much less soluble in these liquids than in carbon disulfide. This finding provides a good example to demonstrate the influence of the solvent on the activity of the solute.[31]

Procedure. Silver foil is roughened with fine emery paper and then thoroughly cleaned. A drop of the test solution is placed on the foil and the solvent (CS_2) allowed to evaporate. A gray or black gray fleck of silver selenide appears.
Limit of Identification: 5 γ selenium

(3) Test by conversion into selenosulfate

Elementary sulfur and elementary selenium are readily soluble in solutions of alkali sulfite. The products are alkali thiosulfate and selenosulfate, respectively:

$$S^0 + SO_3^{-2} \rightarrow S_2O_3^{-2}; \quad Se^0 + SO_3^{-2} \rightarrow SSeO_3^{-2} \qquad (1)$$

In some reactions, thiosulfates and notably selenosulfates behave as though they are loose addition products of sulfite ion and sulfur or selenium. In

References p. 382

contrast to thiosulfate, which remains totally unaltered, selenosulfate reacts with formaldehyde as though it consists of a mixture of selenium and sulfite, i.e., aldehyde-bisulfite is formed and free selenium is precipitated.[32]

$$Na_2SSeO_3 + CH_2O + H_2O \rightarrow CH_2(OH)SO_3Na + NaOH + Se^o \qquad (2)$$

This reaction (2) can serve not only to distinguish between thiosulfate and selenosulfate, but it can be applied to detect free selenium in mixture with free sulfur after the realization of (1).

Procedure.[33] According to the suspected content of free selenium, several milligrams or a gram of the solid sample is warmed with 1–2 ml of 10 % sodium sulfite solution. The suspension is filtered or centrifuged and a drop of the clear liquid is transferred to a micro crucible. One to three drops of 40 % formaldehyde are added and the mixture is warmed. A red precipitate or pink color appears, according to the quantity of selenium.

Limit of Identification: 2.5 γ selenium

(4) Test through catalytic acceleration of the reduction of methylene blue [34]

The reduction of methylene blue to its colorless leuco compound by sulfide ions, which proceeds very slowly in the cold, is catalytically hastened by elementary selenium (compare detection of selenium and selenic acid, page 348.) This catalysis involves $[S...Se^o]^{-2}$ ions, which are produced when free selenium dissolves in alkali sulfide solutions. However, selenosulfide ions are produced also when alkali selenite reacts with alkali sulfide. In this case the redox reaction:

$$SeO_3^{-2} + 2\ S^{-2} + 3\ H_2O \rightarrow Se^o + 2\ S^o + 6\ OH^-$$

is followed by solution of the free sulfur and selenium in excess alkali sulfide solution with production of polysulfide and selenosulfide. Therefore, when the free selenium is to be extracted from a sample through conversion into soluble alkali selenosulfide by digestion with alkali sulfide, the process can be applied only if selenium oxygen compounds are absent. The selenium in such mixtures must be extracted with carbon disulfide, or the oxygenated selenium compounds must be removed beforehand by digesting the sample with dilute mineral acid. Because of the slight solubility of selenium in carbon disulfide, the treatment with mineral acid is preferable.

As a rule, the analytical problem is to detect free selenium in the presence of much free sulfur. In such cases, digestion with alkali sulfide yields considerable (yellow) polysulfide along with selenosulfide. Since polysulfide ions are much more reactive toward methylene blue than sulfide ions (though much less so than selenosulfide ions) it is advisable to convert the alkali

References p. 382

polysulfide to monosulfide. This conversion is accomplished by adding sodium sulfite:

$$Na_2[S...S_x^0] + x\ Na_2SO_3 \rightarrow x\ Na_2S_2O_3 + Na_2S \qquad (1)$$

The sodium sulfite likewise converts selenosulfide into selenosulfate ions:

$$[S...Se^0]^{-2} + SO_3^{-2} \rightleftarrows S^{-2} + SeSO_3^{-2} \qquad (2)$$

In contrast to (1), equation (2) represents an equilibrium reaction which, even in the presence of a large excess of sulfite, does not lead to the complete disappearance of the $[S ... Se^0]^{-2}$ ions that are catalytically effective with respect to the reduction of methylene blue. Consequently, the catalytic effect of a selenosulfide solution is not lessened to any significant extent by the addition of sulfite. As little as 0.08 γ selenium can be detected in one drop of a selenosulfide solution through the hastening of the reduction of methylene blue.

Further information was yielded by the following findings: A solution of 0.2 *M* sodium sulfide was saturated with sulfur by heating it with excess sulfur and filtering. Approximately 12 grams of sulfur per liter went into solution. Two milliliters of this polysulfide solution were treated with 0.05 ml of standard selenium solution in 0.2 *M* sodium sulfide solution (corresponding to 0.5 microgram of selenium). The mixture was decolorized by adding solid sodium sulfite and heating. The solution was cooled, and 2 drops of 0.05 % methylene blue solution introduced. Comparison was made with a blank on 2 ml of 0.2 *M* sulfide containing the same amount of sodium sulfite. The blank retained a distinct blue color for at least 20 minutes, whereas the solution containing selenium became colorless within 1 minute.

This experiment shows that selenium can be detected easily at a ratio of 1 part Se to 48,000 parts of sulfur. Calculated to percentages on a dry weight basis, this corresponds to 0.002 % selenium in sulfur.

10. Free tellurium

Detection through formation of sulfotellurate [35]

Colorless alkali or ammonium sulfide solutions dissolve elementary sulfur and elementary selenium to form polysulfide and selenosulfide, respectively:

$$S^{-2} + x\ S \rightarrow [SS_x]^{-2} \quad \text{or} \quad S^{-2} + x\ Se \rightarrow [SSe_x]^{-2}$$

However, colorless alkali monosulfide is without action on elementary tellurium, in contrast to alkali polysulfide, which dissolves elementary tellurium with production of alkali sulfotellurate:

$$Te^{\circ} + (NH_4)_2S_4 \rightarrow (NH_4)_2TeS_4$$

References p. 382

When tellurium is in mixture with free sulfur, even colorless ammonium sulfide will also suffice to dissolve these elements because polysulfide is formed through the dissolution of the sulfur. The yellow or red-brown solution of alkali polysulfides and polyselenosulfides is decolorized by adding sodium sulfite with production of thiosulfate or selenosulfate, respectively:

$$(NH_4)_2S{\cdot}S_x + x\ Na_2SO_3 \rightarrow (NH_4)_2S + x\ Na_2S_2O_3$$

$$(NH_4)_2S{\cdot}Se_x + x\ Na_2SO_3 \rightarrow (NH_4)_2S + x\ Na_2SSeO_3$$

In contrast, the yellow-red solution of $(NH_4)_2TeS_4$ is not decolorized, but on warming with sulfite there is precipitation of grey-black elementary tellurium: [32]

$$(NH_4)_2TeS_4 + 3\ Na_2SO_3 \rightarrow Te^\circ + (NH_4)_2S + 3\ Na_2S_2O_3$$

Therefore, if samples that are not soluble in water or acid are to be tested for tellurium in mixture with elementary sulfur or selenium, the sample should be warmed first with ammonium polysulfide and then filtered or centrifuged. Solid sodium sulfite is then added to the clear liquid and warmed. If tellurium is present, a black precipitate or a grey to black turbidity will appear.

It should be noted that because of their conversion to $(NH_4)_2TeS_4$, tellurates behave in a manner analogous to that shown by elementary tellurium (compare page 352).

11. Free carbon

(*1*) *Test by oxidation*

Free carbon need be looked for only in solid, dark, insoluble samples. Its presence is indicated if the sample glows, gives off sparks, or burns when heated with access to air. A more sensitive test is to drop a portion into molten alkali chlorate: oxidation to carbon dioxide with decrepitation. Both tests (see page 422 for procedure) are also applicable to organic, (i.e., carboniferous) compounds in general. It is possible to isolate uncombined carbon by treating the sample in turn with acid and alkali, which dissolve basic and acidic inorganic and organic constituents. The residue is washed, dried, and then extracted with ether and carbon disulfide. Graphitic and carbide carbon are too resistant to be revealed by the oxidation procedure.

Limit of Identification: approximately 50 γ carbon

If other modifications of carbon are absent, and if no organic compounds are present, it is possible to detect graphitic carbon (e.g., in iron or steel) and carbide carbon (in silicon carbide, etc.). The method is based on the formation of potassium cyanide, which then is used to produce hydrocyanic acid.[36] The latter can be detected by the procedure given on page 276. Cyanide results if several milligrams of the sample are mixed with 20 times as much potassium azide and cautiously heated in a small hard glass test tube (goggles). The heating should be moderate at first; eventually the mixture is strongly ignited for 2 minutes. The potassium azide decomposes ($2\ KN_3 \rightarrow 2\ K + 3\ N_2$); probably some potassium nitride is also produced and unites directly with the carbon of the sample to form potassium cyanide. The test is more sensitive if potassium (freed of adhering petroleum) is added to the mixture before heating. The metal probably facilitates the formation of nitride and consequently the production of cyanide.

All natural carbons (hard coal, brown coal), as well as all artificial carbons, contain mineral constituents. These are responsible for the basic ash which remains after the carbon has burned away. (See page 507 regarding the identification of basic ashes.)

(2) *Test by heating with molybdenum trioxide* [37]

At temperatures above 250° C, molybdenum^{VI} oxide (molybdic acid) is extremely reactive toward solid, liquid, or gaseous oxidizable substances, including some with which it does not react at room temperature. Partial reduction of MoO_3 to Mo_2O_5 ensues, i.e., to the intensely colored "molybdenum blue" (compare page 362). Organic compounds of almost every variety and state of aggregation can be brought to react with heated molybdenum^{VI} oxide. The same is true of free carbon (charcoal, brown coal, hard coal, but not graphite) and also of carbon in the form that remains after the incomplete combustion of organic and metallo-organic compounds. The reaction, which does not involve atmospheric oxygen, can be represented:

$$4\ MoO_3 + C \rightarrow CO_2 + 2\ Mo_2O_5$$

Solid inorganic oxidizable compounds behave similarly to carbon and carbon compounds when heated with Mo^{VI} oxide. Typical instances are anhydrous alkali sulfites, arsenites, ferrous salts, etc., which thus present impressive instances of solid-solid reactions in the absence of water. The reducing power of hot MoO_3 is so great that it reacts on contact with gaseous ammonia:

$$2\ NH_3 + 6\ MoO_3 \rightarrow N_2 + 3\ Mo_2O_5 + 3\ H_2O$$

Consequently, much molybdenum blue is produced on heating a mixture of ammonium salts and MoO_3 or even ammonium molybdate (see page 399).

When the molybdic acid test for carbon is used, the absence of ammonium

References p. 382

salts and reducing inorganic compounds must be assured. Soluble reducing compounds and likewise ammonium salts can be removed by digestion and treatment with dilute acid. Insoluble inorganic reductants (particularly sulfides, mercurous chloride, acid-insoluble metals) can be dissolved completely by heating with 4 *N* hydrochloric acid to which 6 % hydrogen peroxide is added dropwise during the heating.[38]

The sample, thus freed of reducing inorganic materials, should be separated from the liquid by centrifugation and not by filtration. Otherwise, the dried preparation may contain paper fibres, which on ignition would yield reducing substances, and the latter in turn would react with molybdic acid. The dry residue from one drop of a sugar solution, whose content was equivalent to 1 γ carbon, gave a distinct blue when mixed with molybdic acid and heated.

(*3*) *Test by heating with potassium iodate* [37]

If potassium iodate is heated, there is no release of oxygen: $KIO_3 \rightarrow \rightarrow 3\ KI + 3\ O$ until the fused mass has reached a temperature above 560° C. Accordingly, potassium iodate remains unchanged after several hours at 300–400°. If, however, the salt is mixed with organic matter and then heated, a redox reaction producing potassium iodide and carbon dioxide occurs within a short time (1–5 minutes), and it begins even at 250–300° C. The presence of the resulting potassium iodide in the heated mixture is readily shown by extracting the cooled mass with a little water and adding dilute sulfuric acid to the extract. Iodine is set free:

$$IO_3^- + 5\ I^- + 6\ H^+ = 3\ H_2O + 3\ I_2$$

and yields blue starch-iodine on the addition of freshly prepared starch solution.

With the exception of graphite, all modifications of carbon can be detected by this topochemical reaction with hot potassium iodate. It is assumed of course, that the sample is finely pulverized before being intimately mixed with the potassium iodate. To achieve intimate contact, the solid sample can be moistened in a micro crucible with a concentrated solution of potassium iodate and taken to dryness in an oven before the heating to 300° C. Milligram quantities of the test material are sufficient.

It should be noted that not only free carbon but also all organic compounds, including those that are volatile, show this redox reaction with potassium iodate (compare Volume II).

As in test *2*, the heating procedure with potassium iodate must not be attempted in the presence of inorganic reductants. If need be, they must be removed or destroyed beforehand.

References p. 382

REFERENCES

1. W. Klemm, *Magnetochemie*, Leipzig, 1936.
2. F. Feigl and G. Dacorso, *Chemist-Analyst*, 32 (1943) 28.
3. Compare Feigl, *Chemistry of Specific, Selective and Sensitive Reactions*, New York, 1949, p. 115.
4. C. F. Schönbein, *J. prakt. Chem.*, [1], 79 (1860) 67.
5. F. Feigl and L. Vokac, unpublished studies.
6. R. Kuhn and F. Weygand, *Ber.*, 69 (1936) 1969.
7. F. Feigl, *J. Chem. Education*, 22 (1945) 344; see also G. Gaglio, *Chem. Zentr.*, 1894, II, 452.
8. B. W. Nordlander, *Ind. Eng. Chem.*, 19 (1927) 518.
9. F. Stitt and J. Tomimatsu, *Anal. Chem.*, 23 (1951) 1098.
10. J. Stone, *Ind. Eng. Chem., Anal. Ed.*, 5 (1933) 220.
11. A. C. Vournasos, *Ber.*, 43 (1910) 2264.
12. F. Haber and E. Bruner, *Z. Elektrochem.*, 10 (1904) 706.
13. H. Arnold, *Z. anorg. Chem.*, 88 (1914) 74.
14. O. Frehden and C.-H. Huang, *Mikrochem. ver. Mikrochim. Acta*, 26 (1939) 41.
15. F. Feigl, D. Goldstein and R. A. Rosell, *Z. anal. Chem.*, (1957), in press.
16. B. Gehauf and J. Goldenson, *Chem. Abstracts*, 47 (1953) 6824.
17. R. F. Milton, *Nature*, 164 (1949) 448.
18. F. Feigl and L. Hainberger, *Analyst*, 80 (1955) 807.
19. A. S. Komarowsky and N. S. Poluektoff, *Z. anal. Chem.*, 96 (1934) 23.
20. Comp. Beilstein, *Handb. Org. Chem.*, Vol. II (1922) 550.
21. F. Feigl and N. Braile, *Chemist-Analyst*, 32 (1943) 76.
22. H. Sommer, *Ind. Eng. Chem., Anal. Ed.*, 12 (1940) 388.
23. E. Grünsteidl, *Z. anal. Chem.*, 77 (1929) 283.
24. F. Feigl, studies with L. Weidenfeld, *Dissertation*, Vienna, 1930; comp. also *Mikrochemie* (*Emich Festschrift*) (1930) p. 133.
25. E. Obach, *J. pr. Chem.*, 18 (1878) 258.
26. E. W. Zmaczynski, *Z. anal. Chem.*, 106 (1936) 32.
27. F. Feigl and C. Stark, *Anal. Chem.*, 27 (1955) 1838.
28. F. Feigl and N. Braile, *Chemist-Analyst*, 32 (1943) 76.
29. F. Feigl, unpublished studies.
30. F. Feigl, *J. Chem. Education*, 20 (1943) 300.
31. F. Feigl, Ref. 3, p. 121.
32. A. Brukl and W. Maxymowicz, *Z. anal. Chem.*, 68 (1926) 16.
33. F. Feigl, unpublished studies.
34. F. Feigl and Ph. W. West, *Anal. Chem.*, 19 (1947) 351.
35. F. Feigl, unpublished studies.
36. E. Müller, *J. prakt. Chem.*, 203 (1917) 53.
37. F. Feigl and D. Goldstein, *Mikrochim. Acta* (1956) 1317.
38. Compare A. S. Komarowsky, *Z. anal. Chem.*, 72 (1927) 293; E. Salkowski, *Chem. Ztg.*, 40 (1916) 448.

Chapter 6

The Systematic Analysis of Mixtures by Spot Reactions

The convenience and value of spot tests for identification and detection purposes are well established by the foregoing discussions and procedures. The sensitivity, selectivity, specificity, and particularly the simplicity of many spot reactions make it appear likely that considerable advantage can result if their systematic use is substituted for some of the usual analytical procedures. It will thus be possible to avoid, to a considerable degree, the special separations and operations (precipitation, filtration, solution, etc.) that are so time-consuming when working with larger amounts of the sample. In other words, spot methods make it possible to accomplish qualitative analyses with great saving in time and material.

Very little data are available on the use of spot reactions for the systematic analysis of all cations and anions of a mixture.* At present, not enough is known about the limiting proportions and concentration limits of the spot analysis of mixtures to make a critical evaluation or recommendation of any particular procedure. Statistical evidence and details as to the time and materials consumed, and comparisons of these figures with those of the usual methodes of analysis, are greatly needed. If such particulars were available, they would be extremely valuable in showing how errors can be discovered and avoided. They would also stimulate improvements in the present methods and lead to the development of new spot methods. The aim of an analytical scheme composed entirely of spot tests is not merely to perfect methods for the decisive detection of the constituents of a solution when the concentrations are average and of about the same order. A further objective is the selecting and testing of such spot reactions as can be successfully applied to give unequivocal detection of the materials in the presence of others when the concentrations range over wide ratios. This factor is of paramount importance in actual practice. Spot tests allow a much greater elasticity in this respect than do the usual methods because many sources of error (adsorption of cosolutes on the large bulk of a precipitate, losses by induced precipitation, etc.) are avoided.

* The first reference to this use for spot tests described by the author is by E. Hauser.[1] The first reference to the use of modern spot tests in the systematic detection of cations is by C. J. Van Nieuwenburg.[2]

References pp. 426–427

Several schemes of analysis employing spot methods have been recommended. The procedures described on pages 391, 393 and 404 have been tested by the author and found reliable.

The investigations of Miller [3] on qualitative semimicro analysis with reference to the Noyes and Bray system [4] deserve special attention. The same is true of the scheme found in the recent edition of G. Charlot's book.[5] In the pertinent papers the use of many modern spot reactions is described.

It should be noted that spot reactions are particularly useful as preliminary tests (see page 394) and in the examination of acid-insoluble residues (see page 407).

1. Gutzeit's Analytical Scheme [6]

(1) Test on the original substance

Ammonium: The substance is heated in a crucible with sodium hydroxide. Black filter paper moistened with Nessler's reagent is held over the crucible. The ammonia is revealed by a yellow color.

The spot test can also be carried out using logwood extract. A drop of the reagent is placed on filter paper and held over the reaction vessel (crucible, etc.). A red color indicates ammonia.

Carbonate: The sample is heated with chromic anhydride plus glacial acetic acid. A drop of freshly prepared baryta water is held over the crucible on a glass rod. A white cloudiness indicates carbon dioxide.

Cyanide: The sample is heated with concentrated acetic acid. The vapors are allowed to impinge on filter paper moistened with a solution of hydrocoerulignone $[HOC_6H_2(OCH_3)]_2$ and copper acetate. A purple color indicates cyanide.

Sulfide: The sample is treated with concentrated acetic acid. Any hydrogen sulfide is identified by the orange color on filter paper moistened with $SbCl_3$ held over the crucible.

Chloride: The sample is heated with potassium chromate (permanganate) and concentrated sulfuric acid. A thin glass rod carrying a drop of NaOH is held over the crucible and then dropped into a test tube containing aniline-phenol water. A blue color after boiling indicates chloride.

Bromide: The substance is treated as in the chloride test. Filter paper moistened with a drop of reduced fuchsin solution is held over the crucible. Bromide is revealed by the violet color.

Fluoride: The sample is placed in a lead crucible, moistened with a drop of water and heated with a few drops of concentrated sulfuric acid. Filter paper moistened with Brazil wood extract is laid on the crucible. A red color indicates fluoride.

(2) *Test in the carbonate extract*

Acetic acid is added to the carbonate extract until no further carbon dioxide is evolved.

Iodide: A drop of a 1 % palladium chloride solution, a drop of dilute hydrochloric acid, and finally a drop of the test solution are placed on filter paper. A brown fleck indicates iodide.

Sulfide: A drop of 20 % solution of sodium nitroprusside is placed on filter paper followed by a drop of the test solution. Sulfide sulfur produces a violet color.

Sulfite: A drop of 3 % solution of Solid blue R (Meldola blue) is placed on filter paper and dried. A little powdered cadmium carbonate is placed in the middle of the fleck; then a drop of the substance to be analyzed. The paper is held over fuming hydrochloric acid. A color change from blue to yellow indicates sulfite.

Sulfate: A drop of a fresh solution of sodium rhodizonate is placed on filter paper and treated with barium chloride and then with dilute hydrochloric acid. The red fleck is treated with the test solution. If the color is discharged, sulfate is present.

Nitrite: A drop of naphthylamine-sulfanilic acid reagent is placed on filter paper, treated with dilute sulfuric acid, and then a drop of the test solution added. A red-brown color indicates nitrite.

An alternative method is to spot filter paper with benzidine solution. A drop of the test solution is added, followed by a drop of an aqueous solution of β-naphthol. After development over ammonia, a violet color appears.

Nitrate: A drop of the test solution is placed on a varnished microscope slide, acidified with acetic acid, and treated with a 5 % solution of nitron formate. A precipitate indicates the presence of nitrate.

An alternative test is: a drop of the test solution and a drop of a sulfuric acid solution of 1,5-dihydro-oxyanthraquinone are placed on a watch glass. A violet color indicates nitrate.

Phosphate: A drop of the solution is placed on filter paper, spotted with ammonium molybdate, and dried. After treating with benzidine acetate, a drop of concentrated ammonia is added. A blue color indicates phosphate.

Borate: Filter paper is moistened with turmeric solution. A drop of the test solution and a drop of concentrated acetic acid are added. A red fleck, that turns blue-black with sodium carbonate, indicates borate.

Silicate: Three drops of an ammonium molybdate solution are placed on a watch glass, followed by a drop of the test solution, and a drop of alkali stannite solution. A blue color indicates silicate.

Cyanide: An aqueous solution of pyramidone containing pyridine is

References pp. 426–427

treated with a drop of the test solution. A blue color indicates cyanide.

Ferrocyanide, Thiocyanate, Thiosulfate: A large drop of ferric chloride solution is placed on filter paper and almost dried. The fleck is spotted with diluted hydrochloric acid and then with the test solution. After a short time individual colored areas are observed, due to capillary separation. From the center outward they reveal: ferrocyanide, blue fleck; thiosulfate, white ring; thiocyanate, red ring.

(*3*) *Test in the nitric acid extract*

Silver: A drop of the test solution is spotted on filter paper with a drop of a hydrochloric acid–stannous chloride solution, and then with a solution of chromotropic acid. A black fleck is formed if silver is present.

Alternatively, a drop of the test solution is treated with concentrated ammonia and then spotted with a drop of concentrated potassium chromate solution and a drop of concentrated acetic acid. A red fleck indicates silver.

Lead: A drop of dilute sulfuric acid is placed on filter paper; then a trace of concentrated potassium cyanide solution and a drop of the test solution are added. A 2 % solution of diphenylthiocarbazone in carbon disulfide is placed in the middle of the fleck. A red stain, disappearing on the addition of cobalt nitrate, indicates lead.

Alternatively, a drop of test solution is stirred with a drop of dilute sulfuric acid and developed over ammonia. The middle of the fleck is treated with 3 % hydrogen peroxide, dried, and the edge of the fleck spotted with an alcohol-acetic acid solution of tetramethyl-*p*-diaminodiphenylmethane. A blue color indicates lead.

Mercury: A drop of the test solution is placed on filter paper. A drop of water and then a drop of a 1 % alcoholic solution of diphenylcarbazide are added. After developing over ammonia, a blue color appears. If the stain dissolves in benzene, only univalent mercury is present.

(*4*) *Tests in the aqua regia extract*

Bismuth: (*1*) A drop on filter paper is spotted with tetra-acetyl ammonium hydroxide. A red fleck indicates bismuth.

(*2*) A drop of cinchonine–potassium iodide reagent is placed on filter paper followed by a drop of distilled water and a drop of the test solution. In the presence of Hg^{+2}, Pb^{+2}, and Cu^{+2}, as well as Bi^{+3}, a series of concentric rings develops (capillary separation). From the outside toward the center these are: *copper,* brown; *lead,* canary yellow; *bismuth,* orange; *mercury,* pale yellow.

Copper: (*1*) A drop of the test solution is treated on a varnished microscope slide with a drop of tartaric acid solution and then with a drop of

References pp. 426–427

an alcoholic solution of benzoinoxime. A pale green precipitate indicates copper.

(*2*) One drop is treated on filter paper with 2 % alcoholic solution of 1,2-diaminoanthraquinone-3-sulfonic acid. A dark blue stain indicates copper.

(*3*) A drop on filter paper treated with tincture of guaiacum followed by potassium cyanide solution gives a blue fleck if copper is present.

Cadmium: (*1*) A drop of diphenylcarbazide–thiocyanate reagent is placed on filter paper, treated with a drop of the test solution, and developed over ammonia. A violet stain indicates cadmium.

(*2*) A drop of the test substance is treated with sodium hydroxide and then with a 5 % solution of thiosinamine (allylthiourea). A yellow color indicates cadmium.

Arsenic: (*1*) Two or three drops on a watch glass are rendered alkaline with sodium hydroxide, heated to boiling, and 2 drops of hydrogen peroxide solution added. Two drops of the filtered liquid are transferred with a capillary pipette to filter paper. The moist spot is treated with acetic acid and then with silver nitrate solution. A brick red color indicates arsenic.

(*2*) The moist spot [see (*1*)] is developed over concentrated hydrochloric acid and treated with an alcoholic solution of 8-hydroxyquinoline, and then with ferric chloride solution. A blue-green color indicates arsenic.

Antimony: Two drops are heated to boiling on a watch glass and a few crystals of sodium thiosulfate are added. A vermilion precipitate indicates antimony.

Tin: Two or three drops are heated on a watch glass with magnesium powder and 2 drops of the clear solution are transferred, by means of a capillary, to filter paper and treated with gold chloride. A brown-black color indicates tin.

Iron: On treatment with a 20 % alcoholic solution of 2,4-dinitroso-resorcinol, a pale green color develops. Further tests include treatment with a solution of 1,2,5-sulfosalicylic acid (violet color) or potassium ferrocyanide (blue color).

Chromium: Two or three drops on a watch glass are rendered alkaline with sodium hydroxide, heated to boiling, and treated with hydrogen peroxide or sodium peroxide, and filtered. Three drops of the filtrate are acidified on filter paper with hydrochloric acid, and spotted with an alcoholic solution of orcinol. A brown color develops. Alternatively, spotting with α-naphthylamine causes a violet coloration.

Manganese: A drop of the test solution is treated on filter paper with sodium hydroxide. A drop of tartaric solution is placed in the middle of the moist spot, followed by an acetic acid solution of benzidine. A blue color indicates manganese.

Nickel: A drop of concentrated diammonium phosphate solution is placed on filter paper, treated with the test solution, and the moist spot edged with a 1 % alcoholic solution of dimethylglyoxime, and developed over ammonia. A red stain forms. Other color tests use oxalenediuramineoxime (orange) and α-benzildioxime (red).

Cobalt: Filter paper is moistened with an acetic acid solution of β-nitroso-α-naphthol and dried. A drop of diammonium phosphate solution is placed on the spot followed by the test solution. A rose- red border forms if cobalt is present.

Zinc: A drop of the solution (on a watch glass) is heated with sodium hydroxide, and filtered. The filtrate is acidified on a varnished watch glass with acetic acid, and then treated with antipyrine-thiocyanate reagent. A white precipitate forms if zinc is present.

Aluminum: A drop of a solution of potassium ferrocyanide is placed on filter paper and dried. The test solution is added, followed by dilute ammonia, and finally by an alcoholic solution of alizarin. The fleck is developed over ammonia and dried. A rose-red color encircles the spot if aluminum is present.

Barium: A few drops of the solution are made alkaline with ammonia, heated to boiling, filtered, and the filtrate transferred to filter paper. The moist fleck is treated with a freshly prepared alcoholic solution of sodium rhodizonate and then with dilute hydrochloric acid. A red stain indicates barium.

Strontium: A drop of potassium chromate solution is placed on filter paper, dried, and spotted with a drop of the test solution, and then acetic acid is added. The fleck is treated with sodium rhodizonate and developed over ammonia. A rose-red border around the drop indicates strontium.

Calcium: A drop of the test solution is acidified with hydrochloric acid on a slide, evaporated to dryness, and the residue treated with an ammoniacal solution of phenol. A blue-green color forms if calcium is present.

Magnesium: A drop of the test solution is treated on filter paper with sodium hydroxide, then with tartaric acid and an alcoholic solution of quinalizarin. After drying, a blue fleck is formed with dilute hydrochloric acid. A spot test with dehydrothio-*p*-toluidine sulfonic acid yields an orange-red color if magnesium is present.

Sodium: Three drops of the solution are mixed on a watch glass with 2 drops of ammonia and 2 drops of ammonium carbonate, and filtered. A drop of potassium-cesium-bismuth nitrite reagent is placed on black filter paper and a drop of the filtrate added. A yellow stain forms if sodium is present.

The test solution forms a pale yellow precipitate, visible on a varnished slide, when treated with zinc uranyl acetate.

References pp. 426–427

Potassium: A spot test with sodium bismuth thiosulfate, carried out on black filter paper, produces a yellow stain if potassium is present.

2. Scheme according to Heller and Krumholz [7]

These authors utilize spot methods to identify elements within the familiar groups of the systematic scheme of analysis. Therefore a preliminary separation of the mixture by successive precipitation with HCl, H_2S, and $(NH_4)_2S$ is necessary. These operations can be carried out quickly and on small amounts of material if the centrifuge is used. The identification of the individual ions is carried out by means of spot tests, not only on paper but also on a spot plate.

(1) Cations of the HCl group

Silver: The mixed precipitate of silver, mercurous and lead chlorides thrown down by HCl is covered with 5 % potassium cyanide solution. A drop of the filtrate is treated on a spot plate with an alcoholic solution of *p*-dimethylaminobenzylidenerhodanine. In the presence of silver, a red color forms (see page 59).

(2) Cations of the H_2S group

The sulfide precipitate is treated with yellow ammonium sulfide (residue I, solution I). Residue I is treated with HNO_3 (sp. gr. = 1.2) (residue II, solution II). Solution II is diluted with water and tested for Cu, Pb, Bi, Cd.

Copper: A drop of solution II is mixed on a spot plate with a drop of 1 % zinc acetate solution and a drop of $(NH_4)_2[Hg(CNS)_4]$ solution. In the presence of copper, the precipitate is colored violet (see page 93).

Lead: The authors include no tests for lead. It may be detected as follows: A few drops of solution II are heated to fuming with a drop of concentrated sulfuric acid in a micro crucible; after cooling, water is added. A white precipitate indicates lead. If the contents of the crucible are transferred to a microcentrifuge tube, very small amounts of $PbSO_4$ are visible. The rhodizonate test for lead (page 73) or the dithizone test (page 74) can also be used.

Bismuth: A drop of solution II is treated on a spot plate with a drop of a saturated solution of lead chloride, and 2 drops of a mixture of equal parts of 5 % $SnCl_2$ solution (in 0.5 *N* HCl) and 25 % NaOH. If the solution contains copper, a drop of 2 *N* NaOH and a drop of 5 % potassium cyanide are introduced before adding the stannite. When large amounts of bismuth are present, a black precipitate forms at once. For small amounts, the precipitate appears after a few minutes and a blank test is necessary in this case (see page 77).

Cadmium: Solution II is treated with ammonia. A drop of the filtrate is mixed on a spot plate with a drop of each of the following: 10 % NaOH, 10 % KCN, 0.1 % alcoholic di-*p*-nitrophenylcarbazide, and finally 3 or 4 drops of formaldehyde. In the presence of cadmium, a blue precipitate or color forms, which should be compared with a blank test applied to ammonia (see page 96).

Mercury: Residue II is dissolved in concentrated $HCl+H_2O_2$. A drop of the solution is mixed on the spot plate with a drop of an alcoholic solution of *p*-dimethylaminobenzylidenerhodanine and solid sodium acetate. A red color indicates mercury (see page 67).

Solution I is acidified with HCl. The precipitate of the acid sulfides is treated with HCl (1 : 1) (solution III, residue III).

Antimony: A drop of solution III, diluted with water, is placed on filter paper, treated with phosphomolybdic acid and steamed. A blue color indicates antimony (see page 104).

Tin: A portion of solution III is reduced with metallic magnesium and a drop placed on ammonium phosphomolybdate paper (see page 108). In the presence of tin, a blue color forms.

Arsenic: The authors give no arsenic test; the procedure described on page 99 may be applied.

(3) Cations of the $(NH_4)_2S$ group

The precipitate of sulfides is dissolved in dilute hydrochloric acid plus hydrogen peroxide.

Nickel: A drop of the test solution is mixed on a spot plate with a drop of 3 % H_2O_2, a drop of saturated sodium tartrate solution, 2 drops of a saturated sodium carbonate solution, and a drop of an alcoholic dimethylglyoxime solution. In the presence of nickel, a red precipitate forms; it diffuses to the edge of the liquid.

Cobalt: A drop of sodium phosphate solution, one of the test solution, and one of α-nitroso-β-naphthol, are placed of filter paper, which is then held over ammonia and spotted with 2 *N* H_2SO_4. A permanent brown color indicates cobalt.

Iron: A drop of the solution is treated on a spot plate with a drop of a 1 % KCNS solution. A red color indicates iron.

Chromium: A drop of the test solution is mixed on a spot plate with a drop of bromine water, and 2 or 3 drops of 2 *N* NaOH. A crystal of phenol is added, and a drop of 1 % alcoholic diphenylcarbazide; 2 *N* H_2SO_4 is added drop by drop until the red color of the diphenylcarbazide (due to alkali) disappears. In the presence of chromium, a violet color remains (see page 167).

Manganese: A drop of the test solution is placed on filter paper and spotted in succession with a drop of a KOH solution containing tartrate, and a drop of benzidine solution. A blue color indicates manganese (see page 175).

A portion of the test solution is treated with an excess of 2 *N* NaOH. The filtrate is tested for zinc and aluminum.

Zinc: A drop of the solution, on a spot plate, is rendered just acid with 10 % acetic acid and then stirred with a drop of a 0.1 % copper sulfate solution, and a drop of a solution of 8 g $HgCl_2$ and 9 g NH_4CNS in 100 ml water. In the presence of zinc, a violet precipitate is formed.

Aluminum: A drop of the solution is treated on a spot plate with a drop of a 0.5 % solution of alizarin sulfonic acid and then 2 *N* acetic acid is cautiously added until the violet color disappears. One more drop of acetic acid is then added. A permanent red color indicates aluminum.

Spot tests for the alkali and alkaline earth metals are not included by these authors.

3. Analysis Scheme according to Krumholz [8]

Detection of Hg, Bi, Pb, Cu, Cd, As, Sn, Co, Ni, Fe, Cr, Mn, Al, Zn

About 20 to 50 mg of the solid sample is heated with 0.5 ml concentrated hydrochloric acid and 2 ml bromine water, evaporated to about 0.5 ml, then diluted with 2 ml water, and separated from any insoluble matter by filtering or centrifuging.

The following reactions can be carried out directly on the solution:

Mercury: A drop of the solution is placed on a spot plate and stirred with 5 drops of 10 % sodium phosphate and then a drop of dimethylamino-benzylidenerhodanine is added. A red color indicates mercury (see page 67).

Bismuth: A drop of the solution is mixed on a spot plate with a drop of a 5 % solution of sodium potassium tartrate, a drop of 1 % lead acetate solution, and 3 to 5 drops of sodium stannite (equal volumes of 5 % stannous chloride in 5 % hydrochloric acid and 25 % sodium hydroxide). A brown to black color indicates bismuth (see page 77). A blank test is necessary when testing for small amounts of bismuth.

When mercury is present, a drop of the solution is ignited in a micro crucible; the residue is dissolved in 2 *N* HCl and tested as directed.

Arsenic: A drop of the solution is mixed in a micro crucible with three drops concentrated hydrochloric acid and a little stannous chloride, and warmed. A brown color indicates arsenic (see page 99).

When mercury is present, a drop of hydrogen peroxide and of ammonia are added, then a little magnesium sulfate. The mixture is evaporated to dryness, ignited, and the residue tested as directed.

Nickel: A drop of the solution is mixed on a spot plate with a drop of each of the following: hydrogen peroxide, 5 % sodium potassium tartrate, dimethylglyoxime, sodium carbonate. A red precipitate indicates nickel (see page 149).

Iron: A drop of the solution is mixed on a spot plate with a drop of potassium thiocyanate. A red color indicates iron (see page 164).

Chromium: A drop of the solution is mixed on a spot plate with a drop each of 2 *N* sodium hydroxide, bromine water, or phenol water. The mixture is acidified with 2 drops of 2 *N* hydrochloric acid, and a drop of diphenylcarbazide is added. A violet color indicates chromium (see page 167).

Manganese: A drop of sodium potassium tartrate, one of the test solution, and one of sodium hydroxide are placed on filter paper. The moist fleck is treated with an acetic acid solution of benzidine. A blue color indicates manganese (see page 175).

Tin: Two drops of the solution are mixed with 3 drops of concentrated hydrochloric acid in a microtest tube and a little metallic zinc added. The mixture is boiled for 1 minute, then filtered through a sintered glass filter stick onto a spot plate, and tested with cacotheline. A violet color indicates tin (see page 109).

The remainder of the solution is carefully mixed with 25 % sodium hydroxide until a permanent precipitate forms. Then 5 to 6 more drops of the alkali are added, the suspension is boiled for 1 minute, cooled, and centrifuged. The precipitate is saved. The solution is tested for zinc, aluminum, lead.

Zinc: Two drops of potassium ferricyanide and diethylaniline solution are mixed on a spot plate and a drop of the test solution added. A brown color forms, sometimes only after long standing. A blank test is necessary (see page 178).

Aluminum: A drop of the solution is placed on a black spot plate and 3 drops of 2 *N* acetic acid and a drop of morin solution are added. A green fluorescence develops; it should be compared with a blank test on 2 *N* sodium hydroxide (see page 182).

Lead: A few drops of the solution are treated on a spot plate with a drop of potassium cyanide (to prevent interference from any copper that may have gone into solution) and then a drop of ammonium sulfide is added. A brown color indicates lead.

The *precipitate obtained with sodium hydroxide* is dissolved in a few drops) of concentrated hydrochloric acid. The solution is treated with excess ammonia, centrifuged, and the solution tested for copper, cobalt, cadmium.

Copper: A drop of the solution is placed on spot paper, gently warmed and treated with benzoinoxime. A green color indicates copper.

Cobalt: A drop of the solution is stirred on a spot plate with 2 drops of 2 *N* hydrochloric acid, and 1 drop of 10 % potassium iodide, and then treated with a little solid sodium sulfite. A drop of α-nitroso-β-naphthol and 2 or 3 drops of sodium acetate are added. A brown color indicates cobalt (see page 145).

Cadmium: The rest of the solution is treated with a few drops of 10 % potassium cyanide in a micro test tube. The mixture is heated to boiling and a drop of ammonium sulfide added. A yellow precipitate or turbidity forms if cadmium is present.

4. Spot Analysis of Alloys according to Heller [9]

About 0.01 g of the sample is evaporated to dryness on the water-bath with a few drops of concentrated nitric acid. The residue is taken up with a little 2 *N* nitric acid, filtered, and washed (solution I, residue I). Alloys not attacked by nitric acid are dissolved in bromine-hydrochloric acid, and the solution taken to fumes with nitric acid.

Residue I is tested for tin and antimony. A few drops of concentrated hydrochloric acid are run through the material on the filter, and the test for antimony made on 1 or 2 drops of the filtrate. The rest of the filtrate and the residue are treated with zinc and tested for tin.

A drop of solution I is tested for silver; the rest is heated with a few drops of sulfuric acid to white fumes, then taken up in about 2 ml water, filtered, and washed (precipitate II, solution II). Concentrated potassium hydroxide is poured over the precipitate and the filtrate is tested for lead.

Hydrogen sulfide is led into solution II and the acid sulfides (precipitate III) are separated from solution III. Precipitate III is dissolved in hot 2 *N* nitric acid. The clear filtrate is tested for copper, bismuth and cadmium.

Solution III is oxidized with bromine. Excess bromine is removed by concentrating to 2 ml. This solution is tested for Fe, Ni, Co, Cr, and Mn by suitable spot tests. The rest of the solution is treated with 2 *N* potassium hydroxide. Aluminum and zinc are tested for in the concentrated filtrate.

This separation with potassium hydroxide introduces a high content of alkali salts; they reduce the sensitivity of the tests, both on the spot plate and on paper.

The individual cations can be detected by the following reactions:

Silver: Test with *p*-dimethylaminobenzylidenerhodanine on a spot plate see page 59).

References pp. 426–427

Lead: Test with benzidine, on paper (see page 72).

Tin: Test with ammonium phosphomolybdate paper (see page 108).

Antimony: Test with phosphomolybdic acid paper (see page 104).

Copper: Test by formation of copper zinc mercury thiocyanate, on a spot plate (see page 93).

Cadmium: Test with di-*p*-nitrodiphenylcarbazide, on a spot plate (see page 96).

Bismuth: Test with alkaline stannite solution and lead chloride, on a spot plate (see page 77).

Iron: Test with potassium thiocyanate, on a spot plate (see page 164).

Chromium: Test with diphenylcarbazide (after conversion into chromate), on a spot plate (see page 167).

Cobalt: Test with α-nitroso-β-naphthol, on filter paper (see page 144).

Nickel: Test with dimethylglyoxime, on a spot plate (see page 149).

Manganese: Test with benzidine, on a spot plate. A drop of saturated sodium tartrate solution, one of potassium hydroxide, and one of benzidine are added to a drop of the solution (see page 175).

Zinc: Test by formation of zinc cobalt mercury thiocyanate, on a spot plate (see page 180).

Aluminum: Test with alizarin sulfonic acid, on a spot plate (see page 183).

Table 6 shows the percentages of the constituents of the alloys examined by Heller.

The figures enclosed in parentheses indicate percentages of metals present, but not revealed by this procedure. "0" indicates that a metal was actually found, although it was not reported in the previous analysis. According to Heller, the examination can be accomplished with one-fiftieth of the quantity of sample required for macroanalyses, and the procedure consumes only one-fourth as much time.

5. Ring Oven Method according to Weisz

The ring oven method has proved to be very useful and quite apt to provide developments for the identification of cations in mixtures. The principles of this method are mentioned in Chapter 2. The outstanding value of ring oven methods seems to be that they can be used for solving special problems, mainly for testing minerals and alloys with minimal amounts of test material. For details, refer to the pertinent original works.[10]

6. Application of Spot Tests in Preliminary Examinations

Before proceeding to the systematic wet analysis, it is always advisable to carry out preliminary tests to obtain general information regarding the nature of the sample. The results are often important in establishing the presence or absence of certain substances, in making an intelligent choice of solvents, etc. A small sample, about the size of a millet seed to a grain of corn, is usually sufficient for the preliminary analysis. Solid samples should be very finely pulverized and well mixed; liquid samples should be taken to dryness and the tests carried out on the residue. Loss of volatile materials during the evaporation should be noted. Spot reactions can often be successfully applied directly to the solid sample or to the evaporation residue. Various methods of testing, characteristic of certain groups, will be given.[11]

TABLE 6. ALLOYS ANALYZED BY HELLER

No.	Weight taken, mg	Ag, %	Pb, %	Sn, %	Sb, %	Cu, %	Cd, %	Bi, %	Fe, %	Al, %	Cr, %	Co, %	Ni, %	Mn, %	Zn, %	Other constituents
1	10		63	17	13	7.5										
2	10		64	13	21	1.8										
3	10								33		62					C, Si, S, P
4	10								22					77		C, Si, S, P
5	10		14	75	7	3.3			0							
6	10			100												
7	10								16					83		C, Si, S, P
8	10					64							13		23	
9	10		26	13			12	49								
10	10					99.9										
11	10			7		86			1				2.1		4	P
12	10		2	81	11	6			0.1						0	
13	10		76	10	12				0.1						0.6	
14	10					0.04			32		67			(0.06)		C, Si, S, P
15	10								28		72			(0.1)		C, Si, S, P
16	10								96				3.5			C, Si, S, P
17	10								99					(0.6)		C, Si, S, P
18	10		0			66			0				10		24	
19	10								95					3.5		C, Si, S, P

References pp. 426–427

TABLE 6 *(continued)*

No.	Weight taken, mg	Ag, %	Pb, %	Sn, %	Sb, %	Cu, %	Cd, %	Bi, %	Fe, %	Al, %	Cr, %	Co, %	Ni, %	Mn, %	Zn, %	Other constit-uents
20	10	0.8	99													
21	10								95		1		0.5	(0.5)		C, Si, S, P
22	10		62	36	1.3	0.9									0	
23	10		0.7	21		78			0.2							
24	10		3.8	15	0	78			0.3	0					3.4	
25	10		0.2	11		87			1.3		0				(0.8)	
26	10		0.6	2.2	(0.3)	95			0.3						1.6	
27	10		0.3		99.6	0.08			(0.2)							
28	10					61			0.4				15	(0.2)	24	
29	10		1.2				(0.3)		0						98	
30	20		1.2	0			(0.3)		0						98	
31	10		0.9				(0.5)		0						98	
32	20		0.9				0.5		0						98	
33	10		0.5	22		77			0.8							
34	10				0				94		1.1		3.4	(0.5)		C, Si S, P
35	10		50	(0.1)					0	10					40	
36	10		(0.6)			54			7				2.1		36	
37	10		0.7	1.4		62			1.5						(35)	
38	10		2.5	5.6		0.2				16					75	
39	10		2.6	(8.9)	0.9	82			0.5						4.8	
40	10		3.7	5.9	0.7	84			0.6				0.4		4	P
41	10		12	6.7		79			0.4						2.1	
42	10		0.6			63			0.5						36	
43	10		0.9	22	(0.7)	68			0.1	0.9			4.7			
44	10		3.2	12	0.5	81			4.2						0.6	
45	10		1.1	(0.1)		58			0.2				(0.07)		40	
46	50		1.1	0.1		58			0.2				0.07		40	
47	10		1.9	81	10	5.8			0.1							
48	10		0.7	1.4		69			0.2				0.4		29	
49	10		5.7	5.6		79			4.5				4.4	0.9	1.2	
50	10		(0.03)	(0.15)	(0.2)	99							0.27			As
51	50		0.03	0.15	0.2	99							0.27			As

(1) Heating on a magnesia spoon, in a porcelain dish, or in a micro crucible

This type of preliminary examination serves to establish the alterations, characteristic of certain materials, that are eventually undergone when the solid sample is heated in contact with air. The evaporation residues of liquid samples should be examined in the same manner. The temperature should be raised by slow degrees, finally reaching redness. The material should not be stirred or agitated in any way during the heating; the changes can be observed better under quiet heating. Sometimes it is well to interrupt the heating to retain characteristic effects which disappear on further elevation of the temperature and contact with the air. Examination with a magnifying glass and comparison with a portion of the unheated sample are highly recommended to determine whether changes have extended throughout the material or have affected only certain parts. In this way, it is possible to detect non-uniformities even in fine powders. When mixtures of powders are heated, thermal decomposition and oxidation may bring about characteristic changes in certain constituents of the mixture, and chemical reactions between the constituents may also occur. Water frequently participates in such reactions when moist materials or those carrying "water of crystallization" are heated. Reactions which occur at elevated temperatures, and which often proceed to considerable extents without the participation of water, include: fusion and sintering reactions; reactions of solids with gaseous decomposition products of another solid; direct reactions of two solids.

When mixtures of powders are heated to temperatures far above the boiling point of water, it must be remembered that the reactions may be not only those which occur ordinarily in solution, but also others which are not realizable in solution. Consequently, the composition of mixtures of solids after heating can sometimes be quite different from that of a mixture made up from the constituents that have previously been heated separately.*

Several diagnoses, based on the behavior of solids and mixtures of solids when heated, will be given.

The sample remains unaltered in color and structure: Absence of volatile compounds, i.e., mercury and ammonium salts; carbonates that loose carbon

* A characteristic example may be cited: Solutions containing alkali or alkaline earth nitrates, or nitrites, and alkali formates remain completely unchanged even though boiled. In contrast, a mixture of the solids leaves only carbonates when heated to about 500° C. This product results from the reaction of the nitrate (nitrite) with the hydrogen and carbon monoxide yielded by the pyrolysis of the alkali formate.[12] A like conversion occurs if a mixture of solid nitrate (nitrite) and alkali oxalate is heated. In this instance, the carbon monoxide yielded by the oxalate reduces the nitrate (nitrite) and converts it into carbonate.

References pp. 426–427

dioxide. Absence of organic matter or metallo-organic compounds. Absence of compounds carrying water of crystallization, or much adsorbed water (gels of oxyhydrates). Absence of compounds, which alone or in mixture with suitable compounds are altered on heating because of the occurrence of redox reactions (see later).

The sample melts: Possibility of the presence of nitrates, nitrites, carbonates, formates, chlorates, etc. of the alkali metals; nitrates of alkaline earth metals; silver chloride or bromide; fusible organic and metallo-organic compounds. Quite a few salts carrying "water of crystallization" melt below 100° C. Examples include: alums, vitriols, $Na_2SO_4 \cdot 10\ H_2O$ (32.4° C), $Na_2S_2O_3 \cdot 5\ H_2O$ (48° C), $Zn(NO_3)_2 \cdot 6\ H_2O$ (36.4° C), $Na_3AsO_4 \cdot 12\ H_2O$ (86° C).

The sample chars partially or completely: Presence of organic material; metallo-organic compounds; metal salts of organic acids; metal formates or oxalates, which decompose with evolution of carbon monoxide and yield carbon, particularly when heated rapidly and in contact with solids (establishment of the Boudouard equilibrium: $2\ CO \rightleftarrows CO_2 + C$).

Transient charring which disappears rapidly with accompanying deflagration indicates the presence of organic matter along with nitrate, nitrite, or chlorate.

The sample changes color (without charring): Presence of compounds of certain heavy metals. Darkening indicates formation of oxides from metal salts of volatile or non heat-resistant acids (e.g., $Cu^{+2} \rightarrow CuO$; $Cd^{+2} \rightarrow CdO$; $Fe^{+3}(Fe^{+2}) \rightarrow Fe_2O_3$; $Tl^{+} \rightarrow Tl_2O$; etc.) Lightening of color indicates decomposition of higher oxides (e.g., $PbO_2 \rightarrow PbO$; $NiO_2 \rightarrow NiO$; etc.) or oxidation of sulfides (e.g., $PbS \rightarrow PbSO_4$; $Bi_2S_3 \rightarrow Bi_2O_3$; etc.).

The color of many metal oxides at room temperature is not the same as that of the hot oxide (e.g., ZnO, CdO, NiO, MoO_3, WO_3, etc.). Invariably, the color is deeper at the higher temperature.[13]

Dark finely dispersed metal is deposited from the carbonates, oxides and organic salts of the noble metals. Ignition of copper, cadmium, zinc, etc. salts of organic acids often produces the metal initially; further heating with access of air brings about formation of the metal oxide.

White silver cyanide, thiocyanate, ferro- and yellow ferricyanide are decomposed with production of ignition-resistant silver. All metal ferro- and ferricyanides of base metals leave a residue of ferric oxide and the particular metal oxide. For instance, the white ferrocyanides of Zn, Cd, Mg, Ca, Ba, Sr, Th etc. become yellow-brown; Prussian blue and Turnbull's blue become first dark (Fe_3O_4), later brown (Fe_2O_3). Cupric ferricyanide (brown) and cupric ferrocyanide (violet-brown) are blackened through formation of cupric oxide.

Ammonium salts of molybdic acid, phosphomolybdic acid and tungstic

References pp. 426–427

acid as well as mixtures of ammonium salts with MoO_3 or WO_3 yield blue ignition residues of lower oxides of molybdenum and tungsten, which re-oxidize on continued heating (see page 380 regarding redox reactions of ammonia with molybdic acid, etc.).

Yellow ammonium chromate and orange ammonium bichromate yield finely divided olive green chromic oxide which results from an innermolecular redox reaction, or from the reaction of the products of the thermal cleavage:

$$2\,(NH_4)_2CrO_4 \rightarrow Cr_2O_3 + 2\,NH_3 + N_2 + 5\,H_2O \quad \text{or}$$
$$2\,CrO_3 + 2\,NH_3 \rightarrow Cr_2O_3 + N_2 + 3\,H_2O$$

Yellowish-white silver arsenite becomes brown-black when ignited; metallic silver and silver metarsenate result from the innermolecular redox reaction between cation and anion in the solid state:

$$Ag_3AsO_3 \rightarrow 2\,Ag + AgAsO_3$$

Mixtures of molybdic acid or tungstic acid with anhydrous alkali sulfites and arsenites, as well as with organic compounds, turn blue when heated. These present clear instances of redox reactions between two acids:

$$Na_2SO_3 + 2\,MoO_3 \rightarrow Mo_2O_5 + Na_2SO_4$$
$$Na_3AsO_3 + 2\,MoO_3 \rightarrow Mo_2O_5 + Na_3AsO_4$$

WO_3 reacts analogously. Oxidation to white MoO_3 or yellow WO_3 occurs only on further heating in contact with the air.

When mixtures of anhydrous cobalt or manganese sulfate with alkali nitrates, nitrites, halogenates, or perchlorates are heated they turn brown or black because of the production of higher oxides of Co or Mn.[14] In these cases the fused nitrate, etc., function as oxygen donors to the solid sulfate:

$$2\,CoSO_4 + O \rightarrow Co_2O_3 + 2\,SO_3$$

However, the anhydrous sulfates of cobalt and manganese remain unaltered if heated alone to temperatures of 940 or 820° C respectively.[15]

(2) *Heating in an ignition tube*

This type of preliminary examination will rapidly reveal the formation of certain volatile thermal decomposition products of solid test materials. If liquid samples are being tested, a portion should be taken to dryness on the water bath, with due consideration of the possible volatilization or decomposition of certain materials.

The sample gives off water which condenses on the walls of the tube: Presence of salts carrying water of crystallization; moist substances; metal hydroxides and oxyhydrates; ammonium salts; organic material. The water condensed on the cold part of the tube should be tested with acid-base indicator papers.

Acid vapors (litmus paper): Volatile acids from acid salts; sulfur trioxide from heavy metal sulfates; sulfur dioxide from sulfites, sulfur, or polysulfides with access to air or in the presence of oxidizing materials.

Alkaline vapors (litmus paper): Ammonium salts of heat-stable, non-oxidizing acids; complex metal ammine salts; normal and complex pyridine salts.

A white sublimate: Ammonium salts of volatile acids; As_2O_3, Sb_2O_3; mercury salts of volatile acids; oxalic acid (cautious heating to 150° C).

A yellow sublimate: As_2S_3, As_2S_5; HgI_2; free sulfur or sulfur derived from polysulfides and heavy metal thiosulfates. Yellow mercuric iodide is transformed to the red modification when rubbed with a glass rod.

A gray sublimate: Finely divided metallic mercury from mercury compounds; free arsenic from ter- and quinquevalent arsenic compounds in the presence of reducing substances; free iodine from iodides in the presence of oxidizing substances or acidic oxides.

Yellow or brown vapors: Bromine from bromides in the presence of oxidizing agents; nitrogen oxides from nitrites or nitrates; chromyl chloride, if both bichromate and water-bearing chlorides are present. (Bromine vapors turn starch paper yellow, in contrast to nitrogen oxides which have no visible action.)

Violet vapors: Iodine from iodides and iodates in mixture with acidic substances, mercury or silver iodide.

Colorless vapors: Carbon dioxide from carbonates and salts of carboxylic acids; carbon monoxide from oxalates and formates; cyanogen or hydrocyanic acid from hydrolyzable or thermal decomposable cyanides (e.g. silver, mercuric, and palladium cyanides, ferro- and ferricyanides); hydrogen sulfide from water-bearing sulfides or thiosulfates; ammonia from ammonium salts, thiocyanates and many organic compounds containing nitrogen; oxygen from higher metal oxides or alkali halogenates, persulfates and perchlorates; hydrogen from formates.

In addition to the foregoing tests, the following important fact may be stated: any kind of inorganic (and also organic) nitrogen compounds form N_2O_3 or N_2O_4 when ignited with MnO_2 or Mn_2O_3.[16] These acid nitrogen oxides can be easily detected through the sensitive Griess reaction. Compare page 330.

(3) *Test on the aqueous solution or suspension* [17]

One or two drops of the aqueous solution (or water suspension) of a solid sample are tested with the following reagents on a spot plate, in a micro crucible, or in a micro test tube. Very small quantities of the solid may also be tested directly.

Indicators:[18] An acid reaction with litmus paper indicates free acid, acid salts, or hydrolyzed salts of weak bases. An alkaline reaction indicates alkali or alkaline earth hydroxides, carbonates, sulfides, or hydrolyzed salts of strong bases. If an acid reaction is obtained, the solution is tested with Congo paper. A blue color ($pH \leqq 3$) indicates strong acids. When an alkaline reaction is given with litmus, the solution is further tested with a freshly prepared solution of thymolphthalein. A blue color ($pH \leqq 10.5$) indicates the presence of ammonia, or alkali salts of very weak acids. If tropaeolin O is colored orange-brown ($pH = 12$), free alkali or alkaline earth hydroxides are present.

Nickel dimethylglyoxime equilibrium solution: precipitation of red Ni-dimethylglyoxime indicates the presence of water-soluble and insoluble compounds which consume H^+-ions. The test is specially suited to the detection of insoluble carbonates, oxyhydrates, arsenates, and phosphates, which have no effect on indicator solutions. The chemistry of the test and the preparation of the reagent solution are given on page 508.

Sulfuric acid-potassium iodide solution: Liberation of iodine indicates the presence of oxidizing agents. When hydrogen peroxide or peroxides are suspected, it is advisable to add a small amount of ammonium molybdate (as catalyst) to hasten the liberation of iodine, since normally this action is quite slow.

Acidic solution of ferrous thiocyanate: The colorless solution turns red (formation of ferrithiocyanate) on contact with soluble and insoluble oxidants.[19] Since arsenates do not react at all and ceric salts but weakly, the sensitive color test is more selective than the preceding tests. It can be applied to reveal the presence of oxidizing materials in solutions of arsenic acid or arsenates, which is not possible with acidified potassium iodide solution.

The reagent solution is prepared as follows: 9 g $FeSO_4 \cdot 7\,H_2O$ are dissolved in 50 ml of 1 : 1 hydrochloric acid and some granulated zinc added. After the color has disappeared, 5 g of NaCNS is added and the resulting transient red color is allowed to fade. Then 12 g additional NaCNS is added and the colorless solution is decanted from the undissolved zinc. The solution prepared in this manner keeps for about a day. If it turns red due to autoxidation, it should be decolorized with a little granulated zinc before use.

Alkaline solution of thallous nitrate: The formation of a chocolate-brown precipitate or coloration of TlOOH due to the realization of $Tl^+ + OH^- + O \rightarrow TlOOH$ indicates the presence of oxidants which are effective in basic solution. These include: CrO_4^{-2}, MnO_4^-, $Fe(CN)_6^{-3}$, H_2O_2, $S_2O_8^{-2}$, $HalO^-$. Accordingly, this test is far more selective than the two preceding tests.

Acetic acid-iodine-potassium iodide solution: Decolorization or lightening of the brown solution indicates the presence of reducing substances.

Alkaline suspension of thallic hydroxide: Immediate discharge or lightening of the color of a drop of the yellow-brown suspension * by a drop of the alkaline test solution indicates the presence of compounds which function as energetic reducing agents in basic solution. (Alkali sulfite, arsenite, stannite, hydroxylamine, hydrazine.) Sulfide or high concentrations of cyanide must be absent.

Another spot test for detecting soluble reducing substances employs paper impregnated with manganese dioxide.[20] A drop of the solution, acidified with acetic acid, is placed on the paper, which is then dried in a blast of heated air. Consumption of MnO_2 leaves a lighter area or a white spot on the paper. The reagent paper is prepared by bathing strips of filter paper in a weakly basic dilute solution of potassium permanganate, washing, and drying. The manganese dioxide produced by the reducing action of the cellulose remains homogeneously dispersed in the capillaries of the paper. The color ranges from deep brown to barely visible yellow, depending on the concentration of the permanganate solution and the duration of its action. The reaction of small quantities of reducing materials (use light-colored MnO_2 paper) can be made easily visible if, after reduction, the paper is bathed in an acetic acid solution of benzidine. The unused MnO_2 produces a deep blue color against which the white MnO_2-free areas appear in sharp contrast. The manganese dioxide paper is stable.

Dilute sulfuric acid: Liberation of colorless gases (*a*), or colored gases (*b*). The test should be carried out in a micro test tube, held against white paper.

(*a*) CO_2 from carbonates; HCN from cyanides; H_2S from sulfides; SO_2 from sulfites and thiosulfates; acetic acid from acetates; oxygen from peroxides; hydrogen from base metals; hydrazoic acid from azides.

(*b*) Brown, yellow or violet gas: nitrogen oxides from nitrites; chlorine, bromine, or iodine from halides in the presence of oxidizing agents.

Concentrated sulfuric acid: As with dilute sulfuric acid but more vigorous; also hydrogen chloride from chlorides; oxygen from chromates, permanganates and peroxides; hydrogen fluoride from fluorides and fluosilicates.

Hydrogen sulfide water acidified with hydrochloric acid: A colored precipitate indicates metals of the H_2S-group.

Ammonium sulfide solution: A colored precipitate indicates basic sulfides. When the H_2S test was negative, then sulfides of Co, Ni, Fe, Mn, Tl, and U

* The thallic suspension is prepared by treating an acidic 0.01 % solution of Tl_2SO_4 with excess bromine water, and then removing the unused bromine by adding a little solid sulfosalicylic acid (compare p. 168). A drop of this colorless solution is treated, on a spot plate, with a drop of 2 *N* sodium hydroxide, whereby a small amount of brown TlO(OH) or $Tl(OH)_3$ is precipitated.

may be present. The formation of a yellow or white precipitate indicates the presence of cadmium or zinc.

Alkali hydroxide solution: A black precipitate or color may result from the presence of mercurous ions, or from the simultaneous presence of one or more noble metals (Au^{+3}, Pt^{+2}, Pd^{+2} ions) and a salt of the metals capable of forming higher insoluble oxyhydrates (Tl^{+1}, Fe^{+2}, Co^{+2}, Ce^{+3} ions). In the second case, the finely divided noble metal and the particular higher oxide result (see footnote on this page). A black precipitate or color may also be due to the simultaneous presence of Sn^{+2} and Bi^{+3} salts (deposition of finely divided metallic bismuth, see page 77).

Alkaline hydrogen peroxide: Bleaching indicates sulfide (sulfate formation), or higher oxides of nickel, which are reduced to $Ni(OH)_2$ (see page 355); darkening of color indicates formation of higher oxides of Mn, Co, Fe, Ce, Pb, Tl.

The following procedure, based on a color (redox) reaction with benzidine, can be used to establish the presence of metal ions, which form higher oxides (Mn, Ce, Co, Ni, Pb, Bi, Tl): A drop of freshly prepared NaOBr solution is placed on filter paper. Before the hypobromite has entirely soaked in, a drop of the nearly neutral test solution is added. Higher oxides are formed instantly and are fixed on and in the paper. The excess hypobromite is decomposed by repeated spotting with ammonia water. After washing with water and brief drying, the paper is spotted with acetic acid-benzidine solution. A blue color appears at once if higher oxides are present. Tetrabase may be substituted for the benzidine.

Hydrogen peroxide containing hydrochloric acid: Bleaching or solution of a dark powder indicates higher oxides (PbO_2, MnO_2, etc.) or sulfide (conversion into sulfate), or free metals (oxidative solution).

Ammoniacal silver oxide solution: In the absence of sulfide ions (compare nitroprusside test for sulfides, page 303) a black precipitate or color indicates Tl^{I}, Mn^{II}, Fe^{III}, Co^{II}, Ce^{III}, Cr^{III} salts.* The reagent is prepared by carefully adding sodium hydroxide solution to a 2 % solution of silver nitrate, and then dissolving the brown silver oxide by dropwise addition of ammonia water.

Nitric acid-silver nitrate solution: A white, yellow, or orange precipitate indicates the presence of Cl^-, Br^-, CN^-, CNS^-, I^-, N_3^-, $(FeCN)_6^{-3}$. A black-brown precipitate may be due to the formation of silver sulfide.

* A reaction of the following type occurs:

$$2\,Ag^+ + Mn^{+2} + 4\,OH^- = 2\,Ag^\circ + MnO_2 + 2\,H_2O$$

An adsorption complex of the noble metal and higher metallic oxide is responsible for the black color.

Phosphoric acid–ferrous sulfate solution: A dark color or precipitate indicates noble metals, selenium or tellurium, released by reduction.

Sodium azide-iodine solution: Decolorization of the brown solution after addition of the neutral test material with evolution of gas (nitrogen) indicates the presence of sulfides, thiosulfates, thiocyanates. (Compare the respective tests in Chapter 4.)

Sodium nitroprusside solution: An alkaline test solution turns red-violet when sulfides are present. If a yellow, brown, or dark acid-insoluble sample is to be tested, a few grains of the powdered material are warmed with sodium hydroxide. (Sulfo salts are formed from certain acid sulfides.) After cooling, sodium nitroprusside is added. A red-violet color indicates sulfides of As, Sb, Sn, W, or Mo, since they alone dissolve in sodium hydroxide solution to produce sulfo salts which react with sodium nitroprusside in the same manner as alkali sulfides.

Ferricyanide–ferric chloride solution: A small portion of the sample, freed of soluble reducing materials by thorough extraction with water, is treated on a spot plate with 1 drop each of dilute solutions of $K_3Fe(CN)_6$ and $FeCl_3$. An immediate blue color or precipitate (Prussian blue) indicates the presence of uncombined metals or alloys. (Au, Pt, Pd are exceptions, they react very weakly.) The complete removal of soluble reducing materials is essential because they react with ferric ferricyanide to give Prussian blue.

Silver ferrocyanide paper and ferric sulfate: [21] A green color that turns to blue indicates the presence of Cl^-, Br^-, I^-. The test is based on the formation of Prussian blue by reaction of $Ag_4Fe(CN)_6$ with the halide ions in the presence of a ferric salt. The reagent paper is prepared: photographic silver-nitrate paper is bathed for 10 minutes in 1 % potassium ferrocyanide solution. It is washed with water, bathed in dilute silver nitrate, and again washed with water. It is then immersed in 10 % ferric sulfate solution and dried.

7. Identification of Anions in a Mixture [22]

The systematic analysis of anions in a mixture is much more difficult than that of cations. The reason lies particularly in the great number of anions, especially when the possible presence of metal-containing complex anions must be taken into account. Moreover, there exists a lack of completely satisfactory specific and selective group precipitations and of identification reactions for the constituents of group precipitates. An additional impediment results from the fact that mixtures of salts, which are stable indefinitely, or for some time at least in the solid state, are altered in solution because of redox reactions etc., changes that in great measure are determined by the prevailing pH-value. Because of these difficulties, there is no

References pp. 426–427

good scheme for the systematic search for anions, comparable to the familiar procedures for detecting cations.

However, the analyst is seldom faced with complicated mixtures of anions. Usually the number of anions to be detected is small, especially when technical products are being studied. In many cases, information regarding the origin of the sample and its intended use will indicate which anions are likely to be present.

Nevertheless, the investigation of complicated mixtures of anions is of didactic interest, because the solution of such exercises includes modes of reaction which otherwise will probably not be brought within the bounds of the students' experience. Accordingly, it is advisable to prepare synthetic complicated mixtures of anions, and to use these as test systems on which to try out the following systematic procedure.

No completely satisfactory scheme for the systematic detection of all the inorganic and organic anions has been developed. An orienting, preliminary examination for those anions which occur most commonly is given here. The procedure is based on a division of anions into two groups by the action of zinc nitrate. Spot tests for particular anions are made on the precipitate and with the filtrate.

If the original material does not consist solely of alkali salts, the sample is boiled with strong sodium carbonate solution, or it is fused with sodium potassium carbonate in a platinum spoon. Heavy metal salts are thus converted into water-insoluble carbonates. After adding water and filtering, any ammonium ions are removed by boiling with sodium hydroxide. Nitric acid is then added, but not sufficient to neutralize the solution. (In this way CuO can also be precipitated, if the carbonate solution was blue due to complex copper carbonate.) The weakly alkaline solution is then treated with zinc nitrate, added as a concentrated solution or as solid. The mixture is warmed and filtered. There results:

(*A*) A *precipitate* of the zinc salts of S^{-2}, AsO_4^{-3}, AsO_3^{-3}, PO_4^{-3}, BO_3^{-3}, F^-, $Fe(CN)_6^{-4}$, $Fe(CN)_6^{-3}$, CN^-, VO_3^-, MoO_4^{-2}, WO_4^{-2} (Treatment I).

(*B*) A *solution* containing the anions: CNS^-, Cl^-, Br^-, I^-, ClO_3^-, SO_4^{-2}, $S_2O_3^{-2}$, SO_3^{-2}, CrO_4^{-2}, NO_2^- (Treatment II).

I. Small portions of the well washed precipitate *A* are transferred to filter paper or a spot plate and spotted, or treated as prescribed, with reagents *1* to *10*:

1. Acid lead nitrate solution: black or brown color indicates S^{-2}.

2. Acid solution of ferrous salt: blue color indicates $Fe(CN_6)^{-3}$.

3. Acid solution of ferric salt: blue color indicates $Fe(CN_6)^{-4}$.

References pp. 426–427

4. Stannous chloride solution: yellow to brown color indicates MoO_4^{-2} or WO_4^{-2}.

5. Hydrochloric acid-potassium xanthate solution: violet color indicates MoO_4^{-2} (see page 116).

6. Ammonium molybdate solution, benzidine solution and ammonia: blue color indicates PO_4^{-3} (see page 333).

7. A little of the precipitate is dried and ignited. The ignition residue is subjected to the Gutzeit test for arsenic (page 101). A positive result indicates AsO_4^{-3} or AsO_3^{-3}.

8. A portion on curcuma (turmeric) paper is treated with HCl. On drying, a reddening that turns a dirty blue with KOH indicates BO_3^{-3}. A little of the precipitate may also be warmed in a micro crucible with 3 drops of 0.01 % solution of quinalizarin in concentrated sulfuric acid. A color change (violet to blue) indicates B_2O_3 (see page 339).

9. If no sulfide ions are present, a portion of the precipitate is spotted with mercurous nitrate: a blackening indicates CN^-. When sulfides are present, and complex iron cyanides are absent, the procedure on page 277 may be used. Alternatively, a little of the precipitate may be digested with ammonia water, which dissolves zinc cyanide. The ammoniacal filtrate is tested for cyanide as described on page 276.

10. A small amount of the precipitate is stirred with barium chloride on a spot plate, and a drop of a hydrochloric acid–zirconium–alizarin solution is added. The color soon changes from red to yellow in the presence of F^- (see page 269).

II. The following spot tests on paper or on a spot plate are carried out with single drops of solution *B*:

1. Acid ferric chloride solution: a red color indicates CNS^-.

2. Diphenylcarbazide in sulfuric acid solution: a violet color indicates CrO_4^{-2} (see page 167).

3. Barium rhodizonate suspension: decolorization of the red fleck indicates SO_4^{-2} (see page 313).

4. Aniline sulfate and concentrated sulfuric acid: a blue color indicates ClO_3^-.

5. Palladium chloride solution: blackening indicates I^- (see page 265).

6. Fuchsin-bisulfite solution: blue color indicates Br^- (see page 264).

7. Dilute iodine-potassium iodide solution: decolorization indicates SO_3^{-2} or $S_2O_3^{-2}$. The decolorized solution tested with litmus paper is neutral if $S_2O_3^{-2}$ is present; acid if SO_3^{-2} is present (see page 312).

8. A heated portion of solution *B* is treated with $Pb(NO_3)_2$ and the precipitate spotted with $AgNO_3$: a brown color indicates $S_2O_3^{-2}$.

Nitrates and nitrites may be detected directly in the sodium carbonate or sulfuric acid or plain water extracts of the sample (see pages 326 and 330).

8. Identification Tests for Substances Insoluble or Sparingly Soluble in Acids [23] ("Insoluble Residues")

The examination of certain solids requires consideration of the fact that they are more or less insoluble in the usual solvents (water, dilute and concentrated acids, *aqua regia*). To detect the constituents of acid-insoluble residues, they are usually subjected to attack by special reagents, which convert them into compounds soluble in the usual solvents (water, acids, alkalis). After they are dissolved, they can be identified without difficulty by the aid of suitable reagents.

If the insoluble residue is not a single substance, this method of attack is unsuited to small amounts of material, since it may be necessary, according to the nature of the residues, to use a number of different decomposition agents to reach a decision more quickly.*

The use of suitable spot methods makes it possible to detect quickly and easily numerous compounds that may constitute an insoluble residue. Not more than a small portion is required. One or two milligrams suffices to detect the compounds given here in a mixture (containing 0.01 mol of each constituent). Sometimes even much less of an acid-free, washed sample will be enough.

For practical purposes it often suffices to warm the pulverized sample for about 5 minutes with concentrated hydrochloric acid, and to regard the residue as "insoluble" even though more prolonged action of the acid or the use of *aqua regia* could take more of the sample into solution. The following statements refer to insoluble residues that are left after brief treatment with concentrated hydrochloric acid.

(*1*) *Insoluble sulfides*

Sulfides of mercury (or mercury sulfo salts) and arsenic may be present, also natural sulfides, such as pyrites, molybdenite, galena, etc.

The catalytic acceleration of the reaction between sodium azide and iodine (see page 303) is the best test for sulfides. It gives a positive result with even traces of sulfide sulfur. It is sufficient to place a grain of the powdered sample on a small watch glass and add 1 or 2 drops of iodine-azide solution. The formation of small bubbles of nitrogen on the surface of the solid indicates the presence of sulfides.

* It has been shown [24] that certain insoluble compounds, such as SnO_2, $PbSO_4$, $BaSO_4$, CaF_2, and silver halides, can be converted into water-soluble iodides or double iodides by heating with hydriodic acid (sp.gr. 1.7).

Certain sulfidic minerals, e.g., pyrites, give a weak iodine-azide reaction. In such cases, an activation can be effected [25] by digestion with mercuric chloride solution; HgS which reacts quickly is formed.

Acid-resistant sulfides (As_2S_3, As_2S_5, Sb_2S_3, Sb_2S_5, SnS_2, MoS_3, V_2S_5, WS_3), which dissolve in sodium or potassium hydroxide with production of alkali sulfo salts, can be detected through the fact that their alkaline solutions, like alkali sulfides, give a violet color with sodium nitroprusside (see page 303). The sulfides must be washed free of hydrogen sulfide.

All acid sulfides (including TeS_2 and elementary tellurium) dissolve in yellow ammonium sulfide with production of ammonium salts of the respective sulfo acids; they can be separated in this manner from the basic sulfides. One drop of the resulting solution is enough for the direct detection of:

Tellurium: By formation of a black precipitate or grey coloration after addition of solid sodium sulfite (compare page 352).

Tin: By a positive response to the fluorescence reaction with morin after warming with hydrogen peroxide (compare page 110).

If a drop of the sulfalkaline solution is taken to dryness and the residue ignited gently, it is possible to detect:

Antimony: Through a positive response to the color reaction with rhodamine B (compare page 105).

Molybdenum, tungsten, vanadium: By the procedure given on page 417 for MoO_3, WO_3, V_2O_5.

(2) *Insoluble sulfates*

This category includes $PbSO_4$, $BaSO_4$, $SrSO_4$, $CaSO_4$.

A general test for slightly soluble sulfates is to treat a little of the material on a watch glass with a drop of an acid solution of mercuric nitrate [10 g $Hg(NO_3)_2$ plus 1 ml concentrated HNO_3, in 100 ml water]. Yellow basic mercuric sulfate is formed.[26] $CaSO_4$ and $PbSO_4$ respond at room temperature; $SrSO_4$ and $BaSO_4$ must be warmed to develop the yellow. If sulfides and halogen compounds are absent, the test is decisive. The test material must be colorless; otherwise the yellow cannot be seen.

When neither sulfides or elementary sulfur are present, it is permissible to reduce insoluble sulfates to a soluble sulfide by fusion with metallic potassium (see page 315). The sulfide component may then be detected by the iodine-azide reaction or the sodium nitroprusside reaction. Because of the high sensitivity of the iodine-azide reaction, a fraction of a milligram of sample is sufficient for the detection of sulfate.* The nitroprusside test

* The reduction of insoluble sulfates to alkali sulfides in the alcohol flame has been recommended by Hahn [27]. The procedure in question is described in Vol. II.

is less sensitive but is satisfactory for many purposes. A larger sample is required.

The reduction of sulfate to sulfide may also be accomplished [28] by cautious heating of milligram quantities of the sample along with sodium azide. The sodium resulting from the decomposition: $2\ NaN_3 \rightarrow 2\ Na + 3\ N_2$ is active here.

Lead sulfate may be detected in the presence of alkaline earth sulfates as follows: A sample of the order of 1 milligram is stirred in a micro crucible with a warm mixture of acetic acid and ammonium acetate. The lead sulfate dissolves (formation of slightly dissociated or complex lead acetate). A strip of filter paper is dipped in the suspension and left there for 10 minutes. The lead salt ascends with the solvent. The lead may then be detected in the paper several millimeters above the surface of the liquid, either by sending hydrogen sulfide gas against the paper or by applying a drop of hydrogen sulfide or ammonium sulfide solution.

A quicker, surer, and more sensitive direct test for lead sulfate in an insoluble residue (containing no strontium sulfate) is provided by spotting with a fresh 0.2 % solution of sodium rhodizonate.[29] A blue-violet color appears; it turns scarlet when spotted with acetic acid. The color is due to lead rhodizonate (see page 73).

When lead has been found to be absent from an insoluble sulfate residue, barium sulfate (or strontium or calcium sulfates) are likely to be present. The alkaline earth metals may be detected as follows: A small test portion is mixed with 3 to 4 times its bulk of sodium potassium carbonate and fused in a loop of platinum wire. The melt is heated with water in a microcentrifuge tube and then centrifuged. After decanting the supernatant liquid, the residue is washed several times with water and collected by centrifuging. Finally the solid is dissolved in dilute hydrochloric acid and tested for alkaline earths by the flame color.

It should be noted that the respective water-soluble chlorides are produced if the sulfates of lead, barium, strontium, and calcium are heated with excess solid ammonium chloride. The heating must be conducted so slowly that no fumes of ammonium chloride are evolved.

(3) *Silver halides*

AgI, AgBr, AgCl, AgCN, AgCNS, AgN_3 may be present.

A good and rapid test for silver halides is based on their ready solubility in alkali cyanides, with formation of complex alkali silver cyanides. When the cyanide is used in the form of a solution of potassium nickel cyanide, nickel cyanide is liberated:

$$AgHal + K_2[Ni(CN)_4] = K[Ag(CN)_2] + Ni(CN)_2 + KHal$$

The demasked nickel cyanide can react with an alcoholic solution of dimethylglyoxime with formation of red nickel dimethylglyoxime. At room temperature, the yellow $K_2[Ni(CN)_4]$ solution reacts with dimethylglyoxime hardly at all and consequently a stable mixed solution of the two compounds may be prepared. It produces nickel dimethylglyoxime when brought into contact with silver halides.*

The nickel cyanide formed by the foregoing reaction reacts sluggishly with dimethylglyoxime in the cold, but on addition of ammonia, which dissolves the nickel cyanide, the reaction is quite rapid.** A solution containing $K_2[Ni(CN)_4]$, ammonia, and dimethylglyoxime therefore reacts almost instantaneously with all silver halides (freshly precipitated or aged by drying for several weeks). Red nickel dimethylglyoxime is formed.

To detect a silver halide in an "insoluble residue", it is only necessary to place a bit of the sample on a white spot plate and add 1 or 2 drops of the reagent (see below). A bright red color or precipitate indicates AgCl or AgBr, AgI, AgCNS, AgN_3, AgCN. Silver iodide reacts a little more slowly, owing to its small solubility but the reaction is complete in thirty seconds even in this case.

The *reagent* is best prepared by boiling freshly precipitated and washed $Ni(CN)_2$ with KCN solution, using a deficit of the latter so that not all of the $Ni(CN)_2$ goes into solution. The resulting $K_2[Ni(CN)_4]$ solution (filtrate) keeps. Immediately before the test, a few milliliters of this solution are mixed with a few drops of ammonia, and a few drops of a saturated alcoholic solution of dimethylglyoxime; the mixture is filtered, if necessary.

Silver halides can be detected also through their behavior toward an ammoniacal solution of potassium ferrocyanide containing α,α'-dipyridyl.[31] On warming, the ferrous iron is demasked:

$$Fe(CN)_6^{-4} + 3\ AgHal \rightarrow Fe^{+2} + 3\ [Ag(CN)_2]^{-1} + 3\ Hal^{-1}$$

The ferrous ions set free react with the α,α'-dipyridyl to give a red color (compare detection of iron, page 161). The test is conducted in a microtest tube. A little of the solid is treated with a drop of 1 % potassium ferrocyanide solution, followed by a drop of 2 *N* ammonia, and a drop of a 1 %

* A. R. Ubbelohde, *Analyst*, 59 (1934) 339, was the first to recommend a solution of complex nickel cyanide and dimethylglyoxime as reagent for metal ions that form undissociated cyanides (Ag^+, Pd^{+2}, Hg^{+2}).

** The sluggish reaction of $Ni(CN)_2$ with dimethylglyoxime is due to the fact that it should be viewed as $Ni[Ni(CN)_4]$. Accordingly, only the cationic half of the nickel reacts, and then only slowly because of the slight solubility of the salt. The latter readily dissolves in ammonia to form $[Ni(NH_3)_2][Ni(CN)_4]$ and the $[Ni(NH_3)_2]^{+2}$ ions then react at once with dimethylglyoxime.[30]

solution of α,α'-dipyridyl in alcohol. The mixture is warmed in a water-bath. Depending on the nature and quantity of silver halide, the initial colorless solution becomes pink to deep red within a few minutes.

The silver halides (AgCl, AgBr, AgI, AgCN, AgCNS) can be partially differentiated by their behavior on heating (ignition tube). Silver chloride and bromide melt without decomposition; silver iodide remains unchanged. Silver cyanide and thiocyanate decompose with production of dicyanogen and metallic silver, or a mixture of the metal and silver sulfide is left. Therefore, a positive test for metallic silver after ignition points to the presence of silver cyanide or thiocyanate. Considerable quantities of elementary silver make the ignition residue gray or black. Small quantities of metallic silver can be detected by spotting the ignition residue with phosphomolybdic acid; the resulting blue is due to lower molybdenum oxides (see page 362). The ignition residue can also be digested with warm dilute nitric acid and, after any necessary filtration, the solution is tested with *p*-dimethylaminobenzylidenerhodanine (red precipitate; compare page 59). The ignition test should be applied only when silver azide is known to be absent, since it is explosive. Silver azide can be converted into silver chloride by digestion with sodium chloride solution. Regarding tests for silver azide and silver thiocyanate compare page 286 and 282, respectively.

Silver iodide can be reliably detected by means of palladium chloride.[32] A few milligrams of the sample are spread on filter paper, moistened with a drop of water, and held over boiling water. After 1 or 2 minutes, the spot is treated with a drop of a 1 % $PdCl_2$ solution. On further steaming, the surface of the silver iodide turns brown-black due to the production of palladous iodide. The color is discharged on treatment with ammonia.

Ignition of white silver ferrocyanide and yellow ferricyanide yields a mixture of elementary silver and ferric oxide. In mixture with other silver halides, the color change accompanying this decomposition is characteristic. Compare (*4*) regarding the pyrolysis of ferricyanides and ferrocyanides, and the detection of ignited ferric oxide.

Without exception, the silver salts of those acids which contain the CN-group, i.e. AgCN, AgCNS, $Ag_4Fe(CN)_6$, $Ag_3Fe(CN)_6$, release dicyanogen when heated over a bare flame. This gaseous product can be detected by the procedure given on page 371.

(*4*) *Ferrocyanides and ferricyanides*

Certain ferro- and ferricyanides are not soluble in strong non-oxidizing acids. Instances are: $Ag_4Fe(CN)_6$, $Ag_3Fe(CN)_6$, Prussian blue. When these salts are heated with sodium carbonate solution, the corresponding carbonate and sodium ferro- or ferricyanide results. After removing the metal

References pp. 426–427

carbonate, the filtrate (centrifugate) can be tested for ferro- or ferricyanide ions by the sensitive test described in Chapter 4.

All ferro- and ferricyanides, both soluble and insoluble, have the uniform characteristic of yielding dicyanogen when the dry salt is heated. For example:

$$Ag_4Fe(CN)_6 \rightarrow 4\ Ag^\circ + Fe^\circ + 3\ (CN)_2$$
$$ZnK_2Fe(CN)_6 \rightarrow Zn(CN)_2 + 2\ KCN + Fe^\circ + (CN)_2$$
$$K_4Fe(CN)_6 \rightarrow 4\ KCN + Fe^\circ + (CN)_2$$

These equations are not entirely valid because there is partial reaction between the iron and dicyanogen to give nitrogen and iron carbide, and furthermore the dicyanogen polymerizes in part to paracyanogen $(CN)_x$. An improved yield of dicyanogen can be obtained if the sample is taken to dryness with an alcoholic solution of mercuric chloride and then weakly ignited. For instance:

$$K_4Fe(CN)_6 + 6\ HgCl_2 \rightarrow 4\ KCl + FeCl_2 + 3\ Hg_2Cl_2 + 3\ (CN)_2$$

Analogous reactions occur with other soluble and insoluble ferro- and ferricyanides. The gaseous dicyanogen produced by pyrolysis can be detected as described on page 371 through the blue or red color that appears when the gas is brought into contact with copper-benzidine acetate or potassium cyanide solution containing oxine.

Apart from their differences in color, silver ferrocyanide and silver ferricyanide can be distinguished through their behavior toward dilute ammonia solutions. Silver ferrocyanide is not soluble and consequently it can be separated from other silver halides, with the exception of iodide (and bromide in part) by digestion in ammonia water. If the mixture of silver salts that does not dissolve in ammonia is ignited and the ignition residue then rubbed on a porcelain plate, it is easy to detect the presence of silver ferrocyanide through the brown color (due to Fe_2O_3).

Another sure way to detect silver ferrocyanide in the presence of silver ferricyanide and other silver halides is to suspend the sample in dilute ammonium hydroxide, add a drop of a 1 % alcoholic solution of α,α'-dipyridyl and warm. A red color appears (compare test for silver halides, page 409). If other acid-insoluble ferrocyanides are to detected, a slight quantity of the solid should be treated with a drop of an ammoniacal silver solution and a drop of 1 % alcoholic solution of α,α'-dipyridyl. A red color indicates the presence of ferrocyanide.

(5) *Ferric oxide*

Ignited ferric oxide and iron oxide ores dissolve in acids (in the absence of a reducing agent) with great difficulty. However, a special fusion is not necessary to convert the iron into soluble red ferric thiocyanate. It is sufficient to treat a small portion of the "insoluble" residue with a sulfuric acid solution of potassium or ammonium thiocyanate and warm gently in a micro crucible. The original colorless reagent solution turns blood-red or pink, according to the quantity of iron involved. Even a trace of iron in alumina may be rapidly detected in this way.

Small quantities or even traces of ferric oxide can be detected by fusion with 8-hydroxyquinoline at about 250° C.[33] Depending on the Fe_2O_3 content, the melt turns black-green to dark green, because of the formation of ferric oxinate, which is soluble in fused 8-hydroxyquinoline. For the same purpose the spot test with a solution of α,α'-dipyridyl in thioglycolic acid is recommended.[34] A red color (compare p. 162) is formed.

(6) *Aluminum oxide*

To detect ignited Al_2O_3 in a residue, the sample must be fused with alkali pyrosulfate or bisulfate (in a platinum spoon). These fluxes also serve as SO_3-donors. Soluble $Al_2(SO_4)_3$ results. It is also possible to put alumina into solution by fusing the sample with potassium persulfate. The reaction is carried out in a hard glass test tube.[35] The reaction: $K_2S_2O_8 \rightarrow K_2SO_4 + O + SO_3$ yields the SO_3 essential to convert the aluminum oxide into sulfate.

The following tests may be applied to the aqueous extract of the fusion residue. Any other metal ions present are precipitated with excess caustic alkali and the hydroxides are separated from the aluminate solution by centrifuging. The latter is acidified with acetic acid and mixed with a few drops of a saturated solution of morin in methyl alcohol. A green fluorescence appears in the presence of aluminum (see page 182). Alternatively, the aluminate solution may be treated with a drop of an aqueous solution of sodium alizarin sulfonate and then acidified with acetic acid. The red alizarin lake indicates aluminum (see page 185).

(7) *Titanium^{IV} oxide*

Titanic acid anhydride, $Ti(OH)_4 \cdot aq.$ and basic Ti^{IV} salts dissolve readily in warm dilute sulfuric acid in the presence of hydrogen peroxide and yield a yellow color (pertitanic reaction, see page 195). Ignited TiO_2 must be fused with alkali pyrosulfate or bisulfate, as just described for aluminum oxide, before adding the sulfuric acid and hydrogen peroxide. It is simpler to heat the sample with concentrated sulfuric acid until white fumes appear. After cooling, the residue is treated with solid chromotropic acid (see page 197). A violet color indicates TiO_2.

(8) Zirconium oxide

Strongly ignited zirconium oxide is very difficultly soluble in acids. It can be converted into soluble zirconium sulfate by heating with concentrated sulfuric acid, or more rapidly by fusion with sodium bisulfate, or better with sodium pyrosulfate. In the case of fusion, the water-soluble double sulfate $Na_2Zr(SO_4)_3$ is also formed. Addition of a dilute acetone or an alcohol solution of morin to solutions of these zirconium sulfates produces a strong yellow-green fluorescence in ultraviolet light.[36] The fluorescence is the result of the production of a zirconium-morin adsorption system, which is not destroyed even when concentrated hydrochloric or 1 : 1 sulfuric acid is added (compare page 201). This test is specific for zirconium. Only larger quantities of ferric oxide interfere; the solution of the resulting ferric sulfate absorbs the ultraviolet rays and consequently no fluorescence color appears.

(9) Chromic oxide and anhydrous chromic chloride

These two compounds, which are very difficultly soluble in acids, are not hard to detect even when relatively small amounts are present in mixtures with other colorless substances. Their vivid colors (Cr_2O_3 green, $CrCl_3$ violet) are quite characteristic. To detect tervalent chromium decisively it is sufficient [even with dark gray acid-insoluble chrome iron ore (chromite) [$FeCr_2O_4$] to fuse a little of the powdered sample in a platinum spoon or a porcelain micro crucible with a mixture of sodium carbonate and sodium peroxide (1 : 1). Sodium chromate results.The cooled melt is dissolved in a few drops of concentrated sulfuric acid, and 1 or 2 drops of a 1 % alcoholic solution of diphenylcarbazide solution are added. A violet color indicates chromium (see page 167).

Mercury salts and molybdates in acid solution also give violet colors with diphenylcarbazide. When the insoluble residue is assumed to contain salts of mercury (e.g., HgS), they must be completely eliminated by long ignition previous to the carbonate-peroxide fusion. Alternatively, hydrochloric acid may be used in place of sulfuric acid to dissolve the melt; non-reacting $[HgCl_4]^{-2}$ ions are formed. Any reaction with molybdate may be masked by the addition of a few drops of a saturated aqueous solution of oxalic acid; the complex molybdenum-oxalic acid (see page 171) is formed.

Acid-insoluble Cr^{III} compounds may also be decomposed by fusion (in a platinum spoon) with potassium bifluoride.[37] The resulting potassium chromate can be detected by the diphenylcarbazide color reaction. This procedure eliminates interference by mercury and molybdenum compounds. The

References pp. 426–427

former are thermally decomposed and volatilized; the latter are converted into alkali fluomolybdate, which does not react with diphenylcarbazide.

(*10*) *Lead chromate*

Yellow precipitated lead chromate, either wet or dried, is readily dissolved or decomposed by dilute mineral acids. In contrast, lead chromate that has been fused (m.p. 844° C) is red-brown and is quite resistant to concentrated mineral acids, even in the presence of hydrogen peroxide or sulfurous acid. Accordingly, fused lead chromate may be present in an insoluble residue.*

Fused lead chromate may be detected by the fact that dilute sulfuric acid extracts from it, even at room temperature, sufficient chromic acid to respond to the very sensitive diphenylcarbazide test (page 167). It is better to subject the finely powdered sample to the combined action of 5 *N* sulfuric acid and 1 % alcohol solution of the reagent (2 drops of each). An intense violet color will appear in a few minutes.

Strongly ignited (acid-insoluble) molybdenum trioxide also gives a violet color with diphenylcarbazide. In case its presence is suspected, a saturated solution of oxalic acid should be used in place of the 5 *N* sulfuric acid just prescribed. Oxalic acid masks the MoO_3 by forming complex molybdic-oxalic acid and thus prevents the interfering MoO_3-diphenylcarbazide reaction.

A blue color develops when slight amounts of lead chromate are added to a solution of diphenylamine in concentrated sulfuric acid (compare page 419). Higher metal oxides and large quantities of antimony pentoxide behave likewise. The former can be easily removed by the procedure described on page 367 under "Metallic tungsten and molybdenum".

(*11*) *Manganese dioxide and lead dioxide*

The mineral pyrolusite (MnO_2) is very resistant to even hot strong acids, while PbO_2 is similarly resistant to fairly dilute hydrochloric and nitric acids. Both oxides may, however, readily be brought into solution by treatment with dilute nitric acid and hydrogen peroxide, even at room temperature.** Therefore, when a very small sample of a dark insoluble

* After fusion, lead chromate no longer corresponds exactly to the formula $PbCrO_4$. Part of the salt decomposes when it melts ($4\ PbCrO_4 \rightarrow 4\ PbO + 2\ Cr_2O_3 + 3\ O_2$). Consequently, only precipitated lead chromate dissolves completely in sodium hydroxide ($PbCrO_4 + 4\ NaOH \rightarrow Pb(ONa)_2 + Na_2CrO_4 + 2\ H_2O$) whereas the cold melt always leaves a considerable residue of green Cr_2O_3 on treatment with alkali solutions.

** Whereas neutral hydrogen peroxide is catalytically decomposed on the surface of the manganese dioxide, with evolution of oxygen, the following reaction occurs in the presence of acids: $MnO_2 + 2\ H^+ + H_2O_2 = Mn^{+2} + 2\ H_2O + O_2$. Lead peroxide and other higher metallic oxides behave analogously.

residue is treated on a spot plate with a drop of 2 *N* nitric acid and a drop of 3 % hydrogen peroxide, decolorization indicates the presence of MnO_2 or PbO_2, or a mixture of them.

These oxides also react with an acetic acid solution of benzidine to produce a blue color (see pages 72 and 175). It is sufficient to streak a little of the sample on quantitative paper and treat the streak with a drop of benzidine acetate solution. The blue color appears at once.

Chromate is formed by heating a small sample containing PbO_2 or MnO_2 in a micro crucible with a solution of chromic sulfate in strong sulfuric acid. The product may be identified, after cooling, by the violet color formed on the addition of a drop of a 1 % alcoholic solution of diphenylcarbazide (page 167). The $Cr_2(SO_4)_3$ solution used must be chromate-free (blank test); otherwise it must be treated with a few crystals of sodium sulfite and boiled before the final test is made.

It is easy to distinguish between manganese dioxide and lead peroxide, whose colors (black-brown to black) are identical. A few milligrams or even less of the sample is spotted with freshly prepared sodium rhodizonate solution and dilute acetic acid. Red lead rhodizonate (see page 73) is formed at once, whereas manganese dioxide gives no reaction.

(*12*) *Tungsten*^{VI} *oxide*

Tungstic acid anhydride yields a deep blue when warmed with a solution of stannous chloride in strong hydrochloric acid. Insoluble lower oxides of tungsten are formed. It is advisable, especially in the detection of WO_3 in minerals, first to decompose the sample by fusing with sodium peroxide in a micro crucible, and then to spot the fusion residue with a strongly acid solution of stannous chloride. It should be noted that, under these conditions, a blue color (molybdenum blue) is produced also by MoO_3.

Tungsten^{VI} oxide, in quantities up to 0.2 mg, can be separated from molybdenum^{VI} oxide and also from acid-insoluble basic oxides, by brief fusion with excess alkali bisulfate or pyrosulfate. This process not only converts basic oxides into water-soluble sulfates, but also transforms ignited MoO_3 into water-soluble alkali molybdate. The latter results from the reaction:

$$MoO_3 + Na_2S_2O_7 \rightarrow Na_2MoO_4 + 2\ SO_3$$

Although tungstic acid, WO_3, likewise dissolves in the melt, because of the formation of Na_2WO_4, it is reprecipitated on leaching with water because of the action of excess sodium pyrosulfate, which behaves toward sodium tungstate like free sulfuric acid.* The WO_3.aq. produced in this manner filters and centrifuges well. The isolated product gives tungsten blue if treated with a drop of 1 : 1 hydrochloric acid and a little metallic zinc. Alternatively, the tungstic acid, gathered in the centrifuge tube, may be thoroughly washed, dried, and then heated with ammonium chloride. Tungsten blue is formed: [38]

$$6\ WO_3 + 2\ NH_4Cl \rightarrow 3\ W_2O_5 + N_2 + 2\ HCl + 3\ H_2O$$

If even traces of molybdenum are absent, tungstic acid can also be detected by the procedure just described.

(13) Oxides of molybdenum, tungsten, and vanadium [39]

These oxides, in line with their acidic nature, are soluble in ammonium hydroxide with which they form the ammonium salts of the corresponding metallo acids. If these solutions are treated with hydrogen peroxide, ammonium salts of the per acids result. For example:

$$(NH_4)_2MoO_4 + H_2O_2 \rightarrow (NH_4)_2MoO_5 + H_2O$$

By virtue of the groups (I) and (II) or the corresponding anions (III)

Me –OOH	Me –OO –NH_4	Me –OO^{-1}	
(I)	(II)	(III)	(Me = 1 equivalent Mo,W,V)**

the solutions of the per acids and their alkali salts have a far greater oxidizing action than the equivalent quantity of hydrogen peroxide in acid or ammoniacal solution.[40] Since the formation of (I)–(III) from the metallo acids or their ammonium salts with hydrogen peroxide is immediate, the oxidizing action of hydrogen peroxide in acid or ammoniacal surroundings is enhanced by the addition of MoO_3, WO_3, or V_2O_5.

This activation is strikingly demonstrated by the following effect. Ammoniacal solutions of thallium^{I} salts, which remain unchanged on the introduction of hydrogen peroxide, immediately precipitate brown thallium^{III} oxyhydrate when molybdate is added, because of the formation and activity of MoO_5^{-2} ions. Consequently:

* If MoO_3 or WO_3 is sintered with an excess of sodium sulfate instead of pyrosulfate, these water-insoluble acidic oxides are completely transformed into the corresponding water-soluble sodium salts:

$$WO_3\ (MoO_3) + Na_2SO_4 \rightarrow Na_2W\ (Mo)O_4 + SO_3$$

** Vanadium salts on treatment with hydrogen peroxide yield peroxo-vanadium salts (compare page 123) which can likewise be regarded as derivatives of hydrogen peroxide. With regard to the activation of hydrogen peroxide described and used here, they behave like vanadium per acid. Possibly there is a partial conversion into the latter.

$$Tl^{+1} + NH_3 + H_2O + H_2O_2 \rightarrow \text{no reaction} \quad (1)$$

$$Tl^{+1} + NH_3 + H_2O + MoO_5^{-2} \rightarrow TlO(OH) + MoO_4^{-2} + NH_4^+ \quad (2)$$

Reactions analogous to (2) may be assumed for tungstates and vanadates. Therefore, all of these cases involve catalytic effects effectuated by the formation, action, and regeneration of per compounds.

The foregoing findings have been made the basis of a convenient procedure for the detection of MoO_3, WO_3, and V_2O_5 even in mixture with other oxides. It is essential to use ammoniacal surroundings in this activation of the thallium^{I}-hydrogen peroxide reaction, because in caustic alkali solution hydrogen peroxide by itself will precipitate the brown TlO(OH).

Procedure. A slight quantity of the solid (fractions of mg are enough) is digested warm with several drops of 2 *N* ammonium hydroxide and then centrifuged. One drop of the centrifugate is placed on a spot plate and treated in succession with a drop of 3 % hydrogen peroxide solution and a drop of a 5 % solution of thallous nitrate in 2 *N* ammonia. An immediate brown precipitate or color indicates the presence of MoO_3, WO_3, or V_2O_5.

With respect to the sensitivity of this test, it was found that the *limits of dilution* are around 0.05 γ MoO_3, WO_3, V_2O_5. When very slight amounts of these oxides are involved, the appearance of the color may require several minutes. Under such conditions it is best to make a comparison blank test.

(*14*) *Metallic tungsten*

Molybdenum and vanadium are soluble in strong nitric acid forming the corresponding metallo acids, whereas tungsten is attacked little if at all. This invulnerability is likewise true of highly alloyed tungsten steels and concentrates carrying high proportions of tungsten. Such materials can be decomposed by applying the fact that fusion with alkali nitrite rapidly produces water-soluble alkali tungstate.[41] The fusion may be accomplished in a porcelain crucible. After the melt has cooled, it is leached with water. A drop of the clear filtrate or centrifugate is tested for tungsten by the procedure given in (*13*).

(*15*) *Silica (silicates), silicon (silicides) and borides*

Silica and silicates can be converted into alkali silicates by fusion with sodium–potassium carbonate in a loop of platinum wire or a platinum spoon. The melt is disintegrated with a hot nitric acid solution of ammonium molybdate (preferably in a platinum crucible). A yellow solution of silicomolybdic acid results if silica or silicates are present. A yellow crystalline precipitate in a colorless solution indicates the presence of phosphates.

References pp. 426–427

Traces of silicomolybdic acid (down to 0.1 γ SiO_2) may be detected by the blue color obtained on spotting with an acetic acid solution of benzidine (see page 335).

To convert silicon and silicides to silicate, it is necessary to fuse the sample with a 1 : 1 mixture of sodium carbonate and sodium peroxide on silver or nickel foil. The melt (after cooling) is dissolved in nitric acid in a platinum crucible, and tested for silicomolybdate as just described.

Borides may be decomposed in the same way as silicon compounds. The resulting boric acid may be detected by means of color reactions with hydroxyanthraquinones dissolved in concentrated sulfuric acid (see page 340).

(*16*) *Antimony*^{V} *oxide*

Ignited antimony pentoxide, which is difficultly soluble in acids, may be detected by spotting with a solution of potassium iodide containing a mineral acid and starch. The redox reaction liberates iodine:

$$Sb_2O_5 + 4\,HI \rightarrow Sb_2O_3 + 2\,H_2O + 2\,I_2$$

The reaction is not directly applicable when higher metallic oxides (MnO_2, PbO_2, etc.) are present. They must be removed by means of hydrogen peroxide and diluted mineral acid (page 415) and the test applied to the washed residue.

When higher metal oxides or fused lead chromate are absent, a more reliable test for Sb_2O_5 is based on its behavior toward a sulfuric acid solution of N,N'-diphenylbenzidine (compare page 327). Oxidation of the diphenylbenzidine is involved. After brief contact with solid Sb_2O_5, an intense blue appears.[42] Addition of solid sodium chloride hastens the development of the blue and makes it more permanent.

An excellent method for the detection of antimony pentoxide, even in very small quantities and in the presence of other colorless materials, is based on the production of an insoluble violet salt of the basic dye rhodamine B with H $[SbI_4]$. The procedure is described in detail on page 107.

(*17*) *Niobium oxide* [43]

White amorphous niobium^{V} oxide can be converted into the black-brown Nb^{III} oxide (perhaps even to the metallic state) by heating with reducing agents. This reduction can be accomplished by heating the oxide in the reducing area of a Bunsen flame. In this case, the carbon monoxide of the gas supply serves as the reductant:

$$Nb_2O_5 + 2\,CO = 2\,CO_2 + Nb_2O_3$$

The change in color is best observed if the solid test material is made into a paste or water suspension on a porcelain crucible lid, evaporated to dryness, and then heated strongly. The flame of the burner is then directed on the mass in such fashion that the inner (reducing) cone impinges on the solid. The darkening, which is easily seen with even small particles, disappears completely or partially when the flame is taken away because contact with the air brings about reoxidation of the hot Nb_2O_3. However, the darkening can be restored by repeating the reducing treatment. This test can be utilized to distinguish between tantalic and niobic oxides and also to detect slight amounts of Nb_2O_5 in Ta_2O_5. Titanic oxide is not affected. The test is applicable, of course, only to a colorless or possibly a yellow sample.

The oxides of molybdenum^{VI} (white), tungsten^{VI} (light yellow), and vanadium^{V} (brown) are converted into the intensely colored (dark blue, brown) lower oxides by this reducing thermal treatment. However, these relatively strong acid oxides, in contrast to the faintly acidic Nb_2O_5, can be completely removed from the test material by digestion with warm ammonia (production of water-soluble ammonium salts).

In the absence of oxides of molybdenum and tungsten, Nb_2O_5 can be detected by fusion with zinc chloride.[44] The black product is probably NbO_2; even small amounts are easily seen after the solidified melt is extracted with water. It is likely that anhydrous hydrogen chloride is liberated from the hydrated zinc chloride on fusion and in superheated form enters into the redox reaction with the niobic oxide:

$$ZnCl_2 + H_2O \rightarrow ZnOHCl + HCl$$
$$2\ HCl + Nb_2O_5 \rightarrow 2\ NbO_2 + Cl_2 + H_2O$$

Possibly the reduction does not come to a standstill with the formation of NbO_2 but continues in part to metallic niobium. Oxides of tantalum, titanium, zirconium, and the rare earths are not altered by fusion with zinc chloride. Therefore, this procedure makes possible the detection of slight amounts of niobic oxide in the presence of tantalum oxide.

(*18*) *Insoluble fluorides*

A good reaction for the detection of insoluble fluorides, such as calcium and thorium fluorides, rests on the fact that a violet hydrochloric acid solution of zirconium alizarinate turns yellow in the presence of fluorides, because of the formation of $[ZrF_6]^{-2}$ ions and alizarin (see page 269). A very small portion of the insoluble residue is mixed with 1 or 2 drops of the reagent ($ZrCl_4$ solution in hydrochloric acid mixed with excess sodium alizarin sulfonate), gently heated, and observed for any change from violet to yellow.

Alternatively, the sample may be heated with silica and concentrated sulfuric acid; SiF_4 is liberated. The latter may be hydrolyzed to silicic acid, which is identified by the sensitive silicomolybdate reaction (see page 335).

Since all insoluble fluorides are quantitatively decomposed by warming with concentrated sulfuric acid, with evolution of hydrogen fluoride, the test described on page 274 can be used to detect insoluble fluorides. This test is based on the change in wettability of glass that has been attacked by hydrogen fluoride.

Insoluble fluorides can be removed from other insoluble substances by digestion with an acid solution of beryllium chloride or nitrate through the formation of complex $[BeF_4]^{-2}$ anions.[45] For example:

$$2\ CaF_2 + Be^{+2} \rightarrow 2\ Ca^{+2} + [BeF_4]^{-2}$$

Solution occurs with even gentle warming.

(19) Tin^{IV} oxide and basic tin^{IV} phosphate

Tin dioxide as well as the adsorption compound between stannic hydroxide and phosphoric acid (basic stannic phosphate) are insoluble in strong mineral acids and may therefore be present in an "insoluble" residue. The detection of tin is possible by means of the methods described below.

A little of the insoluble residue is covered with about 6 ml concentrated hydrochloric acid in a evaporating dish. Small pieces of pure zinc rod are added to liberate hydrogen, and the mixture is stirred with a small test tube filled with cold water. The test tube is then removed from the liquid and the portion that was immersed placed in the nonluminous portion of a Bunsen flame. A characteristic blue mantle of flame forms around the tube in the presence of even extremely small amounts of tin. Since this flame coloration (luminescence reaction) is an extremely sensitive test for tin (see page 109), only a trace of sample is necessary for the detection of stannic oxide in an insoluble residue. Glass may be tested for tin by this procedure.

Alternatively, a few milligrams of the insoluble residue may be fused with potassium cyanide in a porcelain micro crucible. The resulting metallic tin is isolated by extracting the melt with water and centrifuging, and then dissolved in hydrochloric acid (caution: hydrogen cyanide). The stannous chloride is then identified by the violet color which develops when a drop of the solution is placed on filter paper impregnated with cacotheline (page 109). The test with ammonium molybdate (page 108) may also be applied.

An excellent method for the identification of tin in SnO_2 and basic stannic phosphate is based on the tin-morin reaction described on page 110 (formation of a fluorescing adsorption compound of morin with tin^{IV} hydroxide).

The product is resistant to acetic acid. The adaptation of this test for the identification of tin in acid-insoluble materials is fully described on pages 439 ff.

If stannic oxide is mixed with about 15 times its weight of ammonium iodide and heated to 500° C. the tin is completely volatilized as yellow-red SnI_2, because of the action of the hydrogen iodide liberated in the thermal dissociation of the ammonium iodide.[46] The oxides of antimony and arsenic likewise form volatile iodides.

(20) Carbon and carbides

Carbon, if present as hard coal, brown coal, charcoal, graphite, or in heavy carbides, is not attacked at all when its mixtures with inorganic materials are treated with hydrochloric acid. Nitric acid and *aqua regia* produce only a slight effect. Consequently, carbon may be present in an insoluble residue, provided the latter is not white or yellow. Materials with these colors can be assumed to contain no free carbon. The presence of carbon in coals is disclosed by the glow which appears when the sample is heated with access to air. The test can be made easily by slowly heating a few milligrams of the insoluble residue on platinum foil.

A more sensitive test for carbon (in coals) is to drop small portions of the insoluble residue into molten potassium chlorate (m.p. 370° C), contained in a porcelain crucible. The carbon oxidizes with decrepitation. No higher oxides (especially MnO_2), should be present, since they cause a violent evolution of oxygen from the chlorate. If the test described on page 415 has given a positive response for higher oxides, the latter must be destroyed beforehand by warming with concentrated hydrochloric acid and hydrogen peroxide. The resulting residue is washed, dried, and then tested with melted chlorate. The carbon of resistant graphite and resistant heavy metal carbides cannot be detected by this procedure.

If reducing compounds are absent or have already been removed, the test described on page 381 can be applied to reveal carbon and organic materials. It is based on the thermal reaction of the sample with potassium iodate (formation of iodide). However, this test fails with graphite and metal carbides.

When carbon and carbonaceous inorganic compounds (even carbides and acetylides) are heated with sodium azide, the heat-resistant sodium cyanide is formed by direct combination:[47]

$$C + Na + N \rightarrow NaCN$$

The reactants for carbon (carbides and acetylides) are formed by thermic decomposition of sodium azide:

References pp. 426–427

$$NaN_3 \rightarrow Na + 3\ N$$

Molten sodium or sodium vapors react with carbon. The resulting sodium cyanide can be identified by the Prussian blue test, i.e., by addition of a mixture of ferric and ferrous salts to the ignition residue and subsequent acidification. Also the cyanide test (page 278) can be used; it is based on the demasking of alkali palladium dimethylglyoxime.

Small amounts (less than 1 mg) of the material to be tested are mixed with a tenfold amount of sodium azide in a micro test tube and cautiously heated. After cooling, one drop of water is added and the cyanide test is carried out. All forms of carbon, including graphite and carbides (e.g., silicium carbide), can be detected in this way.[48]

9. Separation by the Ring Oven Method [49]

A separation scheme, requiring only one drop of test solution, was worked out with the aid of the Weisz ring oven method described on pages 54ff. It comprises the ions: Pb, Bi, Cd, Sb, Sn, Fe, Ni, Co, Al, Zn, Ti.

The test solution must not contain too much hydrochloric acid. One drop (1.5 μl) is placed in the center of a circle of quantitative filter paper. All of the lead, bismuth, copper, cadmium, antimony, tin is precipitated and fixed as sulfide by means of hydrogen sulfide generated from zinc sulfide and dilute sulfuric acid. The apparatus shown in Fig. 35 (page 55) is used. The paper is then placed on the ring oven and the unprecipitated ions are washed out of the fleck by means of 0.1 N hydrochloric acid and collected in a sharply defined ring zone. The paper is dried in an oven and the fleck is cut out of the paper with a 12 mm punch. The rest of the paper now contains (in the ring zone) all of the iron, cobalt, nickel, chromium, manganese, zinc, aluminum, and titanium ("Ring I"). The small punched-out disk is exposed to bromine vapor to oxidize the sulfides and then fumed with ammonia, and then dried briefly to insure the fixing of any antimony and tin. The oxidation is best conducted in the apparatus shown in Fig. 34b (page 55). The disk now contains all of the lead as sulfate, the copper and cadmium as tetrammine complexes, the antimony and tin as hydrous oxides, and the bismuth as basic sulfate. The dried disk is centered on a fresh circle of filter paper, and put into position on the ring oven, and washed there with 1 : 5 ammonium hydroxide to transfer all of the copper and cadmium to the underlying filter paper. The latter is dried and now contains these ions in a sharply defined concentric ring zone ("Ring II").

The disk, on which all of the lead, bismuth, tin, and antimony is present in the initial fleck, is now placed on a second fresh circle of filter paper, and washed on the ring oven with yellow ammonium sulfide to transfer all of the tin and antimony into this new ring zone. The disk and the filter paper are dried. The latter now contains (in "Ring III") all of the antimony and tin as sulfides, while the disk contains all of the lead and bismuth in the original fleck, likewise as sulfides.

The entire separation procedure requires about 15 minutes. The various steps are summed up in the following chart:

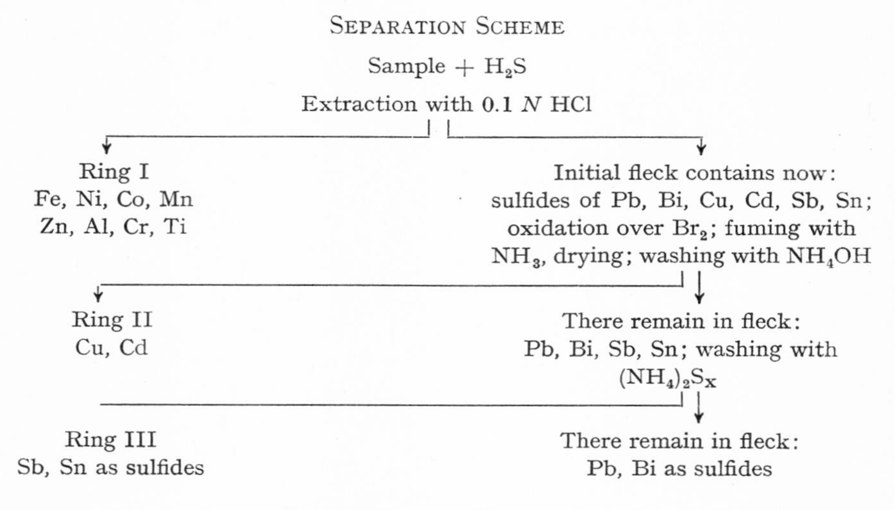

IDENTIFICATION REACTIONS

DISK: The disk is fumed over bromine and ammonia.

Pb: Spray with 0.2 % sodium rhodizonate solution, fume with HCl to disappearance of self-color of the reagent. Violet to red color indicates lead.

Bi: Bismuth is detected in the same fleck. The rhodizonate is destroyed by fuming over bromine and ammonia. The disk is treated on a spot plate with 5 % hydrogen peroxide and dried in an oven to oxidize any remaining sulfide. A drop of saturated lead chloride solution and 2 drops of alkaline stannite solution are then added. If bismuth is present, a black precipitate appears. A blank is advisable.

RING I: The paper is cut into several sectors.

Fe: The sector is fumed over bromine and then sprayed with a solution of potassium ferrocyanide. A blue mark indicates iron.

Co: The sector is fumed over ammonia, and any iron is masked by sodium phosphate. The paper is then sprayed with α-nitroso-β-naphthol. A brown-red streak indicates cobalt.

References pp. 426–427

Ni: The paper is fumed with ammonia, and sprayed with a 1 % solution of dimethylglyoxime. Nickel yields a red mark. If much ferric hydroxide is present, it can be decolorized by bathing the paper in tartrate solution.

Cr: The barely visible streak is followed by a capillary held at an angle and containing 10 % hydrogen peroxide and ammonia. This produces a very narrow moistened line. After drying, the specimen is sprayed with 1 % alcoholic solution of diphenylcarbazide and then with 2 *N* sulfuric acid. Chromium produces a violet streak.

Mn: The sector is sprayed with ammoniacal silver nitrate solution, and warmed. If manganese is present, a dark streak appears due to the formation of metallic silver and manganese oxide.

Zn: The paper is sprayed with mercuric thiocyanate solution and then bathed in 0.2 % cobalt acetate solution. A blue streak indicates zinc.

Al: The specimen is moistened with a methyl alcohol solution of morin, dried, and then bathed in 2 *N* hydrochloric acid. A streak, which fluoresces green in ultraviolet light, signals the presence of aluminum.

Ti: The paper is sprayed with a 5 % solution of chromotropic acid. If much iron is present, the specimen should then be moistened with stannous chloride. Titanium gives a red-brown streak.

RING II: The paper is cut into several sectors.

Cu: A sector is fumed with ammonia, and sprayed with 1 % alcoholic solution of rubeanic acid. An olive-green to black streak indicates copper.

Cd: The paper is moistened with hydrogen sulfide water and then rinsed with water and potassium cyanide solution. The yellow streak due to cadmium sulfide can be made more distinct by subsequent development with 1 % silver nitrate solution. A black-brown streak indicates cadmium.

RING III: The paper is cut into several sectors.

Sn: The paper is treated with hydrogen peroxide and ammonia applied with a capillary held at an angle (see Cr). After this oxidation, the paper is sprayed with 0.02 % morin solution in acetone, followed by bathing in glacial acetic acid-ethyl alcohol (1 : 20). A yellow-green fluorescence under a quartz lamp indicates tin.

Sb: The narrow streak is carefully cut out with scissors, and any antimony sulfide dissolved out on a spot plate with 1 : 1 hydrochloric acid. One drop of potassium iodide solution and one drop of sodium bisulfite solution are added and stirred in. A drop of 0.5 % rhodamine B

References pp. 426–427

solution yields a blue-violet precipitate if antimony is present. A blank is advisable. The original paper should be consulted for additional details.

Alternative Procedure

The filter papers containing the ring zones I, II, III are cut into several sectors and subjected to the appropriate preliminary treatments. The ions of the separate groups are then revealed by spraying the sectors with the prescribed reagents. Obviously, the substances concentrated in the small zone may not be washed out again. The original paper should be consulted for additional details.

Identification Reactions

Ring I: Iron with potassium ferrocyanide.
Nickel with dimethylglyoxime.
Cobalt with α-nitroso-β-naphthol.
Chromium with diphenylcarbazide.
Manganese with ammoniacal silver nitrate.
Zinc with mercuric thiocyanate and cobalt acetate.
Aluminum with morin.
Titanium with chromotropic acid.

Ring II: Copper with rubeanic acid.
Cadmium with hydrogen sulfide water and development with silver nitrate.

Ring III: Tin with morin.
Antimony with rhodamine B.

Disk: Lead with sodium rhodizonate.
Bismuth (after destruction of the sodium rhodizonate with bromine vapor) with sodium stannite and lead chloride.

References

1. E. Hauser, *Z. anal. Chem.*, 60 (1921) 88.
2. C. J. Van Nieuwenburg, *Mikrochemie*, 9 (1931) 199.
3. C. C. Miller, *J. Chem. Soc.* (1940) 1258, 1263; (1941) 72, 786; (1943) 73; (1947) 1347; (1951) 3188; *Metallurgia*, 31 (1944) 39.
4. A. A. Noyes and W. C. Bray, *A System of Qualitative Analysis for the Rare Elements*, New York, 1948.
5. G. Charlot, D. Bezier and R. Gauguin, *Rapid Detection of Cations*, New York, 1954.
6. G. Gutzeit, *Helv. Chim. Acta*, 12 (1929) 829.

7. K. Heller and P. Krumholz, *Mikrochemie*, 7 (1929) 213. The statements of these writers have been amplified and completed in some places. Compare W. C. Davies, *J. Chem. Education*, 17 (1940) 231 regarding a similar scheme of qualitative analysis employing spot tests.
8. P. Krumholz (Vienna), unpublished studies.
9. K. Heller, *Mikrochemie*, 8 (1930) 33.
10. H. Weisz, See references 17, 18, 19, 20, page 56.
11. See the extensive treatment in W. Böttger, *Qualitative Analyse*, Leipzig, 1925, and in C. R. Fresenius, *Qualitative Analysis*, 17th edition (translated by C. A. Mitchell), New York, 1921.
12. F. Feigl and A. Schäffer, *Anal. Chim. Acta*, 7 (1952) 507.
13. Compare E. W. Blank, *J. Chem. Education*, 20 (1943) 171.
14. F. Feigl, *Proc. Int. Symp. on Reactivity of Solids*, Gothenburg, 1952, p. 115.
15. Comp. Cl. Duval, *Inorganic Thermogravimetric Analysis*, Amsterdam, 1953, pp. 189, 211.
16. F. Feigl and J. A. Amarel, *Anal. Chem.*, in press.
17. In part from F. Feigl, unpublished studies.
18. According to W. D. Treadwell, *Tabellen zur qualitativen Analyse*, 13th ed., Vienna, 1932, p. 78.
19. F. Feigl and Cl. Costa Netto, unpublished studies.
20. F. Feigl, *Chemistry & Industry*, 57 (1938) 1161.
21. C. Duval and G. Mozars, *Compt. rend.*, 207 (1938) 862. Statements regarding the differentiation of Cl^-, Br^-, I^-ions are given there.
22. Compare F. Feigl, *Z. anal. Chem.*, 57 (1918) 135.
23. Compare F. Feigl, *Mikrochemie*, 20 (1936) 198.
24. E. R. Caley and M. G. Burford, *Ind. Eng. Chem., Anal. Ed.*, 8 (1936) 63.
25. F. L. Hahn (Mexico), private communication.
26. G. Denigès, *Bull. soc. chim.*, 23 (1918) 36.
27. F. L. Hahn, *Ind. Eng. Chem., Anal. Ed.*, 17 (1945) 199.
28. F. Feigl and D. Goldstein, unpublished studies.
29. F. Feigl and H. A. Suter, *Ind. Eng. Chem., Anal. Ed.*, 14 (1942) 840.
30. Compare F. Feigl and coworkers, *Bull. Lab. Prod. Mineral*, Minist. Agricultura Brazil, 1944.
31. F. Feigl and A. Caldas, *Anal. Chim. Acta*, 3 (1955) 526.
32. N. Orlow, *Chem. Zentr.*, 1906, II, 630.
33. F. Feigl and L. Baumfeld, *Anal. Chim. Acta*, 3 (1949) 15.
34. F. Feigl and A. Caldas, *Anal. Chem.*, 29 (1957) 580.
35. G. Vortmann, *Z. anal. Chem.*, 87 (1932) 193.
36. G. Charlot, *Anal. Chim. Acta*, 1 (1947) 234.
37. A. Guerreiro (Rio de Janeiro), private communication.
38. F. Feigl, unpublished studies.
39. F. Feigl and Cl. Costa Neto, *Monatsh.*, 86 (1955) 336.
40. Comp. F. Feigl, *Chemistry of Specific, Selective and Sensitive Reactions*, New York, 1949, p. 297.
41. H. Arnold, *Z. anorg. Chem.*, 88 (1914) 74.
42. F. Feigl, unpublished studies.
43. A. Caldas (Rio de Janeiro), unpublished studies.
44. F. Feigl and A. Caldas, *Mikrochim. Acta*, (1956) 1310.
45. F. Feigl and A. Schäffer, *Anal. Chem.*, 23 (1951) 351.
46. E. R. Caley and M. G. Burford, *Ind. Eng. Chem., Anal. Ed.*, 8 (1936) 114; see also L. Moser, *Monatsh.*, 53 (1929) 39.
47. J. Mueller, *J. prakt. Chem.*, 95 (1917) 53.
48. F. Feigl and D. Goldstein, unpublished studies.
49. H. Weisz, *Mikrochim. Acta*, (1954) 376.

Chapter 7

Application of Spot Reactions in Tests of Purity, Examination of Technical Materials, Studies of Minerals

Many of the spot tests described in the foregoing sections of this book can be successfully applied, either unaltered or with very little modification, to the solution of problems presented to the practical and technical analyst. In control tests, in tests of purity of technical and pharmaceutical products, in food and water analysis, it is sometimes very important to be able to establish rapidly the presence or absence of certain materials that affect the quality of the material or that are characteristic of a particular origin or method of preparation. The rapid and indisputable detection of definite elements or compounds in samples of minerals and other natural products often has both scientific and technical importance. Spot reactions can often be used successfully in crime laboratories and in juridical chemical studies.

In spot analysis, tests are made directly with solid or liquid samples. However, this does not exhaust the field of application of spot reactions. Frequently, when confronted with solutions so dilute that direct testing by spot reactions is no longer possible, traces can be accumulated on solids which act as "collectors" or "trace catchers". The collector can then be separated from the solution and the traces of the sought material, which it has gathered together from the solution, can then be subjected to sensitive spot reactions right on the collector itself. Collector action [1] may be due to coprecipitation, formation of mixed crystals, and also adsorption. The adsorptive accumulation on solids for the purpose of subsequent detection, which in spot analysis takes the form of capillary separation on filter paper, is used in an extended form in Tswett chromatography and more recently in paper chromatography. The special texts and comprehensive publications should be consulted regarding these applications of spot reactions, which have thus contributed significantly to the solution of difficult microchemical problems.

With no claim to completeness, pertinent examples are described in the following pages. The methods selected have shown in repeated trials that they are especially suitable, and that they can be successfully accomplished

References pp. 512–515

with even small (10–500 γ) amounts of the sample and sometimes with no visible damage to the specimen.

Spot tests for the identification of many components in technical metallic products can be carried out by applying a drop of an appropriate reagent solution to the clean metal surface. References can be found in a paper by Evans and Higgs.[2]

1. Detection of Silver in Alloys and Plating [3]

The sensitive and selective test with *p*-dimethylaminobenzylidenerhodanine (page 59), can be easily applied to the detection of silver in alloys, metal coatings, jewelry, etc.

Procedure. The sample is drawn over a piece of unglazed porcelain (streak plate). The hair-fine scratch on the porcelain is treated with dilute nitric acid, applied with a fine glass rod, and the plate is held over a flame until the acid has evaporated. Filter paper (S & S 589g), moistened with dilute nitric acid, is pressed against the warm plate. After one minute, the paper is spotted with a saturated solution of *p*-dimethylaminobenzylidenerhodanine in acetone. A red violet streak appears if silver is present and remains unaltered after the paper has been bathed in acetone to remove the excess reagent.

This procedure can be conducted so as to obviate any interference from mercury, when amalgamated surfaces are under examination. The mercury can be volatilized by igniting the scratch on the streak plate or by heating the evaporation residue from a drop of the test solution. The silver nitrate will remain for the subsequent nitric acid treatment and is easily identified by the *p*-dimethylaminobenzylidenerhodanine test.

2. Detection of Copper in Alloys [4]

The rubeanic acid spot test for copper (page 87), combined with capillary separation in the paper, is useful in the qualitative examination of alloys, if minute quantities of copper are to be detected.

Procedure. A few grains of the sample are dissolved in 2 drops dilute nitric acid, using a Pavelka tube[5] or a centrifuge tube. For gold alloys, 2 drops of *aqua regia* are used. After solution, the liquid is evaporated to dryness and the residue is warmed gently with a few drops of dilute acetic acid. By tilting the tube, a drop of the acetic acid solution is placed on paper impregnated with a saturated alcoholic solution of rubeanic acid. In the absence of other metals which react with rubeanic acid, Cu rubeanate is formed as a black to olive-green circle or ring.

When the solution also contains nickel, a violet ring of Ni-rubeanate is formed around the copper stain, but separated from it by a white band (see page 88).

A brown central circle may appear, consisting of the silver or gold compounds with rubeanic acid, and surrounded by a narrow band of copper rubeanate.

The formation of these silver and gold compounds, which can seriously interfere with the detection of small amounts of copper, may be prevented by warming the acetic acid solution with a few crystals of potassium bromide or nitrite, or a little hydrogen peroxide, before carrying out the spot test. The silver is thus converted into AgBr or $H[AgBr_2]$ and the gold ions are reduced to the metal.

The reaction (page 93) with ammonium mercury thiocyanate (violet precipitate) is suitable to reveal traces of copper in metallic zinc or cadmium. After solution of 1 g metal in dilute nitric acid, evaporation, and solution of the residue in 2 ml water, addition of excess ammonium mercury thiocyanate, will reveal as little as 0.001 % copper in metallic cadmium, and 0.002 % copper in metallic zinc.[6]

The use of 1,2-diaminoanthraquinone-3-sulfonic acid (see page 85) is recommended [7] if copper is to be detected by a simple procedure and without visible damage to the sample (alloys, plating, etc.).

Procedure. The sample is drawn forcibly across a streak plate or the unglazed bottom of an evaporating dish. Nitric acid or *aqua regia* is applied to the streak with a thin glass rod. The solution is evaporated by cautious fanning with a flame; overheating must definitely be avoided. A strip of reagent paper moistened with 1 *N* NaOH is then pressed against the still lukewarm porcelain. After about 30 seconds, the paper is removed. A blue print (mirror image of the streak) indicates the presence of copper. In this way, as little as hundredths of a per cent of copper can be quickly detected in electro platings, light metals, etc.

The test for copper, based on its catalytic activity [8] (see page 80), can be applied in the form of a drop reaction for the detection of minimal quantities in alloys and pure metals. It is not necessary to prepare solutions of the sample. Minute particles of the solid (1 to 3 mg) can be tested directly on the spot plate. A comparison blank is advisable.

3. Detection of Small Amounts of Copper in Pharmaceutical Products or Foodstuffs

It is often necessary to test for copper when examining pharmaceutical preparations, especially extracts and essential oils. Because of its toxicity, a high copper content is not permissible, but small amounts (less than 0.001 %) are seemingly harmless. However, even small amounts of copper are forbidden in preserves.

A rapid method of detection is to ash the sample and then form the violet Cu-Zn-Hg-thiocyanate (see page 93). Comparison of the intensity of color with that of a precipitate from a known amount of copper, and prepared under the same conditions, can give a fairly accurate estimate of the copper content.[9]

Procedure. About 0.2 to 1 g of the substance is ashed in a small (5 ml) porcelain crucible, fumed with 2 drops concentrated nitric acid, and ignited until the carbon disappears. The ash is then taken to dryness with a drop of concentrated hydrochloric acid, and the residue gently warmed with a further drop of hydrochloric acid, and taken up in 2 ml water. The solution is poured into a test tube, and about 50 mg ammonium fluoride added to mask the effect of iron that is almost always present. The rest of the procedure is given on page 93.

4. Detection of Traces of Copper in Water [10]

If 10–100 milliliters of the water are shaken with several milligrams or centigrams of calcium fluoride or talc, traces of copper are collected through adsorption on the solid and thus completely removed from the sample. (This is one method of producing absolutely copper-free water for this test and other purposes.) The collector is gathered by centrifuging, and the test described on page 81 is applied to it. This test for copper is based on the catalytic effect of copper on the reaction: $2\,Fe^{+3} + 2\,S_2O_3^{-2} \rightarrow 2\,Fe^{+2} + S_4O_6^{-2}$, whose occurrence is revealed by the discharge of the red ferri-thiocyanate color. In the procedure, calcium fluoride (or talc), that has been shaken with copper-free water (see above) is placed in a depression of a spot plate. A like amount of the collector that has been shaken with the suspected water is placed in an adjacent depression. Equal volumes of the reagent solution are added. A slower disappearance of the red color in the case of the copper-free collector indicates the presence of copper in the water being examined.

Another extremely sensitive method [11] for the reliable detection of traces of copper in water is based on the production of copper cupferronate when cupferron (ammonium salt of nitrosophenylhydroxylamine) is added to the water. The product can be taken up in chloroform, the latter evaporated, and the residue tested with alizarin blue (page 92).

Procedure. A drop of dilute hydrochloric acid is added to 100 ml of the water being tested and one or two drops of a 6 % aqueous cupferron [12] solution introduced. (Small amounts of copper may produce no visible precipitate.) The treated water is shaken with 5 ml of chloroform. After separating the two layers, the chloroform is evaporated in a small porcelain crucible, a drop of a saturated pyridine solution of alizarin blue [13] is added to the residue, and the pyridine then volatilized. The residue is treated with a drop or two of acetic anhydride. A blue residue, that adheres firmly to the porcelain, indicates the presence of traces of copper.

In this way, 0.025 γ copper can be detected in 100 ml water. This corresponds to a dilution of 1 : 4,000,000,000.

Addition of cupferron followed by extraction with chloroform provides a method of obtaining water which is absolutely free of copper and iron.

5. Detection of Traces of Copper in Metallic Nickel or Nickel Salts [11]

The method just described for the detection of traces of copper in water by means of alizarin blue can also be employed to reveal traces of copper in metallic nickel or nickel salts.

Procedure. An amount of the material to be tested corresponding to 0.1 g nickel is brought into solution and made up to 100 ml with water. A drop or two of 6 % solution of cupferron is introduced and the remainder of the procedure is carried out as described for the detection of traces of copper in water.

In this way, 0.025 γ copper can be detected in 100 ml of 0.1 % nickel solution, which corresponds to a copper content of 0.000025 %.

6. Detection of Copper in Solutions of Alkali Cyanides [14]

It is impossible to test directly for copper in the colorless aqueous solutions of alkali cuprocyanides. The stability of the cuprocyanide ions is too great to permit an adequate concentration of Cu^{+2} ions for the realization of most analytical reactions. This masking of copper by alkali cyanide is in harmony with the familiar fact that almost all copper salts which are not soluble in water and acids readily dissolve in alkali cyanide with production of alkali cuprocyanide. Therefore, to detect copper in such solutions it has hitherto been necessary to destroy the cyanide by repeated fuming with concentrated mineral acid.

A much simpler procedure is to withdraw the cyanide ions from the complex equilibrium

$$[Cu(CN)_4]^{-2} \rightleftharpoons Cu_2^{+2} + 4\ CN^-$$

by means of formaldehyde. The reaction

$$CN^- + CH_2O \longrightarrow CH_2\begin{matrix}\diagup O^- \\ \diagdown CN\end{matrix}$$

produces ions of glycolic acid nitrile and the copper ions are liberated. The latter can then be detected by the reaction with sodium diethyldithiocarbamate; a brown precipitate is formed:[15]

$$Cu^+ + SC\begin{matrix}\diagup SNa \\ \diagdown N(C_2H_5)_2\end{matrix} \longrightarrow SC\begin{matrix}\diagup SCu \\ \diagdown N(C_2H_5)_2\end{matrix} + Na^+$$

Other reagents for copper may be substituted for sodium diethyldithiocarbamate.

References pp. 512–515

Procedure. A spot plate is used. One drop of the test solution is treated with a drop of the 1 % clear solution of sodium diethyldithiocarbamate and then 1–3 drops of 40 % formaldehyde are added. Depending on the quantity of copper present, a brown precipitate or a red coloration appears. A comparison blank is recommended when slight amounts of copper are suspected.

Limit of Identification: 0.35 γ copper

7. Detection of Lead in Tin Platings [16] and Enamels [17]

The tinned surface to be tested is washed free of grease with alcohol and ether. Then a drop of concentrated nitric or acetic acid is placed on it, and the specimen carefully heated until the acid has evaporated. The gray fleck that remains may contain lead nitrate or acetate as well as metastannic acid. It is treated with a drop of 5 % potassium iodide solution. In the presence of lead (down to 1 %) yellow needles of lead iodide are formed. A slight yellow-green tint indicates the presence of not more than the small amounts of lead which are always present in commercial tin.

When testing enamels for lead, the surface is spotted with concentrated nitric acid, evaporated to dryness, and then moistened with a drop of 10 % potassium iodide solution.

8. Detection of Lead in Alloys and Crude Metals [18]

Lead can be detected quickly in alloys and crude metals by means of the very sensitive reaction with sodium rhodizonate (see page 73). Alloys containing no, or only a little, tin can be tested by merely treating the sample with a buffer solution (pH ca. 3) and sodium rhodizonate. Lead rhodizonate is formed if not less than 0.3 % lead is present. When testing alloys containing much tin, violet stannous rhodizonate may be formed and hide any red lead rhodizonate. Consequently, with such alloys or crude tin, the sample must be treated previously with nitric acid. The resulting $SnO_2 \cdot aq.$ does not react with sodium rhodizonate.

Procedure. *Alloys up to* 12 % *tin.* A few fragments are removed with a fine file. The particles are washed with ether to remove grease. The test is made on a spot plate. Three drops of buffer solution are added, and the solid and liquid kept in agitation for 2 minutes by blowing into the suspension through a pipette. One drop of sodium rhodizonate solution is then added, but not stirred in. A red precipitate or color appears if the alloy contains lead. A blank test, using 3 drops of buffer solution and 1 of drop reagent, should be made when testing alloys very low in lead. The blank becomes completely colorless after a few minutes.

The filings may also be ignited in a micro crucible with access to the air. After cooling, the residue is spotted with buffer solution and sodium rhodizonate.

Either of these procedures will definitely reveal as little as 0.3 % lead.

Procedure. *Alloys rich in tin.* A few fine filings are treated in a micro crucible with several drops of nitric acid, evaporated, and then weakly ignited. The cold residue is treated with 1 drop buffer solution and 1 drop sodium rhodizonate. Red lead rhodizonate appears.

Reagents: *1)* 0.2 % sodium rhodizonate solution
2) Buffer solution: 1.5 g tartaric acid and 1.9 g sodium bitartrate in 100 ml water (pH = 2.79)

Crude and pure metals (Zn, Mg, Al, etc.), that are soluble in sulfuric acid, can be tested for lead as follows: A little of the metal is dissolved in dilute sulfuric acid and the solution evaporated to fumes (SO_3). The residue is taken up in hot water and filtered through a small paper. After washing free of acid, the paper is spread on a porcelain or glass plate, and spotted with buffer solution and sodium rhodizonate. A red fleck indicates the presence of lead sulfate and, consequently of lead.

9. Detection of Lead in Ores, Minerals, and Pigments [18]

Lead can be detected directly in ores, and in minerals which contain no barium or strontium, by means of scarlet-red lead rhodizonate (see page 73). A little (less than 1 mg) of the finely powdered sample is spotted with a drop of buffer solution and a drop of sodium rhodizonate on a spot plate. The surface of the specimen turns red if lead is present. When it is suspected that barium and strontium are present, the sample must be fumed with a drop of sulfuric acid. This transformation into sulfates is necessary because the oxides of these alkali earth metals, as well as barium carbonate, react with sodium rhodizonate. The sulfates of the alkaline earth metals, in contradistinction to lead sulfate, do not react with this reagent.

Lead-bearing sulfidic ores, such as galena, can be tested in the same way as other lead compounds. However, the following **Procedure** has a higher sensitivity. A little of the test material is ignited in a micro crucible, and stirred with a platinum wire. The lead sulfide is thus transformed, in greater part, into lead sulfate. After cooling, the residue is spotted with the reagent and buffer solutions. The sensitivity of the test is raised if the specimen is calcined beforehand, even in the case of non-sulfidic minerals.

The following minerals and ores gave a positive reaction with sodium rhodizonate when this procedure was used:

Galena—PbS	Anglesite—$PbSO_4$
Cerussite—$PbCO_3$	Crocoite—$PbCrO_4$
Stolzite—$PbWO_4$	Lead-bearing zinc blende—ZnS, PbS

Zinc oxide or zinc dross can be tested for lead by direct spotting of the sample with sodium rhodizonate and buffer solution. A rapid method of detecting lead in zinc oxide is especially useful if the material is to be used in the preparation of glass, because lead often enters glass in this way. A little (0.5 to 1 mg) of the sample is fumed with sulfuric acid; the residue is extracted with water; filtered, and the undissolved material on the paper tested with the reagent and buffer solutions.

Lead can be detected in pigments, oils, and varnishes by this reagent. Pigments can be spotted directly; oils, and varnishes must be ashed before making the test.

10. Detection of Traces of Lead in Water, Alkali Salts, and Sulfuric Acid [19]

The decisive detection of traces of lead in water is rather important from the medical and public health standpoint because small amounts of lead accumulate in the human organism and are eliminated at a low rate. There is an extensive literature on the detection and determination of small amounts of lead available.[20]

The test described on page 73 has an identification limit of 0.1 γ and consequently only succeeds with one drop of a solution at a maximal dilution of 1 : 500,000. Lead can be detected in solutions of far greater dilutions if 10 ml of the sample is treated with sodium mercurisulfide solution and solid ammonium chloride. Mercuric sulfide is precipitated:

$$Na_2[HgS_2] + 2\ NH_4Cl \rightarrow 2\ NaCl + (NH_4)_2S + HgS$$

The precipitate functions as a "trace catcher" for lead sulfide. In other words, small quantities of lead, which of themselves will not precipitate as sulfide, are coprecipitated with the mercuric sulfide. If the mixed precipitate is treated with hydrogen peroxide dried, and ignited, the mercuric sulfide is driven off and lead sulfate remains. It can be detected by the formation of red lead rhodizonate when spotted with sodium rhodizonate and buffer solution.

The following procedure will permit the detection of lead in a dilution of 1 : 10,000,000 in 10 ml water. Lead can be detected in 100 ml water at a dilution of 1 : 100,000,000. These dilutions correspond to 0.00001 % Pb and 0.000001 % Pb, respectively. It requires about 20 minutes to carry out this procedure for the detection of traces of lead.

Procedure. Ten ml of the water to be tested is treated with 0.25 ml $Na_2[HgS_2]$ solution. A partial precipitation of HgS ensues because of hydrolysis. The precipitation is completed by addition of 0.6 to 1 g highest purity ammonium chloride and brief digestion. The flocculated precipitate is collected on a porcelain filtering crucible and washed once with water. Five milliliters of 3 % hydrogen peroxide are added, and after 3 minutes the liquid is sucked off. The crucible, together with the wet precipitate, is placed in a porcelain dish and kept on a hot plate until the black mercuric sulfide has disappeared. The crucible is then heated gently over a burner for 1 minute. After cooling, the contents of the crucible are moistened with 5 drops of the rhodizonate solution, the liquid is sucked off, and the excess of the yellow reagent is bleached by spotting with 2 or 3 drops of the buffer solution and then with 3 drops of pure water. If the sample contained lead, red particles of lead rhodizonate will remain. They are quite visible against the white bottom of the crucible.

The quantity of lead rhodizonate formed, and consequently the intensity of the color produced, depends on the lead content of the water. Therefore, it is possible to compare the intensity of the color of the red residue obtained in this test with that produced by residues obtained from various standard solutions. Suitable dilutions are: $1:0.5 \cdot 10^6$; $1:1 \cdot 10^6$; $1:2 \cdot 10^6$; $\cdots 1:1 \cdot 10^7$. It is possible to estimate the lead content by this colorimetric comparison, which is adequate in many cases.

Reagents: *1*) $Na_2[HgS_2]$ solution: 13.6 g $HgCl_2$ are dissolved in 60 ml hot water. Solid Na_2S, and later Na_2S solution, are added until the initial precipitate of HgS has completely dissolved. The solution is then made up to 100 ml.

2) Sodium rhodizonate solution: 2 mg of the salt in 1 ml water. The solution must be freshly prepared. It gradually undergoes oxidative decomposition on standing, but is stable for 2 days if stored in a refrigerator.

3) Buffer solution (pH = 2.79) : 1.5 g tartaric acid and 1.9 g sodium bitartrate in 100 ml water

This method of detecting traces of lead can also be used in the examination of certain "high purity" chemicals. Dissolve 1–5 g of the sample in 10 ml water and apply the procedure just described. Alkali salts of the highest purity were tested for traces of lead by this method. The packages bore labels reading, "Heavy metals as Pb ...0.0003 %," "...0.0005 %," respectively. Colorimetric determination of lead by the procedure given here showed values between $^1/_2$ and $^1/_{10}$ of the indicated lead content. When identical quantities of the same salt, coming from different sources, are tested by this method, it is often possible to discover the origin of the samples by the lead content. It is remarkable that alkali sulfates generally show a higher lead content than other alkali salts. The lead content of acetates usually is less than $^1/_{10}$ of the indicated amount of heavy metals. The same is true for magnesium sulfate of the highest purity. Sulfuric acid can be tested for lead by this method. The sample is evaporated, and the residue treated with sodium rhodizonate and buffer solution.

11. Detection of Lead in Bismuth Compounds [21]

When solutions of lead or bismuth salts are treated with alkali hypobromite, lead peroxide and higher oxides of bismuth are formed. These products react with benzidine acetate to give benzidine blue (see page 72). However, if bismuth salts are heated alone with alkali hydroxides, a yellow precipitate forms, [probably BiO(OH)] and, when once formed, is not converted into higher bismuth oxides on treatment with hypobromite. This resistance is applied in the detection of lead in bismuth salts. The following procedure is recommended in the examination of pharmaceutical bismuth preparations (e.g., bismuth sub-gallate, -tannate, -salicylate).

Procedure. One gram of the sample is ashed in a porcelain crucible and the residue taken up in concentrated nitric acid and evaporated to dryness. Ten drops of water and 20 drops 3 N sodium hydroxide are added, and the mixture boiled for 1 or 2 minutes. The spot test on paper for lead, given on page 72, may be carried out with 1 or 2 drops of the supernatant alkaline liquid, which need not be filtered.

12. Detection of Traces of Metallic Lead in Criminalistic Investigations [22]

In police investigations it is often necessary to detect traces of metallic lead in certain portions of substrates which may have been exposed to the impact of uncoated lead bullets discharged from fire arms. Such substrates, whether they have been penetrated or pierced, may be of various kinds; in general, wood, fabrics, and paper are to be considered. The rhodizonate test as used in the foregoing procedures can be employed.

Slightly colored or colorless materials are spotted at the impact area with a drop of the buffer solution followed by a drop of the reagent solution. If lead is present, a red fleck appears precisely on the spot where traces of lead have been retained. When traces of lead are to be identified or detected on intensely colored substrates, a piece of filter paper wetted with the buffer solution is pressed for several minutes against the surface being tested. A drop of sodium rhodizonate solution is then placed on the filter paper. When lead is present, a red fleck is formed on the paper and indicates exactly the location of the traces of lead on the substrate.

13. Detection of Thallium in Minerals, Alloys, Water, etc.[23]

Thallium can be detected with certainty and sensitivity in technical materials, mineral waters, etc., by means of the reaction of thallium^{III} chloride (bromide) with rhodamine B. The resulting addition compound of thallium^{III} halide with rhodamine B is fluorescent and soluble in benzene (see page 158). When testing minerals and alloys, 1–2 milligrams of the sample are brought to dryness with concentrated nitric acid, the evaporation residue is taken up in several drops of concentrated hydrochloric acid, and the test for thallium described on page 160 is conducted on a drop of the solution. The thallium test can also be used in forensic-medical and criminalistic investigations. Thallium can be detected in dilutions down to 1 : 1,000,000,000 if advantage is taken of the collector action of freshly precipitated hydrous manganese dioxide. For this purpose, 500 ml of the solution are treated with 1–2 milligrams of manganese salt (as chloride, sulfate), excess bromine water added, the solution made basic with caustic alkali, warmed, and filtered. The manganese dioxide will contain all of the thallium. The precipitate is collected by filtration or centrifugation, and dissolved in several drops of concentrated hydrochloric acid containing some alkali sulfate. One drop of the solution can then be subjected to the rhodamine B test.

14. Detection of Bismuth in Alloys [24]

The very sensitive and selective redox reaction between bismuth hydroxide and alkali stannite (page 77) may be applied for the rapid and reliable detection of bismuth in alloys. The material can be dissolved or it may be subjected to attack with bromine vapors. Bismuth, as well as other components of alloys, are thus converted into bromides. This step, as well as the transformation into $Bi(OH)_3$ and the reduction of the latter to dark, finely divided metal, can be carried out without visible damage to the specimen.

Procedure. The sample is drawn over unglazed porcelain (streak plate or the unglazed bottom of an evaporation dish). The scratch is held for 2–3 minutes over bromine water and then spotted with alkali stannite solution. When bismuth is present, the scratch becomes dark and the metal deposited remains in place when washed with water.

In the case of alloys which contain a high percentage of copper or silver, the brominated scratch must be spotted first with a drop of 2 % alkali cyanide solution. After dipping the porcelain into water to remove the water-soluble $K[Ag(CN)_2]$ or $K_2[Cu_2(CN)_4]$ formed, the scratch is spotted with the alkali stannite solution.

Reagents: *a*) Stannous chloride solution: 5 g $SnCl_2$ in 5 ml concentrated HCl diluted to 100 ml with water.

b) 25 % sodium hydroxide.
The stannite solution is made up shortly before use by mixing equal volumes of *a* and *b*.

The excellence of the test is illustrated by the fact that a strong positive response was shown by an alloy containing 3 % Bi, 1 % Ag, 1 % Hg, 80 % Pb, 15 % Sb.

15. Detection of Tin in Minerals, Metallic Objects, etc.

The very sensitive cacotheline reaction, described on page 109, may be applied for the rapid and certain detection of tin in minerals.[25] A few milligrams of the powdered substance, mixed with sodium carbonate and potassium cyanide, are melted on a Wedekind magnesia spoon or on a magnesia rod. Reduction to metallic tin takes place:

$$SnO_2 + 2\ KCN = 2\ KCNO + Sn$$

The melt is dissolved in a little hot concentrated hydrochloric acid (CAUTION: hydrocyanic acid liberated). The solution is tested for tin by treatment with cacotheline, either in a test tube or on paper impregnated with the reagent (see page 109).

The test has been successfully applied to the following minerals:

Cassiterite (SnO_2) from Schlaggenwald
Stannite (Cu_2FeSnS_4) from Zinnwald and also from Cornwall
Teallite ($PbSnS_2$) from Santa Rosa, Bolivia
Cylindrite ($PbSn_6Sb_2S_{21}$) from Poopó
Franckeite ($Pb_3Sn_2Sb_2S_{21}$ with 1.1 % Ge) from Poopó

If tin is to be detected in metals, alloys, etc., a drop of concentrated hydrochloric acid is placed on the surface of the sample, and stirred with a platinum wire to hasten solution of the tin (stannous chloride formed). A drop of the solution if placed on filter paper and treated with a solution of cacotheline. In the presence of tin, a red color forms.

Another test, that can be applied to alloys containing much tin, to tin-plated metals, and to tin itself, is based on the formation of violet stannous rhodizonate.[26] Stannous tartrate is formed when such materials are treated with a tartrate buffer solution (pH ca. 3). The product immediately reacts with the reagent and forms dark violet stannous rhodizonate. Unlike inorganic stannous salts, the rhodizonate does not undergo autoxidation under the test conditions. Consequently, the continued action of the buffer solution and the reagent gradually produces considerable quantities of the colored tin rhodizonate. The test cannot be used for the detection of small amounts of tin in the presence of much lead, because the latter forms red lead rhodizonate (see page 73).

Procedure. One drop of the buffer solution (page 436) and a drop of 0.2 % sodium rhodizonate solution are placed on a flat surface of the sample. No immediate reaction is observed when the solutions are stirred, but after 3–4 minutes, the spotted area becomes slightly gray. The color intensity increases and becomes dark violet after about 6 minutes. If the surface of the metal is scratched with a steel needle and then spot-tested with the two solutions, the furrows turn dark violet within 3 minutes. It is also feasible to treat 1 to 3 mg of cuttings with the solutions. The surface of the particles turns dark violet after a few minutes.

A reliable test for tin in minerals and alloys can be tried after the sample is dissolved in concentrated sulfuric acid, whereby tin^{IV} sulfate is formed. When alkali iodide is added and the acid solution extracted with benzene, tin^{IV} iodide enters the benzene layer. Tin^{IV} hydroxide is formed when a drop of the benzene solution is brought on filter paper and treated with ammonia. The spot of $Sn(OH)_4$ can then be identified by the fluorescence reaction with morin (see page 110).

When antimony is also present in the test material, benzene-soluble SbI_3 is formed, and gives $Sb(OH)_3$ when the benzene fleck is treated with ammonia. This hydroxide, in contrast to $Sb(OH)_5$, also reacts with morin, and with formation of a fluorescing product [27]. The formation of the indifferent antimony^V hydroxyde can be accomplished if the extraction of the iodides is carried out with benzene containing iodine instead of pure benzene. The benzene layer then contains antimony^V iodide, which after subsequent treatment with ammonia is converted into the indifferent antimony^V hydroxide. The excess of iodine can be eliminated by treating the fleck with sodium sulfite.

Procedure. [28] A small quantity of the finely powdered mineral or fine filings of an alloy is brought in a micro test tube; three drops of concentrated sulfuric acid are added, and heated for a few minutes. After dilution with the same volume of water and cooling, solid potassium iodide is added and the mixture is extracted with 10 drops of a 5 % solution of iodine in benzene. One drop of the benzene solution is brought on filter paper and the fleck held over ammonia. Then the fleck is spotted successively with one drop of 0.5 % water solution of sodium sulfite and one drop of 0.05 % morin solution in acetone, and bathed for $^1/_2$ minute in 1 : 1 acetic acid. When tin is present, a blueish green fluorescing fleck (ultraviolet light) remains.

The procedure gave excellent results with the tin-bearing minerals cited on page 439, and also a columbite $Fe(NbO_3)_2$ carrying 0.6 % SnO_2, as well as with various alloys containing tin up to a minimum of 0.02 % (checked by quantitative analysis). The morin test was positive in a series of alloys supposed to be free of tin. The spectrographic examination was in accord with these findings. The alloys, which were spectrographically free of tin, also responded negatively to the morin test.

16. Test for Metallic Tin and Tin in Alloys [24]

When metallic tin and tin in alloys are present, the dissolution of the test material in concentrated sulfuric acid and subsequent conversion of metal sulfates into metal iodides may be replaced by heating the test material with iodine. Tin^{IV} iodide is formed, and also the iodides of other metals in the case of alloys. When the material heated with iodine is digested with benzene, SnI_4, SbI_5 (SbI_3), AlI_3 and some Cu_2I_2 as well as the excess of iodine dissolve in the benzene. When a drop of the benzene solution is brought on filter paper and spotted with alkali hydroxide after evaporation of the benzene, $Sn(OH)_4$, $Sb(OH)_5$ (with copper-containing alloys also $Cu(OH)_2$) remain on the fleck, whereas aluminum is eliminated in the form of aluminum hydroxide.* Elementary iodine, remaining on the fleck after evaporation of the benzene, is converted into water-soluble alkali hypoiodite. After washing the paper in alkali hydroxide, the tin retained on the spot fleck can be detected by the morin reaction (see page 110).

Procedure. A few milligrams of the alloy to be tested, in form of fragments or fine filings, are mixed in a micro test tube with about 0.5 g iodine and held for about 1–3 minutes over a small flame, at the melting temperature of iodine. After cooling, about 0.5 ml benzene is added. One drop of the benzene solution is brought on filter paper and, after evaporation of the benzene, spotted with 1 or 2 drops of 0.5 *N* alkali hydroxide. The filter paper is bathed in alkali hydroxide, the wet paper placed on dry filter paper and spotted with 0.05 % acetone solution of morin solution. Then, the filter paper is bathed for $^1/_2$ minute in 1 : 1 acetic acid. According to the amount of tin, a more or less intensely fluorescing bluish green fleck remains, when exposed to ultraviolet light.

This procedure is somewhat less sensitive than the one described under Section **15**, but it suffices for most technical purposes. If metallic tin or alloys with a tin content of 40–60 % are to be studied, it is enough to shake the test material with a solution of iodine in benzene for about one minute and then bring one drop of the solution on filter paper. The rest of the treatment is as just indicated.

17. Simultaneous Test for Tin and Antimony in Alloys [24]

Sometimes a simultaneous test for tin and antimony instead of separate tests may be advantageous. The following procedure is appropriate. It is based on the fact that $SnBr_4$ and $SbBr_3$ are formed if bromine vapors act on powdered alloys containing tin and antimony. When treated with alkali

* In spite of the solubility of $Sn(OH)_4$ in alkali hydroxide to form stannate, the tin remains on the fleck. Alkali stannate, in contrast to alkali aluminate, is probably retained by the paper through adsorption.

hydroxide, these bromides are converted into the corresponding hydroxides, which can be very sensitively detected with morin by the formation of fluorescing adsorption compounds.

Procedure. The sample is drawn over a piece of unglazed porcelain (streak plate) or over the unglazed bottom of an evaporating dish. The scratched area is held for 2–3 minutes over a vessel containing saturated bromine water. A paper wetted with 0.5 *N* sodium hydroxide is pressed on the site of the scratch. The paper is taken off, washed with dilute sodium hydroxide, spotted with 0.05 % acetone solution of morin and dipped in 1 : 1 acetic acid. When antimony or tin, or both, are present, a fleck appears on the place of the scratch, which shows a bluish green fluorescence when exposed to ultraviolet light.

18. Test for Antimony in Alloys and Mineral Products [24]

When an alloy containing antimony is heated briefly with iodine, antimony^{V} iodide is formed. The latter dissolves in benzene together with iodine and the iodides of the other components of the alloy. After evaporation of the benzene solution and treatment of the residue with aqueous sulfurous acid, Sb^{+3} ions are formed. They unite with iodine ions forming $H[SbI_4]$ provided an excess of iodine and hydrogen ions is present. On addition of the water-soluble rhodamine B, a violet precipitate separates. The precipitate is the salt of the basic dye with the complex acid.

Procedure. A few milligrams of the alloy to be tested (fragments or fine filings) are mixed with about 0.1 g iodine in a micro test tube and heated for about half a minute over a small flame at the melting temperature of the iodine. After cooling, about 0.5 ml benzene is added. A drop or two of the benzene solution is brought into the depression of a spot plate and the benzene evaporated by blowing through a glass capillary. One drop of a saturated sulfurous acid solution is added to the residue, whereby the excess of iodine is completely removed. Then one drop of 5 % potassium iodide solution, one drop of 1 : 1 hydrochloric acid, and one drop of rhodamine B solution are added. When antimony is present, a finely dispersed violet precipitate is formed. A blank test with a solution of iodine in benzene is advised.

Reagent: 0.5 % solution of rhodamine B in 1 : 1 hydrochloric acid

The efficiency of the test for antimony in alloys is illustrated by the fact that it was possible to detect antimony within five minutes in a bronze of the composition: 85 % Cu, 9.96 % Sn, 1.86 % Zn, 1.83 % Pb, 0.24 % Sb, 0.07 % Fe, 0.04 % Ni.

Metallic bismuth if treated as just described likewise produces a violet precipitate. This result is due to the formation of $H[BiI_4]$ which reacts with rhodamine analogously to $H[SbI_4]$. Surprisingly, an alloy containing 20 % bismuth besides 55 % Pb, 12 % Cd and 13 % Sn showed no reaction;

References pp. 512–515

other metals capable of reacting with iodine seemingly prevent the formation of BiI_3.

The formation of the acid-insoluble violet rhodamine B salt of $H[SbI_4]$ also takes place with Sb_2O_5 as starting material. This oxide is completely dissolved without evolution of iodine by a mixture of an acidified solution of potassium iodide and sulfurous acid:

$$Sb_2O_5 + 10\,H^+ + 4\,I^- \rightarrow 2\,Sb^{+3} + 2\,I_2 + 5\,H_2O \quad (1)$$

$$2\,I_2 + 2\,SO_3^{-2} + 2\,H_2O \rightarrow 2\,SO_4^{-2} + 4\,H^+ + 4\,I^- \quad (2)$$

The Sb^{+3} ions formed combine with I^--ions in acid solution to form $H[SbI_4]$ or $[SbI_4]^-$ ions, provided an excess of potassium iodide is employed. The addition of a solution of rhodamine B immediately produces a violet precipitate in solutions containing the complex acid. Even traces of Sb_2O_5 in mixtures with other oxides, or antimony oxide in residues after ignition of sulfidic minerals etc., can be sensitively detected. In this manner all the necessary steps for the identification of antimony^{III} with rhodamine B may be conducted by means of the spot test technique.

Procedure. A minimum amount of the solid test material is placed in a micro crucible or in the depression of a spot plate. When calcination is necessary, a micro crucible is preferable. One drop of 5% potassium iodide solution, one drop of (1 : 1) HCl and one drop of sulfurous acid are added and mixed by blowing through a pipette. Then one drop of the 0.5 % dyestuff solution is added. When antimony is present, a violet precipitate is immediately produced. A blank test with potassium iodide and sulfurous acid is only necessary when very small amounts of antimony are to be detected.

This test is not directly applicable in the presence of bismuth compounds, because under the test conditions $H[BiI_4]$ is formed, and it reacts in the same manner as $H[SbI_4]$ with rhodamine B. When antimony is to be detected in materials containing bismuth, a digestion with ammonium sulfide must be carried out beforehand. Bi_2S_3 is formed, whereas $(NH_4)_3\,SbS_3$ goes into the filtrate. One drop of the latter, after oxidation with hydrogen peroxide, is evaporated, and the residue then tested as previously described.

19. Detection of Arsenic in Minerals [29]

When oxygen compounds of arsenic are heated with sodium formate, arsine is formed.[30] This is a consequence of the fact that sodium formate decomposes when heated over 300° C, i.e. above its melting point (250°C).

$$2\,HCOONa \rightarrow Na_2(COO)_2 + 2\,H$$

The nascent hydrogen formed reacts with As_2O_5 (or with its combinations with metal oxydes) according to:

$$As_2O_5 + 16\,H \rightarrow 2\,AsH_3 + 5H_2O$$

Arsine can be identified through its action on filter paper moistened with silver nitrate by the formation of a brown fleck (compare page 101). When sulfidic minerals are to be examined (test for sulfide-sulfur, see page 491) the pulverized material must be ignited in an open micro crucible, to transform sulfides into sulfates. This treatment is necessary because metal sulfides ignited with sodium formate form hydrogen sufide which reacts with silver nitrate forming black silver sulfide.

Procedure. About one milligram of well powdered and carefully ignited mineral is intimately mixed with 10–15 time excess solid sodium formate, which was previously dried at 120°C. The mixture is transferred into a micro test tube covered with a disk of filter paper moistened with a drop of 10% silver nitrate and heated over a free flame. A black fleck forms on the reagent paper when we are dealing with an arsenic containing mineral.

This method was tested with the following minerals:

$FeAsS$ / $FeS \cdot FeAs_2$ } mispiquel

$FeAsO_4$ – scorodite

$NiAs$ – nicolite (rammelsbergite)

20. Detection of Molybdenum in Tungsten Ores and Technical Materials [31]

The very sensitive and specific color reaction for molybdates with potassium xanthate (page 116), may be applied for the detection of a very small molybdenum content in tungstic acid or tungsten oxides (e.g., in scheelite, $CaWO_4$, an important source of tungstic acid).

Procedure. About 0.1 to 2.0 grams is digested with the least possible volume of concentrated sodium hydroxide in a small crucible and then acidified with a drop of phosphoric acid. Soluble complex phosphotungstic and phosphomolybdic acids are formed. A few crystals of potassium xanthate are then added. A dark violet to red color forms, according to the molybdenum content. The color fades on standing.

Molybdenum can be detected in steels without removing cuttings by subjecting the sample to an oxidizing etch with nitric acid. The resulting molybdic acid is absorbed in filter paper, which is then tested by the xanthate reaction.[32]

21. Differentiation between Palladium and Platinum Foil [33]

A drop of a 2 % alcoholic solution of iodine is placed on the foil being tested and allowed to evaporate in the air. If palladium is present, a brown-black fleck is left due to the formation of palladium iodide. It withstands washing with water, alcohol, acetone or ether. The stain disappears on heating to redness; the palladium iodide decomposes with evolution of iodine. Solutions of alkali iodides likewise discharge the stain formation of $[PdI_4]^{-2}$ ions. Palladium that has been electrolytically deposited on metals can likewise be easily identified by this procedure. No stain is given by platinum on treatment with iodine solution.

22. Detection of Gold in Alloys, Coating, etc. [24]

The sensitive and specific test for gold with rhodamine B (page 129), allows the rapid detection of gold in alloys, metal coatings etc. without visible damage to the specimen.*

Procedure. The sample is drawn over a piece of unglazed porcelain (streak plate) or the unglazed edge bottom of an evaporating dish, and the hair line scratch on the porcelain is spotted with a drop of bromine-hydrochloric acid. After 2 minutes a drop of 10 % water solution of sulfosalicylic acid is added, and the liquid is brought into a micro test tube by means of a small pipette. A drop of 0.2 % aqueous rhodamine B solution and 4–6 drops of benzene are added and the mixture shaken. When gold is present, the red benzene layer shows an orange fluorescence after a few minutes exposure to ultraviolet light.

23. Detection of Nickel Plating and Nickel in Alloys

A nickel content may be rapidly detected by the dimethylglyoxime reaction (see page 149).

Procedures. One or two drops of concentrated nitric acid are placed on the sample and left for 1 or 2 minutes. The solution is taken up on a strip of filter paper and held over ammonia to neutralize the excess acid. A drop of acetic acid is placed on the spot, and the moist portion spotted with a drop of saturated alcohol solution of dimethylglyoxime.[34]

In another method of testing for nickel in alloys, the surface is freed of grease with ether, and a drop of a solution of 0.5 g dimethylglyoxime in 5 ml 98 % alcohol and 5 ml ammonia is added. A pink fleck of nickel dimethylglyoxime forms and may be made clearly visible by rubbing a piece of filter paper on the stain. The test is more sensitive if the portion of the surface to be tested is oxidized by the action of a blowpipe flame before treating it with the reagent.[35]

When a drop of a mixture of equal volumes of 0.1 *N* iodine and saturated alcoholic solution of dimethylglyoxime is placed on the metal surface and allowed to stand, red Ni-dimethylglyoxime is formed.[36]

* Regarding another spot test for the detection of gold in platings compare M. Lerner, *Ind. Eng. Chem., Anal. Ed.*, 15 (1943) 416.

The method of attacking a scratch by bromine (see p. 438) can also be used; this permits the detection of nickel in acid-resistant steel. [37]

24. Detection of Traces of Nickel in Cobalt Salts [38]

The detection of nickel with dimethylglyoxime is quite selective and very sensitive. The test fails, however, in ammoniacal or acetic acid solutions containing small amounts of nickel in the presence of much cobalt, because cobalt ions also react with dimethylglyoxime with formation of soluble brown compounds of bi- and tervalent cobalt. Dimethylglyoxime is thus consumed by the cobalt and not enough may be available to react with the nickel, since the solubility of the reagent is quite limited. Furthermore, it is very difficult to see small amounts of the nickel precipitate in the cobalt-dimethylglyoxime solution, which often is dark brown. Finally, nickel dimethylglyoxime is appreciably soluble in the solution of the cobalt dimethylglyoxime compound. For these reasons, it was thought to be impossible to detect or determine nickel in cobalt salts by the dimethylglyoxime reaction in ammoniacal or acetic acid solutions, when the proportion of nickel to cobalt was less than 1 : 100 for an amount of nickel of 0.03 mg.

These difficulties may be avoided and nickel can nevertheless be detected in the presence of any given amount of cobalt. A simple procedure accomplishes this end. It consists in converting the ions into the complex cyanides, $K_2[Ni(CN)_4]$ and $K_3[Co(CN)_6]$, by the addition of potassium cyanide and hydrogen peroxide. Formaldehyde is added and the dimethylglyoxime reaction is then carried out. The complex $[Co(CN)_6]^{-3}$ ion is not affected by formaldehyde, so that the cobalt is left in a form inactive towards dimethylglyoxime, whereas the $K_2[Ni(CN)_4]$ is decomposed by formaldehyde and alkali, with formation of the potassium compound of glycolic nitrile. Nickel, hydroxide is also produced and reacts at once with dimethylglyoxime.

The chemical changes occurring in this procedure are:

(*a*) Conversion of nickel and cobalt into the complex cyanides:

$$Co^{+2} + 2\,CN^- = Co(CN)_2$$

$$Co(CN)_2 + 4\,KCN = K_4[Co(CN)_6]$$

$$2\,K_4[Co(CN)_6] + H_2O_2 = 2\,K_3[Co(CN)_6] + 2\,KOH$$

$$Ni^{+2} + 2\,CN^- = Ni(CN)_2$$

$$Ni(CN)_2 + 2\,KCN = K_2[Ni(CN)_4]$$

(*b*) Decomposition of the complex nickel cyanide by formaldehyde:

$$K_2[Ni(CN)_4] + 4\ CH_2O + 2\ KOH = Ni(OH)_2 + 4\ H_2C\begin{matrix} OK \\ CN \end{matrix}$$ {potassium salt of the nitrile of glycolic acid*}

(*c*) Formation of nickel dimethylglyoxime:

$$Ni(OH)_2 + 2\ C_4H_8N_2O_2 = Ni(C_4H_7N_2O_2)_2 + 2\ H_2O$$

The following procedure involves the complete course of this series of reactions. It will often reveal nickel in preparations of cobalt which were hitherto taken to be nickel-free, even those labeled "for analysis". When traces of nickel are to be detected, a comparison test should be carried out with an absolutely nickel-free cobalt salt. The preparation of such salts is described below.

Procedure. About 0.05 to 0.06 g of the water-soluble cobalt salt is placed on a spot plate. A similar quantity of a nickel-free cobalt salt is placed in the adjacent depression. Each is dissolved in 1 or 2 drops of water and drops of a saturated solution of potassium cyanide stirred in until the initial precipitate is redissolved. Then 1 or 2 drops of 3 % hydrogen peroxide are added, and the mixture stirred until the solution becomes bright yellow (color of the $[Co(CN)_6]^{-3}$ ions). This occurs in a few minutes. A few milligrams of dimethylglyoxime (tip of knife blade) and then 40 % formaldehyde (a few drops) are added and the mixture stirred. In the presence of nickel, the solution turns orange-red or a red precipitate appears. The nickel-free parallel test remains yellow. Very small color differences, due to traces of nickel, may be distinguished by comparing the tests.

Preparation of cobalt salts practically free of nickel [40]

Ten grams of a cobalt salt containing as little nickel as possible under the circumstances is dissolved in water and treated, while warm, with a concentrated solution of potassium cyanide until the initial precipitate redissolves. The solution is clear and greenish yellow. Then sufficient 3 % hydrogen peroxide is added to produce $K_3[Co(CN)_6]$. The mixture is boiled a few minutes, until the solution turns yellow; otherwise more peroxide should be added. If a slight precipitate has formed, it should be filtered off. The filtrate or solution is boiled down until it becomes viscous. An excess of solid dimethylglyoxime is added, and 40 % formaldehyde is poured into the lukewarm solution, with vigorous stirring. On cooling, nickel dimethylglyoxime and solid dimethylglyoxime precipitate. After an hour, the sus-

* Regarding the mechanism of the precipitation of nickel dimethylglyoxime from alkali cyanide solution containing formaldehyde, see [39].

pension is filtered and the (nickel-free) filtrate is evaporated to dryness over a sand-bath. The residue is cautiously heated further, with continuous stirring, until the mass begins to char and appears quite black. After cooling somewhat, it is mixed to a paste with a little warm water. Then concentrated hydrochloric acid is added and the suspension heated for 1 to 2 hours on a water-bath. The mass is diluted with water and filtered. The filtrate is saved. The residue is dried, ignited, made into a paste with water, treated with concentrated hydrochloric acid, diluted, and filtered. The two filtrates are united. Alkali is added to the warm solution and pink cobalt hydroxide precipitates. It is thoroughly washed, dried, and converted to the oxide by ignition. If desired, reduction in a current of hydrogen produces metallic cobalt. The washed oxide or the metal can be dissolved in acids. Crystallization delivers nickel-free cobalt salts. They are necessary for the parallel tests prescribed in the foregoing procedure.

25. Detection of Cobalt Plating and Cobalt in Alloys [41]

Cobalt in such materials may be rapidly detected by attacking the sample with bromine water ($CoBr_2$ is formed), followed by reaction with α-nitroso-β-naphthol (see page 144).

Procedure. A drop of bromine water is placed on the metal surface and allowed to stand until the bromine color is nearly discharged. Then a drop of a freshly prepared 1 % solution of α-nitroso-β-naphthol in 50 % acetic acid is added. If cobalt is present, a blood red precipitate appears at once.

26. Detection of Zinc in Plated Metals

A drop of 2 *N* sulfuric acid is placed on the surface to be tested, and allowed to react for about a minute. The drop is then taken up on filter paper impregnated with copper sulfate. The fleck on the paper is treated with a drop of ammonium-mercury thiocyanate solution (see page 93). Zinc is revealed by a violet color. The color is due to the formation of a compound containing $Zn[Hg(CNS)_4]$ and $Cu[Hg(CNS)_4]$ which, however, does not have the color of a mechanical mixture of these components (see page 93).

When the fleck is red, owing to ferric thiocyanate from large amounts of iron, a drop of ammonium fluoride solution is added. It discharges the ferrithiocyanate color, and the violet of the zinc-copper-mercury thiocyanate becomes clearly visible.

This test for zinc is not reliable when cobalt and nickel are present because they too form colored double thiocyanates with mercury thiocyanate.

Reagent: Copper sulfate paper: Quantitative filter paper is impregnated with a 0.1 % solution of copper sulfate and dried

27. Detection of Traces of Iron in Fluorides and Phosphoric Acid

The usual thiocyanate reaction (see page 164) is not applicable to the detection of small amounts of iron in fluorides, because the tervalent iron is present in solution as the complex $[FeF_6]^{-3}$. The Fe^{+3} ion concentration delivered by this complex is insufficient for the formation of the red ferri-thiocyanate color.

When solutions contain $[FeF_6]^{-3}$ ions and are thus masked against CNS^- ions, Fe^{+3} ions can be set free by adding a beryllium salt.[42] The reaction is:

$$2\,[FeF_6]^{-3} + 3\,Be^{+2} \rightarrow 3\,[BeF_4]^{-2} + 2\,Fe^{+3}$$

Procedure. (*a*) A drop of the test solution is mixed on a spot plate with several crystals of potassium thiocyanate. A drop of hydrochloric acid is then introduced. If the addition of beryllium chloride or beryllium sulfate is followed by the development of a red or pink color, iron is present.

Reliable tests for traces of iron in alkali fluorides are based on the reduction of iron^{III} to iron^{II} by thioglycolic acid,[43] which occurs even in the presence of much fluoride ion:

$$2\,Fe^{3+} + 2\,HS{-}CH_2{-}COO^- \rightarrow 2\,Fe^{2+} + \begin{array}{l} S{-}CH_2{-}COO^- \\ | \\ S{-}CH_2{-}COO^- \end{array} + 2\,H^+$$

Iron^{II} salts yield a red-violet color[44] with thioglycolic acid and ammonia probably because of the formation of inner complex anions containing iron. This color reaction is more sensitive than the thiocyanate reaction with iron^{III}; it is not impaired by materials which form complexes with iron^{III}.[45] If thioglycolic acid containing α,α'-dipyridyl or α,α'-phenanthroline is used, the sensitivity is still better. The reaction described on page 162 occurs.

Procedure. (*b*): A micro test tube is used. A drop of the test solution, or several grains of alkali fluoride, is treated with one drop of concentrated (80 %) thioglycolic acid (with addition if need be of α,α'-dipyridyl). The reaction mixture is slightly warmed and made alkaline with ammonium hydroxide. A positive response is shown by a blue-violet or red color.

Detection of traces of iron in phosphoric acid

Because of the production of stable complex $[Fe(PO_4)_2]^{-3}$-ions, it is not possible to detect iron in phosphoric acid by means of thiocyanate. Here again, reduction of the iron to the divalent condition by thioglycolic acid is possible in ammoniacal solution, and the violet or red color can then be produced by addition to the reduced solution (acid or ammoniacal) of α,α'-dipyridyl.

Procedure. See Procedure *b*.

28. Detection of Small Amounts of Iron in Mercury Salts [46]

The thiocyanate test, which is usually applicable, is not suitable for the detection of iron in the presence of large amounts of mercury salts. The thiocyanate ion is consumed in the formation of slightly dissociated mercuric thiocyanate, or of double thiocyanates of mercury. The Prussian blue reaction is uncertain, owing to the formation of yellow mercury ferrocyanide, though this test may be improved by adding sodium chloride (solution of the mercury precipitate). The α,α'-dipyridyl test as given here is simpler and more sensitive.

Procedure. A drop or two of the test solution is placed in a micro crucible and treated with a like volume of reagent solution and a little solid sodium sulfite and sodium chloride. A pink color appears at once if small amounts of iron are present. The function of the sodium chloride is to form the soluble complex $Na_2[HgCl_4]$ and thus prevent the precipitation of the white insoluble double salt of $HgCl_2$ and α,α'-dipyridyl. Sometimes a red precipitate of $[Fe(\alpha,\alpha'dip)_3]$ $[HgCl_4]$ is formed instead of a red color.

29. Detection of Traces of Iron in Alumina, Pyrolusite, Titanium Dioxide, Silicates, Colorless Ignition Residues

The sensitive α,α'-dipyridyl test for iron, described on page 161, may be used for the detection of the slight iron content in alumina and pyrolusite, if care is taken to dissolve and reduce the iron to the bivalent state.[46]

Procedure. A quantity of alumina (size of pea) is pulverized and fused with potassium bisulfate in a porcelain crucible. The melt is dissolved in water and sodium sulfite and 1 or 2 drops of an acid solution of dipyridyl are added. When iron is present, a pink color appears.

A similar amount of pyrolusite is powdered and dissolved by warming with a little hydrochloric acid plus 3 % hydrogen peroxide. Sodium sulfite and a little of the solid reagent are added to the solution. A more or less intense pink results, according to the amount of iron present.

A good method for detecting small amounts or even traces of iron in alumina is to heat several milligrams of the finely powdered sample to 250°C in a micro crucible with excess 8-hydroxyquinoline.[47] The molten oxine reacts with ferric oxide to form black-green ferric oxinate, which then dissolves in the colorless melt of the oxine and colors it. Aluminum oxinate is also produced in this fusion reaction; it is light yellow and therefore does not impair the test for iron. The fusion reaction is particularly suited when testing bauxites for iron. It can also be used to detect traces of iron in titanium oxide and in pigments which contain titanium white. Traces of ferric oxide in white ignition residues are also revealed by the oxine fusion method.

Instead of the procedure just described, the following method may be employed.[48] The pulverized sample is placed in a micro test tube along with a drop or two of a 2 % solution of α,α'-dipyridyl or phenanthroline in thioglycolic acid. A red or pink color develops on warming, the shade depending on the iron content of the sample. With ferruginous silicates, the solid sample often turns red. The basis of the test is that thioglycolic acid reduces the tervalent iron and the resulting ferrous iron reacts at once with the α,α'-dipyridyl or phenanthroline (compare page 161).

The procedure affords also a quick means of detecting iron in carbonates (calcite, dolomite, magnesite). The slight iron content of commercial thioglycolic acid gives a pale pink to the reagent solution. Accordingly, a comparison blank is advisable.

30. Detection of Slight Amounts of Iron in Metallic Copper, and in Cobalt- and Nickel-Alloys [48]

The sensitive color reaction of iron^{II} with α,α'-dipyridyl or phenanthroline cannot be employed directly in the presence of much copper because a faint pink would be hidden in the blue solution. Furthermore, Cu^{+2} ions consume these reagents to give ammine ions. The detection of iron in the presence of much Co^{+2} or Ni^{+2} ions is not feasible for the same reasons. Procedure (*a*) given here for the detection of little iron in the presence of much copper makes use of the fact that thioglycolic acid discharges the blue of ammoniacal copper solutions. The redox reaction is:

$$2\,Cu^{+2} + 2\,HS{-}CH_2{-}COO^- \rightarrow \begin{array}{l} S{-}CH_2{-}COO^- \\ | \\ S{-}CH_2{-}COO^- \end{array} + Cu_2^{+2} + 2\,H^+$$

If the solution contains iron^{III}, the latter is likewise reduced by the thioglycolic acid and the resulting iron^{II} then reacts with α,α'-dipyridyl or phenanthroline. The copper^{I} produced by the foregoing reaction probably remains in solution as a complex thioglycolate and gives no reaction with the reagents.

Procedure. (*a*) The test is conducted in a micro test tube. One drop of the test solution is treated with a drop of concentrated ammonium hydroxide, and then with a drop or two of a 2 % solution of α,α'-dipyridyl (or phenanthroline) in 80 % thioglycolic acid. The mixture is warmed. Care must be taken that the solution remains ammoniacal. If iron is present, a red or pink color is obtained. The pink color of the reagent (slight concentration of iron in the commercial product) makes a comparison blank imperative, especially when traces of iron are suspected.

This procedure will reliably reveal 5 γ iron in the presence of 20,000 γ copper. It also serves therefore to test commercial copper for a slight content of iron. A sample is dissolved in nitric acid and the procedure applied to the solution.

Procedure (*a*) cannot be used for the detection of iron in the presence of much cobalt and nickel because these ions form brown complexes with thioglycolic acid in ammoniacal surroundings. Under such circumstances, it is best to employ procedure (*b*), in which traces of iron^{III} hydroxide are gathered by collector action on aluminum hydroxide thrown down in the ammoniacal solution. The purified precipitate gives the characteristic color for iron^{II} on treatment with α,α'-dipyridyl or phenanthroline dissolved in thioglycolic acid.

Procedure. (*b*) The test is made in a centrifuge tube (capacity about 5 mm). A small portion of the alloy is dissolved in a drop or two of concentrated nitric acid by warming in a water-bath. A drop of a 2 % solution of aluminum sulfate is added and, after diluting with 1 ml of water, concentrated ammonium hydroxide is added drop by drop until a deep blue solution is obtained. The liquid is warmed briefly in the water-bath and then centrifuged. The blue solution is decanted, and the precipitate is dissolved in warm dilute hydrochloric acid. Ammonium hydroxide is again added and the suspension is centrifuged once more. If the precipitate of hydrous alumina should still be tinted by adsorbed nickel (cobalt, copper) ammine salts, the solution, reprecipitation, and centrifugation should be repeated. The precipitate, thus purified and gathered together, is treated with a drop or two of the α,α'-dipyridyl-thioglycolic acid reagent solution and warmed in the water-bath. A red or pink color indicates iron. As in procedure (*a*) a blank is urgently recommended.

This procedure revealed definitely the presence of iron in a nitric acid solution prepared from 5 mg of an alloy containing 68.75 % copper, 30.08 % nickel, 1 % manganese, and 0.05 % iron. A sample of nickel (5 mg) containing 0.05 % iron likewise gave a decisive test for iron.

31. Detection of Traces of Iron in Metallic Zinc and Magnesium, and in Zinc- and Magnesium-Salts [48]

When zinc and magnesium containing iron are dissolved in dilute hydrochloric acid, the solution contains not only zinc and magnesium chloride but also iron^{II} chloride. However, under these conditions, the iron does not respond to α,α'-dipyridyl directly unless no more than traces of iron are involved. The sensitive test for divalent iron fails because Zn^{+2} and Mg^{+2}

References pp. 512–515

ions form colorless complex ions with α,α'-dipyridyl and thus consume the reagent. However, this interference can be averted, if concentrated ammonium hydroxide is added *after* the α,α'-dipyridyl has been introduced. The zinc and the magnesium dipyridyl compounds are converted into the respective ammine ions and the reagent is released and made available to the ferrous ions.

Procedure. A micro test tube is used. Several mg or cg of the metal are dissolved in several drops of dilute hydrochloric acid, with gentle warming if need be. The cold solution is treated with a drop of a 2 % solution of α,α'-dipyridyl in dilute hydrochloric acid and then the system is made alkaline with concentrated ammonium hydroxide. A pink to red color indicates traces of iron.

If magnesium or zinc salts are to be tested for iron, and the latter is in the tervalent state, reduction is necessary before the α,α'-dipyridyl reagent is applied. Metallic zinc or magnesium may not be used, because these metals usually contain traces of iron. Hydroxylamine hydrochloride or sulfurous acid should be used. After adding the α,α'-dipyridyl-hydrochloric acid reagent solution, the system should be made basic with ammonium hydroxide.

32. Detection of Traces of Iron in Concentrated Nitric Acid [48]

The thiocyanate test for iron fails or is unreliable when applied to concentrated nitric acid (or after the latter has been diluted). The thiocyanic acid is oxidized and yields orange ether-soluble decomposition products. The usual procedure for testing concentrated nitric acid for traces of iron for their colorimetric determination [49] has been to evaporate 7 ml of the acid, take up the evaporation residue in dilute hydrochloric acid, oxidize with persulfate, and then apply the thiocyanate reaction. The rapid procedure given here is based on the production and fixing of hydrous ferric oxide on alumina (collector) and then the mixed precipitate is subjected to the action of α,α'-dipyridyl or phenanthroline dissolved in thioglycolic acid (compare page 162).

Procedure. A micro centrifuge tube is used (capacity 5 ml). A drop of 2 % solution of aluminum sulfate solution is added to 1 ml of the concentrated nitric acid, and (with cooling) is made alkaline by dropwise addition of concentrated ammonium hydroxide. The precipitate is gathered in the constricted part of the tube by centrifuging and treated there with a drop or two of a 2 % solution of α,α'-dipyridyl or phenanthroline in 80 % thioglycolic acid. The mixture is warmed in the water-bath. A pink or red color indicates iron, the shade depending on the quantity present.

This procedure, accompanied by a comparison blank, definitely revealed 0.001 % iron in concentrated nitric acid. About 3 minutes were required. Traces of iron in hydrochloric, sulfuric, and acetic acid, and likewise in alkali salt solutions can be detected in analogous fashion.

33. Detection of Alumina [50]

The color reaction with the hydroxytriphenylmethane dye Chrome Fast Pure Blue B (see page 189) can be used for the detection of alumina in samples that have been decomposed by fusion with alkali bisulfate or persulfate. A much faster method is to allow an ether solution of this dye to act on the pulverized sample; a red-violet tinting occurs almost at once. Beryllium oxide is made deep blue. These colored products, which are probably adsorption complexes, differ characteristically in that the alumina adsorbate is resistant to dilute acids.

The following procedure is suitable for detecting alumina and for differentiating it from acid-resistant colorless metal oxides. Alkaline earth carbonates are tinted blue-violet by the dye solution, but the colored products are destroyed by acids. Because of their consumption of the reagent, it is best to remove any carbonates by digestion with dilute acid before making the test for alumina.

Procedure. A micro test tube is used. A little of the pulverized sample is treated with a drop or two of the reagent solution. The alumina acquires an intense blue-violet color. After evaporating the ether, two drops of 1 : 1 acetic acid are added and the excess dye is removed by shaking with ether. The test is positive if the solid retains its color.

Reagent: 1 ml of a 1 % water solution of the dye is acidified with dilute hydrochloric acid. The precipitated dye acid is taken up in **10** ml of ether.

This procedure may be applied to the ignition residues of organic aluminum preparations that are used for medical and technical purposes. The ash of leather and paper may likewise be tested for alumina in this way.

34. Detection of Traces of Aluminum in Water [50]

The almost specific test for aluminum with Chrome Fast Pure Blue B (see page 189) can be used for the rapid detection of traces of this element in water. Only 25–50 ml is required. After adding one drop of a 1 % solution of iron^{III} chloride or sulfate, ammonium hydroxide is added and the suspension is centrifuged in portions and the precipitate thus gathered in a conical centrifuge tube. The hydrous ferric oxide acts as collector for the hydrous alumina. The precipitate on the bottom of the tube is dissolved in a drop or two of a mixture of equal volumes of 2 N hydrochloric acid and 80 % thioglycolic acid. One drop of a 5 % water solution of the dye is added and the mixture then made basic with ammonium hydroxide. Finally, the system is acidified with 1 : 1 hydrochloric acid and shaken with ether. The ether layer is pink if the response is positive.

As little as 0.4 γ aluminum was revealed by this procedure in 20 ml of water. This corresponds to a dilution of 1 : 50,000,000.

Aluminum was not detected by this method in sea water. The reason is that the aluminum is not present in ionogenic form but rather in the form of colloidal clay.[51]

35. Detection of Chromium in Rocks and Steel

The sensitive diphenylcarbazide reaction for chromates (page 167) provides a reliable test for even traces of chromium in very small samples of rocks.[52]

Procedure. *(Rocks).* A pin-head sized splinter of the sample is ground to a fine powder in an agate mortar and mixed with four times its bulk of a mixture of equal parts of sodium potassium carbonate and sodium peroxide (or sodium potassium carbonate and potassium chlorate). When a red hot platinum wire is dipped once or twice in this mixture, sufficient material is removed to melt into a transparent bead, in which any chromium is oxidized to chromate. The melt is cooled, placed on a spot plate, and dissolved in 1 or 2 drops of 1 : 1 sulfuric acid, and a drop of 1 % diphenylcarbazide solution added. A violet to pink color forms according to the amount of chromium present. It is necessary to carry out a blank test only when testing for very small amounts of chromium, or when using an old reagent solution, which may itself be slightly colored (yellow to pink).

Chromium was decisively detected in 3 to 4 minutes in:

Peridotite, 0.19 % Cr_2O_3
Olivinic tesinite, 0.04 % Cr_2O_3
Piterite, 0.11 % Cr_2O_3
Smaragdite, 2.30 % Cr_2O_3
Gangue and fuchsite, 0.24 % Cr_2O_3
Pyrope (magnesia-alumina garnet), 1.8–2.4 % Cr_2O_3

Chromium was also detected in the following unanalyzed minerals from various sources: fuchsite, serpentine, chromodiopside, chromochlorite, wavellite.

A very rapid oxidative disintegration of chromium-bearing minerals, rocks and alloys is obtained by fusing or sintering the finely pulverized material with potassium bifluoride (platinum spoon). Potassium chromate results [53] and may be detected by means of the diphenylcarbazide reaction. The fluoride disintegration is particularly recommended for the detection of chromium in steels or special alloys, which are likely to contain molybdenum. The latter, in the form of molybdate ions, reacts with diphenylcarbazide to yield a red-violet color and thus impairs the test for chromium. However, the fluoride method yields no MoO_4^{-2} ions, but instead complex $[MoO_3F_2]^-$ ions, which do not react with diphenylcarbazide.

References pp. 512–515

Procedure. (*Steels*). The specimen is cleaned with emery paper; its surface is treated with 2 or 3 drops of a mixture of equal volumes of nitric acid, 1 : 1 sulfuric acid and water. After 3 minutes, a drop is removed by means of a fine pipette, and evaporated in a porcelain crucible. The residue is ignited gently. Sodium peroxide (tip of knife blade) is added and the mixture fused until no more bubbles appear. The cooled melt is acidified with 1 : 1 sulfuric acid and tested for chromate with diphenylcarbazide solution.

When the steel contains manganese, permanganate is also formed and must be reduced by addition of sodium azide before proceeding to the test for chromium (compare page 345).

Direct oxidative solution can be accomplished by heating several milligrams of the sample (in as small pieces as possible) with 70 % perchloric acid. Steels and cast irons with high carbon content react too violently with the concentrated acid; they should be warmed with dilute acid and the solution then evaporated to fumes of $HClO_4$. The perchloric acid procedure has the advantage that the manganese is not oxidized.[54]

When examining alloys containing molybdenum, it must be remembered that molybdates also give a violet color with diphenylcarbazide. See page 171 regarding the masking of molybdates by oxalic acid.

36. Detection of Manganese in Minerals and Rocks [55]

Page 173 contains a sensitive test for manganese based on the production of permanganate ions. The oxidation of Mn^{II} to Mn^{VII} is accomplished by warming the nitric acid test solution with potassium persulfate in the presence of silver ions as catalyst. This procedure can be employed to reveal small amounts of manganese in minerals and rocks. It is essential to make the manganese accessible beforehand by sintering the sample with potassium persulfate. Manganese^{II} silicates and manganese^{IV} oxide undergo the respective reactions

$$MnO{\cdot}SiO_2 + K_2S_2O_8 \rightarrow MnSO_4 + K_2SO_4 + SiO_2 + O$$

$$MnO_2 + K_2S_2O_8 \rightarrow MnSO_4 + K_2SO_4 + O_2$$

Procedure. Several milligrams of the pulverized sample and several cg of potassium persulfate are heated in a small test tube for several minutes over a free flame. After cooling, two or three drops of 2 *N* sulfuric acid and some solid silver sulfate are added and the liquid brought to boiling. To the hot solution several centigrams of persulfate are added. Depending on the quantity of manganese involved, there appears a more or less intensive violet color.

This procedure revealed manganese in wolframites, columbites, iron ores, and brown glass.

37. Detection of Uranium in Minerals

The potassium ferrocyanide test for uranium (page 205), may be applied to minerals (e.g., broeggerite, samarskite, etc.).

Procedure. [56] A tiny (pin head) sample is finely ground and fused with sodium carbonate or potassium sodium carbonate in a loop of a platinum wire in the blowpipe. On cooling, the melt plus hydrochloric acid is taken to dryness on a platinum crucible lid. The residue is moistened with a drop of the acid and diluted with a little water. The further procedure is described on page 205.

When carrying out the spot test for uranium in minerals, precautions must be taken when titanium is present, which is often true. Titanium gives a yellow stain with ferrocyanide and the product gradually turns brown. In such cases it is advisable to mix a drop of the test solution with ammonia on a watch glass, warm gently, and acidify with acetic acid. The uranyl hydroxide goes into solution, while the titanium hydroxide remains behind. The precipitate is best separated by centrifuging; the titanium-free solution can then be tested for uranium.

The color (fluorescence) reaction (described on page 208) with rhodamine B and benzoic acid is excellent for detecting uranium in minerals. The sample must be fused with sodium carbonate or potassium-sodium carbonate as just described; the objective is to obtain a solution of a complex alkali uranyl carbonate.

38. Detection of Beryllium in Minerals, Ores, and Alloys [57]

Soluble beryllium salts have such great tendency to form a blue water-insoluble compound with quinalizarin (compare page 192) that solutions of even complex alkali beryllium fluoride react with ammoniacal quinalizarin solutions. [58] Beryllium in oxygen and silicate compounds, or bound inter-metallically in alloys, may be readily converted into alkali beryllium fluoride by fusion or sintering with alkali bifluoride. The conversion involves the hydrofluoric acid derived from the thermal decomposition of the bifluoride, and also the alkali fluoride:

$$BeSiO_3 + 6\ HF \rightarrow BeF_2 + 3\ H_2O + SiF_4$$

$$Be^0 + 2\ HF \rightarrow BeF_2 + H_2$$

$$BeF_2 + 2\ KF \rightarrow K_2[BeF_4]$$

If the sintering with potassium bifluoride is carried out under the conditions given here, the fluoridation occurs only on the surface of the material being examined but, because of the extreme sensitivity of the quinalizarin reaction, enough potassium fluoberyllate is produced to insure the specific detection of the beryllium, even though present in small amounts. The products obtained on fusing or sintering magnesium, calcium, ferric, aluminum, titanium, and zirconium oxides with potassium bifluoride do not react with an ammoniacal solution of quinalizarin.

Procedure. A few milligrams of the sample(powder or shavings) are mixed with three times its quantity of potassium bifluoride and fused or sintered for 3–4 minutes in a platinum spoon. The cooled mass is treated with 2 ml of cold water. After 5 minutes, the suspension is filtered or centrifuged. A drop of the clear supernatant liquid is transferred to a spot plate and treated with a drop of quinalizarin solution. A blank test on a drop of water is conducted in an adjacent depression of the spot plate. If beryllium is present, a blue color or precipitate appears, depending on the quantity of beryllium present. The violet of the ammoniacal quinalizarin solution remains unchanged in the blank. If bromine water is added drop by drop, only the blank is decolorized.

Several beryllium-bearing minerals were examined by this procedure, including one whose composition was: 1% BeO; 25% Al_2O_3; 24% CaO; 3% MgO; 5% P_2O_5; 6% Fe_2O_3; 1% Mn_2O_3; 7% SiO_2. The test was invariably confirmative. The same was true of beryllium-bearing alloys.

Reagent: Solution of 50 mg quinalizarin in 10 % ammonia

When beryllium is an essential constituent of tin-free minerals and alloys, its presence can be established by isolating beryllium hydroxide and noting the fluorescence reaction with morin (page 191). Silicate-bound beryllium requires fuming with ammonium fluoride for its release. Solution in dilute hydrochloric acid is sufficient in the case of non-ferrous alloys. Chrysoberyl ($Al_2O_3 \cdot BeO$) must be fused with sodium pyrosulfate.

Procedure.[59] The sample should be divided into as small pieces as possible. One or two milligrams are sintered in a platinum spoon with ammonium fluoride, or fused with sodium pyrosulfate, or dissolved in hydrochloric acid and evaporated. The product is rinsed into a micro crucible with as little water as possible. A drop of the liquid is transferred to a depression of a spot plate and the following are stirred in: three drops of a saturated solution of "Complexone III" (versene) in ammonia (1 : 10); one drop of 0.002 % solution of morin in acetone; then one drop of concentrated ammonia. If beryllium is present, the mixture exhibits a strong yellow-green fluorescence in ultraviolet light.

When very small amounts of beryllium are to be detected, the sample should be decomposed and then the filtration method described on page 192 is applied.

A good procedure for the detection of beryllium in rocks and minerals is based on the fact that alkali beryllate produces a water-soluble yellow-green fluorescent compound with morin.[60]

Procedure. The finely pulverized sample is fused with potassium sodium carbonate. A drop of the melt is transferred to a black spot plate with the aid of a platinum wire. After solution in 3 drops of 5 *N* hydrochloric acid, a drop of a saturated methyl alcohol solution of morin is added, and the solution made alkaline with 4 drops of 5 N sodium hydroxide. A bright green fluorescence in ultraviolet light shows the presence of beryllium.

Careful acidification with 5 N HCl or 30 % acetic acid quenches the fluorescence. If aluminum, tin, or zirconium is present, the fluorescence turns blue.

39. Detection of Tungsten in Minerals

Scheelite is decomposed by evaporation with dilute hydrochloric acid and tungstic acid is precipitated:

$$CaWO_4 + 2\ HCl \rightarrow CaCl_2 + WO_3 + H_2O$$

Analogous decompositions occur with wolframite [$Fe(Mn)WO_4$], stolzite ($PbWO_4$), etc. If the evaporation residue is digested with ammonium hydroxide, water-soluble ammonium tungstate results. The latter reacts with hydrogen peroxide and ammoniacal thallous nitrate solution to yield a precipitate of TlOOH.[61] The chemistry of this test and the procedure for detecting MoO_3, WO_3, and V_2O_5 are given on page 417.

Slight amounts of the pulverized sample are sufficient for the test. The test portion is taken to dryness with a drop or two of hydrochloric acid in a micro crucible. The digestion with ammonium hydroxide and the addition of the reagents can be done in the crucible. When dealing with products containing other minerals it is well to run a comparison test omitting the hydrogen peroxide.

Another reliable test is based on the production of tungsten oxinate.[62]

Procedure.[63] A micro crucible is used. About 5–10 mg of the powdered mineral are fused with three or four times the quantity of sodium peroxide. After cooling, about 0.5 ml of water is added and the suspension warmed several minutes. A drop of the alkaline suspension (colored residues should be centrifuged) is transferred to a spot plate and treated with a drop of a 5 % alcoholic solution of oxine and about 0.05 g solid ammonium chloride. The mixture is stirred and then three drops of concentrated hydrochloric acid are added. A brown precipitate results if the mineral contains tungsten.

40. Detection of Titanium in Minerals

The chromotropic acid reaction of titanium salts, described on page 197, can also be applied for the detection of titanium in minerals (rutile, kaolin, monazite, etc.), after the TiO_2 in them has been converted into water-soluble titanic sulfate by fusion with potassium bisulfate or pyrosulfate.

Procedure.[64] A pin-head size sample is finely powdered and mixed with 3 to 4 times its bulk of potassium bisulfate and fused on platinum foil. The cold melt is transferred to a watch glass, rubbed with a little water, and placed on filter paper impregnated with chromotropic acid. When titanium is present, a red to brown red color appears.

The test is even more sensitive if a drop of a hydrochloric acid solution of stannous chloride is added to the whole melt on the watch glass, and then a few drops of chromotropic acid. In the presence of titanium, the undissolved particles are tinted red as well as the solution.

It is better and simpler to proceed as follows, especially when testing for very small amounts of titanium.[65] The finely powdered sample is mixed with a few drops of concentrated sulfuric acid in a micro crucible or test tube and heated to fuming. After cooling, a small particle of solid chromotropic acid is added and the mixture warmed gently. When titanium is present, a more or less intense violet color appears.

For the detection of traces of titanium in silicates, such as glass, it is advisable to fume a few milligrams of the powdered sample with 2 or 3 drops of hydrofluoric acid in a platinum spoon, and then to heat with 2 or 3 drops of concentrated sulfuric acid. After rinsing the contents of the spoon into a test tube containing a little concentrated sulfuric acid, the test with chromotropic acid may be carried out as described.

41. Detection of Calcium in Silicates [66]

If the evaporation residue of an aqueous zinc chloride solution is heated, hydrogen chloride is given off:

$$ZnCl_2 + H_2O \rightarrow Zn(OH)Cl + HCl \quad \text{(1)}$$

$$Zn(OH)Cl \rightarrow ZnO + HCl \quad \text{(2)}$$

The anhydrous gas is released at temperatures far above the boiling point of concentrated hydrochloric acid (sp. gr. 1.2) which is an aqueous solution containing 41.7 % HCl. Consequently, fusion or sintering of solids with zinc chloride can sometimes accomplish reactions which occur to only a slight extent, if at all, with concentrated aqueous hydrochloric acid. Among such reactions is the decomposition of silicates. For example,

$$CaSiO_3 + 2\ HCl \rightarrow CaCl_2 + H_2O + SiO_2$$

The resulting $CaCl_2$ can be detected by means of sodium rhodizonate through the production of red-violet water-insoluble basic calcium rhodizonate. The presence of excess zinc chloride necessitates alkalization with NH_4OH.

Procedure. The test is conducted in a small quartz crucible. One or two mg of the finely ground sample are mixed with several mg of zinc chloride and a drop of water and taken to dryness. The evaporation residue is then heated to faint redness in the covered crucible. The cold residue is stirred with 1 or 2 drops of hot water and the suspension is transferred to a conical tube and centrifuged. A drop of the clear liquid is treated on a spot plate with a freshly prepared 0.2 % solution of sodium rhodizonate and three drops of concentrated ammonia. The mixing is accomplished by blowing on the liquid. A violet precipitate or color appears if calcium is present. A comparison test is required when tiny quantities of calcium are suspected. The same amount of zinc chloride should be used in both cases since it often gives a slight self-reaction due to slight contamination with calcium and lead.

See page 469 regarding the decomposition of calcium silicates by other methods.

42. Detection of Magnesium in Silicate Rocks

The very sensitive test with *p*-nitrobenzeneazo-α-naphthol (page 225) affords a rapid method for the detection of magnesium in silicates (talc, asbestos, etc.).

Procedure. A few milligrams of the finely powdered sample are mixed with sodium potassium carbonate and fused in a loop of platinum wire. The bead is dissolved in 1 or 2 drops (1 : 1) hydrochloric acid on a spot plate; a drop of 2 *N* NaOH and one of an alkaline solution of the reagent are added. A blue precipitate or color, compared with a blank, indicates magnesium.

The procedure to be used in the presence of large amounts of iron and aluminum is described on page 227.

A very convenient method of decomposing magnesium silicates is to fuse or sinter the test portion with zinc chloride as described in the foregoing section. The sinter residue is stirred with dilute acid and centrifuged. A drop of the clear liquid is placed on a spot plate and the reagent solution is added drop by drop until the violet color is permanent. A pinch (knife blade) of potassium cyanide is added and the solution stirred by blowing on it. A blue precipitate or color indicates magnesium. The potassium cyanide is added because the blue magnesium color lake does not appear in the presence of zinc hydroxide or zincate (see page 227).

43. Detection of Potassium in Silicate Rocks and Glass

The presence or absence of potassium is often characteristic of certain members of the two main classes of silicate rocks. For instance, sodium and potassium felspars are distinguished in the class of the anhydrosilicates. Among the metasilicates, only the leucites contain potassium, and among the orthosilicates, only certain members of the nepheline group contain potassium (or cesium). Likewise, among the hydrosilicates, which include zeolites, mica, serpentine, talcs, only certain varieties contain potassium.

It is not possible to detect potassium directly in silicate rocks since the samples almost always are complex silicates in which the potassium is not bound ionogenically. However, silicates can be decomposed by fuming with hydrofluoric acid plus concentrated sulfuric acid (conversion of SiO_2 into gaseous SiF_4). Subsequent ignition produces sulfate. The residue will contain any potassium. It can then be detected by reaction with dipicrylamine (see page 230).

Procedure. A little (several milligrams) of the rock is pulverized and heated in a platinum crucible or spoon with several drops of hydrofluoric and concentrated sulfuric acid. Solid ammonium fluoride may be used in place of the hydrofluoric acid. When no more acid fumes are evolved, the residue is ignited briefly, allowed to cool, and scraped from the reaction vessel with a nickel spatula. The test portion is placed on orange-red dipicrylamine paper, moistened with 1 drop water, and dried in a current of heated air. The paper is then bathed in 0 1 *N* nitric acid. If potassium is present, a red fleck is left on the paper which is turned bright yellow by the acid. The preparation of the dipicrylamine paper is described on page 232.

A faster and more reliable method of detecting potassium in silicates makes use of a fusion or sintering with hydrated zinc chloride to produce potassium chloride,[66] which yields a white precipitate with sodium tetraphenyl boron (compare page 232). See Section **41** regarding the chemistry of the action of zinc chloride on silicates.

Procedure. The reaction is conducted in a quartz crucible. A small portion of the powdered sample is treated with about 5–10 times the quantity of zinc chloride and a drop of water. After evaporation, the crucible is covered and the contents heated for five minutes over a bare flame. The cooled mass is stirred with a drop or two of hot water, the suspension is transferred to a conical tube, and centrifuged. One drop of the clear liquid is treated with a drop of a 1 % aqueous solution of sodium tetraphenyl boron. If potassium is present, a white precipitate or turbidity appears.

This procedure is especially suitable for testing glass. Fractions of a milligram are adequate.

44. Detection of Ferrous Iron in Silicates and in Acid-resistant, Silica-free Minerals [66]

If silicates containing iron^{II} are fused with zinc chloride, ferrous chloride results and can be easily detected through the color reaction with *α,α'*-dipyridyl (compare page 161). Ferrous iron is likewise liberated by the zinc chloride fusion of silica-free, acid-resistant minerals containing iron^{II}, such as wolframite ($FeWO_4$), ilmenite ($FeTiO_3$), chrome iron ore ($FeCrO_4$), and columbite [$Fe(Mn)Nb_2(Ta)_2O_6$].

Procedure. A little of the pulverized sample is made into a bead by fusion with zinc chloride in a platinum loop. The cold bead is loosened from the loop by immersing it in a drop or two of a hydrochloric acid solution of α,α'-dipyridyl (spot plate or micro test tube). A pink color indicates the presence of iron. Addition of several drops of concentrated ammonium hydroxide makes the color deep red (compare page 453).

The procedure is especially good for detecting slight quantities of iron^{II} in silicates rapidly and with certainty.

45. Detection of Nickel in Silicate Rocks [66]

The decomposition procedure used to break down silicates by heating with zinc chloride (see Sections **41**, **42**, **43**) can be applied to nickel silicates to yield nickel ions which react with dimethylglyoxime. The most important nickel ore in this class is garnierite (hydrated nickel magnesium silicate).

Procedure. A slight quantity of the pulverized sample is fused with zinc chloride in a platinum loop. The cold mass is removed from the platinum wire by stirring it in a drop or two of hot water in a depression of a spot plate. A drop of concentrated ammonium hydroxide and a drop of a 1 % alcoholic solution of dimethylglyoxime are added to the suspension. A red precipitate results if nickel is present.

It is better to make the fusion in a platinum spoon when dealing with silicates that are poor in nickel and to transfer the cold sinter product to a conical tube by means of a drop or two of concentrated ammonium hydroxide. After centrifuging, a drop of the dimethylglyoxime solution is added.

Minerals which also contain iron^{II} will yield ferrous ions when fused with zinc chloride and they too give a red color with dimethylglyoxime and ammonia (compare page 164). Consequently, in such cases it is best to test for ferrous silicates as outlined in Section **44** and if the response is positive the ammoniacal suspension should be warmed with hydrogen peroxide prior to the centrifuging. Ferric hydroxide is precipitated, while the nickel tetrammine ions remain in solution.

The procedure has proved useful as a field test for nickel-poor silicates. In addition, it is easy to detect nickel in chrysoprase (NiNaAlFe silicate) and connarite (NiAl silicate) by this means.

46. Detection of Reducible Metals in Minerals [67]

When hypophosphites are heated they decompose and evolve hydrogen and phosphine. The following thermal decomposition has been assumed for ammonium hypophosphite: [68]

$$7\,NH_4H_2PO_2 \rightarrow 2\,HPO_3 + H_4P_2O_7 + 7\,NH_3 + H_2O + 2\,H_2 + 3\,PH_3$$

If this salt is thermally decomposed in the presence of oxide minerals (which may also contain silica), the hydrogen and phosphorus hydride act as powerful reducing agents; the meta- and pyrophosphoric acids dissolve the metal oxides with production of phosphate, pyrophosphate, or double phosphate. Consequently, it is possible to achieve the same reactions as occur in the so-called phosphate beads in the reducing flame. However, compared with the reactions in the phosphate bead, the ammonium hypophosphite fusion has the following advantages: it can be conducted in a crucible; the reducing fusion occurs at a much lower temperature than the phosphate fusion; there is no danger of a reoxidation, provided an excess of ammonium hypophosphite is used.

The original publication prescribes that approximately 0.1 g of the pulverized mineral be heated for about 2 minutes in a small evaporating dish with 2 g of ammonium hypophosphite. However, the fusions are successful with even a few milligrams of the pulverized sample. The colors of the various melts and their respective indications are:

blue: cobalt, titanium, tungsten	red-brown: molybdenum
green: vanadium, chromium, uranium	black: tellurium, niobium

Characteristic differences in the color of the hot and cooled melts, their behavior on the addition of water, hydrogen peroxide, ammonia, ammonium carbonate, etc. are described in the original paper.

47. Detection of Alkali Metals in Silicates [69]

Chlorides, nitrates, and fluorides of all the metals are converted into carbonates, oxides, or the free metal by evaporating to dryness with oxalic acid, followed by ignition. After taking up the ignition residue in water, the filtrate will only react alkaline to litmus when either calcium oxide or the carbonates of the alkali metals are present. If the ignition with oxalic acid is followed by treatment with ammonium carbonate to convert any calcium oxide to carbonate, then a basic reaction to litmus indicates alkali metals.

Procedure. A finely powdered portion of the silicate is decomposed by fuming with hydrofluoric acid on the lid of a platinum crucible. After the silicon fluoride and excess hydrofluoric acid have volatilized, the metallic fluorides are left. The dry residue is moistened with 2 or 3 drops of hydrochloric acid, and about 1 ml water and a little solid oxalic acid are added. The mixture is evaporated to dryness and then heated to redness. The evaporation and ignition with oxalic acid is repeated. The residue is then moistened with a concentrated solution of ammonium carbonate and evaporated to dryness, but not heated to redness, lest the calcium carbonate revert to the oxide.

References pp. 512–515

The residue is rinsed into a centrifuge tube with a little water. The suspension is warmed and centrifuged. The clear supernatant liquid is evaporated on a platinum lid and then ignited at a dull red heat, to convert any magnesium carbonate into the oxide. The product is moistened with a little water and carefully warmed. The alkali carbonates go into solution, but magnesium oxide adheres to the platinum (usually as a small stain). A drop of the perfectly clear solution is removed with a capillary and placed on litmus paper. If the paper turns blue, the presence of an alkali metal in the silicate is indicated.

When not more than traces of alkali metal are present, the paper does not turn blue at once. The first drop causes no color change, but on adding further drops in the same place, the center of the moist portion remains almost unchanged, with a pale blue circle remaining visible on the periphery. In this way, as little as 0.2 % of alkali may be detected in kaolin.

48. Differentiation of Magnesite and Dolomite or Bräunerite [70]

Numerous slightly soluble magnesium compounds (oxide, hydroxide, carbonate, phosphate, etc.) are colored deep red on heating with an alcoholic-alkaline solution of diphenylcarbazide. The color is stable and persists even after boiling with water. This tinting with diphenylcarbazide is probably due to an adsorption phenomenon similar to the magnesium color reaction described on page 224. The color appears distinctly even in a mixture of 1 part magnesium carbonate and 200 parts calcium carbonate, yet it is not seen at all when the magnesium carbonate is in the form of dolomite which is usually regarded as a double carbonate $CaCO_3{\cdot}MgCO_3$. This masking of the magnesium diphenylcarbazide reaction is clearly due to the fact that dolomite may be regarded as the calcium salt of a magnesium carbonate complex, $Ca[Mg(CO_3)_2]$.[71] The magnesium is thus a constituent of a complex anion, and therefore has lost its normal reactivity just as iron in ferro- or ferricyanides no longer shows its usual ionic reactions. Magnesite and dolomite can therefore be readily differentiated by a simple color reaction which can be carried out as a spot test.

Support for the assumption that the particular structure of dolomite is responsible for the masking of the diphenylcarbazide reaction for magnesium is given by the behavior of the double carbonate $FeCO_3{\cdot}MgCO_3$ (bräunerite). This mineral, which is isomorphous with dolomite, behaves like the latter toward diphenylcarbazide: it gives no reaction for magnesium. However, if dolomite or bräunerite is ignited, the dolomitic linking is destroyed and magnesium can be detected with diphenylcarbazide, in the resulting mixture of oxides.

Procedure. One or two drops of a hot alcoholic alkaline solution of diphenylcarbazide are placed in a depression of a spot plate and a portion (size of pinhead) of the rock added. After 5 minutes, the colored solution is pipetted out and replaced by hot water, and this method of washing is continued until no more coloring matter goes into the water. If the sample is magnesite, or if it contains magnesite particles, a red-violet product remains. When the magnesium is in dolomitic combination (bräunerite), no color results. The magnesium can then be detected by taking a fresh sample and igniting it on platinum. The dolomitic linking is destroyed; the ordinary reactive magnesium hydroxide is formed and can be detected in the manner previously described.

This method of differentiation has been tested on many carbonate and magnesia minerals.[72] In certain cases, the magnesitic or dolomitic nature of the mineral was recognized, although it had been wrongly classified from its crystal habit. The diagnosis from the diphenylcarbazide reaction always agreed with that deduced from the quantitative analysis carried out later.

The procedure also serves to reveal small amounts of non-dolomitic magnesium in chalk. The magnesia content was definitely established in a limestone of the composition: 0.49 % MgO, 64.02 % CaO, 1.77 % Fe_2O_3 + Al_2O_3, 0,23 % SiO_2, and 32.27 % loss on ignition.

Dolomite and bräunerite can be differentiated through the iron^{II} content of the latter. The red color characteristic of Fe^{+2} ions appears if a sample is gently warmed in a micro test tube with a hydrochloric acid solution of α,α'-dipyridyl (compare page 453).

49. Identification of Calcite[73] (Differentiation from Dolomite)

Crystalline or compact calcite and dolomite, which may also occur side by side as a result, for instance, of metasomatosis,[74] are difficult to differentiate because of their similar habit. A rapid differentiation is desired in technical practice in order that the processing and use of these carbonates may be on a sound foundation from the quarry on. The procedure given here is based on the divergent behavior of these materials toward dilute acids. Only calcite is attacked by dilute hydrochloric acid at room temperature with immediate evolution of carbon dioxide, whereas the reaction of dolomite is distinctly more sluggish. If a saturated solution of tartaric acid containing alizarin S (sodium salt of alizarinsulfonic acid) is used as the H^+ ion donor, calcite immediately produces Ca^{++} ions, which react with the alizarinsulfonic acid to yield the purple-red calcium salt which is resistant to tartaric acid. Dolomite remains unchanged. The test for the production of a red color can be conducted on a spot plate with fragments of the crystalline sample or with the powdered material.

50. Differentiation of Hard- and Soft-burned Lime [75]

When calcite is burned, too prolonged and intense heating leads to the so-called hard burned (overburned, deadburned) lime. (The lime obtained from pure calcium carbonate is not deadburned, even though high temperatures are used.) Although soft- and hardburned lime are identical crystallographically, the rate of certain chemical reactions is distinctly slower with the overburned material. The sluggishness may be due to the lower free surface which results in a lesser rate of solution of the hard- or overburned lime as compared with normal burned lime. It is not possible to detect differences in the behavior of these varieties of lime (ignition stages) by means of aqueous solutions of indicators. On the other hand, a solution of thymol blue in nitrobenzene to which a few drops of methyl alcohol are added serves this purpose well. Normal burned lime is tinted blue immediately by this reagent, whereas overburned lime assumes this color very slowly. The color reaction can be made on a hand specimen in a test tube or with the powder on a spot plate.

The reason for the rapid reaction of normal softburned lime is due to the production of $Ca(OH)_2$ on contact with moist air, and the resulting OH^- ions cause the indicator change. This supposition is confirmed by breathing on specimens of the two varieties of lime that have been brushed with the reagent solution. Softburned lime turns blue at once; hardburned lime requires some time before it shows the color change.

51. Differentiation of Caustic Burned and Sintered Magnesite [76]

Magnesium oxide is prepared commercially by igniting magnesite:

$$MgCO_3 \rightarrow MgO + CO_2$$

If the heating is at relatively low temperatures (800–900°C), the product, known as caustic magnesia, sets well with water and is therefore suitable for mortar purposes. If high temperatures (1600–1700°C) are used, the magnesium oxide sinters (primarily because of the Fe_2O_3 contained in commercial raw materials) and the so-called magnesia stone or sinter magnesia no longer hardens with water. These highly calcined products are employed in the preparation of refractory linings, bricks, etc. and for the preparation of laboratory ware.

The two varieties of magnesium oxide can be distinguished by means of an alcohol solution of quinalizarin. If the samples are shaken with the reagent solution, the lower burned material gives a grey-blue turbidity and the sintered magnesite a cornflower blue clear solution.

It is advisable to use test tube reactions for this differentiation, because a higher or lower turbidity can then serve as an index of the degree of calcination of the magnesite. Refractory magnesite products can be tested in this manner for a possible content of caustic magnesia. Obviously, this differentiation reaction is based on the greater free surface of the lower

References pp. 512–515

burned product as indicated by the more rapid and more extensive formation of the magnesium-quinalizarin lake (see page 224).

52. Detection of Lime in Magnesite [77]

Lime (CaO) is a deleterious ingredient in magnesite ($MgCO_3$) and frequently it is desirable to have a rapid method of detecting its presence in rock samples. Natural or unburned specimens can be tested by means of the lake formed by the action of calcium carbonate with alizarin. The color appears very quickly if a weakly acidified reagent solution is used, whereas magnesium carbonate remains unaltered because of its resistance to dilute acids.

Procedure. A particle of the rock, preferably with a polished area, is dropped into boiling reagent solution. The places containing lime turn violet, the magnesite areas become yellow.

Reagent: 0.1 % solution of sodium alizarinsulfonate in *N*/15 hydrochloric acid.

Calcite and dolomite, namely the undecomposed carbonates, may be distinguished in caustic burned magnesite (MgO) by spotting or spraying the sample (grains or surface) with indicators. Phenolphthalein or thymolphthalein may be used. The active earth oxides formed by the burning of the carbonate produce red or blue colors, resp. Magnesite dissociates at 420° C, while calcite and dolomite lose carbon dioxide only at 700° to 900° C, leaving calcium oxide. Accordingly, magnesium oxide can be detected in causticized magnesite with phenolphthalein, while calcite and dolomite are still inactive. The caustic burned magnesite turns red-violet; the harmful ingredients remain uncolored and acquire merely an accessory coloration from the magnesia dust and indicator.

Some of the calcium in caustic magnesite is present, however, as active CaO and it reacts like MgO with the indicators used here. The active lime (CaO) can be detected in caustic burned magnesite in the presence of magnesia (MgO) if a buffer is mixed with the indicator. The alkalinity of the MgO is thus decreased so much that the indicator responds solely to the CaO. A solution of ammonium citrate and thymolphthalein is suitable.

Procedure. The specimen, on a watch glass, is moistened with several drops of *N*/3 ammonium citrate solution and then spotted with thymolphthalein. Within a few seconds CaO grains turn blue, whereas MgO and $CaCO_3$ grains remain unchanged. This procedure can also be used to distinguish CaO from $CaCO_3$.

53. Detection of Calcium in Ashes, Dry Residues, Magnesite, Dolomite, Silicates, etc.

Calcium oxide, which remains after ashing calcareous organic materials as well as after igniting calcium carbonate, can be detected in slight quantities by spotting with sodium rhodizonate. Violet basic calcium rhodizonate is produced at once (compare page 222). Calcium was detected in this manner in the ash of plants, in ignited dolomite, in impure magnesite, and also in the evaporation residue of three drops of alkaline mineral water.

The calcium may be released from silicate bonding by fusion with zinc chloride as described in Section **41**, or the classic procedure of fusing in platinum with sodium-potassium carbonate can be carried out on a micro scale in a spoon. The cold melt is taken up in dilute hydrochloric acid, sodium rhodizonate solution is added, followed by sodium hydroxide.

Calcium can be detected in glass by fuming the powdered sample in a platinum spoon first with ammonium fluoride and then with ammonium sulfate. The residue, containing calcium sulfate, is stirred with a drop or two of 0.5 *N* sodium hydroxide, a drop of the basic solution is transferred to a spot plate, and treated with a drop of sodium rhodizonate solution. Glasses containing calcium give a decided positive response.

54. Differentiation of Calcite and Aragonite [78]

The sensitive test for OH^- ions due to their participation in the reaction

$$Mn^{+2} + 2\ Ag^+ + 4\ OH^- = MnO_2 + 2\ Ag^o + 2\ H_2O$$

(black precipitate), may be used as a good and sensitive method of differentiating between calcite and aragonite. The solubilities in water of these two modifications of $CaCO_3$ differ slightly; aragonite is the more soluble.

When calcium carbonate is suspended in water the reactions are:

$$\underset{\text{solid}}{CaCO_3} \rightleftarrows \underset{\text{dissolved}}{CaCO_3} \rightleftarrows Ca^{+2} + CO_3^{-2}$$

$$CO_3^{-2} + 2\ H_2O \rightleftarrows H_2CO_3 + 2\ OH^- \quad \text{or} \quad CO_3^{-2} + H_2O \rightleftarrows HCO_3^- + OH^-$$

Aragonite gives rise to more OH^- ions than calcite, and so on treating a massive or powdered sample of the more soluble species with an aqueous solution containing Ag^+ and Mn^{+2} ions, the concentration is sufficient to give the OH^- reaction (blackening) after 30–60 seconds. Calcite gives not more than a slight gray color after 10 minutes, and only after several hours does this color deepen to the black developed by aragonite in about 2 minutes.

This differentiation reaction can also be carried out on thin sections. If magnification is used, the slightest occurrence of aragonite, in stalactite sections for example, can readily be detected by the blackening of the aragonite aggregates.

Procedure. The powdered mineral is placed in a watch glass on white paper, or on a spot plate, and treated with a drop of the reagent solution. After about

2 minutes the mass is examined for blackening. If this is apparent, aragonite is present.

Reagent: One gram solid Ag_2SO_4 is added to a solution of 11.8 g $MnSO_4 \cdot 7\,H_2O$ in 100 ml water, and boiled. After cooling, the suspension is filtered and 1 or 2 drops of dilute NaOH solution added. The precipitate is filtered off after 1 to 2 hours. The solution keeps if stored in dark bottles.

The reagent for OH^- ions can also be used to show that basic salts are formed by partial hydrolysis and loss of acid on heating sodium and potassium fluoride. Magnesium chloride behaves analogously. A small sample is melted in a loop of platinum wire and the cooled bead is placed in the reagent solution. The bead is blackened by the action of the basic salts present.

55. Identification of Calcium Sulfate (Gypsum and Anhydrite)

Calcium sulfate can be identified, in the absence of other slightly soluble alkali earth sulfates (and lead sulfate), by its action with an acidified solution of mercuric nitrate (see page 408). Light yellow basic mercuric sulfate is formed. This test [79] requires the absence of all materials which react with the reagent, such as oxides, carbonates, halides, sulfides, etc. Furthermore, it cannot be successfully applied if the color of the sample hides the yellow basic mercuric sulfate.

Another test of wider application [80] is based on the reaction:

$$CaSO_4 + Pb^{+2} \rightarrow PbSO_4 + Ca^{+2}$$

The lead sulfate is then identified by the sensitive specific reaction with sodium rhodizonate (see pages 73 and 434).

Strontium sulfate acts like calcium sulfate. In contrast, barium sulfate, whose solubility product is less than that of lead sulfate, does not react with lead nitrate.

The following procedure can be used for the direct detection of calcium sulfate in the absence of difficultly soluble calcium salts of acids which form slightly soluble lead salts (sulfides, sulfites, phosphates, arsenates). If these are present, the procedure must be modified (see below).

Procedure. A piece of quantitative filter paper is fitted into a Gooch or sintered glass filtering crucible and wetted. Particles of the sample are placed on the paper and spotted with 2 or 3 drops of lead nitrate solution. After a minute, the liquid is sucked off. The excess reagent is removed * by washing

* The washing with dilute nitric acid is essential. If omitted, the paper retains considerable quantities of lead, which react with the rhodizonate even in the absence of $PbSO_4$, produced by $CaSO_4$. The retention of lead by filter paper is due to the hydrolysis of $Pb(NO_3)_2$ solutions followed by adsorption of the hydrolysis product (basic lead salt) by the paper. This basic salt is destroyed and removed by the acid wash liquid.

with 0.2 N nitric acid (2 ml) and then with 3 ml water or alcohol. The acid and water (or alcohol) should be added drop by drop, while gentle suction is being applied. After washing, the paper is placed on a glass plate and spotted with sodium rhodizonate and then with the buffer solution. If the sample contains calcium sulfate, a red spot is left, or the material on the paper turns deep red, due to the formation of lead rhodizonate.

Reagents: *1)* 2 % lead nitrate solution mixed with an equal volume of 0.2 N nitric acid
2) 0.2 % sodium rhodizonate (2 mg in 1 ml water). The solution must be freshly prepared
3) Buffer solution (pH = 2.79) : 1.5 g tartaric acid and 1.9 g sodium bitartrate in 100 ml water

The procedure will reveal the presence of $CaSO_4$ used as filler for paper, provided $Al_2(SO_4)_3$ is absent. The test is carried out on the ash yielded by several milligrams of the paper. Barite ($BaSO_4$) can be tested for $CaSO_4$ in this way. The procedure will reveal gypsum in a mixture of $BaSO_4$ and $CaSO_4$ (1500 : 1). The adulteration of flour with gypsum can be revealed by testing the ash.

Superphosphate of lime can be distinguished from phosphate fertilizers containing no calcium sulfate. The sample (0.1 to 0.5 g) is fumed twice, in a porcelain crucible, with 1 ml concentrated hydrochloric acid. Alcohol is added, and the suspension is filtered. The contents of the filter are thoroughly washed with alcohol and then tested with lead nitrate as described.

If soluble calcium salts are absent, gypsum can be detected (also in the presence of strontium and barium sulfate) by means of violet color obtained when the sample comes into contact with an alkaline solution of sodium rhodizonate. The color is due to calcium rhodizonate.

56. Differentiation of Gypsum and Anhydrite [81]

Calcium sulfate occurs in minerals or technical materials as gypsum ($CaSO_4 \cdot 2\,H_2O$) or as anhydrite ($CaSO_4$). When either is presented in the pure state with no adhering moisture, they may easily be differentiated by the evolution and condensation of water vapor from gypsum when heated in an ignition tube. If the sample is moist, the test is without value of course, or preliminary drying is essential.

A better and more rapid differentiation is based on the fact that gypsum, as the hydrated compound, is more readily soluble in water than anhydrite. At 100° C, the solubility of $CaSO_4 \cdot 2\,H_2O$ is 0.167 %; the corresponding value for $CaSO_4$ is only 0.067 %.[82] The solubilities differ markedly even at room temperature. Accordingly, gypsum reacts much more rapidly with sodium carbonate solution:

References pp. 512–515

$$CaSO_4 + Na_2CO_3 \rightleftarrows CaCO_3 + Na_2SO_4$$

than anhydrite. When a solution of sodium carbonate, reddened with a little phenolphthalein, is used, the gypsum decolorizes it rapidly and the anhydrite much more slowly. This difference can be used to distinguish between these two forms of (powdered) calcium sulfate.

Procedure. A few milligrams of the finely ground samples are placed in adjacent depressions of a spot plate and moistened with 2 or 3 drops of a sodium carbonate solution that is reddened with phenolphthalein. On stirring with a platinum wire, gypsum causes an appreciable lightening of the color in 1 to 2 minutes and complete decolorization in 4 to 5 minutes, compared with 15 minutes and 45 minutes required by powdered anhydrite.

The sample, in bulk, may be tested by spotting it with 1 or 2 drops of the sodium carbonate solution, but the difference is not so pronounced under these conditions.

Reagent: 4 % sodium carbonate solution plus 3 drops of 2 % alcoholic phenolphthalein

A more certain method of distinguishing between gypsum and anhydrite is based on the divergent behavior toward an almost colorless benzene solution of rhodamine B.[83] Gypsum remains as it was, whereas anhydrite yields a scarlet product, which has an intense orange fluorescence in ultraviolet light. Since the benzene solution contains the lacto-form of rhodamine B, the phenomenon used here may involve a chemical adsorption of the dye on the anhydrite. The lacto-form of the rhodamine B (I) is transformed into the red quinoidal zwitter ion (II) as shown in the scheme:

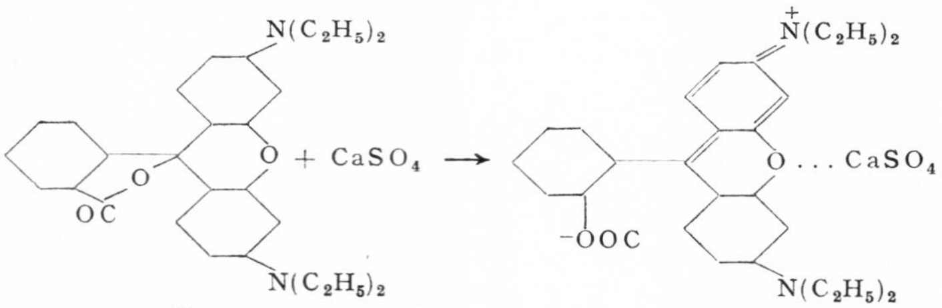

$N(C_2H_5)_2$ ... O ... OC ... O $+ CaSO_4 \longrightarrow$ $\overset{+}{N}(C_2H_5)_2$ ... ^{-}OOC ... $O \ldots CaSO_4$

$N(C_2H_5)_2$ (I) $N(C_2H_5)_2$ (II)

Another plausible explanation is: The lacto-form in the benzene solution is in equilibrium with tiny amounts of the quinoidal form. On contact with anhydrite, the form (II) is adsorbed on the surface of the mineral, but the disturbed equilibrium is instantly restored by transformation of the lacto-form. (Compare in this connection the detection of uranium, page 209.)

Procedure. A slight quantity of the sample (0.1 mg is sufficient) is treated in an Emich conical tube with a drop or two of a saturated benzene solution of rhodamine B and stirred if necessary with a glass thread. The red product collects in the constricted part of the tube on centrifuging. It is readily seen but if need be it can be viewed under the quartz lamp where it exhibits an intense orange-red fluorescence.

Anhydrite and gypsum can be readily distinguished from each other by this method. However, this test is not specific for anhydrite because numerous other anhydrous solids behave in the same fashion as anhydrite toward this dye.

Another method for distinguishing between anhydrite and gypsum is discussed on page 510. It is based on the behavior toward molten potassium thiocyanate.

57. Differentiation of Minerals Containing Barium and Strontium [84]

Two cases come under this heading, namely the differentiation of barite ($BaSO_4$) from celestite ($SrSO_4$), and witherite ($BaCO_3$) from strontianite ($SrCO_3$). These problems can be solved with the aid of the divergent behavior of the alkaline earth sulfates toward a solution of sodium rhodizonate. Strontium sulfate but not barium sulfate is tinted red-violet by contact with a freshly prepared solution of the reagent (formation of strontium rhodizonate, compare page 220). Pulverized barite and celestite are readily distinguished in this way if a milligram or so of the solid is treated in a depression of a spot plate with a drop of a 0.2 % solution of sodium rhodizonate.

If witherite and strontianite are to be differentiated, milligram quantities of the powdered carbonates are mixed in micro crucibles with about 0.5 g of ammonium sulfate and carefully heated until no more fumes of sulfur trioxide appear. The carbonates are transformed to sulfates by this treatment. The cold residues are spotted with a 0.2 % solution of sodium rhodizonate. A red color is formed with the portion originally containing strontianite. This test succeeds in the presence of calcite also, because, like barium sulfate, the calcium sulfate remaining after the fuming with ammonium sulfate does not show any reaction with sodium rhodizonate.

58. Orienting Tests of Glass

(*a*) *Test for earth and heavy metal oxides.*[85] A drop of 10 % hydrofluoric acid immediately frosts glass that is rich in these oxides. The sample can be tested for boron, sodium and potassium by the flame color. The original paper should be consulted regarding methods of testing for other metals after the etching with hydrofluoric acid.

References pp. 512–515

Many of the spot tests described in the present text may be used for the identification of the constituents of glass.

(*b*) *Identification of glass containing bases.*[86] A scratched area of the glass is covered with a drop of a water–ether solution of iodoeosin, $C_{20}H_8I_4O_5$. This acid is converted by the alkali of the glass into its sodium salt, $Na_2C_{20}H_6I_4O_5$, which is insoluble in ether. It sticks to the glass as a distinct red film after the surface is rinsed with ether. This treatment leaves quartz glass colorless, and the red color (water-soluble) indicates a glass containing basic oxides.

Reagent: 0.5 g iodoeosin is dissolved in 13 ml 0.1 *N* NaOH in a separatory funnel, 120 ml water and then 15 ml 0.1 *N* H_2SO_4 are added. The dyestuff precipitates. The yellow-red mixture is vigorously shaken with 1 liter of ether and the dyestuff goes into solution. The ether layer is washed 3 times with 30 ml portions of water. The ether solution of the reagent should be stored in dark bottles

The sensitive reaction of OH^- ions in the test for ammonia (page 237), and also used for distinguishing between calcite and aragonite (page 469):

$$Mn^{+2} + 2\,Ag^+ + 4\,OH^- = MnO_2 + 2\,Ag^o + 2\,H_2O$$

may be applied to powdered glass as a rapid means of detecting free basic oxide.[87] A small heap of the powdered glass, (a few milligrams is enough) is treated with the reagent solution on a spot plate, a watch glass, or a Wedekind magnesia groove, and observed for any development and intensity of gray color. The mixture can be further treated with an acetic acid solution of benzidine, which reacts with the MnO_2 formed (see page 175). A more or less deep blue results, according to the alkalinity of the glass.

Different types of glass gave the following results:[88]

Jena Geräte	light gray
Jena normal 16	gray
Jena-supremax	light gray
Jena-Duran	almost negative
Lead glass	gray
Murano	light gray
Kavalier	dark gray
Pyrex	almost negative
Various optical glasses	gray to dark gray

The microchemical spot test can be applied to reveal alkali given off from different kinds of glass. Small samples (splinters, etc.) are enough. This test is very useful in testing the chemical resistance of glass.[89] It should be noted that the particle size of the glass powder actually used in the test may affect the result. Furthermore, when the molten glass is cooled, there sometimes is an increase in alkali concentration on the surface layers.[90]

(c) *Tests for lead and calcium.* The reactions of lead and calcium ions with sodium rhodizonate (production of red or violet compounds described on page 73 and 222) can be used to classify glass powders.

59. Detection of Free Metals in Oxides, Printing on Paper, etc.[91]

Metal oxides, such as ZnO, CuO, CdO, PbO, Pb_3O_4, that have been prepared by heating the metal in contact with air, often still contain free metal, which was protected against oxidation because its surface was coated with a layer of the oxide. In contrast, metal oxides prepared by calcination of carbonates, nitrates, etc., never contain free metal. Consequently, the detection of free metal in oxides can be used to establish the purity of the sample (particularly of ZnO). The test also will sometimes give information as to the origin of the oxide.

The reaction of free metals with phosphomolybdic acid (see page 362) can be used for rapidly detecting free metal in zinc oxide and other oxides. This test can also be applied to metallic dusts on decorative wrappings, printed paper, etc. The basis of the test is the formation of the highly colored molybdenum blue by the action of phosphomolybdic acid with free metals and alloys.

Procedure. About 10 mg of the zinc oxide is placed in a depression of a spot plate. Without stirring, 3 drops of a saturated water solution of phosphomolybdic acid and 1 drop dilute sulfuric acid are added. The zinc oxide dissolves slowly. Molybdenum blue is formed wherever traces of metallic zinc occur. The blue is quite visible against the undissolved zinc oxide and the white surface of the spot plate. If the oxide contains only a little free metal, blue particles or isolated dots are seen. When the sample is quite impure, no localized reduction is discerned, but the entire liquid turns blue or blue-green. When two specimens are to be compared with respect to their content of metallic zinc, four portions of each are tested in adjacent depressions and the blue dots are counted. Representative data are thus obtained.

Litharge (PbO) and red lead (Pb_3O_4), as well as other metal oxides, can be tested for free metal by this procedure. It is preferable to substitute dilute nitric acid for the sulfuric acid when testing litharge because the sample dissolves faster. It is remarkable that practically all samples of Pb_3O_4 give a positive reaction for metallic lead by this test, despite the fact that they contain quadrivalent lead.

If printing on paper, wrappings, etc., is to be tested for powdered metals, dusts of alloys, etc., it is enough to place a drop of phosphomolybdic acid solution on the paper and to note the formation of a blue coloration.

60. Detection of Barium Sulfate in Pigments, Paper Ash, etc.[92]

Barium sulfate, as such or in the form of lithopones ($BaSO_4$ + ZnS, or + CdS) is used as pigment. It also constitues a weighting material in the manufacture of special papers. It can be easily detected in pigments or the ash of papers because it is completely converted into barium chloride by ignition with ammonium chloride. The conversion is due to the production of hydrogen chloride in the thermal decomposition of ammonium chloride and the gas is formed at about 400°C, in other words at a temperature above the boiling point of sulfuric acid. Under these conditions, it readily reacts with barium sulfate to yield barium chloride, which is easily detected by means of sodium rhodizonate (compare page 216). It should be noted that the barium chloride, which remains when lithopones are heated with ammonium chloride is free of zinc and cadmium. The chlorides of these metals are completely volatilized under the conditions of the test.

Procedure. A micro crucible is used. A few grains of the solid (or ignition residue) is mixed with about 0.5 g of ammonium chloride and heated over a bare flame until fumes are no longer evolved. The residue is transferred to a small test tube and treated with a drop or two of freshly prepared 0.2 % sodium rhodizonate solution. The formation of a red to red-brown precipitate indicates the presence of barium sulfate in the sample.

Gypsum and calcium carbonate do not interfere. The absence of lead compounds must be assured since they react in the same manner as barium sulfate.

61. Ferric Ferrocyanide

The deep blue ferric ferrocyanide is acid-resistant; it conforms to the formula $Fe_4[Fe(CN)_6]_3$. It is widely used as a pigment in water and oil colors, in printing of fabrics, and (dissolved in oxalic acid) as ink. A common use is in laundry bluing. It goes under a variety of names: Berlin blue, Parisian blue, Hamburg blue, Prussian blue, and also Mineral blue.

The compound can be detected easily by ignition; brown Fe_2O_3 remains. When treated with caustic alkali, brown hydrous ferric oxide is produced and the alkaline filtrate responds to the reactions of the $Fe(CN)_6^{-4}$ ion (compare page 287ff). Dicyanogen is produced when the material is ignited (compare page 412). All of these tests can be conducted with milligram quantities of the sample.

62. Procedure for Testing Woods Impregnated with Salts [93]

Spot tests can be applied to investigate the nature of the chemicals used in the preservation of timber (telegraph poles, powerline towers, railway ties, etc.). The mineral salts most commonly used are mercuric chloride, copper sulfate, zinc chloride, sodium fluoride.

Whenever possible, a complete transverse section of the wood should be used for the test. The core of a boring may also be used. When a drop of the reagent solution is placed or sprayed on the section the depth of impregnation may be seen. The spot tests should be carried out on wood that has been half-moistened with a damp cloth or which has been kept for some hours in an atmosphere saturated with water vapor. Sprinkling with water must be avoided lest the impregnating material be washed away.

Procedure. *1) Mercuric chloride:* Spot with ammonium sulfide (mercuric sulfide formation). Wood that has been impregnated for a long time will give the reaction more clearly if it is moistened with dilute hydrochloric acid prior to the test.

2) Zinc chloride: The test depends on the liberation of iodine from potassium iodide solutions by zinc ferricyanide. The reagent solution is a mixture of equal volumes of 1 % potassium ferricyanide solution, 1 % potassium iodide solution, 5 % starch solution. The layer impregnated with zinc chloride turns deep blue when the reagent is sprayed on or if a drop is applied to the specimen.

3) Copper sulfate: Impregnation with copper sulfate, now seldom used, is usually revealed by the grass-green color (due to copper tannates). On adding a drop of 1 % solution of potassium ferrocyanide, a brick red to brown-red color appears (copper ferrocyanide). If the color later turns dirty greenish blue, the copper sulfate contained some iron sulfate.

4) Sodium fluoride: The test depends on the reaction of sodium fluoride with iron^{III} salts to give colorless $Na_3[FeF_6]$. Only an excess of iron salt is revealed by the color reaction with potassium thiocyanate. Higher concentrations of sodium fluoride (from 2 % upward) may be detected by spraying or applying a drop of red, freshly prepared solution of iron^{III} thiocyanate, and observing the decolorization. The reagent consists of 0.6 g KCNS and 0.1 g $Fe_2(SO_4)_3 \cdot 18\ H_2O$ in 200 ml water.

The following reaction, depending on the formation of Prussian blue, is sharper and more permanent. The wood is sprinkled with a 5 % solution of $K_3Fe(CN)_6$, dried, and then a 5 % solution of ferrous ammonium sulfate is added. A dark blue is formed where the wood does not contain sodium fluoride.

When small amounts of the impregnating substance are to be detected, or if the distribution is to be more accurately followed, other sensitive tests for the compounds listed may be used to advantage. (See the appropriate sections.)

63. Mineral Tanning Agents [94]

To determine the method used in mineral-tanned leathers, it is necessary to test the sample for certain inorganic compounds. The most common

References pp. 512–515

mineral tanning compounds are chromium salts; sometimes aluminum salts are used, more rarely, iron salts. They may also be used in combination. Spot tests can be used for the detection of these inorganic tanning agents. The sample required is so small that it may be taken from manufactured products (shoes, small leather articles) without visible damage.

(*a*) *Chromium*

A sample (less than 1 mg) is taken with a razor. It is best cut from the inside of shoes and manufactured articles, since dyes and finishes for leather occasionally also contain chromium. The sample is ashed in a porcelain micro crucible, the operation requires only an instant. Then a little sodium peroxide is added and the mixture heated to complete fusion. The chromium is converted into chromate. After cooling, the mass is dissolved in a few drops of 1 : 5 sulfuric acid and 1 or 2 drops of an alcoholic (about 1 %) solution of diphenylcarbazide are added. A deep violet appears if chromium is present (see page 167).

(*b*) *Iron*

A small piece of the leather (about 1 mg) is ashed in a micro crucible and then evaporated with a drop of nitric acid, ignited again, and fused with a crystal of potassium bisulfate. When cold, the melt is dissolved in 1 or 2 drops of dilute hydrochloric acid and a drop of a solution of potassium or ammonium thiocyanate added. A red color indicates iron (see page 164).

(*c*) *Aluminum*

Aluminum may be detected by the formation of the aluminum-alizarin lake (*1*) or by the fluorescence reaction with morin (*2*) (comp. page 182).

A small sample is ashed in a micro crucible, and the residue fused with a crystal of potassium bisulfate. The melt is dissolved in a few drops of hot water and the solution tested for aluminum by (*1*) or (*2*).

(*1*) *Alizarin test*

A drop of the test solution is placed on dry filter paper that has been impregnated with potassium ferrocyanide. Any iron or chromium is held back in the paper, while the aluminum diffuses through the capillaries and accumulates in a circular zone. A drop of a saturated alcoholic solution of alizarin is added and the paper is held over ammonia. The violet color of ammonium alizarinate then appears. In the presence of considerable amounts of aluminum, the strawberry red of the aluminum alizarinate can be seen at the same time. For smaller amounts the filter paper is warmed (held briefly near a flame or placed in an oven) and the violet ammonium alizarinate decomposes (volatilization of ammonia). The red aluminum lake then becomes clearly visible.

(2) Test with morin

A drop of the solution of the melt is treated in a micro centrifuge tube with 2 drops of a warm, approximately 2 *N* sodium hydroxide solution, centrifuged, and the clear solution is acidified with a few drops 2 *N* acetic acid. Finally a drop of a saturated methyl alcohol solution of morin is added. A green fluorescence appears if aluminum is present.

64. Classification of Inks. Age of Ink Writings [95]

Inks may be divided into three classes according to the nature of the coloring matter:

Iron gallate inks contain ferrous and ferric salts, tannin and gallic acid, free mineral acids (hydrochloric, sulfuric) and acid water-soluble triphenylmethane dyes.

Logwood inks contain logwood extracts, whose coloring material hematoxylin is oxidized by chromates to hematein. The latter gives lake colors with the chromic salt produced in the test procedure.

Dyestuff inks contain various coal tar dyes.

The characteristic constituents of the first two types can be detected by spot reactions on paper to which the ink has been applied.

The principle of the method is based on the conversion of the substance to be identified into suitable insoluble compounds, with simultaneous decomposition of the coloring matter, and fixing of the insoluble products on the paper in the position of the original lines of writing. These colorless, or often barely visible, precipitates adhere firmly to the paper, and can be converted there into colored compounds by suitable spot reactions, and so made visible. The reactions can be carried out as genuine spot tests, or as reactions *in vitro*: a sample of the manuscript is bathed in the reagent solution. This procedure has the advantage that portions of writing of varying intensity can be observed simultaneously. Paper withstands even long treatment with dilute acids, alkalis, or sodium hypochlorite.

To test for *iron*, for example, an alkaline oxidizing medium (hypochlorite) is used; the iron precipitates as ferric hydroxide that is then firmly held on the paper. The excess sodium hypochlorite can be removed with sodium thiosulfate and distilled water. The iron hydroxide can be made visible in the position of the writing by various means, e.g., with thiocyanate, α-nitroso-β-naphthol, potassium ferrocyanide, ammonium sulfide. When a positive reaction for iron is obtained in writing made with logwood and dyestuff inks, it should be remembered that the iron may have been derived from a steel pen. An excellent test for iron in ink is based on spotting the writing with a solution of α,α'-dipyridyl in thioglycolic acid (comp. Section **32**).

To test for *chromium,* the writing is destroyed with a drop of 0.5 N sodium hypochlorite containing 2 % barium chloride. The chromium is thereby converted into insoluble barium chromate, which can be spotted with diphenylcarbazide (see page 167).

The *chloride* and *sulfate* content is an important criterion of the kind of ink used, since iron gallate and logwood inks may be differentiated in this way. The chloride can be detected in iron gallate and logwood inks by the following procedure: 1 % silver nitrate is added to a 10 % solution of sodium nitrite until a large precipitate appears; the latter is dissolved by adding 10 % nitric acid. A strip of the writing is placed in this solution. The chloride in the ink is converted into silver chloride, the body of the ink dissolves, and the coal tar dye is decomposed by the nitrous acid. The ink is decolorized in about 15 minutes. The excess silver nitrate is washed out of the paper with 1 % nitric acid; this takes about 5 minutes. The silver chloride retained on the paper is then reduced to black metallic silver by a mixture of 1 part 35 % formaldehyde and 10 parts of 2 % sodium hydroxide. Finally, the paper is washed several times with distilled water and dried.

When carrying out the chloride test on dyestuff inks, oxidation with nitrite is usually incomplete; permanganate must be used instead, although it gives poorer chloride images than nitrite. The paper is placed in a 1 % silver nitrate solution containing 1 % of free nitric acid, and a few drops of a 2 % potassium permanganate solution are added. After the ink has lost its color, which should require not more than a minute, the manganese dioxide that has precipitated on the paper is dissolved in 2 % nitric acid to which a little solid hydrazine sulfate is added as a reducing agent. The remaining silver nitrate is completely removed by treating with 1 % nitric acid for 5 minutes longer. The rest of the procedure is the same as for gallate inks.

With violet inks, and especially concentrated copying inks, the time required for oxidation with potassium permanganate is too long. The dye is therefore dissolved by immersing the specimen in a mixture containing 10 % perchloric acid, 5 % nitric acid, and 1 % silver nitrate. The violet dye turns yellow. After 5 minutes the reaction is stopped; otherwise the paper is attacked too much. If the dye has not completely dissolved, it can be treated with potassium permanganate. Otherwise the silver nitrate is washed out with 1 % nitric acid, the paper rinsed with water, and the silver chloride reduced with the formaldehyde mixture, and so on.

To test all kinds of ink for sulfate, a strip of the manuscript is treated with a solution containing 4 % lead nitrate, 4 % free perchloric acid, and a few drops potassium permanganate. The sulfate in the ink is converted into lead sulfate, with simultaneous destruction of the dye. The excess lead nitrate is washed out with water saturated with lead sulfate. A few crystals of hydrazine hydrochloride are added to reduce and decompose the precipitated manganese dioxide. The specimen is then washed with distilled water. The lead sulfate is converted into dark brown lead sulfide by treatment with a solution of 2 % sodium sulfide and 2 % sodium hydroxide. To prevent the reaction of the sulfate with any lead originally present in the paper (derived from the size), the paper is immersed in distilled water for 5 minutes before carrying out the test.

If writing containing chlorides is tested in this manner, the script reappears in the original position as metallic silver ("chloride image of the writing").

Conclusions as to the *age of the writing* may be drawn from the form of chloride image or from the distribution of the chloride around the writing. This information may be of considerable importance in legal and criminological work. Details are given in the original paper.

65. Detection of Fixed Alkalis and Ammonium Salts in Alkali Cyanides [96]

Free alkali can be detected in alkali cyanides by means of acid-base indicators only after the cyanide ions have been removed since they cause a basic reaction as a result of the hydrolysis:

$$CN^- + H_2O \rightarrow HCN + OH^-$$

This removal is accomplished by adding mercuric chloride; non-dissociating mercuric cyanide is formed.

Procedure. A tiny bit of the solid or a drop of its solution is stirred on a spot plate with a drop of concentrated mercuric solution, which has been saturated with sodium chloride to repress the hydrolysis of the mercuric chloride. In addition to common ion effect, there is formation of $Na_2[HgCl_4]$. A drop of phenolphthalein solution is added, or the solution is tested with pink litmus paper. If free alkali is present in the sample, a red color will appear or the paper will turn blue.

An approximation of the actual alkalinity can be obtained by means of a pH meter.

Alkali cyanides are often contaminated with ammonium salts.[97] The latter can be detected by the test described on page 236. The production of ammonia due to hydrolysis of the cyanide:

$$KCN + 2\,H_2O = HCOOK + NH_3$$

is prevented by adding a little mercuric chloride to a drop of the test solution in the gas apparatus before the addition of the sodium hydroxide. Non-dissociating mercuric cyanide is formed.

66. Detection of Alkali Earth and Alkali in Ash of Paper, Charcoal, and Coal [98]

The reaction: $2Ag^+ + Mn^{+2} + 4\,OH^- = 2\,Ag^o + MnO_2 + 2\,H_2O$ may be used for detecting very small amounts of free alkali. The extreme sensitivity of the test is due to the ready visibility of the $Ag^o \cdot MnO_2$ adsorption compound (see page 237). This reaction is especially useful in the examination and differentiation of ash derived from different kinds of paper, coal, and charcoal.

In the examination of paper ash, a strip (about 2 × 3 mm) of the specimen is ashed on a crucible lid until the residue has become as near white as possible. The ash is treated with a drop of the reagent solution, and it turns gray or black at once, or in a few minutes, according to the amount of alkali present. The ash of cigarette paper, newspaper, writing paper, and qualitative filter paper turns black immediately owing to the high alkali content, but the ash of quantitative paper shows almost no reaction.

When testing the alkali content of coal ash, a sample (about 0.1 g) of the coal is ashed on a crucible lid. The ash is treated with a drop of the reagent. Results of the examination of coals and charcoals from different sources are given in Table 7.

TABLE 7

Type of coal or charcoal	Color of ash	Degree of reaction
Bone charcoal	Gray	Strongly positive
Activated charcoal, granular (Kahlbaum)	Almost white	Very slight
Activated charcoal, granular (Aussig Chem. works)	Red-gray	Slight
Charcoal for decolorization (Kahlbaum)	Red-gray	Strongly positive
Acid washed blood charcoal	Yellow-gray	Slight
Spongy charcoal	Dark gray	Very strongly positive
Hard coal	White	Strongly positive
Hard coal	Pink	Positive
Brown coal	Gray	Positive
Beechwood charcoal	Gray	Strongly positive

When determining the amount of the water-soluble alkaline constituents of ashes, it is advisable to shake the ash from about 0.1 g of coal or charcoal for several minutes with 10 ml water; allow to stand for 24 hours and then centrifuge. A drop of the solution is then evaporated on a frosted slide and spotted with the reagent. Hard coal and brown coal were found to contain very little soluble alkali, while wood charcoal had a high alkali content. The reaction may be applied to the rapid examination of small samples of ash to indicate its origin, for instance whether from hard or brown coal, or wood charcoal.

Reagent: 2.87 g $Mn(NO_3)_2 \cdot aq$. is dissolved in 40 ml water and mixed with a solution of 3.39 g $AgNO_3$ in 40 ml water, and diluted to 100 ml. Dilute sodium hydroxide is then added drop by drop until a black precipitate separates. It is filtered off. The reagent will keep when stored in dark bottles.

The reaction with nickel dimethylglyoxime equilibrium solution (see page 508) can be applied with excellent results for the detection of alkaline earths and alkalis in ashes.

67. Detection of Ammonium Salts in Chemicals

The testing of the purity of chemicals [99] often prescribes tests for ammonium salts. This can be done reliably by means of the procedure given in detail on page 237. This reaction [ammonia on $AgNO_3$–$Mn(NO_3)_2$ solution] is best carried out on a drop of the saturated solution of the sample, or with several milligrams of the solid.

68. Detection of Ammonium Salts in Filter Papers [100]

Filter papers of low ash content that are satisfactory for qualitative purposes differ from "ashless papers" used in quantitative work in that the latter contain detectable quantities of ammonium salts. This fact probably arises from the practice of removing inorganic compounds from the paper by means of hydrochloric and hydrofluoric acid, reactants that are either not removed completely or rapidly enough by washing with water, and hence the paper is finally subjected to the action of dilute ammonia. The comparatively slight content of ammonium salts is not a harmful impurity since they volatilize completely when the paper is ashed. However, filter paper containing ammonium salts is obviously not suitable for preparing reagent papers intended for moistening with Nessler solution and designed to be used for the detection of ammonia or volatile aldehydes. The detectability of ammonium salts is so typical for ashless papers that it can serve as a basis for distinguishing between qualitative and quantitative filter papers.* Spotting a surface of about 1 cm^2 with Nessler solution is sufficient.

The response to the Nessler reaction of quantitative and qualitative filter papers of various porosities and makes is shown in the following:

Eaton-Dikeman (USA)		615		—
Schleicher & Schüll	(Germany)	589	black ribbon	+++
,, ,,	,,	590		++
,, ,,	,,	589	blue ribbon	+++
,, ,,	,,	507		++
,, ,,	,,	589	white ribbon	+++
,, ,,	(USA)	470		—
,, ,,	,,	470 A		—
,, ,,	,,	598		—
,, ,,	,,	598 YD		+
,, ,,	,,	2043 B		—
,, ,,	,,	597		—
,, ,,	,,	576		—
,, ,,	,,	589	green ribbon	++
,, ,,	,,	589	orange ,,	+
,, ,,	,,	589	black ,,	+
,, ,,	,,	589	white ,,	+
,, ,,	,,	589	blue ,,	+
Mackerey, Nagel & Co.	(Germany)	640 d		++++
,, ,,	,,	640 M		+
J. Green	(England)	802		+++
,,	,,	603		+

69. Imprint and Developing Procedure for Chemical Detection of Heterogeneities in Manufactured Metals, Minerals, etc.[101]

The sensitive tests for metals and metalloids described in previous sections of this book can be applied not only as spot methods employing drops of a test solution or particles of the solid sample, but they also render practicable a special technique, the so-called *imprinting and developing* procedure. This enables the operator to make localized detections of various materials.

The imprinting procedure makes use of gelatin paper which has been impregnated with a reagent that is sensitive and specific for the material to be detected. The prepared paper is pressed moist against a smooth metal surface, and left there for 5 minutes.

* The ash content of filter papers 5.5 cm in diameter is less than 20 γ for quantitative brands and below 100 γ for qualitative varieties. (Compare C. Schleicher & Schuell Co., Catalog 70, New York, 1949.)

The substance being detected reacts with the reagent in the paper, and the colored reaction product forms a definite image. After removing the paper from the surface it can be seen whether the test is positive, and in addition the distribution of certain materials in the test metal itself may be determined from the localized dispersion on the print or image. This procedure is used when the test substance reacts directly with the reagent and gives a colored product. The latter is fixed in the gelatin layer and appears as a mirror image in the finished print, in those places in which the test material resides in the sample. The use of ordinary paper generally gives less satisfactory results, because sometimes the images are indistinct and less permanent. The gelatin layer adds to the clarity of the print, and occasionally inclusions not visible to the normal eye are quite clear in the print. This improvement in visibility is due to the capillary diffusion and spreading of the reaction product from its point of origin into the gelatin layer or filter paper. The resulting magnification in the print is not sufficient to interfere with the localized detection; there is no danger of distortion.

The print is taken on ordinary glossy gelatin paper, like that used in photography (but without silver bromide). The paper should have a glossy surface to give good contrast, and it must be free from the substance tested for. A blank test is necessary to establish this absence.

Gelatin paper of suitable size is immersed in the appropriate reagent solution for about 3 minutes, allowed to drain, and placed on the prepared sample of metal. The paper should be drawn on to the metal rapidly from the side to avoid air bubbles, which lessen the clarity of the print. It is best to press the paper down with stiff cardboard and to leave it in position for about 5 minutes.

Another method of taking a print is to place the moist impregnated gelatin paper on a sheet of glass and press the piece of metal directly on to it, avoiding bubbles. This procedure is used most generally, and particularly when studying pieces of metal of the size used for the usual optical examination.

It should be noted that gelatin paper is especially attacked by strongly acid solutions. In such cases, filter paper with small pores (e.g., S & S 602, hard) may be used successfully.

This method of preparing prints depends on the direct reaction of the solution of the specific reagent on the paper with the substance to be detected. If this method is impracticable, a similar *developing* procedure attains the same end. In the latter, the element to be detected is taken up on the paper with a suitable solvent and, after removing the paper from the metal surface, the image is developed by bathing the paper in an appropriate reagent solution. Gelatin paper or filter paper soaked in the

References pp. 512–515

proper solvent is used in this procedure. Employing the precautions used in taking the ordinary prints, the paper is placed on the prepared metal surface or, inversely, the metal is pressed on the paper, removed after not less than 5 minutes, and immersed in a solution of a specific reagent for the substance to be detected. The reaction quickly takes place in the portions of the paper where, after diffusion in the capillaries, the dissolved element has become fixed, and the image develops.

Sometimes an alkaline pH-range is necessary for the test, whereas solution of the specimen only takes place in an acid medium. Then, between the solution process and the development, the paper is rendered alkaline with ammonia vapor, acid vapors may be used in the opposite case when an acid reaction of the paper is desired.

The imprinting process is useful both for detecting inhomogeneities (undissolved metals, nonmetallic components) and for revealing homogeneously dissolved components of alloys, or for examining a pure metal. Inhomogeneities, segregations, etc., show up as inequalities on the print, whereas prints of alloys containing a homogeneous distribution of the components give an even color over the surface of the paper, the intensity depending, among other factors, on the amount of the element concerned.

A number of examples illustrating the practical application of the method for testing metals are given in the following sections.

(a) Detection of sulfur segregations [102]

Technical irons and steels always contain sulfur and phosphorus, which are the basis of the segregation phenomena. These elements or their compounds (iron and manganese sulfides, iron phosphide), usually are not evenly distributed in the solidified metal but are mainly present in deposits of high phosphorus or sulfur content. The distribution of these metalloids is very important with respect to their effect on the mechanical properties of the material, and hence on its quality. Quantitative chemical analysis gives only the mean content of sulfur and phosphorus, and supplies no detailed information as to distribution. Methods have been worked out to provide a rapid means of locating these segregations.

Metallographic practice employs both the etching [103] and the printing methods. The former uses etching agents, specific in effect, which cause characteristic color changes at the points of segregation. In the imprinting method, the detection is carried out not on the metal itself but on some material pressed against its surface. One procedure uses silver bromide paper,[104] another employs mercuric chloride paper.[105] In both a sharply defined image is obtained, due to reaction of the segregation elements with the silver or mercury salts. With mercuric chloride prints, black and yellow

References pp. 512–515

flecks are formed on the paper. These were assumed to be mercuric sulfide or phosphide. However, a spot test using the iodine-azide reaction (see page 304) showed that mercuric sulfide is present in both the black and the yellow flecks.

Sulfide corrosion in machinery can be detected, even in its first stages, by applying the iodine-azide reaction. For instance, corrosion of distilling apparatus, due to the use of sulfur-bearing oil etc., is easily demonstrated.[106]

Procedure. A strip of gelatin paper of suitable size is immersed for 2 minutes in a hydrochloric acid solution of mercuric chloride. The solution is allowed to drain off and the damp paper is laid on the de-greased test surface. The paper is gently pressed down with a piece of cardboard. The paper is left in contact with the surface for 4 to 5 minutes, then removed, and washed in running water for about half an hour, to remove any excess mercuric chloride. Insufficient washing affects the iodine-azide reaction (formation of mercuric iodide). Filter paper is used to remove most of the water from the gelatin paper, which is then dried at room temperature. A drop of the iodine–azide solution is placed (with a small pipette) on the spot to be tested. When the reaction is positive, bubbles of nitrogen are evolved at once from the segregation imprint, and are easily visible to the naked eye. After 3 to 4 minutes, large bubbles will have formed; they adhere to the paper.

After the reaction has proceeded for 5 minutes, the iodine–azide solution may be treated with a drop of ammonia. This causes the separation of black nitrogen triiodide in all the portions where there was no sulfide. The paper is paler where the iodine–azide reaction occurred, because the iodine was consumed there, and no formation of nitrogen triiodide can occur. When the sulfur content is very high, the drop is sometimes completely decolorized, but the bubbles are clearly visible, none the less.

Reagents: *1*) Iodine–azide solution: 1.3 g sodium azide is dissolved in 100 ml 0.1 *N* iodine

2) Hydrochloric acid-mercuric chloride solution: 10 g mercuric chloride, 20 ml hydrochloric acid (sp. gr. 1.124) and 100 ml water

3) Gelatin paper. A preliminary blank test with the iodine-azide solution is essential to be sure that no sulfur compounds are present in the paper. *

(*b*) *Detection of phosphorus segregations* [107]

The indisputable detection of phosphorus segregations in steel and iron is important; they adversely affect the quality of the metal. The difference in reactivity of the MoO_3 in the complex compound $(NH_4)_3PO_4{\cdot}12MoO_3$

* Gelatin sometimes contains organic sulfur compounds which react with the iodine–azide solution. Any reaction of the paper itself, due to its gelatin content, is so slight, as a rule, that there is practically no danger of confusion with the reaction of the segregation print.

from the behavior of normal molybdates (see page 85) affords a method for the detection of phosphorus segregations. The difference in behavior is especially shown toward reducing agents. Molybdates are reduced by stannous chloride, through the molybdenum blue stage, to the brown or yellow molybdenum^{III} oxide, whereas the reduction of the complex compound stops at molybdenum blue. Thus, after attacking a mixture of molybdates and phosphomolybdates with stannous chloride, molybdenum blue is left. Only the lower molybdenum oxides, which are soluble in acids, are formed when phosphorus is absent.

Procedure. Filter paper, of suitable size, is immersed in a nitric acid-ammonium molybdate solution, allowed to drain, laid on a glass plate, and the test section placed on it. After 3 to 5 minutes, according to the degree of purity of the material, the metal specimen is removed from the paper and placed in a stannous chloride solution (50 ml stannous chloride solution, 50 ml HCl, 100 ml water). This solution dissolves the lower molybdenum oxides and any iron salts taken up by the paper. At first, the imprint is yellow to brown, and after about 3 to 4 minutes, a blue impression is seen. This indicates the presence of phosphorus. Phosphorus-free areas appear colorless in the print. The intensity of color varies from light to dark blue according to the amount of phosphorus present. Since the strongly acid stannous chloride solution vigorously attacks the paper, it is advisable to counteract this effect by adding a little alum, which hardens the paper. When the imprint is sufficiently developed, the excess acid is washed away in running water. The paper is then dried. The appearance of the print shows the distribution of the phosphorus. A uniform blue indicates a homogeneous solid solution, e.g., mixed crystals, whereas segregations show up as localized blue spots on the print.

(*c*) *Detection of copper* [108]

The developing process may be used for the detection of copper in sections of metals. The reaction with rubeanic acid (see page 87) is applied.

Procedure. The prepared metal section is pressed on gelatin paper that has been immersed in 2 : 5 ammonia. After 4 to 5 minutes contact, the paper is placed in an alcohol-water solution of rubeanic acid. Shortly after being swirled in the liquid, the print of the dark (steel-green to black) copper rubeanate appears. The finished developed print is washed for a few minutes in running water and dried.

Reagent: 0.5 g rubeanic acid dissolved in 100 ml alcohol; 5 ml of this solution is diluted to 100 ml with water

The limited solubility of copper in iron causes segregations of free copper when excess copper is present. Their locations are easy to find by the printing method. They show up as dark spots on a uniform gray-green background, whose color is due to the small amount of copper in homogeneous solution.

The method may be used, with some modification, for the detection of small flaws or microcracks in iron. The procedure is to cover the sample with a thin film of copper, e.g., by immersion in copper ammonium chloride. The surface layer of copper adheres loosely and can removed with a little cotton wool. The copper is not so readily removed from the fissures, which can then be located in the copper print.

70. Electrographic Methods

A very interesting procedure for the detection of metals has been worked out by Glazunow[109] and Jirkovský[110]; it is analogous to a method described by Fritz.[111] In this "electrographic" procedure [112] the test substance is used as anode with aluminum foil as cathode, and filter paper moistened with the suitable reagent is placed between these poles. The reagent paper is not laid directly on the aluminum foil, but on another filter paper moistened with potassium chloride or sulfate as electrolyte. This arrangement allows the current to pass more easily. When the circuit is closed, the metal is dissolved anodically and can react directly with the reagent in the paper. The typical colored stains form on the paper.

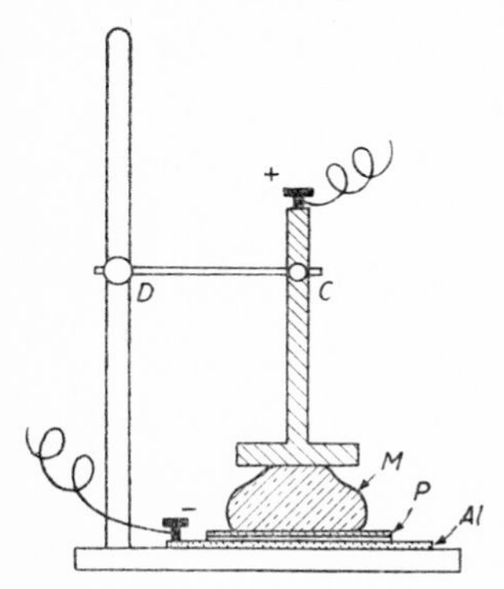

Fig. 40. Set-up for electrographic tests (½ actual size)

A simple apparatus for such electrographic tests is shown in Figure 40. It consists of an aluminum plate as the negative pole (Al), upon which is laid first a layer of filter paper moistened with potassium chloride solution, and then the reagent paper (P) moistened with water or acid. The positive pole is a copper plate with a copper rod (C) soldered to it. A nail or iron wire may also be used to lead in the current. The test substance (M) is placed between the poles; the portion toward the cathode is polished flat. A flashlight battery is the source of current.

Alloys or minerals, which show no appreciable resistance to the passage of the current, may be tested in this apparatus for metals that can be anodically deposited. Sections of metallic products may be tested for the location of constituents without damaging the sample, because in the electrical method, unlike the imprint method, the anodic solution of metals requires no acid reagents. Glazunow detected inclusions of sulfides and phosphides in steel by this procedure.

Jirkovský used the electrographic method primarily for mineralogical purposes. For example, aggregates of PbS and FeS in a mineral section can be detected, after anodic solution, by the sharply defined yellow or blue fleck on chromate or ferrocyanide paper. Jirkovský has been able to detect iron in pyrite, marcasite, pentlandite, sternbergite, and other iron-bearing minerals, with ferrocyanide paper. Similarly, nickel may be detected in minerals with dimethylglyox me paper, cobalt with potassium nitrite and potassium thiocyanate paper; copper with ferrocyanide paper; lead with potassium chromate paper; cadmium with hydrogen sulfide paper; silver with chromate and potassium iodide paper; bismuth with potassium iodide paper; antimony with hydrogen sulfide paper; arsenic with silver nitrate paper; zinc with ferrocyanide paper; aluminum with alizarin paper. Other sensitive tests described in this text may also be used.

This electrographic method is the inverse of the electrolytic (cathodic) deposition of metals. The quantity of the deposit depends on the particular single potential; consequently, in a mixture of metals, the least noble metal is the first to go into anodic solution, the deposition, of course, only affecting the portions lying on the surface. When inhomogeneities are encountered, it must be remembered that such constituents can go into anodic solution only if their resistance is not appreciably greater than that of the adjacent portions.

71. Detection of Fluorine in Rocks and in Mineral Waters[113]

A test for fluorine in minerals is described on page 270 in connection with the zirconium–alizarin reaction. The fluorine test described on page 272, which depends on the volatilization of SiF_4 and its hydrolysis to silicic acid, which is then identified by the molybdate–benzidine reaction, may also be applied to the rapid detection of fluorine in samples of rock or mineral waters. This test is especially recommended when there is much phosphate present, because then the alizarin test is not specific (see page 269).

If the rock contains silica, the powdered sample can be treated directly as described on page 272; otherwise 2 to 4 times its bulk of quartz sand or powdered silica must be added. If the fluorine is combined with silica, the sample must first be fused with sodium carbonate and the cold melt tested. The fusion is best carried out in a bead on platinum wire. Carbonate and sulfide rocks should be ignited beforehand. Mineral waters are tested by evaporating 5 ml samples to dryness with a pinch of quartz sand or powdered silica.

The fluorine was successfully detected using a maximum of 0.1 g of the following rocks:

Ignited zinc blende containing about. . .	0.6 to 1.50 %	fluorine
Andesitoid ,, ,, ...	0.06 %	,,
Granite-gneiss ,, ,, ...	0.02 %	,,
Fleck-amphibolite ,, ,, ...	0.01 %	,,

Mineral waters which bear large amounts of chlorides or iodides must be mixed with 5 ml of hot saturated solution of silver sulfate. The resulting silver halide is filtered off and the filtrate evaporated with quartz sand or silica.

The following mineral waters were tested:

Bilin *Sauerbrunn*	definitely positive
Karlsbad *Sprudel*	,, ,,
Mattoni's *Giesshübler* (H. Mattoni A.-G.) . .	slightly ,,
Tassilo spring (Bad Hall)	negative
Ober-Selters mineral spring	,,
Hunyadi-Janos bitter water	,,
Levico arsenic–iron water	,,

In view of the sensitivity of the fluorine test, it can be assumed that the latter four waters, if they contain any fluorine at all, carry much less than 0.0001 % fluorine.

72. Detection of Free (Unbound) Sulfur in Minerals, Inorganic Mixtures and Polysulfides [114]

Free sulfur in mixture with other inorganic materials can be extracted with carbon disulfide or pyridine. The former dissolves only the crystalline modification of sulfur, whereas pyridine, although less efficient, is a solvent for amorphous sulfur, too.[115] The dissolved sulfur is taken up by the acid-soluble thallous sulfide with formation of acid-insoluble thallous polysulfide. Therefore, when filter paper impregnated with Tl^{I} sulfide is spotted with a drop of a sulfur-containing solution, a fleck of thallous polysulfide is produced, which protects the underlying Tl^{I} sulfide against attack by dilute mineral acids. This protecting layer effect is the basis of a simple test for free sulfur; the details are given on page 372. Elementary selenium behaves in the same way.

Alkali polysulfides can be distinguished from monosulfides by means of this test. It was also possible to show that considerable quantities of free sulfur are present in pyrite, marcasite and galena; only traces of free sulfur could be detected in zinc blende.

A quicker and more reliable test for free sulfur is based on the production of hydrogen sulfide when the solid test material is heated with benzoin at 140–150°C. This fusion reaction is described on page 375.

73. Detection of Sulfide Sulfur in Rocks [116]

The iodine–azide reaction for sulfide sulfur (page 303), in which nitrogen is evolved because of the catalytic action of soluble and insoluble sulfides, may be successfully applied to very small samples to reveal the presence of sulfidic minerals.

Arsenides, antimonides, tellurides, selenides, and free sulfur do not give the iodine-azide reaction. Almost all minerals of these types usually contain small amounts of sulfides and in such cases, the intensity of any evolution of nitrogen should be noted and evaluated before reaching a decision.

The test may be carried out in various ways. The powdered sample is placed in a micro test tube and covered with iodine–azide solution. The production of bubbles of nitrogen is looked for. Alternatively, a grain of the powdered mineral is transferred, as described on page 305, to an Emich microcentrifuge tube with a platinum wire and brought into contact with 1 or 2 drops of iodine–azide solution. Or a drop of the reagent may be placed on the substance *in situ*, or on a mark made with the mineral on a streak plate. The streak should be as broad as possible and any large particles blown away. The advantage of this streak method is that the sample of mineral is hardly damaged.

The catalytic hastening by sulfides of the normally very sluggish reaction: $2\ NaN_3 + I_2 \rightarrow 2\ NaI + 3\ N_2$, can be recognized not only by the production of nitrogen bubbles but also by the disappearance of the free iodine. Accordingly, sulfide sulfur in rocks and minerals can be detected by placing a fragment of the test material or a little of the pulverized specimen on a filter paper moistened with the reagent solution. Pressure is applied through a glass plate. If sulfide is present, a white fleck appears on the paper, which originally is yellow or blue (starch-iodine). The decolorization is especially visible on the reverse side of the paper. In doubtful cases, it is well to run a comparison test, employing an iodine solution of the same concentration but free of sodium azide.

Reagent: Solution of 1 g sodium azide and 1 g potassium iodide and a small crystal of iodine in 3 ml water. The reagent solution may be used warm, if desired.

74. Detection of Selenium in Minerals, Sulfur and Tellurium [117]

The test for selenious acid (page 347), which depends on the oxidation of asym. diphenylhydrazine to the violet quinoneanildiphenylhydrazone, may also be applied to the detection of selenium in minerals. When testing sulfides, it is advisable to evaporate the sample to dryness with concentrated hydrochloric acid and hydrogen peroxide to convert any hydrogen selenide, liberated along with the hydrogen sulfide, into a mixture of selenious and selenic acids. A number of minerals have been examined by the procedure given here, e.g., "cadmium red," clausthalite, zorgite, etc.

References pp. 512–515

Procedure. A little of the finely powdered mineral is boiled in a micro crucible with 3 or 4 drops of 10 % hydrogen peroxide and 4 or 5 drops of concentrated hydrochloric acid; the process is repeated until nearly all the sample has dissolved. The contents of the crucible are then evaporated to dryness to decompose all remaining traces of the hydrogen peroxide.

The residue is taken up in 3 or 4 drops of concentrated hydrochloric acid, a few milligrams of oxalic acid are added, and the mixture evaporated to a damp mass. A drop of dilute hydrochloric acid, 2 drops water, and 4 drops reagent solution are added. A red color appears, either at once or after a few minutes, according to the amount of selenium present.

The same procedure (but without addition of oxalic acid) can be used to detect small amounts of selenium in sulfur or tellurium. An artificially prepared mixture of selenium with sulfur or tellurium gave a light red color when as little as 0.001 % Se was present in a 20 mg test portion.

Examinations of TeO_2 for a possible content of SeO_2 showed that practically all the materials on the market contained selenium. Tellurium dioxide may be purified by long-continued heating at the sintering temperature; the SeO_2 volatilizes. Samples purified by this procedure were mixed with SeO_2 to determine the limiting value of the test for SeO_2 in TeO_2. When 20 mg samples were treated with the reagent in a depression of a spot plate, 0.001 % SeO_2 could still be detected.

75. Detection of Sulfates in Hydrofluoric Acid and Fluorides

When testing the purity of hydrofluoric acid and fluorides, it is important to discover whether any sulfuric acid or sulfate has been left in the product by the manufacturer. The usual test with barium chloride is not feasible here because barium fluoride is sparingly soluble in dilute acid, and barium sulfate is incompletely precipitated from strongly acid solutions. The practice heretofore has been to convert the acid into an alkali salt and then to test for sulfur (sulfate) by the "hepar" method.[118] The sensitive iodine–azide reaction, which is specific for sulfide (see p. 303), can be substituted for the hepar test. It will reveal small quantities of sulfate in one drop of potassium fluoride solution.[119]

Procedure. Several grains of calcined sodium carbonate are placed in the middle of a strip of quantitative filter paper (0.5 × 3 cm), moistened with the test solution, and then dried. The paper is rolled up, laid on a Wedekind magnesia spoon, and heated with the blowpipe flame. An alcohol burner is advisable, to avoid the possible sulfur content of the city gas. Since sulfate is reduced to sulfide by the carbon of the paper, the reducing flame of the blowpipe is not essential. After reducing and fusing, the ash ought to be not quite white but should still contain a little of the carbon residue of the paper. The cooled melt is transferred to a microcentrifuge tube and treated with two drops of water and a drop of iodine-azide solution (preparation see page 305). When sulfide is present, implying sulfate in the sample, distinct bubbles of nitrogen rise through the liquid.

Sodium sulfate (22 γ) was readily detected in the presence of 500 times the amount of potassium fluoride.

Even smaller quantities of sulfate can be revealed if a portion or all of the test solution is treated with barium chloride and the precipitate ($BaSO_4$ and BaF_2) dried and reduced to sulfide with metallic potassium (see page 315). The iodine-azide reaction is used.[120]

76. Detection of Sulfate in Inorganic Fine Chemicals [121]

The detection of sulfates in high grade inorganic chemicals was formerly limited to large samples. About 5 to 10 grams of the test material was dissolved and observed for the formation of a precipitate on the addition of barium chloride. However, there is a considerable possibility of error when small amounts of barium sulfate are expected from concentrated salt solutions since neutral salts can delay or even prevent the separation of barium sulfate. Furthermore, when the dilution is insufficient, precipitates may appear because of concentration effects.

Sulfate can be detected easily in very small amounts of test material if the sulfate is reduced to sulfide by means of metallic potassium and detected by the iodine-azide reaction (see page 303).

Procedure. A few milligrams of the solid and a particle of metallic potassium are placed in a capillary tube and heated until the potassium melts. The tube is then heated strongly and the hot capillary is dropped into a micro test tube containing a few drops of water. A drop of cadmium chloride solution is added, the mixture is acidified with acetic acid, and iodine-azide reagent is introduced. In the presence of sulfide (implying sulfate in the original sample) nitrogen bubbles appear.

Reagents: See page 305.

This procedure revealed sulfate in 3 mg "chemically pure" sodium chloride, oxalate, carbonate, nitrate.

77. Detection of Traces of Nitrate in Lead- and Manganese Dioxide, Antimony pentoxide and Acidic metallic Oxides[122]

Contamination of PbO_2, MnO_2, Sb_2O_5, MoO_3 and WO_3 by nitrate cannot be detected directly through the diphenylamine reaction (see page 327) because these oxides likewise give a blue color with a solution of diphenylamine dissolved in concentrated sulfuric acid. The direct test given here is based on the production of acidic nitrogen oxides when the metal oxides containing nitrate are ignited. If the sample is a nitrate of a heavy metal or a mixture of an oxide and nitrate, the decomposition may be written:

$$Me(NO_3)_2 \rightarrow MeO + N_2O_4 + O$$

In the case of acidic metal oxides, e.g., MoO_3, containing alkali nitrate, the ignition results in the displacement:

$$2\ NaNO_3 + MoO_3 \rightarrow Na_2MoO_4 + N_2O_5\ (= N_2O_3 + O_2)$$

Since N_2O_4 is the mixed anhydride of nitric and nitrous acid, it as well as N_2O_3 will give the characteristic test for nitrous acid with Griess reagent (see page 400).

Procedure. A little of the pulverized sample is placed in a micro test tube and the mouth of the tube is covered with a disk of filter paper moistened with Griess reagent (See page 331). The bottom of the tube is strongly heated with a micro flame. If nitrate is present, the paper shows a pink or red stain, the shade depending on the quantity of nitrate involved.

The sensitivity of this procedure is indicated by the finding that as little as 0.5 γ lead nitrate produces a distinct red coloration.

78. Detection of Traces of Nitrate in Alkali Molybdate, Tungstate and Vanadate [122]

Slight amounts of alkali molybdate, tungstate, and vanadate likewise yield a blue color with a concentrated sulfuric acid solution of diphenylamine (compare the action of the insoluble oxides noted in the preceding section). To detect traces of alkali nitrate in these salts, use may be made of the generation of acidic nitrogen oxides on ignition of such mixtures. The procedure described in **77** can be used. If ammonium salts are being examined, it must be remembered that ammonia is released during the ignition and interferes with the action of the Griess reagent. Accordingly, when the ammonium salts are being examined for the possible presence of nitrate it is necessary to make a preliminary evaporation with dilute caustic alkali.

79. Detection of Traces of Chloride in fine Chemicals [123]

When soluble or insoluble chlorides are warmed with chromic acid-sulfuric acid mixture, free chlorine is given off (along with chromyl chloride, see p. 260). The chlorine can be detected in the gas phase through the blue color it yields on paper impregnated with 4,4′-bis-dimethylamino-thiobenzophenone. The mechanism of the color reaction is discussed in connection with the detection of free halogens (page 368).

Procedure. A micro test tube is used. Amounts ranging from 100 mg to 5 mg of the sample or, if need be, the evaporation residue from one drop of a neutral or weakly alkaline solution, are united with 4 drops of chromic-sulfuric acid mixture. Care must be taken that the upper portion of the test tube remains dry. A disk of reagent paper is placed over the mouth of the tube, which is then immersed to about $^3/_4$ of its length in a boiling water-bath. Depending on the quantity of chloride present, a more or less intense blue color appears on the paper within 1–4 minutes.

Reagents: *1*) Chromic acid-sulfuric acid mixture: A saturated solution of potassium chromate in concentrated sulfuric acid is kept in a boiling water-bath for about 30 minutes to remove traces of chlorine.

2) Thio-Michler's ketone paper: Quantitative filter paper is soaked in 0.1 % benzene solution of 4,4′-bis-dimethylaminothiobenzophenone and dried in the air. The yellow reagent paper is cut into disks. It will keep for weeks if stored in the dark.

The procedure will reveal 0.0005 % chlorine in 100 mg of water-soluble salts: 50 mg is sufficient for chlorine contents of 0.001–0.003 %. Contents as low as 0.02 % chlorine can be detected in a 5 mg sample.

A distinct test is given by 5 drops of concentrated sulfuric acid containing 0.00006 % chlorine.

80. Detection of Bichromate in the Presence of Monochromate [124]

Alkali monochromates, e.g., Na_2CrO_4, are without effect on potassium iodide. The oxidation of iodides by chromates with liberation of iodine requires the presence of hydrogen ions:

$$2\,CrO_4^{-2} + 6\,I^- + 16\,H^+ = 2\,Cr^{+3} + 8\,H_2O + 3\,I_2$$

Bichromate solutions are acidic due to hydrolysis ($Cr_2O_7^{-2} + H_2O \rightleftharpoons 2\,CrO_4^{-2} + 2\,H^+$). The chromium^{III} salts, formed by the reduction of chromate, hydrolyze and therefore also react acid. Consequently, bichromate solutions oxidize iodide solutions and the liberation of iodine serves to indicate bichromate in the presence of chromate. Furthermore, iodine is liberated more rapidly on the addition of a mixture of potassium iodide and potassium iodate to a chromate solution containing bichromate and consequently the hydrogen ions necessary to the reaction:

$$5\,I^- + IO_3^- + 6\,H^+ = 3\,H_2O + 3\,I_2$$

References pp. 512–515

Procedure. A drop of the test solution, or a particle of the solid sample, is mixed with a drop of a 5 % KI-KIO_3 solution, stirred, and a drop of starch solution added. When bichromate is present a distinct blue appears, at most after 5 minutes. The reagent solution should be freshly prepared.

Chromate and bichromate can also be distinguished by a procedure which is less sensitive than the preceding one, but adequate for many purposes. The test is based on the action of sodium sulfite.[125] Only bichromate reacts and yields chromate along with sulfurous acid, and these products then undergo the following redox reaction which likewise requires hydrogen ions:

$$2\ CrO_4^{-2} + 3\ SO_3^{-2} + 10\ H^+ \rightarrow 3\ SO_4^{-2} + 2\ Cr^{+3} + 5\ H_2O$$

The color goes from orange or yellow to green, and the change is quite discernible if spot reactions with the sample and a bichromate-free solution are conducted in adjacent depressions of a spot plate.

81. Detection of Alkali Monochromate in the Presence of Alkali Bichromate [126]

When a neutral solution of the nickel salt of a mineral acid is treated with an excess of dimethylglyoxime, the precipitation of the nickel dimethylglyoxime is incomplete. The clear filtrate is a saturated solution of nickel dimethylglyoxime in which the following equilibrium exists:

$$Ni^{+2} + 2\ DH_2 \rightleftarrows Ni(DH)_2 + 2\ H^+$$

[DH = the radical of dimethylglyoxime (DH_2)]

Since this equilibrium will be disturbed by the introduction of soluble or insoluble materials which consume H^+ ions, red nickel dimethylglyoxime will be precipitated immediately. This fact is the basis of a decisive test for "basic" materials, using the term in its widest sense, i.e., materials which consume H^+ ions (see page 507) [127]

Alkali monochromates consume H^+ ions to form bichromates:

$$2\ CrO_4^{-2} + 2\ H^+ \rightleftarrows Cr_2O_7^{-2} + H_2O$$

Accordingly, the addition of an alkali monochromate to the equilibrium reagent solution results in an immediate precipitation of nickel dimethylglyoxime. Bichromates, which do not consume hydrogen ions, show no action toward this reagent. As little as 2.5 γ K_2CrO_4 can be detected in one drop, after evaporation of the solution.

The diverse behavior of mono- and bichromate toward this equilibrium solution makes it possible to detect quickly the presence of the monochromate in a sample of bichromate. The test is not particularly sensitive, but it affords the only available means of detecting monochromate in the presence of bichromate without making a quantitative analysis, i.e., without determining the ratio K : Cr.

Procedure. A drop of a 10 % solution of the alkali bichromate is taken to dryness in a micro crucible. The residue is treated with a drop of nickel dimethylglyoxime equilibrium solution (for preparation, see page 509). Red nickel dimethylglyoxime appears if monochromate is present. A blank on a pure bichromate solution is recommended when small amounts of monochromate are suspected.

Limit of Identification: 18 γ K_2CrO_4 in the presence of 5000 γ $K_2Cr_2O_7$.

82. Detection of Carbonate in Alkali Cyanides, Sulfites, and Sulfides [128]

The test described on page 337 may be applied for the detection of small amounts of carbonate in alkali cyanides, sulfites, and sulfides. This test depends on the decolorization of a drop of carbonate reddened with phenolphthalein; the carbonic acid forms bicarbonate which gives no color with this indicator.

Liberation of prussic acid, hydrogen sulfide, or sulfur dioxide must be prevented since they too discharge the color of the reagent solution. Cyanides are made non-interfering by adding mercuric chloride; the resulting solution of mercuric cyanide contains practically no cyanide ions. Hydrogen peroxide will convert sulfides and sulfites into sulfate, which does not interfere with the test for carbonate.

Procedure. (*a*) *Test in the presence of cyanides.* About 2 to 3 mg of the solid cyanide, or a drop of a cyanide solution (the concentration should not exceed the equivalent of 5 % potassium cyanide), is placed in the apparatus shown in Figure 23 (page 47) and stirred with four drops of a saturated solution of mercuric chloride. Two drops of 2 *N* sulfuric acid are added and the apparatus closed. Discharge of the color, compared with a blank if necessary, indicates carbonic acid.

(*b*) *Test in the presence of sulfides and sulfites.* A particle of the solid sample or a drop of the solution (the sulfite solution can be saturated, the sulfide solution about 5 %), is stirred with four drops of a 3 % neutral solution of hydrogen peroxide; the further treatment is the same as in (*a*).

The presence of carbonate in a drop of a freshly prepared solution of the following materials, both technical grade and "for analysis", was detected by decolorization of the phenolphthalein drop, after standing about 10 minutes.

Potassium cyanide.	(5 % solution)
Sodium sulfide	(10 % ,,)
Potassium sulfide	(10 % ,,)
Sodium sulfite	(saturated solution)

83. Detection of Silica in Caustic Alkalis and Alkaline Solutions. Vulnerability of Glass and other Siliceous Products to Attack by Alkalis

The sensitive silicic acid test (page 335), which depends on the special reactivity of silicomolybdic acid, may be applied to test for silica in solid caustic alkalis, or in basic solutions which have been kept for a long time in glass. Silica is dissolved from the walls of such containers. This dissolution is very easy to follow in ammonia water; freshly prepared ammonia solutions contain very little silica. On long standing, the silica content increases rapidly, so that old and freshly prepared ammonia solutions may be readily distinguished from each other by this test. Obviously, the vulnerability of the glass makes some difference, but seemingly all kinds of siliceous glass are eventually attacked by alkalis.

Procedure. A drop of concentrated nitric acid is placed in a micro crucible, 1 or 2 drops of the test solution are added with a loop of platinum and stirred with the wire. A small piece of the caustic alkali may also be taken for the test. The addition of molybdate, benzidine etc., is carried out as described on page 335.

The rapidity of the attack on glass or other siliceous materials by caustic alkalis, with resulting removal of silica, affords a means of differentiating the various kinds of glass with respect to their alkali resistance. This may be done by placing 1 or 2 drops of a freshly prepared alkali solution on the test object (e.g., a test tube laid on its side) and allowing the caustic alkali to take effect. The drop is then washed into a micro crucible, using as little nitric acid as possible, and the silica test carried out as described on page 335. By varying the strength of the alkali and the length of attack, data may be obtained for the appraisal of glasses with respect to their alkali resistance.

84. Detection of Silicic Acid in Minerals [129]

(a) After fusion with sodium potassium carbonate

A little of the powdered mineral is decomposed by fusion with sodium-potassium carbonate in a loop of platinum wire; the silicic acid is thus brought into a reactive form. The bead is dissolved (micro crucible) in a drop of 1 *N* nitric acid. Ammonium molybdate solution is added and the rest of the procedure is as described on page 336. Phosphates must be absent.

(b) After conversion into silicon tetrafluoride

Silicates yield silicon tetrafluoride gas when heated with fluorides and concentrated sulfuric acid:

$$SiO_2 + 4\ HF = SiF_4 + 2\ H_2O$$

If the gas is taken up in a drop of water, hydrolysis occurs:

$$3\ SiF_4 + 4\ H_2O = H_4SiO_4 + 2\ H_2SiF_6$$

producing silicic acid and hydrofluosilicic acid. When ammonium molybdate is added, silicomolybdic acid is formed and can be detected with benzidine (pages 272 and 335).

Procedure. A little of the solid is mixed in a small platinum crucible with a few milligrams of calcium fluoride. Two drops concentrated sulfuric acid are added and the crucible covered with a piece of filter paper on which is placed a drop of ammonium molybdate solution. The crucible is heated over a micro-burner for about 1 minute and allowed to cool for 3 to 5 minutes. The ammonium molybdate fleck is then treated with a drop of benzidine and developed over ammonia. Rocks containing carbonate and sulfide should be ignited before the test to prevent frothing (evolution of carbon dioxide) or reduction of the molybdate by hydrogen sulfide.

Limit of Identification: 2.5 γ silicic acid.

Free silica can be distinguished from bound silica by the diverse adsorptive action of finely powdered specimens toward certain dyes. For instance, a solution of safranin O in 30 ml alcohol and 70 ml water colors quartz particles yellow (adsorption), whereas felspar particles are not tinted.[130]

85. Detection of Boron in Rocks and Enamels [131]

The test for boron (page 341) in which a complex boric acid ester of *p*-nitrobenzeneazochromotropic acid is formed, may be applied to detect boron in rocks, enamels, etc., after fusion with potassium hydroxide. The finely powdered sample is fused with potassium hydroxide in a silver crucible. The cold melt is extracted with a few drops of water; the solution is filtered (using the device described on page 42) and the test for boric acid is carried out on the evaporation residue from 1 or 2 drops of the filtrate. When chromate or permanganate is formed in the fusion, the procedure described on page 342 must be used.

The color reaction with hydroxyanthraquinones (see page 340) or the decomposition of methyl borate with alkali fluoride (see page 342), may also be applied to the detection of boron in the melt.

86. Detection of Phosphates in Minerals and Rocks [132]

The test described on page 333 may be applied for the rapid detection of phosphates in rocks and minerals. A few milligrams of the powdered rock or a splinter of the solid are placed on filter paper, moistened with a drop of ammonium molybdate solution, and held over a flame for a short time. A drop of benzidine solution is added and the moist fleck held over ammonia. A brilliant blue forms on the paper either at or near the position of the sample, sometimes in continuous patches, and wherever traces of ammonium phosphomolybdate have been formed. A useful procedure is to carry out the test on a streak plate. The reagents are applied to the streak.

In many instances, the mineral may be treated *in situ* with the nitric acid-molybdate solution. The liquid is taken up on a piece of filter paper and the rest of the test carried out on the paper.

Since arsenates, which resemble phosphates in their geologic habits, do not give the benzidine reaction, no confusion occurs between the two families of compounds. It is therefore possible to detect phosphate in arsenates by this reaction.

The test was applied to 36 minerals from various sources (for details see the original paper). The following powdered rocks, with low apatite content, gave positive results:

Red andesite (Gleichenberg, Styria)	0.57 % P_2O_5
Gray, thick andesite (Gleichenberg, Styria) . . .	0.35 ,, ,,
Granite (Maissau, Lower Austria)	0.26 ,, ,,
Gneiss (Dürnstein, Lower Austria)	0.12 ,, ,,
Fleck-amphibolite (Spitz a. D., Lower Austria) .	0.10 ,, ,,
Granitic gneiss (Waldviertel, Austria)	traces

By varying the test slightly it can also be used to detect or locate apatite in thin sections and on any polished surface of a rock.

Procedure. A small filter paper is impregnated with nitric acid–molybdate solution. The mounted section or the polished area of the rock is warmed (the section should not be heated to the point that the Canada balsam becomes mobile) and pressed at once on the moist paper and left for 1 or 2 minutes. The paper is laid (with the surface of contact upward) on a second paper soaked in benzidine solution, so that the solution rises through the upper paper. The latter is then held over ammonia and an exact blue print is obtained of the crystals of apatite in the section. If the contours of the section have been drawn on the paper, the exact location of the crystals is revealed. Even the smallest crystals of apatite, of whose apatitic nature there may be some doubt, may be identified in this way, thus amplifying or even replacing optical examination.

87. Detection of Chlorine in Minerals and Rocks [133]

Since chlorine is a structural element in relatively few minerals and rocks, the establishment of its presence or absence may be decisive in identification tests. The procedure given on page 495 for detecting traces of chlorine in fine chemicals can be employed here. The basis of the test is the production of free chlorine from chlorides on warming with chromic-sulfuric acid mixture, followed by exposure of the vapors to the yellow paper that has been impregnated with 4,4′-bis-dimethylaminothiobenzophenone. The paper turns blue if the test is positive. Since the color reaction responds to 0.2 γ chlorine, fractions of 1 mg suffice for the detection of chlorine in minerals and rocks. This is shown in Table 8.

TABLE 8. CHLORINE CONTENTS OF MINERALS

Name	Mineral formula	% Chlorine	μg Cl in 0.5 mg
Atacamite	$3\,CuO \cdot CuCl_2 \cdot 3H_2O$	16.6	83
Boracite or Stassfurtite	$Mg_7Cl_2B_{16}O_{30}$	7.9	40
Chlorapatite	$CaFCl \cdot 3Ca_3(PO_4)_2$	3.5	18
Eudialite	$6\,Na_2O \cdot 6(Ca,Fe)O \cdot 20(Si,Zr)O_2 \cdot NaCl$	1.8–1.1	9–6
Hanksite	$9\,Na_2SO_4 \cdot 2Na_2CO_3 \cdot KCl$	2.3	11
Nimetite	$9\,PbO \cdot 3As_2O_5 \cdot PbCl_2$	2.4	12
Pyromorphite	$PbCl_2 \cdot 3Pb_3(PO_4)_2$	2.6	13
Sodalite	$Na_4(AlCl) \cdot Al_2(SiO_4)_3$	7 3	37
Vanadinite	$9\,PbO \cdot 3V_2O_5 \cdot PbCl_2$	2.5	12

The test described here permits the rapid differentiation of apatite [$CaF_2 \cdot 3Ca_3(PO_4)_2$] and chlorapatite [$CaFCl \cdot 3Ca_3(PO_4)_2$]. If about 1 mg of apatite is tested for chlorine, the color reaction is negative or very weak. The same amount of chlorapatite gives an intense blue color.

88. Detection of Free Acids and Basic Compounds in Solutions of Aluminum Salts [134] and of Free Acids in Solutions of Copper and Cobalt Salts [135]

Aqueous solutions of Al salts of strong acids such as $AlCl_3$, $Al_2(SO_4)_3$, react acidic because of hydrolysis. Consequently, any free acid present cannot be detected directly by means of indicators. The same is true of solutions of basic Al salts (e.g., basic aluminum acetate) which give an acid reaction despite their basic constituents.

The hydrolysis of the aluminum salt may be masked by the addition of neutral substances which form complexes. For example, the aluminum ion can be replaced by alkali ions and it can then take a place in compounds whose reaction is neutral. The acid or alkaline reaction of the masked

References pp. 512–515

solution will then reveal the presence of free acids or basic aluminum salts, since the exchange involves the disappearance of only the aluminum ion and not the hydrogen or hydroxyl ions. Alkali oxalates are excellent complex formers with aluminum salts. If sodium oxalate is added to solutions of aluminum salts there is an immediate formation of the sodium salt of aluminum oxalic acid, in which the aluminum is a constituent of a stable complex anion:

$$AlCl_3 + 3\ Na_2C_2O_4 = Na_3[Al(C_2O_4)_3] + 3\ NaCl$$

$$AlCl_3 + HCl + 3\ Na_2C_2O_4 = Na_3[Al(C_2O_4)_3] + 3\ NaCl + HCl$$

$$Al(OH)\ (CH_3COO)_2 + 3\ Na_2C_2O_4 = Na_3[Al(C_2O_4)_3] + 2\ CH_3COONa + NaOH$$

These equations show that solutions of an aluminum salt, after masking with oxalate, react neutral, acid, or alkaline, according to whether a neutral salt, free mineral acid plus neutral salt, or a basic aluminum compound was present originally.

The addition of excess sodium oxalate to solutions of copper or cobalt salts likewise produces complex double oxalates. Consequently the presence of free acid can then be detected by means of indicators in these cases also.

Procedure. A drop or two of the test solution is stirred on a spot plate with a few crystals of sodium oxalate and then tested for free acid or alkali with methyl orange or phenolphthalein.

The presence of small amounts of acid in solutions of copper or cobalt salts can be detected as follows; one drop of the solution is warmed with several crystals of sodium oxalate in a micro crucible. After cooling, the mixture is tested with litmus paper. The presence of metal oxalates or double oxalates does not interfere with the discernment of the color change. As little as 0.05 % free H_2SO_4 can be detected in the test solution prepared from a crystal of the copper sulfate.

89. Detection of Traces of Hydrogen Sulfide in Water [136]

Even slight amounts of soluble and insoluble sulfides catalyze the reaction of sodium azide solution with iodine (page 305). This sensitive reaction can be intensified and applied to the detection of traces of hydrogen sulfide in water if the decomposition of the hydrogen sulfide is localized. This is accomplished by shaking the water to be tested with a drop of metallic mercury. Mercury sulfide is formed and then fixed on the surface of the metal.

Procedure. Ten milliliters of the water are placed in a hard glass test tube or a small stoppered measuring cylinder. A drop of mercury is added. After vigorous shaking, the water is poured off and the mercury placed on a watch glass. If hydrogen sulfide was present in the water, a very thin film of mercury sulfide resides on the surface of the mercury. It is not visible if only small amounts of hydrogen sulfide were present, but on stirring with 1 or 2 drops of a sodium azide–iodine solution, the sulfide reacts at once with evolution of nitrogen. Some of the bubbles of nitrogen adhere to the surface of mercury and are clearly visible with a lens. The use of the mercury dropper (page 39) is recommended.

As little as 0.05 γ of hydrogen sulfide can be detected with certainty in 10 ml water. This corresponds to a dilution of 1 : 200,000,000.

Even smaller amounts of hydrogen sulfide can be detected in a larger volume of water, if the water is shaken, in separate portions, with a drop of mercury, and the latter then tested by the iodine–azide reaction.

90. Hydrogen Ion Concentration of Aqueous Solutions

The hydrogen ion concentration of a solution, i.e. the actual acidity, is expressed by the hydrogen exponent pH. This figure is the negative logarithm of the hydrogen ion concentration. Its value can be determined approximately by means of acid-base indicators. They are organic acids or bases, with two different structural forms, a dissociating (ionogenic) form and a pseudo form. These forms have distinctly different colors.

By using a suitable selection of indicators, with color change intervals at different pH levels (the change is never sudden, but is always gradual) the acidity (pH < 7) or alkalinity (pH > 7) may be found in a drop of liquid on a spot plate or on indicator paper.

The color change and transformation intervals, etc., for the indicators most generally used are given in Table 9.*

TABLE 9. INDICATORS

Name	Color		Change	Color change interval, pH	Indicator solution
	acid	alkaline			
Tropaeolin OO	Red	Yellow	Sharp	1.3 – 3.2	0.1 % aqueous
Methyl orange	Red	Orange	Sharp	3.1 – 4.4	0.1 % ,,
Bromophenol blue	Yellow	Blue	Fairly sharp	3.0 – 4.6	Dilute ,,
Methyl red	Red	Yellow	Sharp	4.2 – 6.3	0.2 % ,,
p-Nitrophenol	Yellow	Colorless	Fairly sharp	5.0 – 7.0	1 % ,,
Litmus	Red	Blue	Sharp	5.0 – 8.0	1 % ,,
Bromothymol blue	Blue	Yellow	Fairly sharp	6.2 – 7.6	Dilute ,,
Neutral red	Yellow	Red	Sharp	6.8 – 8.0	0.1 % ,,
Phenolphthalein	Colorless	Red	Sharp	8.0 –10.0	0.5% alcoholic

* More extensive tables are given in the various handbooks and in texts on quantitative analysis.

A "universal" indicator may be used in an orienting determination of the pH of a solution. This is a mixture of different indicators that give characteristic color changes at different pH values. For example, Bogen's [137] universal indicator is a solution of:

0.1 g phenolphthalein	in 500 ml absolute alcohol
0.3 g dimethylaminoazobenzene	
0.2 g methyl red	
0.4 g bromothymol blue	
0.5 g thymol blue	

A microdrop of the indicator is added to a drop of the test solution on the spot plate, and the following shades appear:

Red	at about	pH	=	2.0
Orange	,,	,,	,,	=	4.0
Yellow	,,	,,	,,	=	6.0
Green	,,	,,	,,	=	8.0
Blue	,,	,,	,,	=	10.0

For the rapid measurement of pH values, it is very advisable to compare the color change in the sample treated with indicator with color charts on which the tint of the indicator is shown with the corresponding pH value. An apparatus suitable for this test is described by Tödt.[138] Various types of pH meters are also in use.

91. Orienting Reactions for Judging Samples of Water

The presence of certain elements and compounds is important when judging water samples. Even very small amounts of such solutes may be detected by spot methods.

Very great dilutions usually have to be considered in water analysis, and it is therefore advisable, when determining the nature of dissolved non-volatile materials, to concentrate the samples or, better still, to evaporate them to dryness and examine the residue.

Other methods are available for concentrating certain constituents. These procedures may be useful in studying minute quantities of dissolved materials. Examples are: Kolthoff's [139] method for Cu^{+2} and Pb^{+2} by shaking with $CaCO_3$; Heller's [140] method for metallic salts by extraction with solutions of dithizone in carbon tetrachloride (accumulation of complex metal dithizonates in CCl_4). There are a number of compounds for which neither an adsorption medium or an extraction system is available. In this connection the work of Schwab and Jockers [141] is likely to be of great importance. They have shown that chromatographic analysis,[142] which has proved so

References pp. 512–515

fruitful in organic and physiological chemistry, may also be applied to inorganic analysis and to the detection of traces of impurities.

Small electrodes ("needle electrodes") for the electrolytic deposition of minute traces of metals were used with success,[143] and it has been found that they are suited to localize traces.[144] Following this electrolytic deposition, the material may be put into solution and identified in various ways. Anodic solution or the printing method (see page 484) may be used to advantage.*

The following is a summary of sensitive tests for some of these dissolved materials, that are worth considering when examining samples of water used for technical purposes.

Ammonia. Test with the manganese-silver reaction, see page 237.

Iron. Test with α,α'-dipyridyl, see page 161.

Fluorine. Test by conversion into silicon tetrafluoride.

A few ml water is evaporated to dryness with a few mg quartz sand and treated as described on page 272. As little as 1 γ of fluorine can be detected in 5 ml water.

Iodide. Test after oxidation to iodate, see page 267.

Silica. Test with ammonium molybdate and benzidine, see page 335.

Small amounts of phosphate in the sample can easily be rendered inactive by adding oxalic acid, after the heating with molybdate (see page 336).

Calcium. Test with ammonium ferrocyanide, see page 220.

Carbon dioxide. Test with sodium carbonate and phenolphthalein (page 337). When waters very low in carbonate are examined, several drops should be used for the test.

Copper. Test by catalytic effect on the ferric-thiosulfate reaction, see page 80.

Magnesium. Test with *p*-nitrobenzeneazo-α-naphthol, see page 225.

Manganese. Test with benzidine, see page 175.

Amounts smaller than 0.15 γ manganese per drop may be detected [145]: 100 to 150 ml water is treated with a few drops of sodium hydroxide solution, boiled, and filtered. The calcium and magnesium carbonates thus precipitated by the carbonate in the alkali coprecipitate any manganese dioxide formed. The mixture is filtered through a quantitative paper and tested on the paper with benzidine solution. A blue color indicates manganese. As little as 1.2 γ manganese (dilution 1 : 25,000,000) may be detected.**

* For example, Alber (*loc. cit.*, Ref. 144, which includes a description of the apparatus used) deposited and identified 0.005 γ Cu from 5 ml of electrolyte. This corresponds to a dilution of 1 : 1,000,000.

** For the detection of manganese by benzidine in water analysis in connection with the ammonia test see W. Olszewski, *Chem. Ztg.*, 47 (1923) 273.

Phosphate. Test with tartaric acid-ammonium molybdate solution, and benzidine,[146] see page 333.

Nitrite. Test with Griess' reagent, see page 330.

Nitrate. Test with brucine, see page 328.

Sulfite. Test with malachite green, see page 311.

Hydrogen sulfide. Test with iodine-azide solution, see page 303, or, for the detection of traces, see page 503.

92. Determination of Hardness of Water; Differentiation of Distilled and Tap Water

The high sensitivity of the reaction:

$$2\ Ag^{+} + Mn^{+2} + 4\ OH^{-} = 2\ Ag^{0} + MnO_2 + 2\ H_2O$$

provides a test for Ag^{+}, Mn^{+2}, and OH^{-} ions (see pages 58, 177 and 237). Since calcium carbonate (aragonite, see page 469) as well as calcium oxide react similarly to free alkali, it is possible to distinguish hard from soft water by this convenient spot test.[147] The residue from the evaporation of a drop of water is treated (after ignition if necessary) with the reagent, and the intensity of the resulting gray or black color is compared with that of a test carried out on water of a known degree of hardness, or on calcium bicarbonate solutions of known concentration. To ensure that the residues for comparison are of the same size and shape, it is advisable to evaporate the samples on slides frosted with hydrofluoric acid. The slides must be scrupulously cleaned with chromic-sulfuric acid and washed with water. Otherwise the water runs about and it is impossible to obtain the drops of equal diameter, which are essential for the comparison with the standards. For this reason, the method of placing the drop on the slide must be the same for every sample and standard solution.

In field work, it is sufficient to evaporate a drop of water on a crucible lid, ignite for a moment, and then add a drop of the reagent to the cold residue. According to the depth of blackening (compared later with a standard treated in the same way), the water may be graded from hard to soft. Distilled water may be distinguished from spring or tap water by this means. The results of the tests may be preserved for permanent reference.

Reagent: for preparation, see page 470.

93. Detection of Basic Inorganic and Organic Materials which React with Mineral Acids [148]

The examination of rocks and minerals, and the chemical testing of a variety of materials, often require a rapid method of detecting soluble or

References pp. 512–515

insoluble products that react with dilute mineral acids to form the corresponding salts. Materials of this kind, which can be considered as basic, using the term in its widest sense, include hydroxides, oxides, carbonates, arsenates, phosphates, fluorides, organic bases and salts of weak organic acids.

A very sensitive test for materials which consume acids, and consequently H^+ ions, is furnished by a nickel dimethylglyoxime equilibrium solution. Nickel dimethylglyoxime can be precipitated completely only from ammoniacal or acetic acid solutions. The precipitation is incomplete from solutions of nickel salts of strong acids [$NiCl_2$, $Ni(NO_3)_2$, $NiSO_4$]. The clear filtrate in these latter cases is a saturated solution of nickel dimethylglyoxime presenting the equilibrium:

$$Ni^{+2} + 2\ DH_2 \rightleftarrows Ni(DH)_2 + 2\ H^+$$
$$(DH_2 = \text{dimethylglyoxime})$$

This equilibrium solution (pH = 1.9) reacts with all materials which consume H^+ ions and which therefore disturb the equilibrium when they are brought into contact with this solution. Red nickel dimethylglyoxime is precipitated and is quite visible even in minute quantities.

The particular advantage of using this equilibrium solution for establishing the presence of basic materials is that it provides a simple rapid method which is also applicable to materials so slightly soluble in water that they do not affect indicators. Furthermore, minimal quantities of solid specimens suffice.

The applications of the procedure for detecting basic compounds and materials that are attacked by acids include:

1. Detection of rocks that are attacked by acids (oxides, carbonates, phosphates, acid-decomposable silicates, etc.).

2. Testing of evaporation and ignition residues (see pages 482 and 507).

3. Detection of alkali chromate in bichromate (see page 497).

4. Detection of free organic bases, and salts of organic bases with weak acids, in mixtures with indifferent materials. (Comp. Vol. II, Chap. 3.)

Procedure. A few grains of the pulverized specimen are placed on a spot plate. One or two drops of the equilibrium solution are added and mixed with the solid by means of a fine glass rod or by blowing through a glass capillary. According to the vulnerability of the specimen to the acid in the equilibrium solution, there will be no change, or a red color will appear immediately, or after several minutes. Very small quantities can be tested on a slide and the color observed under the microscope. Portions of rocks and minerals can be scraped loose with a pen knife and the powder spotted (on the solid specimen) with the reagent.

The production of the red nickel salt can be seen easily only when colorless materials are being tested. With colored specimens, it is well to make blank comparison tests and to use several drops of water in place of the equilibrium solution.

The efficiency of the test can be demonstrated with such "basic" materials ($MgNH_4PO_4$, albumen, etc.) as produce no color change with indicators (phenolphthalein). They react promptly with the equilibrium solution.

Reagent: Nickel dimethylglyoxime equilibrium solution: 2.3 g $NiSO_4 \cdot aq$, (dissolved in 300 ml water) is treated with 2.8 g dimethylglyoxime (dissolved in 300 ml alcohol). The suspension is allowed to stand for 30 minutes and then filtered. The reagent will keep for several weeks if stored in tightly stoppered bottles

94. Detection of Chemically or Adsorptively Bound Water

(1) Detection with molten potassium thiocyanate [149]

When potassium thiocyanate (m.p. 173–179°C) is heated to about 430°C, the molten mass turns blue; after cooling the color disappears.[150] This phenomenon is probably due to the splitting-off of sulphur, which remains in highly dispersed colloidal form [151] in the molten potassium thiocyanate, and which regenerates potassium thiocyanate on cooling:

$$KCNS \rightleftharpoons KCN + S^{\circ} \quad \ldots\ldots (1)$$

An equilibrium between potassium thiocyanate and the isomeric potassium isothiocyanate is also possible:

$$N{\equiv}C{-}S{-}K \rightleftharpoons S{-}C{=}N{-}K \quad \ldots\ldots (2)$$

Sodium thiocyanate (m.p. 287°C) behaves similarly.

Whatever the explanation for the color change, the interesting fact remains that in molten potassium or sodium thiocyanate the sulphur is highly reactive and displays reactions which are not realizable in aqueous solutions of alkali thiocyanates. Among such reactions are: formation of silver sulfide from metallic silver; formation of sodium thiosulfate with sodium sulfite; conversion of metal oxides and sulfates (even lead sulfate) into the corresponding sulfides; the partial reduction of niobic and molybdenum oxides into the respective metals, etc.[152] In some cases, these effects occur even at temperatures below 400°C, i.e., in still colorless melts of potassium thiocyanate, which proves that the sulfur is activated even before the transformations (1) and (2) are discernible. The effects due to the enhanced reactivity of sulfur in molten potassium thiocyanate include the behavior of inorganic materials which liberate water only at higher temperatures and yield hydrogen sulfide in contact with the hot potassium thiocyanate. This

References pp. 512–515

latter reaction doubtless is due to the fact that the water is split off in the form of superheated steam and under such conditions reacts with potassium thiocyanate:

$$KCNS + H_2O \rightarrow KCNO + H_2S$$

Since the resulting hydrogen sulfide is readily detected by lead acetate paper, it thus becomes possible to detect chemically bound or adsorptively held water in certain materials that have been dried at 120°–150°C.

Procedure. About 0.2–0.3 g of well dried potassium thiocyanate is melted in a micro test tube. A small portion of the sample is added to the still colorless melt and the heating is continued until the melt turns blue. Filter paper moistened with 10 % lead acetate solution is held at the mouth of the test tube. The paper develops a black stain, if the sample contains water of crystallization or adsorptively bound water.

Reagent: Potassium thiocyanate (pulverized and dried at 110°C)

The procedure was tried on hydrated sodium sulfate and sodium carbonate, and also on hydrated silicic acid and aluminum hydroxide, both well dried at 120°C. All gave distinct evidence of the formation of hydrogen sulfide. After ignition for 30 minutes, these products were inactive toward fused potassium thiocyanate. Gypsum ($CaSO_4 \cdot 2H_2O$) can easily be distinguished from anhydrite ($CaSO_4$) by this procedure. Likewise, calcined tungstic acid (WO_3) can be differentiated by this procedure from $WO_3 \cdot aq.$ and from WO_3 that has been exposed to moist air.

When carrying out the test for chemically or adsorptively bound water, the following must be absent: salts of metals, which form insoluble sulfides (see above), organic materials, ammonium salts. Solid organic substances which split off water when heated (cellulose etc.) react with fused potassium thiocyanate for the same reasons as inorganic materials containing water. When heated with fused potassium thiocyanate, ammonium salts form thiourea, which decomposes with the formation of hydrogen sulfide (compare Vol. II, Chapters 3 and 4). Organic bases exhibit the same behavior in contact with fused alkali thiocyanates.

(2) *Detection with dipicrylamine-dioxane* [153]

If a drop of water is added to the light yellow anhydrous dioxane solution of dipicrylamine, the color changes to orange. In other terms, the same effect is obtained as by making dipicrylamine solutions basic, i.e. by the formation of alkali salts of dipicrylamine or of its *p*-quinoidal anion.

Water produces this effect probably because dioxane is an acceptor for hydrogen ions thus producing the cations (I). On the other hand, hydroxyl ions transform dipicrylamine into the quinoidal anion (II). Consequently, the addition of water, which is slightly dissociated into H^+ and OH^- ions, results in the formation of the soluble dissociated orange-red salt of (I) and (II):

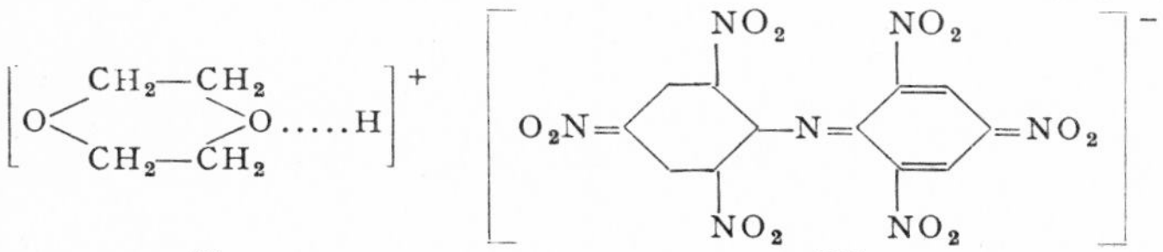

$$\left[O\langle{}^{CH_2-CH_2}_{CH_2-CH_2}\rangle O\cdots H \right]^+ \qquad \left[O_2N{=}C_6H_2(NO_2)_2{-}N{=}C_6H_2(NO_2)_2{=}NO_2 \right]^-$$

(I) (II)

The color change in dioxane-dipicrylamine solution is not limited to the addition of water. Solids which contain chemically bound or adsorptively held water accomplish the same effect.

Procedure. A spot plate is used. A few milligrams of the pulverized specimen are treated with a few drops of the reagent solution. If the sample contains water, the reaction takes place topochemically (on the surface of the specimen) and a light colored material is thus tinted orange-red. A comparison blank, in an adjacent depression of the plate, is advisable. If it seems best to exclude atmospheric moisture, the liquid in the depressions can be shielded by several drops of mineral oil.

Reagent: A 1 % solution of dry dipicrylamine in dry dioxane. The solution should be protected from the entrance of moisture.

Alumina, which had been washed free of alkali and then ignited gave a positive test for water by the procedure described here. This unexpected finding is probably not due to the formation of a salt with the dipicrylamine since strong basic oxides do not show this behavior. Therefore, this test may have some value in identifying dry aluminum oxide.

Hydrated materials, which give off water at room temperature or when gently heated, can be detected in the gas adsorption apparatus shown in Fig. 23, page 47. A drop of the reagent solution is placed on the knob of the stopper after the bulb of the vessel has been charged with a little of the specimen. The closed apparatus is allowed to stand at room temperature or in hot water. A change in the color of the suspended drop is easily seen.

95. Detection of Iodine in Mineral and Sea Water [154]

The test for iodine based on the catalytical acceleration of the redox reaction between tetrabase and chloramine T (see p. 536) can be used for the examination of mineral and sea water.

One drop of mineral water labelled as containing iodine is sufficient to detect this halogen.

When iodine is to be detected in sea water, the presence of oxidizable matter must be taken into consideration, because it consumes chloramine T. This interference may be avoided by the following procedures.

Procedure. (*a*) One drop of the sample is taken to dryness in a small test tube along with $MgCO_3$, and ignited. One or two drops of distilled water are added and the filtered solution is mixed with the reagents in the depression of a spot plate. A blue color indicates the presence of iodine.

Procedure. (*b*) One drop of sea water is added to about 5 ml of distilled water in a test tube, treated with one drop each of the reagent solutions. In the presence of iodine a blue color appears.

Reagents: *1*) Tetrabase: see p. 536

2) Chloramine T: 1% water solution.

REFERENCES

1. E. B. Sandell, *Colorimetric Determination of Traces of Metals*, 2nd. ed., New York, 1950.
2. B. S. Evans and D. G. Higgs, *Spot Tests for the Identification of Certain Metallic Coatings and of Certain Metals in Bulk*, Cambridge, 1943.
3. F. Feigl and H. E. Ballaban, unpublished studies.
4. F. Feigl and H. J. Kapulitzas, *Mikrochemie*, 8 (1930) 242.
5. F. Pavelka, *Mikrochemie*, 4 (1926) 200.
6. F. Feigl, *Mikrochemie*, 7 (1929) 11.
7. H. E. Ballaban, *Mikrochem. ver. Mikrochim. Acta*, 27 (1939) 62.
8. F. L. Hahn and G. Leimbach, *Ber.*, 55 (1922) 3870.
9. F. Feigl and P. Krumholz, *Pharm. Presse*, 5 (1934) 1.
10. F. Pavelka, *Mikrochemie*, 23 (1937) 202.
11. F. Feigl and A. Caldas, *Anal. Chim. Acta*, 8 (1953) 117.
12. For preparation, see K. H. Slotta and K. R. Jacobi, *Z. anal. Chem.*, 80 (1930) 97.
13. For preparation, see J. Auerbach, *J. Chem. Soc.*, 35 (1897) 799.
14. F. Feigl and D. Goldstein, *Mikrochem. ver. Mikrochim. Acta*, 40 (1952) 46.
15. M. Delepine, *Compt. rend.*, 146 (1908) 981; *Bull. Soc. Chim.*, 3 (1908) 625.
16. A. Bobierre, *Compt. rend.*, 80 (1875) 961; J. Fordos, *ibid.*, 80 (1875) 794.
17. J. Grünwald, *Oesterr. Chem. Ztg.*, 14 (1911) 271; see also M. Egorov, *Chem. Abstracts*, 33 (1939) 9187.
18. F. Feigl and H. A. Suter, *Ind. Eng. Chem., Anal. Ed.*, 14 (1942) 840; F. Feigl and N. Braile, *Chemist-Analyst*, 32 (1943) 52.
19. F. Feigl and N. Braile, *Chemist-Analyst*, 69 (1944) 147.
20. American Water Works Association, *Manual of Water Quality and Treatment*, New York, 1940, p. 38.
21. F. Feigl and A. Singer, *Pharm. Presse*, 6 (1935) 37.
22. A. Caldas. *Chemist-Analyst*, 42 (1953) 64.
23. F. Feigl, V. Gentil and D. Goldstein, *Anal. Chim. Acta*, 9 (1953) 393.
24. F. Feigl and V. Gentil, unpublished studies.
25. F. Feigl, studies with R. Nováček.
26. F. Feigl and N. Braile, *Chemist-Analyst*, 32 (1943) 56.
27. H. Gotô, *Chem. Abstracts*, 35 (1941) 1720.
28. F. Feigl, V. Gentil and D. Goldstein, *Mikrochim. Acta*, (1954) 93.

29. F. Feigl and E. Jungreis, unpublished studies.
30. A. C. Vournasos, *Ber.*, 43 (1910) 2269.
31. K. Agte, H. Becker-Rose and G. Heyne, *Z. angew. Chem.*, 38 (1925) 1124.
32. M. T. Petrov, *Chem. Abstracts*, 31 (1937) 2120.
33. E. Merck, *Prüfung der chemischen Reagenzien auf Reinheit*, 4th ed., Darmstadt, 1931. See also J. Rosin, *Reagent Chemicals and Standards*, 2nd ed., New York, 1946.
34. A. Bianchi and E. di Nola, *Chem. Zentr.*, 1910, II, 913.
35. V. Fortini, *Chem. Ztg.*, 36 (1912) 1461.
36. B. S. Evans and D. G. Higgs, Ref. 2, page 10.
37. F. Feigl and D. Goldstein, unpublished studies.
38. F. Feigl and H. J. Kapulitzas, *Mikrochemie (Emich Festschrift)*, 1930, p. 128; *Z. anal. Chem.*, 82 (1930) 417.
39. A. Prins, *Chem. Weekblad*, 27 (1930) 191.
40. F. Feigl and H. J. Kapulitzas, *Z. anal. Chem.*, 82 (1930) 417.
41. B. S. Evans and D. G. Higgs, Ref. 2, page 11.
42. F. Feigl and A. Schaeffer, *Anal. Chem.*, 23 (1951) 353.
43. F. Richter, *Z. anal. Chem.*, 126 (1934) 426.
44. F. Andreasch, *Z. anal. Chem.*, 18 (1897) 601.
45. For literature see F. J. Welcher, *Organic Analytical Reagents*, Vol. IV, New York, 1948, page 161.
46. F. Feigl and H. Hamburg, *Z. anal. Chem.*, 86 (1932) 7.
47. F. Feigl and L. Baumfeld, *Anal. Chem. Acta*, 3 (1949) 15.
48. F. Feigl and A. Caldas, *Anal. Chem.*, 29 (1957) 580.
49. Am. Chem. Soc. Specifications, Washington 1952, p. 227.
50. F. Feigl and D. Goldstein, *Anal. Chem.*, 29 (1957) 456.
51. H. U. Sverdrup, M. W. Johnson and R. H. Fleming, *The Oceans, their Physics, Chemistry and General Biology*, New York, Prentice-Hall Inc., 1949.
52. F. Feigl and R. Nováček, unpublished studies.
53. A. Guerreiro (Rio de Janeiro), private communication.
54. H. H. Willard and J. L. Kassner, *J. Am. Chem. Soc.*, 52 (1930) 2402; H. H. Willard and R. C. Gibson, *Ind. Eng. Chem., Anal. Ed.*, 3 (1931) 88.
55. D. Goldstein, unpublished results.
56. N. A. Tananaeff and G. A. Pantschenko, *Z. anorg. allgem. Chem.*, 150 (1926) 164.
57. F. Feigl and A. Schaeffer, *Anal. Chem.*, 23 (1951) 353.
58. H. Fischer, *Beryllium, its Production and Application*, New York, 1932, pp. 26, 47,
59. F. Feigl and D. Goldstein, unpublished studies.
60. H. L. Zermatten, *Proc. Acad. Sci. Amsterdam*, 36 (1933) 899.
61. F. Feigl and Cl. Costa Neto, *Monatsh.*, 86 (1955) 336.
62. A. de Sousa, *Mikrochem. ver. Mikrochim. Acta*, 40 (1953) 252.
63. F. Feigl and V. Gentil, unpublished studies.
64. N. A. Tananaeff and G. A. Pantschenko, *Z. anorg. allgem. Chem.*, 150 (1926) 163.
65. F. Feigl and H. E. Ballaban, unpublished studies.
66. F. Feigl and A. Caldas, *Mikrochim. Acta*, (1956) 1311.
67. H. B. van Valkenburgh and T. C. Crawford, *Ind. Eng. Chem., Anal. Ed.*, 13 (1941) 459.
68. C. Rammelsberg, *J. Chem. Soc.*, 26 (1873) 1, 13.
69. N. A. Tananaeff, *Z. anorg. allgem. Chem.*, 180 (1929) 80.
70. F. Feigl, *Z. anal. Chem.*, 72 (1927) 113; 74 (1928) 399.
71. Compare F. Ephraim, *Inorganic Chemistry*, 5th ed., London, 1948, p. 819.
72. F. Feigl and H. Leitmeier, *Chem. Abstracts*, 22 (1928) 2903. Compare R. Nováček, *Veda přirodni*, 14 (1933) 257.
73. F. Schwartz, *Berg- u. Hütt. Jahrb.*, 1930, *Sprechsaal*, 1953/1.
74. Compare F. Machatschki, *Vorräte und Verteilung der mineralischen Rohstoffe*, Vienna, 1948, page 71.
75. K. Alberti and F. Schwarz, *Ref. Tonindustrie-Ztg.*, 1956/7.
76. F. Schwartz, private communication.
77. F. Schwarz, *Mikrochim. Acta*, 3 (1938) 126.

78. F. Feigl and H. Leitmeier, *Chem. Abstracts*, 29 (1935) 3263. Compare R. Nováček, *Veda přirodni*, 14 (1933) 257.
79. G. Denigès, *Bull. soc. chim.*, 23 (1918) 36.
80. F. Feigl and N. Braile, *Chemist-Analyst*, 33 (1944) 76.
81. F. Feigl and V. Demant, unpublished studies.
82. A. E. Hill, *J. Am. Chem. Soc.*, 56 (1934) 1071.
83. F. Feigl and V. Gentil, *Anal. Chem.*, 27 (1955) 433, and unpublished studies.
84. F. Feigl and V. Gentil, *Mikrochim. Acta* (1954) 435.
85. F. Mylius and E. Groschuff, *Chem. Abstracts*, 4 (1910) 1798.
86. F. Mylius, *Z. anorg. allgem. Chem.*, 55 (1907) 233.
87. F. Feigl, *Mikrochemie*, 13 (1933) 139.
88. Y. Kondo, *Mikrochim. Acta*, 1 (1937) 157.
89. Compare W. Eitel, M. Pirani, and K. Scheel, *Glasstechnische Tabellen*, Berlin, 1932, p. 509.
90. E. Reser, *Techn. wiss. Abhandl., Osram-Konzern*, 2 (1931) 277.
91. F. Feigl and G. Dacorso, *Chemist-Analyst*, 32 (1943) 28.
92. F. Feigl and V. Gentil, *Mikrochim. Acta*, (1954) 435.
93. See F. Moll, *Z. angew. Chem.*, 38 (1925) 73.
94. K. Klanfer, *Mikrochemie*, 9 (1931) 34.
95. O. Metzger, W. Hess, and H. Rahl, *Beiträge zur kriminalistischen Symptomatologie*, Graz, 1930. Compare also these writers, *Z. angew. Chem.*, 44 (1931) 645.
96. F. Feigl and H. E. Ballaban, unpublished studies.
97. W. Böttger, "Qualitative Analyse anorganischer Verbindungen", in Berl-Lunge, *Chemisch-technische Untersuchungsmethoden*. 8th ed., Berlin, Vol. I, 1932, p. 132.
98. Y. Kondo, *Mikrochim., Acta* 1 (1937) 154.
99. E. Merck, *Prüfung der chemischen Reagenzien auf Reinheit*, 4th ed., Darmstadt, 1931; J. Rosin, *Reagent Chemicals and Standards*, 2nd edition, New York, 1946.
100. F. Feigl, *Z. anal. Chem.*, 152 (1956) 53.
101. M. Niessner, *Mikrochemie*, 12 (1932) 1; G. Gutzeit, M. Gysin, and R. Galopin, *Compt. rend. soc. phys. hist. nat., Genève*, 50 (1934) 192; T. Hiller, *Schweiz. mineralog. petrog. Mitt.*, 17 (1937) 88.
102. M. Niessner, *Arch. Hüttenw.*, 3 (1929) 157; *Mikrochemie*, 12 (1932) 16.
103. E. Heyn, *Mitt. kaiserl. Materialprüfungsamt*, 1906, p. 253; P. Oberhoffer, *Stahl u. Eisen*, 33 (1916) 798.
104. R. Baumann, *Metallurgie*, 3 (1906) 416. Compare K. M. Rauner, *Chem. Abstracts*, 31 (1937 972.
105. E. Heyn, *Stahl u. Eisen*, 26 (1906) 8.
106. K. Weisselberg, *Petroleum*, 31 (1934) 7.
107. M. Niessner, *Mikrochemie*, 12 (1932) 17.
108. M. Niessner, *Mikrochemie*, 12 (1932) 20.
109. A. Glazunow, *Chem. Abstracts*, 26 (1932) 3456; 24 (1930) 5666; *Oesterr. Chem.-Ztg.*, 40 (1938) 217.
110. R. Jirkovský, *Chem. Abstracts*, 25 (1931) 5640; see also S. I. D'yachkovski and T. I. Isaenko, *Chem. Abstracts*, 26 (1932) 48; E. Arnold, *Chem. Abstracts*, 26 (1933) 3671.
111. H. Fritz, *Z. anal. Chem.*, 78 (1929) 418.
112. Comp. H. W. Hermance and H. V. Wadlow, *Electrography and Electro-Spot Testing, Physical Methods in Chemical Analysis*, Vol. 2, New York, 1951.
113. F. Feigl and H. Leitmeier, *Mineralog. petrog. Mitt.*, 40 (1929) 6; see also *Mikrochemie (Pregl Festschrift)*, 1929, p. 83.
114. F. Feigl and N. Braile, *Chemist-Analyst*, 32 (1943) 76; comp. also F. Feigl and C. Stark, *Anal. Chem.*, 27 (1955) 1838.
115. H. S. Sommer, *Ind. Eng. Chem., Anal. Ed.*, 12 (1940) 388.
116. F. Feigl and H. Leitmeier, *Chem. Abstracts*, 24 (1930) 4480.
117. F. Feigl and V. Demant, *Mikrochim. Acta*, 1 (1937) 322.
118. E. Deussen, *Z. anal. Chem.*, 46 (1907) 320.
119. F. Feigl, studies with L. Weidenfeld, *Dissertation*, Vienna, 1930.
120. F. Feigl, studies with L. Badian.

121. F. Feigl and L. Badian, unpublished studies.
122. F. Feigl, unpublished studies.
123. F. Feigl, D. Goldstein and R. A. Rosell, *Z. anal. Chem.* (1957), in press.
124. M. Richter, *Z. anal. Chem.*, 21 (1882) 368.
125. L. Rossi and co-workers, *Chem. Abstracts*, 32 (1938) 8301.
126. F. Feigl and N. Braile, unpublished studies.
127. F. Feigl and C. P. J. Da Silva, *Ind. Eng. Chem., Anal. Ed.*, 14 (1942) 14.
128. F. Feigl and P. Krumholz, *Mikrochemie*, 8 (1930) 131.
129. F. Feigl and P. Krumholz, *Ber.*, 62 (1929) 1138; see also *Mikrochemie (Pregl Festschrift)*, 1929, p. 83.
130. A. L. Engel, *U. S. Bur. Mines Repts. Investigations Tech. Papers*, 3370 (1938) 69.
131. A. S. Komarowsky and N. S. Poluektoff, *Mikrochemie*, 14 (1933-34) 317.
132. F. Feigl and H. Leitmeier, *Mineralog. petrog. Mitt.*, 39 (1928) 224; H. Leitmeier, *Mikrochemie*, 6 (1928) 144.
133. F. Feigl and D. Goldstein, *Mineralog. Petrograph. Mitt.* (1957), in press.
134. F. Feigl and G. Krausz, *Ber.*, 58 (1925) 398.
135. L. Melnik (Ghent), private communication.
136. F. Feigl and L. Weidenfeld, *Mikrochemie (Emich Festschrift)*, 1930, p. 132.
137. E. Bogen, *Süddeut. Apoth. Ztg.*, 68 (1928) 308.
138. F. Tödt, *Chem. Abstracts*, 21 (1927) 1560.
139. I. M. Kolthoff, *Pharm. Weekblad*, 53 (1916) 1739; *comp.* N. Schoorl, *Z. anal. Chem.*, 88 (1932) 328.
140. K. Heller, G. Kuhla, and F. Machek, *Mikrochemie*, 18 (1936) 193.
141. G M. Schwab and K. Jockers, *Naturwissenschaften*, 25 (1937) 44; *Z. angew. Chem.*, 50 (1937) 546.
142. G. Hesse, *Z. angew. Chem.*, 49 (1936) 315, gives a comprehensive review. See also L. Zechmeister and L. v. Cholnoky, *Principles and Practice of Chromatography*, translated by A. L. Bacharach and F. A. Robinson, London, 1941; H. H. Strain, *Chromatographic Adsorption Analysis*, New York, 1942; E. and M. Lederer, *Chromatography, A Review of Principles and Applications*, 2nd ed., Amsterdam, 1957; R. C. Brimley and F. C. Barrett, *Practical Chromatography*, New York, 1953.
143. H. J. Brenneis, *Mikrochemie*, 9 (1931) 385.
144. H. Alber, *Mikrochemie*, 14 (1934) 234.
145. F. Feigl, *Chem. Ztg.*, 44 (1920) 689.
146. See also A. Sulfrian, *Chem. Ztg.*, 56 (1932) 650.
147. Y. Kondo, *Mikrochim. Acta*, 1 (1937) 154.
148. F. Feigl and C. P. J. da Silva, *Ind. Eng. Chem., Anal. Ed.*, 14 (1942) 316.
149. H. E. Feigl (Cambridge), unpublished studies.
150. E. Paternò and A. Mazzucchelli, *Gazz.*, 38 (1908) 137.
151. W. Ostwald, *Kolloid-Beih.*, 2 (1911) 409; P. P. von Weimarn, *ibid.*, 22 (1926) 38; H. Sommer, *Ind. Eng. Chem., Anal. Ed.*, 12 (1940) 368.
152. F. Feigl and A. Caldas, unpublished studies.
153. A. Caldas, *Chemist Analyst*, 43 (1954) 100.
154. F. Feigl and E. Jungreis, *Z. anal. Chem.* (1958), in press.

Tabular Summary of the Limits of Identification Attained by Spot Tests

The limit of identification is given for the most sensitive procedure, if the reaction may be carried out in different ways.

Unless otherwise stated, a macrodrop (about 0.05 ml) is implied.

1. CATIONS

Cation identified	*Reagents or test reactions*	*Limit of identification,* γ	*Page*
Aluminum	Morin	0.2	182
	Alizarin sulfonic acid	0.65	183
	Alizarin	0.15	185
	Quinalizarin	0.005 (in 0.01 ml)	188
	Chrome fast pure blue B	0.1	189
	Pontachrome blue-black R	2 (in 1 ml)	191
	Ammonium aurin tricarboxylate	0.16	191
Ammonium	*p*-Nitrobenzenediazonium chloride	0.67	235
	Litmus paper	0.01	236
	Manganese sulfate and silver nitrate	0.05	237
	8-Hydroxyquinoline and zinc chloride	0.2	237
	Silver nitrate and formaldehyde	0.05	238
	Silver nitrate and tannin	0.1	238
	Nessler's reagent	0.025	238
	Manganese sulfate, hydrogen peroxide and benzidine	0.03	239
Antimony	Zinc on platinum foil	20	103
	Phosphomolybdic acid	0.2	104
	9-Methyl-2, 3, 7-trihydroxy-6-fluorone	0.2	105
	Rhodamine B	0.5	105
	Luminescence	—	107

Cation identified	*Reagents or test reactions*	*Limit of identification, γ*	*Page*
Arsenic	Stannous chloride	1	99
	Reduction to AsH_3	0.5	101
	Silver nitrate	6	102
	Iodine	5	102
	Kairine and ferric chloride	0.005	103
Barium	Sodium rhodizonate	0.25	216
	Induction of precipitation of lead sulfate	0.4	218
	Potassium permanganate and sodium sulfate	2.5	218
	Tetrahydroxyquinone	5	219
	Nitro-3-hydroxybenzoic acid	1.5	219
Beryllium	Morin	0.07	191
	Quinalizarin	0.14	192
	p-Nitrobenzeneazoorcinol	0.2 (in 0.04 ml)	194
	Chromeazurole	0.3	195
	2-(*o*-Hydroxyphenyl)-benzthiazole	0.03	195
Bismuth	Cinchonine and potassium iodide	0.14	76
	Alkali stannite and lead salt	0.01	77
	Luminescence	0.004	79
	Potassium chromithiocyanate	0.4	79
	Alkali stannite	1	80
	Potassium manganese^{II} cyanide	10	80
	Formation of bismuth oxyiodide	25	80
	Quinoline and potassium iodide	1	80
	Thioacetamide	7	80
	Alkali iodide and isobutyl ketone	1	80
Cadmium	Ferrous dipyridyl iodide	0.05	94
	Di-*p*-nitrophenylcarbazide	0.8	96
	Di-*β*-naphthylcarbazone	0.2	98
	Diphenylcarbazide	4	99
	p-Nitrodiazoaminoazobenzene	0.025	99
	Cyanide and sodium sulfide	0.1	99
Calcium	Ammonium ferrocyanide	25	220
	Dihydroxytartaric acid osazone	0.01	221
	Sodium rhodizonate and alkali hydroxide	1	222
	Glyoxal-bis(2-hydroxyanil)	0.05	534

Cation identified	*Reagents or test reactions*	*Limit of identification, γ*	*Page*
Cerium	Hydrogen peroxide and ammonia	0.35	210
	Ammoniacal silver nitrate	1	210
	Benzidine	0.18	211
	Phosphomolybdic acid	0.52	212
	Leuco malachite green	0.03 (in 0.1 ml)	212
	Ammonium naphthoate, anthranilate or salicylate	—	212
Cesium	Gold platinibromide	0.25 (in 0.001 ml)	234
	Potassium bismuth iodide	0.7 (in 0.001 ml)	235
	Auric chloride and palladium chloride	1 (in 0.001 ml)	235
Chromium	Diphenylcarbazide after conversion to chromate		167
	in alkaline solution	0.25	168
	in acid solution	0.8	170
	with sodium peroxide	0.5	169
	Benzidine after conversion to chromate	0.25	171
	2,7-Diaminodiphenylene oxide after conversion to chromate	0.003	171
	Acid alizarin RC	0.6	172
	Silver nitrate after conversion to chromate	6	172
	Pyrrole after conversion to chromate	5	173
	Strychnine after conversion to chromate	0.9	173
Cobalt	Sodium thiosulfate	8	143
	α-Nitroso-β-naphthol	0.05	144
	Rubeanic acid	0.03	146
	Ammonium thiocyanate and acetone	0.5	146
	Diacetylmonoxime *p*-nitrophenylhydrazone	0.1	147
	2-Nitroso-1-naphthol-4-sulfonic acid	0.01	149
	Chromotropic acid dioxime	0.05	149
	Sodium pentacyanopiperidine ferroate	0.05	149

Cation identified	*Reagents or test reactions*	*Limit of identification, γ*	*Page*
Cobalt (contd.)	Hydrogen peroxide and sodium bicarbonate	5	149
	Manganous sulfate and butyraldehyde	1	149
	Sodium azide and *o*-tolidine	0.5	149
Copper	Catalysis of the ferric-thiosulfate reaction	0.02	80
	Benzoin oxime	0.1	82
	Salicylaldoxime	0.5	83
	o-Tolidine and ammonium thiocyanate	0.003 (in 0.015 ml)	84
	Phosphomolybdic acid	1.3	84
	1, 2-Diaminoanthraquinone-3-sulfonic acid	0.02	85
	Rubeanic acid	0.006	87
	Dithizone	0.03	90
	Cuproin	0.05	91
	Alizarin blue	0.004	92
	Benzidine and potassium bromide	0.6	93
	Ammonium mercury thiocyanate and zinc salts	0.1	93
	Dimethylaminobenzylidenerhodanine	0.6	93
	8-Hydroxyquinoline and potassium cyanide	0.4	93
	Hydrobromic acid	0.15	94
	2-Nitroso-1-naphthol-4-sulfonic acid	0.01	94
	Phenetidine chloride and hydrogen peroxide	—	94
	Diphenylcarbazone	0.002	94
	Ammonium bromide and phosphoric acid	0.1	94
	Alkali tartrate, hydrogen peroxide and *o*-hydroxyphenylfluorenone	0.4	94
	Zinc diethyldithiocarbamate	0.002	94
	o-Hydroxyphenylfluorone	1.7	94
	Catalysis of autoxidation of resorcinol	0.005	94

Cation identified	*Reagents or test reactions*	*Limit of identification, γ*	*Page*
Gallium	Potassium ferrocyanide and manganous chloride	5 (in 0.04 ml)	214
	Rhodamine B	0.5	215
Germanium	Increase of acidity due to mannite	2.5	112
	Ammonium molybdate and benzidine	0.25 (in 0.025 ml)	112
	9-Phenyl-2, 3, 7-trihydroxy-6-fluorone	0.13	114
	Diphenylcarbazone	0.01	114
	Hydroxyanthraquinone	5	114
	Ammonium molybdate and hydroxylamine hydrochloride	0.4	115
Gold	*p*-Dimethylaminobenzylidene-rhodanine	0.1	127
	Benzidine	0.02 (in 0.001 ml)	128
	Reduction to metal by heating	1	128
	Rhodamine B	0.1	129
	Reduction to metal by reductants	—	131
	α-Naphthylamine and *n*-butyl alcohol	1	131
	Reduction to metal by ascorbic acid-Complexone III	1	131
Hydrazine	Salicylaldehyde	0.1	239
	— and sodium acetate	0.1	240
	Ammoniacal copper^{II} ferricyanide	0.3	240
	p-Dimethylaminobenzaldehyde	0.001	240
	Alkali copper pyrophosphate and *p*-dimethylaminobenzylidene-rhodanine	0.1	242
	Alkali copper pyrophosphate and alkali thiocyanate	2	242
	Lead^{IV} oxide or thallium^{III} oxide	5	242
	Dinitrobenzene	5	242
Hydroxyl-amine	Diacetylmonoxime and nickel salt	1	242
	Salicylaldehyde and copper salt	1	243
	Formaldehyde and ferric salts	0.5	245

Cation identified	*Reagents or test reactions*	*Limit of identification, γ*	*Page*
Hydroxylamine (contd.)	Conversion to nitrous acid	0.01	245
	Reduction to ammonia	0.1	246
	Ammoniacal cupric ferricyanide	0.2	246
	Potassium copper pyrophosphate and *p*-dimethylaminobenzylidenerhodanine	2	246
	Potassium copper pyrophosphate and alkali thiocyanate	1	246
	Dinitrobenzene	5	246
Indium	Alizarin or quinalizarin	0.05 (in 0.025 ml)	213
Iridium	Condensation of hydrogen	0.18	131
Iron	Potassium ferrocyanide	0.05	161
	α, α′-Dipyridyl	0.03	161
	8-Hydroxyquinoline-7-iodo-5-sulfonic acid	0.5	163
	Potassium thiocyanate	0.25	164
	Dimethylglyoxime	0.4	164
	Disodium-1, 2-dihydroxybenzene-3, 5-disulfonate	0.05	165
	8-Hydroxyquinoline	10	166
	Thioglycolic acid	0.01	167
	Quercetin or quercitrin	3	167
	2-Nitroso-1-naphthol-4-sulfonic acid	0.01	167
	Isonitrosobenzoylmethane	0.02	167
Lead	Benzidine	1	72
	Rhodizonate	0.1	73
	Dithizone	0.04	74
	Sulfuric acid and cadmium tin^{II} iodide	10	75
	Gallocyanine	0.3	75
	Carminic acid	1	75
Lithium	Complex ferric periodate	0.1	233
	Etching of soda glass	1	234
Magnesium	Alkali hypoiodite	0.3	223
	Quinalizarin	0.25 (0.001 γ in 0.001 ml)	224

Cation identified	*Reagents or test reactions*	*Limit of identification, γ*	*Page*
	p-Nitrobenzeneazo-α-naphthol	0.19	225
	p-Nitrobenzeneazoresorcinol	0.5	226
	Titan yellow	1.5	228
	Ammonia and phenolphthalein	0.6	228
	Monoazo dyes	100	229
Manganese	Catalytic oxidation to permanganate in acid solution	0.1	173
	in alkaline solution	2.5	174
	Periodate and tetrabase	0.001	174
	Benzidine	0.15	175
	Ammoniacal silver nitrate	0.05	177
Mercury	Diphenylcarbazone	0.1	64
	Cuprous iodide	0.003	65
	Stannous chloride and aniline	1	66
	p-Dimethylaminobenzylidene-rhodanine	0.33	67
	Catalysis of the reduction of tin^{IV} salts by hypophosphites	0.1	68
	Catalysis of the formation of alumina from aluminum	0.0001	69
	Potassium ferrocyanide	0.002	70
	Dithizone	0.25	71
	Chromotropic acid	100	72
	Potassium ferrocyanide and α, α′-dipyridyl	0.5	72
Molybdenum	Potassium thiocyanate and stannous chloride	0.1	115
	Potassium xanthate	0.01	116
	Phenylhydrazine	0.13	117
	Methylene blue and hydrazine	0.012	117
	o-Hydroxyphenylfluorone	1.7	118
	Tincture of cochineal	0.02	119
	α, α′-Dipyridyl and stannous chloride	0.4	119
	Diphenylcarbazide	2.5	119
	Di-β-naphthylcarbazone	0.3	119
Nickel	Dimethylglyoxime	0.015	149
	— and bromine water	0.12	152
	Rubeanic acid	0.012 (in 0.015 ml)	153

Cation identified	*Reagents or test reactions*	*Limit of identification, γ*	*Page*
Osmium	Induced reduction of nickel salts	0.5	133
	Activation of chlorate solutions	0.005	141
	Benzidine acetate	0.008 (in 0.001 ml)	143
	Potassium ferrocyanide	0.008 (in 0.001 ml)	143
Palladium	Condensation of hydrogen	0.01	131
	Induced reduction of nickel salts	0.0025 (in 1 ml)	133
	Stannous chloride	0.04 (in 0.002 ml)	135
	Catalysis of the reduction of phosphomolybdates by carbon monoxide	0.025	135
	Nickel dimethylglyoxime	0.05	137
	Mercuric iodide	0.08	137
	p-Nitrosodiphenylamine	0.005	138
	Phenoxithine	0.1	139
	3-Hydroxy-1-*p*-sulfonatophenyl-3-phenyltriazine	0.05	140
	Naphthalene-4-sulfonic acid-1-azo-5-*o*-8-hydroxyquinoline	2	140
	p-Dimethylaminobenzylidene-rhodanine	0.004	141
	Mercuric cyanide and diphenyl-carbazide	0.5	141
	Mercaptobenzeneindazole	0.25	141
	Mercuri cyanide and methyl yellow	0.4	141
	p-Fuchsin	0.01	141
	α-Nitroso-β-naphthol	0.5	141
	Potassium ferrocyanide and *a, a'*-dipyridyl	2	141
Platinum	Condensation of hydrogen	0.04	131
	Induced reduction of nickel salts	1.5	133
	Stannous chloride	0.025 (in 0.002 ml)	134
	Alkali iodide	0.5	134
Potassium	Sodium cobaltinitrite	4	230
	– and silver nitrate	1	230
	Dipicrylamine	3	230
	Sodium tetraphenyl boron	1	232

Cation identified	*Reagents or test reactions*	*Limit of identification, γ*	*Page*
Rhodium	Condensation of hydrogen	0.02	131
Ruthenium	Induced reduction of nickel salts	0.5	133
	Activation of chlorate solutions	–	141
	Rubeanic acid	0.2	142
Silver	Manganese nitrate and alkali	2	58
	p-Dimethylaminobenzylidene-rhodanine	0.02	59
	Physical development of silver nuclei	0.005	61
	Catalytic reduction of manganese^{III} and cerium^{IV} salts	0.4	62
	Potassium chromate	2	63
	Stannous chloride	1	63
	Dithizone	0.05	63
	Stannous chloride and chromotropic acid	0.1	64
	Cuprous thiocyanate	–	64
	Phenothiazine	0.54	64
	p-Dimethylaminobenzylidene-thiobarbituric acid	0.2	64
	Ethylenediaminetetraacetic acid	0.05	64
	Potassium ferrocyanide and *α, α'*-dipyridyl	2	64
	Potassium nickelcyanide and dimethylglyoxime	0.5	64
Sodium	Zinc uranyl acetate	12.5	229
	– in ultra violet light	2.5	229
Strontium	Sodium rhodizonate	0.45	220
Thallium	Potassium iodide	0.6	154
	Benzidine	0.5	155
	Phosphomolybdic acid and hydrobromic acid	0.13 (in 0.025 ml)	156
	Dipicryl amine	20	156
	Oxidation to thallium^{III} in acid solution	0.2	157
	Rhodamine B	0.03	158, 161
	Sodium carbonate and ammonium sulfide	0.5	160

Cation identified	*Reagents or test reactions*	*Limit of identification, γ*	*Page*
Thallium (contd.)	Auric chloride and palladium chloride	0.4	161
	Alkaline ferricyanide	0.01	161
	Uranyl sulfate	1	161
	8-Hydroxyquinoline	0.5	161
Tin	Dithiol	0.05	107
	Ammonium phosphomolybdate	0.03	108
	Cacotheline	0.2	109
	Flame color	0.03	109
	Morin	0.05	110
	Dimethylglyoxime and ferric salts	0.04	111
	Diazine green	2	111
	2-Benzylpyridine	1.3 (in 0.01 ml)	111
	Mercuric chloride and aniline	0.6	111
	1,2,7-Trihydroxyanthraquinone	0.2	111
	Anthraquinone-1-azo-4-dimethyl-aniline hydrochloride	0.01	111
	o-Aminophenols	0.06	112
Titanium	Hydrogen peroxide	2	195
	Pyrocatechol	2.7	196
	Chromotropic acid	3	197
	— acid and conc. sulfuric acid	0.1	198
	Morin	0.01	198
	Tannin and antipyrine	0.2	199
	Methylene blue and zinc	0.05	199
	Alizarinsulfonic acid and zinc	0.5	199
	Resoflavine	0.1	199
Tungsten	8-Hydroxyquinoline	15	119
	Stannous chloride	5	120
	Catalysis of Ti^{III}-malachite green reaction	0.1	121
	Diphenyline	6	122
Uranium	8-Hydroxyquinoline	10	204
	Potassium ferrocyanide	0.92 (0.05 γ in 0.001 ml)	205
	Fluorescence	0.001 (in 0.001 ml)	207
	Photolytic decomposition of oxalic acid	2.5	207

Cation identified	*Reagents or test reactions*	*Limit of identification, γ*	*Page*
Uranium (contd.)	Rhodamine B	0.05	208
	Quercetin or quercitrin	3	209
	Sodium phosphate	2.5	209
	Cochineal	2.5	210
	Fluorescein and ammonium chloride	0.12	210
	Phenanthroline and ferric chloride	1	210
Vanadium	Hydrogen peroxide	2.5	123
	Reduction of iron^{III} by vanadium^{IV}	0.1	124
	α-Benzoin oxime	1	124
	8-Hydroxyquinoline	0.5	125
	3,3-Dimethylnaphthidine	0.1	126
	Aniline	3	127
	8-Hydroxyquinoline and sodium tartrate	0.27 (in 0.04 ml)	127
	8-Hydroxyquinoline and chloroform	0.1	127
	Phosphomolybdic acid	0.4 (in 0.04 ml)	127
	Phosphoric acid and alkali tungstate	0.3	127
	Diaminobenzidine hydrochloride	0.25	127
	Diphenylbenzidine	0.05	127
	Quercetin	0.25	127
Zinc	Potassium ferricyanide and diethylaniline	0.001	178
	Dithizone	0.025	178
	Induced precipitation of $Co[Hg(CNS)_4]$	0.2	180
	Potassium ferricyanide and 3,3′-dimethylnaphthidine	0.1	181
	Resorcinol	2	182
	Potassium cobalt^{III} cyanide	0.6 (in 0.002 ml)	182
	Potassium ferricyanide and *p*-phenetidine	0.05	182
	Uranyl ferrocyanide	0.1	182
Zirconium	β-Nitroso-α-naphthol	0.2	199
	Alizarin	0.5	200
	Morin	0.1	201

Cation identified	Reagents or test reactions	Limit of identification, γ	Page
Zirconium (contd.)	*p*-Dimethylaminoazophenylarsonic acid	0.1	202
	Carminic acid	0.5 (in 0.001 ml)	204
	Gallocyanin	—	204
	Chlorobromoamine acid	0.5	204
	Azodyes from mandelic acid	–	204

2. ANIONS

Anion or substance identified	Reagents or test reactions	Limit of identification, γ	Page
Aminosulfonic acid *see* Sulfamic acid			
Boric acid	Tincture of curcuma	0.02	339
	Alizarin S	1	340
	Purpurin	0.6	340
	Quinalizarin	0.06	340
	p-Nitrobenzeneazochromotropic acid	0.08 (in 0.04 ml)	341
	Decomposition of methyl borate with KF	0.01	342
	Increase of acidity with polyhydroxy compounds	0.001	343
	Carminic acid	0.1 (in 0.03 ml)	343
	Sulfuric acid and methyl alcohol	–	343
Bromic acid	Manganese sulfate and sulfuric acid	20	298
Carbonic acid	Sodium carbonate and phenolphthalein	4 (in 0.1 ml)	337
Chloric acid	Manganous sulfate and phosphoric acid	0.05	297
Chromic acid	Diphenylcarbazide	0.5	345
Cyanic acid	Conversion to hydroxyurea	30	285
Dithionous acid *see* Hyposulfurous acid			
Ferricyanic acid *see* Hydroferricyanic acid			
Ferrocyanic acid *see* Hydroferrocyanic acid			

Anion or substance identified	*Reagents or test reactions*	*Limit of identification, γ*	*Page*
Hydrazoic acid	Silver nitrate or ferric chloride	5	286
	Nitrous acid	1	287
Hydriodic acid	Palladous chloride	1	265
	Potassium nitrite and starch	0.025	266
	Bromine water	0.05	267
	Catalytic reduction of $cerium^{IV}$ salts	0.05	267
	Formation of silver iodide	5	268
	Formation of thallous iodide	–	269
	Nitric acid and starch	0.3	269
	Chloramine T and tetrabase	0.0005	535
Hydrobromic acid	Fluorescein	0.3	262
	Fuchsin	3.2	264
	Permolybdate and α-naphthoflavone	1	264
	Oxidation to bromine and action on KI	5	265
	Copper bromide and sulfuric acid	0.02	265
Hydrochloric acid	Silver nitrate in presence of hydrogen peroxide and oxine	2	259
	Conversion to chromyl chloride	0.3	260
	Volatilization and formation of AgCl	1	261
Hydrocyanic acid	Copper acetate and benzidine	0.25	276
	Copper sulfide	1.25	277
	Conversion to thiocyanate	1	277
	Demasking of alkali palladium dimethylglyoxime	0.25	278
	Alkali mercuri chloride	2.5	279
	Picric acid	–	280
	Decolorization of starch-iodine	1	280
	Copper ferricyanide	–	280
	Catalysis of benzoin rearrangement	0.05	280
Hydroferricyanic acid	Benzidine	1	291
	Phenolphthalin	0.5	292

Anion or substance identified	Reagents or test reactions	Limit of identification, γ	Page
Hydroferricyanic acid (contd.)	Induced oxidation of tetrabase	0.1	293
	Trithiourea cuprous chloride	0.9	294
	Thallium sulfate, alkali and benzidine	3.5	294
Hydroferrocyanic acid	Uranyl acetate	0.5	287
	Demasking of ferrous ions	0.2	288
	Ferric chloride	0.07	289
	Titanium tetrachloride	50	294
Hydrofluoric acid	Zirconium alizarinate	1	269
	Zirconium azoarsonate	0.25	271
	Conversion to silicomolybdic acid	1	272
	Etching of glass	0.5	274
	Prevention of formation of oxinates	0.05	275
Hydrogen peroxide	Lead sulfide	0.04	353
	Ferric ferricyanide	0.08	354
	Reduction of higher nickel oxides	0.01	354
	Catalyzed oxidation of phenolphthalin	0.04 (in 0.04 ml)	355
	Auric chloride	0.07	355
	Alkali thiocyanate	0.7	355
	Potassium cerous carbonate	0.1	355
	o-Tolidine	–	356
	Luminol and hemin	0.012	356
	Chromate and diphenylcarbazide	0.5	356
Hydrogen sulfide	Sodium nitroprusside	1	303
	Catalysis of the iodine-azide reaction	0.3	303
	p-Aminodimethylaniline	1	307
	Sodium plumbite	1.8	307
Hypohalogenous acids	Safranin	ClO^- : 0.5	294
		BrO^- : 0.5	294
		IO^- : 2.5	294
	Thallous hydroxide	0.5	295
	Ammonia	0.1	295
	Liberation of halogen	ClO^- : 2	296
		BrO^- : 2.5	296
		IO^- : 2.5	296

Anion or substance identified	*Reagents or test reactions*	*Limit of identification, γ*	*Page*
Hyposulfurous acid	Dinitrobenzene and ammonia	3	320
Iodic acid	Hypophosphorous acid	1	299
	Potassium thiocyanate	4	299
	Pyrogallol	0.25	300
Mercuric cyanide	Silver nitrate and alkali	5	280
	Conversion into HgI_4^{-2} and CN^-	–	281
	Formation of cyanogen iodide	0.25	281
	Potassium ferrocyanide and α, α'-dipyridyl	—	282
	Formation of dicyanogen	—	282
Nitric acid	Ferrous sulfate and sulfuric acid	2.5	326
	Diphenylamine	0.5	327
	Diphenylbenzidine	0.07	327
	Brucine	0.06	328
	Reduction to nitrite	0.05	330
	Sodium salt of chromotropic acid	0.2	330
Nitrous acid	Sulfanilic acid and α-naphthylamine	0.01	330
	1,8-Naphthylenediamine	0.1	331
	Iodine and starch	0.005	332
	Chrysean	0.25	332
	Benzidine	0.7	332
	Ferrous sulfate and acetic acid (ring test)	2	332
	Safranine T	0.5	332
Perchloric acid	Cadmium chloride	1	300
Periodic acid	Manganous sulfate and phosphoric acid	5	301
	Manganese salts and tetrabase	0.5	302
	Formation of complex copper^{III} salts	0.5	302
Permanganic acid	Reduction to MnO_2	0.3	344
Persulfuric acid	Benzidine	0.25	322
	Nickel hydroxide	2.5	323

Anion or substance identified	Reagents or test reactions	Limit of identification, γ	Page
Phosphoric acid	Ammonium molybdate and benzidine	0.05	333
	o-Dianisidine molybdate and hydrazine	0.05	335
Selenious and selenic acids	Hydriodic acid	1 (in 0.025 ml)	346
	Pyrrole	0.5	346
	1,1-Diphenylhydrazine	0.05	347
	Catalysis of the reduction of methylene blue	0.08 (in 0.08 ml)	348
	Thiourea	2	349
	Ferrous sulfate	10	349
	1,8-Naphthylenediamine	1	349
	3,4,3',4'-Tetraaminobenzidine	–	349
	4-Dimethylamino- (or 4-methylthio)-1,2-phenylenediamine	—	349
Silicic acid	Ammonium molybdate and benzidine	0.1	335
Sulfamic acid	Deamidation	8	325
Sulfuric acid (free)	Glucose	8	316
	Methylenedisalicylic acid	2.5	317
	m- or *p*-Hydroxybenzaldehyde	2.5	317
Sulfuric acid (insoluble sulfate)	Reduction to sulfide with potassium	0.5	315
	Reduction to sulfide with carbon	36	315
Sulfuric acid (soluble sulfate)	Barium rhodizonate	5	313
	Barium carbonate and phenolphthalein	5	314
	Precipitation of $BaSO_4$ in presence of permanganate	2.5	314
Sulfurous acid	Sodium nitroprusside	3.2	307
	Zinc nitroprusside	3.5	308
	Induced oxidation of $Ni(OH)_2$	0.4	309
	Induced oxidation of $cobalt^{II}$ azide	0.5	310
	Decolorization of malachite green	1	311
	Formaldehyde	100	312
	Iodine	5	312
	2-Benzylpyridine	7.2 (in 0.1 ml)	313
	Reduction of ferric ferricyanide	5	313

Anion or substance identified	*Reagents or test reactions*	*Limit of identification, γ*	*Page*
Tellurous and telluric acids	Hypophosphorous acid	0.1	349
	Alkali stannite	0.6 (in 0.025 ml)	350
	Formation of complex copper^{III} tellurate	0.2	350
	Ferrous sulfate and phosphoric acid	0.5	351
	Ammonium polysulfide and sulfite	0.5	352
Thiocyanic acid	Catalysis of the iodine-azide reaction	1.5	282
	Permanganate, copper acetate and benzidine	1	284
	Cobalt salts and acetone	1	285
	Ammonium chloride	10	285
Thiosulfuric acid	Catalysis of the iodine-azide reaction	0.025 (in 0.025 ml)	318
	Mercuric chloride	8	319

3. FREE ELEMENTS

Element identified	*Reagents or test reactions*	*Limit of identification, γ*	*Page*
Aluminum	Alkali hydroxide and *o*-dinitrobenzene	—	363
Arsenic	Reduction to arsine	—	366
Carbon	Oxidation by molten $KClO_3$	50	379
	Molybdenum trioxide	1	380
	Potassium iodate	–	381
Dicyanogen	Hydrolysis to hydrogen cyanide	—	370
	Alkali cyanide and oxine	1	371
Halogens	Fluorescein	Cl : 1	368
		Br : 2	368
		I : 6	368
	Thio-Michler's ketone	Cl : 0.2	368
		Br : 0.2	368
		I : 0.5	368
	Formation of polymethine dyes	Cl : 2.5	369
		Br : 2.5	369

Element identified	*Reagents or test reactions*	*Limit of identifications, γ*	*Page*
Lead	Alkali hydroxide and *o*-dinitrobenzene	–	**363**
Mercury	Palladium chloride	–	**364**
	Selenium sulfide	–	**365**
	Cuprous iodide	–	**366**
Molybdenum	Conversion to molybdate	–	**367**
Selenium	Addition to thallous sulfide	1	**375**
	Formation of silver selenide	5	**376**
	Conversion to selenosulfate	2.5	**376**
	Catalysis of the reduction of methylene blue	0.08	**377**
Sulfur	Formation of thallium polysulfide	3	**372**
	Conversion to thiocyanate	–	**373**
	Conversion to thiosulfate or mercury sulfide	–	**374**
	Conversion to hydrogen sulfide	0.5	**375**
Tellurium	Formation of sulfotellurate	–	**378**
Tin	Alkali hydroxide and *o*-dinitrobenzene	–	**363**
Tungsten	Conversion to tungstate	–	**367**
Vanadium	Conversion to vanadate	–	**367**
Zinc	Alkali hydroxide and *o*-dinitrobenzene	–	**363**

Addendum

Calcium

(4) Test for calcium with glyoxal-bis(2-hydroxyanil) [1]

The above mentioned Schiff base (I) forms with metal ions colored inner complex salts.[2] Solutions of calcium, barium and strontium salts when treated with an alcoholic solution of this compound and made alkaline with alkali hydroxide give a red precipitate. Only the calcium compound is stable against alkali carbonate, and may be extracted with chloroform.

It is probable that the formation of the calcium-inner complex salt (II) is involved:

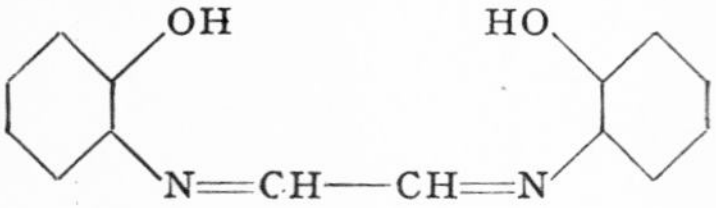

OH HO
N=CH—CH=N
(I)

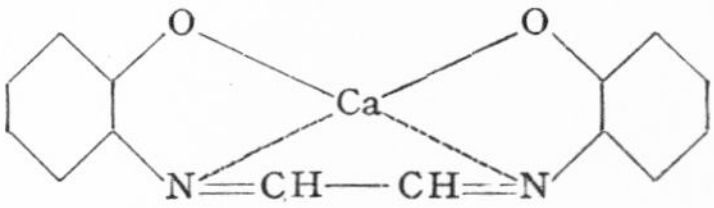

O O
Ca
N=CH—CH=N
(II)

Cobalt and nickel ions show the same behavior as calcium, but in contrast to the latter the red precipitate is insoluble in chloroform. Cadmium ions give a blue precipitate soluble with a blue color in chloroform.

Procedure: A drop of the neutral or acid test solution is treated successively in a micro test tube with four drops of the reagent solution, one drop of a 10% sodium hydroxide solution and one drop of 10% sodium carbonate solution and extracted with 3–4 drops of chloroform. Addition of a few drops of water hastens the separation of the two layers. A red color in the $CHCl_3$ layer indicates the presence of calcium. When testing for small amounts of calcium a blank test is advisable.

Limit of Identification: 0.05 γ calcium
Limit of Dilution: 1:1,000,000
Reagent: 1% solution of glyoxal-bis(2-hydroxyanil) in ethyl alcohol. The reagent can be prepared[2] by dissolving 4.4 g *o*-amino phenol in one liter water at 80° C and adding 3.0 g of 40% water solution of glyoxal. The mixture is kept for 30 minutes at 80° C and after that for 12 hours in the ice box. The precipitate is filtered, washed with water and recrystallized from methanol.

[1] D. Goldstein and C. Stark Mayer (Rio de Janeiro), unpublished studies.
[2] E. Bayer, *Chem. Ber.*, 90 (1957) 2325.

With this procedure it is possible to identify:

20 γ calcium in the presence of 2000 γ strontium
25 ,, ,, ,, ,, ,, ,, 2000 ,, barium
30 ,, ,, ,, ,, ,, ,, 1000 ,, magnesium
20 ,, ,, ,, ,, ,, ,, 500 ,, lead
10 ,, ,, ,, ,, ,, ,, 500 ,, iron
0.5 ,, ,, ,, ,, ,, ,, 500 ,, aluminum

10 γ calcium may even be detected in the presence of 100 γ cadmium if the extraction is made with 0.5 ml chloroform.

Sulfate, oxalate, fluoride and phosphate do not prevent the reaction, although they somewhat lower the sensitivity.

Addendum to p. 268

Hydriodic acid

(5) *Test by catalytical acceleration of the redox reaction between chloramine T and tetrabase*[3]

In water solution, chloramine T (sodium *p*-toluene sulfone chloramide) hydrolizes appreciably according to

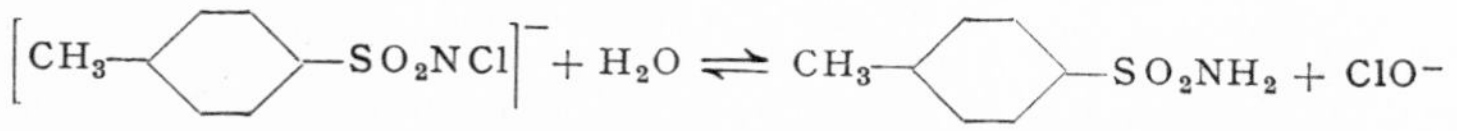

$$[CH_3-C_6H_4-SO_2NCl]^- + H_2O \rightleftharpoons CH_3-C_6H_4-SO_2NH_2 + ClO^-$$

The concentration of hypochlorite is not sufficient to oxidize tetrabase (I) to the blue chinoidal compound (II). Even in slightly acetic acid solution the oxidation proceeds very slowly if the concentration ratios of both partners in the redox reaction (3) are kept within certain boundaries. If, to such a colorless solution, iodide is added, a blue color appears immediately. This means that iodide catalyzes the redox reaction between chloramine T and tetrabase.

The explanation of this catalysis is the following: ClO^- reacts in slightly acid solution with iodide according to (1); the free iodine formed immediately oxidizes tetrabase as shown in (2), whereby iodide is regenerated.

$$ClO^- + 2I^- + 2H^+ \rightarrow Cl^- + H_2O + 2I^\circ \quad (1)$$

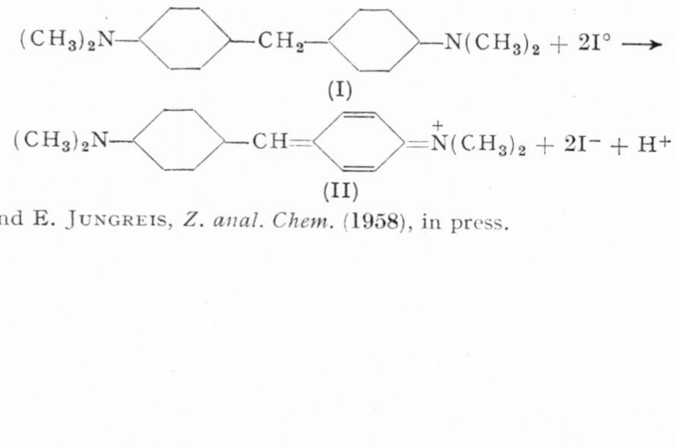

$$\underset{\text{(I)}}{(CH_3)_2N-C_6H_4-CH_2-C_6H_4-N(CH_3)_2} + 2I^\circ \longrightarrow \quad (2)$$

$$\underset{\text{(II)}}{(CH_3)_2N-C_6H_4-CH{=}C_6H_4{=}\overset{+}{N}(CH_3)_2} + 2I^- + H^+$$

[3] F. Feigl and E. Jungreis, *Z. anal. Chem.* (1958), in press.

Summation of (1) and (2) gives the equation of the uncatalyzed reaction in which the catalyst no longer appears.

$$ClO^- + H^+ + (CH_3)_2N-C_6H_4-CH_2-C_6H_4-N(CH_3)_2 \longrightarrow$$

$$Cl^- + H_2O + (CH_3)_2N-C_6H_4-CH=C_6H_4=\overset{+}{N}(CH_3)_2 \quad (3) = (1) + (2)$$

Based on its catalytical action an extremely sensitive test for iodide is possible. Chlorides do not interfere with this test, but bromides when in relatively great quantities (more than 500 γ) also give a pale blue color under the conditions of this test; furthermore, by reasons not yet known, the iodide catalysis is hindered when the amount of bromide is higher than 500 γ.

Procedure: On a depression of a spot plate, a drop of the neutral test solution is mixed with a drop each of the tetra base solution and of a 0.04% freshly prepared water solution of chloramine T. According to the amount of iodide present, a blue color appears immediately or within a few seconds. The blue color is not stable, it changes to green in a few minutes.

Limit of Identification: 0.0005 γ potassium iodide
Limit of Dilution: 1:100,000,000
Reagent: Tetrabase solution. Excess tetrabase is dissolved in N-acetic acid, heated on a water bath, and filtered. Some tetrabase may precipitate, but it is not necessary to filter it.

AUTHOR INDEX

Abrahamczik, E., 35[3]
Agte, K., 444[31]
Aickin, R. G., 263[8]
Ajtai, M., 94[114]
Alber, H. K., 30, 506[144]
Alberti, K., 466[75]
Alexejeff, R., 65
Almássy, G., 198[423], 199[426]
Amis, E. S., 141[269]
Andreasch, F., 449[44]
Andreasch, R., 59[5]
Anft, B., 1[2]
Angerer, E. von, 61[10]
Arnold, E., 489[110]
Arnold, H., 367[13] 418[41]
Artmann, P, 65[34]
Asperger, S., 71[47]
Atack, F. W., 144[280]
Auerbach, J., 93[105], 431[13]
Aufrecht, W., 218[476], 314[118]
Augusti, S., 94[112], 173[359]
Avens, A. W., 326

Babko, A. K., 146[286]
Bacharach, A. L., 505[142]
Badian, L., 68[38], 69[44], 315[121], 342[194], 349[213], 494[120], 494[121]
Baeyer, A. von, 355[225]
Balarew, D., 58[2], 177[373]
Ballaban, H. E., 85[93], 183[392], 198[422], 302[89], 309[89], 429[3], 430[7], 460[65], 481[96]
Balson, E. W., 210[457]
Balston, J. N., 23[71], 49[10]
Bamberger, E., 227[491]
Barannikov, G. J., 327[155]
Barbosa, P. E., 156[311], 230[499]
Barnebey, O. L., 277[44]
Barrett, F. C., 49[9], 505[142]
Basart, J., 270
Baumann, R., 486[104]
Baumfeld, L., 125[216], 161[321], 450[47]
Baumgarten, P., 325[146], 329[161], |330[162], 330[163]
Bayer, E. 534[2]
Beck, G., 109[159], 183[390]
Becker-Rose, H., 444[31]
Beckurts, H., 101[134]
Behr, M., 299[84]
Belcher, R., 24, 30, 126[219], 126[220], 181[381], 216[471]
Bellino, A., 266[21]
Bellucci, J., 199[428]
Belokon, A. N., 212[463]
Bencko, V., 86[94]
Benedetti-Pichler, A. A., 6[13], 182[384]
Berg, R., 237[516], 346[204]
Berman, P., 131[240]
Besthorn, E., 177[377]
Bettendorf, A., 99[132]
Bezier, D., 101[138]
Bhattacharyya, S. C., 140[263]
Bianchi, A., 445[34]
Biehringer, J., 175[369], 294[73], 302[90]
Biltz, W., 210[458], 269[31]
Blancpain, C. P., 105[145]
Blank, E. W., 398[13]
Blau, F., 161[325]
Block, R. J., 23[71], 49[10]
Blohm, H., 118[194]
Blom, J., 245[536], 330[164], 330[166]
Bobierre, A., 433[16]
Bödecker, C., 307[102]
Bogen, E., 505[137]
Borrel, M., 125[218]
Bose, M., 19, 167[345]
Böttger, W., 6[18], 144[279], 236[512], 271[34], 309[105], 395[11], 481[97]
Boussingault, I. B., 236
Braile, N., 12[33], 316[122], 372[21], 375[28], 433[18], 434[18], 435[19], 439[26], 470[80], 491[114], 497[126]
Bran, F., 309[105]
Braun, C. D., 115[183]
Brauner, B., 350[216]
Bray, W. C., 384[4]
Brenneis, H. J., 506[143]
Brimley, R. C., 49[9], 505[142]
Browning, Ph. C., 155[308]
Brukl, A., 299[84], 352, 377[32], 379[32]
Brunck, O., 149[296]
Bruner, E., 366[12]
Brünger, K., 83[83]
Budkewitsch, A. A., 238[520]

Burford, M. G., 407[24], 422[46]
Burgess, L. L., 230[497]
Burkhalter, T., 343[195]
Burkser, E. S., 234[507]

Cacciapuoti, B. N., 144[281]
Cady, L. C., 178[375]
Caldas, A., 64[27], 72[51], 92[104], 141[271], 162[328], 280[50], 288[65], 413[34], 419[43], 420[44], 431[11], 432[11], 437[22], 451[48], 452[48], 453[48], 460[66], 462[66], 463[66], 509[152], 510[153]
Caley, E. R., 41[5], 407[24], 422[46]
Canneri, G., 274[40]
Carboni, P., 332[174]
Carlton, J. K., 45[6], 80[74]
Carvalho, L. F., 207[448]
Cazeneuve, P., 64[29], 167[344]
Čelechovský, J., 86[95a], 93[107]
Černý, P., 8[19]
Chamot, E. M., 339[186]
Chard, S. J., 171[355], 322[140]
Charlot, G., 183[391], 199[427], 201[432], 384[5], 414[36]
Charpentier, G., 355[226]
Cholnoky, K., 49[9]
Cholnoky, L. von, 505[142]
Claeys, A., 94[117], 94[119], 105[147], 114[177], 118[195]
Claisen, L., 84[86]
Clark, R. E. D., 107[154]
Clarke, B. L., 9[26], 22, 94[118], 116[189]
Clarke, L., 339[185]
Claus, A., 163[330]
Clegg, D. L., 49[10]
Clemmsen, E., 318[127]
Cole, H. J., 339[186]
Commanducci, E., 321[135]
Cone, W. H., 178[375]
Conrad, L. J., 124[211], 124[212], 124[213], 127[227]
Conway, E. J., 30[1]
Cornea, J., 11[31], 268[25]
Costa Neto, C., 238[517], 317[125], 321[136], 417[39], 459[61]
Costeanu, R. N., 131[238]
Cozzi, A., 274[40]
Cramer, F., 23[71], 49[10]
Crawford, T. C., 463[67]
Crowell, W. R., 139[261]
Cucuel, F., 69[42]
Cullinane, N. M., 171[355], 322[140]
Curtmann, L. J., 131[241], 133[244]
Cupery, M. E., 325[147]

Dacorso, G., 362[2], 475[91]
Dannenberg, E., 54[16]
Da Silva, C. P. J., 497[127], 507[148]
Daube, F. W., 339[184]
Davidson, D., 285[60], 289[67]
Davies, W. C., 389[7]
De Aguar, A., 331[168]
De Boer, J. H., 200[430], 269[28], 270
Debray, C., 116
De Koninck, L., 230[498]
Delaby, R., 2[9]
Deleo, E., 245[535]
De Leo, F., 266[21]
Delepine, M., 432[15]
Delscheff, P., 125[215]
Demant, V., 246[538], 347[205], 471[81], 492[117]
Denigès, G., 72[52], 103[141], 408[26], 469[79]
Desbassins de Richemont, 326[148]
Deshmukh, G. S., 114[180], 115[182]
De Sousa, A., 119[200], 149[293], 167[339], 204[440], 459[62]
Detscheff, P., 83[82]
Deusse, E., 493[118]
De Vries, G., 99[131]
Diels, O., 243[531]
Dieterich, H., 101[137]
Di Nola, E., 128[232a], 445[34]
Dishen, B. R., 94[116]
Ditz, H., 146[284]
Dobroljubski, O. K., 199[425]
Dodd, A. S., 343[195]
Doi, K., 115[183]
Doležal, J., 64[26], 131[240]
Dolgow, K. A., 128[233], 135[250]
Donau, J., 79[66], 107[153]
Dowzard, E., 101[136]
Droho, O. P., 146[286]
Dromljuk, R. L., 173[358]
Dubnikov, L. M., 274[40]
Dubský, J. V., 86[94], 332[171]
Duckert, R., 18[46], 25, 105[145], 162[326]
Duff, M. A., 72[49]
Dunkelberger, T. H., 332[173]
Dunleavy, R., 204[439]
Duval, C., 25, 128[234], 149[293], 316[123], 404[21]
Duval, R., 64[24]
Duyk, M., 174[365]
Dwyer, F. P., 99[130]
D'yachkovski, S. I., 489[110]

Edge, R., 373
Eegriwe, E., 105[149], 111[167], 111[170], 129[236], 158[315], 178[374], 182[387], 226[490], 308[103]
Ege, Jr., J. F., 170[348]
Egorow, M., 433[17]
Ehrhart, O., 141[273]
Eisleb, O., 84[86]
Eitel, W., 474[89]
Emich, F., 14[43]

Endres, G., 245[536]
Engel, A. L., 500[130]
Engel, E. W., 226[490]
Engelder, C. J., 332[173]
Ephraim, F., 61[12], 83[84], 116, 124[212], 465[71]
Evans, B. S., 68[39], 428[2], 445[36], 448[41]
Evans, M. H., 349[209]

Fairhall, L. T., 159[316]
Falciola, P., 349[208]
Fanstone, R. M., 319[130]
Fauconnier, P., 128[234]
Feigl, F., 1[1], 2[7], 2[8], 3[10], 8[22], 8[25], 10[28], 11[30], 11[32], 12[33], 12[34], 18[46], 19[47], 19[48], 19,[49] 19[50], 19[51], 20[53], 20[57], 24[72], 25, 46[7], 51[13], 59[3], 60[6], 60[8], 62[14], 63[16], 64[27], 64[30], 64[31], 68[38], 69[44], 72[51], 72[54], 73[55], 76[61], 77[62], 77[65], 82[80], 84[88], 85[89], 86[95], 88[99], 92[104], 93[106], 93[109], 94[122], 96[126], 99[129], 104[143], 105[146], 108[155], 110[163], 110[164], 111[165], 112[174], 114[178], 117[191], 118[193], 118[194], 120[202], 122[205], 123[207], 125[216], 125[217], 127[231], 129[235], 132[242], 133[245], 134[249], 135[251], 137[254], 137[255], 138[256], 141[265], 141[266], 141[268], 141[270], 141[271], 141[272], 144[278], 147[289], 149[292], 151[298], 152[299], 152[300], 153[302], 154[305], 155[307], 155[309], 156[311], 157[312], 158[314], 161[321], 161[324], 162[328], 164[334], 165[336], 166[338], 169[346], 170[350], 170[352], 171[353], 172[357], 173[362], 174[363], 175[370], 182[386], 183[392], 185[394], 185[395], 189[397], 191[402], 196[414], 198[422], 202[433], 202[434], 203[436], 205[443], 208[450], 208[451], 211[459], 216[468], 216[469], 217[472], 218[747], 218[475], 218[476], 220[480], 221[481], 222[483], 222[484], 229[496], 230[499], 234[504], 234[506], 236[510], 236[513], 238[517], 239[522], 239[523], 240[524], 240[525], 241[527], 241[528], 242[529], 242[530], 243[532], 244[533], 246[538], 246[539], 246[540], 246[541], 259[2], 261, 264[15], 266[18], 267[22], 270[32], 270[33], 271[35], 272[36], 272[37], 275[42], 278[45], 279[46], 280[48], 280[50], 281[52], 282[55], 285[59], 286[61], 287[62], 287[64], 288[65], 289[66], 291[69], 295[78], 296[79], 297[80], 298[83], 300[87], 301[89], 302[89], 302[91], 303[95], 305[97], 306[98], 307[99], 309[104], 310[106], 312[112], 314[117], 314[118], 315[121], 316[122], 316[124], 317[125], 318[128], 319[129], 320[133], 321[136], 323[142], 323[144], 325[145], 331[167], 333[175], 333[176], 333[177], 333[178], 334[179], 335[181], 335[182], 335[183], 340[187], 340[189], 341[190], 342[194], 343[196], 344[199], 345[200], 347[205], 348[207], 349[213], 350[215], 351[217], 353[218], 354[221], 354[222], 354[223], 355[227], 362[2], 362[3], 363[5], 364[7], 368[15], 371[18], 372[21], 374[24], 375[27], 375[28], 376[29], 376[30], 376[31], 377[33], 377[34], 378[35], 380[37], 381[37], 397[12], 399[14], 400[16], 400[17], 401[19], 402[20], 404[22], 407[23], 409[28], 409[29], 410[30], 410[31], 413[33], 413[34], 417[38], 417[39], 417[40], 419[42], 420[44], 421[45], 423[48], 429[3], 429[4], 430[6], 431[7], 431[11], 432[11], 432[14], 433[18], 434[18], 435[19], 437[21], 437[23], 438[24], 439[25], 439[26], 440[28], 441[24], 442[24], 443[29], 445[24], 446[37], 446[38], 447[40], 449[42], 450[46], 450[47], 451[48], 452[48], 453[48], 454[50], 455[52], 457[57], 458[59], 459[61], 459[63], 460[65], 460[66], 462[66], 463[66], 465[70], 466[72], 469[78], 470[80], 471[81], 472[83], 473[84], 474[87], 475[91], 476[92], 481[96], 489[100], 490[113], 491[114], 491[116], 492[117], 493[119], 494[120], 494[121], 494[122], 495[122], 495[123], 497[126], 497[127], 498[128], 499[129], 500[132], 501[133], 502[134], 503[136], 506[145], 507[148], 509[149], 509[152], 511[145], 535[3]
Ferla, F., 144[281]
Fetkenheuer, B., 274[38]
Fischer, E., 179[377], 307[100]
Fischer, H., 63[20], 74[56], 90[101], 178[376], 192[405], 194[407], 457[58]
Flanders, F., 220
Fleming, R. H., 455[51]
Fleury, P., 174[364]
Flood, H., 22[66]
Fortini, V., 445[35]
Fränkel, E., 62[14], 133[245], 174[363], 309[104], 354[221], 354[222], 355[227]
Frehden, O., 263, 368[14]
Freiser, A., 20
Fresenius, C. R., 103[140], 131[241], 329[159], 395[11]
Fresenius, R., 207[449]
Freytag, H., 111[168], 313[114], 354[220]
Friederici, L., 133[246], 133[247]
Fries, K., 127[221], 182[382]
Fritz, H., 489[111]
Fukutomi, T., 314
Funakoshi, O., 93[110]

Gagliardi, E., 229[494]
Gaglio, G., 364[7]
Gallego, M., 127[225]
Galopin, R., 486[101]
Gambarin, F., 64[25]
Ganassini, D., 262[7]
Gangl, J., 101[137]
Gaptschenko, M. V., 80[72], 199[424]
Gastaldi, E., 276[43]
Gautier, J. A., 2[9]
Gehauf, B., 369[16]
Geilmann, W., 83[83]
Gelei, G., 64[22], 72[50]
Gentil, V., 107[152], 110[164], 120[202], 129[235],

134[249], 137[255], 158[314], 208[451], 222[484], 349[210], 437[23], 438[24], 440[28], 441[24], 442[24], 445[24], 459[63], 472[83], 473[84], 476[92]
German Patent, 171[356]
Gibson, R. C., 456[54]
Gillis, J., 6[14], 18[46], 94[117], 94[119], 105[147], 114[177], 114[179], 118[195], 119[199]
Gilmont, R., 39[4]
Glazunow, A., 21[61], 489[109]
Glemser, O., 85[90], 146[287], 362
Godbert, A. L., 216[471]
Goldenson, J., 369[16]
Goldschmidt, F., 94[116]
Goldstein, D., 147[289], 155[309], 157[312], 158[314], 189[397], 191[402], 208[450], 208[451], 267[22], 287[62], 295[78], 300[87], 323[142], 368[15], 380[37], 409[28], 423[14], 423[48], 437[23], 440[28], 446[37], 454[50], 456[55], 458[59], 495[123], 501[133], 534[1]
Gooch, F. A., 300[88]
Goppelsroeder, Fr., 1, 182[387], 266[19]
Gorbach, G., 36[3a]
Gordon, W. E., 325[147]
Gosh, A. N., 22[67]
Gotô, H., 23[70], 161[319], 161[320], 209[453], 440[27]
Gottlieb, A. L., 300[86]
Greathouse, L. H., 174[366]
Green, W. F., 142
Griess, P., 330[166]
Groschuff, E., 473[85]
Grünsteidl, E., 373[23]
Grünwald, J., 433[17]
Guareschi, I., 264[11]
Guagnini, O. A., 245[534]
Guerreiro, A., 414[37], 455[53]
Gutzeit, G., 21[60], 64[21], 109[157], 140[264], 216[470], 269[26], 280[47], 313[116], 384[6]. 486[101]
Gutzeit, M., 101[133]
Guyard, A., 127[222]
Gysin, M., 486[101]

Haber, F., 309[105], 366[12]
Hagen, S. K., 274[41]
Hahn, F. L., 4[11], 52[14], 80[75], 109[161], 109[162], 132[243], 153, 188[396], 224[487], 225, 262, 327[155], 330[163], 330[166], 331, 343[195], 408[25], 408[27], 430[8]
Hainberger, L., 320[133], 371[18]
Haken, H. L., 133[246]
Halberstadt, S., 119[201]
Halla, F., 261[5]
Hamburg, H., 161[324], 318, 450[46]
Hamilton, W. C., 8[20], 13[42], 54, 106[151]
Hammett, L., 191[399]
Hamy, A., 223[485]
Hauser, E., 383[1]
Hayes, W. G., 202[435]
Hecht, F., 205[442]
Heim, O., 263[9]
Hein, F., 162[327]
Heisig, G. B., 24[72], 141[268], 275[42], 313[115]
Heitman, A. H. C., 318[127]
Heller, K., 4[12], 60[7], 67[37], 96[124], 98[124], 164[333], 183[393], 389[7], 393[9], 505[140]
Henriques, R., 200[429]
Hermance, H. W., 9[26], 22, 50[11], 94[118], 116[189], 489[112]
Hermsdorf, A., 276[43]
Hernegger, F., 207[445]
Hess, W., 479[95]
Hesse, G., 505[142]
Heyn, E., 486[103], 486[105]
Heyne, G., 444[31]
Higgs, D. G., 429[2], 445[36], 448[41]
Hill, A. E., 471[82]
Hill, W. L., 191[401]
Hiller, T., 486[101]
Hinderer, W., 322[139]
Hinsberg, O., 349[211]
Hirsch, G., 12[37], 282[55]
Hirschel, W. N., 242[530]
Hoag, L. E., 207[447]
Hofmann, K. A., 141[273], 197[418], 281[51], 281[52], 303[94]
Holluta, J., 80[76]
Holzbecker, G., 195[411]
Holzer, H., 84[85], 127[231]
Hönel, F., 94[115], 96[125], 97[127], 98[125], 98[128], 119[199]
Hooreman, M., 153[301]
Hopf, P. P., 23[69]
Hoste, J., 91[102], 94[117], 94[119], 105[147], 114[177], 118[195], 127[228], 127[229], 349[212]
Houben, J., 340[188]
Hovorka, V., 167[343]
Howden, R., 170[349]
Huang, C. H., 263, 368[14]

Ibbotson, F., 170[349]
Iijima, Sh., 23[68]
Ilinsky, M., 144[278], 200[429]
Indovina, R., 266[21]
Isaenko, T. I., 489[110]
Ishidate, M., 239[521]

Jackson, C. L., 339[185]
Jacobi, K. R., 65[33], 170[351], 260[3], 298[81], 345[202], 431[12]
Jaeger, G., 224[487], 327[155], 330[166]
Jander, G., 207[449]
Jirkovský, R., 489[110]
Jockers, K., 505[141]
Johnson, M. W., 455[51]
Jolles, A., 336

Jónás, J., 119[196]
Jones, A. L., 165[337]
Jones, E. H., 202[435]
Jost, H., 243[531]
Jungreis, E., 443[29], 511[154], 535[3]

Kahl, S., 317[126]
Kamm, O., 230[497]
Kao, C. H., 139[260]
Kappelmacher, H., 47
Kapulitzas, H. J., 11[30], 88[99], 110[63], 117[191], 151[298], 152[298], 154[305], 281[54], 312[112], 319[132], 429[4], 446[38], 447[40]
Karlik, B., 205, 207[445]
Karrer, P., 166[338]
Kassner, J. L., 456[54]
Kempf, R., 10[29], 353[219]
Kersting, R., 328[158]
Kirchhof, F., 265[17]
Kirk, P. L., 6[17]
Kirmreutter, H., 281[52]
Klanfer, K., 323[144], 354[223], 477[94]
Klemm, W., 361[1]
Klingenberg, J., 204[439]
Kniga, A. G., 292[70]
Knowles, H., 125[214], 158[313]
Kocsis, E. A., 64[22], 72[50], 167[341], 209[452]
Kohn, M., 161[322], 161[323], 291[68], 354
Kolthoff, I. M., 68[40], 72[55], 128[232], 146[284], 146[288], 226, 228[492], 228[495], 267[23], 281[53], 285[58], 327[151], 505[139]
Komarowsky, A. S., 75[58], 93[111], 112[176], 119[197], 127[224], 127[226], 193[406], 194[408], 195[409], 212[460], 213[464], 225[488], 341[192], 342[193], 371[19], 381[38], 500[131]
Kondo, Y., 474[88], 482[98], 507[147]
König, O., 139[261]
Koppel, J., 116[186], 116[187]
Korenman, I. M., 193[406], 225[488], 236[511]
Korenmann, S. M., 20, 212[460]
Koslowsky, M. T., 284[57]
Krafft-Ebing, H., 205[442]
Kràl, H., 303[93]
Kraljic, I., 70[46]
Krauskopf, F. C., 115[183]
Krausz, G., 502[134]
Kreider, D. H., 300[88]
Krings, W., 261[4]
Krulla, R., 1[6]
Krumholz, P., 46[7], 60[7], 60[8], 67[37], 77[62], 94[115], 96[124], 96[125], 97[127], 98[124], 98[125], 98[128], 119[199], 135[251], 127[231], 141[265], 164[333], 170[352], 180[380], 183[393], 202[433], 203[436], 272[36], 272[37], 335[181], 337[182], 340[187], 389[7], 391[8], 431[9], 498[128], 499[129]
Krutman, A., 279[46]
Kucharský, 191[404]
Kuhla, G., 505[140]
Kuhn, R., 320[134], 363[6]
Kul'berg, L., 20, 84[87], 161[318], 212[462], 219[479], 226, 294[77], 356[229]
Kutschment, M. von, 234[507]
Kuzma, B., 350[216]
Kuznetzov, V. J., 20, 106[150], 112[171], 197[420]

Lacroix, S., 215[466]
Lal, M., 259
Lambert, J., 158, 267
Lang, R., 62[15], 63[17], 117[192]
Langenbeck, W., 356[230]
Langer, A., 70[45]
Lapin, L. N., 293[71], 356[231]
Lawson, A., 210[457]
Lecocq, E., 119[198]
Lecoq de Boisbaudran, P. E., 210[456]
Lederer, E., 23[71], 49[9], 505[142]
Lederer, F., 64[31]
Lederer, M., 23[71], 49[9], 505[142]
Lees, H., 287[62]
Léger, E., 76[60]
Lehmann, W., 127[221]
Lehné, M., 146[286]
Leimbach, G., 80[75], 430[8]
Leitmeier, H., 461[72], 469[78], 490[113], 491[116], 500[132]
Lenher, V., 139[260]
Lenz, W., 323[141]
Lerner, A., 313[115]
Lerner, M., 445
Le Strange, R., 23[71], 49[10]
Lettmayr, K., 93[106]
Leuchs, H., 109[158]
Levine, V. E., 349[210]
Lewin, S. Z., 164[331]
Liebermann, C., 105[148]
Lightfoot, J., 127[222]
Lindenbaum, S., 105[148]
Linz, A., 85[91], 104[144], 212[461]
Liokumovich, R. B., 219[479]
Livingstone, R. S., 68[40]
Llacer, A. J., 217[473]
Locher, M., 222[483]
Lochmann, G., 80[71]
Lohmann, W., 182[382]
Longacre, L. A., 149[294]
Lowe, A. J., 129[236], 159[315]
Lowe, C. S., 191[398]
Löwe, J., 266[19]
Lucena Conde, F., 210[455]
Luther, R., 142
Lutz, G., 85[90], 362
Lyons, E., 167[340]

Ma, T. S., 35[2]
Maass, Th., 117[190]
McCay, Th. C., 131[239]
Machatschki, F., 466[74]
Machek, F., 505[140]
McOmie, J. F. W., 49[9]
Mahr, C., 79[68], 80[69]
Makris, G. K., 238[519]
Malaprade, L., 302[92]
Malatesta, G., 128[232a]
Malissa, H., 237[514]
Malitzky, W. P., 284[57]
Malowan, S., 116[186]
Manchot, W., 326[150]
Mannheimer, W. A., 169[346], 240[524], 241[527], 244[533]
Marggraff, I., 325[146], 329[161]
Marins, J. E. R., 141[270], 238[517]
Marquevrol, M., 328[157]
Marshall, H., 173[360], 173[361]
Martini, A., 80[76]
Mate, M., 70[46]
Matwejew, L., 356[229]
Maxymovicz, W., 352, 377[32], 379[32]
Mazzucchelli, A., 509[150]
Meimberg, F., 227[491]
Meissner, H., 109[161]
Melnik, L., 502[135]
Merck, E., 444[33], 489[99]
Mertens, K. H., 232[500]
Metz, L., 303[95]
Metzger, O., 479[95]
Meyer, J., 123[210]
Meyer, R., 332[169]
Michael, I., 127[230]
Michaltschischin, G. T., 134[248]
Milden, H., 175[367]
Miller, C. C., 129[236], 158[315], 384[3]
Milton, R. F., 369[17]
Miranda, L. I., 8[23], 13[41], 94[122]
Moir, J., 276[43]
Moll, F., 476[93]
Molterer, H., 289[66]
Monier-Williams, G. W., 327[153]
Monnier, A., 322[138]
Monnier, R., 140[264]
Montequi, R., 93[108], 127[225]
Montignie, E., 117[190]
Moore, T. E., 158
Moser, L., 422[46]
Moufang, N., 69[41], 109[160]
Mozars, G., 401[21]
Mueller, J., 422[47]
Müller, E., 289[67], 380[36]
Müller, W., 332[169]
Muraour, H., 328[157]
Murray, M. J., 326
Mutnianski, M., 266[21]
Mylius, F., 139[259], 473[85], 477[86]

Nadler, G., 164[332]
Nageswara Rao, M., 210[454]
Nall, W. R., 56[22]
Nazarenko, V. A., 111[166]
Neuber, F., 2[8], 64[30], 72[54], 76[61], 84[88], 99[129], 104[143], 108[155]
Neunhofer, A., 79[67]
Neurath, F., 336
Newell, I. L., 109[158]
Nichols, E. L., 207[446]
Nichols, L. M., 236
Niessner, M., 21[58], 484[102], 486[101], 487[107], 488[108]
Nieuwenburg van, C. J., 383[2]
Noponen, G. E., 327[151]
Nordlander, B. W., 365[8]
Nováček, R., 169[347], 290, 319[129], 439[25], 455[52], 466[72], 469[72]
Noyes, A. A., 384[4]
Nutten, A. J., 126[219], 126[220], 181[381], 232[502]

Obach, E., 374[25]
Oberhoffer, P., 486[103]
Occleshaw, V. I., 64[23]
Oddo, G., 245[535]
Oelke, W. C., 294[75]
Oelschlager, W., 173[361]
Oesper, R. E., 204[439]
Okáč, A., 8[19], 20, 86[95a], 93[107], 153[303], 197[419], 332[171]
Okuma, K., 226[490]
Olszewski, W., 506
Onishi, H., 215[467]
Orelkin, B., 164[334]
Ostwald, W., 61[11], 509[151]
Oudemans, A. C., 81[78]
Overholser, L. G., 138[257], 204[438]
Owetschkin, W., 75[58]

Paal, C., 133[246], 133[247]
Palmer, E., 155[308]
Pannizon, G., 321[137]
Pantschenko, G. A., 115[185], 120[203], 127[223], 198[421], 206[444], 457[56], 459[64]
Papisch, J., 207[447]
Paris, R., 125[218]
Games, E., 509[150]
Paul, S. D., 99[130]
Pavelka, F., 8[24], 10[27], 75[59], 81[79], 201, 204[437], 221, 269[29], 429[5], 431[10]
Pavolini, T., 64[25]
Pawletta, A., 123[210]

Peabock, D., 343[195]
Pertusi, C., 276[43]
Pesez, M., 237[515], 241[526]
Petit, A., 241[526]
Petraschenj, W. J., 223[485]
Petrov, M. T., 444[32]
Piccard, J., 196[415]
Pickholz, S., 141[272], 176[371], 221[481], 298[83]
Pike, N. R., 196[417]
Pincus, H., 287[63], 329[160]
Pinto, C. M., 137[254]
Pirani, M., 474[89]
Pissarjewsky, L., 210[457]
Plank, E., 355[228]
Pollard, F. H., 49[9]
Polster, M., 153[303]
Poluektoff, N. S., 93[111], 112[173], 112[176], 115[181], 119[197], 127[224], 127[226], 156[310], 194[408], 195[409], 213[464], 214[465], 230[499], 341[192], 342[193], 346[203], 349[208], 350[214], 371[19], 500[131]
Porlezza, C., 77[63]
Postowsky, J. J., 347[206]
Potschinok, C. N., 334
Prat, L., 210[455]
Přibil, R., 64[26], 127[230], 191[404]
Prins, A., 447[39]
Procke, O., 233[503]
Prodinger, W., 175[368], 294[72]
Pzribram, E. M., 12[35]

Raacke, I. D., 333[177]
Raghawa Rao, Ph. S. V., 210[454]
Rahl, H., 479[95]
Raibmann, B., 196[416]
Rajmann, E., 60[8], 127[231], 141[265], 201[431], 202[433], 203[436], 267[22], 271[35], 272[36]
Rammelsberg, C., 463[68]
Raquet, D., 143[277]
Raschig, F., 20[54], 245[537], 304[96], 345[201]
Rauner, K. M., 486[104]
Rây, Pr., 87[97], 146[282], 153[304]
Rây, R. M., 87[97], 146[282], 153[304]
Razim, W. W., 12[39]
Reckendorfer, P., 47[8], 310[107]
Reich-Rohrwich, W., 205[442]
Reihlen, H., 166[338], 291[68]
Reith, F. L., 264[14]
Remy-Gennete, P., 224[486]
Reppmann, W., 103[139]
Reser, E., 475[90]
Retter, W., 162[327]
Richter, E., 323[141]
Richter, F., 449[43]
Richter, M., 496[124]
Riegler, E., 235[509]
Riehm, H., 327[154]
Rienäcker, G., 180[378]
Ritter, F., 261[5]
Robinson, F. A., 505[142]
Robinson, J. W., 335[180]
Rogers, L. B., 8[21], 94[121]
Rollet, A. P., 152[300]
Rosell, R. A., 295[78], 368[15], 495[123]
Rosenheim, A., 146[285], 196[416]
Rosenthaler, L., 312[111]
Rosin, J., 444[33], 489[99]
Rossi, L., 280[49], 497[125]
Rothberg, P., 131[241], 133[244]
Ruge, U., 356[230]
Ruigh, W. L., 226[490]
Runge, F., 1
Rupp, E., 104[142]
Rusconi, Y., 25
Rush, R. M., 8[21], 94[121]
Ruttner, T. F., 142
Ruzicka, E., 112[172]

Sachs, F., 349[211]
Salkowski, E., 381[38]
Sanchez, J. V., 180[380]
Sandell, E. B., 6[15], 21[62], 121[204], 130[237], 128[232], 180[379], 267[23], 428[1]
Sander, A., 319[131]
Sarver, L. A., 94[113], 149[290], 167[342]
Savoia, G., 199[428]
Sawicki, E., 349[212a]
Scagliarini, G., 303[94], 307
Schacherl, R., 305[97]
Schaeffer, A., 118[193], 196[414], 333[178], 449[42], 457[57]
Schäffer, A., 341[190], 397[12], 421[45]
Schales, O., 355[224]
Schantl, E., 138[388]
Schapowalenko, A. M., 262[6], 265[16], 276, 294[74], 299[85], 307[101], 332[172]
Scheel, K., 474[89]
Scheinziss, O. G., 80[72], 199[424]
Schenck, R. T. E., 35[2]
Schendel, G., 196[416]
Scheuing, G., 264[12], 311[110]
Schiff, H., 1[3]
Schiff, W., 180[378]
Schiller, W., 332[173]
Schlagdenhauffen, F., 223[485]
Schleicher, C., 484
Schlenk, W., 72[53]
Schlesinger, H. J., 164[331]
Schlumberger, M. E., 339[185]
Schmatolla, O., 109[161]
Schmidt, E., 69[43]
Schmidt, H., 80[77], 123[208]

Schmidt, J., 322[139]
Scholder, R., 133[246]
Scholey, R., 56[22]
Schönbein, C. F., 1[4], 142, 354[221], 363[4]
Schönn, J. L., 195[412]
Schoorl, N., 14[44], 505[139]
Scott, F., 56[23]
Schröder, E., 326[149]
Schulek, F., 264[13]
Schultz, G., 123[208]
Schwab, G., 22[67]
Schwab, G. M., 505[141]
Schwartz, F., 466[73], 466[75], 467[76], 468[77]
Schwarz, G., 62
Schwarz, R., 195[413]
Seboth, R., 261, 351[217]
Seiden, R., 164[331]
Sen, B., 64[28]
Senise, P., 80[74], 149[295], 310[108]
Sensi, G., 182[383]
Sexauer, W., 195[413]
Shapiro, M.Ya., 94[120]
Shemyakin, 212[463]
Sieverts, A., 276[43]
Siewert, M., 116[187]
Silverman, L., 170[348]
Simon, V., 64[26]
Singer, A., 72[54], 437[21]
Skalos, G., 12[38]
Skraup, Z., 1[5]
Slattery, M. K., 207[446]
Slawik, P., 164[334]
Slotta, K. H., 65[33], 170[351], 260[3], 298[81], 345[202], 431[12]
Sogani, N. C., 140[263]
Sommer, F., 287[63], 329[160]
Sommer, H., 373[22], 509[151]
Sommer, H. S., 491[115]
Sommer, L., 197[419]
Sorbini, F., 116[188]
Sorkin, F. P., 343[197]
Sottery, C., 191[399]
Spiegel, L., 117[190]
Stahl, W., 344[198]
Stark, C., 325[145], 375[27], 491[114], 534[1]
Steiger, B., 71[48], 142[274]
Steigmann, A., 22[65], 54[15], 141[267], 149[291]
Steinhauser, M., 240[525], 242[529], 246[540]
Stephen, W. I., 56[21], 126[219], 181[381]
Stěrba-Böhm, J., 115[184]
Stern, R., 2[7], 144[278], 172[357], 185[394], 205[443]
Steward, J., 234[505]
Stewart, O. J., 13[40]
Stitt, F., 365[9]
Stone, J., 227, 269[30], 366[10]
Storfer, E., 294[76]
Strain, H. H., 49[9], 505[142]
Suitsu, K., 226[490]
Sulfrian, A., 507[146]
Suter, H. A., 8[25], 13[41], 19[49], 73[55], 82[81], 217[472], 327[156], 328[156], 344[199], 409[29], 433[18], 434[18]
Suter, P. H., 327[156], 328[156]
Sutthoff, W., 327[151]
Sverdrup, H. U., 455[51]
Swarte, C. E., 115[183]
Szarvas, P., 199[426]
Szebellédy, L., 94[114], 119[196], 182[385], 341[191]

Tafel, J., 69[41], 109[160]
Taimni, I. K., 259
Talbot, B. E., 23[71], 49[10]
Tamchyna, J. V., 218[474], 281[52]
Tananaeff, N. A., 2, 21[63], 50, 58[1], 63[19], 66[36], 75[57], 80[70], 111[169], 115[185], 127[223], 128[233], 134[248], 135[250], 143[276], 161[317], 171[354], 176[372], 198[421], 206[444], 229[493], 235[508], 238[520], 262[6], 265[16], 299[85], 307[101], 332[172], 334, 457[56], 459[64], 464[69]
Tanay, I., 182[385]
Tanay, St. 341[191]
Tartarini, G., 96[123]
Tarugi, N., 116[188]
Teitelbaum, M., 346[204]
Testori, R., 182[383]
Theiss, M., 195[410]
Thomä, E., 309[105]
Thomas, P., 355[226]
Thrun, W. E., 191[400]
Tikhomirov, J. F., 274[40]
Tillmans, J., 175[367], 327[151]
Tjeenk Willink, H. D., 162[327]
Tneiss, M., 229[494]
Tödt, F., 505[138]
Tomanjuk, A. N., 143[276]
Tomimatsu, J., 365[9]
Tomula, E. S., 323[143]
Tornow, E., 69[43]
Treubert, F., 77[64]
Trtílek, J., 332[171]
Tschakirian, A., 112[175]
Tschugaeff, L., 149[296], 164[334]

Ubbelohde, A. R., 410
Uhlenhuth, R., 85[92], 87[96]
Uzel, R., 10[28], 149[292], 233[503], 246[539], 302[91], 350[215]
Uzumasa, Y., 115[183]

Van Atta, A., 63[18]
Van Dalen, E., 66[35], 99[131]
Vanino, L., 77[64]

Van Knorre, G., 122[206]
Van Nieuwenburg, C. J., 18[46], 24
Van 't Riet, B., 66[35]
Van Valkenburgh, H. B., 164[331], 463[67]
Velculescu, A., 11[31], 61[9], 62[13], 268[25]
Veluz, L., 237[515]
Verhoeff, J. A., 242[530]
Verma, M. R., 99[130]
Villiers, A., 313[113]
Virgili, J. F., 303[94]
Vitali, D., 298[82]
Vlácil, F., 167[343]
Vogel, H. W., 146[285]
Vokac, L., 363[5]
Vonesch, E. E., 245[534]
Vortmann, G., 413[35]
Vostřebal, J., 115[184]
Votoček, E., 311[109]
Vournasos, A. C., 366[11], 443[30]
Vrtiš, M., 302[92]

Wadlow, H. V., 50[11], 489[112]
Wagner, H., 281[51]
Walter, E., 166[338]
Wawilow, N. W., 80[73]
Webster, S. H., 159[316]
Weeldenburg, I. G., 150[297], 152[297]
Weidenfeld, L., 218[474], 315[121], 318, 323[144], 354[223], 374[24], 493[119], 503[136]
Weimarn, P. von, 509[151]
Weinland, R., 166[338]
Weisselberg, K., 225[489], 487[106]
Weisz, H., 54[17], 56[18], 56[19], 56[20], 56[23], 259[1], 263[10], 269[27], 283[56], 394[10], 423[49]
Welcher, F. J., 19[52], 65[32], 84[85], 125[217], 164[335], 171[356], 205[441], 449[45]
Wenger, P. E., 18[46], 22[64], 25, 105[145], 162[326]
Werner, A., 83[82], 125[215]
Werner, E. A., 155[306]
Werther, G., 123[209]
West, Ph. W., 1[1], 8[20], 13[42], 20[56], 25, 54, 64[28], 72[49], 80[74], 82[81], 90[100], 106[151], 124[211], 124[212], 124[213], 127[227], 131[239], 141[269], 149[294], 330[165], 335[180], 348[207], 377[34]
Weygand, F., 320[134], 363[6]
White, C. E., 191[398]
Wibaut, J., 92[103], 162[327]
Wicke, C., 309[105]
Wieland, H., 264[12], 311[110], 327[152], 347[206]
Willard, H. H., 174[366], 268[24], 345[201], 456[54]
Williams, H. A., 274[39]
Willits, C. O., 236
Wilson, C. L., 6[16], 6[17a], 266[20], 332[170]
Winckelmann, J., 12[36]
Winkler, L. W., 101[135]
Wittig, G., 232[501]
Wohlers, H. E., 218[477], 315[120]
Wölbling, H., 71[48], 142[274]
Wolf, H., 224[487]
Wollner, A., 146[283]
Wollner, R., 88[98], 143[275]

Yagoda, H., 21[59], 50[12], 218[477], 315[119]
Yoe, J. H., 138[257], 163[329], 165[337], 182[383], 191[401], 204[438], 219[478]
Young, D. W., 13[40]
Young, I. W., 234[505]
Young, Ph., 268[24], 345[201]

Zechmeister, L., 49[9], 505[142]
Zenghelis, C., 108[156], 136[252], 238[518]
Zermatten, H. L., 183[389], 191[403], 458[60]
Ziegler, J. H., 222[483]
Ziegler, M., 146[287]
Zimmermann, F., 210[458]
Zipser, A., 59[5]
Zmaczynski, E. W., 375[26]
Ztack, F. W., 183[393]
Zweig, G., 23[71], 49[10]

SUBJECT INDEX

A

Accutint indicator, 280
Acetate(s), interference in hydrogen peroxide test for titanium, 196
in sodium carbonate-phenolphthalein test for carbonic acid, 338
reaction with cobalt salts, 285
thiocyanate in presence of, 285
Acetic acid, detection of iron in, 453
"ring test" for nitrites, 332
Acetone, cobalt salts reaction with thiocyanates in presence of, 285
thiocyanates in presence of, 285
Acetylides in insoluble residue, 422
Acid(s), free, detection of mercury in presence of, 67
in solutions of aluminum salts, test for, 502
of copper and cobalt salts, detection, 502
substances insoluble in, identification, 407
Acid alizarin RC test for chromium, 172
Acid-insoluble residues, examination by spot tests, 384
Acid radicals, tests for, 258
Acid sulfide group of metals, tests for, 99
Adsorption as supplementary process, 7
Agate mortars, 34
Air bath from nickel crucible, 36 (Fig. 10)
Aldazines, formation, 239
Alizarin, reaction with chromium, iron, manganese, 186
with cobalt, copper, and iron salts, 187
with uranium, 186, 187
test for aluminum, 183
for aluminum in tanned leather, 478
for indium, 213
for mercury, 68
for zirconium, 200
Alizarin blue test for copper, 92
Alizarin paper test for aluminum in minerals, 490
Alizarin S, reaction with calcite, 466
as reagent for boric acid, 340
test for aluminum, 183
Alizarinsulfonic acid, *see also* Alizarin S
as substitute for methylene blue in test for titanium, 199
Alkali(s), in alkali cyanides, detection, 481
and alkali earths in coal ash and in paper, detection, 482
in glass, test for, 474
nickel dimethylglyoxime test for, 483
test for vulnerability of glass and siliceous products to attack by, 499
thallium nitrate test for, in presence of alkali sulfides, 307
Alkali bichromate, detection in presence of alkali monochromate, 496
Alkali chlorate as oxidizing agent for carbon, 379
solutions, activation, as test for osmium and ruthenium, 141
Alkali chloride, masking action in reactions of mercury ions, 279
Alkali copper pyrophosphate test for hydrazine, 242
for hydroxylamine, 246
Alkalicuprocyanide, demasking test for, 16
detection of copper in, 432
Alkali cyanides, carbonate in, detection, 498
fixed alkalis and ammonia in, detection, 481
interference in α-naphthylamine test for gold, 131
reaction with formaldehyde, 97
Alkali ferricyanides as oxidizing agent for diethylaniline, aromatic amines, and monoazo dyes, 178
Alkali fluoride, reaction with boric acid, 342
with ferric salts, 147
Alkali hydrosulfides, test for, in presence of alkali sulfides, 306
Alkali hydroxides, interference in pyrocatechol test for titanium, 196
Alkali hypobromites, reaction with manganese salts, 173

Alkali hypoiodite test for magnesium, 223
Alkali iodide(s), in extraction test for bismuth, 80
interference in diphenylamine test for nitric acid, 327
oxidation to iodine, 327
test for platinum, 134
Alkali mercury chloride test for hydrocyanic acid, 279
Alkali mercury thiocyanates, reaction with cobalt and zinc salts, 180
Alkali metals, in silicates, test for, 464
tests for, 223
zinc in presence of, 182
Alkali molybdates, detection of nitrate in, 495
Alkali monochromates, test for bichromate in presence of, 496
Alkaline earth metals, zinc in presence of, 182
Alkali molybdates, detection of nitrate in, 495
Alkaline earths, nickel dimethylglyoxime test for, 483
Alkaline earth sulfates, behavior toward sodium rhodizonate, 223
detection of lead sulfate in, 409
in presence of each other, 223
Alkaline solutions, detection of silica in, 499
Alkali nitrite(s), fusion with tungsten steels, 418
reaction with molybdenum, tungsten or vanadium metals or oxides, 367
Alkali palladium dimethylglyoxime, demasking, 278, 423
test for hydrocyanic acid, 278
Alkali perchlorates, dry heating of, 300
Alkali periodates, reaction with manganous salts, 174, 301
Alkali peroxides, interference in thallous hydroxide test for hypohalogenites, 295
Alkali pyrophosphate, masking action in diphenylcarbazone test for mercury, 65
Alkali resistance of glasses, 499
Alkali salts, detection of iron in, 453
of lead in, 435
Alkali stannite test for telluric and tellurous acids, 350
Alkali sulfate(s), interference in starch-iodine reaction, 266
in titanium-malachite green test for tungsten, 122
Alkali sulfides, alkali in presence of, thallium nitrate test for, 307
alkali hydrosulfides in presence of, test for, 306
interference in dinitrobenzene test for Rongalite, 320
in nickel hydroxide test for sulfurous acid, 310
Alkali sulfite(s), reaction with molybdenum trioxide and carbon, 380
as substituent for hydroxylamine in test for germanium, 115
Alkali thiocyanates, reaction with hydrogen peroxide, 355
with iodates, 299
Alkali tungstate, detection of nitrate in, 495
test for vanadium, 127
Alkali vanadate, detection of nitrogen in, 495
Alloys, antimony in, detection, 442
beryllium in, detection, 457
bismuth in, detection, 438
cobalt in, detection, 448
copper in, detection, 429
gold in, detection, 445
Heller method of analyzing, 393
lead in, detection, 433
manganese in, detection, 177
metals in, 395, 396
nickel in, detection, 445
ring oven analysis of, 56
silver in, detection, 429
tests for, 361
thallium in, detection, 437
tin in, detection, 439
Alum, interference in starch-iodine reaction, 266
Alumina, detection, 454
by alizarin, 69
differentiation from beryllium oxide and other metal oxides, 454
filter paper impregnated with, 22
fusion with potassium persulfate, 413
identification of, 511
iron in, detection, 413, 450
Aluminon test for aluminum, 191
Aluminum, alizarin S test for, 183
alizarin test for, 185
in alloys, test for, 394
aluminon test for, 191
beryllium in presence of iron and, 193
calcium in presence of, 535
chrome fast pure blue B test for, 189
detection by hydroxytriphenylmethane dyes, 13
in presence of beryllium, cobalt, copper, iron, manganese, nickel, zinc or zirconium, 190
of chromium, 190

in water, 454
indium in presence of, test for, 213
interference in ferric-thiosulfate reaction for copper, 81
metallic, detection by reduction of *o*-dinitrobenzene, 363
in minerals, test with alizarin paper, 490
in mixtures, test for, 388, 391, 392
morin test for, 182
pontachrome blue-black R test for, 191
in presence of cobalt, test for, 188
of magnesium, test for, 188
of metals of ammonium sulfide group, test for, 184, 186
of nickel, test for, 188
of other ions, test for, 184
of other metals, identification limits, 187
of uranium, capillary separation test for, 187
quinalizarin test for, 188
reaction with potassium ferrocyanide, 186
separation of gallium from, 215
solubility in nonoxidizing acids, 362
summary of tests, 516, 532
in tanned leather, test for, 478
test for magnesium in presence of, 228
zinc in presence of, test for, 178
Aluminum activation as test for mercury, 69
Aluminum-chelate compounds, formation, 13
Aluminum chloride, zinc in presence of, 181
Aluminum hydroxide, test for water in, 510
Aluminum oxide in insoluble residue, 413
Aluminum salts, free acids and basic compounds in, detection, 502
interference in alkali hypoiodite test for magnesium, 224
in tests for beryllium, 195
in zinc uranyl acetate test for sodium, 229
reaction with hydroxytriphenylmethane dyes, 189
with 1, 2, 7-trihydroxyanthraquinone, 111
as solvent for calcium fluoride, 270
zinc in presence of, test for, 179
zirconium in presence of, test for, 201
Amalgams, detection, 365
o-Aminophenols, test for tin, 112
Aminosulfonic acid, *see also* Sulfamic acid
reaction with nitrites, 330
removal of nitrites by, 330
Ammonia, in alkali cyanides, detection, 481
and hydrogen peroxide test for cerium, 210
interference in phenoxithine test for palladium, 140
liberation as test for ammonium ions, 236
manganese sulfate and hydrogen peroxide test for, 239
manganese sulfate and silver nitrate test for, 237
Nessler reaction as test for, 238, 384
p-nitrobenzenediazonium chloride test for, 235
in presence of lead, mercury, and silver salts, 238
red litmus paper test for, 236
silver nitrate and formaldehyde test for, 238
silver nitrate and tannin test for, 238
test for hydroxylamine, 246
for hypohalogenites, 295
through formation of zinc oxinate, 237
in water, test for, 506
Ammoniacal silver nitrate, reaction with cobalt, iron, and manganese salts, 211
test for cerium, 210
Ammonium anthranilate, reaction with cerium salts, 212
Ammonium bromide test for copper, 94
Ammonium carbonate group of metals, tests for, 216
Ammonium chloride, reaction with thiocyanates, 285
zinc in presence of, 181
Ammonium ferrocyanide test for calcium, 220
Ammonium hypophosphite test for reducible metals, 463
Ammonium ions, ammonia liberation as test for, 236
interference in phenoxithine test for palladium, 140
in mixtures, test for, 384
summary of tests, 516
Ammonium mercury thiocyanate in zinc salts as reagent for copper, 93
Ammonium molybdate, formation of molybdenum blue from, 380
reaction with hydrogen peroxide, 417
Ammonium molybdate-bendizine test for germanium, 112
for phosphoric acid, 333

for silicic acid, **335**
Ammonium naphthoate, reaction with cerium salts, **212**
Ammonium permolybdate, formation of, **417**
Ammonium pertungstate, formation of, **417**
Ammonium pervanadate, formation of, **417**
Ammonium phosphomolybdate, test for tin, **108**
Ammonium polysulfide and sulfite, test for tellurites and tellurates, **352**
Ammonium salicylate, reaction with cerium salts, **212**
Ammonium salts, in chemicals, detection, **483**
in filter papers, detection, **483**
interference in alkali hypoiodite test for magnesium, **224**
in benzoin oxime test for copper, **83**
in dipicrylamine test for thallium, **157**
in iron periodate test for lithium, **233**
in molybdenum trioxide test for free carbon, **380**
in quinalizarin test for beryllium, **193**
for magnesium, **225**
in test for copper, **86**
in test for hydrocyanic acid, **279**
in test for hydroxylamine, **246**
in test for nitrates in alkali salts, **495**
in test for potassium, **232**
in test for water, **510**
in zinc uranyl acetate test for sodium, **229**
reaction with dipicrylamine, **231**
with molybdenum trioxide, **380**
with sodium cobaltinitrite, **230**
tests for, **235**
Ammonium sulfide group of metals, cobalt in presence of, **143**
indium in presence of, **213**
tests for, **144**
for cations of, **390**
Ammonium thiocyanate, and acetone test for cobalt, **146**
reaction with copper, ferric, and nickel salts, **146**, **147**
and *o*-tolidine test for copper, **84**
Ammonium tungstate, reaction with hydrogen peroxide, **417**
Ammonium vanadate, reaction with hydrogen peroxide, **417**
Amphibolite, phosphates in, detection, **501**
Analysis, of alloys, Heller method, **393**
chromatographic, **505**
of mixtures, Gutzeit method, **384**
Heller and Krumholz method, **389**
Krumholz method, **391**
Andesite, phosphates in, detection, **501**
Anglesite, lead in, detection, **434**
Anhydrite, distinction from gypsum, **471**, **510**
identification, **470**
reaction with rhodamine B, **472**
Aniline, in mercuric chloride test for tin, **111**
oxidation to aniline black by vanadium salts, **127**
reduction of mercuric salts in presence of, **66**
Anions, of acid radicals, tests for, **258**
of metallo acids, tests for, **57**
oxidizing, interference in test for copper, **93**
in mixtures, analysis, **383**, **404**
summary of tests, **527**
Anthraquinone-1-azo-4-dimethylaniline hydrochloride as reagent for metal salts, **111**
Anthraquinones, detection, **340**
Antimonic acid in metastannic acid, detection, **104**
Antimony, in alloys, tests for, **394**, **442**
detection of gallium in presence of, **215**
dithizone test for zinc in presence of, **180**
free, detection of free arsenic in presence of, **367**
gold in presence of, detection, **130**
hydrogen-flame luminescence test for, **79**, **107**
interference in morin test for tin, **440**
9-methyl-2,3,7-trihydroxy-6-fluorone test for, **105**
in minerals, tests for, **442**, **490**
in mixtures, tests for, **387**, **390**
phosphomolybdic acid test for, **104**
in presence of bismuth, detection, **443**
in presence of large amounts of tin, detection, **106**
reduction test for, **103**
rhodamine B test for, **105**, **442**
simultaneous test for tin and, **441**
summary of tests, **516**
tellurium in presence of, detection, **353**
thallium in presence of, detection, **160**
zirconium in presence of, detection, **203**
Antimony compounds, reaction with cacotheline, **109**
Antimony hydride, reaction with silver nitrate, **101**

Antimony oxide, in insoluble residue, 419
in metastannic acid, detection, 104
Antimony pentoxide, detection of nitrate in, 494
diphenylbenzidine test for, 419
interference in diphenylamine test for lead chromate, 415
Rhodamine B test for, 419
Antimony salts, detection of arsenic in presence of, 101
of cadmium in presence of, 96
of tin in, 108
interference in pyrrole test for selenious acid, 346
in test for beryllium, 195
reaction with anthraquinone-1-azo-4-dimethylaniline hydrochloride, 112
with cacotheline, 109
with *p*-dimethylaminoazophenylarsonic acid, 203
with diphenylbenzidine, 327
with xanthone dyestuffs, 105
reduction, of phosphomolybdates by, 156
by stannites, 350
Antimony sulfide paper, 51
Antimony trichloride, reduction of phosphomolybdates by, 108
Antipyrine-tannin test for titanium, 199
Apatite, differentiation from chlorapatite, 502
in rocks, detection, 501
Apparatus, special, for stocking of versatile spot test laboratory, 31
Application of spot tests in preliminary examination, 394
Aqua regia extract, Gutzeit test in, 386
Aqueous solution or suspension, tests on, 400
Aragonite, distinction from calcite, 469
Arsenates, arsenites in presence of, test for, 102
interference in hydroxylamine test for germanium, 115
in potassium thiocyanate test for iron, 164
in test for free arsenic, 367
in test for palladium, 137
in zirconium-alizarin test for fluorides, 269
in zirconium azoarsenate test for fluorides, 272
silver nitrate test for, 102
Arsenic, detection in minerals, 443
in presence of antimony and mercury salts, 101
dithizone test for zinc in presence of, 180
free, detection by formation of arsenic hydride, 366
in presence of free antimony, 367
summary of tests, 532
germanium in presence of, 113
Gutzeit test for, 101, 103
interference in ferric thiosulfate test for copper, 81
in flame color test for tin, 110
kairine test for, 103
in minerals, test with silver nitrate paper, 490
in mixtures, test for, 387, 390, 391
pentavalent, precipitation by hydrogen sulfide, 100
platinum in presence of, detection, 133
in presence of mercury, detection, 100
stannous chloride test for, 99
summary of tests, 517
tellurium in presence of, detection, 353
test by reduction to arsenic hydride and decomposition with silver nitrate or gold chloride, 101
Arsenic acid, catalytic reduction by potassium iodide, 100
interference in molybdate-benzidine test for germanium, 113
phosphates in presence of, 333
phosphoric acid in presence of, detection, 334
reaction with benzidine, 335
with iodides, 102
with molybdates in acid solution, 333
reduction to arsenious acid, 100
Arsenic compounds, in antimony reduction test, 103
reaction with cacotheline, 109
with sodium formate, 443
Arsenic hydride (*see also* Arsine), formation of, 101
as test for arsenic, 366
gold chloride test for, 101
silver nitrate test for, 101
Arsenic pentoxide, test for arsenic trioxide in presence of, 103
Arsenic trioxide in presence of arsenic pentoxide, test for, 103
Arsenimolybdic acid, formation, 333
formation of benzidine blue by, 333
Arsenious acid, reaction with iodine, 102
reduction of arsenic acid to, 100
reduction to arsine, 103
Arsenites, interference in test for free arsenic, 367
in test for palladium, 137
in presence of arsenates, test for, 102

reaction with molybdenum trioxide and carbon, **380**
Arsine (*see also* Arsenic hydride), formation of, **443**
reduction of arsenious acid to, **103**
Asbestos, magnesium in, detection, **461**
Ascorbic acid, reduction of selenites by, **349**
test for gold, **131**
Ashes, detection of calcium in, **468**
Asymmetric diphenylhydrazine test for selenious acid, **347**
Atacamite, detection of chlorine in, **502**
Atomizer head for spraying reagents, 52 (Fig. 32)
Auric chloride test for hydrogen peroxide, **355**
Auric chloride and palladium chloride, reaction with rubidium and thallium salts, **235**
test for cesium, 235
test for thallium, **161**
Auric ions, interference in mercury iodide test for palladium, **138**
in alkali iodide test for platinum, **135**
in pyrrole test for selenious acid, **346**
in test for copper, **93**
Aurintricarboxylic acid, ammonium salt of, *see* Aluminon
Autoradiography and ring oven method, **56**
Azides, carbonates in presence of, **338**
detection in presence of nitrite-consuming materials, **287**
1,8-Aziminonaphthalene, formation, **331**
Azo dyes, adsorptive property of magnesium hydroxide for, 225

B

Barite, calcium sulfate in, detection, **471**
differentiation from celestite, **473**
Barium, detection of calcium in presence of, **223**, **535**
lead sulfate precipitation as test for, **218**
magnesium in presence of, 228
in mixtures, test for, **388**
nitro-3-hydroxybenzoic acid test for, **219**
precipitation of barium sulfate in presence of permanganate as test for, **218**
in presence of strontium, detection, 216, **219**
rhodizonic acid test for, 216
sodium rhodizonate test for, 74, 216, **476**
strontium in presence of, **216**, 220
summary of tests, **517**
tetrahydroxyquinone test for, **219**
Barium carbonate, differentiation from strontium carbonate, **473**
and phenolphthalein test for sulfuric acid, **314**
Barium rhodizonate test for sulfuric acid, **313**
Barium salts, color reaction with glyoxal-bis(2-hydroxyanil), **534**
interference in cerium salt reduction test for iodides, **268**
in rhodizonate test for calcium, **222**
in zinc uranyl acetate test for sodium, **229**
reaction with rhodizonic acid, **313**
with sodium rhodizonate, 220, **313**
Barium sulfate, detection in pigments, paper ash, etc., **476**
of gypsum in presence of, **471**
differentiation from strontium sulfate, **473**
precipitation in presence of permanganate as test for barium, **218**
for sulfuric acid, **314**
reduction to sulfide, **315**
test for soluble sulfates, **314**
Barnes dropping bottle, 32 (Fig. 3)
Basic compounds in solutions of aluminum salts, detection, **502**
Basic sulfide group of metals, tests for, **58**
Bauxite, testing for iron, **450**
Benzaldehyde, reaction with hydrazine, **239**
Benzene as solvent for sulfur, **376**
Benzidine, and ammonium molybdate test for phosphoric acid, **333**
for silicic acid, **335**
2,7-diaminofluorene as substitute for, **322**
oxidation by cerium, cobalt, silver, and thallium salts, **176**
by chromates, **176**
by ferricyanides, **176**
products of, 72
reaction with arsenic and phosphoric acids, **336**
with cobalt, lead and manganese salts, **155**
with chromates, **171**
with copper salts, **276**
with manganese, **298**
with prussic acid, **284**
with vanadates, **172**
test for cerium, **211**

for chromium, 171
for copper, 93, 276
for gold, 128, 131
for lead, 72, 416
for manganese, 8, 175, 416
for nitrites, 332
for persulfuric acid, 322
for thallium, 155
Benzidine acetate, reaction with osmium, 143
test for ferricyanic acid, 291
Benzidine acetate-copper acetate solution, preparation , 276
test for hydrocyanic acid, 276
Benzidine blue, formation, 237, 239, 501
by arsenomolybdic acid, 333
by bismuth oxides, 437
by cerium salts, 211
by chromates, 171
by cobalt salts, 172
by copper salts, 172, 276
by germanium salts, 112
by gold salts, 128
by halides, 284
by hydrated manganese dioxide, 175
by lead peroxide, 72
by lead salts, 172
by metallic oxides, 310
by oxidizing agents, 276, 322
by phosphomolybdic acid, 333
by potassium permanganate, 284
by silicomolybdic acid, 333, 336, 419
by silver salts, 172
by thallic salts, 155
in presence of chlorides, bromides, and iodides, 276, 284
of cyanides, 276
Benzoin, detection by dinitrobenzene, 280
reaction with sulfur, 13, 375, 491
Benzoinoxime, reaction with cupric ions, 82
test for vanadium, 124
Benzoin rearrangement, catalysis of, as test for cyanides, 23, 280
2-Benzylpyridine, reaction with stannous chloride, sulfur dioxide, and sulfites, 313
test for tin, 111
Berlin blue, *see* Ferric ferrocyanide
Beryllium, chromeazurole test for, 195
detection of aluminum in presence of, 190
in minerals, ores, and alloys, 457
2-(*o*-hydroxyphenyl)benzthiazole test for, 195
interference in alizarin or quinalizarin test for indium, 213
morin test for, 191, 458
p-nitrobenzeneazoorcinol test for, 194
in presence of iron and aluminum, 193
of magnesium, detection, 193, 194
quinalizarin test for, 192, 457
summary of tests, 517
Beryllium oxide, differentiation from alumina, 454
reaction with Chrome Fast Pure Blue B, 454
Beryllium salts, for isolation of insoluble, fluorides, 421
reaction with calcium fluoride, 421
with morin, 183
with quinalizarin, 225
as solvent for calcium fluoride, 270
zirconium in presence of, 201
Bettendorf test for arsenic, 100
Bicarbonate in presence of carbonate, test for, 338
Bichromate, interference in tetrabase oxidation test for ferricyanic acid, 293
in presence of alkali monochromates, detection, 496
4,4′-Bis(dimethylamino)-thiobenzophenone, *see also* Thio-Michler's ketone
test for chlorine, 495, 502
for free halogens, 368
Bismuth, in alloys, tests for, 394, 438
cinchonine and potassium iodide test for, 76
detection of copper in presence of, 85
extraction test for, 80
hydrogen-flame luminescence test for, 79
interference in rhodamine B test for antimony, 442
in minerals, test with potassium iodide paper, 490
in mixtures, tests for, 386, 389, 391
potassium chromium thiocyanate test for, 79
potassium manganocyanide test for, 80
in presence of copper, detection, 76, 77, 78
of lead, detection, 76, 77
of mercury, detection, 76, 77, 78
quinoline and potassium iodide test for, 80
reaction with rhodamine B, 107
with thiourea, 349
with 1, 2, 7-trihydroxyanthraquinone, 111
rubeanic acid test for, 90
stannite test for, 77, 80, 438
summary of tests, 517
thioacetamide test for, 80

zinc in presence of, **182**
Bismuth chloride, reaction with rhodamine B, **106**
Bismuth compounds, lead in, detection, **437**
Bismuth iodide, extraction by isobutyl ketone, as test for bismuth, **80**
Bismuth nitrate, in nickel hydroxide test for sulfurous acid, **310**
Bismuth salts, benzidine test for, 72
cadmium in presence of, detection, **96**
interference in benzidine test for lead, 72
in rhodamine B test for uranium, 209
in test for beryllium, 195
reaction with rhodamine B, 208
reduction by stannites, 350
tests for, 80
Bismuth subgallate, lead in, detection, **437**
Bismuth subsalicylate, lead in, detection, **437**
Bismuth subtannate, lead in, detection, **437**
Bisulfites, in presence of sulfites, **313**
Bogen's universal indicator, 505
Boracite, detection of chlorine in, 502
Borate in mixtures, test for, 385
Borax as solvent for calcium fluoride, 270
Boric acid, carminic acid test for, **343**
cochineal fluorescence test for, 119
hydroxyanthraquinone test for, **340**
interference in etching test for fluorides, 274
methyl alcohol flame test for, **344**
methyl borate and alkali fluoride test for, **342**
p-nitrobenzeneazochromotropic acid test for, **341**
in presence of oxidizing agents or fluorides, **342**
reaction with hydroxyanthraquinones, 114
reagents for detecting, 340
summary of tests, 527
test by increase in acidity with organic polyhydroxy compounds, **343**
tincture of curcuma test for, **339**
Borides in insoluble residue, **419**
Boron, in enamel, detection, 500
in glass, test for, **473**
Bräunerite, distinction from magnesite, **465**
Broeggerite, uranium in, detection, 457
Bromates, in hypophosphorous acid test for iodic acid, 299
reaction with diphenylamine, **327**
with manganous sulfate and sulfuric acid, **298**
with pyrogallol, **300**
Bromates, *see* Halogenates
Bromic acid, manganous sulfate and sulfuric acid test for, **298**
summary of tests, 527
Bromides, in chlorides and iodides, detection of small amounts, **264**
cyanides in presence of, **277**
detection in presence of iodides, **263**
fluorescein test for, **262**
interference in cerium salt reduction test for iodides, **268**
in ferrous sulfate-sulfuric acid test for nitric acid, **326**
in hydrogen peroxide test for titanium, **196**
for vanadium, **123**
in prussic acid test for thiocyanates, **284**
in silicomolybdic acid test for fluorides, **273**
in mixtures, test for, **384**
oxidation to bromine, 259, 261, 262
potassium bromide-cupric bromide test for, **265**
in presence of chlorides, fluorescein test for, **263**
of iodides, **263**, **268**
reaction with potassium permanganate, **284**
thiocyanate in presence of, 285
Bromine, detection, **262**
by formation of polymethine dyes, **369**
interference in chromyl chloride test for hydrochloric acid, **261**
oxidation of bromides to, 259, 261, 262
summary of tests, 532
test for hydrobromic acid by oxidation to, **262**
Bromine water, reaction with dimethylglyoxime, **153**
Bromophenol blue as indicator of hydrogen-ion concentration, **504**
Bromothymol blue, as indicator of hydrogen-ion concentration, **504**
as indicator in test for boric acid, **343**
Brown coal, reaction with molybdenum trioxide, **380**
Brown glass, detection of manganese in, **456**
Brucine test for nitric acid, **328**
Butyraldehyde, test for cobalt, **149**

C

Cacotheline, formula of, 109
  reaction for tin salts, 68
  test for stannous chloride, 421
    for tin, 109
Cadmium, in alloys, test for, 394
  calcium in presence of, 535
  copper in, test for, 430
  copper in presence of, detection, 85
  di-β-naphthylcarbazone test for, 98
  di-p-nitrophenylcarbazide test for, 96
  di-p-nitrophenylcarbazone test for, 96
  diphenylcarbazide test for, 99
  dithizone test for zinc in presence of, 181
  ferrous dipyridyl iodide test for, 8, 94
  interference in sodium rhodizonate test for barium, 476
  magnesium in presence of, 227
  in minerals, test with hydrogen sulfide paper, 490
  in mixtures, tests for, 387, 390, 393
  p-nitrodiazoaminoazobenzene test for, 99
  in presence of antimony, bismuth, lead, mercury, silver, thallium, and tin salts, detection, 95, 96
    of copper, detection, 95, 96, 97, 98
    of zinc, detection, 96
  reaction with sodium rhodizonate, 74
  sodium sulfide test for, 99
  summary of tests, 517
Cadmium chloride, reaction with halogenates or nitrates, 301
  test for perchlorates, 300
  zinc in presence of, 181
Cadmium hydroxide, reaction with di-β-naphthylcarbazone, 98
  with di-p-nitrophenylcarbazide, 96
  with p-nitrobenzeneazoresorcinol, 226
Cadmium iodide, dissociation of, 267
  in oxidation test for iodides, 267
    for thallium, 158
Cadmium red, selenium in, 492
Cadmium salts, color reaction with glyoxal-bis(2-hydroxyanil), 534
  interference in zinc uranyl acetate test for sodium, 229
  reaction with anthraquinone-1-azo-4-dimethylaniline hydrochloride, 112
Cadmium stannoiodide test for lead, 75
Cadmium sulfide paper, 51
Calcite, in caustic magnesite, 468
  decomposition by hydrochloric acid, 466
  detection of iron in, 451
  distinction from aragonite, 469
    from dolomite, 466
    of witherite and strontianite in presence of, 473
  reaction with alizarin S, 466
Calcium, ammonium ferrocyanide test for, 220
  detection in ashes, dry residues, magnesite, dolomite, silicates, etc., 468
    in glass, 469
    in presence of aluminum, barium, cadmium, iron, lead, magnesium or strontium, 535
    in silicates, 460
  dihydroxytartaric acid osazone test for, 12, 221
  glyoxal-bis(2-hydroxyanil) test for, 534
  magnesium in presence of, 228
  in mixtures, test for, 388
  reaction with potassium ferrocyanide, 220
  sodium rhodizonate test for, 222, 460, 469, 470, 471
  summary of tests, 517
  test for, in presence of barium and strontium, 223
  in water, test for, 506
Calcium carbonate, reaction with p-nitrobenzeneazo-α-naphthol, 226
Calcium fluoride, detection, 271
  in insoluble residue, 420
  preparation of copper-free water by, 431
  reaction with beryllium salts, 421
  as reagent-purifying adsorbent, 81
  solution of, 270
Calcium hydroxide, reaction with p-nitrobenzeneazo-α-naphthol, 226, 228
  formation of, 222, 460, 469, 470, 471
Calcium salts, color reaction with glyoxal-bis(2-hydroxyanil), 534
  interference in zinc uranyl acetate test for sodium, 229
  reaction with nitro-3-hydroxybenzoic acid, 220
Calcium silicate, reaction with zinc chloride, 460, 469
Calcium sulfate, in barite, detection, 471
  identification, 470
  mercuric nitrate test for, 470
  presence as filler for paper, detection, 471
Capillary effects, 1
Capillary pipette(s), 55 (Fig. 34)
  cleaning of, 44
Capillary separation, of ammine salts as test for nickel, 154

for detection of copper in presence of nickel and cobalt, 88
Carbide carbon, detection, 379
Carbides, carbon in, 422
in insoluble residue, 422
Carbon, free, in insoluble residue, 422
molybdenum trioxide test for, 380
test by oxidation, 379
potassium iodate test for, 381
reaction with sodium and nitrogen, 422
summary of tests, 532
Carbonate(s), in alkali cyanides, sulfites, and sulfides, detection, 498
detection of iron in, 451
of sulfamic acid in presence of, 326
extract, Gutzeit test in, 385
interference in pyrocatechol test for titanium, 196
in mixtures, test for, 384
in presence of azides, cyanides, fluorides, nitrites, sulfides, sulfites, and thiosulfates, 338
sodium carbonate and phenolphthalein test for, 337
test for bicarbonate in presence of, 338
Carbon dioxide, in water, test for, 506
Carbon disulfide, detection of traces of sulfur in, 374
as solvent, 373, 374, 376
Carbonic acid, sodium carbonate and phenolphthalein test for, 337, 498
summary of tests, 527
Carbon monoxide, catalytic reduction of phosphomolybdates as test for palladium and platinum, 135
Carminic acid, reaction with zirconium, 204
as reagent for lead salts, 75
test for boric acid, 343
Carriers in spot test analysis, 54
Cassiterite, tin in, 439
Catalysis, of iodine-azide reaction as test for thiosulfuric acid, 318
of methylene blue reduction test for free selenium, 377
test for selenious acid, 348
Catalytic acceleration, of ferric-thiosulfate reaction as test for copper, 80
of reduction of tin salts as test for mercury, 68
Catalytic reactions, prospects in spot test analysis, 23
utilization in qualitative analysis, 20
Catalytic reduction of cerium salts as test for iodides, 267
Catalytic tests for platinum metals, 131
Catalyzed oxidation of phenolphthalin as test for hydrogen peroxide, 355
Cations in mixtures, systematic analysis, 383, 389
summary of tests, 516
Caustic alkalis, detection of silica in, 499
Celestite, differentiation from barite, 473
Cellophane as medium for spot tests, 12
Cellulose, interference in nickel hydroxide test for persulfates, 324
test for permanganate in presence of chromate, 8, 10, 344
Cementation, 362
Centrifuge tubes, cleaning of, 44
Centrifuging, 42
Ceric salts, interference in test for mercury, 66
Cerium, ammoniacal silver nitrate test for, 210
benzidine test for, 211
hydrogen-flame luminiscence test for, 79
hydrogen peroxide and ammonia test for, 210
interference in phosphomolybdic acid test for vanadium, 127
malachite green test for, 212
manganese in presence of, test for, 176
in mixture of rare earths, 210, 212
of zirconium and thorium salts, 212
phosphomolybdic acid test for, 212
in presence of chromates, cobalt, copper, iron, manganese, silver, and thallium salts, 211
summary of tests, 518
Cerium perhydroxide, formation, 210
Cerium salts, catalytic reduction of, as test for iodides, 267
for silver, 62
interference in benzidine test for lead, 72
in hydrogen peroxide test for vanadium, 123
in *o*-tolidine test for copper, 84
lake formation with quinalizarin, 193
oxidation of benzidine by, 176
reaction with ammonium anthranilate, 212
with ammonium naphthoate, 212
with ammonium salicylate, 212
with quinalizarin, 225
Cerussite, lead in, detection, 434
Cesium, auric chloride and palladium chloride test for, 235
gold platinibromide test for, 234
indium in presence of, 213
lithium in presence of, 233
potassium bismuth iodide test for, 235

summary of tests, 518
Cesium chloride, lithium in presence of, **234**
Cesium salts, interference in test for potassium, 232
reaction with dipicrylamine, 231
Chalk, magnesium in, **466**
Charcoal, carbon in, 422
detection of alkali in ash of, **483**
reaction with molybdenum trioxide, **380**
Charring of organic materials, **316**
Chelate compounds, formation of colored, 13
Chemicals, detection of chloride in fine, **495**
of sulfate in inorganic fine, **494**
testing for purity of, **483**
Chemiluminescence, produced by oxidizing luminol, 356
Chemistry of specific, selective and sensitive reactions, 20
Chloramine-T as oxidizing agent, 262
redox reaction with tetrabase, **511**, **535**
Chlorapatite, detection of chlorine in, **502**
differentiation from apatite, **502**
Chlorate solutions, activation, as test for osmium and ruthenium, **141**
Chlorates, in hypophosphorous acid test for iodic acid, **299**
interference in chromyl chloride test for hydrochloric acid, 261
reaction with brucine, **329**
with diphenylamine, **327**
with manganese sulfate, 297
with manganous sulfate, **301**
Chlorates, *see* Halogenates
Chloric acid, manganous sulfate and phosphoric acid test for, 297
summary of tests, 527
Chloride(s), bromides in presence of, fluorescein test for, **263**
detection, **264**
cyanides in presence of, detection, **277**
detection in fine chemicals, **495**
in presence of other halides, 259
in ink writing, test for, **480**
interference in cerium salt reduction test for iodides, 268
in diphenylcarbazide test for palladium, **141**
in ferrous sulfate-sulfuric acid test for nitric acid, **326**
in hydrogen peroxide test for titanium, **196**
in oxidation test for manganese, **173**
in prussic acid test for thiocyanates, 284
in silicomolybdic acid test for fluorides, **273**
iodides in presence of, **268**
mercury in presence of, detection, **67**
in mixtures, test for, **384**
reaction with potassium permanganate, **284**
thiocyanate in presence of, test for, **285**
Chlorinated water, detection of, **370**
Chlorine, 4,4′-bis(dimethylamino)-thiobenzophenone test for, **495**, **502**
detection, 1
in concentrated sulfuric acid, **496**
by formation of polymethine dyes, **369**
in minerals and rocks, **501**
by thio-Michler's ketone, **300**
development of, in chromyl chloride reaction, **260**
summary of tests, 532
Chlorobromoamine acid test for zirconium, **204**
Chloroplatinic acid, interference in *o*-tolidine test for copper, **84**
Choice of procedure, **28**
Chromate-diphenylcarbazide reaction, discharge by peroxides, **356**
Chromate and potassium iodide paper test for silver in minerals, **490**
Chromates, cellulose test for permanganate in presence of, **344**
cerium in presence of, **211**
conversion of chromium salts to, **167**, **170**, **173**
detection of hypohalogenites in presence of, **296**
in presence of hypohalogenites, **295**
diphenylcarbazide test for, **345**
formation of benzidine blue by, **171**
interference in diphenylcarbazone test for mercury, 65
in hydrogen peroxide test for vanadium, **123**
in morin test for zirconium, **202**
in α-naphthylamine test for gold, **131**
in phosphomolybdic acid test for vanadium, 127
inferricyanide and diethylaniline test for zinc, **178**
in tetrabase oxidation test for ferricyanic acid, **293**
oxidation of benzidine by, **176**
of chromium salts to, **167**, **170**, **173**
in presence of permanganates, **345**
reaction with benzidine, **171**, **211**, **322**
with carbazone, 119

with 3,3′-dimethylnaphthidine, 126
with diphenylamine, 327
with diphenylcarbazide, 415, 416
with ferrous sulfate and sulfuric acid, 326
with hydrogen peroxide, 196
with silver nitrate, 102
reduction by hydrogen peroxide, 356
test for, by formation of silver or lead chromate, 172
test for permanganates in presence of, 344
Chromatographic analysis applied to inorganic analysis, 505
Chromatographic spot tests, 23
Chromatography, analytical separation by, 49
spot reactions in, 428
Chromeazurole test for beryllium, 195
Chrome Fast Pure Blue B, reaction with beryllium oxide, 454
test for alumina, 454
test for aluminum, 189, 454
Chrome iron ore, detection of iron in, 462
Chromic acid, diphenylcarbazide test for, 345
summary of tests, 527
Chromite, detection of chromium in, 414
Chromium, acid alizarin RC test for, 172
in alloys, tests for, 394
aluminum in presence of, 187
ammonium hypophosphite test for, 464
benzidine test for, 171
in chromite, detection, 414
detection of aluminum in presence of, 190
2, 7-diaminodiphenylene oxide test for, 171
diphenylcarbazide test for, 167, 260, 414, 455
indium in presence of, 214
in ink writing, test for, 480
interference in alizarin test for aluminum, 186
in stannous chloride test for arsenic, 100
in test for titanium, 199
iron in presence of, 164
lake formation with alizarin, 186
magnesium in presence of, 228
metallic, interference in test for free molybdenum, tungsten or vanadium, 367
in mixtures, tests for, 387, 390, 392
in presence of cobalt, copper, manganese, and nickel, 169
of cobalt, copper, lead, and silver salts, 172
of manganese, 172
of mercury, 170
of molybdenum, 170
pyrrole test for, 173
in rocks, detection, 455
in steel, detection, 455
strychnine test for, 173
summary of tests, 518
in tanned leather, test for, 478
titanium in presence of, 199
zinc in presence of, 178
Chromium chloride, in insoluble residue, 414
Chromium hydroxide, filter paper impregnated with, 23
Chromium nitrate, zinc in presence of, 181
Chromium oxide in insoluble residue, 414
Chromium salts, conversion into chromates, 167, 170, 173
interference in periodate test for manganese, 175
in potassium thiocyanate test for iron, 164
in test for beryllium, 195
oxidation to chromates, 167, 170, 173
reaction with 1, 2, 7-trihydroxyanthraquinone, 111
as solvent for calcium fluoride, 270
Chromochlorite, chromium in, 455
Chromodiopside, chromium in, 455
Chromotrope 2 B (*see also p*-Nitrobenzene-azochromotropic acid), as reagent in boric acid test, 341
Chromotropic acid, reaction with iron, mercury, silver, and uranyl salts, 198
reaction with metal salts, 198
reaction with titanium salts, 197
and stannous chloride test for silver, 64
test for mercury salts, 72
for nitrates, 330
for Rongalite, 321
for titanium, 197, 198, 459
Chromotropic acid dioxime, test for cobalt, 149
Chromous ions, reaction with diphenylcarbazide, 167
Chromyl chloride, formation as test for hydrochloric acid, 260
Chrysean solution, reaction with nitrites, 332
Chrysoberyl, detection of beryllium in, 458
Cinchonine and potassium iodide test for bismuth, 76
Clausthalite, selenium in, 492

Cleanliness, importance of, 28
Coagulation as supplementary process, 7
Coal, carbon in, 422
Coal ash, detection of alkali and alkali earths in, 482
Cobalt, in alloys, test for, 394, 448
aluminum in presence of, 187, 188, 190
ammonium hypophosphite test for, 464
ammonium thiocyanate and acetone test for, 146
butyraldehyde test for, 149
chromium in presence of, 169
chromotropic acid dioxime test for, 149
copper in presence of, 88, 89
detection of, in presence of ions of ammonium sulfide group, 143
in presence of nickel, 149
diacetylmonoxime *p*-nitrophenylhydrazone test for, 147
hydrogen peroxide test for, 149
indium in presence of, 214
interference in alizarin blue test for copper, 92
in ammonium-mercury thiocyanate test for zinc, 448
in isonitrosobenzoylmethane test for iron, 167
in phosphomolybdic acid test for vanadium, 127
in stannous chloride test for arsenic, 100
iron in presence of, 164
magnesium in presence of, 227
manganese in presence of, 174, 176
in minerals, test with potassium nitrite and potassium thiocyanate paper, 490
in mixtures, tests for, 388, 390, 393
nickel in presence of, 150, 151, 152, 154, 446
in nickel salts, tests for, 147
nitrosonaphthol test for, 17, 144, 448
2-nitroso-1-naphthol-4-sulfonic acid test for, 94, 149
in presence of copper, 144, 145
of iron, 144, 145, 147
of uranium, 144, 145
reaction with dimethylglyoxime, 151
with potassium ferrocyanide, 178
rubeanic acid test for, 88, 146
sodium azide and *o*-tolidine test for, 149
sodium pentacyanopiperidine ferroate test for, 149
sodium thiosulfate test for, 143
solubility in nonoxidizing acids, 362
summary of tests, 518
tervalent, identification by *o*-tolidine, 311
Cobalt alloys, detection of iron in, 451
Cobalt azide, induced oxidation of, as test for sulfites and sulfurous acid, 310
Cobalt^{II} azide, oxidation to Cobalt^{III} azide, 149
Cobalt ions, catalysis of oxidation of manganous ions by, 149
Cobalt hydroxide, reaction with *p*-nitrobenzeneazoresorcinol, 226
Cobalt and iron salts, nickel in presence of, 150, 152
Cobalt mercury thiocyanate test for zinc, 180
Cobaltous hydroxide, reaction with malachite green, 212
Cobalt salts, as catalyst in oxidation test for manganese, 174
cerium in presence of, 211
color reaction with glyoxal-bis(2-hydroxyanil), 531
formation of benzidine blue by, 172
free acids in solutions of, detection, 502
interference in ammonium mercury thiocyanate test for copper, 93
in benzidine test for lead, 72
in cerium salt reduction test for iodides, 268
in dipyridyl test for iron, 162
in ferricyanide and diethylaniline test for zinc, 178
in potassium thiocyanate test for iron, 164
in zinc uranyl acetate test for sodium, 229
nickel in, detection, 446
nickel-free, preparation, 447
oxidation of benzidine by, 176
presence in 1, 2-diaminoanthraquinone-3-sulfonic acid test for copper, 87
reaction with alizarin, 184
with alkali mercury thiocyanates, 180
with ammoniacal silver nitrate, 211
with benzidine, 155, 172, 211
with 1,2-diaminoanthraquinone-3-sulfonic acid, 86
with dimethylglyoxime, 150, 165
with diphenylcarbazide, 168
with nitro-3-hydroxybenzoic acid, 220
with 2-nitroso-1-naphthol-4-sulfonic acid, 167
with rubeanic acid, 142, 153, 154
with sodium azide, 310
with thiocyanates in presence of acetone, 285
with thioglycolic acid, 162

test for thiocyanic acid, 285
Cobalt zincate, formation of, 182
Cochineal fluorescence test, for boric acid, 119
for molybdenum, 119
for uranyl ions, 210
Collectors, of traces, 428
use in preliminary concentration, 21
Colloidal gold solutions, qualitative analysis of, 128
Color intensity of precipitate, 7
Color of metals and alloys, 361
Colored ions, interference in ascorbic acid test for gold, 131
in cuproin test for cuprous ions, 92
in hydrogen peroxide test for cobalt, 149
in methylene blue and hydrazine test for molybdenum, 118
Colorimetric analyses in ring oven method, 56
Color lakes, 86
Color reactions, organic reagents in, 19
Columbite, detection of tin in, 440
of iron in, 462
of manganese in, 456
Columbium, *see* Niobium
Complexing, incorrect synonym for masking, 17
Complexone III, *see* Ethylenediaminetetraacetic acid
Compounds, insoluble, conversion to water-soluble iodides, 407
Concentrating of solutions, aluminum block for, 35 (Fig. 6)
in centrifuge tube, 35 (Fig. 7)
Concentration, influence in spot reactions, 21
Concentration limit, 4
Concentration sensitivity, 4
Conditioning of tests, 5, 18
Confined spot test papers for quantitative analysis, 21
Coordination compounds, 19
Copper, acceleration of ferric-thiosulfate reaction as test for, 80
acid potassium fluoride and *o*-hydroxyphenylfluorone test for, 94
alizarin blue test for, 92
in alloys, detection, 394, 429
ammonium bromide test for, 94
benzidine test for, 93
bismuth in presence of, detection, 77, 78
cadmium in presence of, detection, 95, 96, 97, 98
in cadmium, test for, 430
as catalyst, in oxidation test for manganese, 174
in sodium thiosulfate reduction of iron, 206
catalytic test for, 80, 430
chromium in presence of, 169
cobalt in presence of, 144, 145
detection in alkali cuprocyanide solutions, 432
of aluminum in presence of, 190
of iron in, 451
1,2-diaminoanthraquinone-3-sulfonic acid test for, 85
diphenylcarbazone test for, 94
distinction from gold, 129
dithizone test for, 90
in foodstuffs, detection, 430
hydrobromic acid test for, 94
o-hydroxyphenylfluorone test for, 94
8-hydroxyquinoline test for, 93
interference in test for iron, 167
in thiourea test for selenious acid, 349
iron in presence of, 164
magnesium in presence of, 227
manganese in presence of, 177
mercury in presence of, 67
in metallic zinc, test for, 430
in minerals, test with ferrocyanide paper, 490
in mixtures, tests for, 386, 389, 393
nickel in presence of, 151, 154
2-nitroso-1-naphthol-4-sulfonic acid test for, 94
in pharmaceutical products, 430
phenetidine chloride and hydrogen peroxide test for, 94
phosphomolybdic acid test for, 84
precipitation by antimony sulfide paper, 51
in presence of bismuth, cadmium, lead, and mercury, 85
of cobalt and nickel, detection by capillary separation, 88
of gold, nickel, and silver, 430
of other heavy metals, 90
of hydrochloric acid, 83
reaction with dinitrodiphenylcarbazide, 97
with potassium ferrocyanide, 178
with rubeanic acid, 87, 146, 429, 488
resorcinol test for, 94
salicylaldoxime test for, 83
in sections of metals, developing process for, 488
sodium diethyldithiocarbamate test for, 432

solubility in nonoxidizing acids, 362
summary of tests, 519
test with ammonium mercury thiocyanate in presence of zinc salts, 93
test by ferric thiocyanate formation, 430
o-tolidine and ammonium thiocyanate test for, 84
traces, detection in nickel, 432
uranium in presence of iron and, 206
in water, detection, 431, 506
Copper acetate-benzidine acetate, preparation of solution, 276
test for hydrocyanic acid, 276
Copper benzoinoxime salt, polymer structure of, 82
Copper cupferronate, formation, 431
Copper-free water, preparation, 431
Copper hydroxide, formation, 97
Copper ions, inference in test for cobalt, 149
Copper-manganese solutions as reagent for telluric acid, 351
Copper rubeanate, polymer structure of, 88
Copper salicylaldoxime, formation of, 83, 243
Copper salts, cerium in presence of, 211
complex formation, as test for periodic acid, 302
as test for telluric and tellurous acids, 350
formation of benzidine blue by, 172, 276
free acids in solutions of, detection, 502
interference in cerium salt reduction test for iodides, 268
in chromeazurole test for beryllium, 195
in dithizone test for zinc, 180
in kairine test for arsenic, 103
in potassium thiocyanate test for iron, 164
in zinc uranyl acetate test for sodium, 229
reaction with alizarin, 184
with ammonium thiocyanate, 147
with benzidine, 172, 211, 276
with cacotheline, 68
with dimethylglyoxime, 150, 165
with diphenylcarbazide, 169
with disodium-1,2-dihydroxybenzene-3,5-disulfonate, 166
with nitrosonaphthol, 144
with 2-nitroso-1-naphthol-4-sulfonic acid, 149, 167
with potassium ferrocyanide, 205
with rubeanic acid, 142, 153
reduction, to copper by hypophosphorous acid, 349
of phosphomolybdates by, 156
by stannites, 350
and salicylaldehyde test for hydroxylamine, 243
Copper sulfate, as reagent for telluric acid, 351
solution, detection of free sulfuric acid in, 503
in wood preservatives, detection, 477
Copper sulfide test for hydrocyanic acid, 277
Copper thiocyanate as reagent for silver, 64
Corrosion in machinery metal, detection, 487
Co-solutes, role in spot reactions, 7, 15
Criminalistic investigations, detection of traces of lead in, 437
of traces of thallium in, 438
spot tests in, 23
Crypto fusion reactions, 234
Crystal water, detection of, 510
Cupferron, detection of traces of copper by, 431
Cupric bromide-potassium bromide test for bromides, 265
Cupric ferricyanide test for cyanides, 280
for hydroxylamine, 246
Cupric ferro- and ferri-cyanides, ammoniacal solutions, test for hydrazine by reduction of, 240
Cupric ions, interference in morin test for zirconium, 202
in test for mercury, 71
reaction with benzoinoxime, 82
with potassium iodide, 145
Cupric salts, interference in test for zinc, 178
Cuproin test for cuprous salts, 91
Cupro-tetraiodomercuriate, formation, 366
Cuprous dimethylaminobenzylidenerhodanine, formation, 242, 246
Cuprous ferricyanide, formation, 246
Cuprous halides, interference in tests for free metals and alloys, 362
Cuprous iodide test for mercuric salts, 65
for mercury vapor, 366
Cuprous ions, zinc diethyldithiocarbamate test for, 94
Cuprous salts, cuproin test for, 91
dimethylaminobenzylidenerhodanine reaction with, 59, 93
Cuprous thiocyanate, formation, 230, 242, 246

Curcuma, tincture of, test for boric acid, **339**
Curcumin, conversion to rosocyanine, **339**
reaction with metal compounds, **339**
Cyanates, reaction with hydroxylamine hydrochloride, 285
Cyanic acid, hydroxyurea test for, 285
summary of tests, 527
Cyanides, carbonates in presence of, **338**
copper acetate-benzidine acetate test for, **276**
cupric ferricyanide test for, 280
detection, by copper sulfide, 277
through catalysis of benzoin rearrangement, **23**
dissociation of complex, 97
effect on the starch-iodine reaction, **266**
interference in azide test for hydrazoic acid, **286**
in cerium salt reduction test for iodides, **268**
in 1,2-diaminoanthraquinone-3-sulfonic acid test for copper, **86**
in phenoxithine test for palladium, **140**
in sodium carbonate-phenolphthalein test for carbonic acid, **338**
in test for ammonium salts, **236**
in mixtures, test for, **384**, **385**
nickel in presence of, **153**
picric acid and sodium carbonate test for, 280
in presence of ferro- and ferri-cyanides, iodides, bromides, chlorides, and thiocyanates, 277
of sulfides or sulfites, 277
prussic acid and starch-iodine test for, 280
reaction with ferrous sulfate and sulfuric acid, **326**
with formaldehyde, 97
test for, by catalysis of benzoin rearrangement, 280
Cyanogen, detection by the copper acetate-benzidine acetate reaction, 276
in mercuric cyanide, detection, 281
Cyanogen iodide test for mercuric cyanide, 281
Cyanogen radical, capillary detection of ions containing, **294**
Cylindrite, tin in, 439

D

Decrepitation test for carbon in coal, 422
Demasking of alkali palladium dimethylglyoxime, **278**, **423**
of bivalent iron, 70
of ferrocyanide, 72
of test ions, **16**
Demasking test for ferrocyanic acid, 288
Developer for silver tests, **61**
Developing process for detecting heterogeneities in metals, minerals, etc., **484**
for detecting copper in sections of metals, **488**
Development of spot test analysis, 1
Diacetyldioxime, *see also* Dimethylglyoxime
formation, 242
Diacetylmonoxime and nickel salt test for hydroxylamine, 242
Diacetylmonoxime *p*-nitrophenylhydrazone, reaction with alkali carbonate, **147**
with alkali hydroxide, **147**
with ammonia, **148**
with cobalt ions, **148**
with magnesium oxide, 148
test for cobalt, **147**
1,2-Diaminoanthraquinone-3-sulfonic acid as reagent for copper, **85**
Diaminobenzidine test for vanadium, 127
2,7-Diaminodiphenylene oxide, as substitute for benzidine, **322**
test for chromium, **171**
2,7-Diaminofluorene reagent as substitute for benzidine, **322**
o-Dianisidine molybdate and hydrazine test for phosphoric acid, **335**
reaction with silicates, **335**
Diazine green S(K) test for tin, 111
Diazotization, **332**
Dicyanogen, detection by hydrolysis, **370**
formation, **370**
from ferri- and ferrocyanides, **412**
as test for mercuric cyanide and oxycyanide, **282**
oxine test for, **371**
polymerization of, 412
summary of tests, 532
reaction with iron, **412**
Diethylaniline, oxidation by alkali ferricyanides, **178**
Diethylaniline-ferricyanide test for zinc, 178
Differential diffusion, analytical separation by, **49**
1,2-Dihydroxyanthraquinone sulfonic acid, *see* Alizarin S
1,8-Dihydroxynaphthalene-3,6-disulfonic acid, *see* Chromotropic acid
Dihydroxytartaric acid osazone, test for calcium, 12, 221

for differentiation between tap and distilled water, 222
Dilution limit, 4
p-Dimethylaminoazophenylarsonic acid, reaction with antimony, gold, and thorium salts, 203
with molybdates and tungstates, 203
with tin salts, 204
with titanium salts, 203
test for zirconium, 202, 271
p-Dimethylaminobenzaldehyde test for hydrazine, 240
p-Dimethylaminobenzylidenerhodanine test for cuprous salts, 59
for gold, 59, 127
for mercury, 59, 67, 128
for palladium, 59, 128, 141
for platinum, 59
for silver, 59, 67, 128, 411, 429
p-Dimethylaminobenzylidenethiobarbituric acid test for silver, 64
4-Dimethylamino-1,2-phenylenediamine, reaction with selenious acid, 349
test for selenious acid, 349
Dimethylglyoxime, and potassium nickelcyanide, test for silver, 64
reaction with bromine water, 153
with cobalt salts, 150, 151, 165
with copper salts, 150, 165
with ferrous salts, 111, 150
with gold salts, 150
with palladium, 150
with persulfate, 153
reagent papers impregnated with, 152
test for iron, 164
for nickel, 9, 149, 445
in minerals, 490
in presence of oxidizing agents, 150
in silicate rocks, 463
3, 3′-Dimethylnaphthidine and alkali ferricyanide test for zinc, 181
3, 3′-Dimethylnaphthidine test for vanadium, 126
4-Dimethylthio-1,2-phenylenediamine, reaction with selenious acid, 349
Di-β-naphthylcarbazone, reaction with molybdates, 119
as reagent for cadmium, 98
Dinitrobenzene, color reaction with inorganic reductants, 320
detection of benzoin by, 280
reaction with hydrazine, 242
with hydroxylamine, 242
with sulfides, 242
with sodium hydrosulfite or with Rongalite, 320, 321
reduction of, as test for metallic aluminum, lead, tin and zinc, 363
for magnesium, 364
test for hydrazine, 242
for hydroxylamine, 246
Di-*p*-nitrophenylcarbazide test for cadmium, 96
Di-*p*-nitrophenylcarbazone test for cadmium, 96
Dioxan and dipicrylamine test for chemically or adsorptively bound water, 510
Diphenylamine, reaction with oxidizing agents, 327
test for lead chromate, 415
for nitric acid, 327
Diphenylbenzidine test for antimony pentoxide, 419
for nitric acid, 327
for vanadium, 127
Diphenylcarbadiazone, oxidation of diphenylcarbazide to, 167
Diphenylcarbazide, and mercuric cyanide test for palladium, 141
oxidation to diphenylcarbazone and diphenylcarbadiazone, 167
reaction with magnesium compounds, 465
with mercury salts, 170
with molybdates, 119
with molybdenum trioxide, 415
with vanadates, 170
test for cadmium, 99
for chromates, 345, 415, 416
for chromium, 167, 260, 414, 455
for mercury, 414
for molybdates, 170, 414
Diphenylcarbazone, oxidation of diphenylcarbazide, to, 167
reaction with chromous ions, 167
as reagent for copper, 94
for mercury, 64
test for germanium, 114
Diphenylhydrazine, test for selenious acid, 347, 492
Diphenyline test for tungsten, 122
Diphenylthiocarbazone, *see* Dithizone
Dipicrylamine, and dioxan test for chemically or adsorptively bound water, 510
reaction with ammonium, cesium, rubidium, and thallium salts, 231
test for potassium, 230, 462
for thallium, 156
α,α′-Dipyridyl, ferrous salts of, 15
α,α′-phenanthroline as substitute for, 162

and potassium ferrocyanide, reaction with mercury, 72
with palladium, 64, 72, 141
with silver, 64, 72
test for mercuric cyanide and oxycyanide, 282
for mercury, 72
for palladium, 141
for silver, 64
for silver halides, 410
reaction with ferrous salts, 288
with metallic ions, 162
with molybdates, 119
as reagent for ferrous salts, 111
test for iron, 161, 450, 462, 466
interference of zinc and magnesium ions in, 453
and thioglycolic acid, test for iron, 449, 451, 452, 453
2,2'-Diquinolyl, *see* Cuproin
Discernibility of reaction product, 6
Disodium-1,2-dihydroxybenzene-3,5-disulfonate, reaction with copper and titanium salts, 166
test for iron, 165
Dispersion of precipitate, 7
Dissolved materials, testing for, 40
Distillation tube, 48 (Fig. 30)
Distilled water, differentiation from tap water, 507
Dithioglycolic acid, oxidation of thioglycolic acid to, 107
Dithiol, reduction of stannic salts by, 107
test for tin, 107
Dithionous acid, *see* Hyposulfurous acid
Dithio-oxamide, *see* Rubeanic acid
Dithizone, test for copper, 90
for lead, 74
for mercury, 71
for silver, 63
for zinc, 178
Dolomite, in caustic magnesite, 468
detection of calcium in, 468
of iron in, 451
differentiation from calcite, 466
from magnesite, 465
Dropper pipet, 38 (Fig. 13)
cleaning of, 44
Dropping bottle, 32 (Fig. 1)
Drops, addition and control of, 37
Dry reagents in spot test analysis, 54
Dry residues, detection of calcium in, 468
Drying, as preliminary operation, 35
of impregnated filter paper, 52
of spots, 40
Dyes, adsorptive property of magnesium hydroxide on organic, 225
Dyestuff inks, composition, 479
Dysprosium, hydrogen-flame luminescence test for, 79

E

Earth oxides, test for, 473
Electrodes for electrolytic deposition of traces of metal, 506
Electrographic apparatus, 50 (Fig. 31)
Electrographic detection, 21
analytical separation by, 49
Electrographic methods for detecting metals, 489
Electrographic sampling and ring oven method, 56
Electrolytic depositions, 1
Elements, free, summary of tests, 532
tests for, 361
Enamel(s), detection of boron in, 500
of lead in, 433
Eosin, formation, 262, 297, 368
Equipment, requirements, 28
special, for stocking of versatile spot test laboratory, 31
Erythrosin, formation, 368
Etching test for detecting hydrofluoric acid, 274
for lithium, 234
for sulfur segregations in iron and steel, 486
Ethylenediamine, masking action in rubeanic acid test for copper, 88, 90
Ethylenediaminetetraacetic acid, masking action in morin test for beryllium, 192
test for silver, 64
Eudialite, detection of chlorine in, 502
Evaporating as preliminary operation, 35
Evaporation and ignition residues, testing, 508
Experimental conditions, 7
importance of, 28
Extraction pipet, 45
Extractions, liquid-liquid, 44
liquid-solid, 44

F

Factory controls, spot tests in, 23
Ferric chloride, and phenanthroline, test for uranium, 210
reaction with thiocyanate, 373
zinc in presence of, 181
Ferric ferricyanide, reduction to Turnbull's blue as test for sulfurous acid, 313

test for free metals and alloys, **363**
for hydrogen peroxide, **354**
Ferric ferrocyanide, detection, **476**
formation, **363**
Ferric hydroxide, effect on benzidine test for manganese, **176**
Ferric ions, interference in test for mercury, **71**
reduction by thioglycolic acid, **449, 451, 452, 453**
Ferric oxide, 8-hydroxyquinoline test for, **413**
in insoluble residue, **412**
Ferric oxinate, formation, **413, 450**
Ferric salts, dimethylglyoxime test for ferrous salts in presence of, **165**
ferrous salts in presence of, detection, **162**
and formaldehyde, test for hydroxylamine, **245**
interference in benzidine test for copper, **93**
in *a*-benzoinoxime test for vanadium, **125**
in cuproin test for cuprous salts, **91**
in cuprous iodide test for mercury, **66**
in morin test for titanium, **199**
for zirconium, **202**
in rhodamine B test for uranium, **209**
quercetin or quercitrin test for, **167**
reaction with alkali fluorides, **147**
with ammonium thiocyanate, **147**
with diphenylamine, **327**
with hydroxamic acids, **245**
with nitrosonaphthol, **144**
with potassium ferrocyanide, **205**
with rhodamine B, **208**
with thiosulfates, **80, 143**
reduction as test for tin, **111**
by thioglycolic acid, **162**
rubeanic acid test for, **90**
Ferric thiocyanate, decomposition by mercuric chloride, sodium fluoride, or sodium thiosulfate, **290**
Ferric thiocyanate test for copper, **431**
for thiocyanates in presence of iodides, **283**
Ferricyanic acid (hydroferricyanic acid), benzidine test for, **291**
induced oxidation of tetrabase test for, **293**
phenolphthalin test for, **292**
summary of tests, **528**
trithiourea cuprous chloride test for, **294**
Ferricyanide(s), cyanides in presence of, **277**
and diethylaniline test for zinc, **178**
detection of hypohalogenites in presence of, **296**
ferrocyanides in presence of, **288**
formation of dicyanogen from, **412**
in insoluble residue, **411**
interference in benzidine test for nitric acid, **327**
oxidation of benzidine by, **176**
and *p*-phenetidine test for zinc, **182**
in presence of ferrocyanide, **291**
of other oxidizing materials, test for, **292**
reaction with benzidine, **322**
with safranin, **295**
with silver nitrate, **102**
thallous sulfate test for, **294**
Ferrocyanic acid (hydroferrocyanic acid), demasking test for, **288**
iron chloride test for, **289**
summary of tests, **529**
thiocyanate in presence of, **283**
uranyl acetate test for, **287**
Ferrocyanide(s), catalytic decomposition of, as test for mercury, **70**
cyanides in presence of, **277**
demasking of, **72**
test for, **70**
ferricyanide in presence of, **291**
formation of dicyanogen from, **412**
in insoluble residue, **411**
in mixtures, test for, **386**
in presence of ferricyanides, **288**
of halide ions, **288**
of thiocyanates and iodides, **290**
Prussian blue test for, **9, 289**
reaction with ferrous sulfate and sulfuric acid, **326**
with mercuri amidochloride, **289**
with mercuric ions, **70**
titanium tetrachloride test for, **294**
Ferrocyanide paper test for copper, iron, and zinc in minerals, **490**
Ferrous dipyridyl iodide as reagent for cadmium, **8, 94**
Ferrous ferricyanide, formation, **363**
Ferrous salts of α,α'-dipyridyl and phenanthroline, behavior on filter paper, **15**
Ferrous salt of ferricyanic acid, *see* Turnbull's blue
Ferrous salts, identification by dimethylglyoxime, **111**
by α,α'-dipyridyl, **111**
interference in test for zinc, **178**
in presence of ferric salts, detection, **162**

dimethylglyoxime test for, 165
reaction with dimethylglyoxime, 150
with α,α'-dipyridyl, 288
with molybdenum trioxide and carbon, 380
with 2-nitroso-1-naphthol-4-sulfonic acid, 149
Ferrous silicate, reaction with zinc chloride, 462
Ferrous sulfate and sulfuric acid test for nitric acid, 326
Filter paper(s), capillary action in ring oven method, 55
detection of ammonium salts in, 483
differentiation of qualitative and quantitative, 483
influence of type of, 8
as medium for spot reactions, 1, 7
recommended for spot test use, 54
role of surface in spot reactions, 8
Filtering pipet, 41 (Fig. 15)
Fine chemicals, *see* Chemicals
Fixing baths, qualitative analysis of, 129
Flame color test for boric acid, 344
for tin, 109, 421
Flaws in iron, detection, 489
Fleck, appearance of, 2
Flotation, analytical, 15, 44
Flour, test for adulteration with gypsum, 471
Fluoboric acid, formation, 270
Fluomolybdic acid, behavior with *o*-hydroxyphenylfluorone, 119
Fluorapatite, detection of fluorine in, 271
Fluorescein-potassium bromide paper test for free halogens, 368
Fluorescein test for bromides, 262
for free halogens, 368
for hydrobromic acid, 262
for uranium, 210
Fluorescence, in morin test for aluminum, 182, 479
prospects in spot test analysis, 23
of sodium zinc uranyl acetate, 229
test for aluminum with morin, 182, 479
for uranium, 207
Fluorescing organometallic compounds, 23
Fluoride disintegration of beryllium-bearing alloys, minerals, and ores, 457
of chromium-bearing alloys, minerals, and rocks, 455
Fluorides, boric acid in presence of, 342
carbonates in presence of, 338
etching test for, 274
insoluble, detection by glass test for hydrogen fluoride, 421
by zirconium-alizarin test, 270, 420
tests for, 420
interference in alizarin test for zirconium, 201
in ammonium molybdate test for phosphoric acid, 333
in chromyl chloride test for hydrochloric acid, 261
in *p*-dimethylaminoazophenylarsonic acid test for zirconium, 203
in hydrogen peroxide test for titanium, 195
in morin test for zirconium, 202
in β-nitroso-α-naphthol test for zirconium, 200
in phenoxithine test for palladium, 140
in potassium ferrocyanide test for iron, 161
in potassium thiocyanate test for iron, 164
in sodium carbonate-phenolphthalein test for carbonic acid, 338
in titanium-malachite green test for tungsten, 122
iron in, detection, 449
in mixtures, test for, 384
in presence of nitrates, 273
of oxalates, 271
of phosphates, 271
reaction with *p*-nitrobenzeneazochromotropic acid, 341
sulfates in, test for, 493
titanium in presence of, 195
zirconium azoarsenate test for, 271
Fluorine, in mineral waters, test for, 490
in rocks, test for, 490
in solid substances, detection, 270
in water, test for, 506
zirconium-alizarin test for, 269
Fluorspar, detection of fluorine in, 270
Foods, copper in, detection, 430
spot reactions in analysis of, 428
Forensic investigations, detection of thallium in, 438
Form species of precipitate, 7
Formaldehyde, and ferric salts, test for hydroxylamine, 245
reaction with alkali cyanides, 97
with selenosulfate, 377
with sodium hydrosulfite, 322
with sulfites, 312
and silver nitrate test for ammonia, 238
Formaurindicarboxylic acid, formation, 317
Formhydroxamic acid, formation of, 245

Fractional precipitation of rubeanates, 11
of silver halides, 11
Free carbon, test by oxidation, 379
Free elements, tests for, 361
Free halogens (*see also* Halogens, free), fluorescein test for, 368
starch-potassium iodide test for, 368
tests for, 368
Free hydrochloric acid, test for, 316
Free metals and alloys, ferric ferricyanide test for, 363
phosphomolybdic acid test for, 362
tests for, 361
Free phosphoric acid, test for, 316
Free selenium, reaction with thallous sulfide paper, 372, 375
test by addition to thallous sulfide, 372
by catalysis of methylene blue reduction, 377
by formation of silver selenide, 376
Free sulfur, test by conversion to thiocyanate, 373
by conversion to thiosulfate, 374
by formation of thallium polysulfide, 372
Free sulfuric acid, tests for, 316, 317
Fuchsin, tests for hydrobromic acid, 264
p-Fuchsin, action of sulfurous acid on, 311
test for palladium, 141
p-Fuchsinleucosulfonic acid, formation, 311
Fuchsite, chromium in, 455
Fusing, as preliminary operation, 35
reactions, 13

G

Galena, free sulfur in, 491
Gallium, detection in presence of antimony, gold, thallium, and platinum metals, 215
ferrocyanide and manganese salts test for, 214
rhodamine B test for, 215
separation from aluminum, 215
summary of tests, 520
Gallium salts, reaction with anthraquinone-1-azo-4-dimethylaniline hydrochloride, 112
with morin, 183
Gallocyanine, behavior toward lead sulfate, 10
color reaction with zirconium, 204
as reagent for lead hydroxide, 75
Garnierite, detection of nickel in, 463
Gaseous materials, sampling of, 33
Gaseous reagents, precipitation by, 56
Gases, detection, 46
apparatus for, 47, 48 (Figs. 23–28)
in spot test analysis, 45
Gelatin foils as medium for spot tests, 12
Gelatin paper for imprinting procedure, 484
Germanic acid, reaction with hydroxyanthraquinones, 115
with molybdates in acid solution, 333
with *p*-nitrobenzeneazochromotropic acid, 115
Germanium, ammonium molybdate and benzidine test for, 112
diphenylcarbazone test for, 114
hydroxylamine test for, 115
9-phenyl-2,3,7-trihydroxy-6-fluorone test for, 114
in presence of arsenic, 113
of iron and tin, 113
of selenium, 113
summary of tests, 520
test by formation of a complex with mannite, 112
Germanimolybdic acid, formation, 112, 333
Glass, alkali in, detection, 474
alkali resistance of, 499
calcium in, detection, 475
chemical resistance of, test for, 474
containing bases, identification, 474
differentiation as to alkali resistance, 499
detection of calcium in, 469
lead in, detection, 475
medium for spot tests, 12
metal oxides in, detection, 473
tin in, detection, 421
titanium in, test for, 460
vulnerability to attack by alkalis, test for, 499
Glass ware for stocking of versatile spot test laboratory, 30
Glucose, combination of germanic acid with, 112
Glycerol, combination of germanic acid with, 112
Glyoxal-bis(2-hydroxyanil), color reactions with, 534
test for calcium, 534
Gneiss, phosphates in, detection, 501
Gold, in alloys, detection, 445
ascorbic acid test for, 131
benzidine test for, 128
copper in presence of, 429
detection, in presence of palladium and platinum, 131

of gallium in presence of, 215
p-dimethylaminobenzylidenerhodanine reaction for, 59, 127
distinction from copper, 129
interference in cuprous iodide test for mercury, 66
in phenoxithine test for palladium, 140
α-naphthylamine test for, 131
palladium in presence of, test for, 136
in presence of antimony, mercury, and thallium, 130
reaction with rhodamine B, 107
reduction of gold salts to metallic, 128
rhodamine B test for, 129, 445
silver in presence of, 60
summary of tests, 520
test with reducing substance, 131
thallium in presence of, 160
Gold alloys, qualitative analysis of, 129
Gold chloride, reaction with rhodamine B, 106
test for arsenic hydride, 101
Gold ions, as catalysts in pentacyanoaquoferroate formation, 71
Gold platinibromide, reaction with rubidium salts, 234
test for cesium, 234
Gold salts, interference in test for palladium, 136
in *o*-tolidine test for copper, 84
reaction with cacotheline, 68
with *p*-dimethylaminoazophenylarsonic acid, 203
with *p*-dimethylaminobenzylidenerhodanine, 141
reduction to metallic gold, 128, 349
by stannous chloride, 134
Granite, phosphates in, detection, 501
Granitic gneiss, phosphates in, detection, 501
Graphite, carbon in, 422
in insoluble residue, 423
Graphitic carbon, detection, 379
Griess reaction of nitrites, 246, 330, 400
Grinding of solids, 33
Gutzeit procedure, in analysis of mixtures, 384
test for arsenic, 101, 103
Gypsum, as adulterant of flour, test for, 471
detection in presence of barium or strontium sulfate, 471
distinction from anhydrite, 471, 510
identification, 470
in a mixture of barium and calcium sulfates, 471

H

Halide ions, ferrocyanides in presence of, 288
Halides, interference in silver ion catalysis of diphenylcarbazide test for chromium, 170
reaction with cobalt salts in presence of acetone, 285
with potassium permanganate, 284
silver carbonate and phenolphthalein preliminary test for, 261
Halogenates, detection in presence of hypohalogenites, 295
interference in benzidine test for nitric acid, 327
in nitrous acid test for hydrazoic acid, 287
in pyrrole test for selenious acid, 346
in test for perchlorates, 301
reaction with cadmium chloride, 301
Halogen hydracids, detection of hydrazoic acid in presence of, 286
preliminary test for, 261
Halogens, free, 4,4′-bis(dimethylamino)-thiobenzophenone test for, 368
fluorescein test for, 368
reaction with potassium cyanide, 370
starch-potassium iodide test for, 368
summary of tests, 532
Hamburg blue, *see* Ferric ferrocyanide
Hanksite, detection of chlorine in, 502
Hard coal, reaction with molybdenum trioxide, 380
Hardness, of metals and alloys, 361
of water, determination, 507
Heating of spot plates, 12
Heller and Krumholz method for analysis of mixtures, 389
Heller method, for analysis of alloys, 393
of water, 505
Hemin, oxidation of luminol in presence of, as test for hydrogen peroxide, 356
Hepar reaction for sulfate in hydrofluoric acid, 493
Hexanitrodiphenylamine, *see* Dipicrylamine
Hood, arrangement, 37 (Fig. 11)
special section for spot tests, 29
Hydrazine, alkali copper pyrophosphate test for, 242
color reaction with dinitrobenzene, 320
detection of hydroxylamine in, 245
p-dimethylaminobenzaldehyde test for, 240
dinitrobenzene test for, 242

lead^{IV} oxide test for, 242
thallium^{III} oxide test for, 242
in presence of hydroxylamine, 243–246
reaction with benzaldehyde, 239
with salicylaldehyde, 239
salicylaldehyde test for, 23, 239
summary of tests, 520
test by reduction of ammoniacal cupric solutions, 240
Hydrazine salts, interference in nitrous acid test for hydrazoic acid, 287
Hydrazoic acid, nitrous acid test for, 287
as reducing agent of permanganate in test for chromate, 345
removal of nitrites, 329
summary of tests, 528
test by formation of silver or iron azide, 286
Hydriodic acid, for converting insoluble compounds into water-soluble iodides, 407
palladous chloride test for, 265
reaction with tellurous acid, 346
summary of tests, 528
test for selenious acid, 346
Hydrobromic acid, fluorescein test for, 262
fuchsin test for, 264
permolybdate and α-naphthoflavone test for, 264
and phosphomolybdic acid test for thallium, 156
potassium permanganate test for, 265
as reagent for copper, 94
summary of tests, 528
test by oxidation to bromine, 262
Hydrochloric acid, copper in presence of, detection, 83
detection of iron in, 453
formation of chromyl chloride as test for, 260
free, test for, 316
summary of tests, 528
test for, by precipitation of silver chloride in presence of other halide ions, 259
volatilization test for, with formation of silver chloride, 261
Hydrocyanic acid (*see also* Prussic acid), alkali mercury chloride test for, 279
alkali palladium dimethylglyoxime test for, 278
copper acetate-benzidine acetate test for, 276
copper sulfide test for, 277
formation, 380
of thiocyanate as test for, 277
summary of tests, 528
Hydroferricyanic acid, *see* Ferricyanic acid
Hydroferrocyanic acid, *see* Ferrocyanic acid
Hydrofluoric acid, etching test for, 274
sulfates in, detection, 493
summary of tests, 529
test by conversion into silicomolybdic acid, 272
by prevention of formation of oxinates, 275
zirconium-alizarin solution test for, 269
zirconium azoarsenate test for, 271
Hydrogen, condensation, detection of platinum metals by, 131
Hydrogen chloride group, tests for cations of, 389
Hydrogen cyanide (*see also* Prussic acid), detection of dicyanogen by hydrolysis to, 370
interference in hydrolysis test for dicyanogen, 370
Hydrogen-flame luminescence test for antimony, 79, 107
for bismuth, 79
for manganese, 79
for rare earths, 79
Hydrogen-ion concentration, determination, 504
indicators for, 504, 505
Hydrogen peroxide, activation of, 417
and ammonia test for cerium, 210
auric chloride test for, 355
decomposition by oxides, 415
discharge of chromate–diphenylcarbazide reaction by, 356
ferric–ferricyanide test for, 354
interference in ammonium molybdate test for phosphoric acid, 333
in nitrous acid test for hydrazoic acid, 287
in thallous hydroxide test for hypohalogenites, 295
lead sulfide test for, 10, 353
and manganese sulfate test for ammonia, 239
and nitric acid test for lead dioxide, 415
for manganese dioxide, 415
persulfates in presence of, detection, 324
and phenetidine-HCl test for copper, 94
reaction with alkali thiocyanates, 355
with ammonium molybdate, tungstate or vanadate, 417
with chromates, 196
with iron salts, 196

with molybdates, 196
with nickel oxide, 323
with *o*-tolidine solution, 356
with vanadates, 196
with vanadium salts, 417
reduction of chromates by, 356
of nickel oxides as test for, 354
summary of tests, 529
test by catalyzed oxidation of phenolphthalin, 355
for cobalt, 149
by oxidizing luminol in presence of hemin, 356
with potassium cerous carbonate, 327
for titanium, 195, 413
for vanadium, 123
Hydrogen phosphide, reaction with silver nitrate, 101
Hydrogen sulfide, catalysis of the iodine-azide reaction as test for, 303
detection of free sulfur by conversion to, 375, 491
formation, 13
as test for thiocyanates, 285
group, tests for, 58
tests for cations of, 389
interference in silicomolybdic acid test for fluorides, 273
paper test for antimony and cadmium in minerals, 490
reaction with cacotheline, 109
with silver nitrate, 101
set-up, 46 (Fig. 22)
sodium nitroprusside test for, 303
summary of tests, 529
in water, detection, 503, 507
Hydroxamic acids, reaction with ferric salts, 245
Hydroxy acids, organic, interference in alizarin test for zirconium, 201
Hydroxyanthraquinones, reaction with boric and germanic acids, 114
test for boric acid, 340
Hydroxybenzaldehyde test for free sulfuric acid, 317
o-Hydroxybenzaldehyde, *see* Salicylaldehyde
Hydroxylamine, ammonia test for, 246
color reaction with dinitrobenzene, 320
cupric ferricyanide test for, 246
detection in presence of hydrazine, 245
diacetylmonoxime and nickel salts test for, 242
in *p*-dimethylaminobenzaldehyde test for hydrazine, 242
dinitrobenzene test for, 246
formaldehyde and ferric salts test for, 245
in presence of hydrazine, 243–246
reaction with dinitrobenzene, 242
salicylaldehyde and copper salts, test for, 243
sodium copper pyrophosphate test for, 246
summary of tests, 520
test by conversion to nitrous acid, 245
test for germanium, 115
Hydroxylamine hydrochloride, reaction with cyanates, 285
2-(*o*-Hydroxyphenyl)benzthiazole test for beryllium, 195
o-Hydroxyphenylfluorone (*see also* 9-Phenyl-2,3,7-trihydroxy-6-fluorone), and acid potassium fluoride, test for copper, 94
color reactions by, 114, 118
test for molybdenum, 118
for copper, 94
8-Hydroxyquinoline (*see also* Oxine), test for copper, 93
for ferric oxide, 413
for iron in alumina, 450
for tungsten, 119
for vanadium, 125
8-Hydroxyquinoline-7-iodo-5-sulfonic acid test for iron, 163
3-Hydroxy-1-*p*-sulfonatophenyl-3-phenylhydrazine test for palladium, 140
3-Hydroxy-1-*p*-sulfonatophenyl-3-phenyltriazine test for palladium, 140
Hydroxytriphenylmethane dyes, detection of aluminum by, 13
reaction with aluminum salts, 189
Hydroxyurea, formation of, as test for cyanic acid, 285
Hypobromites, detection by formation of eosin, 297
Hypochlorites, detection of, 297
Hypohalogenites, ammonia test for, 295
detection of other oxidants in presence of, 295
in presence of chromates, ferricyanides or permanganates, 296
differentiation of, 296
interaction of, 297
general test for, 296
reaction with benzidine, 322
safranin test for, 294
summary of tests, 529
thallous hydroxide test for, 295
Hypoiodites, detection by zinc chloride and starch, 297

Hypophosphorous acid, in extraction test for bismuth, 80
reduction of metallic salts by, 349
test for iodic acid, 299
for telluric and tellurous acids, 349
Hyposulfurous acid, *see also* Sodium hydrosulfite
alkali salts of, 320
as reductants, 320
summary of tests, 530

I

Identification limit, 4
Igniting as preliminary operation, 35
Ignition, and evaporation residues, testing, 508
residues, detection of iron in, 450
test for silver salts, 411
tube, heating of sample in, 399
Ilmenite, detection of iron in, 462
Impermeable surfaces as reaction medium, 1
Impregnated reagent paper, preparation, 22
spotting on, 51
Imprinting procedure for detecting heterogeneities in metals, minerals, etc., 484
Increase of sensitivity of spot tests, 8, 9, 11, 12, 15, 44, *see also* Sensitivity, increase
Indicators, for measuring hydrogen-ion concentration, 504, 505
Indium, alizarin test for, 213
in presence of aluminum, 213
of ammonium sulfide group of metals, 213
of cesium, 213
of chromium, 214
of cobalt, iron, manganese, nickel, and zinc, 213
quinalizarin test for, 213
summary of tests, 521
Indium salts, reaction with morin, 183
Infrared heating of spot plates, 12
Inhomogeneities in metals and rocks, 21
Inks, classification, 479
Ink writing, chemical identification, 479
determination of age, 479, 481
Inorganic materials, and organic (basic), reacting with mineral acids, detection, 507
"Insoluble residue", detection of mercury in, 305
detection of sulfides in, 304, 305
identification of constituents, 407
Insoluble sulfates, identification, 408
Insoluble sulfides, identification, 407
Interfaces, role in spot reactions, 8
Intermetallic mercury compounds, detection, 365
Iodates, interference in test for zinc, 178
oxidation of pyrogallol by, 300
reaction with alkali thiocyanates, 299
with diphenylamine, 327
with ferrous sulfate and sulfuric acid, 326
starch-iodine reaction for detecting, 299
test for iodides by conversion to, 267
Iodates, *see* Halogenates
Iodic acid, hypophosphorous acid test for, 299
potassium thiocyanate test for, 299
pyrogallol test for, 300
summary of tests, 530
Iodide(s), bromides in, detection, 264
bromides in presence of, 263
catalytic reduction of cerium salts as test for, 267
conversion to iodates as test for, 267
cyanides in presence of, 277
detection of bromides in presence of, 263
ferrocyanides in presence of, 290, 291
interference in butyraldehyde test for cobalt, 149
in chromyl chloride test for hydrochloric acid, 261
in ferrous sulfate-sulfuric acid test for nitric acid, 326
in hydrogen peroxide test for vanadium, 123
in oxidation test for manganese, 173
in prussic acid test for thiocyanates, 284
in silicomolybdic acid test for fluorides, 273
in mixtures, test for, 385
oxidation to iodine, 259
palladous chloride as reagent for, 265
in presence of bromides, 268
of chlorides, 268
reaction with arsenic acid, 102
with cobalt salts, in presence of acetone, 285
with potassium permanganate, 284
reduction of iodine to, 102
silver nitrate test for, 268
starch test for, 269
test by catalysis of tetrabase-chloramine T reaction, 511, 535
by oxidation to free iodine, 266
for small amounts of, 267, 268

thallium nitrate test for, 269
thiocyanate in presence of, 283, 285
in water, test for, 506
Iodide ions, interference in test for mercury, 71
Iodine, adsorption on magnesium hydroxide, 225
detection in mineral and sea water, 511
oxidation of alkali iodides to, 327
of iodides to, 259
of pyrogallol by, 300
reaction with arsenious acid, 102
reduction to iodide, 102
release on starch filter paper as test for nitrous acid, 332
summary of tests, 532
test for iodides by oxidation to, 266
Iodine–azide reaction, catalysis of, as test for alkaline earth sulfates, 315
for detecting sulfide corrosion in machinery metal, 487
for hydrogen sulfide, 303
for insoluble sulfates, 315
for insoluble sulfides, 305
for lead sulfate, 315
for soluble sulfides, 304
for sulfates, 408
for sulfides, 303, 314, 407, 493
for sulfide sulfur, 20
in rocks, 491
for thiocyanates, 304, 305
for thiocyanic acid, 282
for thiosulfates, 304, 305, 374
for thiosulfuric acid, 318
Iodoeosin, formation, 263
test for basic oxides in glass, 474
Ions containing cyanogen radical, capillary detection, 294
Iridium, detection by condensation of hydrogen on, 131
summary of tests, 521
Iridium salts, reaction with anthraquinone-1-azo-4-dimethylaniline hydrochloride, 112
Iron, in alloys, tests for, 394
in alumina, detection, 413, 450
aluminum in presence of, 187
beryllium in presence of aluminum and, 193
calcium in presence of, 535
cerium in presence of, 211
cobalt in presence of, 145, 147
detection in acetic acid, 453
in alkali salts, 453
of aluminum in presence of, 190
in calcite, 451
in carbonates, 451
in cobalt alloys, 451
in concentrated nitric acid, 453
in dolomite, 451
in hydrochloric acid, 453
in magnesite, 451
in magnesium, 452
in magnesium salts, 452
in metallic copper, 451
in nickel alloys, 451
in sulfuric acid, 453
of uranium in presence of, 209
in zinc, 452
in zinc salts, 452
dimethylglyoxime test for, 164
α, α'-dipyridyl test for, 161, 450, 462, 466
disodium-1,2-dihydroxybenzene-3,5-disulfonate test for, 165
dithizone test for zinc in presence of, 180
ferrous, demasking of, 70
detection in silicates and minerals, 462
in fluorides, test for, 449
germanium in presence of, 113
8-hydroxyquinoline-7-iodo-5-sulfonic acid test for, 163
indium in presence of, 213, 214
in ignition residues, detection, 450
in ink writing, test for, 479
interference in cerium salt reduction test for iodides, 268
in fluorescence test for uranium, 207
in oxine test for vanadium, 126
in phosphomolybdic acid test for vanadium, 127
isonitrosobenzoylmethane test for, 167
lake formation with alizarin, 184
magnesium in presence of, 228
manganese in presence of, 176
in mercury salts, test for, 450
in minerals, test with ferrocyanide paper, 490
in mixtures, tests for, 387, 390, 392
nickel in presence of, 151, 154
2-nitroso-1-naphthol-4-sulfonic acid test for, 94, 167
oxine test for, 166
in phosphoric acid, test for, 449
phosphorus segregations in, detection, 487
potassium ferrocyanide test for, 161
potassium thiocyanate test for, 164
in potassium thiocyanate test for molybdenum, 116

in presence of chromium, cobalt, copper, and nickel, detection, 164
in pyrolusite, test for, 450
reaction with curcumin, 339
with dicyanogen, 412
with potassium ferrocyanide, 178
solubility in nonoxidizing acids, 362
sulfur segregation in, detection, 486
summary of tests, 521
in tanned leather, test for, 478
testing of bauxite for, 450
thiocyanate reaction for, 283
thioglycolic acid test for, 167
titanium in presence of, 199
in titanium dioxide, detection, 450
uranium in presence of, 205, 206
in water, test for, 506
Iron azide, formation, as test for hydrazoic acid, 286
Iron chloride test for ferrocyanic acid, 289
Iron and cobalt salts, nickel in presence of, 150, 152
Iron^{II} dimethylglyoxime test for vanadium, 124
Iron gallate inks, composition, 479
Iron ions, interference in test for cobalt, 149
Iron ores, detection of manganese in, 456
Iron^{III} oxinate, formation of, 166
Iron periodate complex test for lithium, 233
Iron salts, (II), in presence of (III), detection, 162
interference in ammonium mercury thiocyanate test for copper, 93
in tests for beryllium, 195
in test for palladium, 136
in *o*-tolidine test for copper, 84
reaction with alizarin, 184
with ammoniacal silver nitrate, 211
with chromotropic acid, 198
with hydrogen peroxide, 196
with pyrocatechol, 196
with 1,2,7-trihydroxyanthraquinone, 111
reduction of phosphomolybdates by, 156
as solvent for calcium fluoride, 270
Iron sulfate as reagent for tellurites and tellurates, 351
Iron ware for stocking of versatile spot test laboratory, 31
Isobutyl ketone test for bismuth, 80
Isonitrosobenzoylmethane, test for iron, 167

J

Jena glass, alkali in, 474

K

Kairine test for arsenic, 103
Kalignost, *see* Sodium tetraphenyl boron
Kaolin, detection of titanium in, 459
Kavalier glass, alkali in, 474
Kolthoff's method of water analysis, 505
Krumholz method of analysis of mixtures, 391

L

Laboratory requirements, 28
Lanthanum, hydrogen-flame luminescence test for, 79
Lanthanum salts, lake formation with quinalizarin, 193
reaction with quinalizarin, 225
Lead, in alkali salts, detection, 435
in alloys, tests for, 394, 433
benzidine test for, 72
in bismuth compounds, test for, 437
bismuth in presence of, detection, 76, 77
cadmium stannoiodide test for, 75
copper in presence of, detection, 85
in criminalistic investigations, detection of traces, 437
detection of calcium in presence of, 535
in glass, 475
dithizone test for, 74
for zinc in presence of, 180
interference in sodium rhodizonate test for barium, 476
magnesium in presence of, 227
metallic, detection by reduction of *o*-dinitrobenzene, 363
in metals, detection, 433
in minerals, ores, and pigments, detection, 434
in minerals, test with potassium chromate paper, 490
in mixtures, tests for, 386, 389, 392
in neutral solution, detection by gallocyanine, 10, 75
in oils, pigments, and varnishes, detection, 435
in pharmaceutical preparations, detection, 437
in presence of other metals, detection, 74
reaction with potassium iodide, 155
sodium rhodizonate test for, 73, 416, 433, 437

solubility in nonoxidizing acids, 362
in sulfuric acid, detection, 435
summary of tests, 521, 533
thallium in presence of, test for, 156
in tin platings and enamels, 433
in water, detection, 435
in zinc dross or zinc oxide, detection, 434
zinc in presence of, 182
Lead chloride, silver in a mixture of mercurous and silver chlorides and, 60
Lead chromate, decomposition by fusion, 415
diphenylamine test for, 415
in insoluble residue, 415
interference in diphenylbenzidine test for antimony pentoxide, 419
Lead dioxide, decomposition of hydrogen peroxide by, 415
detection of nitrate in, 494
Lead glass, alkali in, 474
Lead hydroxide, detection by gallocyanine, 75
Lead nitrate, zinc in presence of, 181
Lead oxide test for hydrazine, 242
Lead oxides, detection of free metal in, 475
Lead peroxide, conversion of lead salts to, 72
formation of benzidine blue by, 72
hydrogen peroxide–nitric acid test for, 415
in insoluble residue, 415
and manganese dioxide, distinction, 416
Lead salts, in alkaline stannite test for bismuth, 77
ammonia in presence of, 238
cadmium in presence of, detection, 95
carminic acid as reagent for, 75
conversion to lead peroxide, 72
formation of benzidine blue by, 172
interference in cuproin test for cuprous ions, 91
in rhodizonate test for calcium, 222
in kairine test for arsenic, 103
reaction with benzidine, 172, 416
with potassium chromium thiocyanate, 79
Lead sulfate, induced precipitation of, as test for barium, 218
in insoluble residue, sodium rhodizonate test for, 409
iodine-azide reaction for, 315
in presence of alkaline earth sulfates, 408
reaction with activated sulfur, 509
reduction to sulfide, 315
Lead sulfide, from sulfate, formation by activated sulfur, 509
test for hydrogen peroxide, 10, 353
Leather, detection of alumina in ash of, 454
testing of, for tanning minerals, 477
Leuco nitro diamond green test for thallium, 161
Lime, differention of hard- and soft-burned, 466
in magnesite, detection, 468
soft-burned, color reaction with thymo blue, 467
Limit of concentration, 4
of dilution, 4
of identification, 4
Limiting proportion, 14
Liquid materials, sampling of, 33
Litharge, detection of free metal in, 475
Lithium, complex iron periodate test for, 233
etching test for, 234
in presence of cesium, potassium, rubidium, and sodium, 233
of alkali chlorides, 234
specific test for, 13
summary of tests, 521
Lithium salts, interference in zinc uranyl acetate test for sodium, 229
reaction with sodium cobaltinitrite, 230
Litmus, as indicator of hydrogen-ion concentration, 504
paper test for ammonia, 236
Lithopones, detection in pigments, paper ash, etc., 476
Logwood inks, composition, 479
Luminescence test for antimony, 79, 107
for bismuth, 79
for manganese, 79
for tin, 421
Luminol, oxidation in presence of hemin, as test for hydrogen peroxide, 356
Luster of metals and alloys, 361

M

Macrochemical tests, choice of, 57
Macro methods, 6
Magnesia spoon, heating of sample on, 397
Magnesite, caustic, calcite and dolomite in, 468
detection of calcium in, 468
of iron in, 451
differentiation of caustic burned and sintered, 467

distinction from bräunerite, 465
from dolomite, 465
lime in, detection, 468
Magnesium, alkali hypoiodite test for, 223
aluminum in presence of, quinalizarin test for, 188
in asbestos and talc, 461
beryllium in presence of, 194
calcium in presence of, 535
in chalk, detection, 466
detection of iron in, 452
metallic, detection by reduction of *o*-dinitrobenzene, 364
in mixtures, test for, 388
p-nitrobenzeneazo-α-naphthol test for, 225, 461
p-nitrobenzeneazoresorcinol test for, 226
phenolphthalein paper test for, 228
in presence of other metals, tests for, 227
quinalizarin test for, 193, 224
in silicate rocks, detection, 461
summary of tests, 521
in tap water, detection, 227
test for, with monoazo dyes, 229
titan yellow test for, 228
Magnesium compounds, reaction with diphenylcarbazide, 465
Magnesium hydroxide, adsorption of iodine on, 225
Magnesium ions, interference in α,α'-dipyridyl test for iron, 453
Magnesium salts, detection of iron in, 452
interference in dipicrylamine test for thallium, 157
in test for copper, 86
in zinc uranyl acetate test for sodium, 229
reaction with ammonium ferrocyanide, 221
with *p*-nitrobenzeneazoorcinol, 194
Magnesium silicate, reaction with zinc chloride, 461
Magnesium sulfate, interference in starch-iodine reaction, 266
in titanium-malachite green test for tungsten, 122
Magnetic properties of metals and alloys, 361
Malachite green, decolorization of, as test for sulfurous acid, 311
reaction with cobaltous hydroxide, 212
with manganese hydroxide, 212
with silver oxide, 212
with thallium hydroxide, 212
test for cerium, 212
Malachite green–titanium reaction, catalysis of, as test for tungsten, 121
Malonic acid, masking action in rubeanic acid test for copper, 88, 90
Mandelic acid azo dyes test for zirconium, 204
Manganates, interference in pyrrole test for selenious acid, 346
Manganese, in alloys, tests for, 177, 394
aluminum in presence of, 187
benzidine test for, 8, 175, 298, 416
chromium in presence of, 168, 172
detection of aluminum in presence of, 190
in minerals and rocks, 456
dithizone test for zinc in presence of, 180
hydrogen-flame luminescence test for, 79
indium in presence of, 213, 214
interference in diphenylcarbazide test for chromium, 456
lake formation with alizarin, 186
magnesium in presence of, 228
in mixtures, tests for, 387, 391, 392, 394
nickel in presence of, 151
periodate and tetrabase test for, 174
in presence of cerium, test for, 176
of cobalt, 174, 176
of copper, test for, 177
of iron, test for, 176
of nickel, 174
of silver, test for, 177
of small amounts of silver salts, 173
of thallium, test for, 177
reaction with potassium ferrocyanide, 178
silver ammine salts test for, 177
in special steels, test for, 177
summary of tests, 522
test by catalytic oxidation to permanganate, 173, 174, 456
titanium in presence of, 199
in water, test for, 506
Manganese dioxide, in boric acid, detection, 342
decomposition of hydrogen peroxide by, 415
detection of nitrate in, 494
hydrogen peroxide-nitric acid test for, 415
in insoluble residue, 415
and lead peroxide, distinction, 416
reaction with potassium persulfate, 456
Manganese hydroxide, conversion to hydrated manganese dioxide, 175

reaction with malachite green, 212
Manganese nitrate and alkali test for silver, 58
Manganese oxides, reaction with nitrogen compounds, 400
Manganese salts, catalytic reduction of, as test for silver, 62
cerium in presence of, 211
and ferrocyanide test for gallium, 214
interference in benzidine test for lead, 72
in cerium salt reduction test for iodides, 268
in fluorescence test for uranium, 207
in *o*-tolidine test for copper, 84
in zinc uranyl acetate test for sodium, 229
oxidation to permanganates, 170
presence in 1,2-diaminoanthraquinone-3-sulfonic acid test for copper, 87
reaction with alkali hypobromites, 174
with ammoniacal silver salts, 211
with benzidine, 155, 211
with diphenylcarbazide, 168
with nitro-3-hydroxybenzoic acid, 220
with persulfates, 302
with potassium persulfate, 456
and tetrabase test for periodic acid, 302
Manganese sulfate, and hydrogen peroxide test for ammonia, 239
reaction with chlorates, 297
and silver nitrate test for ammonia, 237
zinc in presence of, 181
Manganous ions, oxidation to manganese dioxide, 149
Manganous salts, interference in test for zinc, 178
reaction with alkali periodates, 174, 301, 302
Manganous sulfate, and phosphoric acid test for chloric acid, 297
for periodic acid, 301
reaction with chlorates, periodates, and persulfates, 301
and sulfuric acid test for bromic acid, 298
Mannite, combination with germanic acid, 112
test for germanium, 112
Marcasite, free sulfur in, 491
Masking of test ions, 16
Mercaptobenzeneindazole test for palladium, 141
Mercuri amidochloride, reaction with ferrocyanides, 289
Mercuric chloride, activation of iodine-azide reaction of pyrites, 408
decomposition of ferric thiocyanate by, 290
diphenylcarbazone test for, 65
interference in tests for free metals and alloys, 362
in wood preservatives, detection, 477
Mercuric cyanide, differentiation from mercuric oxycyanide, 282
and diphenylcarbazide test for palladium, 141
potassium ferrocyanide and α,α'-dipyridyl test for, 282
starch–iodine test for, 281
summary of tests, 530
test by conversion to mercuric iodide and cyanide ion, 281
by formation of cyanogen iodide, 281
by formation of dicyanogen, 282
by precipitation with silver nitrate, 280
Mercuric iodide and cyanide ion, conversion test for mercuric cyanide, 281
Mercuric ions, as catalysts in pentacyanoaquoferroate formation, 71
reaction with ferrocyanide, 70
Mercuric nitrate test for calcium sulfate, 470
Mercuric oxycyanide, differentiation from mercuric cyanide, 282
potassium ferrocyanide and α,α'-dipyridyl test for, 282
reaction with potassium iodide, 282
test for, by formation of dicyanogen, 282
Mercuric salts, cuprous iodide test for, 65
detection by chromotropic acid, 72
interference in cerium salt reduction test for iodides, 268
in pyrrole test for selenious acid, 346
reaction with alkali and manganese salts, 58
Mercuric sulfide, test for free sulfur, by conversion to, 374
Mercurous chloride, interference in molybdenum trioxide test for free carbon, 381
silver in a mixture of lead and silver chlorides and, 60
test for gold, 131
Mercurous ions, interference in phenoxithine test for palladium, 140
Mercurous salts, detection by chromotropic acid, 72
interference in cuproin test for cuprous ions, 91
in α-naphthylamine test for gold, 131

in cuprous iodide test for mercury, 66
reaction with alkalis, 58
reduction of phosphomolybdates by, 156
rubeanic acid test for, 90
Mercury, arsenic in presence of, detection, 100
bismuth in presence of, detection, 77, 78
catalytic acceleration of the reduction of tin salts as test for, 68
chromium in presence of, detection, 170
copper in presence of, detection, 85
cuprous iodide test for, 65
detection in acid solution, 70
in neutral solution, 69
p-dimethylaminobenzylidenerhodanine test for, 59, 67, 128
diphenylcarbazone test for, 64
dithizone test for, 71
electrolytic precipitation, 69
free, summary of tests, 533
gold in presence of, 130
in "insoluble residue", detection, 305
in mixtures, tests for, 386, 390, 391
potassium ferrocyanide test for, 72
precipitation by antimony sulfide paper, 51
in presence of chlorides, detection, 67
of copper, detection, 67
of free acids, detection, 67
reaction with diphenylcarbazide, 414
with potassium iodide, 155
with rhodamine B, 107
with thiourea, 349
silver in presence of, detection, 60
solubility in nonoxidizing acids, 362
stannous chloride and aniline test for, 66
summary of tests, 522
test by activation of aluminum, 69
by catalytic decomposition of ferrocyanide, 70
by reduction of palladium chloride, 364
thallium in presence of, tests for, 155, 160
vapor, selenium sulfide as reagent for, 365
test by reduction of palladium chloride, 364
Mercury chloride, reaction with rhodamine B, 106
Mercury dropper, 39 (Fig. 14)
Mercury iodide test for palladium, 137
Mercury salts, ammonia in presence of, 238
cadmium in presence of, 96
detection of arsenic in presence of, 101
interference in ascorbic acid test for gold, 131
in dimethylaminobenzylidenerhodanine test for gold, 128
in dithizone test for zinc, 180
in kairine test for arsenic, 103
in potassium thiocyanate test for iron, 164
for molybdenum, 115
in silver nitrate test for arsenic hydride, 101
in test for chromium, 168
in test for copper, 86
in test for palladium, 136
in *o*-tolidine test for copper, 84
in zinc uranyl acetate test for sodium, 229
iron in, detection, 450
reaction with anthraquinone-1-azo-4-dimethylaniline hydrochloride, 112
with chromotropic acid, 198
with *p*-dimethylaminobenzylidene-rhodanine, 141
with diphenylcarbazide, 170
with potassium chromithiocyanate, 79
reduction by stannites, 350
by stannous chloride, 100
Mercury sulfides, detection by sodium azide, 305
Mercury vapor, test with cuprous iodide, 366
Metallic ions, interference in titanium–malachite green test for tungsten, 122
reaction with α,α'-dipyridyl, 162
Metallic objects, tin in, detection, 439
Metallic oxides, detection of nitrate in, 494
heavy, test for, 473
Metallic salts, interference in hydrogen peroxide test for vanadium, 123
Metallic samples, search for traces in, 21
Metal(s), ammonium carbonate group, tests for, 216
ammonium sulfide group, aluminum in presence of, 183
indium in presence of, 213
tests for, 143
copper in, developing process for detection, 488
electrodes for electrolytic deposition of traces of, 506
electrographic test for, 489
free, in oxides, detection, 475
in printed paper, detection, 475
reaction with phosphomolybdic acid, 475

test for, 361
group arrangement, 58
group separations, 58
hydrogen sulfide group, acid sulfide group, tests for, 99
basic sulfide group, tests for, 58
imprinting and developing procedure for detecting heterogeneities in, 484
inhomogeneities in, 21
lead in, detection, 433
oxides, higher, decomposition of hydrogen peroxide by, 415
in diphenylamine test for lead chromate, 415
in diphenylbenzidine test for antimony pentoxide, 419
phosphorus segregations in, test for, 487
plated, test for zinc in, 448
sulfur segregations in, test for, 486
tests for, 57
utensils for stocking of versatile spot test laboratory, 30
Metastannic acid, detection of antimony oxide or antimonic acid in, 104
in insoluble residue, 421
Methyl alcohol, flame test for boric acid, 421
Methyl borate and alkali fluoride test for boric acid, 342
4-Methyl-1,2-dimercaptobenzene, *see* Dithiol
Methylene blue, catalysis of reduction, test for free selenium, 377
for selenious acid, 348
test for sulfides, 307
and hydrazine test for molybdenum, 117
and zinc test for titanium, 199
Methylene white, formation by molybdate, 117
Methylenedisalicylic acid test for free sulfuric acid, 317
Methyl orange as indicator of hydrogen-ion concentration, 504
Methyl red as indicator of hydrogen-ion concentration, 504
4-Methylthio-1,2-phenylenediamine test for selenious acid, 349
9-Methyl-2,3,7-trihydroxy-6-fluorone test for antimony, 105
Methyl yellow test for palladium, 141
Micro beaker, aluminum support for, 36 (Fig. 9)
water bath for, 36 (Fig. 8)
Microcentrifuge tube, 42 (Fig. 17)
Microchemical apparatus, standardization, 30
Microchemical equipment, 30
Microchemical tests, choice of, 57
Microchemistry, non-existence of, 6
Micro crucibles, 12
heating of sample in, 397
Micro distillation apparatus, 48 (Fig. 29)
Microdrops, 5
Microfiltration, 42 (Fig. 16)
Microgram analysis, 6
Micro methods, 6
Micro mortar, 34 (Fig. 4)
Micro pipet-buret, 39
Micro sieve, 34 (Fig. 5)
Micro test tubes, 12
Milligram analysis, 6
Minerals, arsenic in, detection, 443
beryllium in, detection, 457
chlorine in, detection, 501
ferrous iron in, detection, 462
imprinting and developing procedure for detecting heterogeneities in, 484
lead in, detection, 434
manganese in, detection, 456
phosphates in, detection, 500
reducible metals in, detection, 463
selenium in, test for, 491, 492
silicic acid in, detection, 499
spot reactions in studies of, 428
sulfur in, detection, 491
tungsten in, detection, 459
thallium in, detection, 437
tin in, detection, 439
titanium in, detection, 459
uranium in, detection, 457
used for leather tanning, test for, 477
Mineral acids, detection of inorganic and organic (basic) materials that react with, 507
detection of selenites in solutions of, 349
Mineral blue, *see* Ferric ferrocyanide
Mineral tanning agents, 477
Mineral water, detection of iodine in, 511
fluorine in, tests for, 490
Mixing of solids, 33
Mixtures, analysis of anions in, 404
of common metals, detection, 57
systematic analysis by spot tests, 383
Molybdates, interference in alizarin test for zirconium, 201
in diphenylcarbazone test for mercury, 65
in dithiol test for tin, 108
in ferrous sulfate test for nitric acid, 326
in oxine test for tungsten, 120
in pyrrole test for selenious acid, 346

in cuprous iodide test for mercury, 66
in tests for vanadium, 123
reaction with α-benzoinoxime, 125
with *p*-dimethylaminoazophenylarsonic acid, 203
with diphenylamine, 327
with diphenylcarbazide, 170, 414
with α,α'-dipyridyl, 119
with ferrous sulfate, 326
with hydrogen peroxide, 196
with oxine, 125
with rhodamine B, 106, 107
with thiocarbazide, 120
with 1,2,7-trihydroxyanthraquinone, 111
reduction by stannous chloride, 108
Molybdenum, ammonium hypophosphite test for, 464
chromium in presence of, 170
cochineal fluorescence test for, 119
detection of tungsten in presence of, 120
di-β-naphthylcarbazone test for, 119
diphenylcarbazide test for, 119
o-hydroxyphenylfluorone test for, 118
interference in diphenylcarbazide test for chromium, 456
in disodium-1,2-dihydroxybenzene-3,5-disulfonate test for iron, 166
in test for titanium, 199
metallic, detection by conversion to metallo acid salts, 367
reaction with alkali nitrite, 367
methylene blue and hydrazine test for, 117
phenylhydrazine test for, 117
potassium thiocyanate and stannous chloride test for, 115
potassium xanthate test for, 116, 444
reaction with curcumin, 339
reduction of molybdenum oxides to, 509
in steel, detection, 444
summary of tests, 522, 533
in technical materials, detection, 444
in titanium-malachite green test for tungsten, 122
in tungsten ores, detection, 444
tungsten in presence of, 120
zirconium in presence of, 203
Molybdenum blue, constituents, 362
formation by alloys, 362, 475
by antimony salts, 104
by benzidine, 273
by carbon monoxide and palladium salts, 136
by cuprous iodide, 66
by diphenylcarbazide, 170
by germanium molybdic acid, 112
by hydrazine hydrate, 335
by metals, 362, 475
by phosphomolybdic acid, 333
by potassium cuprocyanide, 85
by reductants, 380
by silicomolybdic acid, 273, 336
by silver, 411
by stannous chloride, 108, 416
by thallium salts, 156
by vanadium salts, 127
Molybdenum oxalic acid, formation, 171
Molybdenum oxide(s), in insoluble residue, 417
interference in cacotheline test for tin, 109
reaction with alkali nitrite, 367
reduction to molybdenum, 509
test for, by hydrogen peroxide activation, 417
Molybdenum salts, interference in test for chromium, 168
reaction with anthraquinone 1-azo-4-dimethylaniline hydrochloride, 112
Molybdenum trioxide, detection of nitrate in, 494
interference in reduction test for niobium pentoxide, 420
reaction with ammonium salts, 380
with diphenylcarbazide, 415
reducing power, 380
separation from tungsten trioxide, 416
test for free carbon, 380
Monazite, detection of titanium in, 459
Monoazo dyes, test for magnesium, 229
Morin, reaction with beryllium, gallium, indium, scandium, thorium and zirconium salts, 183
test for aluminum, 182
in tanned leather, 479
for beryllium, 191, 458
for tin, 110, 421, 440, 441
for titanium, 198
for zirconium, 201, 414
Munktell filter paper, 54
Murano glass, alkali in, 474

N

Naphthalene-4-sulfonic acid-1-azo-5-*o*-8-hydroxyquinoline test for palladium, 140
α-Naphthoflavone and permolybdate test for hydrobromic acid, 264
α-Naphthylamine, and sulfanilic acid test for nitrous acid, 330
test for gold, 131

1,8-Naphthylenediamine, reaction with selenites, **349**
test for nitrites, **331**
Neodymium, hydrogen-flame luminescence test for, 79
Neodymium salts, lake formation with quinalizarin, **193**
reaction with quinalizarin, 225
Nessler reaction, **236**
as spot test for ammonia, **238**, **384**
Neutral red as indicator of hydrogen-ion concentration, 504
Nickel, in acid-resistant steel, detection, **446**
in alloys, test for, **394**, **445**
aluminum in presence of, **188**
chromium in presence of, **169**
in cobalt salts, detection, **446**
copper in presence of, **88**, **429**
detection of aluminum in presence of, **190**
of cobalt in presence of, **149**
of copper traces in, **432**
in silicate rocks, **463**
dimethylglyoxime test for, **9**, **149**, **445**, **463**
in presence of oxidizing agents, **152**
indium in presence of, **214**
interference in alizarine blue test for copper, 92
in ammonium mercury thiocyanate test for zinc, 448
in dimethylglyoxime test for iron, **165**
in ferric-thiosulfate test for copper, **81**
in isonitrosobenzoylmethane test for iron, **167**
in stannous chloride test for arsenic, 100
in thiocyanate test for cobalt, **146**
iron in presence of, **164**
magnesium in presence of, 227
manganese in presence of, test for, **174**
in minerals, test with dimethylglyoxime, **490**
in mixtures, tests for, 388, 390, 392
plating, detection, **445**
in presence of cobalt, **150**, **151**, **152**, **154**, **446**
of cobalt and iron salts, 150, 152
of copper, **152**, **154**
of cyanides, **153**
of iron, **151**, **154**
of manganese, **151**
reaction with potassium ferrocyanide, **178**
rubeanic acid test for, **88**, **146**, **153**
summary of tests, 522
Nickel alloys, detection of iron in, **451**
Nickel dimethylglyoxime, formation, **242**
precipitation from alkaline cyanide solutions in presence of formaldehyde, **446**
reaction with oxidant, **153**
as reagent for acid-consuming materials, **508**
test for palladium, **11**, **137**
Nickel-free cobalt salts, preparation, **447**
Nickel hydroxide, induced oxidation of as test for sulfurous acid, 309
test for persulfuric acid, **323**
reaction with *p*-nitrobenzeneazoresorcinol, 226
Nickel oxides, reduction of, as test for hydrogen peroxide, **354**
Nickel^{IV} oxyhydrate, formation and decomposition, **323**
Nickel salt and diacetylmonoxime test for hydroxylamine, 242
Nickel salts, as catalyst in oxidation test for manganese, **174**
cobalt in, test for, **147**
color reaction with glyoxal-bis(2-hydroxyanil), **534**
induced reduction of, as test for platinum metals, **133**
interference in ammonium mercurithiocyanate test for copper, **93**
in benzidine test for lead, 72
in cerium salt reduction test for iodides, 268
in dipyridyl test for iron, 162
in potassium thiocyanate test for iron, **164**
in test for zinc, 178
in zinc uranyl acetate test for sodium, 229
presence in 1,2-diaminoanthraquinone-3-sulfonic acid test for copper, **87**
reaction with ammonium thiocyanate, **147**
with 1,2-diaminoanthraquinone-3-sulfonic acid, 86
with diphenylcarbazide, 169
with rubeanic acid, **142**, **153**
with sodium hypophosphite, 133
with thioglycolic acid, 162
zinc in presence of, 179
Nickel silicate, reaction with zinc chloride, **463**
Nimetite, detection of chlorine in, 502
Niobium, ammonium hypophosphite test for, **464**
fluorescence test for, 207

reaction with curcumin, 339
reduction of niobium pentoxide to, 509
Niobium pentoxide, detection in presence of tantalum oxide, 420
distinction from tantalic oxide, 420
fusion with zinc chloride, 420
reduction to niobium, 509
reduction test for, 419
Niobium salts, interference in cacotheline test for tin, 109
in cuprous iodide test for mercury, 66
Niobium tetroxide, formation of, 420
Niobium trioxide, formation by reduction of pentoxide, 419
Nitrates, chromotropic acid test for, 330
conversion to carbonates, 397
detection in alkali molybdates, tungstates or vanadates, 495
in metallic oxides, 494
in presence of oxidants, 328
fluorides in presence of, 273
interference in cerium salt reduction test for iodides, 268
in chromyl chloride test for hydrochloric acid, 261
in hydrogen peroxide test for titanium, 196
in methylene blue and hydrazine test for molybdenum, 118
in pyrrole test for selenious acid, 346
in silicomolybdic acid test for fluorides, 273
in test for perchlorates, 301
in titanium–malachite green test for tungsten, 122
in mixtures, test for, 385
in nitrites, test for, 329
reaction with cadmium chloride, 301
with pyrogallol, 300
reduction to nitrites, 330
in water, test for, 507
Nitric acid, brucine test for, 328
concentrated, detection of iron in, 453
diphenylamine or diphenylbenzidine test for, 327
extract, Gutzeit test in, 386
ferrous sulfate and sulfuric acid test for, 326
summary of tests, 530
Nitrites, acetic acid "ring test" for, 332
benzidine test for, 332
carbonates in presence of, 338
conversion to carbonates, 397
Griess reaction of, 246, 330
hydroxylamine in presence of, 246
interference in butyraldehyde test for cobalt, 149
in chromyl chloride test for hydrochloric acid, 261
in cobalt salt test for thiocyanates, 285
in ferrous hydroxide test for hydroxylamine, 246
in potassium thiocyanate test for molybdenum, 115
in sodium carbonate-phenolphthalein test for carbonic acid, 338
in thiourea test for selenious acid, 349
in mixtures, test for, 385
1,8-naphthylenediamine test for, 331
nitrates in, test for, 329
reaction with aminosulfonic acid, 329
with brucine, 329
with chrysean, 332
with diphenylamine, 327
with potassium thiocyanate, 164
reduction of nitrates to, 330
removal by hydrazoic acid, 329
by sodium azide or aminosulfonic acid, 330
sulfanilic acid test for, 325, 330
test by release of iodine on starch filter paper, 332
toluosafranine test for, 332
in water, test for, 507
p-Nitrobenzeneazochromotropic acid, reaction with germanic acid, 115
test for boric acid, 341
p-Nitrobenzeneazo-α-naphthol, reaction with calcium carbonate, 226
with calcium hydroxide, 226, 228
test for magnesium, 225, 461
p-Nitrobenzeneazoorcinol, reaction with magnesium salts, 194
test for beryllium, 194
p-Nitrobenzeneazoresorcinol, reaction with cadmium, cobalt, and nickel hydroxides, 226
test for magnesium, 226
p-Nitrobenzenediazonium chloride, preparation, 236
test for ammonia, 235
p-Nitrodiazoaminoazobenzene as reagent for cadmium, 99
Nitrogen, reaction with carbon and sodium, 422
Nitrogen compounds, reaction with manganese oxides, 400
Nitrogen tetroxide, formation of, 495
reaction with thio-Michler's ketone, 369
Nitro-3-hydroxybenzoic acid, reaction

with calcium, cobalt, manganese, strontium and zinc, 220
test for barium, 219
p-Nitrophenol as indicator of hydrogen-ion concentration, 504
p-Nitrophenylnitrosamine ammonium salt, formation of, 235
Nitroso compounds, identification by alkali pentacyanoaquoferroates, 71
in test for mercury, 70
p-Nitrosodiphenylamine test for palladium, 138
α-Nitroso-β-naphthol, reaction with copper and uranyl salts, 144, 145
with ferric and palladium salts, 144
reagent for cobalt, 144
test for cobalt, 17, 144, 448
for palladium, 141
β-Nitroso-α-naphthol test for zirconium, 199
2-Nitroso-1-naphthol-4-sulfonic acid, reaction with copper salts, 149, 167
with ferrous salts, 149
test for cobalt, 94, 149
for cobalt salts, 167
for copper, 94
for iron, 94, 167
Nitroso-nitrobenzene, alkali salts of aci-form, formation of, 320
Nitrosylthiocyanate, formation, 164
Nitrous acid, conversion of hydroxylamine to, 245
1,8-naphthylenediamine test for, 331
reaction with sulfamic acid, 325
with thio-Michler's ketone, 369
sulfamic acid test for, 325
sulfanilic acid and α-naphthylamine test for, 330
summary of tests, 530
test for hydrazoic acid, 287
test by release of iodine on starch filter paper, 332
Noble metals, reaction with alkali and manganese salts, 58
Noble metal salts, reduction by stannous chloride, 100
Non-uniformities in fine powders, detection, 397
Nuclear growth as supplementary process, 7

O

Off-print process, 21
Oils, lead in, detection, 435
Olivinic tesinite, chromium in, 455
Optical glasses, alkali in, 474
Ores, beryllium in, detection, 457
lead in, detection, 434
Organic acids, interference in *p*-dimethylaminoazophenylarsonic acid test for zirconium, 203
in stannous chloride test for tungsten, 121
reaction with molybdates, 115, 116
Organic hydroxy acids, interference in alizarin test for zirconium, 201
Organic materials, charring, 316
Organic reagents, 19
Organic substances, interference in test for water, 510
Original substance, Gutzeit test on, 384
Orthoperoxyvanadic acid, formation of, 123
Osmium, activation of chlorate solutions as test for, 141
detection by induced reduction of nickel salts, 134
detection of palladium in presence of, 136
interference in phenoxithine test for palladium, 140
in catalytic test for palladium, 136
reaction with benzidine acetate, 143
with potassium ferrocyanide, 143
ruthenium in presence of, detection, 143
summary of tests, 523
Osmium salts, and catalytic action of mercury, 68
interference in cerium salt reduction test for iodides, 268
reduction by stannous chloride, 134
Osmium tetroxide, action on potassium iodide, 142
Oxalates, fluorides in presence of, 271
interference in ammonium molybdate test for phosphoric acid, 333
in hypoiodite test for magnesium, 224
in phenoxithine test for palladium, 140
in potassium ferrocyanide test for iron, 161
in potassium thiocyanate test for iron, 164
in zirconium-alizarin test for fluorides, 269
Oxalic acid, in extraction test for bismuth, 80
test for uranium, 207
for uranyl ions, 208
thiocyanate in presence of, 282
Oxidants, boric acid in presence of, 342
chloramine-T as, 262

detection of nitrates in presence of, 328
in presence of hypohalogenites, 295
dimethylglyoxime test for nickel in presence of, 152
formation of benzidine blue by, 322
interference in hydroxylamine test for germanium, 115
in 8-hydroxyquinoline-7-iodo-5-sulfonic acid test for iron, 163
in α-naphthylamine test for gold, 131
in thallous sulfate test for ferricyanides, 294
reaction with *p*-nitrobenzeneazochromotropic acid, 342
Oxidation, 8
test for free carbon, 379
for thallium, 157
Oxides, basic, in glass, detection, 474
free metals in, detection, 475
Oxidizable material, *see* Reductants
Oxidizing cations, reaction with thio-Michler's ketone, 369
Oxinates, prevention of formation, test for hydrofluoric acid, 275
Oxine (*see also* 8-Hydroxyquinoline), halogenated, formation of, 259
test for dicyanogen, 371
for iron, 166
for thallium, 161
for tungsten, 459
for uranium, 204
for vanadium, 125, 127
Oxine tungstate, formation of, 119
Oxine uranate, formation of, 205

P

Palladium, detection by condensation of hydrogen on, 131
of gold in presence of, 131
by induced reduction of nickel salts, 133
by mercuric cyanide and diphenylcarbazide, 141
in presence of osmium, 136
in presence of platinum or rhodium, 140
p-dimethylaminobenzylidenerhodanine reaction for, 59, 128, 141
distinction from platinum foil, 444
ferrocyanide and α,α'-dipyridyl test for, 141
p-fuchsin test for, 141
3-hydroxy-1-*p*-sulfonatophenyl-3-phenyltriazine test for, 140
interference in dimethylaminobenzylidenerhodanine test for gold, 128
in cuprous iodide test for mercury, 65
mercaptobenzeneindazole test for, 141
mercury iodide test for, 137
methyl yellow test for, 141
naphthalene-4-sulfonic acid-1-azo-5-*o*-8-hydroxyquinoline test for, 140
nickel dimethylglyoxime test for, 11, 137
p-nitrosodiphenylamine test for, 138
nitrosonaphthol test for, 141
phenoxithine test for, 139
in platinum salts, test for, 136
in presence of gold and platinum, stannous chloride test for, 135
reaction with dimethylglyoxime, 150
with potassium ferrocyanide and α,α'-dipyridyl, 64, 72, 141
rubeanic acid test for, 90
silver in presence of, 60
summary of tests, 523
test by catalysis of the carbon monoxide reduction of phosphomolybdates, 135
Palladium chloride, and auric chloride, reaction with rubidium and thallium salts, 235
test for cesium, 235
reaction with silver iodide, 411
reduction, as test for mercury, 364
Palladium dimethylglyoxime, formation of, 128
Palladium ions, as catalysts in pentacyanoaquoferroate formation, 71
Palladium salts, interference in alkali iodide test for platinum, 135
in test for bismuth, 80
reaction with cacotheline, 68
with nitrosonaphthol, 144
with rubeanic acid, 143
reduction by stannous chloride, 134
Palladous chloride test, for hydriodic acid, 265
Paper ash, detection of alkali in, 482
of alumina in, 454
of barium sulfate or lithopones in, 476
Papers, recommended for spot test use, 54
Paper chromatography, prospects in spot test analysis, 23
Paracyanogen, formation of, 412
Paraffined porcelain dishes as medium for spot tests, 12
Parisian blue, *see* Ferric ferrocyanide
Pentacyanoaquoferroates, formation of, 70
identification of nitroso compounds by, 71
Peracids, interference in benzidine test for nitric acid, 327

Perchlorates, cadmium chloride test for, 300
detection in presence of chlorate or nitrate, 301
summary of tests, 530
Percompounds, reaction with nickel oxide, 323
Peridotite, chromium in, 455
Periodates, interference in pyrrole test for selenious acid, 346
reaction with manganous sulfate and phosphoric acid, 301
and tetrabase test for manganese, 174
Periodic acid, formation of complex copper salts as test for, 302
formation of stable copper salts by, 174, 302
manganese salts and tetrabase test for, 302
manganous sulfate and phosphoric acid test for, 301
as preventive of copper catalysis, 174
summary of tests, 530
Permanganate(s), chromates in presence of, 345
detection of hypohalogenites in presence of, 296
interference in tetrabase oxidation test for ferricyanic acid, 293
in test for zinc, 178
oxidation of manganese salts to, 170, 173, 456
in presence of chromate, 8, 344
cellulose test for, 8, 10, 344
reaction with benzidine, 322
with 3,3′-dimethylnaphthidine, 126
with diphenylamine, 327
with sodium azide, 170
Permanganic acid, summary of tests, 530
test by oxidation of cellulose, 344
Permolybdate and α-naphthoflavone test for hydrobromic acid, 264
Peroxide reaction for titanium, 195, 413
Peroxides, discharge of chromate-diphenylcarbazide reaction by, 356
interference in benzidine test for nitric acid, 327
reaction with diphenylamine, 327
Peroxodisulfatotitanic acid, formation, 195
Peroxo-vanadium salts, formation of, 417
Persulfates, detection in presence of hydrogen peroxide, 324
interference in test for zinc, 178
reaction with manganese salts, 302
with manganous sulfate, 301
and phosphoric acid, 298
with pyrogallol, 300
Persulfuric acid, benzidine test for, 322
nickel hydroxide test for, 323
summary of tests, 530
pH, rôle in spot reactions, 16
Pharmaceutical materials, lead in, detection, 437
spot reactions in examination of, 428
test for copper in, 430
Phenanthroline, and ferric chloride, test for uranium, 210
ferrous salts of, 15
α,α′-Phenanthroline, as substitute for α,α′-dipyridyl, 162
and thioglycolic acid, test for iron, 449, 451, 452, 453
p-Phenetidine and ferricyanide test for zinc, 182
Phenetidine-HCl and hydrogen peroxide test for copper, 94
Phenolphthalein, and barium carbonate test for sulfuric acid, 314
as indicator of hydrogen-ion concentration, 504
paper test for magnesium, 228
reduction to phenolphthalin, 292
and silver carbonate, preliminary test for halides, 261
and sodium carbonate, test for carbonic acid, 337
Phenolphthalin, catalyzed oxidation of, as test for hydrogen peroxide, 355
reduction of phenolphthalein to, 292
solution, preparation of, 293
test for ferricyanic acid, 292
Phenothiazine test for silver, 64
Phenoxithine test for palladium, 139
Phenylhydrazine, test for molybdenum, 117
9-Phenyl-2,3,7-trihydroxy-6-fluorone test for germanium, 114
see also *o*-Hydroxyphenylfluorone
Phosphate(s), fertilizers, distinction from superphosphate, 471
fluorides in presence, of, 271
interference in alizarin test for zirconium, 201
in butyraldehyde test for cobalt, 149
in *p*-dimethylaminoazophenylarsonic acid test for zirconium, 203
in hydroxylamine test for germanium, 115
in hypoiodite test for magnesium, 224
in potassium thiocyanate test for iron, 164

in zirconium-alizarin test for fluorides, 269
in zirconium azoarsenate test for fluorides, 272
in minerals, detection, 500
in mixtures, test for, 385
in presence of arsenic and silicic acids, 333
reaction with molybdates, 115
in acid solution, 333
in rocks, detection, 500
in water, test for, 507
Phosphomolybdates, catalytic carbon monoxide reduction of, as test for palladium, 135
interference in pyrrole test for selenious acid, 346
platinum as catalyst in reduction of, 136
reduction by antimony, copper, iron, mercurous, and tin salts, 156
by antimony trichloride, 108
by stannous chloride, 108
Phosphomolybdic acid, formation, 333
formation of benzidine blue by, 333
of molybdenum blue by, 333
and hydrobromic acid test for thallium, 156
reaction with free metals, 475
with stannous chloride, 108
test for antimony, 104
for cerium, 212
for copper, 84
for free metals and alloys, 362
for vanadium, 127
Phosphoric acid, ammonium molybdate and benzidine test for, 333
o-dianisidine molybdate and hydrazine test for, 335
effect on quinalizarin test for magnesium, 225
free, test for, 316
interference in methylenedisalicylic acid test for free sulfuric acid, 317
in molybdate–benzidine test for germanium, 113
in stannous chloride test for tungsten, 121
iron in, detection, 449
and manganous sulfate test for chloric acid, 297
for periodic acid, 301
as medium for pyrrole test for selenious acid, 346
in presence of arsenic acid, 333
of silicic acid, 334
reaction with benzidine, 336
with molybdates, 115
silicic acid in presence of, 336
summary of tests, 530
thiocyanate in presence of, 282
Phosphorus segregations in iron and steel, detection, 487
Photo-reactions, prospects in spot test analysis, 23
Picric acid, and sodium carbonate, reaction with sulfides, 280
test for cyanide, 280
Pigments, barium sulfate and lithopones in, detection, 476
lead in, detection, 434
Pipet bottle, 32 (Fig. 2)
Platinum, alkali iodide test for, 134
as catalyst in reduction of phosphomolybdates, 136
detection by condensation of hydrogen on, 131
of gold in presence of, 131
of palladium in presence of, 140
p-dimethylaminobenzylidenerhodanine reaction for, 59
foil, distinction from palladium, 444
interference in *p*-nitrosodiphenylamine test for palladium, 139
in cuprous iodide test for mercury, 66
metallic, ferric ferricyanide test for, 363
palladium in presence of, 135
in presence of arsenic, detection, 133
of other noble metals, stannous chloride test for, 134
rubeanic acid test for, 90
silver in presence of, 60
summary of tests, 523
Platinum metals, catalytic tests for, 131
detection of gallium in presence of, 215
by induced reduction of nickel salts, 133
Platinum salts, interference in test for bismuth, 80
reaction with *p*-dimethylaminobenzylidenerhodanine, 141
reaction with rubeanic acid, 143
reduction to platinum by hypophosphorous acid, 349
by stannous chloride, 135
Platinum spoon fused into glass tube, 37 (Fig. 12)
Polyethylene containers for storage of reactive reagents, 31
Polymethine dyes, detection of chlorine and bromine by formation of, 369
Polyselenides, detection of free selenium in, 491
Polysulfides, detection of free sulfur in, 491

distinction from monosulfides, 491
formation of, 378
reaction with free tellurium, 378
Pontachrome blue-black R test for aluminum, 191
Porcelain, dish, heating of sample in, 397
spot plates, 12
ware for stocking of versatile spot test laboratory, 30
Portable kits for spot test work, 30
Potassium, detection in presence of sodium, 232
dipicrylamine test for, 230, 462
in glass, test for, 473
lithium in presence of, 233
in mixtures, test for, 389
in presence of sodium, 231
in silicate rocks, detection, 461
sodium cobaltinitrite and silver nitrate test for, 230
sodium tetraphenyl boron test for, 232, 462
summary of tests, 523
Potassium bichromate, detection of potassium chromate in, 508
Potassium bifluoride, decomposition of chromium compounds by, 414
Potassium bismuth iodide, reaction with thallium, 235
test for cesium, 235
Potassium bromide-cuprous bromide test for bromides, 265
Potassium cerous carbonate test for hydrogen peroxide, 355
Potassium chlorate, decrepitation test for carbon in coal, 422
detection of perchlorate in presence of, 301
Potassium chloride, lithium in presence of, 234
Potassium chromate, in potassium bichromate, detection, 508
test for lead in minerals, 490
for silver, 63
Potassium chromium thiocyanate as reagent for bismuth, 79
Potassium cobalticyanide test for zinc, 182
Potassium cuprocyanide, reduction of molybdenum in phosphomolybdates by, 85
Potassium cyanide, formation, 380
reaction with free halogens, 370
Potassium ferricyanide, detection of ferrocyanide in presence of, 289
Potassium ferrocyanide, detection in presence of ferricyanide, 289
and α,α'-dipyridyl, reaction with mercury, 72
with palladium, 64, 72, 141
with silver, 64, 72
test for mercury, 72
for mercuric cyanide and oxycyanide, 282
for palladium, 141
for silver, 64
for silver halides, 410
interference in iodine–azide test for thiocyanate, 283
and manganese salts test for gallium, 214
reaction with aluminum, 186
with calcium, 220
with cobalt, copper, iron^{II}, manganese, and nickel, 178
with copper and ferric salts, 205
with osmium, 143
with silver chloride, 262
test for iron, 161
for uranium, 205, 457
Potassium fluoride, acid, and *o*-hydroxyphenylfluorone, test for copper, 94
detection by zirconium alizarinate, 269
Potassium fluoroberyllate, formation, 457
Potassium iodate test for free carbon, 381
Potassium iodide, action of osmium tetroxide on, 142
as catalytic reducing agent for arsenic acid, 100
paper test for bismuth in minerals, 490
potassium thiocyanate in, test for, 283
reaction with cupric salts, 145
with lead, mercury, and silver, 155
with mercuric oxycyanide, 282
starch paper, detection of chlorine by, 1
test for thallium, 154
Potassium manganocyanide as reagent for bismuth, 80
Potassium nickelcyanide and dimethylglyoxime, test for silver, 64
Potassium nitrate, detection of perchlorate in presence of, 301
Potassium nitrite and potassium thiocyanate paper, test for cobalt in minerals, 490
Potassium perceric carbonate, formation, 355
Potassium permanganate, formation of benzidine blue by, 284
precipitation of barium sulfate in presence of, as test for barium, 218
test for hydrobromic acid, 265
Potassium persulfate, detection by benzidine, 322

fusion with alumina, 413
reaction with manganese silicate or dioxide, 456
Potassium salts, interference in dipicrylamine test for thallium, 157
in zinc uranyl acetate test for sodium, 229
Potassium silicate, reaction with zinc chloride, 462
Potassium sulfate, zinc in presence of, 181
Potassium thiocyanate, formation by activated sulfur, 509
molten, heating, 509
in potassium iodide, test for, 283
in presence of sodium thiosulfate, 284
and stannous chloride test for molybdenum, 115
test for chemically or adsorptively bound water, 509
for iodic acid, 299
for iron, 164
α,α′-dipyridyl test as alternative for, 162
Potassium xanthate, test for molybdenum, 116, 444
Praseodymium, hydrogen-flame luminescence test for, 79
Praseodymium salts, reaction with quinalizarin, 225
Precipitate, color intensity, 7
dispersion of, 7
form species of, 7
formation of, 43
washing of, 44
visible beginning of, 7
Precipitation reactions, organic reagents in, 19
Preliminary operations, 28
Preliminary tests, for acid radicals, 259
for halogen hydracids, 261
of sample, 57
spot reactions in, 57, 384, 394
Prospects of spot test analysis, 23
Protective layer effect, 11
Proteins, prevention of starch-iodine reaction by, 266
Prussian blue, detection of, 476
distinction from Turnbull's blue, 291
formation, 161, 283, 289, 294, 354, 363
test for ferrocyanide, 9, 289, 290
for sodium cyanide, 423
Prussic acid, *see also* Hydrocyanic acid
benzidine reaction for, 284
copper acetate–benzidine acetate reaction for, 276
and starch–iodine test for cyanides, 280
test for thiocyanates by conversion to, 284
Pulverized mixtures, heating, 397
Purity, of reagents, importance of, 28
spot reactions in tests of, 428
Purpurin, as reagent for boric acid, 340
Purpurogallin, formation, 300
Pyrex glass, alkali in, 474
Pyridine, as solvent, 373, 376
Pyrite, detection of free sulfur in, 491
iodine-azide reaction of, 407
Pyrocatechol, iron salts of, 166
reaction with antimony salts, 105
with iron salts, 196
test for titanium, 196
Pyrogallol, oxidation by chromate and iodine, 300
reaction with antimony salts, 105
with bromate, nitrate, and persulfate, 300
test for gold, 131
for iodic acid, 300
Pyrolusite, test for, 415
test for iron in, 450
Pyromorphite, detection of chlorine in, 502
Pyrope, chromium in, 455
Pyrotungstic acid, oxine ester, formation of, 120
Pyrrole, test for chromium, 173
for selenious acid, 346
Pyrrole blue, formation, 346

Q

Quantitative determinations by spot reactions, 21
Quantity sensitivity, 4
Quercetin or quercitrin reaction with uranium salts, 167
test for ferric salts, 167
for uranium, 209
for vanadium, 127
Quinalizarin, for differentiation of magnesites, 467
reaction with beryllium salts, 225
with cerium, lanthanum, neodymium, praseodymium, thorium and zirconium salts, 193, 225
with magnesium, 193
test for aluminum, 188
for beryllium, 192, 457
for boric acid, 340
for indium, 213
for magnesium, 224
Quinoline and potassium iodide as reagents for bismuth, 80

Quinoneanildiphenylhydrazone, formation, 347

R

Radioactive materials, ring oven analysis of, 56
Rare earths, cerium in mixture of, 210-212
Réaction à la goutte, 2
à la touche, 2
Reagents, for use in preliminary examination of a mixture, 401–406
stocking of, 31
Recrystallization as supplementary process, 7
Red lead, detection of free metal in, 475
Reducible metals, detection in minerals, 463
Reductants, interference in alkali hypoiodite test for magnesium, 224
in catalytic test for iodine, 512
in hydroxylamine test for germanium, 115
in test for metallic aluminum, lead, tin and zinc, 364
inorganic, interference in potassium iodate test for free carbon, 381
Region of uncertain reactions, 14
Requirements for successful application of spot tests, 28
Residue, "insoluble", detection of substances in, 407
Resoflavine test for titanium, 199
Resorcinol, acceleration of autoxidation of, as test for copper, 94
as test for silver, 94
prevention of starch-iodine reaction by, 266
test for zinc, 182
Rhodamine B, reaction with anhydrite, 472
with antimony, 105
with bismuth, 107
with gold, 107
with mercury, 107
with molybdates, 106, 107
with tungstates, 106, 107
test for antimony, 105, 442
for antimony pentoxide, 419
for gallium, 215
for gold, 129, 445
for thallium, 158, 161, 437
for uranium, 208, 457
Rhodanine dyes, behavior on filter paper, 15
Rhodium, detection by condensation of hydrogen on, 131
of palladium in presence of, 140
summary of tests, 524
Rhodium salts, interference in cacotheline test for tin, 109
Rhodizonate test for calcium, 222
for lead, 437
Rhodizonic acid, reaction with barium salts, 313
test for barium, 216
Ring oven method, 54, 394, 423
identification reactions, 424, 426
separation scheme, 424
"Ring test" for nitrites, 332
Rinmann's green, formation, 182
Rocks, basic, detection, 508
boron in, detection, 500
chlorine in, detection, 501
chromium in, detection, 455
fluorine in, detection, 490
inhomogeneities in, 21
manganese in, detection, 456
nickel in, detection, 463
phosphates in, detection, 500
sulfide sulfur in, iodine–azide reaction for, 491
Rongalite, detection by chromotropic acid, 321
by thermal decomposition, 321
differentiation from sodium hydrosulfite, 321
oxidation of, 322
reaction with dinitrobenzene, 320, 321
stability of, 320
Rosocyanine conversion of curcumin to, 339
Rubeanates, formation of polymers of, 153
fractional precipitation of, 11
Rubeanic acid, reaction with copper, cobalt, and nickel salts, 142, 153
with copper and nickel, 146
test for bismuth, 90
for cobalt, 89, 146
for copper, 87, 429
for ferric salts, 90
for mercurous salts, 90
for nickel, 89, 153
for palladium, 90
for platinum, 90
for ruthenium, 142
for silver, 90
Rubidium, lithium in presence of, 233
Rubidium chloride, lithium in presence of, 234
Rubidium salts, interference in test for potassium, 232
reaction with auric chloride and palladium chloride, 235

with dipicrylamine, 231
with gold platinibromide, 234
Ruthenium, activation of chlorate solutions as test for, 141
detection by induced reduction of nickel salts, 134
interference in phenoxithine test for palladium, 140
in test for palladium, 136
in presence of osmium, test for, 143
rubeanic acid test for, 143
summary of tests, 524
Ruthenium salts, reaction with rubeanic acid, 143
Rutile, titanium in, detection, 459

S

Safranin, reaction with ferricyanides, 295
test for hypohalogenites, 294
Safranine T, *see* Toluosafranine
Salicylaldazine, formation, 239, 244
Salicylaldehyde, and copper salts test for hydroxylamine, 243
reaction with hydrazine, 239
test for hydrazine, 239
Salicylaldoxime, as reagent for copper, 83, 243
test for hydrazine, 23
Samarium, hydrogen-flame luminescence test for, 79
Samarskite, uranium in, detection, 457
Samples, preliminary tests, 57
Sampling, 32
Sapamine, for preparation of impregnated papers, 22
Scandium salts, reaction with morin, 183
Scheelite, decomposition of, 459
molybdenum in, 444
Schiff bases of glutaconic aldehyde, formation of, 369
Schleicher and Schüll filter paper, 54
Sea water, detection of iodine in, 511
Selective reactions, designation, 18
Selective reagents, designation, 18
Selenates, reduction by iron sulfate, 351
Selenic acid, distinction from selenious acid, 346
summary of tests, 531
Selenious acid, catalysis of methylene blue reduction, test for, 348
4-dimethylamino-1,2-phenylenediamine test for, 349
diphenylhydrazine test for, 347, 492
distinction from selenic acid, 346
hydriodic acid test for, 346
interference in tellurium test, 351
4-methylthio-1,2-phenylenediamine test for, 349
pyrrole test for, 346
reaction with 4-dimethylamino (or thio)-1,2-phenylenediamine, 349
reduction to selenium, 346, 349
summary of tests, 531
Selenites, in mineral acid solutions, detection, 347
reaction with 1,8-naphthylenediamine, 349
with 3,4,3′,4′-tetraaminobenzidine, 349
with thiourea, 349
reduction by ascorbic acid, 349
by iron sulfate, 351
Selenium, free, detection, of free tellurium in presence of, 379
detection in presence of free sulfur, 375
in minerals, inorganic mixtures and polyselenides, detection, 491
reaction with thallous sulfide paper, 12, 372, 375
summary of tests, 533
test by addition to thallous sulfide, 375
by conversion to selenosulfate, 376
by formation of silver selenide, 376
germanium in presence of, 114
in minerals, test for, 492
in presence of sulfur, detection, 376, 378
of tellurium, test for, 346
reaction with silver, 376
reduction of selenious acid to, 346, 349
stannous chloride test for, 100
in sulfur, detection, 492
in tellurium, detection, 492
tellurium in presence of, tests for, 349–352
Selenium oxygen compounds, interference in methylene blue test for selenium, 377
lenium sulfide as reagent for mercury
Sevapor, 365
Selenosulfates, distinction from thiosulfates, 377
formation of, as test for free selenium, 376
reaction with formaldehyde, 377
Selenosulfides, formation of, 378
Semimicroanalysis, 6
Semimicrochemical tests, choice of, 57
Sensitivity of spot reactions, 4
data, validity, 58
increase by formation of colored products, 8
by insoluble reagents, 9

by protective layer effect, 11
by shaking out with organic liquids, 12, 15, 44
measurement, 13
numerical values, 6
Separation methods, 39
Sequestering of test ions, 16
Serpentine, chromium in, 455
Sieving solid samples, 34
Silica, in caustic alkalis and alkaline solutions, 499
in insoluble residue, 418
in water, test for, 506
Silicate rocks, potassium in, detection, 461
test for magnesium in, 461
Silicates, alkali metals in, test for, 464
conversion of silicon and silicides to, 419
detection of calcium in, 460, 468
ferrous iron in, detection, 462
fusion reactions with zinc chloride, 460, 461, 462, 463
in insoluble residue, 418
interference in test for phosphates, 335
in mixtures, test for, 385
reaction with *o*-dianisidine molybdate, 335
titanium in, 460
Siliceous products, vulnerability to attack by alkalis, test for, 499
Silicic acid, ammonium molybdate and benzidine test for, 335
hydrated, detection of water in, 510
interference in etching test for fluorides, 274
in molybdate–benzidine test for germanium, 113
in minerals, test for, 499
phosphates in presence of, 333
phosphoric acid in presence of, 334
in phosphoric acid, test for, 336
reaction with molybdates in acid solution, 333
silicomolybdate reaction for, 418
summary of tests, 531
Silicides, conversion to silicate, 418
in insoluble residue, 418
Silicofluorides, etching test for, 274
Silicomolybdate reaction for silicic acid, 418
Silicomolybdic acid, formation, 333, 334
formation of benzidine blue by, 333, 336, 419
of molybdenum blue by, 336
test for hydrofluoric acid by conversion into, 272
Silicon, conversion to silicate, 419
in insoluble residue, 418
Silicon dioxide, interference in fluorescence test for uranium, 207
Silicon tetrafluoride, conversion to, as test for silicic acid in minerals, 499
formation, 272
Silk threads as medium for spot tests, 12
Silver, in alloys, test for, 393, 429
catalytic reduction of manganese and cerium salts as test for, 62
chromotropic acid and stannous chloride test for, 64
copper in presence of, 429
copper thiocyanate test for, 64
deposition test for, 61
p-dimethylaminobenzylidenerhodanine test for, 59, 67, 128, 411, 429
p-dimethylaminobenzylidenethiobarbituric acid test for, 64
dithizone test for, 63
for zinc in presence of, 180
ethylenediaminetetraacetic acid test for, 64
manganese in presence of, 177
manganese nitrate and alkali test for, 58
metallic, ferric ferricyanide test for, 363
in minerals, test with chromate and potassium iodide paper, 490
in a mixture of lead, mercurous and silver chlorides, detection, 60
in mixtures, test for, 386, 389
phenothiazine test for, 64
potassium chromate test for, 63
potassium ferrocyanide test for, 64
potassium nickelcyanide test for, 64
precipitation by antimony sulfide paper, 51
in presence of gold, palladium, and platinum, detection, 60
of mercury, detection, 60
of thallium, catalytic test, 63
reaction with activated sulfur, 509
with potassium iodide, 155
with selenium, 376
with sodium rhodizonate, 74
resorcinol test for, 94
rubeanic acid test for, 90
in silver plating, detection, 429
stannous chloride test for, 63
summary of tests, 524
thallium in presence of, 155
Silver ammine salts, test for manganese, 177
Silver azide, danger in ignition test, 411
formation, as test for hydrazoic acid, 286
Silver carbonate and phenolphthalein, preliminary test for halides, 261

Silver chloride, identification, 262
reaction with potassium ferrocyanide, 262
silver in a mixture of lead and mercurous chlorides and, 60
Silver ferricyanide, detection of silver ferrocyanide in presence of, 412
Silver ferrocyanide, conversion to silver ferricyanide, 262
detection in presence of silver ferricyanide or silver halides, 412
distinction from silver ferricyanide, 412
Silver halides, demasking test for, 16
detection of silver ferrocyanide in presence of, 412
differentiation by heating, 411
α,α'-dipyridyl and potassium ferrocyanide test for, 410
fractional precipitation of, 11
in insoluble residue, 409
Silver iodide, reaction with palladium chloride, 411
Silver ions, as catalysts in pentacyanoaquoferroate formation, 71
Silver nitrate, and formaldehyde test for ammonia, 238
and manganese sulfate test for ammonia, 237
paper test for arsenic in minerals, 490
precipitation in alkaline solution as test for mercuric cyanide, 280
as reagent for arsenic hydride, 101
and sodium cobaltinitrite test for potassium, 230
and tannin test for ammonia, 238
in test for persulfate in presence of hydrogen peroxide, 325
test for arsenates, 102
for iodides, 268
Silver oxide, reaction with malachite green, 212
Silver salts, ammonia in presence of, 238
cadmium in presence of, 95
and catalytic action of mercury, 68
cerium in presence of, 210
formation of benzidine blue by, 172
ignition test for, 411
interference in ascorbic acid test for gold, 131
in benzidine test for lead, 72
in cerium salt reduction test for iodides, 268
in cuproin test for cuprous ions, 91
in cuprous iodide test for mercury, 66
in dimethylaminobenzylidenerhodanine test for gold, 128
in mercuric iodide test for palladium, 138
in *o*-tolidine test for copper, 84
manganese in presence of, 173
in manganese sulfate test for ammonia, 237
oxidation of benzidine by, 176
in oxidation of chromium salts to chromates, 170
reaction with benzidine, 172, 211
with chromotropic acid, 198
with *p*-dimethylaminobenzylidenerhodanine, 141
with potassium chromium thiocyanate, 79
by ethylenediaminetetraacetic acid, 64
reduction to silver by hypophosphorous acid, 349
by stannites, 350
in sodium cobaltinitrite test for potassium, 230
Silver selenide, formation as test for free selenium, 376
Silver sulfide, formation by activated sulfur, 509
Sintering reactions, 13
Small volumes of liquid, concentration and drying, 35
Smaragdite, chromium in, 455
Sodalite, detection of chlorine in, 502
Sodium, detection of potassium in presence of, 232
in glass, test for, 473
lithium in presence of, 233
in mixtures, test for, 388
molten or vapor, reaction with carbon, 423
potassium in presence of, 232
reaction with carbon and nitrogen, 422
summary of tests, 524
zinc uranyl acetate test for, 229
Sodium azide, decomposition, 422
reaction with cobalt salts, 310
with metallic sulfides, 304
with permanganates and chromates, 170
with sulfo salts, 305
removal of nitrites by, 329, 330
and *o*-tolidine, test for cobalt, 149
test for mercury sulfide, 305
Sodium carbonate, crystal water in, detection, 510
extracts in tests for acid radicals, 258
hydrated, test for water in, 510
and phenolphthalein test for carbonic acid, 337

as solvent for thallium, 157, 158, 160
Sodium chloride, lithium in presence of, 233
solution, detection of iodine and bromine in, 265
Sodium cobaltinitrite, reaction with ammonium, lithium, and thallium salts, 230
and silver nitrate test for potassium, 230
Sodium cyanide, Prussian blue test for, 423
Sodium diethyldithiocarbamate test for copper, 432
Sodium fluoride, decomposition of ferric thiocyanate by, 290
in wood preservatives, test for, 477
Sodium formate, decomposition of, 366
reaction with arsenic compounds, 443
Sodium hydrosulfite, differentiation from Rongalite, 321
reaction with dinitrobenzene, 320, 321
with formaldehyde, 322
stability of, 320
Sodium hypophosphite, reaction with nickel salts, 133
Sodium hyposulfite, *see* Sodium hydrosulfite
Sodium nitroprusside, detection of sulfides by, 408
reaction with sulfides, 303
test for hydrogen sulfide, 303
for sulfide, 315
for sulfurous acid, 307
Sodium pentacyanopiperidine ferroate test for cobalt, 149
Sodium phosphate, interference in iodine-azide test for thiocyanate, 283
test for uranyl ions, 209
Sodium plumbite test for sulfides, 307
Sodium potassium carbonate, fusion with, as test for silicic acid in minerals, 499
Sodium rhodizonate, behavior of alkaline earth sulfates toward, 223
differentiation of barium and strontium minerals by, 473
reaction with barium, cadmium, silver, tellurium, and tin, 74
reaction with barium salts, 220, 313
with strontium salts, 216,
test for barium, 216, 476
for calcium, 222, 460, 469, 470, 471
for lead, 73, 416, 433
for lead sulfate in insoluble residue, 409
for strontium, 220
Sodium salts, interference in dipicrylamine test for thallium, 157
sulfate in "pure", detection, 494
Sodium sulfate, crystal water in, detection, 510
Sodium sulfide, detection, 303, 305
test for cadmium, 99
Sodium sulfite, in presence of sodium thiosulfate, detection, 312
reaction with activated sulfur, 509
Sodium sulfoxylate-formaldehyde, *see* Rongalite
Sodium tetraphenyl boron, test for potassium, 232, 462
Sodium thiosulfate, decomposition of ferric thiocyanate by, 269
potassium thiocyanate in presence of, 284
in presence of sodium sulfide, 319
as reagent for molybdenum, 115
sodium sulfite in presence of, 312
test for cobalt, 143
Sodium zinc uranyl acetate, formation, 229
Sodium zirconium sulfate, formation, 414
Sol formation as supplementary process, 7
Solid(s), choice of solvent for, 41
extraction with organic solvents, 33
fusion with disintegrating agents, 33
grinding and mixing, 33
metal sulfides, reaction with sodium azide, 305
preparation of solutions of, 33
reactions with reagent solution, 27
sampling of, 33
sieving of, 34
testing by spot reactions, 40
volatilization of components, 33
Solid and liquid phases, separation, 41
Solid-solid reactions in absence of water, 380
Solubility, determination, 40
of metals in nonoxidizing acids, 361
Solutions, concentrating, 35
Special techniques, 50
Specific reactions, designation, 18
Specific reagents, designation, 18
Spot analysis, applications, 7, 23
Spot colorimetric determinations, 21
Spot colorimetry, 14, 50
prospects in spot test analysis, 23
Spot nephelometry, 51
Spot reactions, influence of concentration, 21
on solid materials, 21

Spot test analysis, value in other provinces of chemistry, 3
Spotting, 27
Spraying of reagents on filter paper, 52
Standard products, 29
Standardization of microchemical apparatus, 30
Stannic salts, interference in test for beryllium, 195
  reduction by dithiol, 107
    by thioglycolic acid, 107
Stannite(s), color reaction with dinitrobenzene, 320
  reduction of metallic salts by, 350
  test for bismuth, 77, 80, 438
  tin in, 439
Stannous chloride, and aniline test for mercury, 66
  reaction with 2-benzylpyridine, 313
    with cacotheline, 421
    with phosphomolybdic acid, 108
  reduction of gold, osmium, palladium, and platinum salts, 134
    of mercuric salts, 66
    of molybdates, 108
    of phosphomolybdates, 108
  test for arsenic, 99
    for gold, 131
    for palladium in presence of gold and platinum, 135
    for platinum in presence of other noble metals, 134
    for selenium, 100
    for silver, 63
    for tellurium, 100
    for tungsten, 120, 416
Stannous salts, interference in phenoxithine test for palladium, 140
  reaction with alkali and manganese salts, 58
Starch, filter paper impregnated with, 23
  test for iodides, 269
Starch–iodine reaction, detection of iodates by, 299
  sensitivity, 266
  test for mercuric cyanide, 281
Starch–potassium iodide test for free halogens, 368
Stassfurtite, detection of chlorine in, 502
Steamer (treating paper with gases), 46 (Fig. 21)
Steel, acid-resistant, detection of nickel in, 446
  chromium in, detection, 455
  manganese in, detection, 177
  molybdenum in, detection, 444
  phosphorus segregations in, detection, 487
  ring oven analysis of, 56
  sulfur segregations in, detection, 486
Stilliréaction, 2
Stirrer for centrifuge tube, 43 (Fig. 19)
Stolzite, decomposition of, 459
  lead in, detection, 434
Strontianite, differentiation from witherite, 473
Strontium, barium in presence of, 216, 219
  detection of calcium in presence of, 223, 535
  magnesium in presence of, 228
  in mixtures, test for, 388
  in presence of barium, 217, 220
  sodium rhodizonate test for, 220
  summary of tests, 524
Strontium carbonate, differentiation from barium carbonate, 473
Strontium salts, color reaction with glyoxal-bis(2-hydroxyanil), 534
  interference in cerium salt reduction test for iodides, 268
    in rhodizonate test for calcium, 222
  in zinc uranyl acetate test for sodium, 229
  reaction with nitro-3-hydroxybenzoic acid, 220
    with sodium rhodizonate, 216
Strontium sulfate, detection of gypsum in presence of, 471
  differentiation from barium sulfate, 473
Strychnine test for chromium, 173
Submicrogram methods, 6
Suction, removing liquid by, 43 (Figs. 18 and 20)
Sulfamic acid, deamidation test for, 325
  detection in presence of sulfite, sulfates or carbonates, 326
  interference in nitrous acid test for hydrazoic acid, 287
  reaction with nitrous acid, 325
  summary of tests, 531
  test for nitrous acid and nitrites, 325
Sulfanilic acid, and α-naphthylamine test for nitrous acid, 330
Sulfates, barium rhodizonate test for, 313
  catalysis of iodine–azide reaction as test for, 408
  detection of sulfamic acid in presence of, 326
  in fluorides, detection, 493
  in hydrofluoric acid, test for, 493
  in ink writing, test for, 480
  in inorganic fine chemicals, test for, 494

insoluble, identification, 408
reduction to sulfides, 315
summary of tests, 531
tests for, 315
interference in alizarin test for zirconium, 201
in cerium salt reduction test for iodides, 268
in fluorescence test for uranium, 207
in β-nitroso-α-naphthol test for zirconium, 200
in zirconium-alizarin test for fluorides, 269
in zirconium azoarsenate test for fluorides, 272
in mixtures, test for, 385
soluble, summary of tests, 531
test by formation of barium sulfate, 314
Sulfide(s), carbonate in, detection, 498
carbonates in presence of, 338
catalysis of iodine–azide reaction as test for, 303, 407
as catalyst of iodine–azide reaction, 318
color reaction with dinitrobenzene, 320
corrosion of machinery metal, detection, 487
cyanide in presence of, 277
detection of free sulfur in presence of, 375
distinction from polysulfides, 491
insoluble, identification, 407
in "insoluble residue", detection, 305
interference in azide test for hydrazoic acid, 286
in molybdenum trioxide test for free carbon, 381
in sodium carbonate–phenolphthalein test for carbonic acid, 338
in tests for free metals and alloys, 362
for phosphates, 335
iodine–azide test for, 314
methylene blue test for, 307
in mixtures, test for, 384, 385
presence of, in iodine–azide test for thiocyanid acid, 282
reaction with dinitrobenzene, 242
reduction of methylene blue by, test for selenious acid, 348
reaction with picric acid and sodium carbonate, 280
with sodium nitroprusside, 303, 307
with triphenylmethane dyestuffs, 311
reduction of insoluble sulfates to, 315
selenium in, test for, 492
sodium nitroprusside test for, 315
sodium plumbite test for, 307
solid, test for, 305
soluble, test by catalysis of iodine-azide reaction, 304
sulfites in presence of, 308
sulfur, iodine–azide reaction for, 20
in rocks, 491
thiocyanates in presence of, 284
thiosulfates in presence of, 319
Sulfite(s), bisulfites in presence of, 313
carbonate in, detection, 498
carbonates in presence of, 338
cyanide in presence of, 277
detection of sulfamic acid in presence of, 326
interference in azide test for hydrazoic acid, 286
in sodium carbonate–phenolphthalein test for carbonic acid, 338
in thioglycolic acid test for iron, 167
in mixtures, test for, 385
nickel hydroxide test for, 309
in presence of sulfides and thiosulfates, 308
reaction with 2-benzylpyridine, 313
with cacotheline, 109
with ferrous sulfate and sulfuric acid, 326
with fuchsin, 311
with triphenylmethane dyestuffs, 311
sodium nitroprusside test for, 307
test for, by induced oxidation of $cobalt^{II}$ azide, 310
in thiosulfate solutions, tests for, 312
in water, test for, 507
4-Sulfo-2,2′-dihydroxyazonaphthalene, zinc salt of, *see* Pontachrome blue-black R
Sulfo salts, reaction with sodium azide, 305
Sulfotellurate, detection of free tellurium by formation of, 378
Sulfur, activation by heating of molten potassium thiocyanate, 509
detection of traces in carbon disulfide, 374
free, detection by conversion to hydrogen sulfide, 375, 491
of free tellurium in presence of, 379
in presence of free selenium, 375
of sulfides, 375
in minerals, inorganic mixtures and polysulfides, detection, 491
reaction with benzoin, 375, 491
with thallous sulfide paper, 12
summary of tests, 533

test by conversion to mercuric sulfide, 374
by conversion to thiocyanate, 373
by conversion to thiosulfate, 374
by formation of thallium polysulfide, 372
thallous sulfide test for, 491
reaction with benzoin, 13
with thallous sulfide paper, 12, 372
segregations in iron and steel, tests for, 486
selenium in, detection, 492
selenium in presence of, detection, 376, 377
tellurium in presence of, detection, 352
Sulfur dioxide, benzidine blue test for, 310
interference in silicomolybdic acid test for fluorides, 273
reaction with 2-benzylpyridine, 313
with zinc nitroprusside, 308
Sulfuric acid, barium carbonate and phenolphthalein test for, 314
barium rhodizonate test for, 313
concentrated, detection of chlorine in, 496
in copper sulfate solution, dectection, 502
detection of iron in, 453
and ferrous sulfate test for nitric acid, 326
free, hydroxybenzaldehyde test for, 317
methylenedisalicylic acid test for, 317
summary of tests, 531
tests for, 316, 317
insoluble sulfate, summary of tests, 531
lead in, detection, 435
and manganous sulfate test for bromic acid, 298
soluble sulfate, summary of tests, 531
test by precipitation of barium sulfate in presence of permanganate, 314
Sulfurous acid, action on *p*-fuchsin, 311
autoxidation of, in test for cobalt, 149
decolorization of malachite green as test for, 311
detection by reduction of ferric ferricyanide to Turnbull's blue, 313
induced oxidation of nickel hydroxide as test for, 309
sodium nitroprusside test for, 307
summary of tests, 531
test for, by induced oxidation of cobalt^{II} azide, 310
Superphosphate, distinction from phosphate fertilizers, 471
Supersaturation as supplementary process, 7
Supplementary processes, 7

T

Talc, magnesium in, detection, 461
preparation of copper-free water by, 431
as reagent-purifying adsorbent, 81
Tannin–antipyrine test for titanium, 199
Tannin–silver nitrate test for ammonia, 238
Tanning minerals, testing of leather for, 477
Tantalum, reaction with curcumin, 339
Tantalum oxide, detection of niobium pentoxide in presence of, 420
Tantalum pentoxide, distinction from niobic oxide, 420
Tap water, differentiation from distilled water, 507
Tartaric acid, masking action in cuproin test for cuprous ions, 92
thiocyanate in presence of, 282
Tartrates, interference in potassium thiocyanate test for iron, 164
Teallite, tin in, 439
Technical materials, molybdenum in, detection, 444
spot reactions for examination of, 428
Techniques, special, 50
of spot tests, 27
Tellurates, detection of, 379
reduction to tellurium, 349, 352
Telluric acid, alkali stannite test for, 350
ammonium polysulfide and sulfite test for, 352
complex copper salt formation as test for, 350
formation of stable copper salts by, 174, 303
hypophosphorous acid test for, 349
as preventive of copper catalysis, 174
summary of tests, 532
Tellurites, reduction to tellurium, 349, 352
Tellurium, ammonium hypophosphite test for, 464
free, detection by formation of sulfotellurate, 378
in presence of free sulfur or selenium, 379
formation, 352, 353
reaction with polysulfides, 378
summary of tests, 533
in presence of antimony, arsenic and tin, 353
of selenium, test for, 350
of selenium and sulfur, detection, 352
reaction with thiourea, 349
reduction of tellurites and tellurates to, 349, 350, 352

selenium in, test for, 492
selenium in presence of, test for, 346
stannous chloride test for, 100
Tellurium compounds, reaction with cacotheline, 109
Tellurium dioxide, detection of selenium dioxide in, 348
Tellurium sulfosalts, formation, 353
Tellurous acid, alkali stannite test for, 350
ammonium polysulfide and sulfite test for, 352
complex copper salt formation as test for, 350
hypophosphorous acid test for, 349
reaction with hydriodic acid, 346
summary of tests, 532
Testing of materials, prospects in spot test analysis, 23
3,4,3′,4′-Tetra-aminobenzidine test for selenites, 349
Tetrabase, induced oxidation as test for ferricyanic acid, 293
and manganese salts test for periodic acid, 302
redox reaction with chloramine T, 511, 535
Tetraborates, interference in phenoxithine test for palladium, 140
Tetrabromofluorescein, *see* Eosin
Tetraethyl rhodamine, *see also* Rhodamine B
test for antimony, 105
1, 2, 5, 8-Tetrahydroxyanthraquinone, *see* Quinalizarin
Tetrahydroxyquinone test for barium, 219
Tetraiodofluorescein, *see* Erythrosin
Tetramethyl-*p*-diaminodiphenylmethane, *see* Tetrabase
Thallium, in alloys, detection, 437
auric chloride and palladium chloride test for, 161
benzidine test for, 155
catalytic test for silver in presence of, 63
detection of gallium in presence of, 215
dipicrylamine test for, 156
gold in presence of, 130
interference in cuproin test for cuprous salts, 91
leuco nitro diamond green test for, 161
manganese in presence of, 177
in minerals, detection, 437
oxine test for, 161
phosphomolybdic acid and hydrobromic acid test for, 156
potassium iodide test for, 154
in presence of antimony, test for, 152
of gold, test for, 160
of lead, tests for, 155, 156
of mercury, tests for, 155, 160
of silver, test for, 155
reaction with sodium rhodizonate, 74
with potassium bismuth iodide, 235
rhodamine B test for, 158, 161, 437
sodium carbonate as solvent for, 160
summary of tests, 524
in technical materials, detection, 437
test by oxidation, 157
uranyl sulfate test for, 161
Thallium carbonate, solubility, 258
Thallium chloride, reaction with rhodamine B, 106
Thallium hydroxide, preparation of suspension, 402
reaction with malachite green, 212
Thallium nitrate, as reagent for alkali, 307
test for iodides, 269
Thallium oxide test for hydrazine, 242
Thallium^{III} oxyhydrate, formation from thallium^{I} salts, 417, 459
Thallium polysulfide formation as test for free sulfur, 372
Thallium salts, cerium in presence of, 211
detection of cadmium in presence of, 95
interference in benzidine test for lead, 72
in *o*-tolidine test for copper, 84
oxidation of benzidine by, 176
reaction with anthraquinone-1-azo-4-dimethylaniline hydrochloride, 112
with auric chloride and palladium chloride, 235
with benzidine, 211
with dipicrylamine, 231
with potassium chromium thiocyanate, 79
with sodium cobaltinitrite, 230
Thallous hydroxide test for hypohalogenites, 295
Thallous polysulfide, formation, 491
Thallous selenide, formation, 375
Thallous salts, formation of thallium^{III} oxyhydrate from, 417, 459
Thallous sulfate test for ferricyanides, 294
Thallous sulfide, test for free selenium by addition to, 375
test for free sulfur, 491
Thallous sulfide paper, reaction with free selenium, 12, 372, 375
with sulfur, 12, 372
Thioacetamide test for bismuth, 80
Thiocarbazide, reaction with molybdates, 120

Thiocyanates, catalysis of iodine–azide reaction as test for, 303, 305
as catalyst of iodine–azide reaction, 283, 318
conversion to prussic acid as test for, 284
cyanides in presence of, 277
ferric thiocyanate test for, in presence of iodides, 283
ferrocyanides in presence of, 290
formation as test for hydrocyanic acid, 277
interference in butyraldehyde test for cobalt, 149
in test for cyanates, 285
in mixtures, test for, 386
in presence of acetate, bromide, chloride, iodide, and thiosulfate, 285
of ferrocyanic, oxalic, phosphoric, and tartaric acid, 282
of halides, detection, 284
of iodides, 283
of sulfides and thiosulfates, test for, 284
reaction with ammonium chloride, 285
with ferric chloride, 373
with ferrous sulfate and sulfuric acid, 326
as test for iron, 283
test for, by formation of hydrogen sulfide, 285
for free sulfur by conversion to, 373
Thiocyanic acid, catalysis of the iodine–azide reaction as test for, 282
cobalt salts test for, 285
in presence of sulfides and thiosulfates, 282
summary of tests, 532
Thiocyanides, interference in phenoxithine test for palladium, 140
Thioglycolic acid, reaction with cobalt and nickel ions, 162
reduction of ferric ions by, 449, 451, 452, 453
of ferric salts by, 162
of stannic salts by, 107
test for iron, 167
Thio-Michler's ketone, *see also* 4,4′-Bis-dimethylamino-thiobenzophenone
detection of chlorine by, 300
reaction with free halogens, 368
with oxidizing cations, nitrous acid or nitrogen tetroxide, 369
Thiosulfates, carbonates in presence of, 338
catalysis of iodine–azide reaction as test for, 303
distinction from selenosulfates, 377
interference in azide test for hydrazoic acid, 286
in butyraldehyde test for cobalt, 149
in sodium carbonate–phenolphthalein test for carbonic acid, 338
in zirconium–alizarin test for fluorides, 269
in zirconium azoarsenate test for fluorides, 272
iodine–azide reaction for, 374
in mixtures, test for, 386
presence of, in iodine–azide test for thiocyanic acid, 282
in presence of sulfides, test for, 319
reaction with cacotheline, 109
with cobalt salts, 285
with ferric salts, 80, 143
with ferrous sulfate and sulfuric acid, 326
sulfites in presence of, 308
test for free sulfur by conversion to, 374
thiocyanates in presence of, 284, 285
Thiosulfate solutions, test for sulfites in, 312
Thiosulfuric acid, summary of tests, 532
test by catalysis of iodine–azide reaction, 318
Thiourea, reaction with bismuth, mercury, and tellurium, 349
with selenite, 349
Thorium, interference in alizarin or quinalizarin test for indium, 213
zirconium in presence of, 203
Thorium fluoride in insoluble residue, 420
Thorium nitrate, in test for uranium, 210
Thorium salts, cerium in a mixture of zirconium salts and, 212
interference in fluorescence test for uranium, 207
in morin test for titanium, 199
lake formation with quinalizarin, 193
reaction with *p*-dimethylaminoazophenylarsonic acid, 203
with morin, 183
with quinalizarin, 225
zirconium in presence of, 201
Thulium, hydrogen-flame luminescence test for, 79
Thymol blue, color reaction with soft-burned lime, 467
Tin, in alloys, test for, 394, 433, 441
o-aminophenol test for, 112
ammonium phosphomolybdate test for, 108
anthraquinone-1-azo-4-dimethylaniline hydrochloride test for, 111

antimony in presence of large amounts of, 106
in antimony salts, detection, 108
2-benzylpyridine test for, 111
cacotheline test for, 109
diazine green S(K) test for, 111
dithiol test for, 107
flame color test for, 109
germanium in presence of, 113
interference in phosphomolybdic acid test for vanadium, 127
luminescence reaction for, 421
magnesium in presence of, 228
mercuric chloride and aniline test for, 111
metallic, detection by reduction of *o*-dinitrobenzene, 363
in metallic objects, detection, 439
in minerals, detection, 439
in mixtures, tests for, 387, 390, 392
morin test for, 110, 421, 440, 441
in presence of antimony, detection, 108
reaction with sodium rhodizonate, 74
simultaneous test for antimony and, 441
summary of tests, 525, 533
tellurium in presence of, 353
1,2,7-trihydroxyanthraquinone test for, 111
zinc in presence of, 183
zirconium in presence of, 204
Tin compounds in antimony reduction test, 103
Tin phosphate in insoluble residue, 421
Tin-plated metals, tin in, detection, 439
Tin platings, lead in, detection, 433
Tin salts, detection of cadmium in presence of, 96
detection by reduction of ferric salts, 111
identification by cacotheline reaction, 68
reaction with *p*-dimethylaminoazophenylarsonic acid, 204
reduction by hypophosphites accelerated by mercury salts, 68
of phosphomolybdates by, 156
Titanium, ammonium hypophosphite test for, 464
chromotropic acid test for, 197, 198, 459
hydrogen peroxide test for, 195, 413
methylene blue and zinc test for, 199
in minerals, detection, 459
morin test for, 198
in presence of chromium, iron, manganese, and uranium, 199
of fluoride, 196
of other elements, test for, 198
pyrocatechol test for, 196
reaction with curcumin, 339
resoflavine test for, 199
in silicates, test for, 460
summary of tests, 525
tannin–antipyrine test for, 199
uranium in presence of, 457
zirconium in presence of, 203
Titanium dioxide, interference in fluorescence test for uranium, 207
iron in, detection, 450
Titanium–malachite green reaction, catalysis of, as test for tungsten, 121
Titanium oxide in insoluble residue, 413
Titanium salts, conversion to titanium fluoride, 123
interference in cacotheline test for tin, 109
in test for beryllium, 195
reaction with chromotropic acid, 197
with *p*-dimethylaminoazophenylarsonic acid, 203
with disodium-1,2-dihydroxybenzene-3,5-disulfonate, 166
with 1,2,7-trihydroxyanthraquinone, 111
zirconium in presence of, 201
Titanium tetrachloride test for ferrocyanide, 294
Titan yellow test for magnesium, 228
Titrimetric analysis, 1
o-Tolidine, and ammonium thiocyanate test for copper, 84
identification of tervalent cobalt by, 311
and sodium azide, test for cobalt, 149
solution as reagent for hydrogen peroxide, 356
Toluene as solvent for sulfur, 376
Toluosafranine test for nitrites, 332
Toning baths, qualitative analysis of, 129
Trace analysis, prospects in spot test analysis, 23
Trace catchers, 428
Traces, search for, 21
Transformation of amorphous into crystalline forms as supplementary process, 7
1,2,7-Trihydroxyanthraquinone, as reagent for aluminum, chromium, iron, titanium, and zirconium salts, 111
as reagent for bismuth, 111
for molybdates, 111
test for tin, 111
Triphenylmethane dyestuffs, reaction with sulfites, mono- and polysulfides, 311

Trithiourea cuprous chloride test for ferricyanic acid, 294
Tropaeolin OO as indicator of hydrogen-ion concentration, 504
Tschugaeff reaction for nickel, 149, 445
Tungstates, interference in alizarin test for zirconium, 201
in cuprous iodide test for mercury, 66
in dipyridyl test for molybdenum, 119
in ferrous sulfate test for nitric acid, 326
in minerals, detection, 416
reaction with α-benzoinoxime, 125
with *p*-dimethylaminoazophenylarsonic acid, 203
with ferrous sulfate, 326
with oxine, 125
with rhodamine B, 106, 107
Tungsten, ammonium hypophosphite test for, 464
detection in minerals, 459
in presence of molybdenum, 120
diphenyline test for, 122
8-hydroxyquinoline test for, 119
in insoluble residue, 416, 418
interference in methylene blue and hydrazine test for molybdenum, 118
in potassium thiocyanate test for molybdenum, 115
in test for titanium, 199
metallic, detection by conversion to metallo acid salts, 367
reaction with alkali nitrite, 367
ores, molybdenum in, detection, 444
in presence of molybdenum, test for, 120
stannous chloride test for, 120, 416
summary of tests, 525, 533
test by catalysis of titanium–malachite green reaction, 121
zirconium in presence of, 203
Tungsten blue, formation, 66, 416, 417
Tungsten oxide, detection of nitrate in, 494
in insoluble residue, 416, 417
interference in cacotheline test for tin, 109
in reduction test for niobium pentoxide, 420
properties, 120
reaction with alkali nitrite, 367
separation from molybdenum trioxide, 416
test for, by hydrogen peroxide activation, 417, 459
Tungsten oxinate, formation of, as test for tungsten, 459
Tungsten steels, fusion with alkali nitrite, 418
Tungstic acid, calcined, differentiation from WO_3·aq., 510
Tüpfelreaktion, 2
Turmeric, *see* Curcuma, tincture of
Turnbull's blue, distinction from Prussian blue, 291
formation, 294, 354, 363
reduction to ferro-ferrocyanide, 161
unsuitability for spot tests, 291

U

Ultramicro methods, 6
Uncertain reactions, region of, 14
Uranium, aluminum in presence of, 187
ammonium hypophosphite test for, 464
cobalt in presence of, 144, 145
detection in presence of iron, 209
ferric chloride and phenanthroline test for, 210
fluorescein test for, 210
fluorescence test for, 207
lake formation with alizarin, 186
in minerals, detection, 457
oxine test for, 204
potassium ferrocyanide test for, 205, 457
in presence of iron, 205
of iron and copper, test for, 206
of titanium, 457
oxalic acid test for, 207
quercetin or quercitrin test for, 209
rhodamine B test for, 208, 457
summary of tests, 525
titanium in presence of, 199
Uranium salts, interference in cacotheline test for tin, 109
reaction with alizarin, 187
with anthraquinone-1-azo-4-dimethylaniline hydrochloride, 112
with quercetin or quercitrin, 167
Uranyl acetate test for ferrocyanic acid, 287
Uranyl ferrocyanide test for zinc, 182
Uranyl ions, interference in test for mercury, 71
Uranyl salts, cochineal test for, 210
interference in cerium salt reduction test for iodides, 268
in oxine test for vanadium, 126
oxalic acid test for, 207
reaction with chromotropic acid, 198

with nitrosonaphthol, 144, 145
with rhodamine B, 208
sodium phosphate test for, 209
Uranyl sulfate test for thallium, 161

V

Vanadates, interference in morin test for zirconium, 202
in oxine test for tungsten, 120
in pyrrole test for selenious acid, 346
in test for zinc, 178
reaction with benzidine, 172
with diphenylamine, 327
with diphenylcarbazide, 170
with hydrogen peroxide, 196
with oxine, 125
Vanadic acid as catalyst for alkali chlorate solutions, 142
Vanadinite, detection of chlorine in, 502
Vanadium, alkali tungstate test for, 127
ammonium hypophosphite test for, 464
α-benzoinoxime test for, 124
diaminobenzidine test for, 127
3,3'-dimethylnaphthidine test for, 126
diphenylbenzidine test for, 127
hydrogen peroxide test for, 123
interference in test for titanium, 199
$iron^{II}$ dimethylglyoxime test for, 124
$iron^{II}$-α,α'-dipyridyl test for, 124
metallic, detection by conversion to metallo acid salts, 367
reaction with alkali nitrite, 367
oxine test for, 125, 127
phosphomolybdic acid test for, 127
in presence of MoO_3 or WO_3, 126
quercetin test for, 127
summary of tests, 526, 533
Vanadium oxide in insoluble residue, 417
test for, by hydrogen peroxide activation, 417
Vanadium pentoxide, interference in reduction test for niobium pentoxide, 420
reaction with alkali nitrite, 367
Vanadium salts, catalytic oxidation of aniline to aniline black by, 127
interference in test for chromium, 168
reaction with hydrogen peroxide, 417
Vanadyl salts as catalysts for alkali chlorate solutions, 142
Vapors in spot test analysis, 45
Varnishes, lead in, detection, 435
Versatile spot test laboratory, 30
Versene, *see* Ethylenediaminetetraacetic acid
Volatile compounds, detection, 46
apparatus for, 47, 48 (Figs. 23–28)
Volatilization test for hydrochloric acid, 261

W

Water, chemically or adsorptively bound, detection, 509
chlorinated, detection of, 370
detection of aluminum in, 454
of copper in, 431
of crystallisation, heating of materials carrying, 397
differentiation between tap and distilled, 222, 507
hardness, determination, 507
hydrogen sulfide in, detection, 503, 507
lead in, detection, 435
magnesium in, 227, 506
mineral, detection of thallium in, 437
spot methods for judging samples of, 505
Wavellite, chromium in, 455
Whatman filter paper, 54
Witherite, differentiation from strontianite, 473
Wolframite, decomposition of, 459
detection of iron in, 462
of manganese in, 456
Wood, spot tests for chemicals used in preserving, 476
Working methods in spot test analysis, 31

X

Xanthate test for molybdenum, 116, 444
Xantho dyestuffs, reaction with antimony salts, 105

Y

Yttrium, hydrogen-flame luminescence test for, 79

Z

Zinc, in alloys, tests for, 394
aluminum in presence of, 187
cadmium in presence of, 96
detection of aluminum in presence of, 190
of iron in, 452
3,3'-dimethylnaphthidine and alkali ferricyanide test for, 181
dithizone test for, 178
ferricyanide and diethylaniline test for, 178
and p-phenetidine test for, 182
indium in presence of, 213, 214

interference in ferric thiosulfate test for copper, 81
in sodium rhodizonate test for barium, 476
magnesium in presence of, 227
metallic, copper in, test for, 430
detection by reduction of *o*-dinitrobenzene, 363
and methylene blue test for titanium, 199
in minerals, test with ferrocyanide paper, 490
in mixtures, tests for, 388, 391, 392
in plated metals, detection, 448
potassium cobalticyanide test for, 182
in presence of alkali and alkaline earth metals, 182
of aluminum, test for, 178
of aluminum salts, 179
of bismuth, test for, 182
of chromium, test for, 178
of lead, test for, 182
of nickel salts, 179
of tin, test for, 182
resorcinol test for, 182
solubility in nonoxidizing acids, 362
summary of tests, 526, 533
test by coprecipitation with cobalt mercury thiocyanate, 180
uranyl ferrocyanide test for, 182
Zinc blende, free sulfur in, detection, 491
lead in, detection, 434
Zinc chloride, detection of hypohalogenites by, 297
fusion with niobium pentoxide, 420
fusion reactions with, 460, 461, 462, 463, 469
in wood preservatives, test for, 477
Zinc diethyldithiocarbamate test for cuprous salts, 94
Zinc dross, lead in, detection, 434
Zinc ions, interference in α,α'-dipyridyl test for iron, 453
Zinc nitroprusside, reaction with sulfur dioxide, 308
Zinc oxide, free metal in, detection, 475
lead in, detection, 434
Zinc oxinate, formation, as test for ammonia, 237
Zinc salts, detection of iron in, 452
interference in chromeazurole test for beryllium, 195
in thioglycolic acid test for iron, 167
in zinc uranyl acetate test for sodium, 229
reaction with alkali mercury thiocyanates, 180
with anthraquinone-1-azo-4-dimethylaniline hydrochloride, 112
with nitro-3-hydroxybenzoic acid, 220
Zinc sulfide paper, 51
Zinc uranyl acetate test for sodium, 229
Zirconium, alizarin test for, 200
detection of aluminum in presence of, 190
p-dimethylaminoazophenylarsonic acid test for, 202, 271
interference in alizarin or quinalizarin test for indium, 213
mandelic acid azo dyes test for, 204
morin test for, 201, 414
β-nitroso-α-naphthol test for, 199
in presence of aluminum, beryllium, thorium, and titanium salts, 201
of antimony, test for, 203
of molybdenum, tungsten, and titanium, test for, 203
of thorium, 203
of tin, test for, 204
reaction with carminic acid, 204
with chlorobromoamine acid, 204
with curcumin, 339
with gallocyanin, 204
summary of tests, 526
Zirconium alizarinate test for fluorides, 420
Zirconium–alizarin lake decomposition, 271
Zirconium–alizarin paper, preparation, 269
Zirconium–alizarin solution as reagent for fluorine, 269, 490
Zirconium–alizarin test for hydrofluoric acid, 269
Zirconium azoarsenate paper, preparation, 272
test for fluorides, 271
Zirconium oxide, ignition test, 414
Zirconium salts, cerium in a mixture of thorium salts and, 212
interference in morin test for titanium, 199
in test for beryllium, 195
lake formation with quinalizarin, 193
reaction with morin, 183
with quinalizarin, 225
with 1,2,7-trihydroxyanthraquinone, 111
Zorgite, selenium in, 492

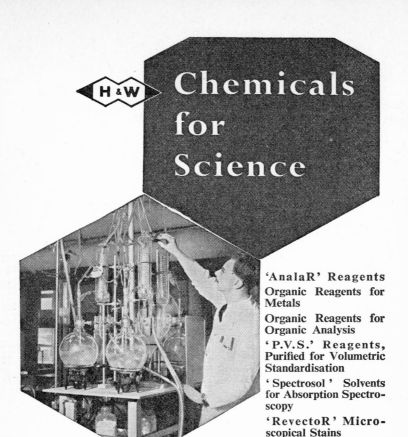